Proceedings of the Third International Symposium on Materials and Sustainable Development

Benmounah Abdelbaki · Brahim Safi
Mohammed Saidi
Editors

Proceedings of the Third International Symposium on Materials and Sustainable Development

Springer

Editors
Benmounah Abdelbaki
Faculty of Engineering Sciences
M'hamed Bougara University of Boumerdes
Boumerdés, Algeria

Mohammed Saidi
Faculty of Engineering Sciences
M'hamed Bougara University of Boumerdes
Bourmedés, Algeria

Brahim Safi
Faculty of Engineering Sciences
M'hamed Bougara University of Boumerdes
Bourmedés, Algeria

ISBN 978-3-030-07837-9 ISBN 978-3-319-89707-3 (eBook)
https://doi.org/10.1007/978-3-319-89707-3

Printed on acid-free paper

This Springer imprint is published by the registered company Springer Nature Switzerland AG
The registered company address is: Gewerbestrasse 11, 6330 Cham, Switzerland

Preface

The healthy environment, the key of the vaulted against pollution, depends mainly on the manufacture processes of the materials, their quality, and durability. Indeed, the interest granted to these by the previous editions of ISMSD2017 Symposium, was an undeniable factor which corroborated our strategy aimed at the continuous improvement of elaborated products whose properties are, above all, functions of their structure and their chemical composition. This is confirmed by the large number of papers that will be on display. This symposium, the third of its kind, organized by the Research Unit of Materials, Process and Environment, in collaboration with others University, will certainly contribute to reinforcing the links and cooperation between our institutions. During the ISMSD2017 Symposium which has deal with the following topics:

Theme 1: Advanced materials and nanotechnology
Theme 2: Chemistry of building materials and materials
Theme 3: Waste management, recycling, and environment
Theme 4: Materials processing technology

The scientific committee and the organizing committee have made every effort to ensure that the scientific aspect is beneficial (respect for the duration of presentation, explicit answers to the questions asked, clarification of unresolved points, by the moderators and specialists present, continuation discussions during coffee breaks, in-depth discussions on the work presented in the form of posters and oral presentations, etc.) for all participants. Thus, the audience of the ISMSD2017 meeting will widen which was be a potential catalyst for the holding of the next edition.

Objectives of the Third International Symposium on Materials and Sustainable Development

Given the quick development of materials, and therefore the interest they generate, the research unit: Materials, Processes, and Environment of University M'hamed Bougara of Boumerdes is working to make this symposium a tradition that will gather every two years, researchers from different universities and industrialists, in order to know, among other things, studies carried out on materials, to foster links between researchers, to strengthen scientific exchanges between them and to create partnerships that will consolidate relations between laboratories.

Moreover, Ph.D. students will not fail to find, in this colloquium, one of the privileged spaces to collect information that will help them to carry out their thesis. Finally, contacts will be made between academics and industry to eventually lead to mutually profitable agreements.

Benmounah Abdelbaki
Brahim Safi
Mohammed Saidi

Organization

Research Unit: Materials, Processes, and Environment

University M'hamed Bougara of Boumerdes

Scientific and Advisory Committee

M. T. Abadlia	Univ. Blida, Algeria
A. Benmounah	UMBB, Algeria
A. Zerizer	UMBB, Algeria
A. Tairi	UMBB, Algeria
A. Bali	ENP, Algeria
B. Bezzazi	UMBB, Algeria
L. Alex	Univ. Reims, France
A. Ponton	Univ. Paris 7, France
M. Bououdina	Univ. Bahrain, Bahrain
K. Boumchedda	UMBB, Algeria
M. Saidi	UMBB, Algeria
M. Al Bakri Abdullah	Univ. Perlis, Malaysia
M. Hamiane	UMBB, Algeria
C. Smith	CNRS-Univ. Limoges, France
M. Hachemi	UMBB, Algeria
J. M. Vallet	CICRP-Marseille, France
L. Bleron	ENSTIB, France
K. Mohammedi	UMBB, Algeria
A. Yahia	Univ. Sherbrooke, Canada
Y. Ghernouti	UMBB, Algeria

T. H. Daouadji	Univ. Tiaret, Algeria
A. Oudia	Univ. Toulouse, France
M. Sonebi	Queens Univ. Belfast, UK
K. Louhab	UMBB, Algeria
A. Bezazi	Univ. Guelma, Algeria
D. Achoura	Univ. Annaba, Algeria
S. Kenai	Univ. Blida, Algeria
M. Mouli	ENP-Oran, Algeria
A. Iratni	UMBB, Algeria
M. Bederina	Univ. Laghouat, Algeria
R. Chaid	UMBB, Algeria
Y. Leprince-Wang	Univ. UPEM, France
A. Diouri	Univ. Rabat, Maroc
B. Chemani	UMBB, Algeria
H. Chemani	UMBB, Algeria
A. Nouani	UMBB, Algeria
O. Limam	Univ. El Manar, Tunis
F. Charrier	Univ. Pau, France
K. Ait Tahar	Univ. Bouira, Algeria
S. Azem	UMMTO, Algeria
D. Miloš	Acad. Sciences- ITAM, Czech Republic
A. Amara	C. Univ. Ain Temouchent, Algeria
M. S. Ghembaza	Univ. Sidi Bel-Abbes, Algeria
Y. Abdelaziz	Univ. Bechar, Algeria
H. Essawy	National Research Centre, Egypt
N. Dokhane	UMBB, Algeria
S. Simon	National Museums Berlin, Germany
M. Z. Boureghda	UMBB, Algeria
D. Aboutaleb	UMBB, Algeria
H. Aknouche	UMBB, Algeria
B. Benotmane	UMBB, Algeria
R. Tala Ighil	UMBB, Algeria
A. Daoui	UMBB, Algeria
F. Bensouici	UMBB, Algeria
B. Safi	UMBB, Algeria
B. Rabehi	UMBB, Algeria
R. Kheribet	UMBB, Algeria
L. Legendre	Université de Rennes 1, France
R. Benzerga	Université de Rennes 1, France
T. Bouziani	Univ. Laghouat, Algeria
C. Franzen	IDK, Germany
S. Lecheb	UMBB, Algeria
A. Chellil	UMBB, Algeria
K. Chahour	UMMTO, Algeria
H. Mechakra	UMBB, Algeria

Contents

Advanced Materials and Nanotechnology

Investigation of Structural, Optical and Electrical Properties of Al Doped SnO_2 Thin Films Synthesized by Sol-Gel 3
A. Boumezoued, K. Guergouri, C. Azizi, A. Aberkane, and A. Khial

Formation Energies of Mn Doped $ZnSnAs_2$ 10
H. Bouhani-Benziane, K. Hebali, A. Bakhti-Siad, O. Sahnoun, M. Baira, and M. Sahnoun

The Improvement of Tandem a-Si:H/μc-Si:H Solar Cells Perfomance by Optimized the Front Contact Barrier Height 15
Abbas Belfar, Tarik Kadi, and Hocine Ait-Kaci

Realization of Diene Dienophile Interface Reaction in Oil/Water Emulsion 25
Abdallah Cheraitia and Nawal Brahimi

A New Four Variable Refined Shear Deformation Theory for Buckling and Vibration of Functionally Graded Plates 34
Ahmed Bakora, Fouad Bourada, Abdelouhed Tounsi, and Adda Bedia El Abbas

Optical and Structural Properties of ZnS:La Thin Films Elaborated by Sol-Gel Method 44
A. Tounsi, D. Talantikite-Touati, R. Khalfi, H. Merzouk, and H. Haddad

Theoretical and Experimental Studies of Micropores Catalysts 52
Atallah Dehbi, Nadia Toumi, Rawia Imene Bahoussi, Meryem Imene Amrane, Affaf Tabti, Imene Benchikh, Zoubida Lounis, Farouk Hamza Reguig, and Abdelkader Bengueddach

Influence of Deposition Time on the CdS Thin Films Prepared by a Chemical Bath Deposition (CBD) 67
D. Haouanoh, R. Talalghil, F. Bensouici, and M. Toubane

Theoretical Investigation of Structural, Elastic and Electronic Properties of $SnHfO_3$ Compound: Ab Initio Calculations 75
D. Chenine and Z. Aziz

Al-Doping Effects on Structural and Morphological Properties of ZnO Aerogels Synthesized in Supercritical Ethanol 81
F. Mouzaia, D. Djouadi, A. Chelouche, and L. Hammiche

High Quality Nano Thin Layer Silicon Transfer Using Plasma Hydrogenation 89
F. Okba

Enhacing Pre-stressed Concrete Beam's Capacity Using Externally Bonded Pre-stressed Composite Plate 95
Hanane Mebsout, Baghdad Krour, and Mohamed Bachir Bouiadjra

Artificial Metamaterials for High Efficiency Silicon Solar Cells 105
Houria Hamouche and Mohammed M. Shabat

A New Technique for the Evaluation of the Parameters of GaAlAsSb(p)/GaAlAsSb(n)/InAsSb Double Interface from Current-Voltage Curve Analysis 116
K. Mahi, B. Messani, and H. Ait-kaci

Controlling Drug Release Through Poly(2-Hydroxyethylmethacrylate-co-acrylic Acid) Grafted Sodium Alginate 126
L. Bounabi, N. Bouslah Mokhnachi, N. Haddadine, and A. Benaboura

Theoretical Study of Eu^{3+} and Li^{+} Co-doped Ca_2SnO_4 134
M. Baira, A. Siad-Bekhti, N. Siziani, H. Bouhani-Benzian, and M. Sahnoun

Synthesis Cordierite Materials Starting from Algerian Kaolin 141
M. Kheloui, K. Boumchedda, S. Debbakh, and B. Rabehi

Nonlinear Behavior of FGM Plate in Cylindrical Bending Under Uniform Loading 147
Mohamed Bourada, Fouad Bourada, Abdelhakim Kaci, Mohamed Bouremana, and Abdelouahed Tounsi

Template Optimization of Block Copolymer Thin Films by Self-Assembly Process 157
M. Loucif Seiad, V. K. MKuppuswamy, M. Ferhat, and R. Gronheid

Hydroxy Ethyl Cellulose-Graft-Acrylamide Superabsorbant Hydrogel for Controlled Drug Delivery of Cetirizine 166
N. Bouslah Mokhnachi, L. Bounabi, N. Baa, F. Ouazib, and N. Haddadine

Improving the Power Efficiency of Organic Photovoltaic Cells with P3HT:ICBA Used as Active Layer 175
O. Ourahmoun

Solar Cell Based on p-Si/n-Zno:Al Thin Films: Simulation and Parameters Extraction 181
Safia Kerour, Meryem Lasladj, and Abdeslam Bouloufa

Material Fracture of Dynamic VAWT Blade 189
S. Lecheb, A. Chellil, H. Mechakra, B. Safi, Y. Beggah, and Y. Lounici

Effect of Dopant Material on the Performance UV Photodetector Based SnO_2 Thin Films Deposited by Sol-Gel Dip-Coating Method 197
Kaour Selma and Rechem Djamil

Non Local Vibration of Damped Viscoelastic—FGM-Beam 206
Souad Hamzi, Ismail Mechab, and Yassine Senhadji

Chemistry of Building Materials and Materials

Characterization of "Float Glass" Eroded by Sandblasting 219
Djamila Aboutaleb, Brahim Safi, Ahmed Hammouda, and Mohammed Hadj Ammar

Effect of Chemical Activation on the Surface Properties of Apricot Stones Based Activated Carbons and Its Adsorptive Properties Toward Aniline 228
Zoubida Kecira, Asma Benturki, Mounir Daoud, and Oumessaâd Benturki

Porosity Effect of a FG Porous Plate on the Behavior of Bonded Beam and Strengthened Structure Instability 241
A. Chedad and N. Elmeiche

Evaluation of the Impact Sound Transmission Through Concrete Floor Using Numerical Simulation and Laboratory Experiment 254
Abdelouahab Bouttout and Mohamed Amara

An Overview of Functionally Graded Materials «FGM» 267
Adda Hadj Mostefa, Slimane Merdaci, and Noureddine Mahmoudi

Experimental and Numerical Damaged Characteristics in Composite Plate with Crack Growth by Vibration Analysis 279
A. Chellil, S. Lecheb, H. Mechakra, S. Chellil, and B. Safi

Tensile Behavior of Bio-Composites: Date Palm Rachi/Epoxy, Date Palm Pits/Epoxy 288
A. Dakhli, R. Benzidane, R. Zehaf, L. M. Bennegadi, and Z. Sereir

Fabrication and Characterization of TiO_2 Nanoparticles 296
A. Bessi, B. Boudine, and C. Boudaren

In Vitro Corrosion Study by EIS of Stainless Steel for Orthopaedic Applications 301
Faiza Kakaa and Mosbah Ferkhi

Application of the Rietveld Method in the Study of Phases in High Chromium Cast Irons 306
Fatma Zohra Boulassel, Benabbas Abderrahim, and Rekaik Mouna

New Materials Rigid Obtained by the Acid Hydrolysis of Starch Followed by a Grafting of Acrylamide with and Without Bis-acrylamide 313
Karima Boumerdassi and Aicha Serier

Copolymerization of B-Pinene with Limonene Using Heterogeneous Catalyst 330
K. Bekhemas, A. Hachemaoui, and A. Yahiaoui

Synthesis, Characterization, Electrochemical of Ligational Behavior of Curcumin Drug Towards Some Transition Metal Ions 337
L. Abane Merzouk and A. Adkhis

Zinc Based Materials as Efficient Catalysts for a Clean Synthesis of Biginelli Product 345
L. Bennini, T. Mazari, C. Khiar, S. Menad, M. Halouane, and C. Rabia

Study of the Electronic and Elastic Properties of ZrN and Zr_3N_4 353
M. Azibi and N. Saoula

Degradation Studies on PVC Plasticized Submitted to Gamma Radiation up to 50 KGy 358
N. Boutouchent-Guerfi, A. Benaboura, and A. Dib

Electronic and Thermoelectric Properties of $CdSnAs_2$ 365
N. Si Ziani, H. Bouhani-Benziane, M. Baira, A. E. K. Belfedal, and M. Sahnoun

Milligels Synthesis and Characterization: Mebeverine Hydrochloride Uptake and Release 372
N. Baït, B. Grassl, and A. Benaboura

Dynamic Rheology Study of In-situ Gelation Process of Polyacrylamide-Montmorillonite Composite Hydrogels 378
N. Baït, B. Grassl, and A. Benaboura

Temperature Effects on Corrosion Inhibition of Mild Steel in 3% NaCl Solution by EDTA and Sodium Benzoate 385
N. Brixi, L. Sail, A. Bezzar, O. Sebouai, and S. Benmesmoudi

Dynamic Behavior of the Composite Rotor Blade Using an Adaptive Damper 398
S. Chellil, A. Nour, A. Chellil, S. Lecheb, H. Mechakra, and A. T. Settet

Micro-Meso Mechanics Based Modeling of Damage Evolution in Cross Ply Laminates Composites 408
S. Benbelaid and B. Bezzazi

Thermoelectric Properties of Single Filled Skutterudites $Yb_xCo_4Sb_{12}$ 420
Z. Benhalima, A. Bekhti-Siad, and M. Sahnoun

Waste Management, Recycling and Environment

Recycling of Plastic Waste and Valorization of Clay Material in the Road Area 427
Oum Essaad Ouaaz and Bachir Melbouci

Experimental Study of an Inclined-Plane Electrostatic Separator 439
A. Nadjem, M. Kachi, K. Rouagdia, and M. Remadnia

Effect of Liquid Type on the Hydraulic Characteristic of Compacted Local Geomaterials for Use as Hydraulic Barriers 451
Abdellah Demdoum, M. K. Gueddouda, I. Goual, and H. Berkak

Application of Factorial Plan for the Recovery of Phosphate Rejects +15 mm by Flotation Method Case of the Mine of Djebel Onk, Tebessa - Algeria 465
M. Ali, I. Abdelaziz, O. H. Malek, and M. Mohamed

Nanofiltration Performance for Synthetic and Natural Water Defluorination: Application to South-Algeria Groundwater 481
Amina Ramdani, Sana Gassara, André Deratani, and Safia Taleb

Treatment of Colored Waters by Beads Chitosan, Extracted from Shrimp Waste 492
D. Aili, W. Arbia, and L. Adour

Use of the Potato Peels and the Bentonite as Additives of Flocculation in the Wastewater Treatment of the STEP of Saïda . . . 506
Fatima Zohra Choumane, Fatma Kandouci, and Bouhana Maachou

Demineralization Effect on the Physicochemical and Interfacial Properties of Sweet Whey 516
K. Acem, A. Choukri, N. Saoudet, S. Tergou, and H. Trari

Treatment of Medical Waste Using Electrocoagulation Process 527
K. Madi-Azegagh, F. Aissani-Benissad, and I. Yahiaoui

Biosorption of the Anionic Dye Direct Red 89 by the Aquatic Plant *Callitriche obtusangula* 540
Khaled Benabbas, Imene Hocini, and Nabila Khellaf

Extraction of Propionic Acid from Waste Water Using Rosmary 549
M. Timedjeghdine, A. Djellouli, and A. Hasseine

Investigate Method to Reduce Methane Emission During Maintenance Operations from Algerian Natural Gas Transportation 556
M. Louhibi–Bouiri and M. Hachemi

Kinetics and Isotherms Adsorption of Reactive Dye by Thermally Treated Dolomite 568
Samira Ziane and Amine Khelifa

Degradation of a Pharmaceutical Pollutant by Coupling Photo-Fenton and Adsorption Processes 580
S. Cherif, M. R. Boudraa, A. Moussa, R. Maachi, N. Nasrallah, and N. Guendouz

Synthesis and Structural Characterization of Glass Foam Composites for Electromagnetic Application 593
Younes Lamri, Ratiba Benzerga, Azzedine Ayadi, Laurent Le Gendre, and Fayrouz Benhaoua

Materials Processing Technology

Bonded Composite Repairs Analysis in Pipes Under Internal Pressure Using Finite Element Technics 605
A. Achour, B. Bachir Bouiadjra, and D. Ouinas

Elaboration and Characterization of Ceramic Materials from Kaolin-Phosphate Mixture 613
Chouia Fateh and Belhouchet Houcine

Optimization of Tool Geometry Parameters for Rotary Peeling Veneer Process Based on TLBO Algorithm 620
Bourab Fatma-zohra, Aknouche Hamid, and Hamrani Abderrachid

Light Scattering Applied to the Study of Biological Tissues: To an Optical Biopsy 637
H. Chorfi, K. Ayadi, and L. Boufendi

The Effect of Doping with N and Cu Atoms on the Hydrogen Sensing Properties of the ZnO$(10\bar{1}0)$ **Surface and ZnO Nanowires: A First-Principles Study** 644
M. A. Lahmer

Macromolecular Complex Architectures: Synthesis and Characterization 657
N. Cherifi, A. Benaboura, and L. Billon

Study of COD Adsorption on Deferent Activated Carbon Prepared from the Date Stones of the South of Algeria 665
Ouahiba Belaid, Ahmed A. Bebba, Zineb Barhoum, and Imane Raache

Study of Structural, Morphological and Optical, Properties of Fe Doped SnO_2 Semiconductor Thin Films Prepared by Sol-Gel Technique 676
S. Benkara, H. Ghamri, and M. Zaabat

Effect of a Thermal Degradation on the Mechanical Behavior of a 5-Harness Satin Weave Carbon–PEEK 683
T. Saoudi and M. A. Belouchrani

Author Index 699

Advanced Materials and Nanotechnology

Investigation of Structural, Optical and Electrical Properties of Al Doped SnO_2 Thin Films Synthesized by Sol-Gel

A. Boumezoued(✉), K. Guergouri, C. Azizi, A. Aberkane, and A. Khial

Laboratory of Active Components and Materials, Faculty of Exact Sciences and Science of Nature and Life, University Larbi Ben M'Hidi of Oum El Bouaghi, Oum El Bouaghi, Algeria
amal.boum4@gmail.com

Abstract. Among materials well known to be used as Transparent Conducting Oxide (TCO) and in the field of solar cells, SnO_2 is one of the best. In this work we investigate the doping effect with aluminum on some physical properties of SnO_2. Al doped SnO_2 thin films have been synthesized by sol–gel route using the dip coating method. The films were deposited on glass substrates with different Al concentrations (1%, 3%, 5% and 10%) and heated at 500 °C. The used precursor is $SnCl_3$ dissolved in ethanol. The obtained thin layers were characterized by X-ray diffraction, UV-visible spectroscopy, by AFM microscopy and electrical measurements. The XRD spectra exhibit a tetragonal crystalline structure of SnO_2 with a preferential direction along (101) for all samples, with grain sizes varying between 10.66 and 20.08 nm. On the other hand the UV-visible spectra show that: all samples of pure and Al doped SnO_2 films have a good transmittance with values between 70–95% in visible region, and an optical gap varying between 3.28 to 3.64 eV. The electrical properties (resistivity, mobility and bulk concentration of charge carriers) obtained by means of Hall Effect indicate that the SnO_2 n-type is inverted to p-type.

Keywords: Nanomaterials · Thin films · Sol-Gel · SnO_2 · Solar cells

1 Introduction

Nanoscale materials have received great interest in recent years due to their many technological applications, very promising in various fields. Among these materials, tin oxide (SnO_2) is considered as one of the most important, due to its attractive properties like optical transmittance, low resistivity, stability to heat treatment and piezoelectric behavior [1, 2]. These properties make SnO_2 a suitable semiconductor for devices such as the gas sensors, the Li-ion batteries, the Light emitting diodes, the solar cells and in photovoltaic [3, 4]. SnO_2 thin films are prepared by various techniques, such as electron beam evaporation [5], sol-gel coating [6], chemical vapor deposition [7], etc.… The widely used, among these methods, is the Sol-Gel technique, because of its low cost and the good quality of the obtained products. In literature Al [8], Sb [9], Ln [10], Ga [11] and Li [12] are often, cited as dopants to obtain p type.

B. Abdelbaki et al. (Eds.): SMSD 2017, *Proceedings of the Third International Symposium on Materials and Sustainable Development*, pp. 3–9, 2018.
https://doi.org/10.1007/978-3-319-89707-3_1

In this study, pure and Al doped SnO_2 thin films have been synthesized by sol-gel technique with different concentrations: 1%, 3%, 5% and 10%. The films are deposited on ordinary glass. The structural, optical and electrical properties of different samples are discussed. The crystal orientation of the SnO_2 thin films doped by different Al concentrations was identified by XRD, using Cu Kα (0.154 nm) line. The optical transmittance spectra were revealed by JASCO V-630 UV-vis spectrometer. The electrical properties were studied by HMS-3000 Hall effect measurements.

2 Experimental Procedure

2.1 Sample Preparation

Undoped and doped thin oxide with Al has been synthesized by dip coating method on the glass substrates. The used precursor $SnCl_3.2H_2O$ (>99.00% in purity) is dissolved in absolute ethanol. Small amounts of Al using $AlCl_3$ (>99.8% in purity) are added to the solution to obtain the desired ratio [Al]/[Sn]. The solution is then stirred at 70 °C for 2 h to obtain a homogeneous and transparent solution. The dip-coated SnO_2: Al glass assembly was dried at 200 °C for 10 min. The processing step was repeated that way for 15 times to obtain the desired thickness. After the multilayer coating, the samples were finally calcined at 500 °C for 2 h.

2.2 Results and Discussion

2.2.1 Structural Properties

Figure 1 shows the XRD patterns of pure and Al doped SnO_2 thin films synthesized by sol-gel route. The spectra exhibit the presence of three most intense peaks of the tetragonal rutile structure at 26.74°, 34.28° and 51.96° corresponding to (110), (101) and (211) planes. The tetragonal structure was preserved with (110) preferred crystal orientation on the glass substrates, and only tetragonal SnO_2 phase is present with large variation of two theta. The crystallite size of pure and Al doped SnO_2 is calculated using Scherrer's formula [13]:

$$D = 0.9\lambda/\Delta\theta \cos\theta \quad (1)$$

Where D is the grain size, λ the X-ray wave length (λ = 1.5418A°), Δθ the full width at half-maximum (FWHM) and θ the Bragg angle.

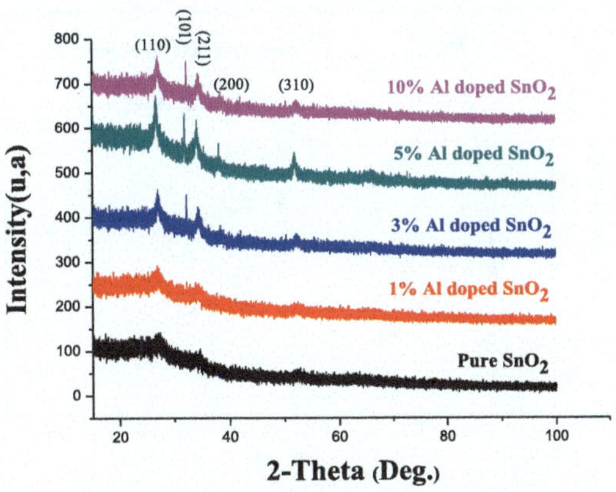

Fig. 1. X-ray diffraction patterns of Al doped SnO_2 thin films as function of Al mol% (0%, 1%, 3%, 5% and 10%).

As observed in Fig. 2, the average values of crystalline size and strain of the samples were calculated on the (110) crystal plane as they are shown in Table 1. We notice that the crystalline size decreases with increasing Al concentration; the values vary from 10.66 to 20.08 nm.

Table 1. XRD data of pure and Al doped SnO_2

Sample	2θ(deg)	Δθ(Rad)	D(nm)	σ(Gpa)
Pure SnO_2	76.35	0.016	10.66	+0.233
1% Al:SnO_2	26.74	0.010	20.08	+0.145
3% Al:SnO_2	26.99	0.012	19.25	+0.172
5% Al:SnO_2	26.84	0.014	15.23	+0.208
10% Al:SnO_2	26.86	0.015	14.43	+0.211

The Al^{3+} ions can occupy the interstitial sites as donors or the Sn^{4+} sites as acceptors. It should be noted that the efficient doping effects are obtained when the ionic radius of dopant is the same as or smaller than the host ion (Sn^{4+} (r = 0.71 A°) and Al^{3+} (r = 0.51 A°)).

The calculated strain values show a decreasing trend for Al doped SnO_2 in comparison with pure samples, which indicates that doped samples are subjected to a strain dilatation.

2.2.2 Optical Properties

The optical transmittance and the direct band gap properties of SnO_2 films as a function of Al concentration in thin films are presented in Figs. 3 and 4, respectively. The optical transmittance was averaged over the wavelength range of 200 to 1100 nm. The absorption coefficient (α) was determined by using Eq. (2):

$$T = e^{\alpha t} \tag{2}$$

Where, t is the thickness of the film.

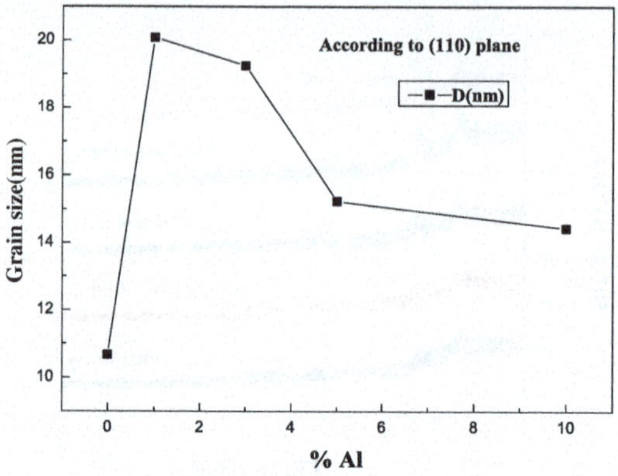

Fig. 2. SnO_2 thin films grain size as a function of Al concentration.

The absorption refers to the excitation of an electron from the valence band to the conduction band. The absorption coefficient was then used to estimate the direct optical energy band-gap (Eg) using the relation (3):

$$(\alpha h\upsilon)^2 = A(h\upsilon - Eg) \tag{3}$$

Where h is the Planck's constant, ν is the frequency of the incident photon, A is a constant for direct transition and α is the absorption coefficient. The direct optical energy band-gap (Eg) was estimated by extrapolating the linear portion of the curve $(\alpha h\nu)^2$ as a function of (hν) for direct allowed transition to the point where $\alpha h\nu = 0$ [14].

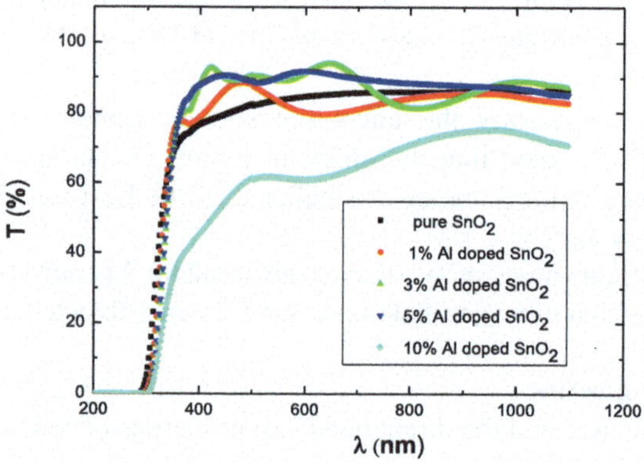

Fig. 3. Optical transmittance of SnO_2 thin films as a function of Al concentration.

Figure 3 shows the variation of transmittance in the range $200 \leq \lambda \leq 1100$ nm of $Al_xSn_{1-X}O_2$ thin films. Our films exhibit a high transparency (94.80%) in the visible region. It was found that the highest value of the band gap was 3.64 eV, found for pure SnO_2, which is equal to the typical value (3.6 eV). The value of the band gap decreases with increasing Al concentration as shown in Fig. 4. The decrease in the band gap (Table 2) is attributed to the presence of holes in $Al_xSn_{1-X}O_2$.

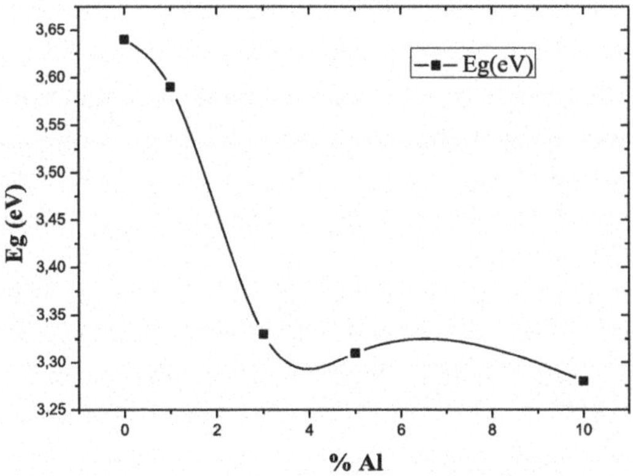

Fig. 4. Variation of SnO_2 band gap as a function of Al concentration.

Table 2. Transmittance and optical band gap of Al doped SnO_2 thin films annealed for different concentration.

Sample	Transmittance (%)	Band gap (Ev)
Pure SnO_2	86.28	3.64
1% Al:SnO_2	89.65	3.59
3% Al:SnO_2	93.27	3.33
5% Al:SnO_2	94.80	3.31
10% Al:SnO_2	77	3.28

2.2.3 Electrical Properties

Hall measurements were performed in order to determine the resistivity, mobility, carrier type, and the probable inversion from n to p conductivity. Table 3 shows the charge carrier concentration, the resistivity as a function of Al concentration. The pure SnO_2 layer is n-type semiconductor with resistivity around 90 Ω.cm. The table below indicates that the electrical parameters were influenced by aluminum incorporation in tin oxide (SnO_2).

Table 3. Electrical parameters of SnO_2:Al thin films heated at 500 °C.

Sample	N (cm^{-1})	ρ (Ω.cm)	R_H (cm^3/C)	Type
Pure SnO_2	$-9.51.10^{+3}$	$1.97.10^{+6}$	$-6.55.10^{+5}$	n
1% Al:SnO_2	$-1.02.10^{+14}$	$8.04.10^{+4}$	$-6.93.10^{+4}$	n
3% Al:SnO_2	$-2.31.10^{+13}$	$2.99.10^{+4}$	$-2.69.10^{+5}$	n
5% Al:SnO_2	$+1.66.10^{+14}$	$1.2.10^{+4}$	$+3.74.10^{+4}$	p
10% Al:SnO_2	$+2.56.10^{+14}$	$1.08.10^{+2}$	$+1.12.10^{+7}$	p

Figure 5 shows the charge carrier concentration as a function of Al content; the charge carrier was found to be high around $2.56.10^{+14}$ cm^{-1}, the samples have p and n-type conductivity as listed in Table 3; when the Al content increases the charge carrier increases as seen in Fig. 5.This fact is explained in terms of the change in charge carrier concentration at the surface film.

As mentioned in literature, if the Sn tetravalent ions were substituted by Al trivalent ions, one broken bond (hole) is produced, which acts as an acceptor energy level near the valence band and accepts an electron from it, thus increasing the hole concentration or the p-conductivity [15].

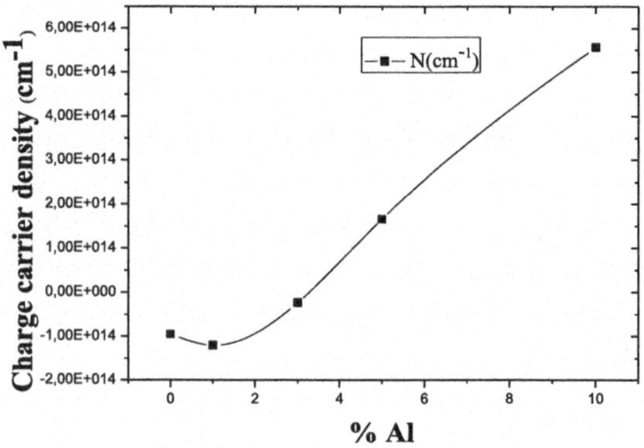

Fig. 5. The charge carrier of pure and Al doped SnO_2 thin films.

3 Conclusion

Nanostructures of pure and Al (1%, 3%, 5%, and 10%) doped SnO_2 have been synthesized by sol-gel route. A (110) oriented tetragonal structure was confirmed by X-ray patterns analysis. According to this orientation the obtained grain sizes of Al doped SnO_2 are nanometric and vary between 10.66 and 20.08 nm.

The sample of pure and Al doped SnO_2 have a good transmittance in the visible with values between 77–95%, and the optical gap varies between 3.28 to 3.64 eV.

Electrical measurements of Al doped SnO_2 nanostructure films permitted us to conclude that Al^{3+} is the suitable dopant for acceptor doping process, with the substitution of the lower valency of cation in Sn^{4+} position of tin oxide structure one or a few hole may be produced and then decreases the n-type conduction or increases hole density and thus p-conductivity.

Moreover a very low resistivity $1.08.10^{+2}$ (Ω.cm) and a high transmittance in visible spectrum (95%) give Al doped SnO_2 the behavior of a good quality TCO for photovoltaic devices.

References

1. Jin, Z.C., Hamberg, J., Granqvist, C.G.: Opticalproerties of sputter-deposition ZnO:Al thin films. J. Appl. Phys. **64**, 5117 (1988)
2. Rusu, R.S., Russia, G.I.: On the electrical and optical characterization of CdO thin films. J. Optoelectron. Adv. Mater **7**, 823 (2005)
3. Penza, M., Cozzi, S., Taglient, M.A., Mirenghi, L., Martucci, C., Quirini, A.: Characterization of transparent and conductive electrodes of indium tin oxide thin films by sequential reactive evaporation. Thin Solid Films **71**, 349 (1999)
4. Ishibashi, S., Higuchi, Y., Ota, Y., Nakamura, K.: Low resistivity indium–tin oxide transparent. In: 6th Vacuum and Surface Sciences Conference of Asia and Australia (VASSCAA-6). IOP Publishing (1998)
5. Khan, A.F., Mehmood, M., Rana, A.M., Bhati, M.T., Mahmood, A.: Optical characterization of rf-magnetron sputtered nanostructured SnO2 thin films China. Phys. Lett. **7** (2009). 77803
6. Cavicchi, R., Semancik, S.: The growth of thin, epitaxial SnO_2 films for gas sensing application. Thin Solid Films **206,** 81
7. Jiang, J.C., Lian, K., Meletis, E.I.: Influence of oxygen plasma treatment on the microstructure of SnO_x thin films. Thin Solid Films **411**, 203 (2002)
8. Mehdi, M., Mohagheghi, B., Saremi, M.S.: The influence of Al doping on the electrical, optical and structural properties of SnO_2 transparent conducting films deposited by the spray pyrolysis technique. J. Phys. D Appl. Phys. **37**, 1248–1253 (2004)
9. Ni, J., Zhao, X., Zheng, X., Zhao, J., Liu, B.: Electrical, structural, photoluminescence and optical properties of p-type conducting, antimony-doped SnO_2 thin films. J. Acta Mater. **57**, 278–285 (2009)
10. Ji, Z.G., He, Z.J., Song, Y.L., Liu, K., Ye, Z.Z.: Fabrication and characterization of indium-doped p-type SnO_2 thin films. J. Cryst. Growth **259**, 282–285 (2003)
11. Huang, Y.X., Ji, Z.G., Chen, C.: Preparation and characterization of p-type transparent conducting tin-gallium oxide films. Appl. Surf. Sci. **253**, 4819–4822 (2007)
12. Mehdi, M., Mohagheghi, B., Saremi, M.S.: Electrical, optical and structural properties of Li-doped SnO_2 transparent conducting films deposited by the spray pyrolysis technique: a carrier-type conversion study. Semicond. Sci. Technol. **19**, 764–769 (2004)
13. Wurst, J.C., Nelson, J.A.: Lineal intercept technique for measuring grain size in two-phase polycrystalline ceramics. J. Amer. Cer. Soc. **55**, 109–111 (1972)
14. Tauc, J., Grigorovici, R., Vancu, A.: Optical properties and electronic structure of amorphous germanium. Phys. Status Solidi (b) **15**, 627–637 (1966)
15. Bagheri Mohagheghi, M.M., Shokooh Saremi, M.: J. Phys. D: Appl. Phys. **37**, 1248–1253 (2004)

Formation Energies of Mn Doped $ZnSnAs_2$

H. Bouhani-Benziane[1(✉)], K. Hebali[1], A. Bakhti-Siad[2], O. Sahnoun[1], M. Baira[1], and M. Sahnoun[1]

[1] Laboratoire de Physique Quantique de la Matière et Modélisation Mathématique (LPQ3M), University of Mascara, Mascara, Algeria
bouhanibenziane_hamida@yahoo.fr
[2] Laboratoire de Matériaux Applications et Environnement (LMAE), Mascara, Algeria

Abstract. The defect calculations are firstly performed by replacing a single cation (namely Zn and Sn) with a single Mn atom in the pure chalcopyrite ZnSnAs2 supercell, and their corresponding formation energies show that the substitution of a Sn atom (rather than Zn) by Mn is strongly favored. Thereafter, a comparison of total energy differences between ferromagnetic (FM) and antiferromagnetic (AFM) are given. Surprisingly, the exchange interaction between a Mn pairs is found to oscillate with the distance between them. Consequently, the AFM alignment is energetically favored in Mn-doped ZnSnAs2 compounds, except for low impurity concentration associated with lower distances between neighboring Mn impurities, in this case the stabilization of FM increases. Moreover, the ferromagnetic alignment in the Mn-doped ZnSnAs2 systems behaves half-metallic; the valence band for majority spin orientation is partially filled while there is a gap in the density of states for the minority spin orientation. This semiconducting gap of ~ 1 eV opened up in the minority channel and is due to the large bonding-antibonding splitting from the p-d hybridization. Our findings suggest that the Mn-doped ZnSnAs2 chalcopyrites could be a different class of ferromagnetic semiconductors.

Keywords: Chalcopyrite · Ferromagnetic · Exchange interaction Impurities · Half-metallic

1 Introduction

Quite recently, high-temperature ferromagnetism was found to exist in chalcopyrite-type compounds by introducing ferromagnetic (FM) property at room temperature via doping them by transition materials (TM) impurities. Among them manganese-doped II-IV-V2 chalcopyrite's such as ZnGeP2:Mn, [1] CdGeP2:Mn [2], ZnGeAs2:Mn, ZnSiAs2:Mn, CdGeAs2:Mn [3] and ZnSnAs2:Mn [4] have demonstrated Curie temperatures of up to 367 K. The general consensus among researchers in this area is that ferromagnetism in these materials originates from the interactions between itinerant electrons or holes and localized electrons. In this context Choi et al. [5] reported experimentally the discovery of ferromagnetism in Mn-doped ZnSnAs2 single crystal with Curie temperature of 329 K. They found that the magnetic moment from saturation magnetization per Mn atom is around 3.63 μB at 5 K. However, from theoretical point of view, Yi et al. [6] point out that

B. Abdelbaki et al. (Eds.): SMSD 2017, *Proceedings of the Third International Symposium on Materials and Sustainable Development*, pp. 10–14, 2018.
https://doi.org/10.1007/978-3-319-89707-3_2

ZnSn1- xMnxAs2 for x = 0.125 prefers the AFM ground state rather than the FM state. In addition, the local structures around Mn atoms in the ZnSnAs2:Mn studied by XAFS measurements demonstrated that the Mn atoms are substituted into both Zn and Sn cation sites [7].

However, the physical origin of FM or AFM states in Mn-doped II–IV–V2 chalcopyrites is still in debate [8, 9] and more theoretical and experimental studies are required. This situation motivates us to study the electronic and magnetic properties of Mn-doped ZnSnAs2chalcopyrite, using the full potential linearized augmented plan wave plus orbital's local (FPLAPW + lo), in order to give more understand about the properties of herein materials. In the following discussions, the possible HM-AFM states are highlighted.

2 Method of Calculation

All calculations presented in the present work were performed using the full potential linearized augmented plane-wave (LAPW) method as implemented in the Wien2k package [10]. The electronic exchange-correlation is described within the generalized gradient approximation (GGA) of Perdew-Burke-Ernzerhof (PBE) [11]. The wave function of the charge density and potential in the interstitial region is expanded up to lmax = 10, as well as the energy cut-off of RMT.Kmax = 8.5 were kept fixed in our calculation, where RMT denotes the smallest atomic sphere radius and Kmax gives the magnitude of the largest K vector in the plane wave expansion.

3 Formation Energy of the Mn-Doped $ZnSnAs_2$ Chalcopyrite

In order to simulate magnetic properties we have modeled the situation when manganese atoms replace either the site of Zn (MnZn) or the site of Sn (MnSn) atoms. The Mn doping has been modeled in a sufficiently large and fully relaxed supercell, which is a $2 \times 2 \times 1$ translated 64-atoms simple tetragonal (Fig. 1).

The defect calculations are performed by placing a single Mn atom in a supercell, containing 64 atoms and having lattice vectors (2,0,0)a, (0,2,0)a, and (0,0, ηu)a; the lattice parameters a and c = 2aηu are chosen as for the pure bulk chalcopyrite. In this case, the systems are in a ferromagnetic configuration. We consider substitutional defects, i.e., replacing a single cation (namely Zn and Sn) with a Mn atom in the pure chalcopyrite ZnSnAs2 cell. The formation energy of a defect is estimated as follows [12].

$$E_f = \frac{1}{64}\left[E((Zn,Sn)_{1-x}Mn_x, As_2) - E(ZnSnAs_2) - \mu(Mn) + \mu(Zn,Sn)\right]$$

where E((Zn,Sn)1-xMnxAs2) is the total energy of the supercell containing the Mn dopant, E(ZnSnAs2) is the total energy of the pure host having the same dimension to supercell, and μMn and μ(Zn,Sn) are the chemical potentials of bulk Mn, Zn and Sn, respectively. Our calculated formation energies of a Mn defect in ZnSnAs2 are 0.248 eV and 0.216 eV for Mn substituting Zn and Sn, respectively. In the chosen limit of the chemical potentials, this implies that the substitution of a Sn atom (rather than Zn) by Mn is strongly favored. Generally, when 3d TM ions are substituted for the cations of the host, their electronic structure is influenced by the strong 3d orbitals of the magnetic ion and the p orbitals of the neighboring anions.

Fig. 1. Mn doped ZnSnAs2

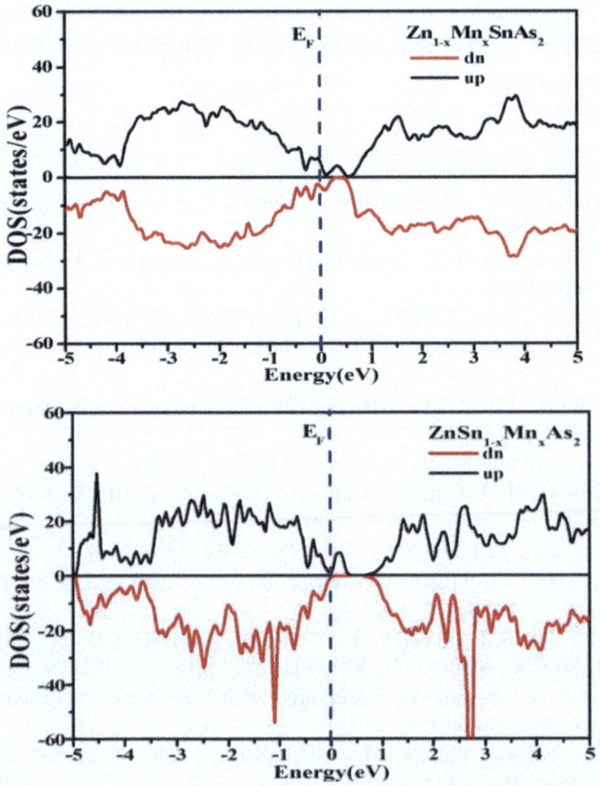

Fig. 2. Total density of states for Mn doped ZnSnAs2. Only a single Mn atom is substituted into Zn or Sn cation sites in the supercell of 64 atoms (x = 0.0625).

In our case, the hybridization between Mn dopant and its neighboring host atoms results in the splitting of the energy levels near the EF, which shifts the majority spin states downward and minority spin states upward. Density of states in Fig. 2 show spin splitting between the majority-spin and minority-spin channel near EF, which implies that Mn doping can result in magnetism in the ZnSnAs2 host. When Mn ions substitute on Zn and Sn sites, the obtained magnetic moment per Mn is 4.00 μB and 3.90 μB, respectively, which is in reasonable agreement with the experimental moment (3.85 μB/Mn) for Mn-doped ZnSnAs2 (x = 0.056) [8] assumed due to substitution on both Zn and Sn sites.

4 Conclusions

Using the first-principles density functional theory within the GGA, we have studied the electronic structure of ZnSnAs2 doped with 12.5% Mn impurities for various possible geometries to replicate the situation where the Mn atom would appear either to cluster or be separated. The calculated total magnetic moment is about 4 μB per unit cell and mostly located in Mn atom. Our results show that the substitution of Mn leads to the AFM alignment is energetically favored in Mn-doped ZnSnAs2 compounds

References

1. Medvedkin, G.A., Hirose, K., Ishibashi, T., Nishi, T., Voevodin, V.G., Sato, K.: J. Cryst. Growth **609,** 236 (2002)
2. Medvedkin, G.A., Ishibashi, T., Nishi, T., Hayata, K., Hasegawa, Y., Sato, K.: Jpn. J. Appl. Phys. **10**A (2000). L949
3. Morozov, A.S., Koroleva, L.A., Zashchirinskii, D.M., Khapaeva, T.M., Marenkin, S.F., Fedorchenko, I.V., Szymczak, R.A., Krzymanska, B.: Solid state phenomena **168–169,** 31 (2011)
4. Choi, S., Cha, G.B., Hong, S.C., Cho, S., Kim, Y., Ketterson, J.B., Jeong, S.Y., Yi, G.C.: Solid State Commun. **122,** 165 (2002)
5. Yi, H., Lee, S.: J. Magn. Magn. Materials **272**, e243 (2004)
6. Oomae, H., Toyota, H., Emura, S., Asahi, H., Uchitomi, N.: EPJ Web of Conferences, vol. 75 (2014). 09004
7. Freeman, A.J., Zhao, Y.-J.: J. Phys. Chem. Solids **64**, 1453 (2003)
8. Yu-Jun, Z., Picozzi, S., Continenza, A., Geng, W.T., Freeman, A.J.: Phys. Rev. B **65** (2002). 094415
9. Galanakis, I., Özdöğan, K., Şaşoğlu, E., Aktaş, B.: Phys. Rev. B **75**, (2007). 092407
10. Blaha, P., Schwarz, K., Madsen, G., Kvasicka, D., Luitz, J.: WIEN2K, an augmented plane wave local orbitals program for calculating crystal properties, Technical University of Vienna, Vienna (2001)
11. Ozdogan, K., Upadhyay Kahaly, M., Sarath Kumar, S.R., Alshareef, H.N., Schwingenschlogl, U.: J. Appl. Phys. **111**, (2012). 054313
12. Zhang, C.-W., Yan, S.-S.: Appl. Phys. Lett. **95**, (2009). 232108

The Improvement of Tandem a-Si:H/μc-Si:H Solar Cells Perfomance by Optimized the Front Contact Barrier Height

Abbas Belfar(✉), Tarik Kadi, and Hocine Ait-Kaci

Laboratoire de Physique des Plasmas Matériaux Conducteurs et leurs Applications (LPPMCA), Département de Physique Energétique, Faculté de Physique, USTO-MB, Université des Sciences et de la Technologie d'Oran – Mohamed Boudiaf, BP 1505, Oran, Algeria
a_belfar@hotmail.com, abbas.belfar@univ-usto.dz

Abstract. In this study we have examined by simulation the role of the front contact barrier heights Φ_{b0} (barrier at TCO/P window layer interface), on the performances of a-Si:H/μc-Si:H based tandem solar cell. The numerical simulation has been carried out by using AMPS-1D (Analysis of Microelectronic and Photonic Structures) program. The output parameters of tandem solar cell, like open circuit voltage (V_{OC}), fill factor (FF), short circuit current (J_{SC}) and efficiency (Eff) are numerically calculated for different values of Φ_{b0}. The simulation results shown that the efficiency value increases from 9.07% until 11.60% when the front contact barrier height (Φ_{b0}) value increase from 1.1 eV to 1.4 eV. However, we achieved the best conversion efficiency of 12.33% with Φ_{b0} equal to 1.9 eV and with a values of 60 nm and 1200 nm for thicknesses of top and bottom absorber layers, respectively.

Keywords: Tandem solar cell · a-Si:H · μc-Si:H · Barrier height Simulation

1 Introduction

A disadvantage related to the a-Si: H technology is the relatively low photovoltaic efficiency and the light-induced degradation phenomenon, which leads to a decrease in the initial efficiency (Deng and Schiff 2003). For simple junction module of a-Si: H a stabilized efficiencies the order of 10% are obtained (Cythelia Sarl and Alain Riacaud 2011). In order to improve the efficiency of solar cells based on hydrogenated amorphous silicon (a-Si:H) and microcrystalline hydrogenated silicon (μc-Si:H), tandem (micromorph) cells were manufactured. The idea of multi-junction cells produced by the juxtaposition of a a-Si: H upper cell and a μc-Si: H base cell makes it possible to reinforce the stability of the thin-film cells. The highest stabilized photovoltaic yield 12% for tandem solar cells was obtained for an a-Si: H/μc-Si: H configuration at the University of Neuchâtel (Meier et al. 1998). Experimental and computer modeling works showed that hole barrier at the front contact TCO (Transparent Conducting Oxide)/p-layer interface can also influence the solar cell performances (Rubinelli et al. 1992; Sanchez Sinencio and Williams 1983; Guha et al. 1986). It has been shown in single junction a-Si:H n-i-p

B. Abdelbaki et al. (Eds.): SMSD 2017, *Proceedings of the Third International Symposium on Materials and Sustainable Development*, pp. 15–24, 2018.
https://doi.org/10.1007/978-3-319-89707-3_3

solar cell that an increase of the work function of the p-layer which increases the built-in voltage (Vbi) of the device and leads to a V_{OC} enhancement. This also necessitates a using of a front electrode with a higher work function (Biron et al. 2012; Belfar and Ait Kaci 2013). In this work, we have analyzed the effect of the front contact barrier heights (Φ_{b0}) on the performances of p-i-n (a-Si: H)/p-i-n (μc-Si: H) tandem solar cell by using the one-dimensional device simulation program AMPS-1D (Analysis of Microelectronic and Photonic Structure). For these reasons, two tandem solar cells, one with back reflection (RB = 0.8) and the other without back reflection (RB = 0) are simulated and output parameters, like short circuit current (J_{SC}), open circuit voltage (V_{OC}), fill factor (FF) and efficiency (Eff); for both tandem structures are numerically calculated and compared by using different values Φ_{b0}.

2 Amps-1D Model for Amorphous Silicon and Tandem Solar Cell Structure

2.1 About AMPS-1D

The simulations carried out using one-dimensional device simulation program AMPS-1D for the analysis were based on solves the Poisson's equation and the continuity equations of electrons and holes by using finite differences and the Newton-Raphson technique (Fonash et al. 1997) and used to analyze the carriers transport behavior of semiconductor electronic and optoelectronic device structures including solar cells.

The Poisson's equation is given by:

$$\frac{d}{dx}\left(-\varepsilon(x)\frac{d\Psi}{dx}\right) = q * \left[p(x) - n(x) + N_D^+(x) - N_A^-(x) + p_t(x) - n_t(x)\right] \quad (1)$$

Where, ε is the dielectric permittivity of the material. Ψ, n and p represent the electrostatic potential, the free electrons density and the free holes densities, respectively. nt and pt are the trapped electrons and holes. In Eq. (1) ionized concentrations of donors and acceptors are given by N_D^+ and N_A^- , respectively, and q the free charge of electron.

Free electrons and holes continuity equations are:

$$\frac{1}{q}\left(\frac{dJ_n}{dx}\right) = -G_{op}(x) + R(x) \quad (2)$$

$$\frac{1}{q}\left(\frac{dJ_p}{dx}\right) = -G_{op}(x) - R(x) \quad (3)$$

Jn represents the electron's current density and Jp the hole's current densities. In this model, direct and indirect Schokley-Read-Hall (SRH) recombination phenomena are taken into account through the recombination rate R(x). Numerical simulation requires a model for the trap density of states (DOS) in the sample. For the density of localized states in the mobility gap of amorphous silicon and microcrystalline silicon it has been assumed that there are both acceptor-like states and donor-like states modeled by exponential band tails (Urbach tails) and Gaussian mid-gap states (associated to silicon dangling bonds).

2.2 Tandem Solar Structure

The use of microcrystalline silicon (μc-Si:H) as a complement to amorphous silicon (a-Si:H) is one of the most promising ways of creating a tandem solar cells. This leads to the "microcrystalline/amorphous" or "micromorph" tandems, introduced first under the name of "mixed stacked a-Si:H/μc-Si:H cells" by IMT Neuchatel in 1994 (Meier et al. 1998). The structure of the tandem a-Si:H/μc-Si:H cells simulated in this work is presented in Fig. 1. To make the connection between top cell and bottom cell we have used a very thin defective layer with a very reduced gap equal to 1 eV (RL).

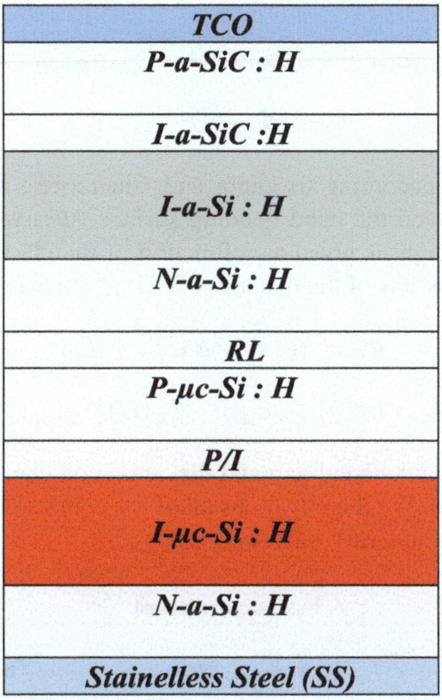

Fig. 1. Schematic diagram of the tandem a-Si:H/μc-Si:H solar cell.

The simulation requires input parameters such as surface recombination velocities, barrier heights, power density of the radiation, and characteristics of the layers forming the structure to simulate. In this calculation, we assumed that the front contact barrier height φ_{bO} (TCO/p-layer) was varied between 1.1 eV and 1.9 eV the back contact barrier height φ_{bL} (n layer/metal) is chosen equal to 0.21 eV. The surface recombination speeds of electrons and holes were both 1×10^7 cm/s. As a source of illumination, an AM 1.5 solar radiation with a power density of 100 mW/cm^2 was used.

The light absorption coefficient, for the different layers was already incorporated in the AMPS-1D program. We also assumed that the back contact has a reflection (RB) coefficient value of 0.80 in the case of structure with back reflection and a value of 0. in the other structure without back reflection. However, the reflection of light at the front face (RF) was set at 0.2. The other main electric parameters employed in the simulation were mainly based on (Poissant et al. 2003; Belfar and Mostefaoui 2011).

3 Results and Discussions

In AMPS, barriers at ITO/p interface and n/metal interface are described by φ_{b0} and φ_{bL} respectively. The barrier heights φ_{b0} and φ_{bL} are depicted in Fig. 2, which shows the band diagram of a n-i-p simple junction solar cell structure in thermodynamic equilibrium. In one hand, and according to Arch et al. (Arch 1991) description the barrier heights are related to the work functions ($\Phi_{w,front}$ and $\Phi_{w,back}$) of the contacts by:

$$\varphi_{b0} = \Phi_{w,front}(TCO) - \chi_e(p)|_{x=0} \tag{4}$$

$$\varphi_{bL} = \Phi_{w,back} - \chi_e|_{x=L} \tag{5}$$

On the other hand, according to Dutta and Chatterjee's (Dutta and Chatterjee's 2004), the relation between the band bending surface (Φh) which is a barrier for the holes and Φ_{bo} barrier height is given as a function of the electronic affinity χ of the p layer and the output work Φw of the material (TCO) at the front contact by the relation:

$$\Phi_h = Eg(p) - Ea\,(p) - \Phi_{b0} \tag{6}$$

$$\Phi_h = Eg(p) - Ea\,(p) - \Phi_w(ITO) + \chi_e(p) \tag{7}$$

Here $E_g(p)$ and $E_{ac}(p)$ are the band gap and the activation energy of the p layer. $E_g(p)$, $\chi_e(p)$ and $E_{ac}(p)$ are fixed. We thereby focus all of the contact effects into the front TCO (ITO)/p nc-Si:H contact and vary only φ_{bo}.

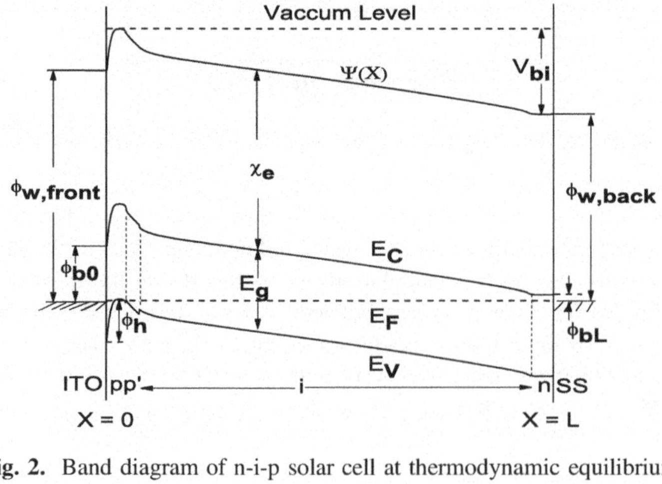

Fig. 2. Band diagram of n-i-p solar cell at thermodynamic equilibrium.

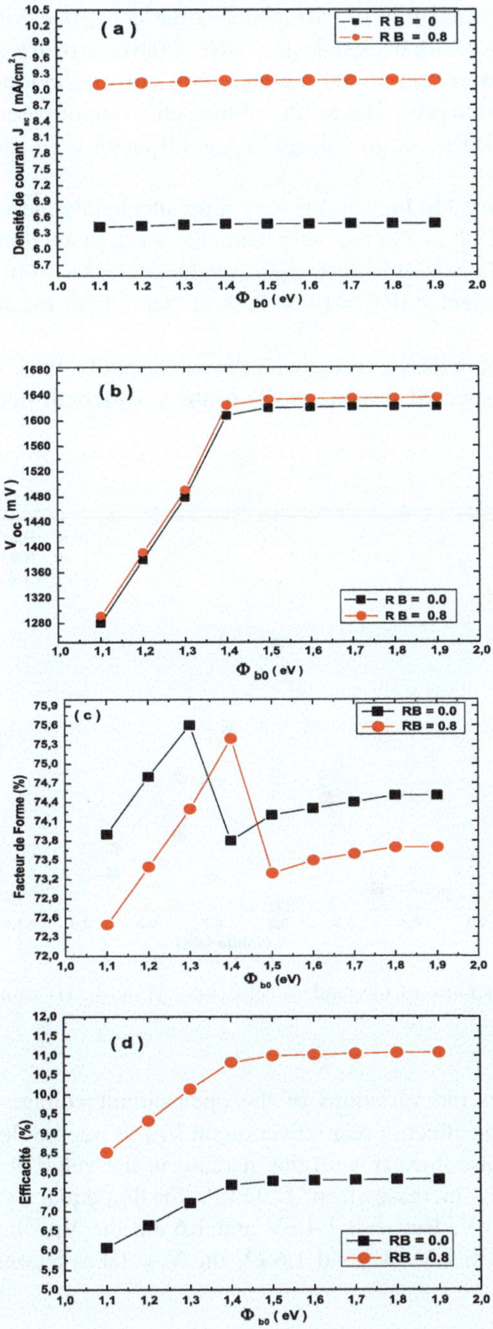

Fig. 3. The output parameters of the tandem a-Si:H/μc-Si:H solar cell for different values of Φ_{b0}.

In this work the simulation of two tandem solar cells, one without back reflection (RB = 0) and the other with back reflection (RB = 0.8) are realized. The thicknesses of the absorbent layers were set at 100 nm and 1200 nm for the top sub-cell and for the bottom sub-cell, respectively. The results of the cell's output parameters, namely short circuit current (J_{SC}), open-circuit voltage (V_{OC}), fill factor (FF) and efficiency (Eff) are shown in Fig. 3.

For the J_{SC} presented in Fig. 3(a) is very little affected by the variations of the front contact barrier height (Φ_{bO}). On the other hand, there is a marked improvement with the addition of a back reflector. However, the J_{SC} value increases from 6.47 mA/cm^2 in the absence of a back reflector (RB = 0) to 9.18 mA/cm^2 with the presence of the back reflector (RB = 0.8).

This improvement in the J_{SC} value with the presence of a back reflector is due to the improvement of the spectral response (SR) in the wavelength range from 0.53 μm to 1 μm (Fig. 4).

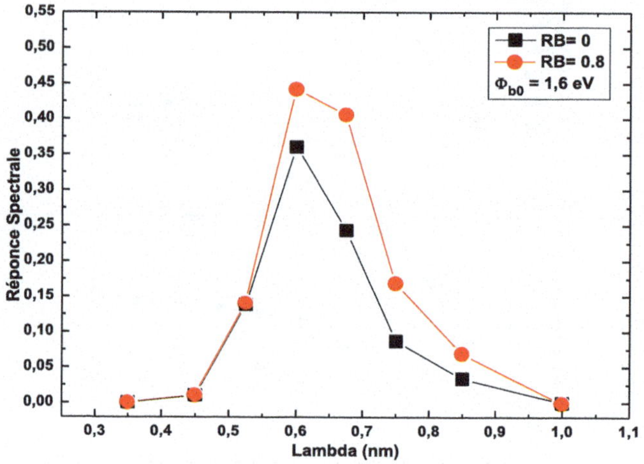

Fig. 4. The spectral response of the tandem cells (a-Si: H/μc-Si: H) simulated as a function of the back reflection (RB).

Figure 3(b) shows the variations of the open circuit voltage (V_{OC}). On the one hand, it is clear that the effect of rear reflection on V_{OC} is negligible. On the other hand, the V_{OC} values increase linearly with the increase in the value of Φ_{b0} from 1.1 eV to 1.4 eV. However, V_{OC} increases from 1292 mV for Φ_{b0} equal to 1.1 eV to 1625 mV for Φ_{b0} equal to 1.4 eV. Between 1.4 eV and 1.6 eV the V_{OC} increases very slowly from 1625 mV to 1636 mV. Beyond 1.6 eV the V_{OC} takes a constant value.

For the fill factor (FF) presented in Fig. 3(c), it is found affected by the two parameters (RB and Φ_{b0}). In the case of cell without back reflector (RB = 0), and when Φ_{b0} varied between 1.1 eV and 1.3 eV the FF increases from 73.9% to 75.6%, then decreased from 75.6% to 73.8% between 1.3 eV and 1.4 eV. Beyond 1.4 eV, the value of FF increases slowly. In the case of cell with back reflector (RB = 0.8), when Φ_{b0} varied between 1.1 eV and 1.4 eV the FF increases from 72.5% to 75.4%, then between 1.4 eV and 1.5 eV it decreased from 75.4% to 73.3%. Beyond 1.5 eV it increases very slowly to reach the value of 73.7%.

Figure 3(d) shows the variations of the efficiency (Eff). Firstly, it is clear that the effect of the back reflection is important on the efficiency. Secondly, the Eff values increase linearly with increasing the value of Φ_{b0} from 1.1 eV to 1.4 eV. For RB = 0, and when Φ_{b0} varie from 1.1 eV to 1.4 eV the efficiency increases from 6.07% to 7.68%, then between 1.4 eV and 1.5 eV Eff increases slowly from 7.68% to 7.78% and beyond 1.5 eV Eff becomes constant. However, for RB = 0.8, and when Φ_{b0} varie from 1.1 eV to 1.4 eV the efficiency increases from 8.52% to 10.83%, then between 1.4 eV and 1.5 eV Eff increases slowly from 10.83% to 11.08% and beyond 1.5 eV Eff becomes constant.

The simulation results presented previously showed that the best performances of the tandem cells are obtained with large values of the front contact barrier (Φ_{b0}).

The simulation calculations allowed us to represent the energy band diagrams for different values Φ_{b0} (Fig. 5). According to Figs. 2 and 5, it is clear that the increase of Φ_{b0} reduces the value of Φ_h, which acts as a barrier for photo-generated holes that point to the front contact. Experimentally, and from the relation 7, in order to increase the value of Φ_{b0}, when the values of $E_g(p)$ $E_{ac}(p)$, and $\chi_e(p)$ where set constant, it is necessary to choose for the front contact a TCO (Transparent Conducting Oxide) which have a high output.

The output work function of TCO, like ITO (Indium Thin Oxide) can vary between 4.3 eV and 5.3 eV, depending on the stoichiometry and the deposition method of the ITO layer (Marks et al. 2000).

To examine the simultaneous effect of the top absorber layer thickness (d_1) and the front contact barrier height (Φ_{b0}) we have varied d_1 from 50 nm to 140 nm and Φ_{b0} from 1.1 eV to 1.9 eV. The simulation results of the efficiency were presented in Fig. 6. In our case, the maximum efficiency of 12.33% was obtained at Φ_{b0} = 1.9 eV in the case of the tandem solar structure with back reflector (RB = 0.8).

Knowing that the efficiency is expressed following the relation:

$$E_{ff} = \frac{J_{sc} V_{oc} FF}{P_{in}} \tag{8}$$

In conclusion, on one hand, and contrary to the J_{SC} (Fig. 3-a), the V_{OC} (Fig. 3-b) and the FF (Fig. 3-c) are very little affected by the back reflection. On the other hand, it was found that the J_{SC} is very little affected by the Φ_{b0} variations.

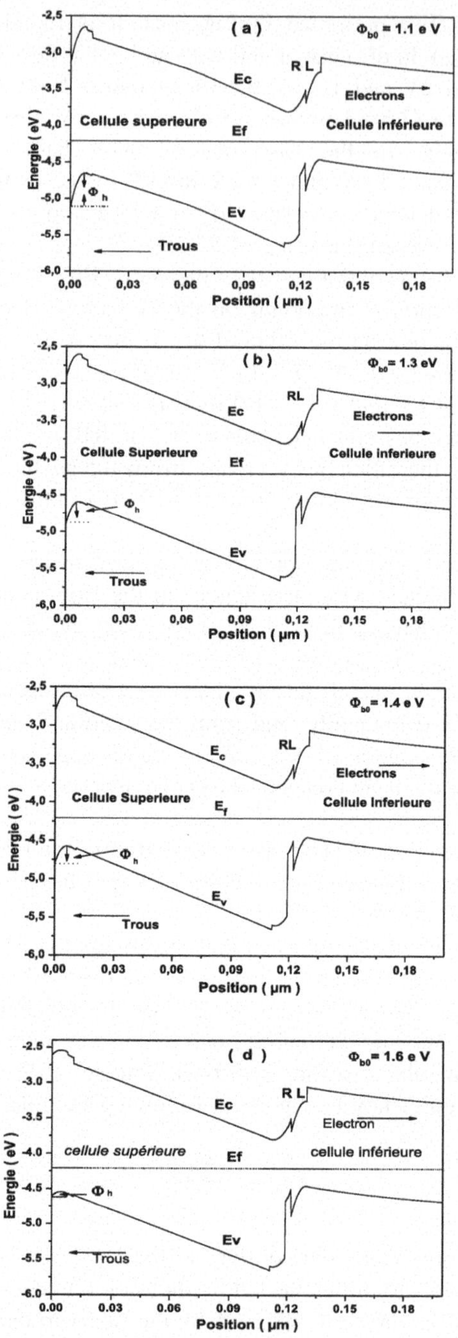

Fig. 5. Energy band diagram of the tandem a-Si:H/µc-Si:H solar cell for different values of Φ_{b0}.

Contrary to the FF and especially the V_{OC} are improved with the increase of Φ_{b0} from 1.1 eV to 1.6 eV. According to the Eq. (8), the improvement in the efficiency (Fig. 3-d) is mainly due to the improvement of J_{SC} values with back reflection and also due to the improvement of the open-circuit voltage (V_{OC}) values with Φ_{b0} high values.

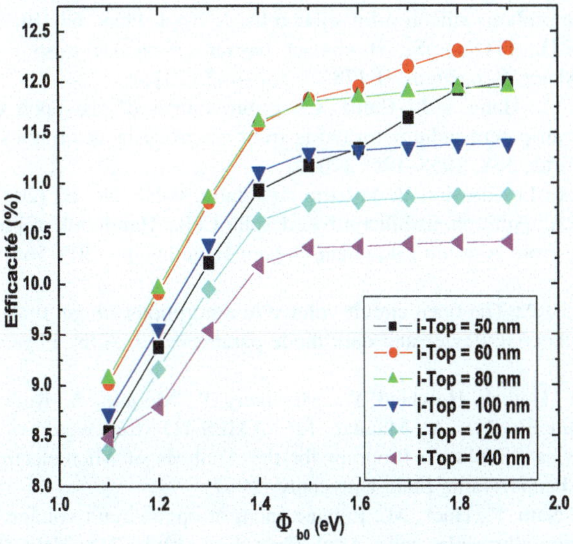

Fig. 6. The efficiency variations of the tandem solar cell with back reflector as a function of the top absorber layer thickness i-a-Si: H and the front contact barrier height (Φ_{b0}).

4 Conclusions

The main objective of this work was to study by simulation the performances of tandem solar cells a-Si:H/μc-Si:H based on hydrogenated amorphous silicon (a-Si: H) and hydrogenated microcrystalline silicon (μc-Si: H. For this we used a one-dimensional numerical simulation code AMPS-1D to study the improvement of two tandem structures, one without back reflector (RB = 0) and the other with back reflector (RB = 0.8). The influence of the front contact barrier height (φ_{b0}) for electrons (or barrier height φ_h for holes) on the performances of our studied devices was examined. The results of simulation showed that the best output parameters were obtained in the case of tandem device with back reflection and for high values of front contact barrier height (Φ_{b0}). However, the maximum efficiency of 12.33% was obtained at Φ_{b0} = 1.9 eV in the case of the tandem solar cell with back reflector (RB = 0.8).

References

Belfar, A., Mostefaoui, R.: Simulation of n1–p2 microcrystalline silicon tunnel junction with AMPS-1D in a-Si:H/μc-Si: H Tandem solar cells. J. Appl. Sci. **11**(16), 2932–2939 (2011)

Arch, J.K., Rubinelli, F.A., Hou, J.Y., Fonash, S.J.: Computer analysis of the role of p-layer quality, thickness, transport mechanisms, and contact barrier height in the performance of hydrogenated amorphous silicon p-i-n solar cells. J. Appl. Phys. **69**(10), 7057 (1991)

Belfar, A., Ait, K.H.: ITO/pnc-Si: H contact barrier effects on n-i-p′-p silicon solar cell performances. Mater. Sci. Engin. B **178**(7), 438–442 (2013)

Biron, R., Pahud, C., Haug, F.J., Ballif, C.: Improvement of the open circuit voltage by modifying the transparent indium–tin oxide front electrode in amorphous n–i–p solar cells. J. Non-Cryst. Solids **358**, 1958–1961 (2012)

Sarl, C., Riacaud, A.: Les dossiers de la lettre du solaire, vol 1, No. 02 (2011)

Deng, X., Schiff, E.A.: Amorphous Silicon-based Solar Cells. Handbook of Photovoltaic Science and Engineering from Antonio Luque and Steven Hegedus, pp. 505–565. Wiley, Chichester (2003)

Dutta, U., Chatterjee, P.: The open circuit voltage in amorphous silicon *p*-i-n solar cells and its relationship to material, device and dark diode parameters. J. Appl. Phys. **96**(4), 2261–2271 (2004)

Fonash, S., Arch, J., Hou, J., Howland, W., Mcelheny, P., Moquin, A., Rogosky, M., Tran, T., Zhu, H., Rubinelli, F.: A Manual for AMPS-1D for windows 95/NT a One-DimensionalDevice Simulation Program for the Analysis of Microelectronic and Photonic Structures. The Pennsylvania State University (1997)

Guha, S., Yang, J., Nath, P., Hack, M.: Enhancement of open-circuit voltage in high efficiency. Amorphous silicon alloy solar cells. Appl. Phys. Lett. **49**(4), 218–219 (1986)

Marks, T.J., Veinot, J.G.C., Cui, J., Yan, H., Wang, A., Edleman, N.L., Ni, J., Huang, Q., Lee, P., Armstrong, N.R.: Progress in high work function TCO OLED anode alternatives and OLED nanopixelation. Synth. Met. **127**, 29–35 (2000)

Meier, J., Keppner, H., Dubail, S., Droll, U., Torres, P., Pernet, P., Ziegler, Y., Selvan, J., Cuperus, J., Fischer, D., Shah, A.: Mater. Res. Soc. Symp. Proc. **507**, 139–144 (1998)

Poissant, Y., Chatterjee, P.: Roca iCabarocas, P.: Analysis and optimization of the performance of polymorphous silicon solar cells: experimental characterization and computer modeling. J. Appl. Phys. **93**(1), 170–174 (2003)

Rubinelli, F.A., Arch, J.K., Fonash, S.J.: Effect of contact barrier heights on a-Si: H p-i-n detector and solar-cell performance. J. Appl. Phys. **72**(4), 1621–1630 (1992)

Sanchez, S.F., Williams, R.: Barrier at the interface between amor- phous silicon and transparent conducting oxides and its influence on solar cell performance. J. Appl. Phys. **54**(5), 2757–2760 (1983)

Realization of Diene Dienophile Interface Reaction in Oil/Water Emulsion

Abdallah Cheraitia(✉) and Nawal Brahimi

Industrial Analysis and Engineering of Materials Laboratory (LAIGM), University May 8th, 1945, Guelma, Guelma, Algeria
cheraitiaabdellah@yahoo.fr

Abstract. The aim of this work is the realization of the Diels-Alder reaction at the interface of an oil/water system and its application to micro encapsulation in developing heat-sensitive membrane microcapsules. This synthesis occurs in three steps:

- The synthesis of hydrophilic dienophile using polyethylene glycol with different molecular weights according to alcohol - maleic anhydride reaction.
- The synthesis of hydrophobic diene from either hydrophobic oligomer grafted to a pyrrole molecule, or by modifying octanoic acid with furan derivatives.
- The Diels-Alder reaction occurs between a diene and a dienophile hydrophobic.

Several combinations of only adducts were prepared from modified triethylene glycol (TEG) and polyethylene glycol (PEG) by action of maleic acid (MA), as a dienophile and (pyrrole/Octanol and furan derivative/octanoic acid) as diene.

Under vigorous stirring, other adducts have been elaborated in an oil/water emulsion system by adding an aqueous solution of TEG-MA in an organic solution of Pyrrole-Octanol (in Octanol) or in water/oil emulsion system by adding an organic solution of Pyrrole-Octanol in Octanol in an aqueous PEG_{6000}-MA solution.

The products obtained were characterized with different techniques such as chromatography TLC thin layer, FTIR infrared spectroscopy, thermal analysis by differential scanning calorimetry DSC and optical microscopy with image acquisition system.

Keywords: Diels · Alder reaction · Diene · Dienophile · Polyethylene glycol

1 Introduction

As early as 1906, Diels observed the addition reaction of cyclopentadiene with parabenzoquinone. The 1928 publication made known the systematic study which he carried out with the help of his student Alder on this type of reaction, known as the cycloaddition of Diels-Alder (Diels and Alder 1928).

B. Abdelbaki et al. (Eds.): SMSD 2017, *Proceedings of the Third International Symposium on Materials and Sustainable Development*, pp. 25–33, 2018.
https://doi.org/10.1007/978-3-319-89707-3_4

This reaction is the process in which conjugated dienes and alkenes, called dienophiles (Scheme 1), combine to form substituted cyclohexenes, called cycloadducts.

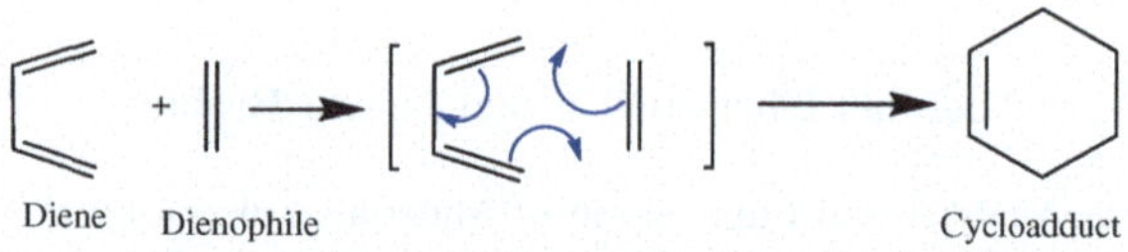

Scheme 1. Mechanism of the Diels-Alder reaction

In the Diels-Alder reaction, an assembly of four conjugated atoms having four π electrons reacts with a double bond containing two π electrons. For this reason, the reaction is called a cycloaddition [4+2] (Chen et al. 1998). Were a new simultaneous and stereospecific bonds are formed (Scheme 2).

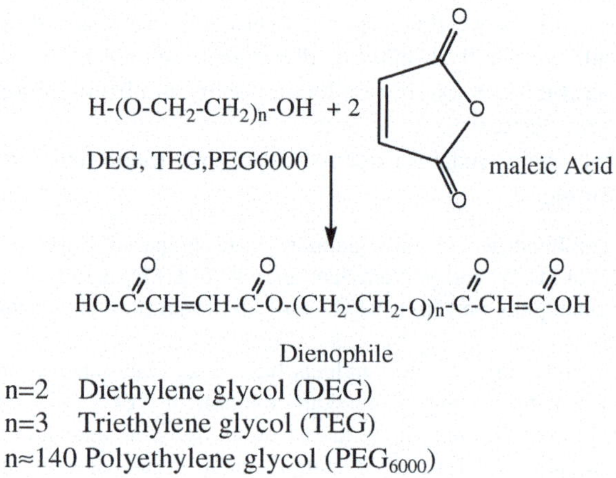

Scheme 2. Synthesis of dienophiles using maleic anhydride (MA) with DEG, TEG and PEG_{6000}

In general, the Diels-Alder reaction is carried out at room temperature and in the presence of a solvent. Hydrocarbons are often the best solvents. Water, which is a highly unlikely solvent for most organic reactions, has a great accelerating effect on the Diels-Alder reaction.

Various dienes and dienophiles were used to realize this reaction (Fringuelli 2002). Aromatic hetero-cycles: pyrrole and furan as electron donor are used as diene in Diels Alder cycloaddition (Fringuelli 1990). Furans can also act as dienophiles (Chen et al. 1998). As dienophile Maleic anhydride presents a good reactivity toward cyclopentadiene in the Diels-Alder Reaction (Sauer et al. 1964).

Modified pyrrole and furan are also used in the Diels Alder reaction (Fringuelli 2002).

2 Experimental Procedure

2.1 Synthesis of Dienophiles

The syntheses of dienophilic molecules have been developed according to the alcohol-maleic anhydride (MA) reaction using diethylene glycol (DEG) as a model.

In a two-necked flask equipped with a condenser, from (diethylene, triethylene and polyethylene) glycol and by adding an equivalent amount of maleic acid according to the above reaction, with stirring and at a temperature of 60 °C. After 15 h of stirring a viscous dienophiles solution, are obtained for the DEG and TEG and a yellow powder, for the PEG_{6000}.

2.2 Synthesis of Dienes

The dienes have been prepared according to the electrophilic substitution reaction (by the action of octanol on pyrrole) or by the amine-acid reaction (furan-2-yl methanamine-octanoic acid).

In a two-necked flask equipped with a condenser, distilled pyrrole is added with an equivalent amount of octanol at 80 °C. (Scheme 3), or octanoic acid with an equivalent amount of furan-2-ylmethanamine at 0 °C (Scheme 4). After 24 h with stirring at

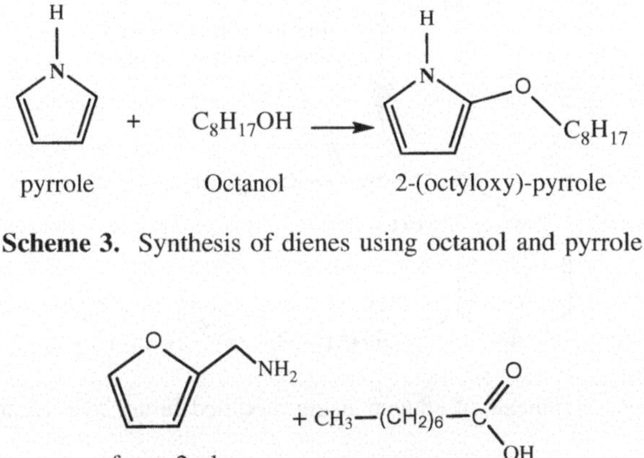

Scheme 3. Synthesis of dienes using octanol and pyrrole

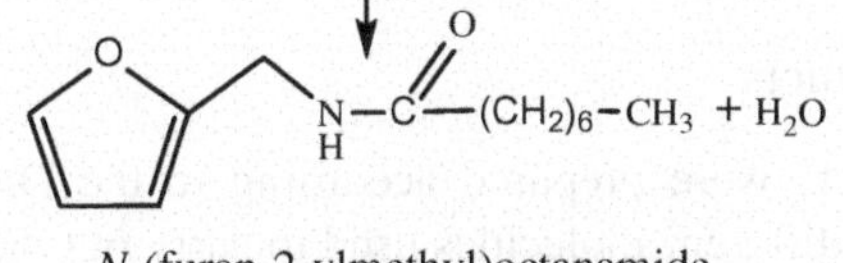

Scheme 4. Synthesis of dienes using furan-2-yl methanamine and octanoic acid

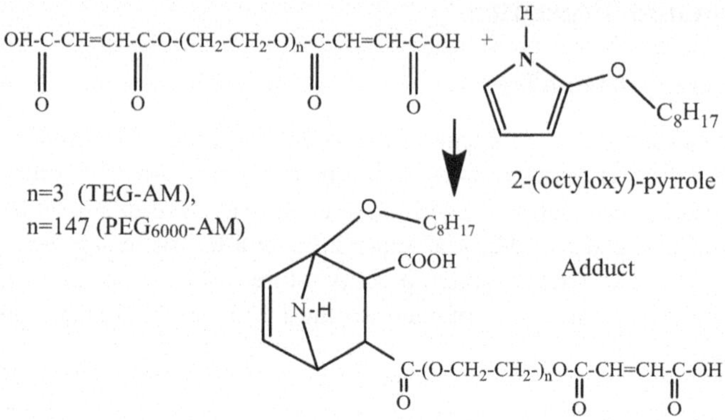

Scheme 5. Synthesis of adducts using modified pyrrole

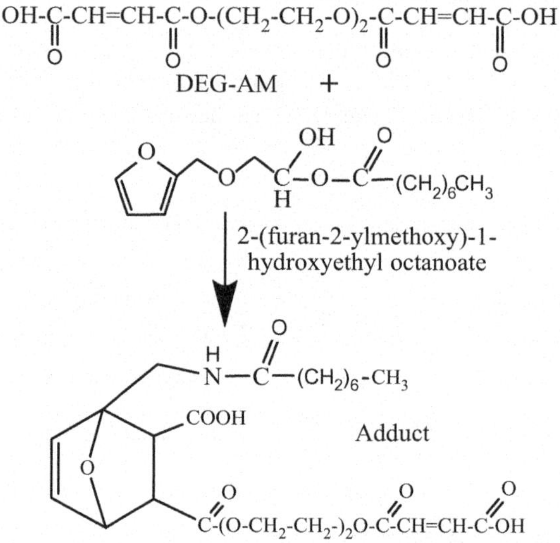

Scheme 6. Synthesis of adducts using modified furan-2-ylmethanamine

ambient temperature, a water-insoluble brown liquid is recovered (for pyrrole) and a powder is obtained after drying (for furan-2-ylmethanamine).

2.3 Synthesis of Adducts

The syntheses of adducts were prepared according to the Diels-Alder reaction. The following table lists the different molecules used to carry out these syntheses (Table 1).

Table 1. Different adducts synthetized

Dienophile	Diene	Adduct
Modified TEG	Pyrrole-Octanol	TEG-AM/Pyrrole-Octanol (Scheme 5)
Modified PEG_{6000}	Pyrrole-Octanol	PEG_{6000}-MA/Pyrrole-Octanol (Scheme 5)
Modified DEG	furan-2-ylmethanamine–octanoic Acid	DEG-MA furan-2-yl methanamine–octanoic Acid (Scheme 6)

3 Synthesis of Adducts in Emulsion

Diels-Alder reactions were carried out in emulsion from different dienophiles and synthesized dienes.

3.1 Synthesis of Water/Oil Emulsion Adducts

Three adducts were synthesized in water/oil emulsion as shown in Table 2:

Table 2. Different adducts synthetized in water/oil emulsion

Dienophile	Diene	Adduct
Modified TEG	Pyrrole-Octanol	TEG-MA/Pyrrole-Octanol
Modified DEG	Pyrrole-Octanol	DEG-MA/Pyrrole-Octanol
Modified DEG	furan-2-ylmethanamine–octanoic Acid	DEG-MA furan-2-yl methanamine–octanoic Acid

In order to have a water/oil emulsion, the adducts prepared from the diene Pyrrole-Octanol and that prepared from the DEG-MA/furan-2-yl methanamine– octanoic Acid dienes are solubilized in an organic solvent (Octanol), with stirring at a rate of 750 rpm and at room temperature for 15 min, then is added dropwise, the aqueous solution of the dienophile “TEG-MA” solubilized in distilled water (Fig. 1).

The mixture passes from a transparent yellow liquid product to a turbid liquid, after one hour a very fine emulsion of micro-droplets (about 20 microns for TEG-AM and 20–40 microns for DEG-MA) (Fig. 2).

3.2 Synthesis of Oil/Water Emulsion Adduct

Only one adduct is synthesized in Oil/ Water emulsion, under the same conditions of the previous experiments PEG_{6000}-AM/ Pyrrole-Octanol is prepared.

The mixture, at the beginning, passes from a white transparent liquid product to a turbid liquid white. After ½ h spherical micro-droplets are observed and after an hour the micro-droplets change shape (oval). A non-stable emulsion is obtained a transparent yellow (micro-droplets between 20 and 1200 μm).

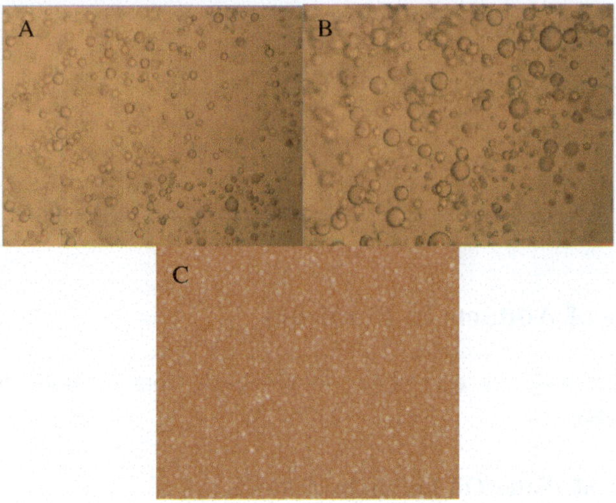

Fig. 1. Water/oil emulsion TEG-MA/Pyrrole–Octanol (A) and DEG-MA/Pyrrole–Octanol (B) and TEG-AM furan-2-yl methanamine–octanoic Acid (C) Adducts

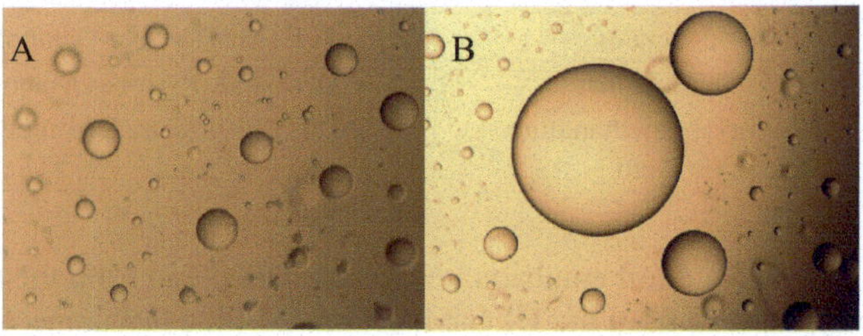

Fig. 2. Oil/ Water PEG_{6000}-AM/Pyrrole–Octanol Adduct after ½ h (A) and after 1 h (B).

4 Characterizations

The techniques used for the characterization of the synthesized products are generally FTIR spectroscopy and calorimetry DSC, in addition to optical microscopy with image acquisition system for observing emulsions.

The FTIR spectrum of the dienophiles (Fig. 3) did not show maleic anhydride vibration of =CH bond at 3100 cm^{-1} and the appearance of the vibration band of the ester function at 1740 cm^{-1}, vibrating band of the ether function at 1150-1050 cm^{-1} and the acid OH bond at 3400-3200 cm^{-1} and other functions.

For the diene (Pyrrole-Octanol), the vibration of C-O bond is observed at 1150-1050 cm^{-1}. Whereas for the diene N - (furan-2-ylmethyl octanamide), the appearance of the vibration of C=O bonds of the amide at 1639 cm^{-1}, NH at 1590 cm^{-1} and C≡N at 1400-1000 cm^{-1} which indicates that the reaction has occurred (Fig. 4).

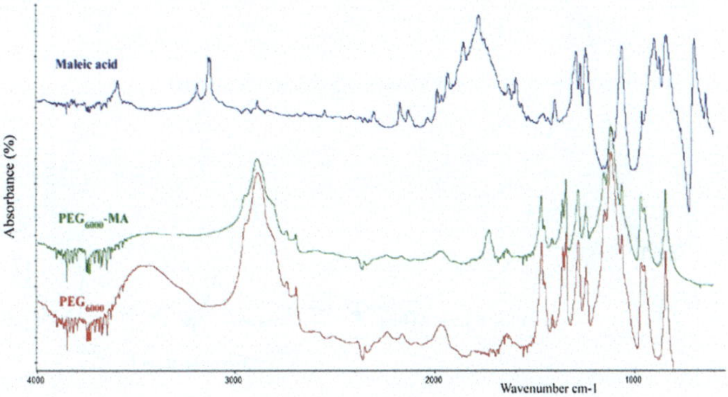

Fig. 3. FTIR Spectrum of the PEG_{6000}-MA dienophile

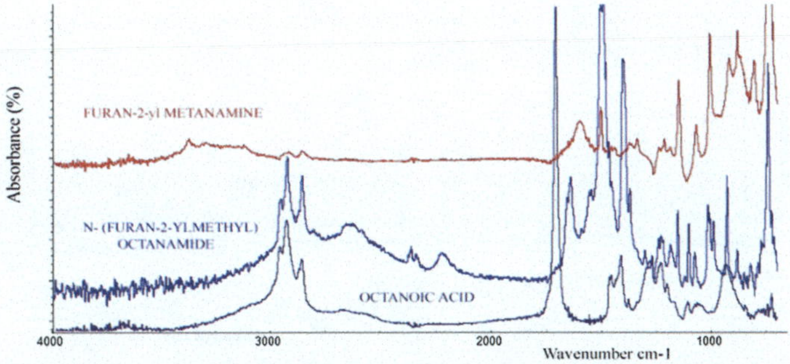

Fig. 4. FTIR Spectrum of the diene N-(furan-2-ylmethyl) octanamide,

The FTIR spectra of the PEG_{6000}-MA/Pyrrole-Octanol (Fig. 5) and TEG-MA/Pyrrole-Octanol adducts, show the preservation of the vibration band of the ester at 1730 cm^{-1} as well as the vibration band of the ether at 1150-1100 cm^{-1}. The vibration band of the CH at 1500-1470 cm^{-1}, the NH at 900-600 cm^{-1} and the decrease in vibration of the C=C bond to 1540-1505 cm^{-1}.

The adduct DEG-MA/furan-2-yl methanamine-octanoic acid FTIR spectrum (Fig. 6), show's a decrease in the C=C bands at 1505 cm^{-1} and =CH at 857-736 and 600 cm^{-1} of the diene with the disappearance of =CH bond at 3100 cm^{-1} for the dienophile and the conservation of the other bands. The DSC spectra show the absence of the melting peak at 65 °C of the diene. An endothermic peak is observed at 76 °C (probably due to the retro Diels Alder). Beyond 125 °C, the endothermic "bump" observed on the diene thermogram.

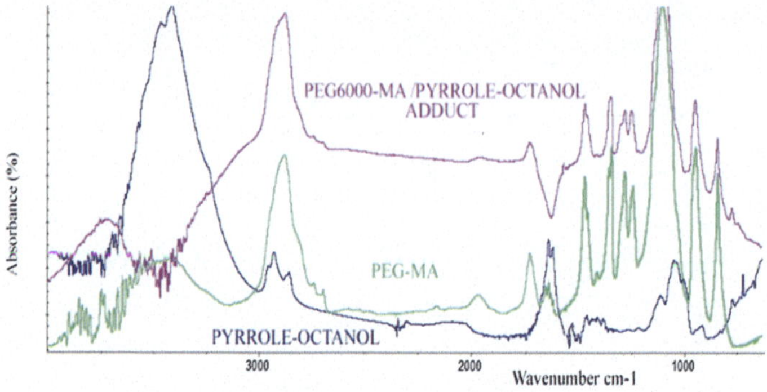

Fig. 5. FTIR Spectrum of the PEG_{6000}-MA/Pyrrole-Octanol adduct

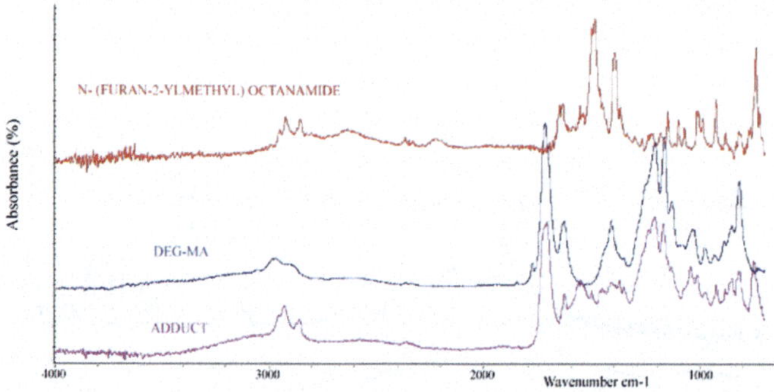

Fig. 6. FTIR of DEG-AM/furan-2-yl methanamine octanoic acid adduct

5 Conclusions

In the first place, new hydrophilic dienophiles have been synthesized by the action of maleic anhydride on oligomers such as diethylene glycol, triethylene glycol and polyethylene glycol. Thus, hydrophobic dienes can be synthesized from the hydrophobic oligomers grafted to a Pyrrole molecule, or by modifying the octanoic acid by furan-2-yl methanamine.

In the second step, the dienes and the modified dienophiles are reacted in order to synthesize the adducts in mass and then in emulsion.

The Diels-Alder reactions between different dienes were carried out without solvent and in water-in-oil or oil-in-water emulsion. We have observed that the adduct obtained from octanoic acid furfurylamine and diethylene glycol maleic anhydride is hydrophilic.

The Diels-Alder reactions between a hydrophobic diene and a dienophile appear to take place at the interface of the emulsions but contrary to what is expected the adduct formed is not amphiphilic but hydrophilic.

References

Chen, C.-H., Rao, P.D., Liao, C.-C.: Furans Act as Dienophiles in Facile Diels−Alder Reactions with Masked o-Benzoquinones. J. Am. Chem. Soc. **120**, 13254–13255 (1998)

Diels, O., Alder, K.: Synthesen in der hydroaromatischen Reihe. I. Justus Liebigs Ann. Chem. **460**, 98–122 (1928)

Fringuelli, P., Taticchi, A.: The Dieh-Alder Reaction: Selected Practical Methods. Wiley Interscience, New York (2002)

Fringuelli, P., Taticchi, A.: Diene in the Dieh-Alder Reaction. Wiley Interscience, New York (1990)

Sauer, J., Wiest, H., Mielert, A.: A study of the DIELS-ALDER Reaction, I. Reactivity of Dienophiles with Cyclopentadiene and 9,10-Dimethyl–anthracene. Chem. Ber. **97**, 3183–3207 (1964)

A New Four Variable Refined Shear Deformation Theory for Buckling and Vibration of Functionally Graded Plates

Ahmed Bakora(✉), Fouad Bourada, Abdelouhed Tounsi, and Adda Bedia El Abbas

Material and Hydrology Laboratory, Faculty of Technology, Civil Engineering Department, University of Sidi Bel Abbes, Sidi Bel Abbes, Algeria
bakora2010@live.fr

Abstract. A new four variable shear deformation theory is developed for buckling, and vibration of functionally graded plates. The theory accounts for parabolic distribution of transverse shear stress, and satisfies the free transverse shear stress conditions on the top and bottom surfaces of the plate without using shear correction factor. Unlike the conventional shear deformation theory, the proposed four variables refined shear deformation theory contains only four unknowns and has strong similarities with classical plate theory in many aspects such as equations of motion, boundary conditions, and stress resultant expressions. The material properties of plate are assumed to vary according to power law distribution of the volume fraction of the constituents. Equations of motion are derived from the Hamilton's principle. The closed-form solutions of simply supported plates are obtained and the results are compared with those of first-order shear deformation theory and higher-order shear deformation theory. It can be concluded that the proposed theory is accurate and efficient in predicting the buckling, and vibration responses of functionally graded plates.

Keywords: Buckling · Vibration · Functionally graded plate · Plate theory

1 Introduction

Functionally graded materials (FGMs) are a new class of composite structures that is of great interest for engineering design and manufacture. FGMs are heterogeneous composite materials, in which the material properties vary continuously from one interface to the other. This is achieved by gradually varying the volume fraction of the constituent materials. Rectangular thick plates made of FGMs are often employed as a part of engineering structures. Analytical studies on FG rectangular thick plates are, however, rare whereas several studies have been numerically performed to analyze the mechanical, thermal or the thermo-mechanical responses of FG rectangular plates. When the plate is thin, the classical plate theory (CPT) is used to analyze FG rectangular plate problem. In an outstanding work on the free vibration analysis of thin rectangular plates with a pair of opposite edges simply supported, exact solutions were presented by Leissa (1973) for all possible combinations of classical boundary conditions along the other

B. Abdelbaki et al. (Eds.): SMSD 2017, *Proceedings of the Third International Symposium on Materials and Sustainable Development*, pp. 34–43, 2018.
https://doi.org/10.1007/978-3-319-89707-3_5

edges. Due to ignoring the effect of shear deformation through the plate thickness, the CPT is valid only for thin plates and gives proper results for lower frequencies. In addition, the CPT underestimates deflections and overestimates frequencies. In order to eliminate the deficiency of the CPT for moderately thick plates, the first order shear deformation theory (FSDT), including the effects of transverse shear deformation and rotary inertia, was employed by many research groups using analytical and numerical methods. Hosseini Hashemi et al. (2010) presented an analytical method for free vibration analysis of moderately thick FG rectangular plates supported by either Winkler or Pasternak elastic foundations The application of the Reissner Mindlin theory to plate problem has attracted the attention of many researchers. Various methods have been applied to compute natural frequencies and buckling loads for thick rectangular plates with different boundary condition, namely the analytical Navier (Reddy and Phan (1985)) and Levy (Liew et al. (1996); Zenkour (2001)) solutions, the Rayleigh–Ritz method (Liew et al. (1995), (1998); Wang et al. (1994); Kitipornchai et al. (1993); Cheung and Zhou (2000)). A new formula for the shear correction factors, used in the Mindlin plate theory, was obtained for FG plates. Although many studies on the vibration analysis of FG rectangular Mindlin plates have been carried out, the FSDT requires shear correction factor, depending on geometric parameters, boundary conditions, loading of the plate and the materials. This is due to the fact that the FSDT overestimates frequencies for thick FG plates, whereas third order shear deformation plate theory (TSDT), higher-order shear deformation plate theory (HSDT) and the three dimensional (3-D) elasticity solution not only require no shear correction factor but also model a plate with smaller displacements and higher rigidity. Reddy and Phan (1985) have used the Navier solution in order to analyze the free vibration and buckling of isotropic, orthotropic and laminated rectangular plates with simply supported edge condition according to the HOPT of Reddy (1999). Hanna and Leissa (1994) have developed a completely higher order shear deformation plate theory, including energy functional, equation of motion and boundary condition. They have used Rayleigh–Ritz method for free vibration solution of fully free rectangular plate.

The objective of this paper is to propose a new exact closed form approach for free vibration and buckling analysis of thick rectangular FG plates based on the third-order shear deformation theory. The material properties are assumed to be graded through the thickness in accordance with a power-law distribution. Hamiltonian principle is used to extract the equations of dynamic equilibrium and natural boundary conditions of the plate.

2 Theory and Formulation

2.1 Functionally Graded Plates

Consider a ceramic–metal FGM plate of length a, width b and thickness h resting on an elastic foundation. A coordinate system (x, y, z) is established in which (x, y) plane on the middle surface of the plate and z is thickness direction $(-h/2 \leq z \leq h/2)$ as shown in Fig. 1.

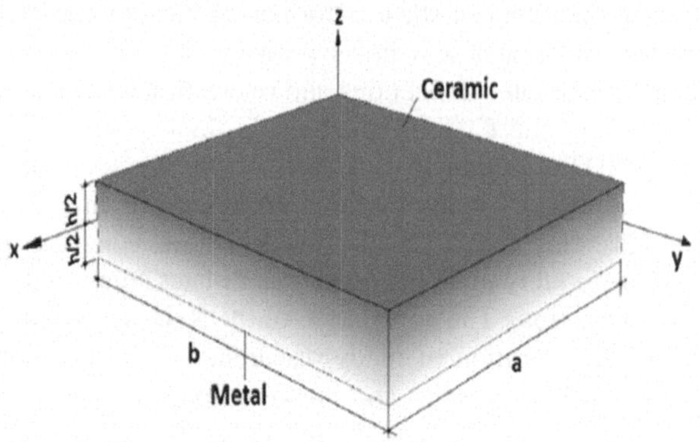

Fig. 1. Geometry and coordinate system of an FGM plate.

Effective properties (modulus of elasticity E) of FGM plate are determined by linear rule of mixture as

$$E(z) = E_m + (E_c - E_m)\left(\frac{1}{2} + \frac{z}{h}\right)^p \tag{1}$$

Where P is volume fraction index $(0 \leq p \prec \infty)$ and Poisson ratio v is assumed to be constant. Subscripts m and c stand for the metal and ceramic constituents, respectively. It is evident from Eq. (1), that the upper surface of the plate $(z = h/2)$ is ceramic-rich, while the lower surface $(z = -h/2)$ is metal-rich.

2.2 Kinematics and Strains

The present study uses higher order shear deformation plate theory to establish governing equations and determine the buckling loads and natural frequencies

$$\begin{aligned} u(x, y, z, t) &= u_0(x, y, t) - z\frac{\partial w_0}{\partial x} + k_1 f(z)\int \theta(x, y, t)\, dx \\ v(x, y, z, t) &= v_0(x, y, t) - z\frac{\partial w_0}{\partial y} + k_2 f(z)\int \theta(x, y, t) dy \\ w(x, y, z, t) &= w_0(x, y, t) \end{aligned} \tag{2}$$

$$f(z) = \frac{5}{4}\left(z - \frac{4z^3}{3h^2}\right)$$

Where u_0, v_0, w_0, θ are four unknown displacements of the midplane of the plate, $f(z)$ denotes shape function representing the variation of the transverse shear strains and stresses within the thickness. The constants k_1 and k_2 depends on the geometry.

The nonzero strains associated with the displacement field in Eq. (2) are

$$\begin{Bmatrix} \varepsilon_x \\ \varepsilon_y \\ \gamma_{xy} \end{Bmatrix} = \begin{Bmatrix} \varepsilon_x^0 \\ \varepsilon_y^0 \\ \gamma_{xy}^0 \end{Bmatrix} + z \begin{Bmatrix} k_x^b \\ k_y^b \\ k_{xy}^b \end{Bmatrix} + f(z) \begin{Bmatrix} k_x^s \\ k_y^s \\ k_{xy}^s \end{Bmatrix}, \tag{3a}$$

$$\begin{Bmatrix} \gamma_{yz} \\ \gamma_{xz} \end{Bmatrix} = g(z) \begin{Bmatrix} \gamma_{yz}^0 \\ \gamma_{xz}^0 \end{Bmatrix}, \; g(z) = \frac{df(z)}{dz} \tag{3b}$$

$$\begin{Bmatrix} \varepsilon_x^0 \\ \varepsilon_y^0 \\ \gamma_{xy}^0 \end{Bmatrix} = \begin{Bmatrix} \frac{\partial u_0}{\partial x} \\ \frac{\partial v_0}{\partial x} \\ \frac{\partial u_0}{\partial y} + \frac{\partial v_0}{\partial x} \end{Bmatrix}, \; \begin{Bmatrix} k_x^b \\ k_y^b \\ k_{xy}^b \end{Bmatrix} = \begin{Bmatrix} -\frac{\partial^2 w_0}{\partial x^2} \\ -\frac{\partial^2 w_0}{\partial y^2} \\ -2\frac{\partial^2 w_0}{\partial x \partial y} \end{Bmatrix},$$

$$\begin{Bmatrix} k_x^s \\ k_y^s \\ k_{xy}^s \end{Bmatrix} = \begin{Bmatrix} k_1\,\theta \\ k_2\,\theta \\ (k_1 A + k_2 \mathbf{B}) \frac{\partial^2 \theta}{\partial x \partial y} \end{Bmatrix}, \; \begin{Bmatrix} \gamma_{yz}^0 \\ \gamma_{xz}^0 \end{Bmatrix} = \begin{Bmatrix} k_2 B' \frac{\partial \theta}{\partial y} \\ k_1 A' \frac{\partial \theta}{\partial x} \end{Bmatrix},$$

For elastic and isotropic FGMs, the constitutive relations can be expressed as

$$\begin{Bmatrix} \sigma_x \\ \sigma_y \\ \tau_{xy} \\ \tau_{yz} \\ \tau_{xz} \end{Bmatrix} = \begin{bmatrix} C_{11} & C_{12} & 0 & 0 & 0 \\ C_{12} & C_{22} & 0 & 0 & 0 \\ 0 & 0 & C_{66} & 0 & 0 \\ 0 & 0 & 0 & C_{44} & 0 \\ 0 & 0 & 0 & 0 & C_{55} \end{bmatrix} \begin{Bmatrix} \varepsilon_x \\ \varepsilon_y \\ \gamma_{xy} \\ \gamma_{yz} \\ \gamma_{xz} \end{Bmatrix} \tag{4}$$

$$C_{11} = C_{22} = \frac{E(z)}{1-\nu^2}, \; C_{12} = \frac{\nu E(z)}{1-\nu^2}, \; C_{44} = C_{55} = C_{66} = \frac{E(z)}{2(1+\nu)},$$

Where $(\sigma_x, \sigma_y, \tau_{xy}, \tau_{yz}, \tau_{xz})$ and $(\varepsilon_x, \varepsilon_y, \gamma_{xy}, \gamma_{yz}, \gamma_{xz})$ are the stress and strain components.

2.3 Equation of Motion

Hamilton's principle is herein utilized to determine the equations of motion

$$0 = \int_0^t (\delta U + \delta V - \delta K)\, dt \tag{5}$$

Where δU is the variation of strain energy; δV is the variation of the external work done by external load applied to the plate; and δK is the variation of kinetic energy which their expressions can be found on Hebali et al. (2016).

The following equations of motion are obtained:

$$\delta u_0: \frac{\partial N_x}{\partial x} + \frac{\partial N_{xy}}{\partial y} = I_0 \ddot{u}_0 - I_1 \frac{\partial \ddot{w}_0}{\partial x} + k_1 A' J_1 \frac{\partial \ddot{\theta}}{\partial x} \tag{6a}$$

$$\delta v_0: \frac{\partial N_{xy}}{\partial x} + \frac{\partial N_y}{\partial y} = I_0 \ddot{v}_0 - I_1 \frac{\partial \ddot{w}_0}{\partial y} + k_2 B' J_1 \frac{\partial \ddot{\theta}}{\partial y} \tag{6b}$$

$$\begin{aligned} \delta w_0: & \frac{\partial^2 M_x^b}{\partial x^2} + 2\frac{\partial^2 M_{xy}^b}{\partial x \partial y} + \frac{\partial^2 M_y^b}{\partial y^2} + N_x^0 \frac{\partial^2 w_0}{\partial x^2} + 2N_{xy}^0 \frac{\partial^2 w_0}{\partial x \partial y} \\ & + N_y^0 \frac{\partial^2 w_0}{\partial y^2} = I_0 \ddot{w}_0 + I_1 \left(\frac{\partial \ddot{u}_0}{\partial x} + \frac{\partial \ddot{v}_0}{\partial y} \right) - I_2 \nabla^2 \ddot{w}_0 \\ & + J_2 \left(k_1 A' \frac{\partial^2 \ddot{\theta}}{\partial x^2} + k_2 B' \frac{\partial^2 \ddot{\theta}}{\partial y^2} \right) \end{aligned} \tag{6c}$$

$$\begin{aligned} \delta \theta: & -\mathrm{k}_1 M_x^s - \mathrm{k}_2 M_y^s - (k_1 A + k_2 B) \frac{\partial^2 M_{xy}^s}{\partial x \partial y} + k_1 A' \frac{\partial S_{xz}^s}{\partial x} \\ & + k_2 B' \frac{\partial S_{yz}^s}{\partial y} = -J_1 \left(k_1 A' \frac{\partial \ddot{u}_0}{\partial x} + k_2 B' \frac{\partial \ddot{v}_0}{\partial y} \right) \\ & - K_2 \left((k_1 A')^2 \frac{\partial^2 \ddot{\theta}}{\partial x^2} + (k_2 B')^2 \frac{\partial^2 \ddot{\theta}}{\partial y^2} \right) + J_2 \left(k_1 A' \frac{\partial^2 \ddot{w}_0}{\partial x^2} + k_2 B' \frac{\partial^2 \ddot{w}_0}{\partial y^2} \right) \end{aligned} \tag{6d}$$

Where dot-overscript convention indicates the differentiation with respect to the time variable t. The stress resultants N_i, M_i, Q_i and the inertias $I_i, j_i, K_i (i = 1, 2, 3)$ are defined by

$$\left(N_i, M_i^b, M_i^s\right) = \int_{-h/2}^{h/2} (1, z, f)\ \sigma_i dz, (i = x, y, xy) \text{ and } \left(S_{xz}^s, S_{yz}^s\right) = \int_{-h/2}^{h/2} g\left(\tau_{xz}, \tau_{yz}\right) dz$$

$$(I_0, I_1, I_2) = \int_{-h/2}^{h/2} (1, z, z^2) \rho(z) dz$$

$$(J_1, J_2, K_2) = \int_{-h/2}^{h/2} (f, zf, f^2) \rho(z) dz$$

N_x, N_y, N_{xy} are transverse and in-plane applied loads.

The equations of motion can be expressed in terms of displacements (u_0, v_0, w_0, θ) and the appropriate equations take the form Hebali et al. (2016).

2.4 Analytical Solution for Simply-Supported FG Plates

The Navier solution method is employed to determine the analytical solutions for which the displacement variables are written as product of arbitrary parameters and known trigonometric functions to respect the equations of motion and boundary conditions

$$\begin{Bmatrix} u_0 \\ v_0 \\ w_0 \\ \theta \end{Bmatrix} = \sum_{m=1}^{\infty}\sum_{n=1}^{\infty} \begin{Bmatrix} U_{mn} & e^{i\omega t} & \cos(\alpha x)\sin(\beta y) \\ V_{mn} & e^{i\omega t} & \sin(\alpha x)\cos(\beta y) \\ W_{mn} & e^{i\omega t} & \sin(\alpha x)\sin(\beta y) \\ X_{mn} & e^{i\omega t} & \sin(\alpha x)\sin(\beta y) \end{Bmatrix} \tag{7}$$

Where ω is the frequency of free vibration of the plate, $\sqrt{i} = -1$ the imaginary unit. With

$$\alpha = m\pi/a, \ \beta = n\pi/b$$

Considering that the plate is subjected to in-plane compressive loads of form: (here γ_1 and γ_2 are non-dimensional load parameters).

$$N_x^0 = \gamma_1 N_{cr},\ N_y^0 = \gamma_2 N_{cr},\ N_{xy}^0 = 0 \tag{8}$$

Substituting Eq. (7) into Eq. (6a), (6b), (6c) and (6d) the following problem is obtained

$$\left(\begin{bmatrix} S_{11} & S_{12} & S_{13} & S_{14} \\ S_{12} & S_{22} & S_{23} & S_{24} \\ S_{13} & S_{23} & S_{33}+N & S_{34} \\ S_{14} & S_{24} & S_{34} & S_{44} \end{bmatrix} - \omega^2 \begin{bmatrix} m_{11} & 0 & m_{13} & m_{14} \\ 0 & m_{22} & m_{23} & m_{24} \\ m_{13} & m_{23} & m_{33} & m_{34} \\ m_{14} & m_{24} & m_{34} & m_{44} \end{bmatrix} \right) \begin{Bmatrix} U_{mn} \\ V_{mn} \\ W_{mn} \\ X_{mn} \end{Bmatrix} = \begin{Bmatrix} 0 \\ 0 \\ 0 \\ 0 \end{Bmatrix} \tag{9}$$

Where the expressions of $\left(S_{ij}, m_{ij}, N, i=j=1,2,3,4\right)$ are found on Hebali et al. (2016).

3 Numerical Examples and Discussions

In this section, various numerical examples are presented and discussed to check the accuracy of present HSDT in investigating the bending, buckling, and vibration behaviors of simply supported FG plates. For proposed examples, an Al/Al_2O_3 plate fabricated of aluminum (as metal) and alumina (as ceramic) is examined. The Young's modulus and density of aluminum are $E_m = 70$ GPa and $\rho_m = 2702$ kg/m^3, respectively, and those of alumina are $E_c = 380$ GPa and $\rho_c = 3800$ kg/m^3 respectively.

For validation purpose, the computed quantities are compared with those reported utilizing various existing plate models. For convenience, the following dimensionless quantities are employed in illustrating the numerical results in graphical and tabular form

$$D = \frac{Eh^3}{12(1-\nu^2)} \quad \overline{\overline{N}} = \frac{N_{cr}b^2}{\pi^2 D} \quad \overline{N} = \frac{N_{cr}a^2}{E_m h^3}$$

$$\overline{\overline{\omega}} = \omega h\sqrt{\rho_c/E_c} \quad \overline{\omega} = \omega\frac{a^2}{h}\sqrt{\rho_c/E_c}$$

The computed data are compared with those calculated by and Hosseini-Hashemi et al. (2011a) based on FSDT and Hosseini-Hashemi et al. (2011b) based on TSDT. It is concluded from (Table 1) that there is an excellent agreement between the results determined by present model, FSDT (Hosseini-Hashemi et al. 2011a), and TSDT (Hosseini-Hashemi et al. 2011b). From this investigation can be observed that the resulting curves (Fig. 2) are very close to the curves plotted by employing a TSDT (Reddy 2000) and the CPT overestimates the results of thick plate. Thus, in general, the present model is successfully validated.

Table 1. Comparison of nondimensional fundamental frequency $\overline{\overline{\omega}}$ of square plate.

a/h	Method	Power law index (*P*)				
		0	0.5	1	4	10
5	FSDT[a]	0.2112	0.1805	0.1631	0.1397	0.1324
	HSDT[b]	0.2113	0.1807	0.1631	0.1378	0.1301
	Present	0.2113	0.1807	0.1631	0.1378	0.1301
10	FSDT[a]	0.0577	0.0490	0.0442	0.0382	0.0366
	HSDT[b]	0.0577	0.0490	0.0442	0.0381	0.0364
	Present	0.0577	0.0490	0.0442	0.0381	0.0364
20	FSDT[a]	0.0148	0.0125	0.0113	0.0098	0.0094
	HSDT[b]	0.0148	0.0125	0.0113	0.0098	0.0094
	Present	0.0148	0.0125	0.0113	0.0098	0.0094

[a]Taken from Hosseini-Hashemi (2011a)
[b]Taken from Hosseini-Hashemi (2011b)

Table 2 presents the values of the non-dimensional buckling loads $\overline{\overline{N}}$ of isotropic plate ($P = 0$) under various loading cases for various values of aspect ratio a/b and thickness ratio h/b. The computed values are compared with the results reported by Shufrin and Eisenberger (2005) based on FSDT and TSDT. An excellent agreement is proved for all types ranging from moderately thick to very thick plates. It is demonstrated that the present novel four variable refined plate theory and TSDT predict almost the same values, and CPT over-estimates the buckling loads of plate due to neglecting transverse shear deformation influences. The difference between CPT and shear deformation models diminishes when the side-to-thickness ratio a/h increases (Fig. 3).

Table 2. Comparison of nondimensional critical buckling load $\overline{\overline{N}}$ of isotropic plate under different loading types $P = 0$

h/b	Method	Loading type (γ_1, γ_2)		
		(1, 0)	(0, 1)	(1, 1)
0.1	FSDT[a]	3.7865	3.7865	1.8932
	TSDT[a]	3.7866	3.7865	1.8933
	Present	3.7866	3.7865	1.8933
0.2	FSDT[a]	3.2638	3.2637	1.6319
	TSDT[a]	3.2653	3.2653	1.6327
	Present	3.2653	3.2653	1.6327
0.3	FSDT[a]	2.6533	2.6533	1.3266
	TSDT[a]	2.6586	2.6586	1.3293
	Present	2.6586	2.6586	1.3293
0.4	FSDT[a]	1.9196	1.9196	1.0513
	TSDT[a]	1.9550	1.9550	1.0567
	Present	1.9550[a]	1.9550[a]	1.0567

[a]Taken from Shufrin and Eisenberger (2005)

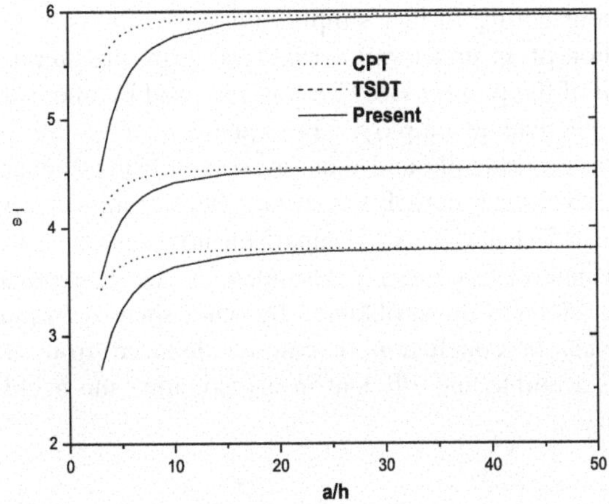

Fig. 2. Comparison of the variation of nondimensional fundamental frequency $\overline{\omega}$ of square plate versus thickness ratio a/h.

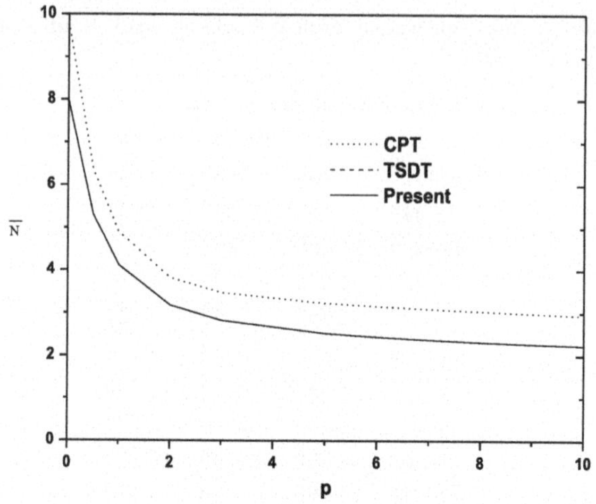

Fig. 3. Comparison of the variation of nondimensional critical buckling load $\overline{N}$ of square plate under biaxial compression versus power law index P $(a = 5h)$.

4 Conclusions

A novel higher-order shear deformation theory is developed for buckling, and vibration of FG plates. By considering further simplifying suppositions to the existing HSDT, with the incorporation of an undetermined integral term, the number of variables and equations of motion of the present HSDT are diminished by one, and hence, make this model simple and efficient to employ. The equations of motion are determined by utilizing the Hamilton's principle and then are solved using Navier's procedure. The exactitude of the developed model has been checked for the buckling, and free vibration responses of FG plates. All comparison investigations demonstrate that the buckling load, and natural frequency determined by the developed model with four variables are almost close to those obtained by other shear deformation theories containing five variables. In conclusion, it can be deduced from this work that the developed theory is accurate and efficient in investigating the buckling, and vibration behaviors of FG plates.

References

Leissa, A.W.: The free vibration of rectangular plates. J. Sound Vib. **31**, 257–293 (1973)

Hosseini-Hashemi, S., Rokni Damavandi Taher, H., Akhavan, H., Omidi, M.: Free vibration of functionally graded rectangular plates using first-order shear deformation plate theory. Appl. Math. Model. **34**(5), 1276–1291 (2010)

Reddy, J.N., Phan, N.D.: Stability and vibration of isotropic, orthotropic and laminated plates according to a higher-order shear deformation theory. J. Sound Vib. **98**(2), 157–170 (1985)

Liew, K.M., Xiang, Y., Kitipornchai, S.: Analytical buckling solutions for mindlin plates involving free edges. Int. J. Mech. Sci. **38**(10), 1127–1138 (1996)

Zenkour, A.M.: Buckling and free vibration of elastic plates using simple and mixed shear deformation theories. Acta Mech. **145**, 183–197 (2001)
Liew, K.M., Xiang, Y., Kitipornchai, S.: Research on thick plate vibration: a literature survey. J. Sound Vib. **180**(1), 163–176 (1995)
Liew, K.M., Wang, C.M., Xiang, Y., Kitipornchai, S.: Vibration of Mindlin Plates. Elsevier, Amsterdam (1998)
Wang, C.M., Xiang, Y., Kitipornchai, S., Liew, K.M.: Buckling solutions for mindlin plates of various shapes. Eng. Struct. **16**(2), 119–127 (1994)
Kitipornchai, S., Xiang, Y., Wang, C.M., Liew, K.M.: Buckling of thick skew plates. Int. J. Numer. Method Eng. **36**, 1299–1310 (1993)
Cheung, Y.K., Zhou, D.: Vibrations of moderately thick rectangular plates in terms of a set of static Timoshenko beam functions. Comput. Struct. **78**(6), 757–768 (2000)
Reddy, J.N.: Theory and Analysis of Elastic Plates. Taylor & Francis, Philadelphia (1999)
Hanna, N.F., Leissa, A.W.: A higher order shear deformation theory for the vibration of thick plates. J. Sound Vib. **170**(4), 545–555 (1994)
Hebali, H., Bakora, A., Tounsi, A., Kaci, A.: A novel four variable refined plate theory for bending, buckling, and vibration of functionally graded plates. Steel Compos. Struct. **22**(3), 473–495 (2016)
Hosseini-Hashemi, S., Fadaee, M., Atashipour, S.R.: A new exact analytical approach for free vibration of Reissner-Mindlin functionally graded rectangular plates. Int. J. Mech. Sci. **53**(1), 11–22 (2011a)
Hosseini-Hashemi, S., Fadaee, M., Atashipour, S.R.: Study on the free vibration of thick functionally graded rectangular plates according to a new exact closed-form procedure. Compos. Struct. **93**(2), 722–735 (2011b)
Shufrin, I., Eisenberger, M.: Stability and vibration of shear deformable plates-first order and higher order analyses. Int. J. Solids Struct. **42**(3–4), 1225–1251 (2005)

Optical and Structural Properties of ZnS:La Thin Films Elaborated by Sol-Gel Method

A. Tounsi[1(✉)], D. Talantikite-Touati[1], R. Khalfi[1], H. Merzouk[2], and H. Haddad[1]

[1] Department of Chemistry, Abderrahmane Mira University, Bejaïa, Algeria
amel89sm@yahoo.fr

[2] Department of Physic, Abderrahmane Mira University, Bejaïa, Algeria

Abstract. The thin layers of undoped ZnS and ZnS doped La were deposited on glass substrates using sol-gel and dip-coating methods. The structural characterization of these samples was carried out by the X-rays diffraction (XRD), scanning electron microscopy (SEM) and atomic force microscopy (AFM). UV-visible spectrophotometry and Fourier Transform Infrared spectroscopy (FTIR) have been used to study the effect of deposition layers on the optical properties of ZnS doped La thin films. X-ray diffraction measurements show that these films are amorphous. Atomic force microscopy images of the films have revealed homogeneous and granular structure and the SEM micrographies show that the amorphous deposit films with uniform and porous structure. The optical transmission spectra in the UV - visible range have shown that the doped film present a good optical transmission ranging between 50% and 70% in the visible. The calculated values of gap were between 3.68 and 2.77 eV. Doping with lanthanum decrease the optical gap of ZnS. Analysis FTIR of the samples confirm the presence of the vibration band of Zn-S at 668 cm^{-1}.

Keywords: ZnS · ZnS doped La · Sol-gel method · XRD and UV-visible

1 Introduction

Recently, researches on properties of nanocrystalline semiconductors have attracted much attention [1]. II-VI semiconductors have an immense scientific and technological importance. Among them, Zinc sulfide (ZnS) is the most important with a wide optical band gap (3.6 eV). Because of its favorable electronic and optical properties, ZnS have a many applications such as solar cells [2], electro-luminescent and optoelectronic devices [3]. Various physical and chemical methods are used to prepare ZnS thin films such as RF magnetron sputtering [4], hydrothermal method [5], spray pyrolysis [6], chemical bath deposition [7] and sol-gel [8, 9].

The sol-gel technique is a simple technique based on the series of hydrolysis and condensation reactions to form a wet-gel structure. This method has been extended to the fabrication of thin films or coatings on different substrates.

B. Abdelbaki et al. (Eds.): SMSD 2017, *Proceedings of the Third International Symposium on Materials and Sustainable Development*, pp. 44–51, 2018.
https://doi.org/10.1007/978-3-319-89707-3_6

Several studies were reported in literature on doping ZnS with 3d transition metal and rare earth (RE) [10, 11] ions, in order to improve luminescent properties of zinc sulfide.

We have not found in literature, a studies about La-ZnS thin films elaborated by sol-gel technique. For this reason, we have chosen to present in this work the doping of ZnS by the lanthanum. We studied the effect of La-doping on the structural and optical properties of ZnS films and we present the preparation and the characterization of ZnS and La doped ZnS thin films. So, the thin layers of La doped ZnS with two compositions, ZnS and 0.6La + 0.94ZnS, were deposited on glass substrates by a dip-coater using sol-gel method. The structural properties were carried out by the X-ray diffraction (XRD), scanning electron (SEM) and atomic force (AFM) microscopies. The Fourier transform infrared measurements were used to confirm the formation of ZnS bonds and identify any adsorbed species on the crystallite surface. UV-visible spectroscopy at room temperature were used for the optical investigation.

2 Experimental

ZnS doped La thin films with two concentrations (0 and 6%) were prepared by following steps: zinc acetate dihydrate $Zn(CH_3COO)_2.2H_2O$ and lanthanum III nitrate hexahydrate ($La(NO_3)_3.6H_2O$) were firstly dissolved in propanol-1. Thiourea CH_4N_2S was dissolved in propanol-1 separately by stirring at room temperature. Triethanolamine was added to the first solution drop by drop. Then, a solution of NaOH was added to adjust the pH at 12 and to obtain a homogeneous transparent solution. Finally, the solution of thiourea was added to this solution and mixed in the same flask. A constant agitation is maintained during 2 h. Thin films of ZnS doped La were deposited on properly cleaned glass substrates by dip-coating.

The deposited films were dried at 423 K to remove the mixed solvent and organic substances such as C, H and N.

The structural characterization of these samples was carried out by the X-rays diffraction (XRD) using a (XPert PRO PANatycal) with Cu Kα radiation ($\lambda = 1.54051A°$), scanning electron microscopy (SEM) (SEM Quanta 200, FEI France) and atomic force microscopy (AFM) with three-dimensional (3D) and two dimensional (2D). UV-visible using UV/visible spectrophotometry (SHIMADZU UV 1800 PC) in the range of 200–1100 nm at room temperature and Fourier Transform Infrared spectroscopies (FTIR) (Iraffinity-1 Shimadzu) between 400 and 4000 cm^{-1} have been used to study the effect of deposition layers on the optical properties of ZnS doped La thin films. All the characterizations have been recorded at room temperature.

3 Results and Discussion

3.1 XRD Analysis

Figure 1 shows the XRD patterns of $ZnS:La^{3+}$ (0 and 6%) thin films deposited on glass substrate.

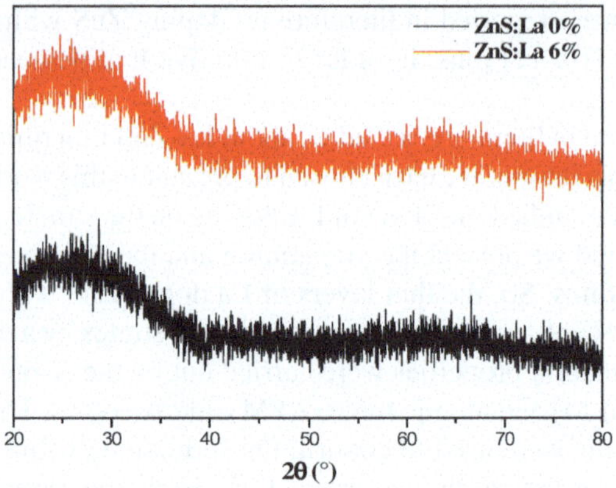

Fig. 1. X-ray diffraction of ZnS:La^{3+} thin films (0-6%).

The analysis present amorphous structures for all deposits thin layers, and no peak were detected. This result was expect. Indeed, thin films produced by sol-gel method and dip-coating give an amorphous compound [12] and it has been explained by highly disordered present within a deposit materials and a very small grain sizes [13]. A sintering of La-ZnS thin films carried out at 500 °C in air during 2 h show a characteristic phase of ZnO. This behavior has been observed by many authors [14, 15] which report the oxidation of ZnS to ZnO in the presence of air oxygen.

3.2 AFM Analysis

Figure 2 shows the surface morphology of the thin layers of pure ZnS and ZnS:La^{3+} (6%), deposited on glass substrate characterized by the atomic force microscopy (AFM) with two dimensional (2D).

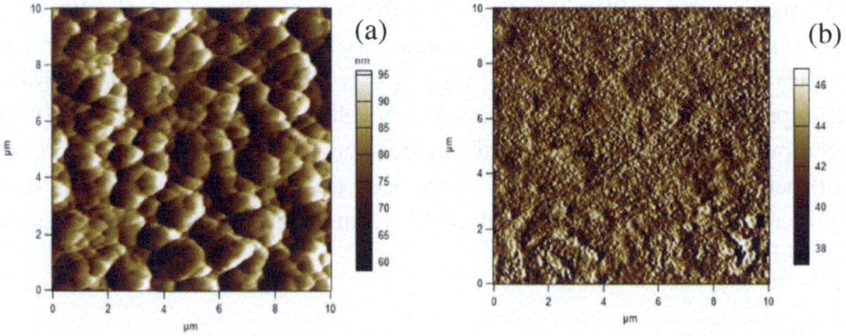

Fig. 2. AFM images 2D of the surface of thin films (1-x)ZnS-xLa (x = 0, 6%): (a) x = 0%, (b) x = 6%.

Thin film of pure ZnS shows no planar surface with a very important roughness (RMS = 90.505 nm). Doping greatly decreases roughness and the surface sample of

ZnS:La is more homogeneous and less rough compared with a pure ZnS. The obtained values of RMS were consigned in Table 1.

Table 1. The values of the RMS (the surface roughness) of the samples ZnS:La^{3+} (0–6%).

Compositions	RMS (nm)
ZnS:La^{3+} 0%	90.505
ZnS:La^{3+} 6%	12.331

3.3 SEM Analysis

Figure 3 shows micrographies of the different compositions of ZnS:La^{3+} (0 and 6%) at room temperature. Pure ZnS has some bright spots scattered on the surface. Clusters of different sizes and shapes were given by the sample of 6%.

(a)

(b)

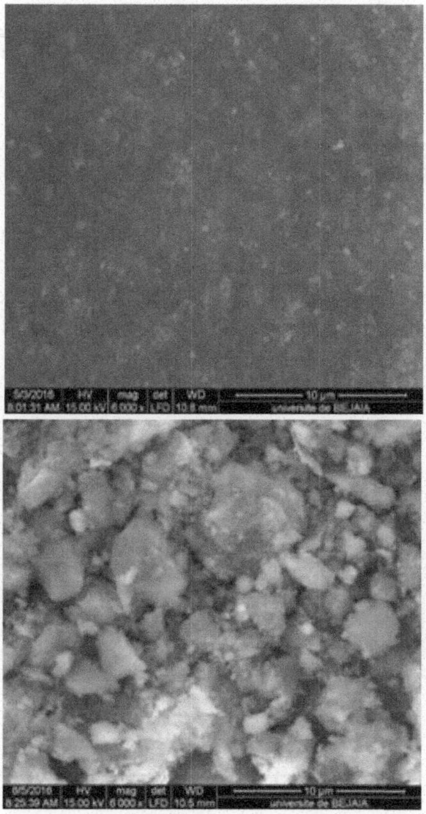

Fig. 3. The SEM micrographs of the different compositions of ZnS:La^{3+}: (a) x = 0%, (b) x = 6%.

3.4 UV-Visible Analysis

The optical properties of the zinc sulfide doped lanthanum (ZnS:La^{3+}) thin films were determined from transmittance measurements using UV/visible spectrophotometry in the range of 200–1100 nm at room temperature. Figure 4 shows the transmittance spectra of ZnS:La^{3+} thin films according to the wavelength.

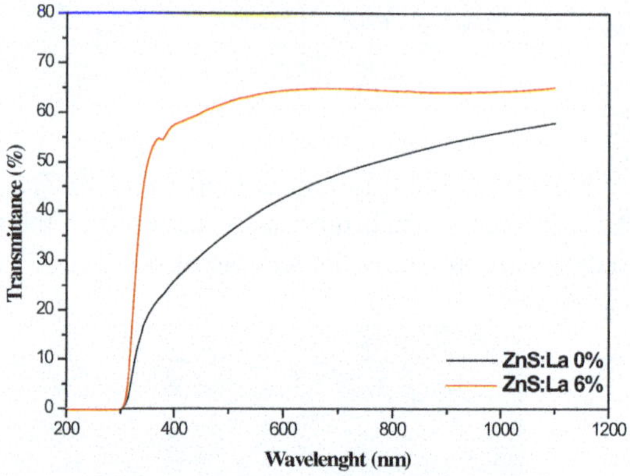

Fig. 4. Optical transmittance spectra of ZnS:La^{3+} thin films (0 and 6%).

The behavior of curves shows a good homogeneity of deposits ZnS:La thin films. The compositions of ZnS and ZnS:La^{3+} 6% have a good transparency in the visible domain. In general, the doping of ZnS with La improves the transmittance of ZnS.

The UV-visible absorption curves of zinc sulfide doped lanthanum thin films were represented in Fig. 5. We observed a considerable increase on the wavelength absorption edge for the ZnS:La^{3+} (6%) composition compared to pure ZnS.

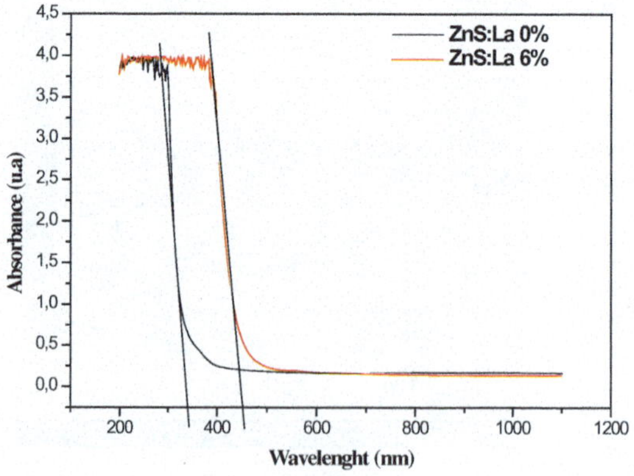

Fig. 5. Optical absorption spectra of ZnS:La^{3+} thin films (0–6%) on glass substrates.

In order to understand the performance of thin films ZnS with dopant concentration, it's necessary to determine the direct band gap (Eg). The gap was calculated using the formula (1):

$$\mathbf{Eg} = \mathbf{h}\nu = \mathbf{hc}/\boldsymbol{\lambda} \quad (1)$$

Table 2 summarizes the gap values and the maximum transmittance of ZnS thin films doped with lanthanum (0 and 6%). The gap values decrease with doping. These values are in agreement with those found by other authors [11, 16].

Table 2. The absorption threshold, the gap values and the maximum transmittance of $ZnS:La^{3+}$ thin films (0–6%).

Compositions	Absorption thresholds λ (nm)	Optical gap Eg (eV)	Maximal transmittance Tmax (%)
$ZnS:La^{3+}$ 0%	337	3.68	56
$ZnS:La^{3+}$ 6%	448	2.77	65

3.5 FTIR Analysis

Figure 6 shows the transmittance spectra of $ZnS:La^{3+}$ thin films (0 and 6%).

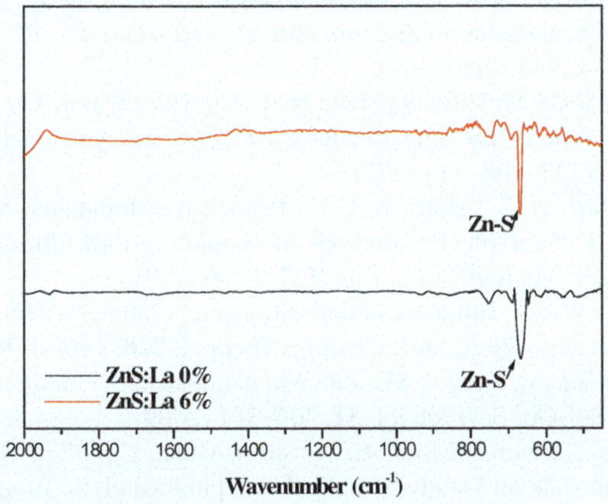

Fig. 6. FTIR spectra of $ZnS:La^{3+}$ thin films (0–6%).

The FITR analyses confirmed the presence of ZnS. The two curves observed on the Fig. 5 of the transmittance spectra of ZnS:La show only one intense peak at 668 cm^{-1} which is attributed by several authors at the Zn-S vibration band [11, 17].

4 Conclusion

Zinc sulfide doped lanthanum ($ZnS:La^{3+}$) with two concentrations (0 and 6%) thin films were deposed on glass substrates by a dip-coater using sol-gel method.

X-ray diffraction analysis has revealed that the thin layers for all compositions are amorphous. Atomic force microscopy images of the films have revealed a homogeneous and granular structure. Doping decreases roughness. The optical transmission spectra in the UV - visible range have shown that all the doped films present a good optical transmission in the visible. The values of gap were between 3.68 and 2.77 eV. Analysis FTIR of the samples confirms the presence of the vibration band of Zn-S at 668 cm^{-1}.

References

1. Sahraei, R., Darafarin, S.: Preparation of nanocrystalline Ni doped ZnS thin films by ammonia-free chemical bath deposition method and optical properties. J. Lumin. **149**, 170–175 (2014)
2. Liu, W.-L., Chen, W.-J., Hsieh, S.-H., Yu, J.-H.: Growth behavior of nanocrystallinezns thin films for solar cell using CBD technique. Procedia Eng. **36**, 46–53 (2012)
3. Akhtar, M.S., Malik, M.A., Riaz, S., Naseem, S., O'Brien, P.: Optimising conditions for the growth of nanocrystalline ZnS thin films from acidic chemical baths. Mater. Sci. Semicond. Process. **30**, 292–297 (2015)
4. Gayou, V.L., Salazar-Hernandez, B., Constantino, M.E., Andres, E.R., Dıaz, T., Macuil, R.D., Lopez, M.R.: Structural studies of ZnS thin films grown on GaAs by RF magnetron sputtering. Vacuum **84**, 1191–1194 (2010)
5. Hoa, T.T.Q., The, N.D., McVitie, S., Nam, N.H., Vu, L.V., Canh, T.D., Long, N.N.: Optical properties of Mn-doped ZnS semiconductor nanoclusters synthesized by a hydrothermal process. Opt. Mater. **33**, 308–314 (2011)
6. Filipovic, L., Selberherr, S., Mutinati, G.C., Brunet, E., Steinhauer, S., Köck, A., Teva, J., Kraft, J., Siegert, J., Schrank, F.: Methods of simulating thin film deposition using spray pyrolysis techniques. Microelectron. Eng. **117**, 57–66 (2014)
7. Zhou, L., Tang, N., Wu, S.: Influence of deposition temperature on ZnS thin film performance with chemical bath deposition. Surf. Coatings Technol. **228**, 146–149 (2013)
8. Yang, P., Szatkowski, A., Bredol, M.: ZnS:Mn nanoparticles conjugated to sol–gel-derived silica matrices. J. Sol-Gel SciTechnol. **51**, 306–314 (2009)
9. Gómez-Barojas, E., Sánchez-Mora, E., Castillo-Abriz, C., Flores-Rodríguez, E., Silva-González, R.: Synthesis and study of optical and photocatalytic properties of Mn and Sm doped ZnS grown by sol-gel. J. Supercond. Nov. Magn. **26**, 2337–2340 (2013)
10. Liang, Z., Mu, J., Han, L., Yu, H.: Microbe-assisted synthesis and luminescence properties of monodispersed Tb^{3+}-doped ZnS nanocrystals. J. Nanomater. **2015**, 1–6 (2015)
11. Varughese, G., Aswathy, S.R., Usha, K.T., Kumar, A.S.: Optical band gap of wurtzite zinc sulphide doped with lanthanum ions. Adv. Mater. Res. **622–623**, 752–756 (2013)
12. Goktas, A., Aslan, F., Yasar, E., Mutlu, I.H.: Preparation and characterization of thickness dependent nano-structured ZnS thin films by sol–gel technique. J. Mater. Sci. Mater. Electron. **23**, 1361–1366 (2012)
13. Luque, L., Gómez-Gutiérrez, C.M., Lastra, G., Carrillo-Castillo, A., Quevedo-López, M.A., Olivas, A.: Role of zinc source in chemical bath deposition of zinc sulfide thin films on Si_3N_4. J. Electron. Mater. **43**(11), 4317–4321 (2014)

14. Wahab, R., Ansari, S.G., Kim, Y.S., Dhage, M.S., Seo, H.K., Song, M., Shin, H.S.: Effect of annealing on the conversion of ZnS to ZnO nanoparticles synthesized by the sol-gel method using zinc acetate and thiourea. Met. Mater. Int. **15**(3), 453–458 (2009)
15. Bu, I.Y.Y.: Sol–gel synthesis of ZnS(O, OH) thin films: influence of precursor and process temperature on its optoelectronic properties. J. Lumin. **134**, 423–428 (2013)
16. Chen, Y., Huang, G.F., Huang, W.Q., Zou, B.S., Pan, A.: Enhanced visible-light photoactivity of La-doped ZnS thin films. Appl. Phys. A **108**, 895–900 (2012)
17. Reddy, D.A., Liu, C., Vijayalakshmi, R.P., Reddy, B.K.: Effect of Al doping on the structural, optical and photoluminescence properties of ZnS nanoparticles. J. Alloys Compd. **582**, 257–264 (2014)

Theoretical and Experimental Studies of Micropores Catalysts

Atallah Dehbi[1(✉)], Nadia Toumi[1], Rawia Imene Bahoussi[2], Meryem Imene Amrane[3], Affaf Tabti[1], Imene Benchikh[1], Zoubida Lounis[1,4], Farouk Hamza Reguig[5], and Abdelkader Bengueddach[1]

[1] Laboratory of Chemistry of Materials LCM, Ahmed Ben Bella University of Oran 1, Oran, Algeria
atallah160@yahoo.fr

[2] Laboratory of Technology and Solid Properties LTPS, Abdelhamid Ibn Badis University, Mostaganem, Algeria

[3] Laboratory of Organic Chemistry, Physical and Macromolecular, University Djillali Liabes, Sidi Bel Abbes, Algeria

[4] Laboratory of Engineering of Industrial Safety and Suitable Development LISIDD, Mohamed Ben Ahmed University of Oran 2, Oran, Algeria

[5] Laboratory of Physical Chemistry Macromolecular LCPM, Ahmed Ben Bella University of Oran 1, Oran, Algeria

Abstract. The main objective of this presentation is to study the oxidation of allylic alcohols with the use of zeolites of type MFI and more particularly the ZSM-5 by the combination of experimental techniques and molecular modeling. This type of zeolite (MFI) is generally used in separation and catalysis processes. This is due to the relatively medium pore size, its high specific volume, its low Si/Al ratio, which makes it possible to prepare materials rich in metal ions, and also its low cost of synthesis. In a first step, an experimental study was carried out by syntheses of ZSM-5 zeolites with low transition metal contents, as we have done a series of catalytic tests on the oxidation of allylic alcohols by these zeolite products. The various physicochemical characterization techniques give us information on the structure and texture of the products obtained and confirm the effectiveness of the catalytic tests. We also used molecular modeling, which is an essential tool for predicting the reactivity of these materials. Molecular modeling techniques have been carried out using Gaussian 09 software used in specific computer programs to evaluate physicochemical, energy properties, determine reaction pathways, identify reaction intermediates and transition states of molecules and clusters (zeolite, oxidizing agent, substrate). All the results obtained showed the bearing capacity of the experimental study (in vitro) with a computational study (in silico) to predict all the scientific information on the zeolite-substrate system.

Keywords: ZSM-5 · Zeolite · Catalysis · Characterization · Reactivity

B. Abdelbaki et al. (Eds.): SMSD 2017, *Proceedings of the Third International Symposium on Materials and Sustainable Development*, pp. 52–66, 2018.
https://doi.org/10.1007/978-3-319-89707-3_7

1 Introduction

In the field of chemistry, nanoscience is omnipresent, among the materials nanoporous, the zeolites which occupy a very important place. It is crystalline materials, generally aluminosilicates, of which there are several families ordered according to the morphology that their networks can have. The insufficiency of the charges (Z^-) induced by the isomorphic substitution of a few silicon atoms by aluminum atoms and balanced by a compensation of cations (M^+), called extra-lattice cations, which may be transition metals, alkaline, and alkaline-earth, or protons (H^+). The percentage of Si atoms on Al, the presence of other atoms in the lattice, the nature of the extraneous cations: M^+, Fe^+, Ni^+,…) (Le Dred 1982) and structural (the positions of the extra-cations cations, etc.) give these materials specific physicochemical properties making it possible to use them in very wide applications, such as catalysis, adsorption, ion exchange and gas separation.

In order to establish relations between their chemical compositions and their physicochemical adsorption and catalytic properties, on the other hand, a complete characterization of the zeolites appears to be involved.

There are a large number of spectroscopies (NMR spectroscopy, X-ray diffraction, ultraviolet and infrared spectroscopy (IR)). These techniques allow us to extract quantitative and qualitative information on the nature of the surface: acid or basic, number of active sites, the nature and size of elementary cells, etc.…).

Experimental results of oxidation of allyl alcohols with the methods of molecular modeling help us to predict many properties such as structures, energies, reaction mechanisms and spectroscopic observables. The determination of these properties enabled us to identify the transition states of the molecules during an oxidation reaction by M-ZSM-5, in order to improve the performance of our catalysts.

2 Materials and Methods

2.1 Synthesis of ZSM-5 in an Alkaline Fluoride Medium

The syntheses of zeolites of MFI type in fluoride medium proposed by (Guth et al. 1987). Crystallization of silicalite-1 without the incorporation of hetero-elements in the structure. The solubility of the silica is ensured by the presence of the fluoride ion introduced in the form of ammonium hydrogen difluoride (NH_4 HF_2). The hydrothermal synthesis of ZSM-5 zeolite in an alkaline fluoride medium produces smaller crystallites with faster crystallization rates.

The composition of the reaction mixture: 100 SiO_2, $A1_2OI_3$, 27 Na_2O, x KF, 40 TPA-Br, 7000 H_2O.

2.2 Preparation of ZSM-5 Zeolites Exchanged with Transition Metals

Cation exchange is an important operation since it makes it possible to modulate the adsorption and catalysis properties of the zeolites. Several studies on the thermodynamic and kinetic aspects as well as the parameters of the zeolite exchange process are published in the bibliography, the passage from the sodium or proton form to the MZ forms where M is a given cation and Z is the zeolite (usually metals of transition).

The previously synthesized ZSM-5 is exchanged with sodium as follows: two grams of synthesized zeolite are contacted with 200 ml of NaCl (one mole) solution for one day at room temperature with gentle stirring under the same conditions as previously (Lounis et al. 2006). The Na-ZSM-5 samples are later exchanged with metal ions by bringing two grams of the solid material into contact with a 40 ml solution of metal salts, the salts used are nitrates of metals. This cation exchange is carried out at 80 °C for two hours. The resulting M-ZSM-5 was filtered and washed several times. The final product is dried at 80 °C.

3 Characterizations of M-ZSM-5

3.1 Characterization by X-Ray Diffraction

The analysis was carried out using a Philips X-ray diffractometer PW-1130/00, working on the monochromatic Kα1 radiation of Copper (1.5406 Å). A counter (PM 8203) records a signal proportional to the intensity of the diffracted ray (Fig. 1).

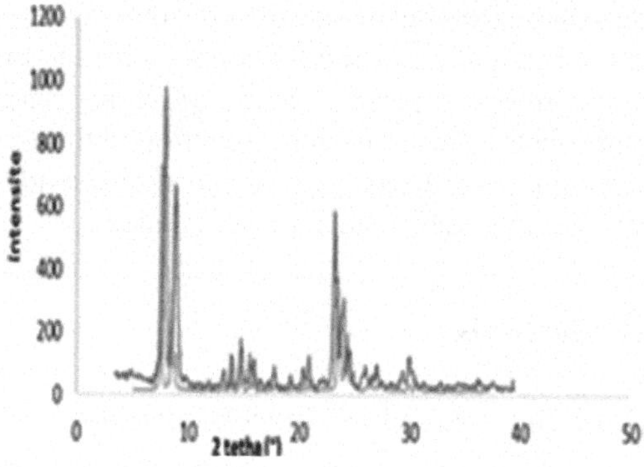

Fig. 1. X-ray diffractograms of zeolites (black) ZSM-5, (red) Ni-ZSM-5

3.2 Characterization by BET

The zeolites before being subjected to this analysis are degassified at a temperature of 350 °C. for a duration of 16 h, in order to eliminate all traces of water. Nitrogen adsorption-desorption isotherms at 77 K in a relative pressure range (P/P_0) varying from 0.0–1, allow to characterize the textural properties of microporous materials, i.e. the size of pores, surface area and pore volume. ZSM-5 is a zeolite of medium porosity because it has tubular channels with a maximum aperture of 0.56 nm, it is good to know that the active sites of these catalysts are located both on the external surface of the crystallites, then in micropores (Figs. 2 and 3) (Csicsery 1982).

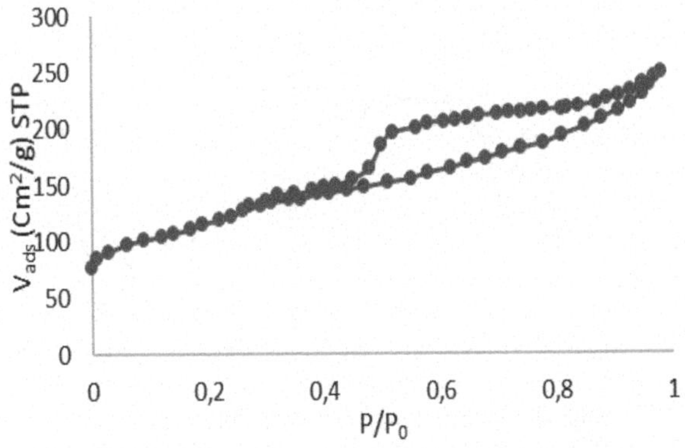

Fig. 2. Adsorption isotherms at 77 K of zeolites ZSM-5

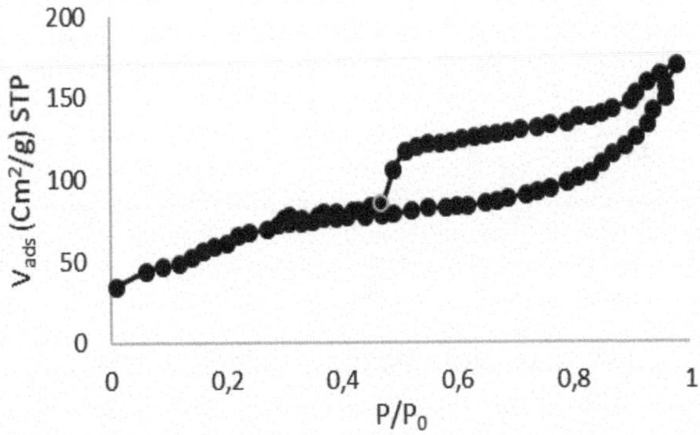

Fig. 3. Adsorption isotherms at 77 K of zeolites Ni-ZSM-5

3.3 Characterization by IR Spectroscopy

Fourier transform infrared spectroscopy is based on the absorption of infrared radiation by the catalysts analyzed. Through the detection of the characteristic vibrations of the chemical bonds, it makes it possible to analyze the chemical functions present in the materials. For this purpose, we performed this technique using a Perkin-Elmer FT 1720 spectrometer in a range of 400 to 4000 cm^{-1} with a resolution of 2 cm^{-1}. The samples were packaged as dispersions in KBr pellets (Figs. 4 and 5).

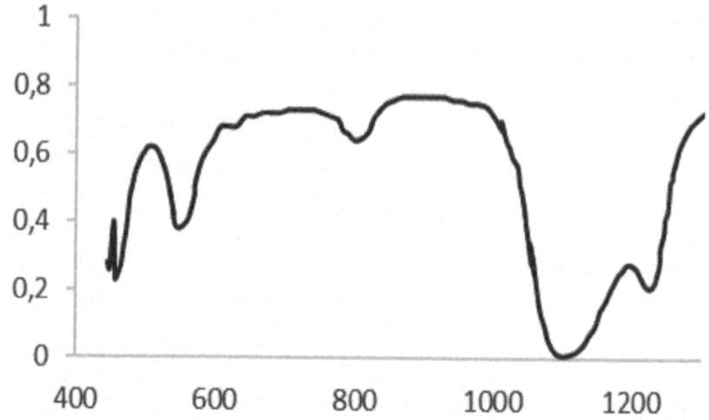

Fig. 4. Infrared analysis of ZSM-5 zeolites

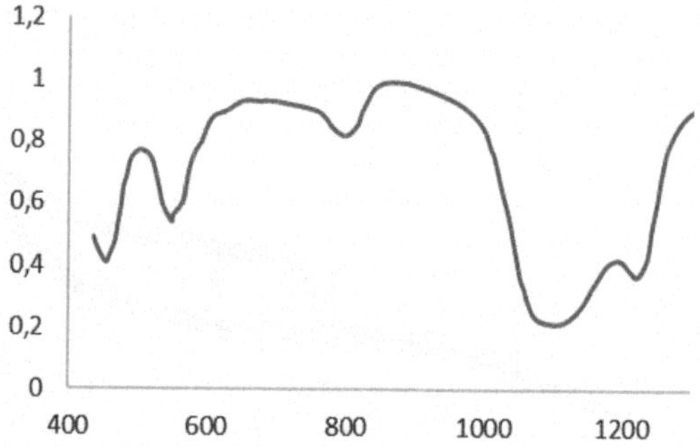

Fig. 5. Infrared analysis of Ni-ZSM-5 zeolites

3.4 Characterization by SEM Microscopy and TEM

Detailed analysis of the images provided by scanning electron microscopy and transmission electron microscopy of the ZSM-5 and Ni-ZSM-5 shown in Figs. 6, 7, 8 and 9 confirms the presence of metal elements in these zeolites.

Fig. 6. SEM images of ZSM-5

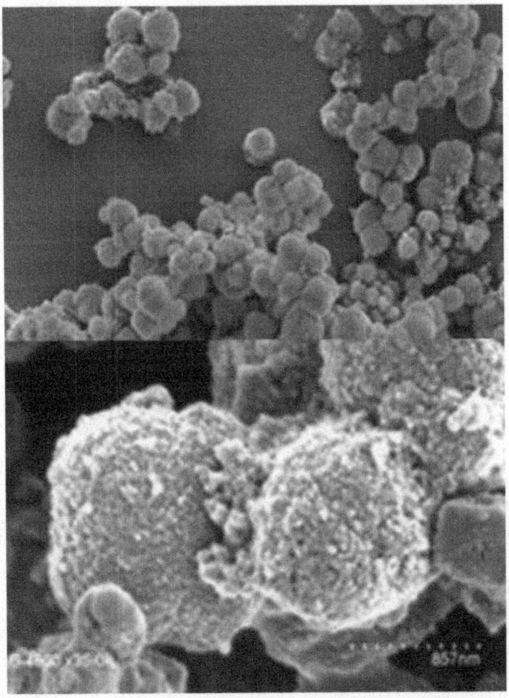

Fig. 7. SEM images of Ni-ZSM-5 exchanged

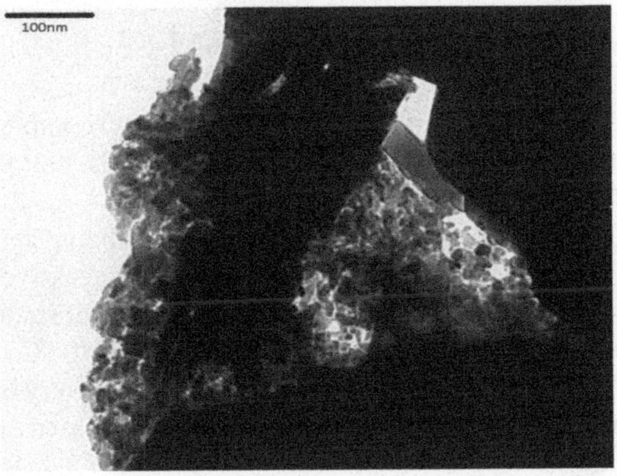

Fig. 8. TEM images of ZSM-5

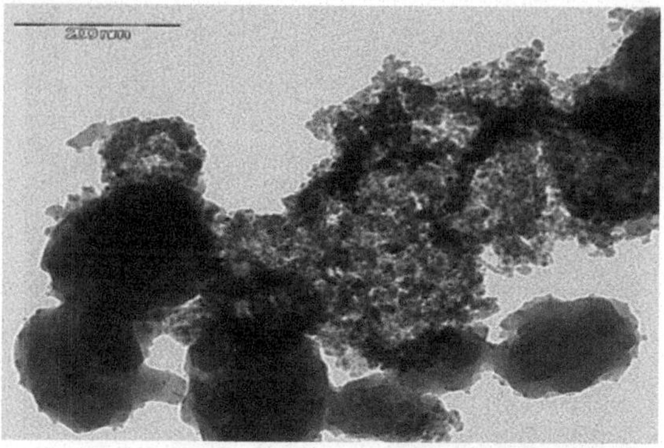

Fig. 9. TEM images of Ni-ZSM-5

3.5 Characterization by X-Ray Fluorescence

Analysis by X-ray fluorescence spectrometry highlights the chemical composition of our samples (Table 1) and confirms the presence of its constituents.

Table 1. Chemical analysis of ZSM-5 and the form exchanged with Ni

Compounds	Si %	Al %	Na %	O %	K %	Ni %	Si/Al %
ZSM-5	44.4	1.47	1.25	52.5	0.23		30.20
Ni-ZSM-5	44.5	1.3	0.24	52.3	0.53	3.5	34.23

4 Oxidation of Allyl Alcohols by M-ZSM-5

The oxidation of the alcohols to their corresponding carbonyl compounds is an important reaction in chemistry (Hudlicky 1999), for this purpose the following protocol with different oxidants (Belarbi et al. 2014):

ZSM-5 (3 mg, 0.002 mmol of Ni), 5 ml of toluene, oxidizing agent (4 mmol of 35% aqueous H_2O_2) or SPC (Na_2CO_3.1.5 H_2O_2) are introduced successively into a 25 ml flask equipped with a refrigerant, after stirring for 15 min, the substrate (1 mmol) is then added. The reaction mixture is brought to a temperature of 80 °C. in an oil bath, with magnetic stirring. The evolution of the reaction is carried out every half hour by analysis on TLC plates. Developer: UV lamp (546 nm) and developer based on phosphomolybdic acid diluted in ethanol.

At the end of the reaction, the solution is filtered through a small column of alumina eluted with ether to reduce the remaining oxidant. The solvents are removed on a rotary evaporator. The concentrated filtrate is purified by flash chromatography or by preparative plates. The eluent used for both types of chromatography is petroleum ether/ethyl acetate (90/10).

Gas chromatographic (GC) analyzes coupled with mass spectrometry were carried out on a Schimadzu GC-14 B chromatogram equipped with an Apiezon L column and an FID flame ionization detector. The results of the oxidation reactions of

5alphacholestanol in the presence of Ni-ZSM-5 H_2O_2, (Na_2CO_3.1.5 H_2O_2) (Chen et al. 1996) were tabulated in Table 2.

Table 2. Results of oxidation reactions of 5alphacholestanol in the presence of Ni-ZSM-5, H_2O_2 and (Na2CO3.1.5 H_2O_2)

Sample	Substrat	Oxidants	t (h)	Formed product and yield (%)
1	5alphacholestanol	H_2O_2	6	5alphacholestanone (50)
2		SPC	6	5alphacholestanone (81)

5 Oxidation of Allyl Alcohols by M-ZSM-5

5.1 Modèle de Cluster et Détails de Calculs

All calculations optimizations, vibration frequency and fixed point energy (i.e., reagents, products, reaction intermediates and transition state) were performed with the gaussian 09 program package, using a functional hybrid B3LYP which is an element of approximation of the interactions of exchange-correlation in order to model molecular orbitals and to reach the structure of the most stable intermediate state. The morphological architecture of the zeolite represents a major role in the catalytic activity and in selectivity.

In particular, the frame of the ZSM-5 elementary cell is crystallographically composed of five tetrahedral (T), the distinct sites of the 128 T according to the theoretical works (van Koningsveld et al. 1987), which relates to the intersection of the right and zigzag channels (giving the formation of small pores) having free diameters equal to 5.4 × 5.6 Å and 5.1 × 5.4 Å, respectively, with the Pnma group, the ZSM-5 zeolite is of orthorhombic form and with parameters of (Å), b = 19.8243 (Å), c = 13.4034 (Å). And those of Ni-ZSM-5 are: a = 20.0792 (Å), b = 19.9425 (Å), c = 13.4420 (Å). due to the existence of channels and to these crystallographic properties, the MFI molecular sieve is an essential element in the catalysis domain.

For the efficiency of the calculations, we used ONIOM (ONIOM3) method with DFT (Kohn et al. 1975) (B3LYP) and base 6-31G (d, p) for the high level at 5T which covers the window of 10 membranes of zigzag channels and which represent the active site of the zeolite and B3LYP with a base split-valence set. The Los Alamos LanL2DZ (lanL2DZ) is used for nickel atoms. To represent the van der Waals interactions and the influence of zeolitic networks on the substrate and the oxidant; the universal force field (UFF) was applied and also for the low level calculations (hydrogen atoms). For the case of five tetrahedra, the base 6-31G (d, p) is applied for the whole cluster.

On the other hand, to generate the Bronsted acid sites (H^+) by an isomorphic substitution, the silicon atoms of the framework can be partially substituted by other elements such as aluminum Al^{+3} (the T5 site in the mesh element is the most favorable position for the substitution of the Si^{+4} atom by Al^{+3}). The negative charge created and compensated by the H^+ cations. The resulting models are shown in Fig. 10. These Bronsted sites can be exchanged by Ni^+ nickel atom or nickel NiO^+ oxides. In order to obtain neutral clusters, we have saturated the oxygen atoms with hydrogen atoms. The final models of the clusters are shown in Fig. 11.

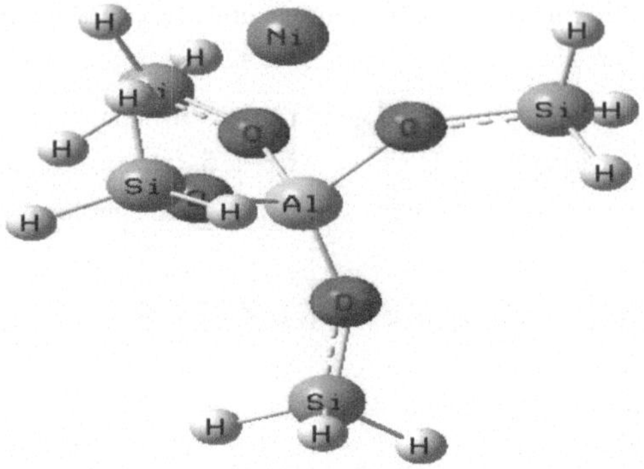

Fig. 10. Optimized cluster geometry of $[Ni]^{+1}$ ZSM-5^{-1}

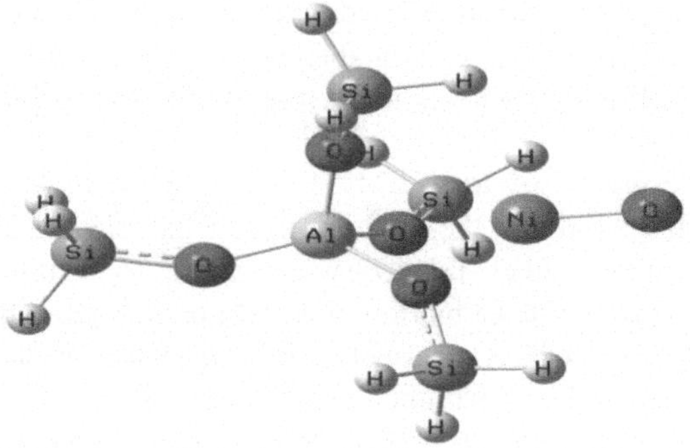

Fig. 11. Optimized cluster geometry of $[NiO]^{+1}$-ZSM-5^{-1}

The first step for the oxidation of 5alphacholestanol to 5alphacholestanone by SPC on the Ni-ZSM-5 and NiO-ZSM-5 catalyst consists in optimizing the geometry of 5T, substrate and oxidant since the latter were allowed to relax at the same time as the terminal hydrogen atoms that are geometrically fixed for the forced to rotate in the TO direction of the next Si site. The vibration frequencies were calculated at a temperature of 298°K due to the lack of experimental and thermochemical information in the literature.

The zero-point energy corrections (ZPE) obtained after the frequency calculations were added to the total load, the vibration frequencies and the free energies of Gibbs are obtained by vibration analysis. To simulate the experimental experiment, all frequency values were scaled by the factor (0.9613) (Pantu et al. 2007). Mulliken atomic charges and Mulliken atomic spin (Jakkapan et al. 2013) densities were acquired by Mulliken population analysis.

To obtain electronic configurations of the nickel atoms, we used the natural orbital regime (NBO). The convergence criteria take into account the maximum force and root-mean-square strengths (rms), the maximal displacement and rms displacement values in the Gaussian software 09 are: 0.000450, 0.000300, 0.001800 and 0.001200 respectively.

The theoretical calculation protocol used in this study is as follows: Zero-point energies (ZPE's) were calculated with many SM numbers for the cluster system to obtain the spin multiplicity (SM) and single point energy (SPE). The cluster and the adsorbed molecules, SPC and 5alphacholestanol, were then carefully geometrically optimized by means of geometric equilibrium (EG) calculations.

For a preset distance, we have located the adsorbed molecules, SPC and 5alphacholestanol closest to the active site of the catalyst. The calculation of intrinsic coordinates of the reaction (IRC) gives us the energy profile, also to find the transition points TS (the highest values of energy) and the stationary points EGs (the minimum energy values). to check whether the reaction links the right products with reagents, The SPE singular energy point is calculated when SPC and 5alphacholestanol are in the vicinity of the zeolite.

The relative energy is illustrated by the following formula:

$$\Delta E = E_{System-}\left(E_{Cluster} + E_{adsorbate}\right)$$

Where E_{System} is the calculated energy of the optimized structure, $E_{Cluster}$ includes the energy of the entire cluster (zeolite-alcohol), and $E_{Adsorbat}$ includes substrate-adsorbent. Finally, we re-optimize the geometry with the minimum energy on the energy profile (Fig. 12), without fixing the coordinates of the reaction. After the cluster optimization

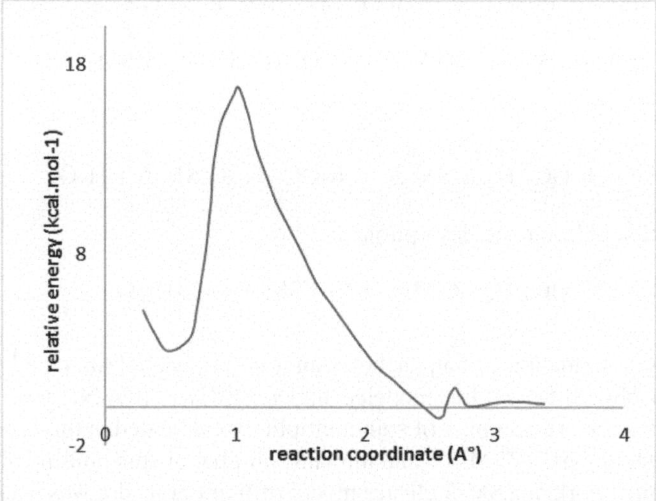

Fig. 12. Energy profile for the transfer of protons from CH_3O of grafted species $OC_{28}H_{47}$ to form 5alphacholestanone and SPC (step 5)

calculations, the highest energy of the energy profile is the starting point of a transition state, the transition structures have been acquired using Berny's algorithm implemented in the software.

The single point energy values for all SM numbers of the Ni-ZSM-5 cluster are calculated, the lowest SPE was determined to be 6 for the Ni-ZSM-5 and the Ni-ZSM-5 cluster comprising 5alphacholestanol and SPC molecules, which means that there are five odd electrons. This last lowest energy was accepted for all calculations. The optimized geometries for the Ni-ZSM-5 cluster modeled as $[(SiH_3)_4\text{-}AlO_4\text{-}Ni]$ and calculated with a neutral load, see Fig. 10. The values of the silica-oxygen distances of the Ni-ZSM-5 cluster were calculated to be 1.54 Å.

Selective oxidation of 5alphacholestanol to 5alphacholestanone by SPC on Ni-ZSM-5:

Step 1: adsorption of SPC (oxidant)

$$\text{SPC} + \text{Ni-ZSM-5} \rightarrow \text{Ni-SPC-ZSM-5} \tag{1}$$

Step 2: Dissociation of SPC and desorption of water

$$\text{Ni-SPC-ZSM-5} \rightarrow \text{Ni-O-ZSM-5} + H_2O \tag{2}$$

Step 3: adsorption of 5alphacholestanol

$$\text{Ni-O-ZSM-5} + C_{27}H_{48}OH \rightarrow \text{O-Ni-OH}C_{27}H_{47}\text{-ZSM-5} \tag{3}$$

Step 4: Transfer of protons from OH to form grafted species

$$\text{O-Ni-OH}C_{27}H_{47}\text{-ZSM-5} \rightarrow \text{HO-Ni-O}C_{27}H_{47}\text{-ZSM-5} \tag{4}$$

Step 5: The transfer of protons from $OC_{27}H_{47}$ to form 5alphacholestanone and water

$$\text{HO-Ni-O}C_{27}H_{47}\text{-ZSM-5} \rightarrow H_2\text{O-Ni-O}C_{27}H_{46}\text{-ZSM-5} \tag{5}$$

Step 6: Water desorption

$$H_2\text{O-Ni-O}C_{27}H_{46}\text{-ZSM-5} \rightarrow \text{Ni-O}C_{27}H_{46}\text{-ZSM-5} + H_2O \tag{6}$$

Step 7: 5alphacholestanone desorption

$$\text{Ni-O}C_{27}H_{46}\text{-ZSM-5} \rightarrow \text{Ni-ZSM-5} + C_{27}H_{46}O \tag{7}$$

The catalytic oxidation of 5alphacholestanol to 5alphacholestanone by SPC at the $[Ni]^{+1}$ site in Ni-ZSM-5 zeolite modeled under $[(SiH_3)_4\text{-}AlO_4\text{-}Ni]$ was theoretically studied in this work. The number of spin multiplicity calculated in this work is equal to six (6) for clusters $[Ni]^{+1}\text{-ZSM-5}^{-1}$ and the same number of spin multiplicity used in the literature. In this work, a ZSM-5 cluster model of five tetrahedra was used to generate 5alphacholestanone from 5alphacholestanol with SPC, also obtained very similar structural and energy characteristics of many 5T (Soltanali et al. 2015) adsorption complexes, as mentioned above, Table 3 shows the spin densities and the Mulliken atomic charges of the nickel atom of the cluster for all steps for equilibrium geometry and transition state geometries. As can be seen from this table, the nickel atoms have a high spin density, which means that a number of odd electrons are centralized on the nickel atoms. Moreover, the atomic charges of the nickel atom for the clusters after the reactions are slightly superior to those of the initial clusters of ZSM-5, this gives a more positive sign on the nickel atoms in the cluster throughout the reactions by analogy with the initial ZSM-5 clusters. The activation barriers for the reaction path involved in the oxidation of 5alphacholestanol to 5alphacholestanone are also tabulated in Table 3. Prior to the oxidation of 5alphacholestanol, the [Ni] sites promote the dissociation of SPC leading to the formation of [Ni-O] sites. The extra-oxygen species formed on the ZSM-5 clusters are the reactive centers for the oxidation of 5alphacholestanol to 5alphacholestanone. Dissociation of the SPC molecule at $[Ni]^{+1}$ sites has been studied on the same cluster models and discussed in previous theoretical studies. The most important reaction for the oxidation of 5alphacholestanol is the activation of the 5alphacholestanol O-H bond. In this work, the calculation of the value of the activation barrier for the transfer of protons from OH of 5alphacholestanol adsorbed on the [NiO] site to form a grafted species (step 4) is 10.28 $kcal.mol^{-1}$. This value is essentially lower than the value of the dissociation energy of the C-H bond of the grafted species $OC_{27}H_{47}$. Also indicate that the Ni-ZSM-5 catalyst has a positive impact on the oxidation of 5alphacholestanol to 5alphacholestanone. In the next step, activating grafted $OC_{27}H_{47}$ to form 5alphacholest-anone and water on the surface (step 5) required a very high energy barrier (E_{act} = 39,

Table 3. Mulliken atomic charges, spin densities and activation barriers for the EG and TS structures of the 5alphacholestanol selective oxidation mechanism to 5alphacholestanone by SPC on the Ni-ZSM-5 catalyst.

Sample	Geometry	Charge	Spin density	Activation barriers ($kcal.mol^{-1}$)
Ni-ZSM-5	EG	+0.65	5.832	0
Step 1: adsorption of SPC	EG	+1.32	5.255	0
Step 2: dissociation of SPC	EG	+1.22	5.146	23
Step 3: adsorption 5alphacholestanol	EG	+1.41	5.346	0
Step 4: transfer of OH proton to the grafted species	EG	+1.31	5.688	41
	TS	+1.19	5.576	
Step 5: transfer of protons from $OC_{28}H_{47}$ to 5alphacholestanone and water	EG	+0.91	4.664	201
	TS	1.12	4.854	
Step 6: Water desorption	EG	+0.57	4.971	59
Step 7: desorption of 5alphacholestanone	EG	+0.76	4.856	168

19 kcal.mol^{-1}), this is possibly due to the fact that the formation of the [HO-Ni-$OC_{27}H_{47}$] grafted species (step 4) is a reaction step which is strongly thermodynamically favored at the site ($\Delta E = -134{,}08$ kcal. mol^{-1}). After formation of 5alphacholestanone and H_2O on the catalyst, the water desorbs first with a desorption barrier value equal to 17.45 kcal. mol^{-1}, then desorption of 5alphacholest-anone from the catalyst surface exhibited a higher desorption barrier (36.56 kcal.mol^{-1}).

6 Conclusions

In recent years, a relatively widespread approach to research projects involves combining theoretical modeling with experimental techniques for the identification and characterization of study objects. More particularly, in the field of heterogeneous catalysis, to improve the physico-chemical properties of the synthesized products on the one hand and to solve the environmental and economic problems on the other hand.

Physico-chemical analytical techniques were carried out after the synthesis of MFI zeolites and on catalytic tests in order to understand the effectiveness of these catalysts. X-ray diffraction tests the crystallinity and purity of the zeolitic materials obtained, this technique shows that the characteristic peaks of the synthesized products located in the region $2\theta = (22 - 24°)$ present higher intensities with respect to the region $2\theta = (6 - 10°)$, this may be due to the fact that the organic structuring agent is occluded in this region at higher inter-reticular distances. Scanning and transmission electron microscopy (SEM, TEM) allowed us to observe the morphology and the distribution of the structural elements. We have sphere sizes ranging from 1 to 3 microns resulting from working in alkaline fluoride medium. The analysis showed a distribution of the constituents almost identical in different measuring zones. Nitrogen adsorption/desorption manometry confirmed the presence of crystal agglomerates constituting the layers of the catalysts (absence of an amorphous phase). The optimized zeolite analysis allowed us to calculate the values of the specific surface area. It is of the order of 350 m^2/g. Analysis by infrared spectroscopy of the solid showed two bands of vibrations close to 550 cm^{-1} and 1225 cm^{-1} confirm the structure of the ZSM-5 zeolites.

Analysis of the results obtained shows that the oxidation of the allylic alcohols provides the corresponding ketones with very high yields and complete conversions.

In the second part of this work and after calculations of energy minimization (the stable state of the molecules), molecular modeling makes it possible to visualize the organic species used during this work in order to determine their different characteristics (shape, size, maximum congestion, etc.). It also makes it possible to visualize the position of the structuring species inside a microporous framework. The various molecular modeling calculations were performed using a BullX computing station with gaussian9 software.

The essential reaction steps involved in the selective catalytic oxidation of 5alphacholestanol to 5alphacholestanone by SPC at nickel sites on Ni-ZSM-5 zeolite represented by [$(SiH_3)_4$-AlO_4-Ni] were studied using the ONIOM3 (B3LYP/6-31G (d,p), LANL2DZ,UFF) calculations. The optimized structures are in excellent agreement with the mesh parameters of the experimental structures. The choice of the reactive center was based on an earlier study of the dissociation of SPC on the exchanged metal ZSM-5. The activation and interaction energies of the reaction intermediates are presented. It is found that the reactive cation Ni^{+1} interacts more strongly with alcohol than Bronsted H^{+1} and that step 5 (5alphacholestanone formation) is the most difficult reaction between seven steps with high activation energy ($E_{act} = 39.19$ kcal.mol^{-1}) for the decomposition of the grafted species. These characteristics show that this type of catalyst can have good performance and efficiency for the selective oxidation of the steroidal alcohol to bioactive molecules. We have mentioned that the theoretical calculations are in agreement with the experimental data.

The DFT methods prove their effectiveness in determining the structures molecular (intermediate and transitions), frequencies of vibration and energy variation. They provide a very effective theoretical tool for practical applications for comparing and completing experimental studies. The studies we carried out on the Ni-ZSM-5-substrate clusters showed this clearly.

References

Yahia, A.: Effect of solid concentration and shear rate on shear-thickening response of high-performance cement suspensions. Constr. Build. Mater. **53**, 517–521 (2014)

Lounis, Z., Riahi, A., Djafri, F., Muzart, J.: Chromium-exchanged zeolite (CrE-ZSM-5) as catalyst for alcohol oxidation and benzylic oxidation with t-BuOOH. Appl. Catal. Gen. **309**, 270–272 (2006)

Chen, J.D., Lempers, H.E.B., Sheldon, R.A.: Synthesis, characterization and catalytic oxidation of alcohols with chromium-substituted aluminophosphates. J. Chem. Soc., Faraday Trans. **92**, 1807 (1996)

Le Dred, R.: Les surfaces des solides Minéraux. Soc. Fr. Miner., Crist. 297 (1982)

Colloids and Surfaces a Physicochemical and Engineering Aspects 453, 86–93

Csicsery, S.M.: In: Rabo, J.A. (ed.) Zeolite Chemistry and Catalysis, vol. 12, p. 171. American Chemical Society (1976)

Hudlicky, M.: Oxidations in Organic Chemistry. ACS Monograph 186 (1999)

van Koningsveld, H., van Bekkum, H., Jansen, J.C.: On the location and disorder of the tetrapropylammonium (TPA) ion in zeolite ZSM-5 with improved framework accuracy. Acta Crystallogr. B **43**, 127 (1987)

Jakkapan, S., Jumras, L.: Structures and mechanisms of the dehydration of benzaldoxime over Fe-ZSM-5 zeolites: a DFT study. J. Struct. Chem. **24**, 1307–1318 (2013)

Soltanali, S., Halladj, R., Ektefa, F.: A DFT exploration of the adsorption of methanol on the brönsted acid site of nanostructured H-ZSM-5 catalyst: probing the effect of cluster size on the O-H···O hydrogen bond based on NMR and NQR parameters. J. Clust. Sci. **26**, 565–579 (2015)

Pantu, P., Boekfa, B., Limtrakul, J.: The adsorption of saturated and unsaturated hydrocarbons on nanostructured zeolites (H-MOR and H-FAU): an ONIOM study. J. Mol. Catal. Chem. **277**, 171 (2007)

Belarbi, H., Lounis, Z., Hamacha, R., Bengueddach, A., Trens, P.: Textural properties of ZSM-5 nanocrystals prepared in alkaline potassium fluoride medium (2014)

Kessler, H., Chezeaua, J.M., Guth, J.L., Strub, H., Coudurier, G.: Nmr and ir study of B and BAl substitution in zeolites of the MFI-structure type obtained in non-alkaline fluoride medium. Zeolites **7**(4), 360–366 (1987)
Kohn, W., Sham, L.: Self-consistent equations including exchange and correlation effects. J. Phys. Rev. **137**, A1133 (1965)

Influence of Deposition Time on the CdS Thin Films Prepared by a Chemical Bath Deposition (CBD)

D. Haouanoh[1(✉)], R. Talalghil[1,2], F. Bensouici[1,3], and M. Toubane[1,3]

[1] URMPE Unit, University M'Hamed Bougara Boumerdes, 35000 Boumerdes, Algeria
haouanoh.djedjiga@yahoo.com

[2] Institute of Electrical and Electronic Engineering, University M'Hamed Bougara Boumerdes, 35000 Boumerdes, Algeria

[3] Physics Department, Faculty of Sciences, University M'Hamed Bougara Boumerdes, 35000 Boumerdes, Algeria

Abstract. Cadmium sulfide thin layers are frequently used in photovoltaic, especially in the realization of Cu2ZnSnS4 (CZTS) based solar cells as buffer layer. In this work, cadmium sulphide (CdS) thin films have been carried out by using chemical bath deposition (CBD) method. They were deposited onto glass substrates from chemical bath containing cadmium sulfate, thiourea and ammonia at pH = 12. The temperature of the bath was fixed at 60 °C, and under 150 RPM stirring, the time deposition was varied (10 mn–40 mn). X-ray diffraction revealed the cubic and hexagonal structures formation with secondary phase CdO. AFM images of CdS thin films show that the substrate is covered with granular structures of different size, the root mean square (Rms) value prepared at different time (10 mn, 20 mn, 30 mn and 40 mn) are, 26.05, 25.08, 19.93 and 38.55 nm. High transmission is obtained in the visible and near infra-red.

Keywords: CdS · CBD · Deposition time · AFM · DRX

1 Introduction

Cadmium sulfide (Oladeji et al. 2000) de is a chemical compound which has the formula CdS. It has a yellow color and is an n–type semiconductor with an optical band gap of 2.40 eV at room temperature (Mathew et al. 1995), (Hayashi e al. 1988).

These last years, the CdS thin layers have received intensive attention due to their very important role in the photovoltaic technology (Bätzner et al. 2001). It has been used in several types of thin film solar cells, such as, Cu (In, Ga)Se_2 (Repins et al. 2008), CdTe (Contreras et al. 2002), and Cu_2ZnSnS_4 CZTS (Mahajan et al. 2017). Specifically, optoelectronic devices (Chu et al. 1991), photo-detectors (Pal et al. 1997), optical filter and multi-layer light emitting diodes (Calixto and Sebastian 1999) and field effect transistors (Schön et al. 2001), (Levinson et al. 1982).

B. Abdelbaki et al. (Eds.): SMSD 2017, *Proceedings of the Third International Symposium on Materials and Sustainable Development*, pp. 67–74, 2018.
https://doi.org/10.1007/978-3-319-89707-3_8

There are several methods for depositing CdS thin films like: vacuum evaporation (Chu and Chu 1995), laser evaporation (Wang et al. 1992), electro-deposition (Edamura and Muto 1993), spray pyrolysis (Chamberlin and Skarman 1966), (Subbaramaiah and Raja 1994), (Feigelson et al. 1977) and chemical bath deposition (CBD) (Ramaiah et al. 2001), (Feitosa et al. 2004), (Cortes et al. 2004), (Prabahar and Dhanam 2005). Among these various techniques, CBD is a simple and slow process, no requirement for sophisticated instruments, inexpensive and safe (Ramaiah, et al. 2001), and has been used to procure large area homogeneous, adherent and reproducible thin films (Hiie et al. 2006), it requires a low deposition temperature (Çetinörgü et al. 2006), and is considered an excellent technique for depositing nano-crystals. The CBD method is based on controlled precipitation of the material; the slow release of sulfide ions through controlled thiourea hydrolysis and the presence of the cadmium salt and NH_3 facilitate uniform precipitation on the substrate (Ximello-Quiebras et al. 2006), (Ximello-Quiebras et al. 2004). In this paper, CdS thin films were deposited by chemical bath deposition technique onto glass substrates. The influence of the time deposition on the structure, morphology and optical properties CdS layers were investigated.

2 Experimental Procedure

The CdS thin films have been deposited by CBD technique onto glass substrates after being cleaned in acetone, ethanol and methanol ultrasonically. The chemical bath containing 8 ml of cadmium sulfate ($3CdSO_4$, $8H_2O$) (0.02M), 16 ml of thiourea ($CS(NH_2)_2$) (1M), 8 ml ammonia (NH_4OH) and 68 ml distillation water. The pH of the solution was 12. The glass substrates were kept vertically in the chemical bath, the temperature of the deposition process was fixed at 60 °C while the deposition duration is varied between 10 and 40 min.

The solution was stirred during the deposition in order to ensure homogenous distribution of the reagents. After finalizing the deposition process, the CdS films were washed with distilled water ultrasonically for 1 min in order to remove the loosely adhered CdS particles on the film and dried in air. Finally, the all films were annealed in the air atmosphere at the temperature fixed at 400 °C.

3 Film Characterization

The crystal structure of the films was determined by X-ray diffraction (XRD) with a Philips X'Pert diffractometer with theta-2theta geometry equipped with a Cu-Kα radiation source. The film morphology and Film roughness were determined by Atomic Force Microscope (Pacific Nanotechnology, advancing nanotechnology). Optical measurements were carried out with UV-Visible type Optizen 2120 UV.

4 Results and Discussion

4.1 Structural Properties

Figure 1 shows the patterns of the four CdS thin films deposited with different deposition time (10–40 min). We note the presence of nine peaks. The peaks located at 26.5° and 52.1° correspond to the CdS cubic planes (111) and (311). (According to the JCPDS card Nº 01-098-0440), or the hexagonal plans (002) and (112). (According to the JCPDS card Nº. 01-080-0006). The peaks attributed at 24.9° and 28.3° correspond to the reflection from (100) and (101) present the CdS hexagonal planes. The peak present at 44.0° corresponds to the CdS cubic plane (220). The five peaks confirm the mixture of two structures CdS hexagonal and cubic with results reported by other groups (Enríquez and Mathew 2003), (Metin and Esen 2003) There are four peaks attributed to the plans (111), (200), (220) and (311) attributed to 2θ = 33.0°, 38.3°, 55.3° and 65.9º respectively, these peaks as assigned to the CdO cubic structure (according to the JCPDS card Nº. 01-075-0592). We can also notice from the diffraction spectra that the peaks intensity increases with enhance of time deposition, and this is due to the growth of material involved in the diffraction process.

The crystallites size ($D = k\lambda/B\cos\theta$) for all films were calculated by means of Scherrer's equation (Cullity 1978), where k is the poly-crystals, λ is the X-ray wavelength, B is the full-width at half maximum (FWHM) of the respective diffraction peak in radians and θ is the Bragg angle.

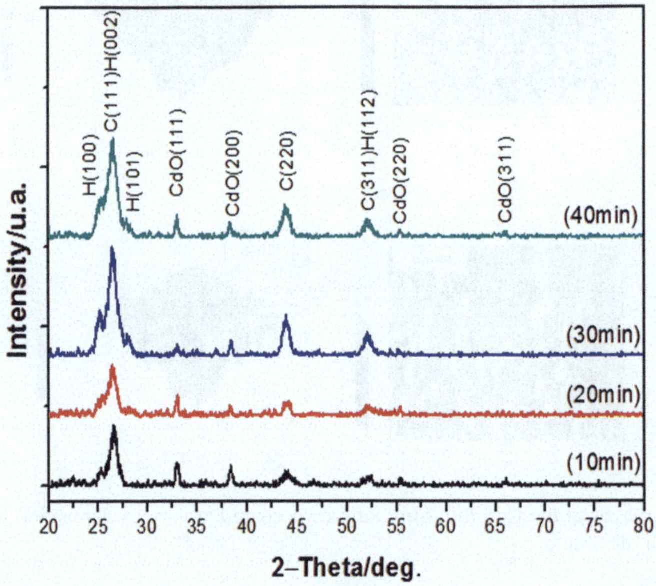

Fig. 1. (XRD) spectra for CdS thin films prepared at different deposition times (10 min, 20 min, 30 min, 40 min).

Fig. 2. AFM images of the CdS thin films surface prepared at various deposition times (10 min, 20 min, 30 min, 40 min).

The crystallite size is estimated from the most intense peak which corresponds to the plane (111) for CdS cubic, the most intense peak which corresponds to the plane (101) for CdS hexagonal and the most intense peak which corresponds to the plane (111) for CdO cubic. The obtained results are reported in Table 1. It can be seen the crystallites size' decrease of the CdS cubic and the crystallites size' increase of the CdS hexagonal with the deposition time increase. The results can be explained by the crystal structure transformation from cubic to hexagonal. The size of CdO crystallites is stable for all films.

Table 1. Grain size of CdS cubic, CdS hexagonal and CdO structures for CdS thin films deposited with various deposition times.

Deposition time		10 min.	20 min.	30 min.	40 min.
Crystallites size D (A°)	CdS C	373.39	287.50	219.23	111.65
	CdS H	/	297.05	369.30	426.68
	CdO	132.37	132.37	132.37	132.24

4.2 Surface Morphology

Figure 2 shows the 2D and 3D AFM images in 3 μm × 3 μm scale for CdS thin films prepared at different deposition time, from these images we can observe compact surface with the bigger grains. The R_{ms} values of CdS thin films deposited at various deposition times 10, 20, 30 and 40 min, are 26.07, 25.09, 19.93 and 28.38 nm respectively. It can be concluded that the R_{ms} roughness decrease with deposition time increase from 10 min to 30 min. The result could be made clearer by grain density enhancement.

Beyond 30 min, it can be seen an increase in the R_{ms} with increase deposition time this result can be explained by the grains' agglomeration.

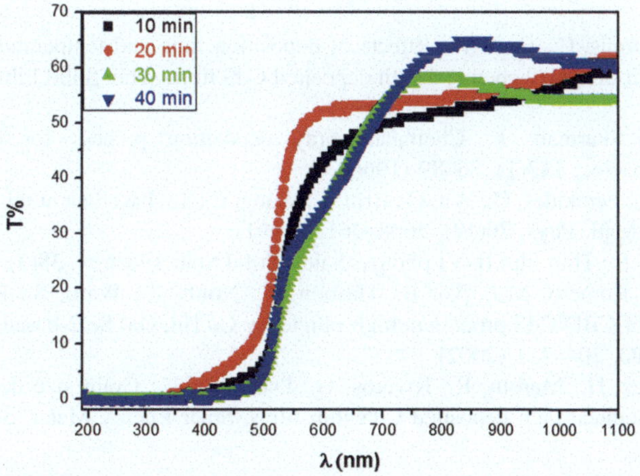

Fig. 3. CdS transmission spectra for CdS thin films deposited at different times (10 min, 20 min, 30 min, 40 min).

4.3 Optical Properties

CdS thin films prepared with different deposition times in the wavelength range of 190–1100 nm transmittance spectra are shown in Fig. 3. The absorption edge between 480 nm and 500 nm for all films to (close to CdS room-temperature energy gap 2.40–2.50 eV) (Bhargava 1997), (Pankove 1971). With the clean yellow- orange color of the films. The average transmission in the wavelength range 500–1100 nm has values between 50% and 65%. We can see the transmittance increase with deposition time in the wavelength range 650 nm and 950 nm.

5 Conclusions

In this report CdS layers were deposited on glass substrates by CBD with varied deposition times. The obtained results show that increasing time yields to a cubic and hexagonal CdS structures, with the presence of CdO phase. The crystallite size of the CdS cubic structure decreases with deposition time increase, contrary to CdS hexagonal structure. The AFM images showed a compact surface with bigger grains. The R_{ms} values decrease with deposition time from 10 min to 30 min. Beyond 30 min the R_{ms} roughness increase with the deposition time increase. Transmittance spectra illustrated that the absorption edge is between 480 nm and 500 nm, and the transmission increases with deposition time in the wavelength range 650 nm and 950 nm.

References

Bätzner, D., Romeo, A., Zogg, H., Wendt, R., Tiwari, A.: Development of efficient and stable back contacts on CdTe/CdS solar cells. Thin Solid Films **387**(1), 151–154 (2001)

Bhargava, R.: Properties of wide bandgap II-VI semiconductors. IET (1997)

Calixto, M., Sebastian, P.: A comparison of the properties of chemical vapor transport deposited CdS thin films using different precursors. Solar Energy Mater. Solar Cells **59**(1), 65–74 (1999)

Çetinörgü, E., Gümüş, C., Esen, R.: Effects of deposition time and temperature on the optical properties of air-annealed chemical bath deposited CdS films. Thin Solid Films **515**(4), 1688–1693 (2006)

Chamberlin, R., Skarman, J.: Chemical spray deposition process for inorganic films. J. Electrochem. Soc. **113**(1), 86–89 (1966)

Chu, T., Chu, S.S., Ferekides, C., Wu, C., Britt, J., Wang, C.: 13.4% efficient thin-film CdS/CdTe solar cells. J. Appl. Phys. **70**(12), 7608–7612 (1991)

Chu, T.L., Chu, S.S.: Thin film II–VI photovoltaics. Solid State Electron. **38**(3), 533–549 (1995)

Contreras, M.A., Romero, M.J., To, B., Hasoon, F., Noufi, R., Ward, S., Ramanathan, K.: Optimization of CBD CdS process in high-efficiency Cu (In, Ga) Se 2-based solar cells. Thin Solid Films **403**, 204–211 (2002)

Cortes, A., Gomez, H., Marotti, R., Riveros, G., Dalchiele, E.: Grain size dependence of the bandgap in chemical bath deposited CdS thin films. Solar Energy Mater. Solar Cells **82**(1), 21–34 (2004)

Cullity, B.: Elements of X-Ray Diffraction, 2nd edn., pp. 197–356. Addision-Wesley Pub. Co. Inc., Reading (1978)

Edamura, T., Muto, J.I.: Electrodeposition and properties of copper-doped cadmium sulphide. Thin Solid Films **235**(1–2), 198–201 (1993)

Enríquez, J.P., Mathew, X.: Influence of the thickness on structural, optical and electrical properties of chemical bath deposited CdS thin films. Solar Energy Mater. Solar Cells **76**(3), 313–322 (2003)

Feigelson, R.S., N'Diaye, A., Yin, S.Y., Bube, R.H.: II-VI solid-solution films by spray pyrolysis. J. Appl. Phys. **48**(7), 3162–3164 (1977)

Feitosa, A., Miranda, M., Sasaki, J., Araujo-Silva, M.: A new route for preparing CdS thin films by chemical bath deposition using EDTA as ligand. Braz. J. Phys. **34**(2B), 656–658 (2004)

Hayashi, T., Nishikura, T., Suzuki, T., Ema, Y.: Formation and properties of In-doped high-conductivity CdS film. J. Appl. Phys. **64**(7), 3542–3550 (1988)

Hiie, J., Dedova, T., Valdna, V., Muska, K.: Comparative study of nano-structured CdS thin films prepared by CBD and spray pyrolysis: annealing effect. Thin Solid Films **511**, 443–447 (2006)

Levinson, J., Shepherd, F., Scanlon, P., Westwood, W., Este, G., Rider, M.: Conductivity behavior in polycrystalline semiconductor thin film transistors. J. Appl. Phys. **53**(2), 1193–1202 (1982)

Mahajan, S., Sygkridou, D., Stathatos, E., Huse, N., Kalarakis, A., Sharma, R.: Enhancement in the efficiency of crystalline Cu 2 ZnSnS 4 thin film solar cell by using various buffer layers. Superlattices Microstruct. **109**, 240–248 (2017)

Mathew, S., Mukerjee, P., Vijayakumar, K.: Optical and surface properties of spray-pyrolysed CdS thin films. Thin Solid Films **254**(1–2), 278–284 (1995)

Metin, H., Esen, R.: Annealing studies on CBD grown CdS thin films. J. Cryst. Growth **258**(1), 141–148 (2003)

Oladeji, I.O., Chow, L., Ferekides, C.S., Viswanathan, V., Zhao, Z.: Metal/CdTe/CdS/Cd 1 − xZnxS/TCO/glass: a new CdTe thin film solar cell structure. Solar Energy Mater. Solar Cells **61**(2), 203–211 (2000)

Pal, U., Silva-Gonzalez, R., Martinez-Montes, G., Gracia-Jimenez, M., Vidal, M., Torres, S.: Optical characterization of vacuum evaporated cadmium sulfide films. Thin Solid Films **305** (1–2), 345–350 (1997)

Pankove, J.: Optical Processes in Semiconductors, p. 92. Prentice-Hall, Englewood Cliffs (1971)

Prabahar, S., Dhanam, M.: CdS thin films from two different chemical baths—structural and optical analysis. J. Cryst. Growth **285**(1), 41–48 (2005)

Ramaiah, K.S., Pilkington, R., Hill, A., Tomlinson, R., Bhatnagar, A.-K.: Structural and optical investigations on CdS thin films grown by chemical bath technique. Mater. Chem. Phys. **68**(1), 22–30 (2001)

Repins, I., Contreras, M.A., Egaas, B., DeHart, C., Scharf, J., Perkins, C.L., Noufi, R.: 19.9%-efficient ZnO/CdS/CuInGaSe2 solar cell with 81.2% fill factor. Progr. Photovolt. Res. Appl. **16**(3), 235–239 (2008)

Schön, J., Schenker, O., Batlogg, B.: RETRACTED: Solution Processed CdS Thin Film Transistors. Elsevier (2001)

Subbaramaiah, K., Raja, V.S.: Preparation and characterization of all spray-deposited p-CuIn (S0. 5Se0. 5) 2/n-CdZnS: in thin film solar cells. Solar Energy Mater. Solar Cells **32**(1), 1–6 (1994)

Wang, X., Spitulnik, F., Campell, B., Noble, R., Hapanowicz, R., Condrate, R., Petrou, A.: Continous wave Nd: YAG laser deposition of CdS thin films. Thin Solid Films **218**(1–2), 157–160 (1992)

Ximello-Quiebras, J., Contreras-Puente, G., Aguilar-Hernandez, J., Santana-Rodriguez, G., Readigos, A.A.-C.: Physical properties of chemical bath deposited CdS thin films. Solar Energy Mater. Solar Cells **82**(1), 263–268 (2004)

Ximello-Quiebras, J., Contreras-Puente, G., Rueda-Morales, G., Vigil, O., Santana-Rodríguez, G., Morales-Acevedo, A.: Properties of CdS thin films grown by CBD as a function of thiourea concentration. Solar Energy Mater. Solar Cells **90**(6), 727–732 (2006)

Theoretical Investigation of Structural, Elastic and Electronic Properties of $SnHfO_3$ Compound: Ab Initio Calculations

D. Chenine[1,2](✉) and Z. Aziz[1]

[1] Laboratory of Technology and Solid Properties, Faculty of Sciences and Technology, Abdelhamid Ibn Badis University, BP 227, 27000 Mostaganem, Algeria
d.chenine@yahoo.ca

[2] Laboratory of Structure, Elaboration and Molecular Materials Applications (S.E.A.2M.), Département de Chimie, Université de Mostaganem, B.P. 981, R.P.27000 Mostaganem, Algeria

Abstract. Perovskite materials are intensively investigated due to their potential used in diverse applications. In This study structural, electronic, elastic and magnetic properties of cubic $SnHfO_3$ material have been investigated using the gradient generalized approximation GGA within the framework of Density Functional Theory (DFT). We have calculated the structural properties such as the equilibrium lattice constant, the bulk modulus and it is pressure derivative and they are in good agreement with the available data. Additionally, the results show that our material is stable in non-magnetic phase via the ferro- magnetic one. Furthermore, to confirm the stability for material of title, the elastic properties have been studied in detail and it is concluded that $SnHfO_3$ is elastically stable. The magnetic moment is also discussed in our study; we find that the main contribution to the total magnetic moments comes from the oxygen atom, while the contributions of Sn and Hf atoms are very small. Finally, the electronic band structures reveal that this material exhibits a direct band gap (X-X) semiconductor by using GGA approximation.

Keywords: Structural · Perovskite materials · Electronic structure Magnetic properties

1 Introduction

Perovskites are a family of compounds having crystal structures related to the mineral $CaTiO_3$. In the ideal cubic perovskite structure ABX_3, the coordination number of A cation is 12 and for cation B is 6. The cations A and B are coordinated by X anions. The X anions are coordinated by two B cations. The compounds which adopt the Hafnat-perovskite structures ABX_3, where A is alkali metal and B is alkaline earth or transition metal are very interesting materials due to their unique properties (ferroelectricity, semiconductivity and optical properties). This class of compounds used in many technological devices such as a lens material (Daniels et al. (1983) and Frederikse et al. (1964)), electro-optical applications (Yamada and Kanemitsu 2015). Among these

B. Abdelbaki et al. (Eds.): SMSD 2017, *Proceedings of the Third International Symposium on Materials and Sustainable Development*, pp. 75–80, 2018.
https://doi.org/10.1007/978-3-319-89707-3_9

compounds, the $SnHfO_3$ which crystallizes in an ideal cubic perovskite structure at ambient conditions and no structural transitions was observed. The crystal structure of $SnHfO_3$ is cubic and belongs to the space group Pm3m (No. 221). The cubic unit cell contains one molecule with the Sn sits at the origin (0, 0, 0), the Hf at the body center (1/2, 1/2, 1/2) and the three oxygen atoms at the three face centers (0, 1/2, 1/2), (1/2, 0, 1/2), and (1/2, 1/2, 0). Figure 1 shows the crystal structure of the $SnHfO_3$.

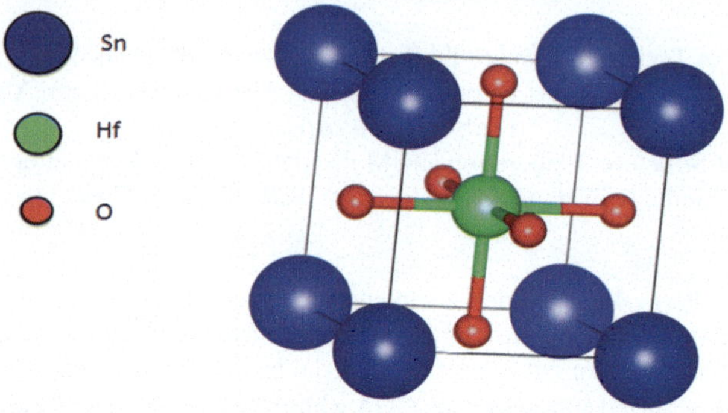

Fig. 1. Crystal structure of $SnHfO_3$ obtained with XCrysDens.

In this paper, we present a first-principles study of the structural, elastic and electronic properties of cubic perovskite $SnHfO_3$ using full potential linearized augmented plane wave (FP-LAPW) method with the generalized gradient approximation in the scheme of Perdew-Burke-Ernzerhof (GGA-PBE) (Perdew and Schmidt 2001).

2 Computational Methodology

The calculations reported here were performed using the Full Potential Linearized Augmented Plane Wave (FP-LAPW) method (Hohenberg and Kohn 1964) within the frame work of the density functional theory (DFT) (Perdew et al. (1996)) as implemented in the wien2k software (Blaha et al. (2001)). The exchange-correlation functional was calculated by the generalized gradient approximation in the scheme of Perdew-Burke-Ernzerhof (PBE) and the Tran-Blaha modified becke-johnson potential (TB-mBJ). So in this method, the space is partitioned into non-overlapping muffin-tin (MT) spheres separated by an interstitial region (IR). The region of nonoverlapping atomic spheres is treated as a linear combination of radial functions times spherical harmonics, whereas in the interstitial region a plane wave expansion is used. In all configurations studied the non-magnetic and ferromagnetic, we use a parameter Rmt X Kmax equal to 7 (where Rmt is the smallest of all atomic sphere radii and K_{max} is the maximum value of the wave vector K). The value of G_{max} =12, where Gmax is defined as the magnitude of largest vector in charge density Fourier expansion, while the expansion of the wave functions was set up to $l_{max} = 10$ inside of muffin-tin. We have chosen the muffin-tin radii to be 2.50, 1.98 and 1.71 a.u. for Sn, Hf and O, respectively. The integration of the full Brillouin zone is performed with $11 \times 11 \times 11$ k-points mesh, which corresponds to 56 k-points in the irreducible wedge.

The self-consistent cycles calculations are considered to be converged when the total energy is stable within 0.1mRy (10^{-4} Ry). The cut off energy was chosen as −6 Ry, which defines the separation of valence and core states.

3 Results and Discussions

3.1 Structural Properties

The calculated total energy versus reduced volume by using GGA approximation for $SnHfO_3$ in their ferromagnetic (FM) and non-magnetic (NM) configurations is shown in Fig. 2. From this fig, it is obvious that the performed calculations predict the non-magnetic (NM) to be the more stable (lower energy). To calculate the relax ground state properties, optimizations are performed for each unit cell and the structural parameters are evaluated by using the fitted Birch–Murnaghan equation of state. The lattice constants (a_0) and ground state energies (E_0) for this compounds are presented in Table 1. The calculated lattice parameters for cubic $SnHfO_3$ using the GGA approximation are 3.88Å, which show a good agreement with experimental value of 3.82Å.

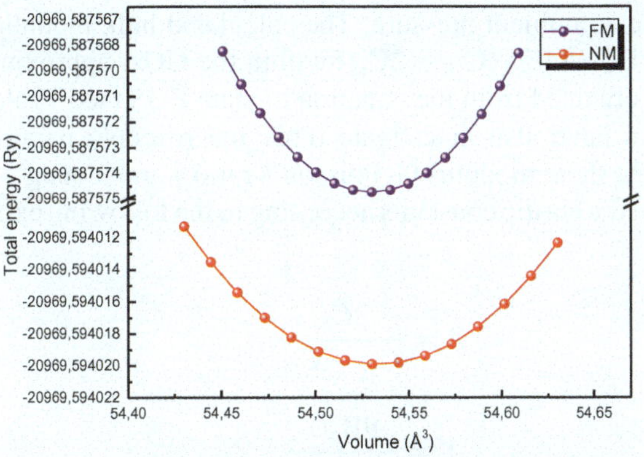

Fig. 2. Total energy as a function of volume for $SnHfO_3$ in Ferromagnetic and Non-magnetic phase.

Table 1. The Chemical Composition

Compounds	Configuration	a_0 (Å)	B_0 (GPa)	B'_0	E_0(Ry)
$SnHfO_3$	FM	3.87	167.70	4.46	−20969.5876
	NM	3.88	173.86	4.94	−20969.5940

3.2 Elastic Properties

To confirm the stability of the studied compound, the elastic properties have been investigated by using the GGA approximation. These properties define the abilities of materials that recover and return to its original shape after stress ceases. And they are essential for many technological applications related to the mechanical properties of solids, for example, thermo elastic stress, sound velocities, internal strain and fracture toughness.

Considering the cubic system, there are only three independent elastic constants, namely, C_{11}, C_{12} and C_{44}. By calculating the total energy as a function of strain, the elastic constants (C_{11}, C_{12} and C_{44}) are determined using the volume conserving tetragonal and orthorhombic strain. The calculated elastic constants at P = 0 GPa and T = 0 K are listed in Table 2. To the best of our knowledge; no experimental or theoretical data are available in the literature for comparison. Therefore, we hope that our results could provide reference data for future investigations.

Table 2. Calculated elastic constants C_{ij} and the bulk modulus B_0 (in GPa), anisotropy factor A, shear modulus G (in GPa), Young's modulus Y (in GPa) and the Poisson's ratio γ of the cubic $SnHfO_3$.

Material	Method	C_{11}	C_{12}	C_{44}	B_0	A	G	Y	B_0/G	γ
$SnHfO_3$	GGA	349.74	87.54	68.46	175.76	0.52	93.52	238.9	1.87	0.27

The obtained values satisfy the above stability criteria, indicating that our material is elastically stable at ambient pressure. The calculated bulk modulus from the elastic constants such as $B_0 = (1/3)(C_{11} + 2C_{12})$ within the GGA approximations has nearly the same value as obtained from the equation of state EOS (see Table 1).

In addition, we have also investigate other macroscopic parameters such as the anisotropy factor A, shear modulus G, Poisson's ratio γ and Young's modulus Y. They are calculated from the elastic constants according to the following expressions (Wallace 1972):

$$A = \frac{2C_{44}}{C_{11} - C_{12}} \tag{1}$$

$$Y = \frac{9B_0G}{3B_0 + G} \tag{2}$$

$$\gamma = \frac{3B_0 - Y}{6B_0} \tag{3}$$

And

$$G = \frac{1}{5}\left(C_{11} - C_{12} + 3C_{44}\right) \tag{4}$$

where A is the anisotropy factor, Y is the young's modulus, ν is the Poisson's ratio and G is the shear modulus. We find that the (B_0/G) value is equal to 1.87 which allow us to say that our compound is ductile in nature, we know that there is a limit value for B_0/G so if it is higher than 1.75 the material is ductile otherwise it is brittle (Kanoun et al. (2004)). Other important quantity is the anisotropy factor A, which indicates if the material is anisotropic or isotropic. If the value of A is equal to unity, the compound is isotropic and any other value indicates that the compound is anisotropic.

3.3 Electronic Properties

The electronic band structure of $SnHfO_3$ compound are calculated at the theoretical lattice constant along high-symmetry directions in the first Brillouin zone, as shown in Fig. 3. We have employed GGA (PBE) without spin orbit configuration. From this fig we can see clearly that the valence band maximum (VBM) is 0 eV at the M point but the conduction band minimum (CBM) is 1.86 eV at the X point. Hence, $SnHfO_3$ is direct band gap material of value 1.86 eV and acts like a semiconductor material which is interesting fro optoelectronic and thermoelectric application.

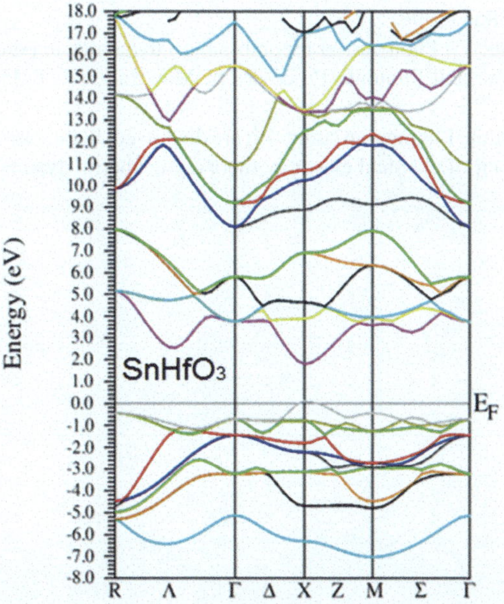

Fig. 3. The calculated band structures for $SnHfO_3$ by using GGA for non-spin polarized state.

4 Conclusions

In summary, we have used first-principles, FP-LAPW method to evaluate the structural,elastic and electronic properties of $SnHfO_3$ perovskite. This compound has a cubic crystal structure with space group Pm3m. the calculated results indicate that the non-magnetic ordering is energetically more favorable than the ferro-magnetic phase. In finally, the calculated elastic constants obey to the cubic stability conditions and our material is ductile in nature. So, we are not aware of any published data for the elastic properties of $SnHfO_3$ so our results may be considered as predictions of these properties.

References

Blaha, P., Schwarz, K., Madsen, G.K.H., Kvasnicka, D., Luitz, J.: WIEN2 K. An augmented plane wave + local orbitals program for calculating crystal properties (2001)

Daniels, R.R., Margaritondo, G., Heaton, R.A., Lin, C.C.: Experimental study of the electronic structure of KMg F3. Phys. Rev. B **27**(6), 3878 (1983)

Frederikse, H.P.R., Thurber, W.R., Hosler, W.R.: Electronic transport in strontium titanate. Phys. Rev. **134**(2A), A442 (1964)

Hohenberg, P., Kohn, W.: Inhomogeneous electron gas. Phys. Rev. **136**(3B), B864 (1964)

Kanoun, M.B., Merad, A.E., Cibert, J., Aourag, H., Merad, G.: Properties of strained zinc-blende GaN: first-principles study. J. Alloys Compd. **366**(1), 86–93 (2004)

Perdew, J.P., Schmidt, K.: Jacob's ladder of density functional approximations for the exchange-correlation energy. AIP Conf. Proc. **577**(1), 1–20 (2001)

Perdew, J.P., Burke, K., Ernzerhof, M.: Generalized gradient approximation made simple. Physi. Rev. Lett. **77**(18), 3865 (1996)

Yamada, Y., Kanemitsu, Y.: Photocarrier recombination dynamics in perovskite semiconductor $SrTiO_3$. In: Excitonic and Photonic Processes in Materials, pp. 1–35. Springer, Singapore (2015)

Wallace, D.C.: Where finite Lagrangian strains ηij are discussed. In the case of infinitesimal strains these reduce to our εij of classical elasticity theory. In: Thermodynamics of Crystals (1972). Chap. 1

Al-Doping Effects on Structural and Morphological Properties of ZnO Aerogels Synthesized in Supercritical Ethanol

F. Mouzaia(✉), D. Djouadi, A. Chelouche, and L. Hammiche

Laboratoire de Génie de l'Environnement (LGE), University of Bejaia, Route Targa Ouzemour, 06000 Bejaia, Algeria
fatahmz@gmail.com

Abstract. Aerogels of pure and aluminium doped ZnO (Al-ZnO) were synthesized by sol-gel route followed by supercritical drying in ethanol. The atomic percentage of the dopant is fixed at 1, 3, 5 and 10 at.%. The elaborated aerogels were characterized by X-ray diffraction (XRD), scanning electron microscopy (SEM), X-ray energy dispersive analysis (EDAX) and Fourier transform infrared spectroscopy (FTIR). The DRX results revealed good crystalline quality of the synthesized samples with hexagonal wurtzite crystallographic structure. No crystalline phase related to aluminum ions or aluminum based compounds has been detected. With increasing Al content in aerogels, the crystallinity deteriorates and the cell parameters are found to be slightly changed. SEM images showed that the presence of Al atoms in the ZnO lattice strongly alters grain morphology, especially for higher doping levels (≥ 5 at.%). The EDAX spectra showed well-defined peaks related only to Zn, O and Al and confirmed the absence of peaks related to other impurities. FTIR measurements have shown that the introduction of the A1 ions in the aerogels greatly reduces the intensity of the Zn-O bond-vibrating band and shifts it towards low wavenumbers.

Keywords: ZnO aerogels · Supercritical ethanol · DRX · Aluminum ions Structural properties

1 Introduction

Zinc oxide (ZnO) attracts a great deal of attention from researchers in laboratories all over the world (Chen *et al.* 2014). It is a most attractive field of research, in view of these qualities either in optoelectronics and electroluminescence diode (LED) (Saleem *et al.* 2016), varistors (Liu *et al.* 2007), gaz sensors (Ryzhikov *et al.* 2015) or in the biomedical (Tallósy *et al.* 2014, Gopikrishnan *et al.* 2010, Madhumitha *et al.* 2016). It is also known for its large optical gap (3.37 eV) and its high exciton binding energy (60 meV) at room temperature.

ZnO can be synthesized as either a nanoscale powder, thin layers or monoliths. Several methods have been used to elaborate ZnO powders, such as spray pyrolysis (Dounia *et al.* 2016), radiofrequency sputtering (Bidmeshkipourand and Shahtahmasebi

B. Abdelbaki et al. (Eds.): SMSD 2017, *Proceedings of the Third International Symposium on Materials and Sustainable Development*, pp. 81–88, 2018.
https://doi.org/10.1007/978-3-319-89707-3_10

2013), chemical vapor deposition (CVD) (Singh *et al.* 2008), and sol-gel (Kwon *et al.* 2013, Meddouri *et al.* 2016, Djouadi *et al.* 2014). The sol-gel method is a method, which has several advantages, because of its simplicity and low cost, and the possibility to obtain various sample forms with good crystalline qualities. In recent years, research has focused on doping semiconductors to improve their optical, electronic and structural properties (Panda and Tseng 2013; Narula *et al.* 2015). The dopant atoms generally used are transition metals as Al, Ga, Cd, Cu ... etc (Cunha et al. 2013, Slimi et al. 2018).

2 Experimental Procedures

Pure ZnO and Al-doped ZnO aerogels were obtained by mixing, in adequate volume and mass proportions, zinc acetate dihydrate ($Zn(C_2H_3O_2)_2$, $2H_2O$), methanol and aluminum nitrates ($Al(NO_3)_3$, $9H_2O$) under continuous magnetic stirring. The obtained solution underwent a drying at supercritical conditions of ethanol (critical temperature Tc = 243 °C and critical pressure Pc = 63.6 bars) in a one-liter reactor (autoclave), in order to obtain pure and Al-doped ZnO nanoparticles in the form of powder aerogels. Once the supercritical conditions reached, the solvent was removed and the autoclave was allowed to cool spontaneously to room temperature. The obtained aerogels were characterized without any chemical or heating treatments.

The crystalline structure of the as-prepared aerogels was studied using a PanAlytical diffractometer where the X-rays are produced from a radiation source CuKα (λ = 0.154 nm), with an acceleration voltage of 40 kV and a current of 30 mA. Infrared spectra (FT-IR) were recorded with a Shimadzu's IRAffinity-1 spectrometer by KBr pellet technique. SEM micrographs were recorded by a Quanta 2000 microscope with energy dispersive X-ray (EDS) spectroscope. EDS was used to identify the composition of the obtained products.

3 Results and Discussion

Figure 1 shows the XRD patterns of pure and Al-doped ZnO aerogels powders synthesized in supercritical ethanol with different Al contents. All the observed peaks [(100), (002), (101), (102), (110), (103) and (112)] are those of the ZnO hexagonal wurtzite structure (JCPDS 36-1451). Peaks have been found to be quite sharp and intense which implies high crystallinity of the aerogels. No crystalline phases related to Al or Al based compounds have been detected by the XRD measurements which indicates that the doping atoms are introduced in ZnO crystal lattice. The most important XRD peaks of the aerogels are shown in Fig. 2 to investigate the effect of Al atoms on the structural properties of the samples. It is observable that the most intense peaks are those of pure ZnO. This means that the introduction of Al atoms in ZnO lattice deteriores the crystalline quality of the samples. Compared to the undoped aerogel peaks, it can be noted that the peaks position of the aerogels doped with 1 and 10 at.% shifts towards lower diffraction angles more than those of the aerogels doped with 3 and 5 at.%. In addition, it can be show that the peak (101) of ZnO doped with 5 at.% is the most intense one and indicate that the introduction of Al atoms ameliorates the crystalline quality in the direction (101).

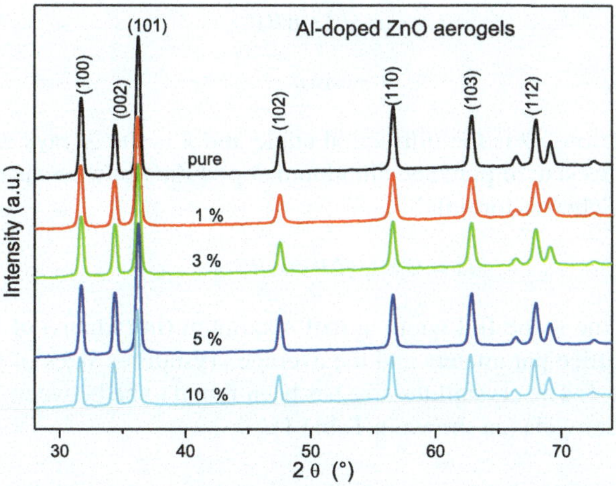

Fig. 1. XRD patterns of pure and Al-doped ZnO aerogels synthesized in supercritical ethanol.

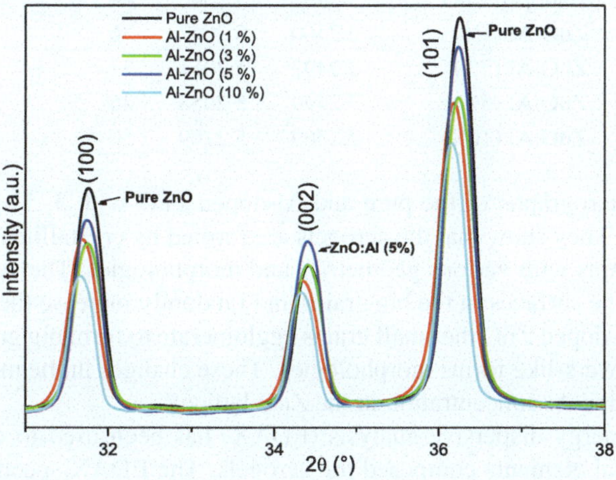

Fig. 2. XRD patterns of the most important peaks of the Fig. 1.

In addition, compared to the pure ZnO peaks, the peaks of the doped aerogels shift toward low angle indicating the existence of tensile stress introduced by the lattice distortion. This result confirms the largening of the lattice parameters and the peaks moving to small-angles. Normally, due to the smaller radius of Al^{3+} ions compared to Zn^{2+} ions, the incorporation of Al^{3+} will lead to the decrease of lattice constants (Serier *et al.* 2011).

The lattice parameters of ZnO cell a and c were calculated respectively from the positions of (101) and (002) using the following equations:

$$a = \lambda/(\sqrt{3}\sin\theta) \tag{1}$$

$$c = \lambda/\sin\theta \tag{2}$$

In these equations, θ is the diffraction angle and λ is the X rays wavelength. The average crystallites size of pure and aluminum doped ZnO powders is calculated using the well-known Scherrer formula:

$$D = 0.89/(\Delta\theta\cos\theta). \tag{3}$$

Where $\Delta\theta$ is the is the full width at half maximum (in radians) of Miller's planes. The calculated lattice parameters and the average crystallites sizes of the aerogels are resumed in Table 1. The crystallites size has been found to be between 36 and 50 nm in all the prepared aerogels (as shown in Table 1).

Table 1. The lattice parameters and crystallites size of the pure and Al-doped ZnO aerogels

Aerogel	$a = b$ (Å)	c (Å)	D (nm)
Undoped ZnO	3.2482	5.2040	44
ZnO:Al (1%)	3.2560	5.2047	36
ZnO:Al (3%)	3.2497	5.2051	37
ZnO:Al (5%)	3.2499	5.2058	45
ZnO:Al (10%)	2.2569	5.2169	50

The SEM micrographs of the pure and Al-doped ZnO with 3, 5 and 10 at. % are shown in Fig. 3. They show that the aerogels are formed by crystallites which agglomerate to form grains with various geometries and morphologies. The very small grains are attached on the surfaces of the big grains and randomly increase their sizes. For the 5 and 10 at.% Al-doped ZnO the small grains agglomerate to form big grains with torus-like and cauliflowers-like forms morphologies. These changes in the morphologies are probably due to the Al concentration in the ZnO lattice.

The X-ray energy dispersive analysis (EDAX) has been used to characterize the different chemical elements composed the aerogels. The EDAX spectra are shown in Fig. 4. The EDAX data of these aerogels show the presence only O, Zn, Al and C atoms. The C atoms come from the grid used in the experimentation. The results indicate that the percentage of Al in the samples did not strongly deviates from their stoichiometry.

Fig. 3. SEM images of pure and Al-doped ZnO (3, 5 and 10 at.%).

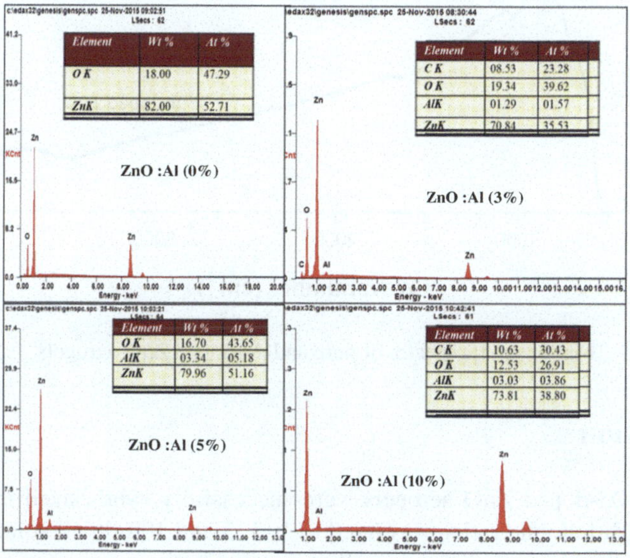

Fig. 4. EDAX spectra of pure and Al-doped ZnO (0, 1, 5 and 10 at.%).

Recorded FT-IR spectra of pure and Al-doped ZnO aerogels are shown in Fig. 5. The vibrational bands with different intensities are observed in all spectra approximately at the same positions (3429, 2370, 1624, 1416 and 1030 cm^{-1}). No bonds related to Al or Al-based compounds are found in the FT-IR curves. The small intensities of the lines testify the high chemical purity of the elaborated aerogels. The band at 2370 cm^{-1} can be assigned to CO_2 in trace and the one at 1416 cm^{-1} is assigned to C-C asymmetric stretching bond vibration. The peak at 1036 cm^{-1}can be attributed to aromatic C = C stretching mode. The absorption bands at 3429 cm^{-1} is due to the stretching mode of O-H group which reveals the existence of small amount of water absorbed by the ZnO aerogels (Djouadi *et al.* 2012). The band at 1624 cm^{-1} is due to the vibration of H-O-H bond (Vigneshwaran *et al.* 2006). A very strong band between 400 and 460 cm^{-1} due to the vibration of the Zn-O bond characterizes the spectra. The intensity of the band related to Zn-O bond decreases after the introduction of Al atoms in the aerogels. It can be noted that the presence of Al atoms in ZnO lattice shifts the absorption band of Zn-O.

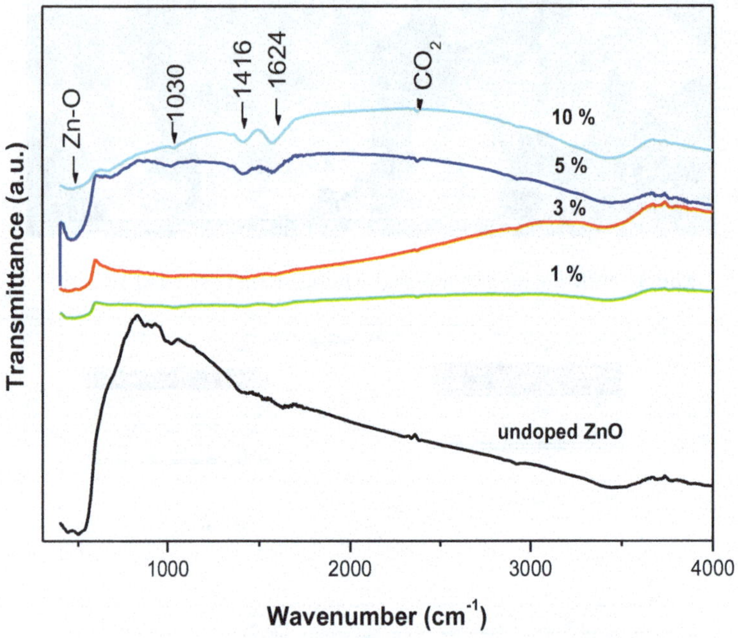

Fig. 5. FTIR spectra of pure and Al-doped ZnO aerogels

4 Conclusion

Undoped and Al-doped ZnO aerogels were successfully synthesized in supercritical ethanol. [Al]/[Zn] atomic ratios was fixed to 1, 3, 5 and 10% in all solutions. The Al-doping caused the changes in the crystallite size and lattice parameters of ZnO lattice. XRD characterizations manifest that the aerogels have high crystallinity and shared the presence of wurtzite structure. Al-doping strongly deteriorates the crystalline quality of the samples. EDAX studies confirm the high purity of the aerogels and the incorporation of Al atoms in ZnO lattice. FTIR spectra confirm that the presence of Al ions in ZnO lattice induces a shift of Zn-O bond vibration band toward low wavenumbers and strongly decreases its intensity. The introduction of dopant elements in the ZnO lattice shifts the absorption The morphology of the aerogels grains was found to be strongly dependent on the Al concentration.

References

Bidmeshkipourand, S., Shahtahmasebi, N.: Different properties of aluminum doped zinc oxide nanostructured thin films prepared by radio frequency magnetron sputtering. Semiconductors **47**, 787–790 (2013)

Chen, J., Pan, J.Y., Qingguo, D., Alagappan, G., Lei, W., Li, Q., Xia, J.: Highly efficient white quantum dot light-emitting diode based on ZnO quantum dot. Appl. Phys. A **117**, 589–591 (2014)

Cunha, D.M., Ito, N.M., Xavier, A.M., Arantes, J.T., Souza, F.L.: Zinc oxide flower-like synthesized under hydrothermal conditions. Thin Solid Films **537**, 97–101 (2013)

Djouadi, D., Meddouri, M., Chelouche, A.: Structural and optical characterizations of ZnO aerogel nanopowder synthesized from zinc acetate ethalonic solution. Opt. Mater. **37**, 567–571 (2014)

Djouadi, D., Chelouche, A., Aksas, A.: Amplification of the UV emission of ZnO: Al thin films prepared by Sol-Gel Method. J. Mater. Envir. Sci. **3**, 585–590 (2012)

Dounia, R., Migalska-Zalas, A., Addou, M., Bernede, J.C., Outzourhit, A., Benbrahim, M.: Preparation and characterization of highly transparent and conductive indium-zinc oxide thin films deposited by pyrolysis spray technique. Opt. Quantum Electron. **48**, 339 (2016)

Gopikrishnan, R., Zhang, K., Ravichandran, P., Baluchamy, S., Ramesh, V., Biradar, S., Ramesh, P., Pradhan, J., Hall, J.C., Pradhan, A.K., Ramesh, G.T.: Synthesis, characterization and biocompatibility studies of zinc oxide (ZnO) nanorods for biomedical application. Nano-Micro Lett. **2**, 31–36 (2010)

Kwon, B.J., Woo, H.J., Park, J.Y., Jang, K.: Optical properties of ZnO powder prepared by using a proteic sol-gel process. J. Korean Phys. Soc. **62**, 739–742 (2013)

Liu, H.Y., Kong, H., Ma, X.M., Shi, W.Z.: Microstructure and electrical properties of ZnO-based varistors prepared by high-energy ball milling. J. Mater. Sci. **42**, 2637–2645 (2007)

Madhumitha, G., Elango, G., Roopan, S.M.: Biotechnological aspects of ZnO nanoparticles: overview on synthesis and its applications. Appl. Microbiol. Biotechnol. **100**, 571–581 (2016)

Meddouri, M., Hammiche, L., Slimi, O., Djouadi, D., Chelouche, A.: Effect of cerium on structural and optical properties of ZnO aerogel synthesized in supercritical methanol. Mater. Sci. Pol. **34**, 659–664 (2016)

Narula, C., Kaur, I., Kaur, N.: Investigation of optical properties of mixed ligand directed ZnO luminescent nanoparticles for application in light emitting diodes. J. Mater. Sci. Mater. Electr. **26**, 8167–8175 (2015)

Panda, D., Tseng, T.Y.: One-dimensional ZnO nanostructures: fabrication, optoelectronic properties, and device applications. J. Mater. Sci. **48**, 6849–6877 (2013)

Ryzhikov, A., Jonca, J., Kahn, M., Fajerwerg, K., Chaudret, B., Chapelle, A., Menini, P., Shim, C.H., Gaudon, A., Fau, P.: Organometallic synthesis of ZnO nanoparticles for gas sensing: towards selectivity through nanoparticles morphology. J. Nanoparticles Res. **17**, 280 (2015)

Saleem, M., Manzoor, A., Zaffar, M., Hussain, S.Z., Anwar, M.S.: Tailoring of ZnO with selected group-II elements for LED materials. Appl. Phys. A **122**, 589–595 (2016)

Serier, H., Demourgues, A., Majimel, J., Gaudon, M.: Infrared absorptive properties of Al-doped ZnO divided powder. J. Solid State Chem. **184**, 1523–1529 (2011)

Singh, P., Kumar, A., Kaushal, A., Kaur, D., Pandey, A., Goyal, R.N.: In situ high temperature XRD studies of ZnO nanopowder prepared via cost effective ultrasonic mist chemical vapour deposition. Bull. Mater. Sci. **31**, 573–577 (2008)

Slimi, O., Djouadi, D., Hammiche, L., Chelouche, A., Touam, T.: Structural and optical properties of Cu doped ZnO aerogels synthesized in supercritical ethanol. J. Porous Mater. **25**, 595–601 (2018)

Tallósy, S.P., Janovák, L., Ménesi, J., Nagy, E., Juhász, A., Balázs, L., Deme, I., Buzás, N., Dékány, I.: Investigation of the antibacterial effects of silver-modified TiO2 And ZnO plasmonic photocatalysts embedded in polymer thin films. Environ. Sci. Pollut. Res. **21**, 11155–11167 (2014)

Vigneshwaran, N., Kumar, S., Kathe, A.A., Varadarajan, P.V., Prasad, V.: Functional finishing of cotton fabrics using zinc oxide-soluble starch nanocomposites. Nanotechnology **17**, 5087–5095 (2006)

High Quality Nano Thin Layer Silicon Transfer Using Plasma Hydrogenation

F. Okba(✉)

University Ferhat Abbas Sétif 1, Sétif, Algeria
fouad.okba@univ-setif.dz

Abstract. In order to transfer thin layers of silicon onto Si substrate, we report the effect of a thin epitaxial layer of SiGe on the nucleation and the growth of hydrogen platelets during plasma hydrogenation. SiGe strained layer was purposely used to introduce H trapping centers into Si wafers. Transmission electron microscopy has been used to quantitatively study the evolution of both density and sizes of these platelets during plasma hydrogenation. Upon hydrogenation, it is shown near SiGe layer that the platelets grow in size, reduce their density and the overall volume they occupy increase. This phenomenon is due to a non-conservative Oswald ripening. At the depth of the SiGe layer, a continuous (100) oriented cracks, which are parallel to the Si surfaces are formed.

Keywords: Nanomaterial · Thin layer · Silicon · Plasma hydrogenation Epitaxial

1 Introduction

In the conventional ion-cut technology for SOI production, hydrogen platelets are produced by both ion implantation and thermal annealing technical. The thickness of the transferred layer is controlled by adjusting the energy of hydrogen ion implantation.

However, in plasma hydrogenation, a stress is purposely introduced by a buried strained layer in Si substrate in order to trapping H platelets at the depth of the strain layer, which determines the cleavage location (Chen and Lau 2005), (Shao et al. 2008). Strain can facilitate (100) oriented platelets formation and cracking (Shao et al. 2005). The agglomeration of vacancies is kinetically and thermodynamically preferred in compressively strained SiGe layers (Shao et al. 2005). High resolution electron microscopy of electron irradiated strained Si has shown preferential aggregation of vacancies on (001) planes within the SiGe layer (Fedina et al. 2000). The interaction of H with those vacancies which are agglomerated along (100) planes facilitate the formation of (100) oriented platelets. Different studies have proposed that vacancies play a central role in platelets formation (Nastasi et al 2005), (Cherkashin et al. 2015). The cracking along (100) planes is facilitated by the presence of an interface shear strain. Once (100) oriented platelets are formed, the shear strain can facilitate cracking a long the interface (Shao et al. 2006). It is well known that platelets only form on {111} planes during plasma hydrogenation (Nickel et al. 2000). In this work, we report the platelets evolution induced by plasma hydrogenation on Si with presence of stress introduced by a buried

B. Abdelbaki et al. (Eds.): SMSD 2017, *Proceedings of the Third International Symposium on Materials and Sustainable Development*, pp. 89–94, 2018.
https://doi.org/10.1007/978-3-319-89707-3_11

SiGe stained layer in a Si substrate in order to transfer thin film of silicon. Furthermore, (111), (100) and (010) oriented platelets are observed between the Si surface and the stain layer. In both sample, continuous (100) oriented cracks, which are parallel to the Si surfaces, are formed at the depth of the SiGe layer.

The kinetics growth study (Grisolia et al. 2000), permit us to see the evolution of the density and diameter of platelets during plasma exposure.

2 Experimental

In order to transfer a thin layer of silicon, a molecular bean epitaxial growth of monocrystalline (100) Si sample, containing a 5-nm –thick Si0.80Ge0.20 layer at the depth of 150-nm, under the surface. The samples were subjected to plasma hydrogenation at a temperature of 300 °C and from 30 min to 120 min by step of 30 min.

All sample were prepared for cross section (CS) imaging along both the [110] and [001] direction using the standard techniques involving mechanical grinding by tripod followed by an ion milling by PIPS until electron transparency. Low energy and currents (3 keV and 7 μAm) were applied to avoid local heating of the simples. A transmission electron microscopy (TEM) Jeol 2010 operating at 200 keV was used to characterize the samples.

Each sample was characterized in terms of types of defects and their size, density and depth location evolution with plasma hydrogenation time.

3 Results and Discussion

Bright Field (BF) off-Bragg under focused images of all hydrogenated samples are shown in Fig. 1. A {111}, (001) and very few (010) oriented platelets are observed in all hydrogenated sample except the sample of 120 min that no platelets are observed between the Si surface and the SiGe layer. In all samples, there are a continuous (100) oriented cracks that has formed at the depth of SiGe layer.

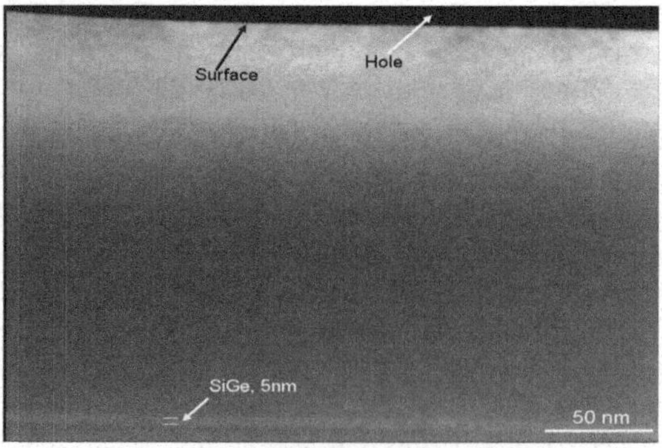

Fig. 1. TEM micrographs from Si/SiGe/Si before plasma hydrogenation taken in CS (110), it shows the position and the thickness of the SiGe layer

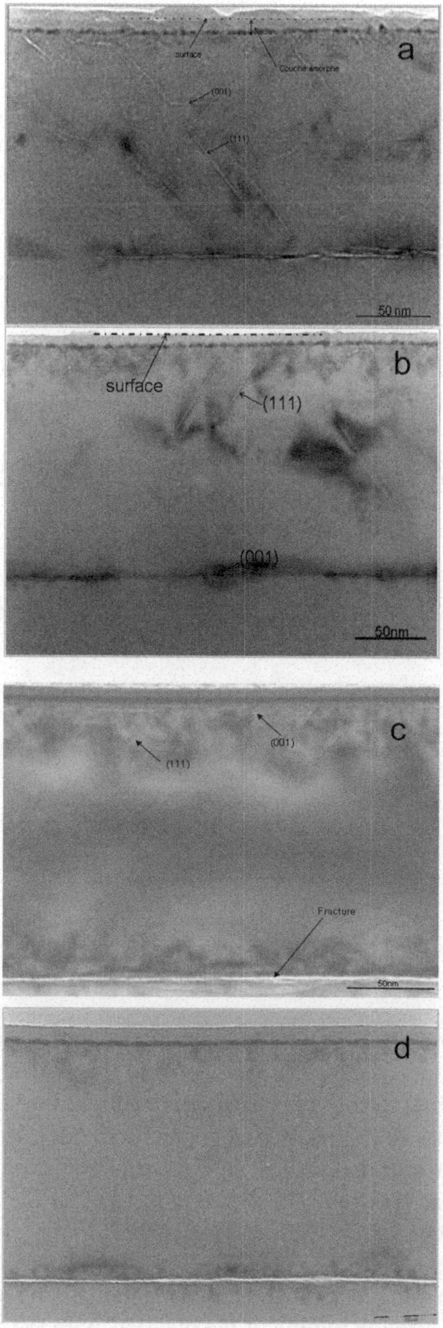

Fig. 2. TEM micrographs from Si/SiGe/Si after plasma hydrogenation at different times: (a) 30 min, (b) 60 min, (c) 90 min, (d) 120 min.

Note that only two variants of 4{111} platelets and only one variant of 3(100) variants are visible in CS (110) geometry. Also no variants of 4{111} platelets and two variants of 3(100) variants are visible in CS (010) geometry.

All four TEM micrographs in Fig. 2 show the formation of amorphous layer around 10 nm thick on the surface, which is due to plasma hydrogenation treatment.

For the sample of 30 min (Fig. 2a), we have observed that the (111) and (001) platelets exist from the surface to the SiGe layer, but the (111) platelets density is superior than (001) platelets. A similar repartition in depth of platelets than the sample of 30 min platelets was observed for the sample of 60 min (Fig. 2b). However, for the sample of 90 min (Fig. 2c), the platelets are concentred near the surface and the thickness of the crack is larger than the precedent samples.

No platelets are observed for the sample of 120 min (Fig. 2d), from the surface to SiGe layer, except the presence of a large and continuous crack are in the SiGe layer position.

Also, we have observed the formation of cavities connected to the fracture line in SiGe (Fig. 3).

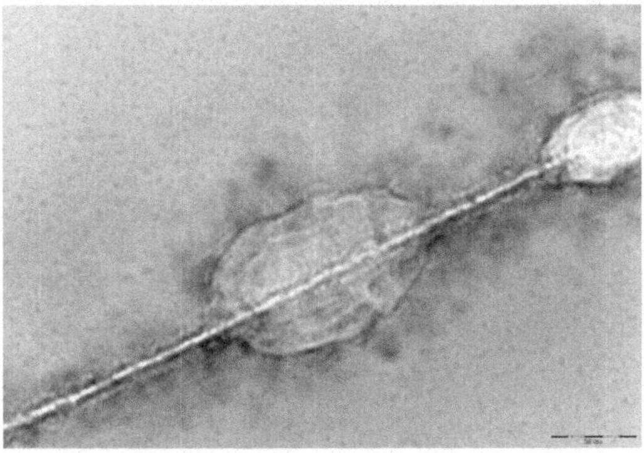

Fig. 3. TEM micrographs from Si/SiGe/Si after plasma hydrogenation at 120 min. Images taken in CS (110).

Figures 4 and 5, shows the time evolution during plasma exposure of the density and the diameter of all variants of platelets existing from the Si surface and SiGe layer.

The Fig. 4 shows that, during plasma hydrogenation, the mean density decreases with exposure time, thus, after 30 min, there is no more nucleation of platelets.

An apparent shift towards the surface: the width of «depletion region» increases: and disappears completely at 120 min. The platelets located on SiGe dissolve the existing platelets in Si and prevent the nucleation of new ones. A massive H transfer from H in Si and from the surface.

The Fig. 5 shows that the mean diameter of platelets increases with exposure time. Thus, this confirms the Growth regime of the platelets with the time of hydrogenation.

Also, the diameter increases when distance towards SiGe layer decreases. Already true after 30 min: preferential nucleation on SiGe. Being larger these platelets are traps

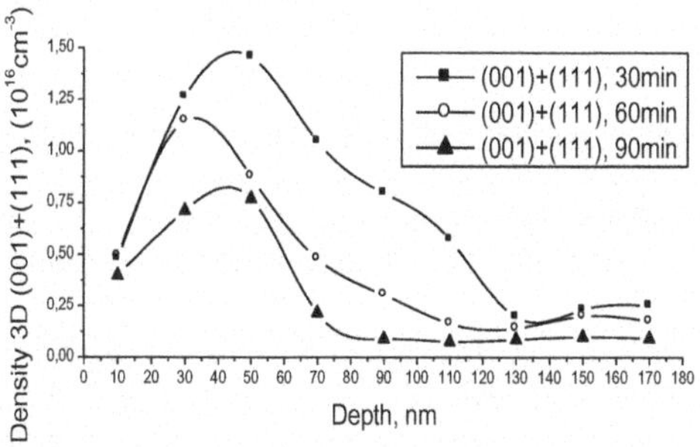

Fig. 4. Evolution of platelets density at 30 min, 60 min, 90 min

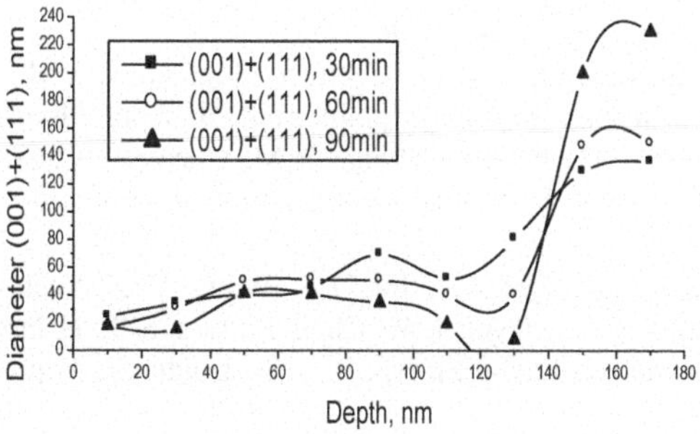

Fig. 5. Evolution of diameter at 30 min, 60 min, 90 min

for H emitted by smaller platelets in Si. Hydrogen diffuses from small platelets to large platelets due to Oswald ripening.

4 Conclusion

In summary, we have investigated the evolution of platelets and the role of stress for H-platelet formation during plasma-hydrogenation treatment on Si/SiGe/Si structure. The buried SiGe layer can act as H trapping centers during hydrogenation and the H-platelet are formed at the depth of SiGe layer. The platelet orientation is strong dependent on the in-plane compressive stress introduced by the SiGe layer, which determines the final cleavage of Si in the plane of stress and parallel to the substrate surface. The formation of platelets on SiGe not only helps in localizing the platelets at some selected depth BUT ALSO «cleans» the top layer from pre-existing defects.

References

Chen, P., Lau, S.S.: Silicon layer transfer using plasma hydrogenation. Appl. Phys. Lett. **87**, 111910 (2005)

Shao, L., Wang, Y.Q., Swadener, J.G., Nastasi, M., Thompson, P.E., Theodore, N.D.: Cracking in hydrogen ion-implanted Si/$Si_{0.8}Ge_{0.2}$/Si heterostructures. Appl. Phys. Lett. **92**, 061904 (2008)

Shao, L., Lin, Y., Swadener, J.G., Lee, J.K., Jia, Q.X., Wang, Y.Q., Nastasi, M., Thompson, P.E., David Theodore, N., Chu, P.K., Alford, T.L., Mayer, J.W., Chen, P., Lau, S.S.: Plasma hydrogenation of strained Si/SiGe/SiSi/SiGe/Si heterostructure for layer transfer without ion implantation. Appl. Phys. Lett. **87**, 091902 (2005)

Fedina, L., Lebedev, O.I., Van Tendeloo, G., Van Landuyt, J., Mironov, O.A., Parker, E.H.C.: In situ HREM irradiation study of point-defect clustering in MBE-grown strained $Si_{1-x}Ge_x$/(001) Si structures. Phys. Rev. B **61**, 10336 (2000)

Nastasi, M., Höchbauer, T., Lee, J.K., Misra, A., Hirth, J.P., Ridgway, M., Lafford, T.: Nucleation and growth of platelets in hydrogen-ion-implanted silicon. Appl. Phys. Lett. **86**, 154102 (2005)

Cherkashin, N., Daghbouj, N., Darras, F.-X., Fnaiech, M., Claverie, A.: Cracks and blisters formed close to a silicon wafer surface by He-H co-implantation at low energy. J. Appl. Phys. **118**(24), 245301 (2015)

Shao, L., Lin, Y., Swadener, J.G., Lee, J.K., Jia, Q.X., Wang, Y.Q., Nastasi, M., Thompson, P.E., Theodore, N.D., Alford, T.L., Mayer, J.W., Chen, P., Lau, S.S.: H-induced platelet and crack formation in hydrogenated epitaxial Si/$Si_{0.98}B_{0.02}$/Si structures. Appl. Phys. Lett. **88**, 021901 (2006)

Nickel, N.H., Anderson, G.B., Johnson, N.M., Walker, J.: Nucleation of hydrogen-induced platelets in silicon. Phys. Rev. B **62**, 8012 (2000)

Grisolia, J., Ben Assayag, G., Claverie, A.: A transmission electron microscopy quantitative study of the growth kinetics of H platelets in Si. Appl. Phys. Lett. **76**, 852 (2000)

Enhacing Pre-stressed Concrete Beam's Capacity Using Externally Bonded Pre-stressed Composite Plate

Hanane Mebsout[1,2], Baghdad Krour[1,3(✉)], and Mohamed Bachir Bouiadjra[3]

[1] University Mustapha Stambouli of Mascara, BP 763, Route de Mamounia, 29000 Mascara, Algeria
kr_bag@yahoo.fr

[2] Laboratoire des Sciences et Techniques de l'Eau, University Mustapha Stambouli of Mascara, BP 763, Route de Mamounia, 29000 Mascara, Algeria

[3] Laboratoire des structures et matériaux avancés en génie civil et travaux publics, University Dijllali Liabes of Sidi Bel Abbes, BP 89, 22000 Sidi-Bel-Abbès, Algeria

Abstract. The present study is carried out in order to highlight the influence of the use of new advanced materials like composite plates on the behavior of a pre-stressed concrete beams.

In fact, a composite plate, due to its high stiffness, when bonded to a pre-stressed concrete beam may increase its bearing capacity and change its behavior by reducing the tensile stress level over the cross section.

In this study, a composite plate is bonded to a pre-stressed concrete beam in aim to test its influence on the stress distribution in the cross section. The composite plate reduce the tensile stresses but does not eliminate them and the cross section remains in its class.

However, when a pre-stressed composite plate is applied, the tension stresses may be eliminated and the cross section passes from class 2 to class 1 for example, where only compressive stresses are applied. Consequently, the cracks due to tensile stresses are avoided and the pre-stressed beam will be able to support more loads than before.

Keywords: Advanced materials · Pre-stressed composite plate
Pre-stressed concrete beam · Stress distribution · Cross section

1 Introduction

The reinforcement of reinforced concrete structures by bounding composite plates is a technique that has grown in the field of civil engineering. Because of their high rigidity and light weight, the composite plates are considered as a good solution for the preservation of existing structures and for the improvement of their performance.

Thanks to researchers and industrialists working in the field of composite materials and polymers as well as those working in the field of civil engineering who constantly find applications relevant to this technique, the reinforcement and repair of civil engineering structures using composite materials are made easy, efficient and metrizable.

B. Abdelbaki et al. (Eds.): SMSD 2017, *Proceedings of the Third International Symposium on Materials and Sustainable Development*, pp. 95–104, 2018.
https://doi.org/10.1007/978-3-319-89707-3_12

Although in the past this reinforcement method was often applied to reinforced concrete and steel structures, today it has extended to pre-stressed concrete structures in order to improve their performance, particularly with regard to tension stresses which could affect their good behavior. This technique is used for several reasons, for example:

- Non-performing or inadequate building materials.
- No accordance with good practice or rules.
- Aggressive environment not envisaged during the design.
- Increased loads due to changes in the use of the structure.
- Improvement of the durability of old structures.
- Improvement of service conditions such as deflection reduction, reduction of internal stresses and reducing risk of cracking.
- Change of design rules requiring the conformity of the structure.

The use of composite materials in the reinforcement of structures is due to the following advantages:

- Low sensitivity to corrosion.
- Very high mechanical performance.
- Excellent durability.
- Ease of transport due to its lightness and its packaging in roll.
- Excellent fatigue behavior.

Several studies based on theoretical and experimental approaches have been conducted in this research field in order to investigate the various aspects of this type of reinforcement.

Aslam et al. (2015) published a review paper on the strengthening of RC beams using pre-stressed fiber reinforced polymers. This study had as goal to show advantages and disadvantages of many reinforcement techniques as the near surface mounted (NSM), externally bonded reinforcement (EBR) and externally post-tensioned techniques (EPT) by reviewing several works in this field.

Hosseini et al. (2016), conducted and experimental an numerical study on flexural strengthening of reinforced low strength concrete slabs using pre-stressed NSM-CFRP laminates in which the use of the pre-stressed CFRP laminates is tested and the results show an interesting increasing of the bearing capacity of the concrete slabs.

Emdad and Al-Mahaidi (2015) tested the influence of a pre-stressed CFRP laminates on the crack growth of steel plates. In this experimental and numerical study, it is shown that the pre-stressed CFRP laminates decreases greatly the deformation in the steel plates specimens.

Gao et al. (2016) curried out a new experimental and numerical study on flexural behavior of preloaded reinforced concrete beams strengthened by pre-stressed CFRP laminates. The main contribution in this study is a new technique for tensioning and application of pre-tensioned CFRP laminates strengthening system.

Rezazadeh et al. (2015) proposed an analytical approach for the flexural analysis of RC beams strengthened with pre-stressed CFRP. This Approach considers the ultimate flexural capacity of NSM-CFRP strengthened beams when concrete cover delamination is the governing failure mode. Based on three linear branches corresponding to the pre-cracking, post-cracking, and post-yielding stages, a moment–curvature relationship is developed.

Benachour et al. (2008) conducted an analytical study of the interfacial stresses of RC beam reinforced by a pre-stressed CFRP plate by using laminates theory and varying fibers orientation.

All these studies has been conducted for RC beams strengthened with pre-stressed CFRP laminates. The present work aims to extend the use of this strengthening technique to pre-stressed concrete beams. It is question to show the positive effect of a pre-stressed CFP plate externally bonded on a pre-stressed concrete beams.

2 Theoretical Basis

In this work we will try to show the effectiveness of the reinforcement of a pre-stressed concrete beam by a composite plate. Indeed, the pre-stressed concrete beams are dimensioned according to Eurocode2 (1991) on the basis of a section class.

We remind that the class I section does not tolerate any tension, the class II section allows tension but restricted to the tensile strength of the concrete and in the class III section there is no tensile limits in the concrete but those of the passive and the active steels are limited. The compression stress is limited to a certain level depending on the action combinations for the three classes.

The bonding of a sheet of rigid and strong composite materials could well increase the bearing capacity of the pre-stressed concrete section and then completely change its behavior since the pre-stressed concrete section with the composite material plate constitute a monolith body capable to change the class of pre-stressed concrete section.

In aim to show the effectiveness of this type of reinforcement, we conduce our analysis on the following assumptions:

- All materials are considered homogeneous and elastic.
- The adhesion between the concrete and the composite material plate is considered as perfect.
- The adhesive is considered sufficiently rigid to transmit the stresses without deformation.
- The planar section and perpendicular to the neutral axis before deformation remains plane and perpendicular to the neutral axis after deformation (Bernoulli hypothesis).
- The influence of passive and active steels in the stiffness of the section is neglected.

Let us consider a simply supported pre-stressed concrete beam subjected to transverse loading and reinforced by a plate made of pre-stressed composite material as shown in Fig. 1.

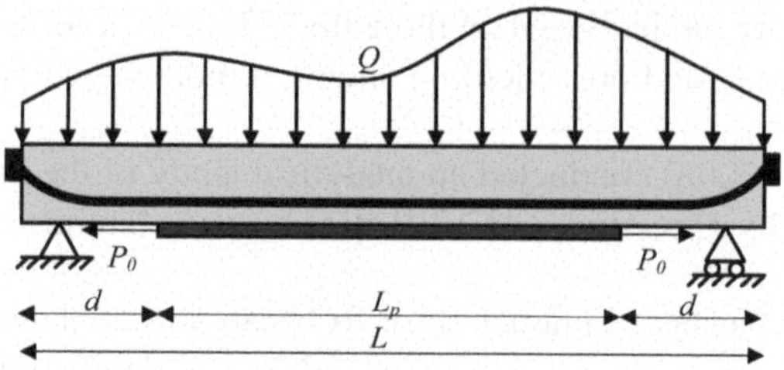

Fig. 1. Simply supported pre-stressed concrete beam with bonded pre-stressed CFRP plate.

The cross section shape is presented in Fig. 2.

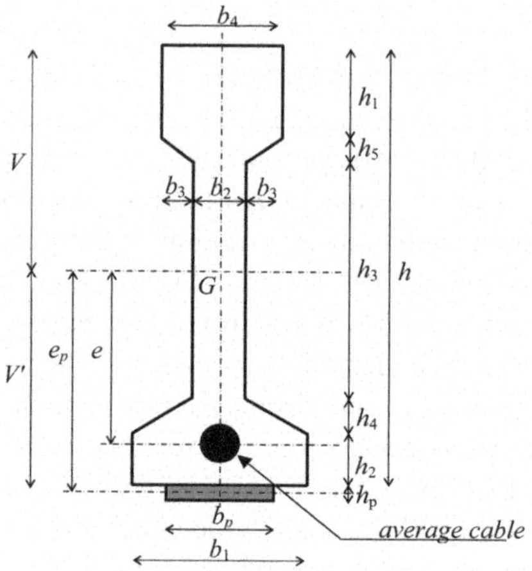

Fig. 2. Cross section shape.

In order to demonstrate the influence of composite plate reinforcement, it is necessary to calculate the distribution of normal stresses at the cross-section. This calculation will allow us to measure the stress level before and after reinforcement.

The application of the composite plate on the concrete gives us a new concrete-plate composite section. To calculate the stresses at this cross section, it is necessary to transform it into a homogeneous concrete section.

Based on the assumption of the perfect adhesion between the concrete and the composite plate we can write the following relations:

$$B_{eq} = B_c + nB_p \quad (1)$$

$$I_{eq} = I_c + nI_p \quad (2)$$

$$n = \frac{E_p}{E_c} \quad (3)$$

With:

B_c: area of the concrete section.
B_p: area of the composite plate section.
B_{eq}: area of the homogeneous concrete section.
I_c: quadratic moment of inertia of the concrete section.
I_p: quadratic moment of inertia of the composite plate section.
I_{eq}: quadratic moment of inertia of the homogeneous concrete section.
n: concrete-composite equivalence coefficient.
E_c: elastic modulus of the concrete.
E_p: elastic modulus of the composite plate.

The normal stress a cross the homogeneous cross-section is due to the external bending moment, to the compression stress of the pre-stress caused by pre-tensioned steel and also to the moment due to the eccentricity of the pre-stressing of the steel as well as that due to the pre-stressing of the composite plate.

It should be mentioned that the composite plate is pre-stressed by means of a tension which is translated into compression in the concrete once it has been bonded thereto. it is practically the same principle of pre-stressing with steels.

So we can write:

$$\sigma(y) = \frac{M_z(x)}{I_{eq}}y + \left(\frac{P + P_0}{B_{eq}}\right) + \left(\frac{Pe + P_0 e_p}{I_{eq}}\right)y \tag{4}$$

In the case of a pre-stressed concrete beam without reinforcement, the Eq. (4) becomes:

$$\sigma(y) = \frac{M_z(x)}{I_c}y + \frac{P}{B_c} + \frac{Pe}{I_c}y \tag{5}$$

In the case of a pre-stressed concrete beam with no pre-stressed composite plate reinforcement, the Eq. (4) becomes:

$$\sigma(y) = \frac{M_z(x)}{I_{eq}}y + \frac{P}{B_{eq}} + \frac{Pe}{I_{eq}}y \tag{6}$$

Where:

y: is the ordinate at which the normal stress is calculated with respect to the centroid of the homogeneous section.
P: is pre-stressing effort due to pre-tensioned steel.
P_0: is pre-stressing effort due to pre-tensioned CFRP plate.
e: eccentricity of the pre-stressing effort due to pre-tensioned steel with respect the centroid of the homogeneous concrete section.
e_p: eccentricity of the pre-stressing effort due to pre-tensioned CFRP plate with respect the centroid of the homogeneous concrete section.

Thus, it is necessary to calculate above all, the position of centroid of the new homogeneous section using the following equation:

$$y_G = \frac{\sum_{i=1}^{2} y_i B_i}{\sum_{i=1}^{2} B_i} = \frac{B_{cc} y_G + B_p \left(h + \frac{h_p}{2} \right)}{B_{eq}} \tag{7}$$

Where:

B_{cc}: is the area of the compressed section of the concrete. This implies that the tensioned part of the concrete section is neglected.
h: height of the concrete section.
h_p: height of the CFRP composite plate section.

The Eqs. (1) to (7) allow us to calculate the normal stress distribution across the homogenous concrete section.

In this study, let consider a simply supported pre-stressed concrete beam reinforced by a pre-stressed CFRP plate having the following geometrical proprieties listed in Table 1.

Table 1. Geometrical proprieties of studied beam.

h (mm)	h_1 (mm)	h_2 (mm)	h_3 (mm)
1300	270	150	700
b_1 (mm)	b_2 (mm)	b_3 (mm)	b_4 (mm)
500	150	100	350
h_4 (mm)	h_5 (mm)	h_p (mm)	d (mm)
100	80	variable	300
b_p (mm)	V (mm)	V' (mm)	e_p (mm)
150	674	626	627.2
L (m)	L_p (m)	d (mm)	
26.00	25.40	300	

The elastic modulus of both concrete and CFRP plate are respectively $E_c = 36$ GPa and $E_p = 165$ GPa and the compressive strength of concrete f_{c28} is equal to 30 MPa.

The beam is subjected to transversal load causing a maximum bending moment equal to 2 000 KN.m pre-stressed with compressive effort equal to 2.36 KN corresponding to 17 tendon of T15 with 25% of lose at the middle of the beam and designed as class II section according to Eurcode2 (1991).

3 Results and Discussion

This study is conducted to show the efficiency of the pre-stressed CFRP plate bounded at the bottom of the concrete beam. Thus, we have to evaluate the normal stress distribution across the pre-stressed concrete section before the reinforcement.

Using Eq. (5) we obtain the following diagram:

Figure 3 shows that the concrete section is well designed in class II according to Eurocode2 (1991) since the tensile stress is less than the tensile strength of concrete f_{t28} equal to 2.4 MPa and the compressive stress in lower than $0.6f_{c28}$ equal to 18 MPa.

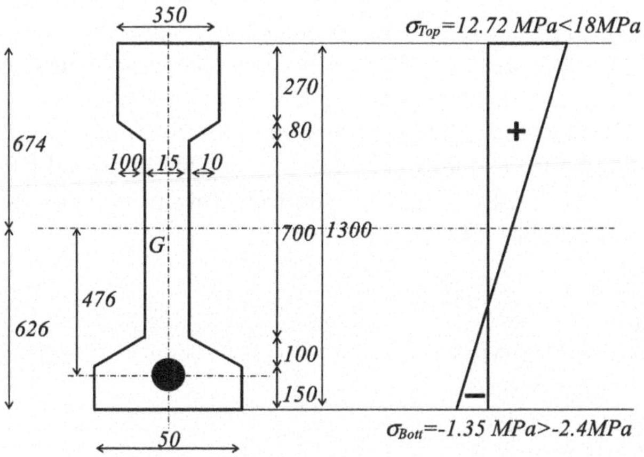

Fig. 3. Normal stress distribution before reinforcement.

It is now question to apply a externally CFRP plate on the bottom of the concrete section without any pre-stressing of the CFRP plate. The thickness of CFRP plate is

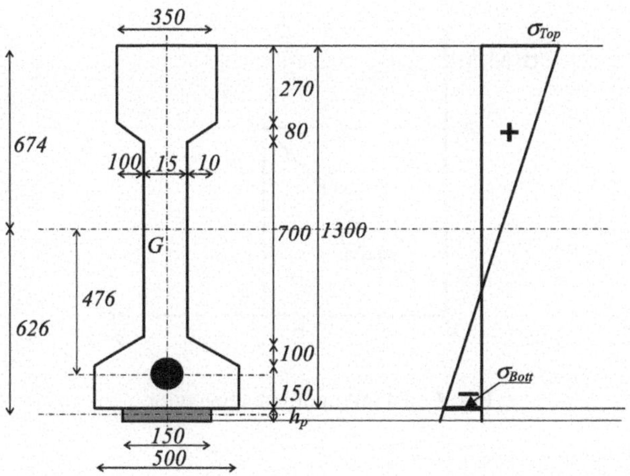

Fig. 4. Normal stress distribution with no pre-stressed CFRP plate reinforcement.

varied in order to test its influence on the stress distribution across the concrete section (Fig. 4).

Tables 2 shows that the use of CFRP plate without pre-stressing does not change the class of the section.

Table 2. Normal stresses across the concrete section with respect CFRP plate Thickness.

h_p (mm)	0	1	2	4
σ_{Bott} (MPa)	−1.35	−1.34	−1.34	−1.32
σ_{Top} (MPa)	12.72	12.68	12.64	12.54

However, we need a plate of 340 mm of thickness to eliminate all the tension stress at the bottom of the cross section. this thickness is so large to be realized in practice.

It is now obvious that the use of non pre-stressed CFRP plate for the reinforcement does not change the behavior of the pre-stressed concrete cross section.

In the following part it is question to use an available industrial CFRP plate called Sika® CarboDur®, S1512/240* witch has the following proprieties, Notice technique Sika® Carbodur® Lamelles (2009):

- Thickness 1.2 mm.
- Width: 150 mm.
- Tensile strength = 3100 MPa.
- Failure deformation: 1.80%
- Elastic modulus: 165 GPa.
- Density 1,60
- Fiber volume ratio > 68%.

A pre-tension effort is applied to the CFRP plate in order to create a pre-stressing effect to the concrete pre-stressed beam.

The CFRP plate behaves elastically until failure. However, in this study, the CFRP plate is pre-stressed up to less than 50% of its strength (Fig. 5).

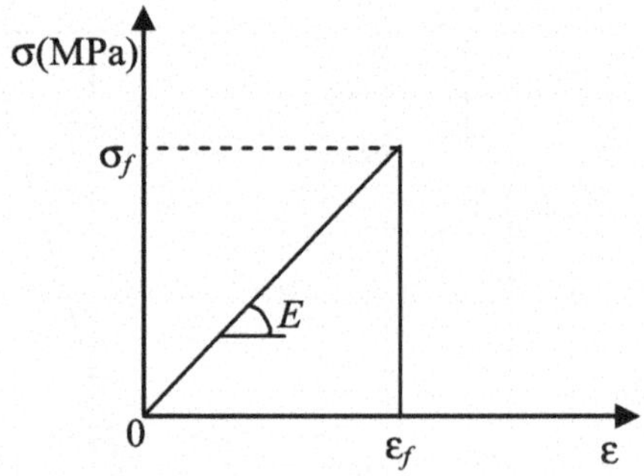

Fig. 5. CFRP laminates behavior.

This table shows clearly that the use of pre-stressed CFRP plate change the behavior of the pre-stressed concrete section and all tensile stresses are eliminated and the concrete cross section is subjected only to compressive stresses (Table 3).

Table 3. Normal stresses across the concrete section with respect pre-stress load of CFRP.

h_p (mm)	0	1.2	1.2	1.2
P_p (KN)	0	100	150	200
σ_{Bott} (MPa)	−1.35	−0.43	0.02	0.47
σ_{Top} (MPa)	12.72	12.97	13.13	13.28

According Eurocde2 (1991), the pre-stressed concrete cross section is now classified as class I since it supports only compressive stresses and cracks risk can be avoided.

In the following section, we will test the reinforcing of a pre-stressed concrete cross section of class III witch admits tensile stresses superior than the tensile strength of concrete.

We obtain this stresses by increasing the bending moment taken equal to 2200 KN.m The results, are summarized in Table 4:

Table 4. Normal stresses across the concrete section with respect pre-stress load of CFRP.

h_p (mm)	0	1.2	1.2	1.2
P_p (KN)	0	100	150	200
σ_{Bott} (MPa)	−3.31	−2.37	−1.91	−1.46
σ_{Top} (MPa)	14.82	15.07	15.23	15.38

The results presented in Table 4 show that it is possible to transform a pre-stressed concrete section from class III to class II where tensile stresses are limited to tensile strength of concrete, consequently we can prevent cracks occurrence.

4 Conclusions

The efficiency of the use of pre-stressed CFRP laminates in the strengthening of pre-stressed concrete beams is demonstrated in this work. in fact, this type of reinforcement changes the behavior of the concrete cross section by reducing considerably or completely the tensile stress and prevents against cracks. Therefore, the durability of this type of structure is assured.

It is also shown that using CFRP laminates without pre-stressing reduces tensile stresses but does not change the concrete cross section behavior since the tensile stress is not eliminated.

This technique of strengthening is capable to increase the bearing capacity of pre-stressed concrete beams and allows them to carry more loads than expected in the design, particularly when the structures are note in conformity with the rules.

References

Aslam, M., Shafigh, P., Jumaat, M.Z., Shah, S.N.R.: Strengthening of RC beams using prestressed fiber reinforced polymers – a review. Constr. Build. Mater. **82**, 235–256 (2015)

Hosseini, M.M., Dias, S.J.E., Barros, J.A.O.: Flexural strengthening of reinforced low strength concrete slabs using prestressed NSM CFRP laminates. Compos. Part B **90**, 14–29 (2016)

Emdad, R., Al-Mahaidi, R.: Effect of prestressed CFRP patches on crack growth of centre-notched steel plates. Compos. Struct. **123**, 109–122 (2015)

Gao, P., Gu, X., Mosallam, A.S.: Flexural behavior of preloaded reinforced concrete beams strengthened by prestressed CFRP laminates. Compos. Structures **157**, 33–50 (2016)

Rezazadeh, M., Barros, J., Costa, I.: Analytical approach for the flexural analysis of RC beams strengthened with prestressed CFRP. Compos. B **73**, 16–34 (2015)

Benachour, A., Benyoucef, S., Tounsi, A., Adda Bedia, E.A.: Interfacial stress analysis of steel beams reinforced with bonded prestressed FRP plate. Eng. Struct. **30**, 3305–3315 (2008)

Eurocode2: Calcul des structures en béton NF-EN 1992-1-1. ISSN 0335-3931

Notice technique, Edition juillet 2009, Numéro 9.14, Version n° 137.2009, Sika® Carbodur® Lamelles

Artificial Metamaterials for High Efficiency Silicon Solar Cells

Houria Hamouche[1(✉)] and Mohammed M. Shabat[2]

[1] Centre de Développement des Energies Renouvelables, CDER, BP 62, Route de l'Observatoire, 16340 Bouzaréah, Algiers, Algeria
hamouche_houri@yahoo.fr

[2] Departement of Physics, The Islamic University of Gaza, P.O. Box 108, Gaza Strip, Palestine

Abstract. The reduction of the significant optical losses due to the reflection and the increase of the penetration of the solar photons into the silicon initiated the development of new photoelectrical techniques for silicon solar cells. Insertion of metamaterials in solar cells seems to be one of the interesting approaches owing to the promising properties of these news materials. Metamaterials are a class of man-made subwavelength structured composite materials. Metamaterials with their unusual electromagnetic properties allow unprecedented guiding of the incident light. In this paper, we undertake a numerical study of the optical response of a multilayer planar waveguide structure based on metamaterials for silicon solar cells. The structure comprises three layers embedded in air. The metamaterial layer is sandwiched between a glass layer and a silicon layer. We first analyze the electromagnetic parameters of the metamaterial layer. We then determine the optical parameters of this structure, namely the reflection, the transmission and the absorption. This study is based on the use of the transfer matrix method which is a very useful algorithm for reflection and transmission calculations of multilayer structures. According to the law of conservation of energy, the absorption coefficient is inferred from the transmission and reflection coefficients. Numerical simulations by using MAPLE software show a weak reflection of light at the air/glass front interface for optimal thicknesses of the structure's layers. These results are compared with those of a conventional cell structure without metamaterial. Additionally, the effect of the incidence angles for both Transverse Electric (TE) and Transverse Magnetic (TM) polarizations is discussed.

Keywords: Optical losses · Metamaterials · Transfer matrix method · Thickness · Incidence angles

1 Introduction

The need of high efficient silicon solar cells is the mean reason for the development of new photoelectrical techniques that bring a reduction in both the energy losses and costs that restrict the conversion cell efficiency. High efficiency requires the increase of the low penetration of the solar photons into the silicon that lead to an improvement in the light-generated current. Moreover, the efficiency of silicon solar cells is very limited by the significant optical losses due to the reflection of light at the air/glass front interface and the

B. Abdelbaki et al. (Eds.): SMSD 2017, *Proceedings of the Third International Symposium on Materials and Sustainable Development*, pp. 105–115, 2018.
https://doi.org/10.1007/978-3-319-89707-3_13

unwanted transmission from the rear side of the cell (Goetzberger and Hoffmann 2005). Optical confinement techniques are then required to overcome these optical limitations (Shabat et al. 2016). In the past, optical performance of silicon solar cells have been largely studied by many design concepts, such as antireflection coatings (Bouhafs et al. 1998), light trapping (Markvart 2000) and thermal oxide surface passivation (Biro et al. 2009). Dielectric planar solar waveguide structures in which light is confined by total internal reflection, have been the first realized structures (Okamoto 2006). During the last few years, insertion of metamaterials in dielectric solar waveguides has attracted an increasing attention. Metamaterials are a class of man-made subwavelength structured composite materials with unusual electromagnetic properties which allow unprecedented guiding of incident light (Noginov and Podolskiy 2012; Ubeid et al. 2011; Ubeid et al. 2012). In dielectric-metamaterials waveguides, the light is trapped by surface bound electromagnetic waves that propagate along dielectric-metamaterials interfaces named Surface Plasmon Polariton ((Hosseini and Massoud 2006); Lavoie et al. (2012)).

In this paper, we undertake a numerical study of the optical response of a multilayer planar waveguide structure based on metamaterials for silicon solar cells. The structure comprises three layers embedded in air. We first analyze the electromagnetic parameters of the metamaterial layer. We then determine the optical parameters of this structure, namely the reflection, the transmission and the absorption. This study is based on the use of the transfer matrix method which is a very useful algorithm for reflection and transmission calculations of multilayer structures. According to the law of conservation of energy, the absorption coefficient is inferred from the transmission and reflection coefficients. The reflection and the transmission coefficients of this structure were compared with those of a structure without metamaterial. Additionally, the effect of the incidence angles for both Transverse Electric (TE) and Transverse Magnetic (TM) polarizations is discussed.

2 Electromagnetic Parameters of the Metamaterial Layer

The relative electrical permittivity $\varepsilon_r(\omega)$ and the relative magnetic permeability $\mu_r(\omega)$ are the fundamental electromagnetic parameters describing light-matter interactions. The propagation of the light through a metamaterial medium is characterized by the refractive index

$$N = -\sqrt{\varepsilon_r(\omega) \cdot \mu_r(\omega)} = n_r + i \cdot n_i \tag{1}$$

n_r and n_i are the real and the imaginary part of the metamaterial's refractive index, respectively. In this study, the relative magnetic permeability of the metamaterial is taken equal to -1 $\left(\mu_r = -1\right)$ and the Drude model is used to express the relative permittivity $\varepsilon_r(\omega)$

$$\varepsilon_r(\omega) = 1 - \frac{\omega_p^2}{\omega^2 + i\gamma\omega} \tag{2}$$

The plasma frequency ω_p and the collision frequency γ are given by (Jichun et al. (2011))

$$\gamma = \frac{\omega \cdot \varepsilon_2}{1 - \varepsilon_1}, \quad \omega_p = \sqrt{(1 - \varepsilon_1) \cdot (\omega^2 + \gamma^2)} \tag{3}$$

with

$$\begin{aligned} \varepsilon_1 &= -1.1 \cdot \lambda^3 - 39 \cdot \lambda^2 - 12 \cdot \lambda + 12 \\ \varepsilon_2 &= 7.3 \cdot \lambda^8 - 100 \cdot \lambda^7 + 580 \cdot \lambda^6 - 1900 \cdot \lambda^5 \\ &\quad + 3700 \cdot \lambda^4 - 4400 \cdot \lambda^3 + 3200 \cdot \lambda^2 \\ &\quad - 1300 \cdot \lambda + 210 \end{aligned} \tag{4}$$

The real and imaginary parts of the metamaterial's refractive index is presented in Fig. 1.

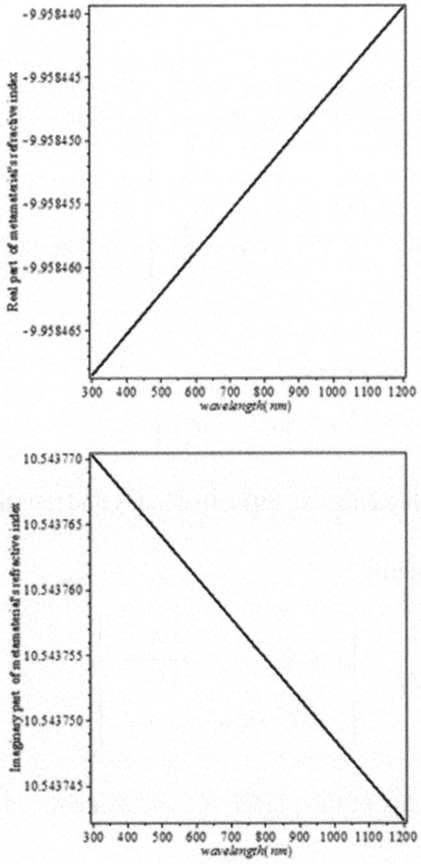

Fig. 1. Complex refractive index of metamaterial versus the wavelength

Figure 1 shows that the real and imaginary parts of the metamaterial's refractive index are practically constants. For our calculations, we have used a complex refractive index N equal to $-9.96 + i \cdot 10.54$

3 Derivation of the Optical Parameters

The optical parameters were determined using Transfer Matrix Method. Transfer Matrix Method is a powerful approach for the analysis of periodic structures. The Transfer Matrix Method employs continuity boundary conditions between layers of material and wave equations to describe the reflectance and transmittance values across each layer. The complete transfer matrix M for a stack of N layers $l_1, l_2, \ldots l_N$ between two infinite layer l_0 and l_s is obtained by multiplying the individual transfer matrices (Hamouche and Shabat (2016)).

$$M = M_0^{-1}\left[\prod_1^N M_j\right]M_s = \begin{bmatrix} M_{11} & M_{12} \\ M_{21} & M_{22} \end{bmatrix} \tag{5}$$

with

$$M_0^{-1} = \frac{1}{2}\begin{bmatrix} 1 & \frac{1}{n_{cla}^{eff}} \\ 1 & -\frac{1}{n_{cla}^{eff}} \end{bmatrix} \tag{6}$$

and

$$M_s = \begin{bmatrix} 1 & 1 \\ n_{sub}^{eff} & -n_{sub}^{eff} \end{bmatrix} \tag{7}$$

n_{cla}^{eff} and n_{sub}^{eff} are the cladding effective refractive index and the substrate effective refractive index, respectively.

M_j is the jth layer's matrix

$$M_j = \begin{bmatrix} \cos\delta_j & \frac{i\sin\delta_j}{n_j^{eff}} \\ i n_j^{eff}\sin\delta_j & \cos\delta_j \end{bmatrix} \tag{8}$$

n_j is the refractive index of the j layer. n_j^{eff} and δ_j are the effective refractive index and the phase thickness, respectively.

n_j^{eff} and δ_j are defined as

$$n_j^{eff} = \begin{cases} n_j \cos\theta_j \ \textit{for TE polarization} \\ \dfrac{n_j}{\cos\theta_j} \ \textit{for TM polarization} \end{cases} \tag{9}$$

and

$$\delta_j = \frac{2\pi}{\lambda} n_j d_j \cos\theta_j \tag{10}$$

λ and d_j are the wavelength and the thickness of the j layer, respectively.

The angle of refraction θ_j is related to the incidence angle θ_{cla} by the Snell's law

$$n_{cla} \sin\theta_{cla} = n_j \sin\theta_j \tag{11}$$

The reflection and transmission amplitudes can be written in terms of the transfer matrix coefficients as

$$r = \frac{M_{21}}{M_{11}}, \ t = \frac{1}{M_{11}} \tag{12}$$

The reflection coefficient R is the square of the reflection amplitude $R(\lambda) = |r|^2$ and the transmission coefficient T is given by $T(\lambda) = \frac{n_{sub}^{eff}}{n_{cla}^{eff}} |t|^2$.

The absorption coefficient A is given by application of the law of conservation of energy A = 1-R – T.

4 Numerical Simulation Results

The studied structure comprises three layers embedded in air. The metamaterial layer of 1 nm thickness $(d_{LHM} = 1\text{ nm})$ is sandwiched between a glass layer with refractive index 1.47 ($n_{Glass} = 1.47$) and a silicon layer with refractive index $3.6 - i \cdot 0.00015$ $(n_{Si} = 3.6 - i \cdot 0.00015)$. For optimum antireflection effect, the thickness of glass and silicon layers are taken 100 nm ($d_{Glass} = d_{Si} = 100$ nm). Air is considered as the cladding and the substrate of the structure ($n_{cla} = n_{sub} = 1$).

4.1 Numerical Simulation Results for Normal Incidence

The structure has firstly been examined under normal incidence. In Fig. 2, the reflection, transmission and absorption coefficients are plotted as a function of the wavelength.

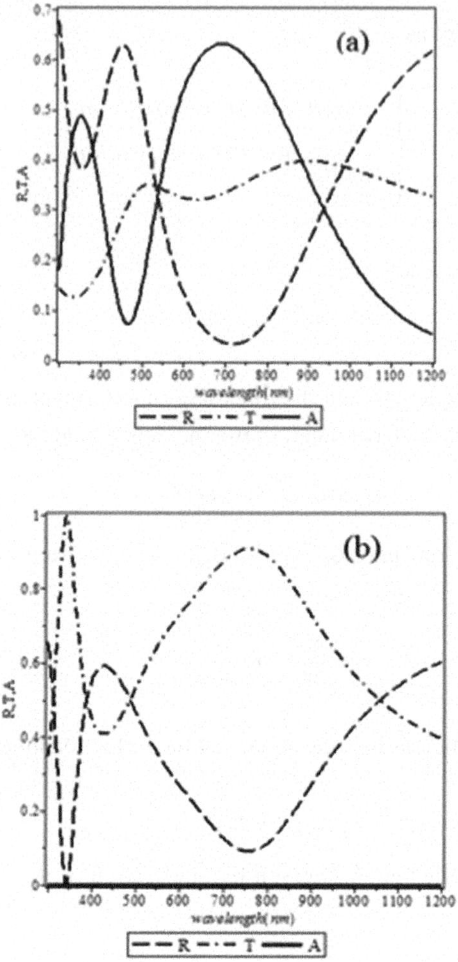

Fig. 2. Reflection, transmission and absorption coefficients at normal incidence versus the wavelength (a) with metamaterial layer, (b) without metamaterial layer.

Figure 2 shows a strong enhancement of absorption in the structure (Fig. 2(a)) in comparison with the absorption of a structure of air/glass/silicon/air (Fig. 2(b)). A reduction in the losses by reflection and transmission is illustrated in Fig. 3. The reflection of the structure in the wavelength range 530–1050 nm, is remarkably weaker than the reflection of the structure without the metamaterial layer. Moreover, the unwanted transmission from the rear side of the structure is considerably minimized.

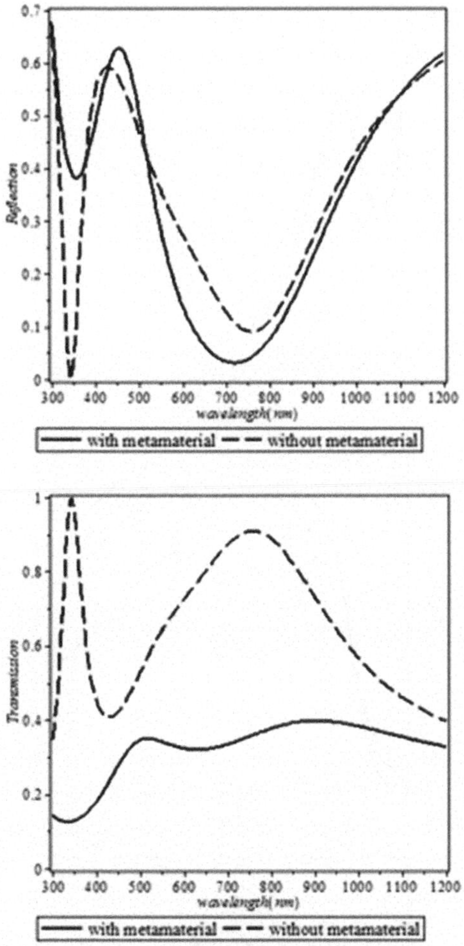

Fig. 3. Reflection and transmission coefficients at normal incidence versus the wavelength.

4.2 Numerical Simulation Results for Oblique Incidence

The dependence of the optical parameters on the incidence angles is considered in the two modes of polarization. As it can be seen in Fig. 4, the reflection and the absorption of the Transverse Electric (TE) polarization increase with the increasing incident angles in the wavelength range of 800 nm $< \lambda <$ 1200 nm and 530 nm $< \lambda <$ 850 nm, respectively. The reflection and the absorption are weakly dependent on the incidence angles in the range of 300 nm $< \lambda <$ 800 nm and 300 nm $< \lambda <$ 530 nm, respectively. The transmission decreases with the increasing incident angles in the wavelength range of 500 nm $< \lambda <$ 1200 nm.

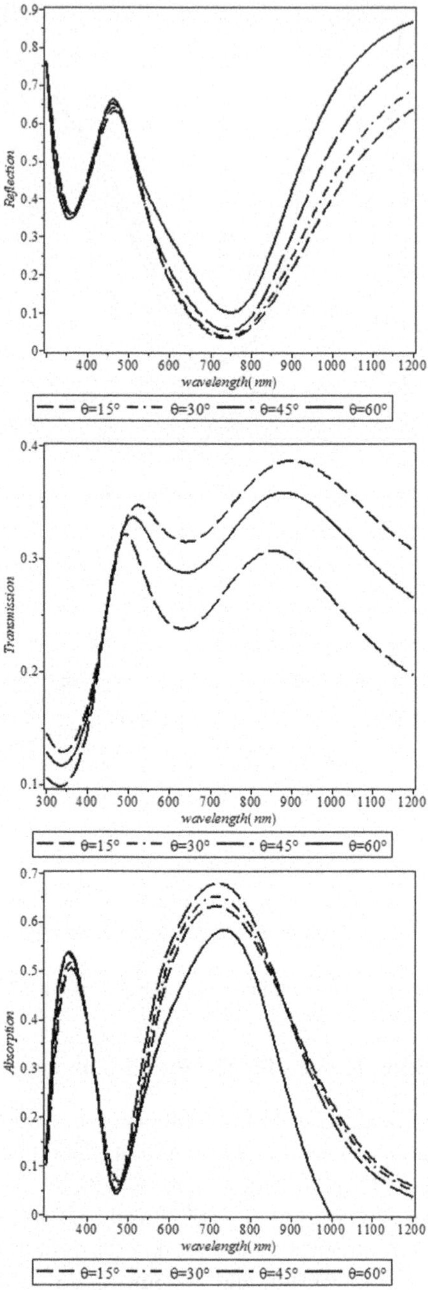

Fig. 4. Reflection, transmission and absorption coefficient of TE polarization as a function of wavelength under different incident angles.

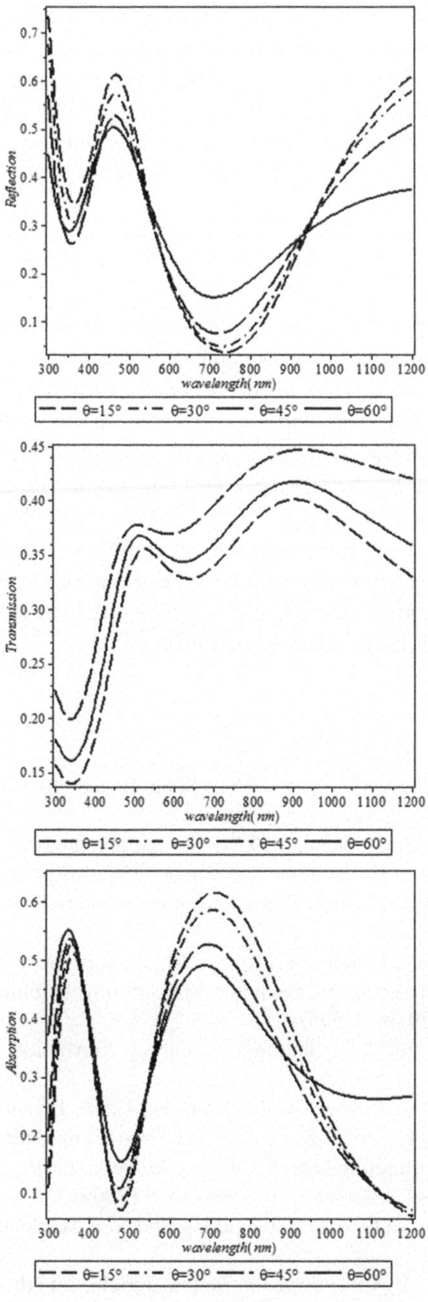

Fig. 5. Reflection, transmission and absorption coefficient of TM polarization as a function of wavelength under different incident angles.

In Fig. 5, it shows that the optical parameters behavior is inversed for the TM polarization. The reflection and the absorption decrease with the increasing incident angles in the wavelength range of 950 nm $< \lambda <$ 1200 nm and 550 nm $< \lambda <$ 950 nm, respectively. The transmission increases with the increasing incident angles in the wavelength range of 500 nm $< \lambda <$ 1200 nm. The same graph is obtained for the transmission of the two polarization modes at the angles of 30 and 60.

5 Conclusion

The numerical analysis presented in this paper shows that the insertion of a nanolayer metamaterial in a conventional silicon solar cell structure of air/glass/silicon/air can achieve an improvement in the light absorption and a minimization of the reflection and transmission losses. From the simulations carried out in this study it has been found that for suitable thicknesses of the structure's layers the optical characteristics of this structure are better than those of the structure without the metamaterial layer. Indeed, the absorption of the conventional cell at normal incidence is obtained nearly zero and a weak reflection for each polarization of incident angles is observed in the range of 530–900 nm. In the same range, the absorption of TE polarization is found higher than the absorption of TM polarization. In conclusion, the studied structure turned out to be very interesting for guiding light in silicon solar cells.

References

Goetzberger, A., Hoffmann, V.U.: Photovoltaic Solar Energy Generation. Springer Series in Optical Sciences. Springer, New York (2005)

Shabat, M.M., El-Amassi, D.M., Schaadt, D.M.: Design and analysis of multilayer waveguides containing nanoparticles for solar cells. Sol. Energy **137**, 409–412 (2016)

Li, J., Chen, Y., Liu, Y.: Mathematical simulation of metamaterial solar cells. Adv. Appl. Math. Mech. **3**, 702–715 (2011)

Bouhafs, D., Moussi, A., Chikouche, A., Ruiz, J.M.: Design and simulation of antireflection coating systems for optoelectronic devices: application to silicon solar cells. Solar Energy Mater. Solar Cells **52**, 79–93 (1998)

Markvart, T.: Light harvesting for quantum solar energy conversion. Prog. Quantum Electron. **24**, 107–186 (2000)

Biro, D., Mack, S., Wolf, A., Lemke, A., Belledin, U., Erath, D., Holzinger, B., Wotke, E.A., Hofmann, M., Gautero, L., Rentsch, J., Preu, R.: Thermal oxidation as a key technology for high efficiency screen printed industrial silicon solar cells. In: Proceedings of the 34th IEEE Photovoltaic Specialists Conference, pp. 1594–1599 (2009)

Okamoto, K.: Fundamentals of Optical Waveguides, 2nd edn. Academic Press, Elsevier, London (2006)

Noginov, M.A., Podolskiy, V.A.: Tutorials in Metamaterials, 1st edn. Taylor & Francis Group, LLC, Boca Raton (2012)

Ubeid, M.F., Shabat, M.M., Sid-Ahmed, M.O.: Effect of negative permittivity and permeability on the transmission of electromagnetic waves through a structure containing left-handed material. Nat. Sci. **3**, 328–333 (2011)

Ubeid, M.F., Shabat, M.M., Sid-Ahmed, M.O.: Numerical study of a structure containing left-handed material waveguide. Indian J. Phys. **86**, 125–128 (2012)

Hosseini, A., Massoud, Y.: A low-loss metal-insulator-metal plasmonic bragg reflector. Opt. Express **14**, 11318–11323 (2006)

Lavoie, B.R., Leung, P.M., Sanders, B.C.: Low-loss surface modes and lossy hybrid modes in metamaterial waveguides. Photonics Nanostruct. Fundam. Appl. **10**, 602–614 (2012)

Hamouche, H., Shabat, M.M.: Enhanced absorption in silicon metamaterials waveguide structure. Appl. Phys. A **122**(1–7), 685 (2016)

A New Technique for the Evaluation of the Parameters of GaAlAsSb(p)/GaAlAsSb(n)/InAsSb Double Interface from Current-Voltage Curve Analysis

K. Mahi(✉), B. Messani, and H. Ait-kaci

Laboratoire de Physique des Plasmas, Matériaux Conducteurs et leurs Applications, Faculté de physique, Université d'Oran des Sciences et de la Technologie Mohamed Boudiaf, Boite Postale 1505 El-M'naouer, 31000 Oran, Algeria
mahikhalidou@yahoo.fr

Abstract. Parameters characterizing a p-n junction based device like ideality factor n, saturation current Is, shunt resistance Rsh and series resistance Rs are very important since they can give a first idea on conduction processes occurring across the interface of the junction. They also inform about the device performances and the possibilities to optimize them. For a good understanding of the device operation and its performances, these parameters are to be determined as precisely as possible. In this work, a new and simple method is proposed for the analysis the current-voltage (I-V) characteristic of a p-n junction diode. The method involves performing a mathematical operation on the experimental data that allows to calculate the parameters at values of forward current smaller than the reverse saturation current, is firstly applied to a simulated p-n diode. This technique was applied successfully to a heterojunction using antimonide semiconductors and the results were compared to those obtained by using two other methods encountered in the literature.

Keywords: I-V characteristics · Diode · Ideality factor · Saturation current GaAlAsSb

1 Introduction

The ideally current-voltage characteristics of the p-n junction or Schottky diode is given by the following expression (Sze 1981):

$$I = I_S\left(\exp(\frac{\beta V}{n}) - 1\right) \tag{1}$$

where I is the measured current, V the external applied voltage, Is the saturation current and n the ideality factor. The term β in Eq. (1) represents the thermal voltage q/kT. Generally, the current-voltage (I-V) characteristic for a semiconductor p-n junction

B. Abdelbaki et al. (Eds.): SMSD 2017, *Proceedings of the Third International Symposium on Materials and Sustainable Development*, pp. 116–125, 2018.
https://doi.org/10.1007/978-3-319-89707-3_14

based device can be described by a single exponential expression, which takes into account the effects of parasitic series resistance Rs and shunt resistance Rsh. When Rs is sufficiently important, it's effect Rs can be clearly observed on forward semi-logarithmic plot of the I-V characteristic and causes a deviation from linearity of this plot (see Fig. 1). The effect of Rsh appears at low forward or reverse current-voltage characteristic (Lee et al. 1998). This resistance is used to model the excess currents, generally due to surface or bulk defects. At low forward or reverse bias, the relation between the current across the p-n junction and the external applied voltage is given by the well known formula (Aubry and Meyer 1994):

$$I = I_S\left(\exp(\frac{\beta V}{n}) - 1\right) + \frac{V}{R_{sh}} \tag{2}$$

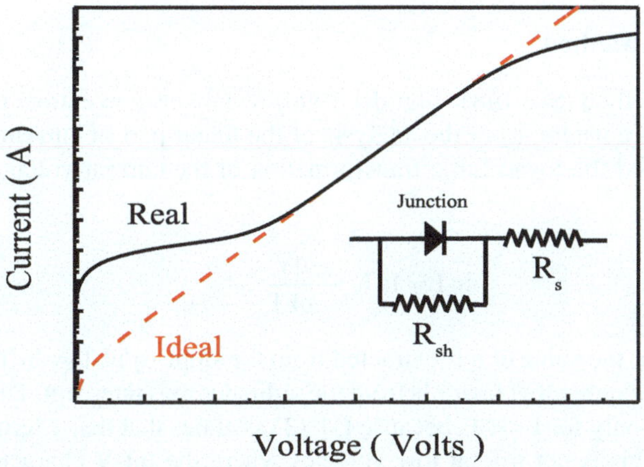

Fig. 1. Semi-logarithmic representation of the ideal and real I-V characteristics of the homojunction diode.

The value of the saturation current Is and the ideality factor n are very important since they can give a first idea on conduction processes occurring across the interface of the junction. They also inform about the device performances and the possibilities to optimize them. For example, n = 1 when the forward current is limited by thermionic emission. Shockley-Read-Hall model assumes that recombination in the space charge region taking place via a single trap level within middle of the gap of low doped side of the junction provides n = 2 and for an exponential defect distribution, the value of n may lie between 1 and 2 (Radziemska 2005; Jain and Kapoor 2005; Bayhan 2004). Different methods have been proposed to extract these parameters from the experimental current-voltage (I-V) curve (Kavasoglu et al. 2009; Mikhelashvili et al. 2001; Slavica et al. 2015; Evangelou et al. 1993; Ferhat-Hamida et al. 2002; Chan and Phang 1987; Bashahu and Nkundabakura 2007; Arcipiani 1985). Standard method which requires the presence of linear region in the ln(I) versus V plot (Sze 1981; Shroder 1990). This relatively simple analysis fails when large series resistance Rs is present. Methods using some auxiliary functions that allow the separation of the effect of Rs (Norde 1979;

Cheung and Cheung 1986). Method using the derivative of the current with respect to voltage (Lee et al. 1998; Werner 1988). Method using the integration of the current with respect to voltage (Ortiz-Conde et al. 1995; Kaminski et al. 1999). In this paper, we present a simple and generalized analytical method for the determination of the ideality factor n and the saturation current Is of a p-n junction based device from forward current-voltage characteristics, our proposed method is based on the use of first derivative of the current with respect to voltage. The technique is applied successfully to both simulated and experimental data, measured on a InAsSb based heterojunction. The results obtained on both simulated and measured characteristics are, finally, compared to those obtained by using other methods encountered in the literature.

2 Determination of Diode Parameters

2.1 Standard Method

The standard method (Sze 1981; Shroder 1990) widely used to extract the p-n junction based device parameters, from the analysis of the linear part of current-voltage curve. This method used the logarithmic transformation of the current-voltage characteristic based in Eq. (1):

$$\ln I = \ln I_S + \frac{qV}{nkT} \tag{3}$$

Using Eq. (3) the value of n is extracted from the slope ($q/nkT = \Delta(\ln I)/\Delta V$) and the value of the pre-exponential factor Is from the ordinates axis intercept. This conventional method is valid only for I << Is because Eq. (3) assumes that the -1 term in Eq. (1) can be neglected. This is not true at low voltages where the lnI-V characteristics deviate from linearity due to the non-exponential behavior of the junction near the origin.

2.2 Method Using the Integral (Ranuarez et al.)

Ranuarez et al. proposed the following procedure to extract the ideality factor and the saturation current. This technique is based on the numerical integration of the forward current from zero up to the maximum voltage value of interest. Integrating Eq. (1) (Ranuarez et al. 1999):

$$\int_0^V I.dV = \frac{nkT}{q} I - V.I_S \tag{4}$$

A quantity A, which corresponds to a voltage, can be obtained when dividing Eq. (4) by the current (I). Another quantity B, corresponding to a current, is obtained by dividing this same equation by the voltage (V) term. A and B are given by:

$$A = \frac{\int_0^V I.dV}{I} = \frac{nkT}{q} - \left(\frac{V}{I}\right) I_S \tag{5}$$

$$B = \frac{\int_0^V I.dV}{V} = \frac{nkT}{q}\left(\frac{I}{V}\right) - I_S \tag{6}$$

According to Eq. (5) a plot of A as a function of V/I should produce a straight line with slope –Is and ordinates axis intercept of nkT/q. Similarly, according to Eq. (6) a plot of B as a function of I/V should produce a straight line with slope nkT/q and ordinates axis intercept –Is. The use of either one of these equations represents the basis of the proposed method. The numerical integration can be obtained by the means of the trapezoidal method. In that case, since we consider that the curve can be approximated by a straight line between two successive I-V pairs of data, the voltage hop must be as small as possible to have the best accuracy. The results obtained by this method of the simulated data correspond only to an ideal diode (Eq. (1)). However, in the case where a structure present a leakage currents, has the technique of Ranuarez is applicable?

2.3 Our Method

Our proposed method for the extraction of the saturation current Is and the ideality factor n is based on the use of first derivative of the current with respect to voltage. Differentiating Eq. (1) with respect to voltage gives:

$$\frac{dI}{dV} = \frac{q}{nkT} I_S \exp\left(\frac{qV}{nkT}\right) \tag{7}$$

by addition and subtraction of the quantity qIs/nkT in Eq. (7) we can obtain:

$$\frac{dI}{dV} = \frac{q}{nkT} I_S \left[\exp\left(\frac{qV}{nkT}\right) - 1\right] + \frac{qI_S}{nkT} \tag{8}$$

$$\frac{dI}{dV} = \frac{q}{nkT} I + \frac{qI_S}{nkT} \tag{9}$$

Equation (9) can be rewritten as

$$I = \frac{nkT}{q}\frac{dI}{dV} - I_S \tag{10}$$

Our proposed method is based on the plot of the current I versus conductance dI/dV. It is clear from Eq. (10) that the plot A should produce a straight line with slope nkT/q. The intercept with the y axis gives directly the value of saturation current –Is of the device.

3 Test and Comparison

To illustrate the approach, a diode characteristic corresponding to a single exponential expression and a leakage current term of the type of Eq. (2), is simulated. The values used for Is and n are 0.6 pA, 2.0 respectively and several different values of Rsh. The result of the simulation is given on Fig. 2.

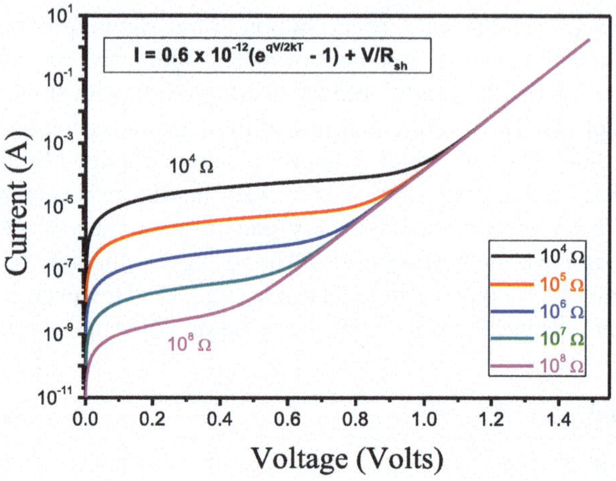

Fig. 2. Simulated I-V characteristics using the parameters with Is = 0.6 pA, and n = 2 and several different values of Rsh.

The plot obtained by using our method as calculated by Eq. (10) is presented in Fig. 3. It is clearly that all the curves corresponding to the different quantity of leakage current, are superposed lines for the arbitrary chosen shunt resistance. The values of n and Is, extracted from Fig. 3, corresponding to the different values used for Rsh, are given on Table 1. The results show that both n and Is do not depend on the value used

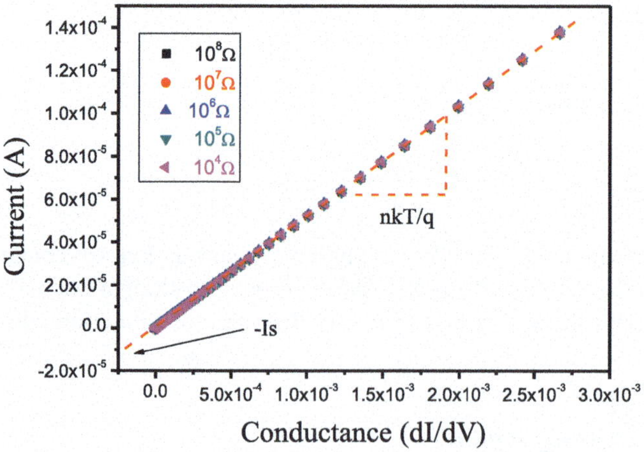

Fig. 3. Plot of Eq. (10) using simulated data of Fig. 2. The slope gives an extracted value means of n = 1.19 ± 0.016, and the ordinates axis intercept gives an extracted value means of Is = 0.599 ± 0.012pA.

for the shunt resistance. Our method gives results in very good agreement with n and Is, used initially for the simulation of the current-voltage characteristic.

Table 1. Comparison of various parameter extraction methods for different values of the shunt resistance.

Method	Parameters	Shunt resistance (Ω)				
		10^8	10^7	10^6	10^5	10^4
Our method	n	2.00	1.99	2.00	2.01	2.01
	Is(pA)	0.60	0.60	0.59	0.61	0.61
Standard method	n	1.99	1.77	1.59	1.44	1.20
	Is(pA)	0.62	1.89	2.90	4.72	8.42
Integration method	n	0.88	0.56	0.11	0.09	0.02
	Is(pA)	3.25	6.66	11.2	56.1	89.5

From data of Fig. 2, the results for the plot A proposed by Ranuarez et al. are regrouped into one graph; Fig. 4. The proposed techniques, which use integration procedure, suffer from numerical errors. They are very sensitive to the measurement's noise. These methods also fail in the presence of leakage.

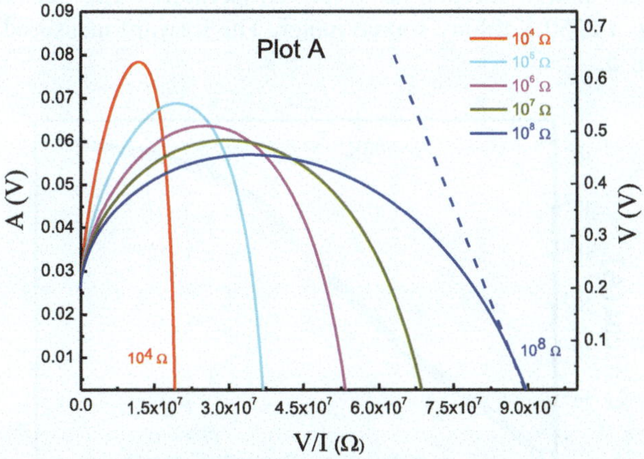

Fig. 4. Ranuarez Plot A of Eq. 5 using simulated data of Fig. 2.

The simulated current-voltage characteristic was analyzed using the methods proposed by Ranuarez et al. and standard method. The results obtained are compared to those obtained by our method, in Table 1. It is clear from Table 1 that values of Is and n obtained by integration or standard method are very sensitive to the leakage current (Rsh). Accurate results are extracted for very higher value of shunt resistance. For small shunt resistance, values of Is and n are erroneous. In contrary when using our proposed method, values extracted for Is and n are very close and in good agreement with those used for the simulation, for any chosen quantity of leakage current.

4 Application to an InAsSb Based Double Heterojunction Diode

Antimonide based materials, such as InAsSb have a strong potential for the development of mid-infrared devices such as lasers or detectors operating between 2 and 5 μm. Applications in this spectral range are found in various domains such as pollutant detection, infrared thermal imaging, lidars, or optical countermeasures. It has been shown that the ternary InAsSb alloy grown on the GaSb or InAs substrate is a promising candidate for the development of room temperature mid-infrared photodetectors (Rogalski 1989; Kim and Razeghi 1998). This material offers a number of advantages such as wide spectral range, low electron effective mass, and high mobility at room temperature.

The studied structure was fabricated on a Varian Gen II Solid-source molecular beam epitaxy machine (Aït-Kaci. et al. 2005). A 1 μm thick and unintentionally doped $InAs_{0.9}Sb_{0.1}$ layer was deposited on n-type, Te-doped ($4 \times 10^{17} cm^{-3}$) and (100) oriented GaSb substrate, at 430 °C with a growth rate of 0.5 ML/s. To form a p-n junction, as a transparent window, a 30 nm thick and Te-doped $Ga_{0.6}Al_{0.4}As_{0.034}Sb_{0.966}$ and a 30 nm thick Be-doped $Ga_{0.6}Al_{0.4}As_{0.034}Sb_{0.966}$ layers were deposited on InAsSb, with a growth rate of 1 μm/h and a V/III ratio of about 2.

In this section, using our method described above, we studied an experimental forward current-voltage characteristic, obtained on GaAlAsSb(p)/GaAlAsSb(n)/InAsSb(n) double heterojunction, at room temperature. The measurements were performed using a 2250 Keithley source meter. The forward measured characteristics are given on Fig. 5.

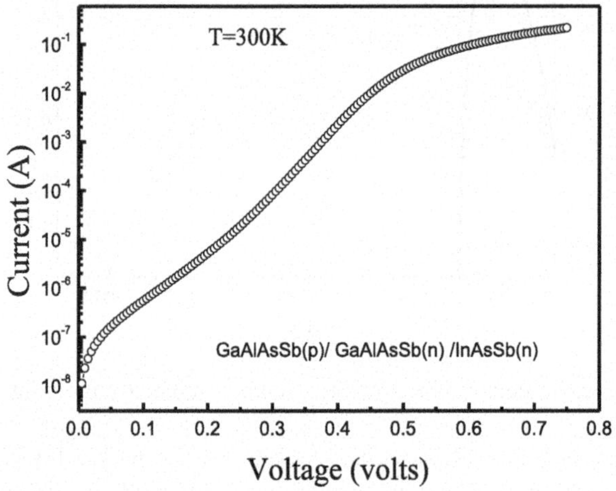

Fig. 5. Forward current-voltage characteristic measured on a GaAlAsSb(p)/GaAlAsSb(n)/InAsSb(n) based diode, at room temperature.

To extract saturation current (Is) and the ideality factor (n) of the studied diode, we used plot of Eq. (10), described in the previous section. The results are represented on Fig. 6. The results obtained, using our method on experimental data, are also compared to those obtained the two other techniques on Table 2. Saturation current and ideality factor, extracted by our method or standard method are more homogenous. For integration method the values of Is and n are erroneous.

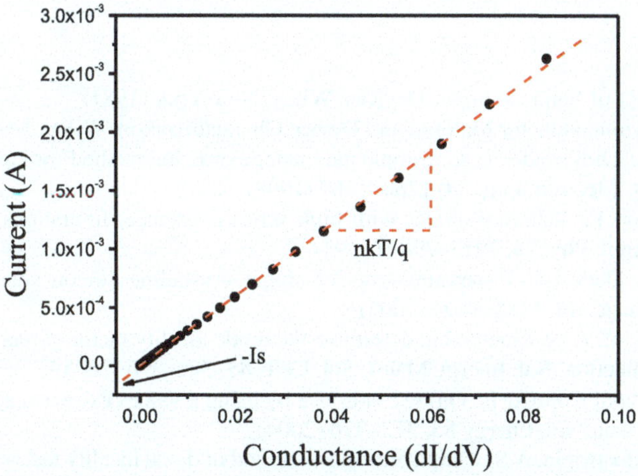

Fig. 6. Plot of Eq. (10) extracted from the measured data of Fig. 5. The slope gives an extracted value of n = 1.17, and the ordinates axis intercept gives an extracted value of Is = 1.57×10^{-8} A.

Table 2. Extracted parameters values of a diode using various methods

Parameter	Standard method	Ranuarez et al.	Our method
Is(A)	3.56×10^{-7}	1.33×10^{-4}	1.57×10^{-8}
n	1.33	0.08	1.17

The results show that the device studied is characterized by small saturation current close to 1.57×10^{-8}A and an ideality factor around 1.17. This later value indicates that the current flow across the interface, when the device is under forward polarizations, is due to the minority carrier diffusion phenomena (Sze 1981).

5 Conclusions

A new and simple method is proposed for the extraction of the ideality factor and the saturation current of an InAsSb based heterojunction device. First, the technique was applied to a synthetic current voltage characteristic with a saturation current and ideality factor equal to 0.6 pA and 2, respectively and several different values of shunt resistance. The results obtained were in excellent agreement with the data used for the simulated characteristic. Our results were also compared to those obtained by a integration based and standard methods as well as the method proposed by Ranuarez et al., with different values of the shunt resistance (Rsh), and on the same synthetic I-V curve. These two procedures seem to suffer from the leakage current. When applying our method to the measured currents voltage characteristic, n and Is, corresponding to our InAsSb based diode are around 1.17, 157 nA.

References

Sze, S.M.: Physics of Semiconductor Devices. Wiley, New York (1981)

Shroder, D.K.: Semiconductor Material and Device Characterization. Wiley, New York (1990)

Lee, J.I., Brini, J., Dimitriadis, C.A.: Simple parameter extraction method for non-ideal schottky barrier diodes. Electron. Lett. **34**, 1268–1269 (1998)

Aubry, V., Meyer, F.: Schottky diodes with high series resistance: limitations of forward I-*V* methods. J. Appl. Phy. **76**, 7973–7984 (1994)

Radziemska, E.: Dark I-U-T measurements of single crystalline silicon solar cells. Energy Convers. Manage. **46**, 1485–1494 (2005)

Jain, A., Kapoor, A.: A new method to determine the diode ideality factor of real solar cell using Lambert W-function. Sol. Energy Mater. Sol. Cells **85**, 391–396 (2005)

Bayhan, H.: Study of CdS/Cu(In, Ga)Se2 interface by using n values extracted analytically from experimental data. Sol. Energy **83**, 372–376 (2004)

Kavasoglu, N., Kavasoglu, A.S., Oktik, S.: A new method of diode ideality factor extraction from dark I-V curve. Curr. Appl. Phys. **9**, 833–838 (2009)

Mikhelashvili, V., Eisenstein, G., Uzdin, R.: Extraction of schottky parameters with a bias dependent barrier height. Solid-State Electron **45**, 143–148 (2001)

Slavica, M.P., Milena, D.D., Tatijana, D., Danilo, N., Martin, P.C.: Concerning a novel mathematical approach to the solar cell junction ideality factor estimation. Appl. Math. Model. **39**, 3248–3264 (2015)

Evangelou, E.K., Papadimitriou, L., Dimitriades, C.A., Giakoumakis, G.E.: Extraction of schottky diode (and p-n junction) parameters from I-V characteristics. Solid-State Electron **36**, 1633–1635 (1993)

Ferhat-Hamida, A., Ouennoughi, Z., Hoffmann, A., Weiss, R.: Extraction of schottky diode parameters including parallel conductance using a vertical optimization method. Solid-State Electron **46**, 615–619 (2002)

Chan, D.S.H., Phang, J.C.H.: Analytical methods for the extraction of solar-cell single- and double-diode model parameters from I-V characteristics. IEEE Trans. Electron Dev. **34**, 286–293 (1987)

Bashahu, M., Nkundabakura, P.: Review and tests of methods for the determination of the solar cell junction ideality factors. Sol. Energy **81**, 856–863 (2007)

Arcipiani, B.: Generalization of the area method for the determination of the parameters of a non-ideal solar cell. Rev. Phys. Appl. **20**, 269–272 (1985)

Norde, H.: A modified forward I-V plot for schottky diodes with high series resistance. J. Appl. Phys. **50**, 5052–5053 (1979)

Cheung, S.K., Cheung, N.W.: Extraction of schottky diode parameters from forward current-voltage characteristics. Appl. Phys. Lett. **49**, 85–87 (1986)

Werner, J.H.: Schottky barrier and pn-junction I/V plots-small signal evaluation. Appl. Phys. **47**, 291–300 (1988)

Ortiz-Conde, A., García Sánchez, F.J., Liou, J.J., Andrian, J., Laurence, R.J., Schmidt, E.P.: A generalized model for a two-terminal device and its applications to parameter extraction. Solid-State Electron **38**, 265–266 (1995)

Kaminski, A., Marchand, J.J., Laugier, A.: I-V methods to extract junction parameters with special emphasis on low series resistance. Solid-State Electron. **43**, 741–745 (1999)

Ranuarez, J.C., Garcia Sanchez, F.J., Ortiz-Conde, A.: Procedure for determining diode parameters at very low forward voltage. Solid-State Electron. **43**, 2129–2133 (1999)

Aït-Kaci, H., Nieto, J., Rodriguez, J.B., Grech, P., Chevrier, F., Salesse, A., Joullié, A., Christol, P.: Optimization of InAsSb photodetector for non-cryogenic operation in the mid-infrared range. Phys. Status Solidi. **202**, 647–651 (2005)

Rogalski, A.: $InAs_{1-x}Sb_x$ infrared detectors. Prog. Quant. Electron **13**, 191–231 (1989)

Kim, J.D., Razeghi, M.: Investigation of InAsSb infrared photodetectors for near-room temperature operation. Opto-Electron. Rev. **6**, 217–230 (1998)

Controlling Drug Release Through Poly(2-Hydroxyethylmethacrylate-co-acrylic Acid) Grafted Sodium Alginate

L. Bounabi(✉), N. Bouslah Mokhnachi, N. Haddadine, and A. Benaboura

Faculty of Chemistry, Laboratory of Synthesis Macromolecular and Thio-Organic Macromolecular, University of Sciences and Technology Houari Boumediene, 16111 Algiers, Algeria
Leila_bounabi@yahoo.fr

Abstract. The present paper reveals the potential uses of grafted hydrogel of Alginate-g-poly(2-hydroxyethyl methacrylate-co-acrylic acid) in controlled drug delivery. The graft copolymerization reaction was carried out in a homogeneous medium, in the presence of potassium persulfate (KPS) as initiator and N, N'-methylenebisacrylamide (MBA) as crosslinker. The grafting Efficiency was calculated and found to be 77.47%.

To investigate the structure of the obtained hydrogel, Fourier transform infrared spectroscopy (FTIR), thermogravimetric analysis (TGA) and Differential scanning calorimetry (DSC) were used. Swelling behavior of the obtained hydrogel in different pH media was investigated. It was found that the synthesized hydrogel has a high pH sensitivity. Drug loading and controlled release properties of the grafted hydrogel, in both stimulated gastric (pH 1.2) and stimulated intestinal (pH 7.4) media at 37 °C, were also evaluated using Tramadol chlorhydrate as model drug. The kinetics of drug release from hydrogel matrices has been examined using several mathematical models. The results showed that the best fitted curves were obtained with Higuchi model. The in vitro release suggests that this hydrogel can be considered as an excellent candidate to design novel drug delivery systems.

Keywords: Alginate · Hydrogel · Graft copolymer · Swelling Controlled release

1 Introduction

Hydrogels are defined as three-dimensional networks of hydrophilic polymers that swell in water or biological fluids. This property of hydrogels made them as important materials in a variety of applications such, agriculture (Abd El-Rehim 2006), wastewater treatment (La et al. 2011), biomedical area, tissue engineering (Daniele et al. 2014), and drug delivery system (Nehls et al. 2016).

Stimuli-responsive hydrogels that can respond to environmental physical and chemical stimuli, such as temperature, pH, electric field, chemical environments, etc., have

B. Abdelbaki et al. (Eds.): SMSD 2017, *Proceedings of the Third International Symposium on Materials and Sustainable Development*, pp. 126–133, 2018.
https://doi.org/10.1007/978-3-319-89707-3_15

become an important area of research and development in the field of medicine, pharmacy and biotechnology (Wang and Wang 2010).

In recent years, much interest has been paid to the development of natural-based hydrogels. Sodium alginate is a polyanionic natural polymer, that has received considerable attention in industrial and medical fields (AL-Kahtani and Sherigara 2014) due to its good biocompatibility and biodegradability.

The objective of the present work was to combine the advantages of a natural and a synthetic polymer to obtain hydrogels with improved properties such as pH responsivity and absorbent properties. After characterization of the synthesized hydrogels by different techniques, the release profile, of Tramadol hydrochloride (TH) from these drug carrier systems was investigated.

2 Materials and Methods

2.1 Materials

2-hydroxyethylmethacrylate (HEMA), Acrylic acid (AA), N,N'-methylenebisacrylamide (MBA), Sodium alginate (Alg), were purchased from Sigma-Aldrich. Potassium persulfate (KPS) was purchased from Fluka. All agents were analytical grade and were used without further purification and all solutions were prepared with distilled water.

2.2 Synthesis of Alg-g-P(HEMA-co-AA)

Sodium alginate solution (3%, w/v) was prepared in distilled water under constant stirring. After being purged with nitrogen gas for 20 min to get rid of dissolved oxygen in the solution, and heating to 60 °C in a water bath. the KPS solution as initiator and 1 g followed by HEMA and AA monomers were added. Afterwards MBA solution was introduced in to the flask dropwise. The reactional mixture was heated to 70 °C and kept for 6 h to complete polymerization. The resulting gel product was dried at 50 °C and then it was taken in water and kept for a week with occasional shaking to remove the water soluble part from the hydrogel. The water insoluble gel sample was further dried to a constant weight. The percentage grafting efficiency *GE* (%) were calculated as

$$GE(\%) = \left(\frac{w_1 - w_0}{w_2}\right) \times 100 \tag{1}$$

where w_0, w_1, and w_2 denote the weights of alginate, graft hydrogel, and monomers, respectively.

2.3 Characterization of Graft Hydrogel

The synthesized hydrogels were characterized by the following methods

2.3.1 Fourier Transform Infrared (FTIR) Spectroscopy

FTIR spectra of the hydrogels were recorded on a FTIR spectrometer (Perkin Elmer, model-Spectrum-Two) using KBr pellet made by mixing KBr with fine powder of the polymer gel samples.

2.3.2 Differential Scanning Calorimetric Analysis (DSC)

The alginate, hydrogel, and drug-loaded hydrogel beads were subjected to DSC analysis. All the samples were heated from 30–300 °C at a heating rate of 10 °C/min under nitrogen atmosphere using a Jade, Perkin-Elmer DSC apparatus.

2.3.3 Thermogravimetry Analysis (TGA)

TGA/DTG studies were performed on TGA Mettler-Toledo under nitrogen atmosphere at a heating rate of 10 °C/min in the temperature range from ambient to 600 °C.

2.3.4 Swelling Measurements

The swelling properties of the as-prepared hydrogels were investigated by gravimetric analysis. Briefly, the dehydrated samples were immersed in buffer solution at room temperature. At regular intervals, the hydrogels were drawn out from the medium, weighed after wiping off the excess solution with filter paper, and then taken back to the original solution. The measurements were taken again and again until the weight of hydrogels remained stable at a constant value. The swelling ratio (SR, %) was calculated using the following equation:

$$SR(\%) = \left(\frac{w_i - w_0}{w_0}\right) \times 100 \quad (2)$$

Here, w_i and w_0 are the weights of the swollen and dry hydrogels, respectively.

2.4 Study of Drug Loading and Release of the Hydrogel

TH was loaded in the hydrogel network by diffusion method. Alg-g-P(HEMA-co-AA) hydrogels was added to the drug solution.

The release rate experiments were performed in simulated gastric fluid (SGF) and in a simulated intestinal fluid (SIF) at 37 ± 05 °C with shaking the solutions during experiments. At different time intervals, Aliquot samples were withdrawn and replaced with the same volume of fresh solution. The amount of drug released was analyzed using UV–visible spectrophotometer at 271 nm after suitable dilutions.

3 Results and Discussion

The Alg-g-P(HEMA-co-AA) hydrogels was synthesized by free radical polymerization using potassium persulfate as free radical initiator under nitrogen atmosphere. The grafting efficiency was found to be 77.47%.

3.1 FTIR Spectroscopy

FTIR spectroscopy was used to investigate the structure of the hydrogels. The FTIR spectra of alginate, Tramadol chlorhydrate (TH), P(HEMA-co-AA), alginate-g-P(HEMA-co-AA), and alginate-g-P(HEMA-co-AA)/TH gels are shown in Fig. 1. In alginate spectrum, the absorption band at 3441 cm^{-1} is assigned hydrogen bonded OH group. Asymmetric and symmetric stretching vibrations of $-COO^{-}$ groups of alginate are observed at 1619 and 1419 cm^{-1}, respectively. The characteristic peak of polysaccharide structure corresponding to C–O stretching vibration appears at 1031 cm^{-1}.

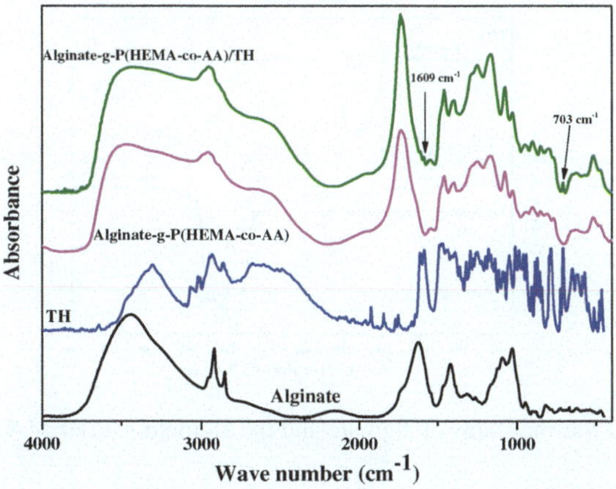

Fig. 1. FTIR spectra of a sodium alginate, TH, P(HEMA-co-AA) alginate-g-P(HEMA-co-AA) and alginate-g-P(HEMA-co-AA)/TH

The alginate-g-P(HEMA-co-AA) spectrum (Fig. 1), showed a broad band at about 3464 cm^{-1} corresponding to the resulting of stretching band of -OH groups of the alginate and -OH groups of the copolymer P(HEMA-co-AA), in addition, we can also see the absorption bands at 1729 cm^{-1} corresponding to the stretching C = O group of P(HEMA-co-AA).

Compared to the FTIR spectrum of alginate-g-P(HEMA-co-AA) hydrogel, the drug loaded hydrogel alginate-g-P(HEMA-co-AA)/TH spectrum, shown in the same figure, had revealed new peaks appearing at 1609 cm^{-1} and 703 cm^{-1} corresponding to the C = C, OH stretching vibration of Tramadol, respectively. These bands confirmed the presence of the TH in the alginate-g-P(HEMA-co-AA)/TH hydrogels.

3.2 TGA Analysis

Thermal stability of hydrogel was investigated by comparing the TGA curves of sodium alginate and Alginate-g-P(HEMA-co-AA) Fig. 2.

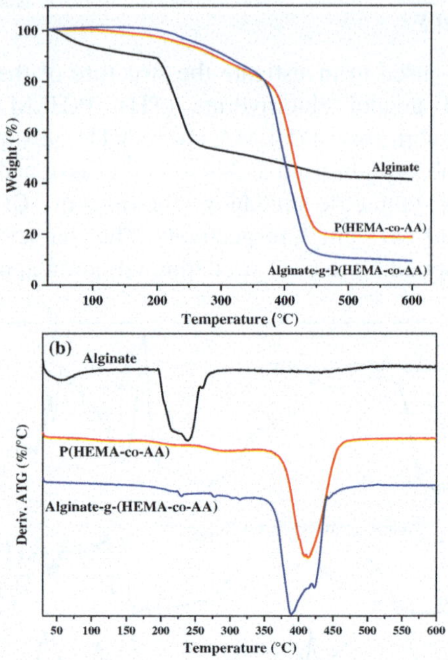

Fig. 2. TGA/DTG thermograms of alginate and the alginate-g-P(HEMA-co-AA) hydrogel.

As illustrated in Fig. 2, alginate thermogram shows a first mass loss at around 86 °C due to moisture elimination and a second mass loss from 193 to 284 °C corresponding to polymer chains decomposition. The decomposition of Alginate-g-P(HEMA-co-AA) occurs from 341 to 477 °C.

The examination of the derivative curves reveals the absence of the alginate decomposition peak from the grafted hydrogel thermogram indicating that the P(HEMA-co-AA) copolymer was able to protect Alg from further degradation.

These observations have clearly indicated that grafting of poly(HEMA-co-AA) has improved the thermal stability of alginate.

3.3 DSC Analysis

DSC further confirmed the formation of graft copolymers. The Tg of the alginate disappears in the thermogram of grafted hydrogel as a result of the restricted chain motion of the macromolecules (Fig. 3).

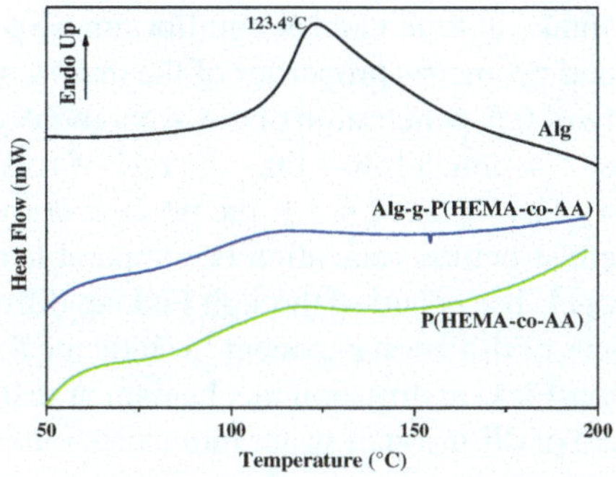

Fig. 3. DSC of Alginate, P(HEMA-co-AA), Alginate-g-P(HEMA-co-AA).

3.4 Swelling Study

The hydrogels based on the P(HEMA-co-AA) copolymers exhibit pH-sensitive behavior. Figure 4 shows that at pH 7.4 the swelling rate increases considerably compared to the values obtained in pH 1.2. This phenomenon is attributed mainly to the effect of carboxylic groups of acrylic acid comonomer.

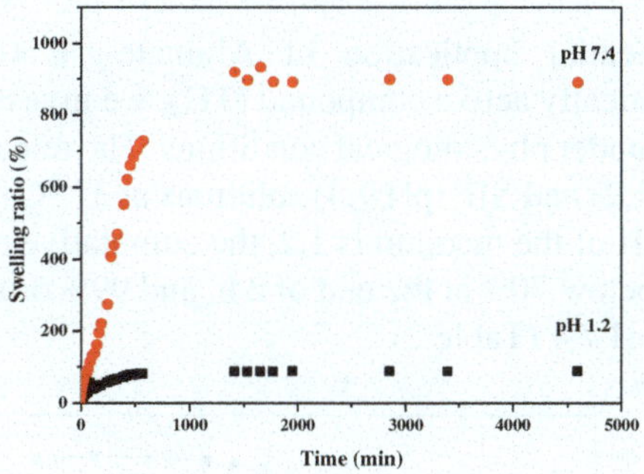

Fig. 4. Swelling behaviour of hydrogels at pH 1.2 and 7.4

At high pH, the carboxylic groups are converted to carboxylate anions with dilatation of the gel network due to electrostatic repulsion causing increased swelling. (Mandal and Ray 2013). In contrast, at low pH, hydrogen bonds between the COOH groups of the copolymer can easily be developed. This results in the formation of additional physical networks (Wang and Wang 2010). Thus, the hydrogels shrink at low pH.

To describe the mechanism transport of water molecules from polymeric matrices, Fick's law of diffusion was used.

$$\frac{w_t}{w_\infty} = kt^n \tag{3}$$

w_t and w_∞ are the water intake at time t and at equilibrium respectively. k is a constant indicating the structural and geometric properties of the matrix, n is a number indicative of the type of diffusion. If $n \leq 0.5$, penetration of the water molecules follows the Fickian law and the rate of diffusion is much lower than the rate of relaxation. In cases, where n = 1, the diffusion is very fast. If $0.5 < n < 1$, the process corresponds to non-Fickian anomalous diffusion where structural relaxation is comparable to diffusion.

The swelling of hydrogels has occurred through Fickian diffusion mechanism at low pH (Table 1), but the values of diffusion exponent 'n' indicate that swelling of the polymers occurred through non-Fickian diffusion mechanism at distilled water and pH 7.4. In this mechanism, the rate of diffusion of water molecules into the polymer matrix and rate of polymer chains relaxation are comparable.

Table 1. Swelling kinetic parameters of the hydrogel samples.

	n	k	R^2	Swelling mechanism
pH 1.2	0.45	2.91	0.99	Fickian
pH 7.4	0.60	4.68	0.99	non-Fickian
Distilled water	0.47	2.73	0.99	non-Fickian

3.5 TH Loading and Releasing

To determine the potential application of Alginate-g-P (HEMA-AA) hydrogel containing a pharmaceutically active compound (TH), we investigated the drug release behavior of this system under physiological conditions. The release rate experiment was performed in SFG (pH 1.2) and SIF (pH 7.4) solutions at 37 °C (Fig. 5). As can be seen from Fig. 5, when the pH of the medium is 1.2, the cumulative release rate of TH from the tested hydrogels is below 70% at the end of 2 h, and 99% drug release was recorded in alkaline medium of pH 7.4 (Table 2).

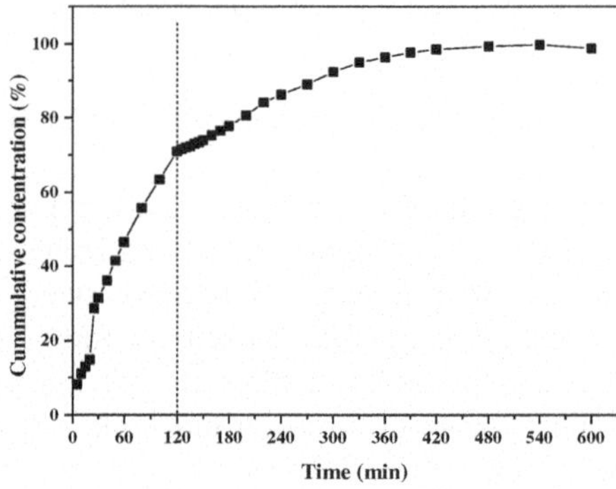

Fig. 5. Release of TH from Alginate-g-P(HEMA-co-AA) as a function of time in pH 1.2 and pH 7.4 at 37 °C

Table 2. Kinetic parameters of drug release by using different models

Zero order		First order		Higuchi		Krosmery-Peppas		
K_0	R^2	K_1	R^2	K_h	R^2	n	k_{KP}	R^2
0.04	0.97	0.54	0.89	0.09	0.998	0.49	0.96	0.996

4 Conclusions

Alginate-g-P(HEMA-co-AA) hydrogel, were synthesized through grafting of AA and HEMA monomers onto Alginate using potassium persulfate as an initiator and methylene bis acrylamide as a crosslinking agent under an inert atmosphere. The results of FTIR spectra showed -OH of alginate participated in graft polymerization with acrylic acid and hydroxyethylmethacrylate.

The pH-sensitive Alg-g-P(HEMA-co-AA) copolymer was successfully synthesized by free radical polymerization,

The Tramadol chlorhydrate release was significantly increased when pH of the medium was changed from acidic to alkaline.

References

Abd El-Rehim, H.A.: Characterization and possible agricultural application of polyacrylamide/sodium alginate crosslinked hydrogels prepared by ionizing radiation. J. Appl. Polym. Sci. **101**, 3572–3580 (2006)

AL-Kahtani, A.A., Sherigara, B.S.: Semi-interpenetrating network of acrylamide-grafted-sodiumalginate microspheres for controlled release of diclofenac sodium, preparation and characterization. Colloids Surf., B **115**, 132–138 (2014)

Daniele, M.A., Adams, A.A., Naciri, J., North, S.H., Ligler, F.S.: Interpenetrating networks based on gelatin methacrylamide and PEG formed using concurrent thiol click chemistries for hydrogel tissue engineering scaffolds. Biomaterials **35**, 1845–1856 (2014)

La, Y.H., McCloskey, B.D., Sooriyakumaran, R., Vora, A., Freeman, B., Nassar, M., Hedrick, J., Nelson, A., Allen, R.: Bifunctional hydrogel coatings for water purification membranes: improved fouling resistance and antimicrobial activity. J. Membr. Sci. **372**, 285–291 (2011)

Mandal, B., Ray, S.K.: Synthesis of interpenetrating network hydrogel from poly(acrylic acid-co-hydroxyethyl methacrylate) and sodium alginate: Modeling and kinetics study for removal of synthetic dyes from water. Carbohydr. Polym. **98**, 257–269 (2013)

Nehls, E.M., Rosales, A.M., Anseth, K.S.: Enhanced user-control of small molecule drug release from a poly(ethylene glycol) hydrogel *via*azobenzene/cyclodextrin complex tethers. J. Mater. Chem. B **4**, 1035–1039 (2016)

Wang, W., Wang, A.: Synthesis and swelling properties of pH sensitive semi-IPN superabsorbent hydrogels based on sodium alginate-g-poly(sodium acrylate) and polyvinylpyrrolidone. Carbohydr. Polym. **80**, 1028–1036 (2010)

Theoretical Study of Eu^{3+} and Li^{+} Co-doped Ca_2SnO_4

M. Baira[1(✉)], A. Siad-Bekhti[2], N. Siziani[1], H. Bouhani-Benzian[1], and M. Sahnoun[1]

[1] Laboratoire de Physique Quantique de la Matière et Modélisation Mathematique (LPQ3M), Faculté des Sciences Exactes, Université Mustapha Stambouli, Mascara, Algeria
bnaima64@yahoo.fr

[2] Laboratoire de Matériaux Applications et Environnement (LMAE), Faculté des Sciences Exactes, Université Mustapha Stambouli, Mascara, Algeria

Abstract. The electronic structures of the non-doped and Li^{+}, Eu^{3+} co-doped Ca_2SnO_4 materials were calculated employing the density functional theory with the WIEN2k package. The later is based on the full potential linearized augmented plane wave method (FP-LAPW), an approach which is among the most precise and reliable ways to calculate the electronic structure of solids. The generalized gradient approximation (GGA) method was applied to the non-doped material. In order to describe better the strongly correlated 4f electrons of the Eu^{3+} the GGA+U method was used. The GGA+U method requires as the input the Coulomb repulsion strength (Hubbard parameter U) and the exchange parameter J (magnetic coupling constant), which are related to the Slater integrals. In this study, the effects of Eu^{3+} and Li^{+} co-doping on the lattice parameters, and also on the electronic structures are presented and analyzed in details.

Keywords: Ca_2SnO_4:Eu^{3+} · Li^{+} · Electronic properties
GGA+U · DFT

1 Introduction

Rare earths (RE) based luminescent materials also called phosphors, are recently widely applied in emissive displays and fluorescent lamps [1], but a special attention has recently been paid to develop highly-luminescent phosphors having superior spectral qualities that can be activated in near UV region. The rare-earth elements can readily be used as the dopant, impurities called activators, added to the crystal of the material giving excellent luminescence properties. The electronic configuration and the distribution of the activator in the crystal are of high importance for the light emission. Among all the RE activators, trivalent europium (Eu^{3+}) is the most commonly used for many practical purposes in the lighting and display field. The two fundamental provition for any phosphor material are the high luminescence yield and long-term stability [2]. Recent research reports reveal that the doping of trivalent europium ions along with the alkali metal ion (Li^{+}) emerged as an effective way to enhance the photo-luminescent brightness

B. Abdelbaki et al. (Eds.): SMSD 2017, *Proceedings of the Third International Symposium on Materials and Sustainable Development*, pp. 134–140, 2018.
https://doi.org/10.1007/978-3-319-89707-3_16

and efficiency of the phosphors [3]. With the removal of the charge unbalance problem with a less defective crystal structure, alkali metal ions not only enhance the efficiency of trivalent europium ions, but also play a crucial role to control the crystallinity and morphology of the phosphors [4].

As specific examples of widely used rare earth phosphors, one could mention alkaline earth metal stannates M_2SnO_4 (M = Ca, Sr, and Ba) [5]. Calcium stannates (Ca_2SnO_4), which has an inverse spinel structure, was adopted as the host material of emitting phosphor [6].

But so far, to the best of our knowledge, there is no research report available on the photo-luminescence properties of $Ca_2SnO_4:Eu^{3+}$ phosphors with alkali metal ion Li^{+} co-doping. Accordingly, in this paper we have reported the effect of alkali metals ions co-doping on the structural, electronic and optical properties of $Ca_2SnO_4:Eu^{3+}$ investigated by first principles calculations. Furthermore, the addition of alkali metals ions makes Eu^{3+} doping in Ca_2SnO_4 energetically favorable. It is interesting to observe that alkali metal ions act as co-dopants in the host lattice and serves only as a charge compensator. This is confirmed by comparing the optical peak intensity in both samples; with and without alkali metal ions doping. The obtained interesting results make this phosphor suitable for next generation solid state lighting as well as field emission display applications.

2 Method of Calculation

The first principles calculations are performed by using the full potential linearized augmented plane-wave approach (FL-LAPW+ lo) [7] basis set as implemented in the WIEN2k code [8]. The exchange-correlation potential is calculated within the generalized gradient approximation (GGA) of Pradew-Burke-Ernezerhof (PBE) [9], and Tran and Blaha modified Becke-Johonson potential (TB-mBJ) [10]. The plane wave expansion cutoffs were 7.0 for expanding the wave function ($R_{MT}K_{max}$) which determines the matrix size (convergence), where K_{max} is plane wave cut off and R_{MT} is the smallest muffin tin sphere radii. For simulate optical properties we replace the Eu^{3+} and alkali metals ion Li^{+} in the site of Ca^{2+} in Ca_2SnO_4 structure. In the present work, a $1 \times 2 \times 2$ supercell was constructed for the Ca_2SnO_4 co-doped by Eu^{3+} and alkali metal ion (Li^{+}) translated 14-atoms simple orthorhombic.

3 Results and Discussion

3.1 Structural Properties

As an olivine structure mineral, Ca_2SnO_4 crystallizes in the orthorhombic structure with space group **no. 55** (*Pbam*) having four formula units in each unit cell which is shown in Fig. 1. This system containing 1 Ca, 1 Sn and 2 O atoms are inequivalent with atomic sites Ca1 (0.0703, 0.3183, 0.5), Sn1 (0, 0, 0), O1 (0.8067, 0.136, 0),

O2 (0.2293, 0.0479, 0.5). The calculation of the total energy as a function of the volume was used for determination of theoretical lattice constant and bulk modulus of the compound. The equilibrium lattice constant, bulk modulus, and its pressure derivative are calculated by fitting the calculated total energy into the Murnaghan's equation of state. The calculated lattice parameters of Ca_2SnO_4 are a = 5.803 Å, b = 9.841 Å and c = 3.307 Å. The lattic parameters a, b and c are consistent with the experimental results of a = 5.753 Å, b = 9.703 Å and c = 3.267 Å [11].

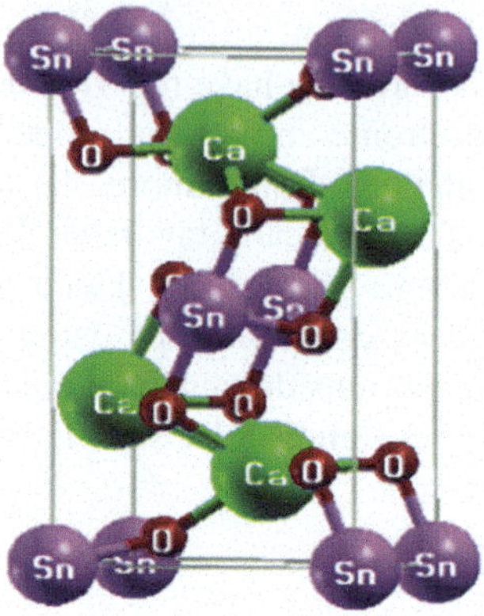

Fig. 1. Schematic crystal structure of Ca_2SnO_4.

3.2 Electronic Properties

In this work we discuss our results of the electronic properties of Ca_2SnO_4 by the density of states and band structure are shown in Fig. 2 using the GGA-PBE and TB-mBJ functional. It is seen this compound shows an indirect optical band gap. The gap between the lowest energy of the conduction band and the highest energy of the valence band is given in Table 1, along with available experimental and theoretical

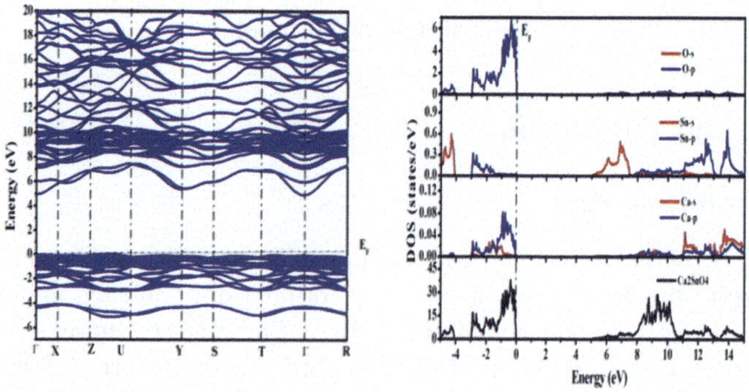

Fig. 2. Band structure of Ca_2SnO_4 with density of states using the TB-mBJ functional.

values. It indicated that there was difference between the experimental and theoretical results. This difference was normal because the program GGA often underestimated the size of the band gap [12]. But the TB-mBJ which gives the band gap values corresponding exactly to the experimental result. It can be concluded that this compound belongs to the category of materials, with large band gap.

The DOS and PDOS of Ca_2SnO_4 (Fig. 2) conveyed the useful information that: The top of the valance band is dominated by O 2p and Sn 5p states. While the bottom of the conduction band is mainly composed of the Ca 4s(3p) and Sn (5s). Valence band spread around the energy region 0 to −2.98 eV. the VB width being more is an essential property and it is discussed in detail while explaining optical properties in the later section. The core valence band is situated around −4.2. to −5.0 eV and this state is mainly due to O-p and Sn-$_S$ state for the compound.

Table 1. Calculated band gaps, in eV of Ca_2SnO_4 compound using GGA and TB-mBJ functional, along with available experimental values.

Ca_2SnO_4	GGA-PBE	TB-mBJ	Exp
Band gap (eV)	2.99 [PW]	5.05 [PW]	5.05 [12]

3.3 Optical Properties

The calculation optical properties at the equilibrium lattice constant are presented in Figs. 3 and 4, for the energy range up to 45 eV. Figure 3 shows the real and imaginary parts of the dielectric function for Ca_2SnO_4. In order to describe the optical anisotropy, the $\varepsilon_1(\omega)$ and $\varepsilon_2(\omega)$ are calculated for incident radiations with the electric field vector $\vec{E}$ polarized along the crystallographic a, b, and c axes, i.e. E_{xx}, E_{yy}, E_{zz}. The imaginary part of dielectic function stats at particular energy called threshold energy, and this

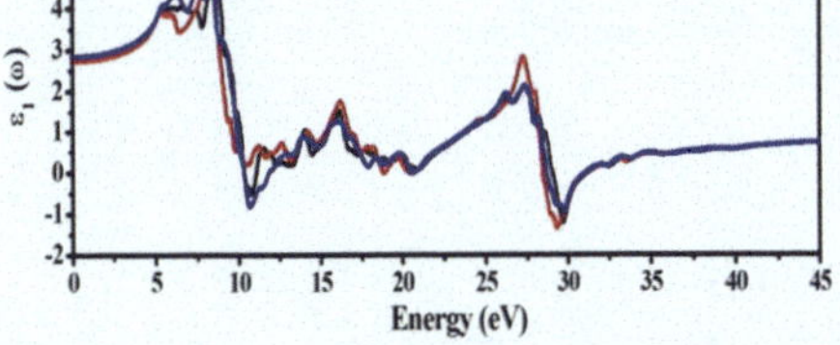

Fig. 3. Calculated real $\varepsilon_1(\omega)$ part and imaginary $\varepsilon_2(\omega)$ part of the dielectric function for Ca_2SnO_4.

threshold energy for Ca_2SnO_4 is 5.05 eV indicating the transition of electrons from valance band to the bottom of conduction band from O 2p and Sn 5p states to Ca 4s (3p) states. The peak in higher energy region of dielectric function indicates the transition of electrons from core valance band to conduction band creating hole in the core valance band, this hole recombines with the electrons in the valance band giving rise to the cross luminescence with fast decay.

The calculation linear absorption spectrum is displayed in Fig. 4. This coefficient $\alpha(\omega)$ increases rapidly when the photon energy is higher than the absorption edge, then it increase to reach its maximum at approximately **28** eV ([010]).

3.4 Electronic Properties of Ca_2SnO_4 Co-doped by $Eu^{3+}$and Li^+

In our calculations, the *4f* states of minority spin of Ca_2SnO_4:Eu^{3+}, Li^+ (Fig. 5) interact with delocalized energy level that defines the conduction band bottom, where the 4f states are located (1,2, 2,1, 2,3, 2,4, 5 eV) above the Femi level. For Eu-doped Ca_2SnO_4, Canimoglu et al. [13] using the GGA + U with Hubbard parameter $U_{eff} = 7$ eV [14, 15], have reported that the 4f states are located between the valence and conduction bands. This is probably due to the differently used Hubbard U parameter and the application of the Hubbard corrections to the Eu *4f* states.

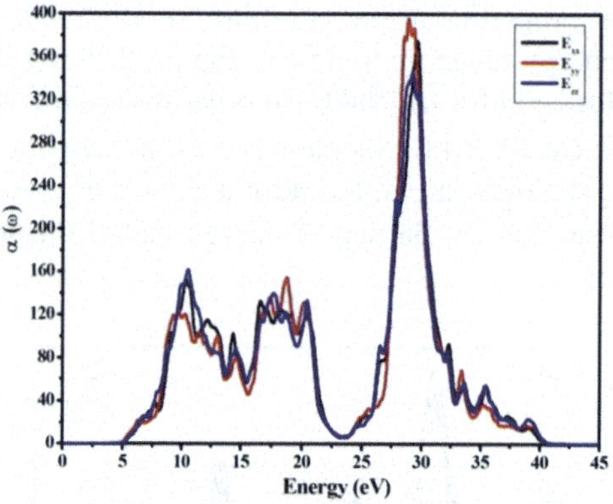

Fig. 4. Calculated absorption coefficient $\alpha(\omega)$ for Ca_2SnO_4.

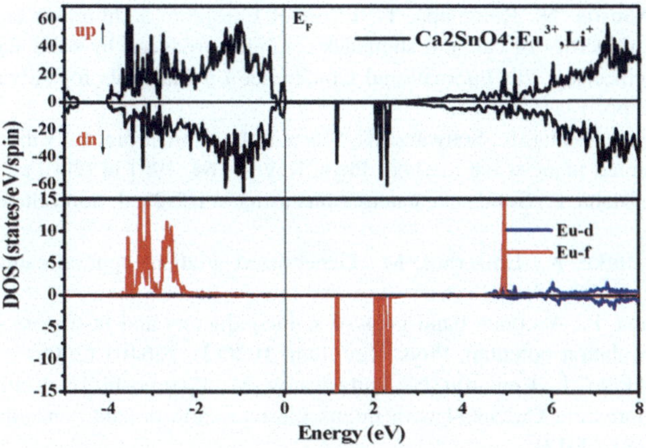

Fig. 5. Total and partial densities of states for Ca_2SnO_4 co-doped by Eu^{3+} and Li^{+}

4 Conclusions

In conclusion, the red phosphors, Eu^{3+} doped Ca with charge compensation, one approach to achieve charge compensation is: $2Ca^{2+} \rightarrow E^{3+} + M^{+}$, where M^{+} is a monovalent cation such as Li^{+}, Na^{+} and K^{+}.

Therefore, Ca_2SnO_4:Eu^{3+} phosphors with charge compensation are potential red candidates for red emitting phosphors of white LEDs.

Hence, this study can light the way for more investigation of this popular oxide material for various chemical and physical applications. The calculated structural parameters and band gap are in excellent agreement with the experimental findings.

References

1. Voncken, J.H.L.: Recycling of rare earths. In: The Rare Earth Elements. Springer Briefs in Earth Sciences. Springer, Cham (2016)
2. Diaz Torres, L.A., De la Rosa Cruz, E.: Spectroscopic Characterization of Yb^{3+} doped Barium Zirconate León Guanajuato, México (2008)
3. Subhajit, S., Swati, D., Uttam, K.G., Nilesh, M., Bipin, K.G., Kalyan, K.C.: Charge compensation assisted enhanced photoluminescence derived from Li-codoped $MgAl_2O_4$:Eu^{3+} nanophosphors for solid state lighting applications. Dalton Trans. **42**, 12965 (2013)
4. Monika, D.L., Bhushana, N., Krishna, R.H., Nagabhushana, B., Sharma, S., Thomas, T.: Synthesis and photoluminescence properties of novel Sr2CeO4:Dy3+ nanophosphor with enhanced brightness by Li+ co-doping. RSC Adv. **4**, 38655–38662 (2014)
5. Andrius, S., Artūras, K., David, E., Danuta, D., Simas, S., Marlies, V.B., An, H., Aivaras, K., Thomas, J.: Luminescence properties of Sm^{3+}-doped alkaline earth ortho-stannates. Opt. Mater. **36**, 1146–1152 (2014)

6. Jintara, P., Anuson, N., Rungnapa, T.: Effect of Europium addition on crystallization and luminescent properties of calcium stannates ceramics produced by solid-states reaction. In: Procedia Engineering. 7th International Conference on Materials for Advanced Technologies, vol. 94, pp. 14–17 (2014)
7. Madsen, G.K.H., Blaha, P., Schwarz, K., Sjöstedt, E., Nordström, L.: Efficient linearization of the augmented plane-wave method. Phys. Rev. B. **64**, 195134 (2001)
8. Schwarz, K., Blaha, P.: Solid state calculations using WIEN2k. J. Solid State Chem. Fr. **102**, 22640 (2003)
9. Perdew, P., Burke, K., Ernzerhof, M.: Generalized gradient approximation made simple. Phys. Rev. Lett. **77**, 3865 (1996)
10. Tran, F., Blaha, P.: Accurate band gaps of semiconductors and insulators with a semilocal exchange-correlation potential. Phys. Rev. Lett. **102**(22), 226401 (2009)
11. Tadashi, I., Akiko, T., Kentarou, N., Puttaswamy, M., Kazuyoshi, U., Kenji, T., Mineo, S.: Long phosphorescent Ca2SnO4 with minuscule rare earth dopant concentration. J. Dalton Trans. **42**, 4781 (2013)
12. Rogers, E.G., Dorenbos, P.: Vacuum energy referred $Ti^{3+/4+}$ donor/acceptor states in insulating and semiconducting inorganic compounds. J. Lumin. **153**, 40–45 (2014)
13. Canimoglu, A., Garcia-Guinea, J., Karabulut, Y., Ayvacikli, M., Jorge, A., Can, N.: Appl. Radiat. Isot. **99**, 138–145 (2015)
14. Steeneken, P.G., Tjeng, L.H., Elfimov, I., Sawatzky, G.A., Ghiringhelli, G., Brookes, N.B., Huang, D.-J.: Exchange splitting and charge carrier spin polarization in EuO. Phys. Rev. Lett. **88**, 047201 (2002)
15. Antonov, V.N., Harmon, B.N., Yaresko, A.N.: Electronic structure of mixed-valence and charge-ordered Sm and Eu pnictides and chalcogenides. Phys. Rev. B **72**, 085119 (2005)

Synthesis Cordierite Materials Starting from Algerian Kaolin

M. Kheloui(✉), K. Boumchedda, S. Debbakh, and B. Rabehi

Department Engineering of Materials, URMPE, FSI, University of Boumerdes, Boumerdès, Algeria
kheloumeriem@yahoo.com

Abstract. The objective of this work is to study the synthesis of the cordierite phase from Algerian kaolin known as DD kaolin. To do this, the kaolin is washed and sieved to prepare a particle size fraction <20 μ, rich in clay particles, and precipitated magnesium hydroxide is prepared to give magnesium and finally silica fume is added to the mixtures experimental, to fill the silica deficit. The weighing of the materials is carried out in such a way as to produce mixtures close to the ideal phase of the cordierite (2MgO.2Al2O3.5SiO2). The mixing is carried out in a large quantity of water with a metal stirrer at high speed. The mixture is heated during stirring until the sludge is obtained. The final drying is continued in an oven, the coarse powder obtained is finely ground in a porcelain mortar.

The results of investigations carried out by simultaneous thermal analysis (ATD-TG) show a first exothermic peak at 945 °C. which is attributed to μ-cordierite, a second exothermic peak at 1200 °C. which indicates the crystallization of μ-cordierite in Indialite, and finally an endothermic peak above 1400 °C which indicates the fusion of the indialite.

The results of investigations of the X-ray diffraction analysis on the calcined powders at high temperature, shows that only the Indialite phase is present from 1300 °C.

Keywords: Cordierite · Kaolin · Synthesis · Ceramic · Sintering

1 Introduction

Cordierite ceramic is composed mainly of a phase corresponding to the formula ($2MgO.2Al_2O_3.5SiO_2$), it is one of the main ternary compounds in the MgO-Al_2O_3-SiO_2 system.

Cordierite materials still interests research development for two main reasons; (I) for its remarkable properties; Thermal shock resistance, chemical stability, and stability of dielectric properties. Cordierite is well known in catalytic converters of vehicles, and refractory applications, but also as a substrate material for integrated circuit boards. Recently cordierite showed of real performances in the filtration applications.

(II) Low cost manufacturing from natural raw materials available throughout the world. Natural raw materials often used are Kaolin [2], talc [3], stevensite [3], sepiolite

B. Abdelbaki et al. (Eds.): SMSD 2017, *Proceedings of the Third International Symposium on Materials and Sustainable Development*,, pp. 141–146, 2018.
https://doi.org/10.1007/978-3-319-89707-3_17

[4], gibbsite [5], and magnesite. To respect the stoichiometric composition formula of cordierite ($2MgO.2Al_2O_3.5SiO_2$), it's added magnesium hydroxyde, magnesium compounds, alumina [4], silica powder [5].

This study presents our research on the development of cordierite materials from abundant Algerian kaolin. This kaolin known as DD is located in the mountain "Djebel Debbagh" in eastern Algeria; it has a blackish coloring that limits its use in paper, luxury porcelain and sanitary articles where brightness and whiteness are required. The precipitated magnesium hydroxide is added to the starting mixtures to compensate for the magnesium deficiency, thus approaching the stoïchiometric formula of the cordierite.

2 Experimental Procedure

2.1 Preparation and Characterization of Materials

The kaolin DD is dissolved in water for several days to separate the clay particles from the other particles (sand, limestone, etc.), then the barbotine is sieved under 20 microns under tap water to give fraction of kaolin after decanting and drying. The characterization show DD kaolin is composed mainly by kaolinite and halloysite (> 80 wt.%), the other minerals are represented by gibbsite which is bearer of alumina, and by manganese compounds which gives blackish color to that raw material (Fig. 1).

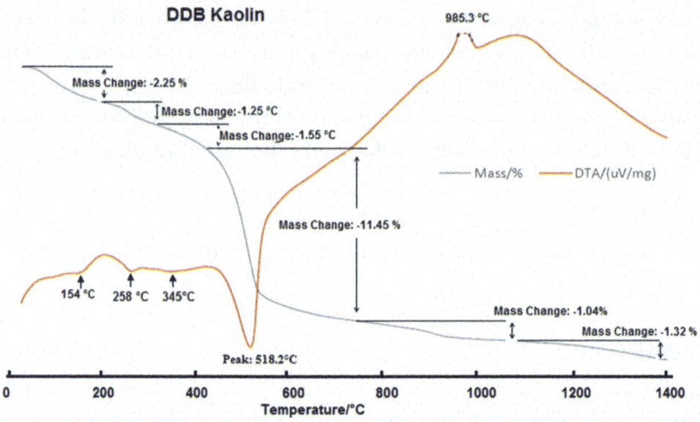

Fig. 1. ATD-TG of DDN kaolin

The chemical composition (Table 1) show that kaolin is very rich in alumina (Al_2O_3 > 46 wt.% at calcined state). The ignition loss exceed 17 wt.%, it's constituted mainly by H_2O because DD kaolin is poor in carbonates (CaO + MgO < 0.50%).

The simultaneous thermal analysis (TG/DTA) shows several transformations of DD kaolin. (1) First weight loss (−2.25 wt.%) corresponds to elimination of interlayer molecules H_20 of halloysite and todorokite. (2) Dehydratation of gibbsite is observed at 311.1 °C (−0.71 wt.%). (3) this weight loss may be attributed to the second dehydratation of todorokite. (4) Dehydroxylation of metahalloysite happens at 518 °C

Table 1. The chemical composition

Compounds	SiO_2	Al_2O_3	MnO	MgO	Other oxides	I. loss
DD black kaolin	53.60	46.25	1.68		<0.50	16.80
Magnesium hydroxyde	–	–	–	66	7	27
Amorphous silica	98	–	–	–	–	1

(−11.45 wt.%). The last weight losses (5, and 6) can be attributed to the reduction of manganese oxides. Exothermic peak observed at 985.3 °C can be attributed to the formation of spinel γ-Al_2O_3 and or mullite starting from metakaolinite.

The magnesium hydroxide ($Mg(OH)_2$) is precipitated from magnesium chloride solution by adding ammonia. Ammonia is added in excess (+5%) to guarantee the totality of the chemical reaction, and then the suspension is washed with distilled water several times until close to pH 7, for eliminate excess from ammonia. After decantation the magnesium hydroxide is dried at 100 °C and characterized.

STA analysis shows clearly shows the decomposition of $Mg(OH)_2$ at 390 °C, the associated loss of weight is 27.50% which gives a content of 88% $Mg(OH)_2$. The exothermic peak (680 °C) may be attributed to the crystallization of MgO (periclase). All peaks in the XRD analysis are attributed to periclase (Figs. 2 and 3).

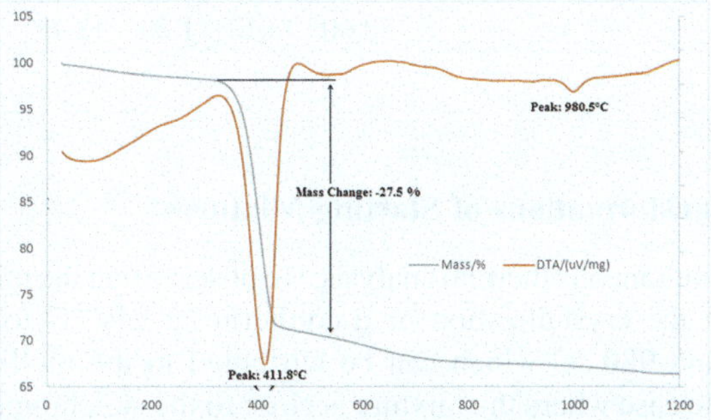

Fig. 2. DTA-TG curve of precipitated magnesium hydroxide

2.2 Preparation of Experimental Mixtures

The weighing of the starting materials is carried out in such a way as to produce mixtures close to the ideal phase of the cordierite (13.78%MgO.34.86%Al2O3.51.36% SiO2). The final calculation gives 72 g of kaolin, 20 g of magnesium hydroxide, and 12 g of silica fume (Table 2).

The mixing is carried out in a large quantity of water with a metal stirrer at high speed. The mixture is heated during stirring until the sludge is obtained. The final drying is continued in an oven, the coarse powder obtained is finely ground in a porcelain mortar.

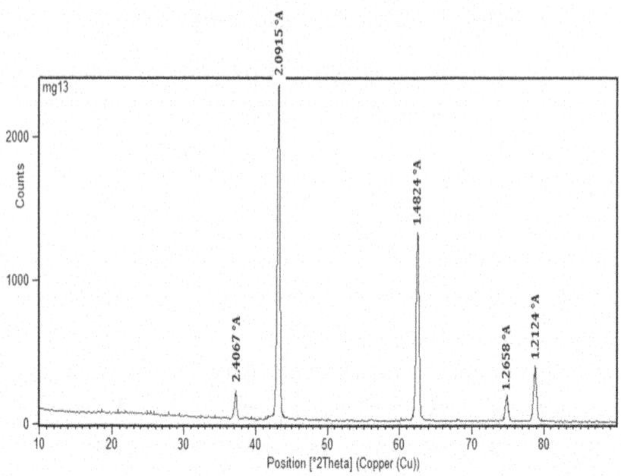

Fig. 3. XRD pattern of precipitated magnesium hydroxide calcined at 1000 °C

Table 2. Weight mixture of starting materials

Starting materials	SiO_2	Al_2O_3	MnO	MgO	OH
72% DD kaolin	38.60	33.12	1.20	0.08	8.24
20% Magnesium hydroxide	—	—	—	13.40	5.60
12% Amorphous silica	12	—	—	—	——
Total	50.60	33.12	1.20	13.48	

3 Results

3.1 Thermal Transformations of Starting Mixtures

According to the simultaneous thermal analysis, we observe that the mixtures with a MgO deficiency result in the crystallization of µ-cordierite at 940 °C followed by another crystallization around 980 °C which can be attributed to the mullite ($3Al_2O_3.2SiO_2$) (Fig. 4). Then, in the case where the mixture is close to the stoichiometric mixture of the cordierite, we observe the 2 exothermic peaks that of µ-cordierite at 940 °C, and that of

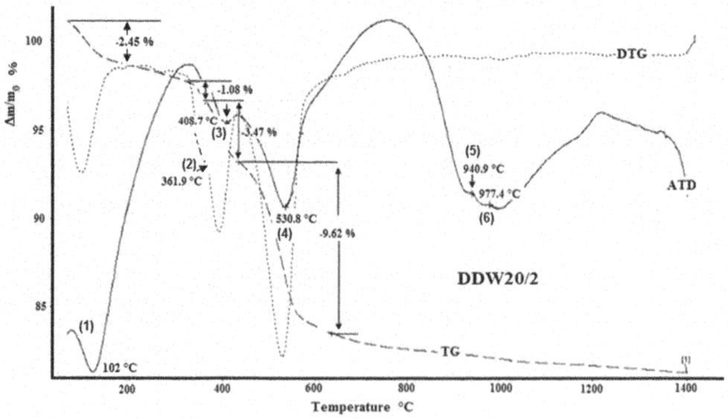

Fig. 4. ATD-TG of firstly mixture

the indialite from μ-cordierite at 1200 °C. The peak of the mullite has completely disappeared (Fig. 5).

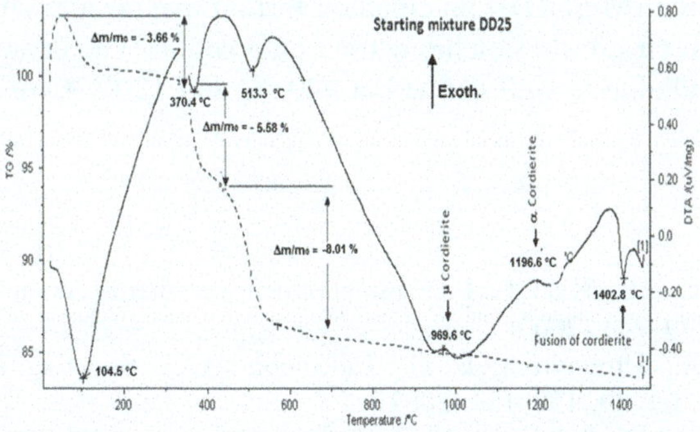

Fig. 5. ATD-TG of stoichiometric mixture

It appears that stoichiometric mixtures give only cordierite, while the other mixing compositions make it possible to synthesize cordierite-mullite composites.

The investigations of the thermal transformations carried out on the calcined stoichiometric mixture at 1000 °C, 1200 °C and 1300 °C for 2 h, show that the appearance of the indialite phase begins at 1200 °C with the presence of cristobalite. From 1300 °C, cristobalite decreases in intensity whereas the crystallization of indialite is perfectly observable (Fig. 6).

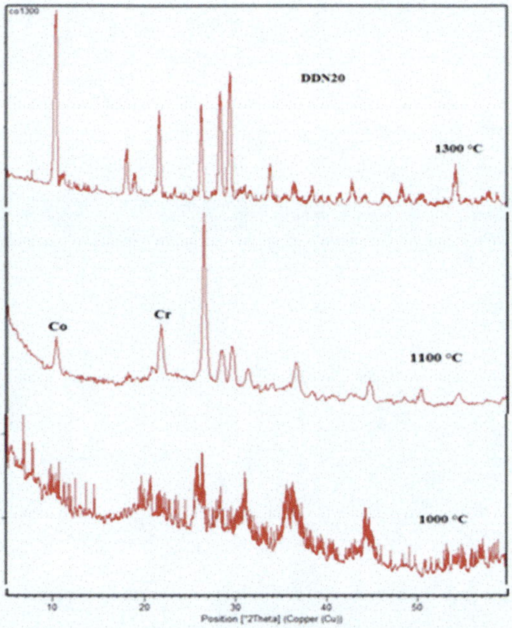

Fig. 6. XRD analysis of stoichiometric mixture calcined at 1000 °C, 1100 °C, and 1300 °C

4 Conclusion

In this study the cordierite ceramic was synthesized from a mixture of magnesium kaolin and hydroxide ($Mg(OH)_2$) precipitated from a magnesium chloride solution by adding ammonia to ensure its stoichiometric composition. The crystallization of the μ cordierite and α Indialite is well defined at 940 °C and 1200 °C respectively.

References

1. Kobayashi, Y., Sumi, K., Kato, E.: Preparation of dense cordierite ceramics from magnesium. Ceram. Int. **26**, 739–743 (2000)
2. Cameruccia, M.A., Urretavizcayab, G., Cavalieria, A.L.: Sintering of cordierite based materials. Ceram. Int. **29**, 159–168 (2003)
3. Goren, R., Gocmez, H., Ozgur, C.: Synthesis of cordierite powder from talc, diatomite and alumina. Ceram. Int. **32**, 53–56 (2006)
4. Zhou, J., Dong, Y., Hampshire, S., Meng, G.: Utilization of sepiolite in the synthesis of porous cordierite ceramics. Appl. Clay Sci. **52**, 328–332 (2011)
5. Bejjaoui, R., Benhammou, A., Nibou, L., Tanouti, B., Bonnet, J.P., Yaacoubi, A., Ammar, A.: Synthesis and characterization of cordierite ceramic from Moroccan stevensite and andalusite. Appl. Clay Sci. **49**(3), 336–340 (2010)

Nonlinear Behavior of FGM Plate in Cylindrical Bending Under Uniform Loading

Mohamed Bourada[1,3](✉), Fouad Bourada[2], Abdelhakim Kaci[1,3], Mohamed Bouremana[1], and Abdelouahed Tounsi[1]

[1] Laboratoire des Matériaux et Hydrologie, Universite Djillali Liabes, Sidi Bel Abbes, Algeria
med_bourada@yahoo.fr

[2] Département de Génie civil, Centre Universitaire Belhadj Bouchaib, Ain Témouchent, Algeria

[3] Laboratoire de modélisation et simulation multi-échelles, Université de Sidi Bel Abbes, Sidi Bel Abbes, Algeria

Abstract. In this work, the nonlinear behaviour of an exponential functionally graded plate (E-FGP) is studied. The plates are subjected to uniform loads and their geometric nonlinearity is introduced in the strain–displacement equations based on Von-Karman assumptions. The material properties of functionally graded plates are assumed to vary continuously across the thickness of the plate in accordance with the exponential law distribution; the Hamilton's principle is used herein to derive the solution of the system. Several numerical results for exponential functionally graded plates are given in the form of explicit graphs; and the effects of material properties on deflections and normal stresses across the thickness are determined.

Keywords: Exponential functionally graded plate · Nonlinear behaviour The Hamilton principle

1 Introduction

The mechanical properties of the functionally graded materials vary discretely on one surface to the other. This is achieved by gradually varying the volume fraction of the constituent materials. the new property functions are adapted and the performance of materials in harsh environments could be improved.

Several studies of Functionally graded plates on linear thermal bending of FGM plates are presented such as (Yamanoushi et al. 1990; Koizumi 1993; Tanigawa et al. 1996; Reddy et al. 1999; Praveen et al. 1999; Mizuguchi and Ohnabe 1996).

The response of functionally graded (FG) ceramic-metal plates using a plate finite element that accounts for the transverse shear strains is analysed by Praveen and Reddy (1998). Reddy (2000) developed solutions for rectangular FG plates based on the (TSDT) third-order shear deformation plate theory. A nonlinear cylindrical bending analysis of FG plates based on the Classical Plate Theory (CPT) has been performed (Sun and Chin 1988, 1991; Navazi et al. 2006).

B. Abdelbaki et al. (Eds.): SMSD 2017, *Proceedings of the Third International Symposium on Materials and Sustainable Development*, pp. 147–156, 2018.
https://doi.org/10.1007/978-3-319-89707-3_18

In this investigation, the displacements and stresses in exponential functionally graded plates in cylindrical bending subjected to uniform pressureIt are accurately determined. The governing equations are obtained using the classical plate theory (CPT). The nonlinear equilibrium equations are obtained By using the von Karman nonlinear strains and then reduced to a linear differential equation. This equation is solved for simply supported BCs and the effects of several parameters such as material index, loading type and plate dimensions are studied. It is shown that the results show that the linear analysis yields wrong results not only quantitatively but also qualitatively.

2 Theoretical Formulations

2.1 Properties of the E-FGM Constituent Materials

In this investigation, consider a rectangular E-FG plates having uniform thickness h and length $l = 2a$. Plates are made of a mixture of ceramics and metals; and it is assumed that their composition is gradual and that they are smoothly varied from the ceramic-rich top surface of the plate ($z = +h/2$) to the metal rich bottom surface ($z = -h/2$). As a consequence, the material properties of FGM plates, such as the Young's modulus E is function of depth z, measured from the middle plane of plate.

The functional relationship between E and z for the ceramic–metal FGM plates is given in the below equation; as explained by Sallai et al. (2009)

$$E(z) = A\,e^{B\,(z+h/2)} \tag{1a}$$

with

$$A = E_2 \text{ and } B = \frac{1}{h}\ln\left(\frac{E_1}{E_2}\right) \tag{1b}$$

where $E(z)$ indicates a typical Young's modulus and E_1 and E_2 denote Young's modulus of the bottom (i.e. $z = +h/2$) and top surface (i.e. $z = -h/2$), respectively.

2.2 Governing Equations for Nonlinear Cylindrical Bending of E-FGM Plates

The classic plate theory (CPT) is applied throughout the work. As a result of the CPT assumptions, the Kirchhoff normalcy condition is incorporated, therefore

$$u(x,y,z) = u_0(x,y) - z\frac{\partial w_0}{\partial x}, \tag{2a}$$

$$v(x,y,z) = 0, \tag{2b}$$

$$w(x,y,z) = w_0(x,y), \tag{2c}$$

where (u, v, w) are the displacements in the (x, y, z) directions, respectively. Also, (u_0, v_0, w_0) are the displacements of the mid-plane in the (x, y, z) directions, respectively.

The nonlinear von Karman strain–displacement relations are used as follows:

$$\varepsilon_x = \frac{\partial u_0}{\partial x} + \frac{1}{2}\left(\frac{\partial w_0}{\partial x}\right)^2 - z\frac{\partial^2 w_0}{\partial x^2}, \tag{3a}$$

$$\varepsilon_y = \varepsilon_z = 0, \tag{3b}$$

$$\gamma_{xy} = \left(\frac{\partial u_0}{\partial y} + \frac{\partial v_0}{\partial x} + \frac{\partial w_0}{\partial x}\frac{\partial w_0}{\partial y}\right) - 2z\frac{\partial^2 w_0}{\partial x \partial y}; \quad \gamma_{xz} = \gamma_{yz} = 0, \tag{3c}$$

The constitutive relations for nonzero strains are given by

$$\begin{Bmatrix} \sigma_x \\ \sigma_x \\ \tau_{xy} \end{Bmatrix} = \begin{bmatrix} Q_{11} & Q_{12} & 0 \\ Q_{12} & Q_{22} & 0 \\ 0 & 0 & Q_{66} \end{bmatrix} \begin{Bmatrix} \varepsilon_x \\ \varepsilon_x \\ \gamma_{xy} \end{Bmatrix}, \tag{4}$$

Using the material properties defined in (1), stiffness coefficients, Q_{ij}, can be expressed as

$$Q_{11} = Q_{22} = \frac{E(z)}{1 - \nu^2}, \tag{5a}$$

$$Q_{12} = \frac{\nu\, E(z)}{1 - \nu^2}, \tag{5b}$$

$$Q_{66} = \frac{E(z)}{2(1 + \nu)}, \tag{5c}$$

Use of Hamilton's principle yields the Euler–Lagrange equations as

$$N_{x,x} = 0, \tag{6a}$$

$$Q_{x,x} + q + N_x w_{,xx} = 0, \tag{6b}$$

$$M_{x,x} - Q_x = 0, \tag{6c}$$

where q is the transverse load. The stress, shear force and moment resultants are defined as follows:

$$(N_x, Q_x) = \int_{-h/2}^{h/2} (\sigma_x, \tau_{xy}) dz \text{ and } M_x = \int_{-h/2}^{h/2} \sigma_x z\, dz \tag{7}$$

From (6a) one can conclude that

$$N_x = N_x^0 = const. \tag{8}$$

Consequently, Eqs. (6b) and (6c) become

$$M_{x,xx} + q + N_x^0 w_{,xx} = 0, \tag{9}$$

By substituting of (3) into (4) and the subsequent results into (7), the generalized stress resultants in terms of displacement components will be obtained which can be presented as follows:

$$N_x = A_{11}\left(u_{,x} + \frac{1}{2}w_{,x}^2\right) - B_{11}w_{,xx}, \tag{10}$$

$$M_x = B_{11}\left(u_{,x} + \frac{1}{2}w_{,x}^2\right) - D_{11}w_{,xx}, \tag{11}$$

A_{11}, B_{11} and D_{11} are called extension, bending–extension coupling, and bending stiffness coefficients, respectively, and are defined as follows:

$$(A_{11}, B_{11}, D_{11}) = \int_{-h/2}^{h/2} Q_{11}(1, z, z^2)dz \tag{12}$$

Substituting Eq. (11) into Eq. (12) yields

$$M_x = \frac{B_{11}}{A_{11}}N_x^0 + \left(\frac{B_{11}^2}{A_{11}} - D_{11}\right)w_{,xx}, \tag{13}$$

By substituting (13) into (9), one can obtain

$$w_{,xxxx} - k^2 w_{,xx} = q_0, \tag{14}$$

where

$$k^2 = \frac{N_x^0}{\left(D_{11} - \frac{B_{11}^2}{A_{11}}\right)}, \tag{15a}$$

$$q_0 = \frac{q}{\left(D_{11} - \frac{B_{11}^2}{A_{11}}\right)}, \tag{15b}$$

3 General Solution

In this study it is assumed that an E-FG plate is subjected to a uniform transverse load q on its top surface. It is intended here to obtain analytical solutions for the non-linear bending of the E-FG plate.

3.1 Nonlinear Analysis

(14) is a linear fourth-order ordinary differential equation whose solution is readily available. The general solution is as follows:

$$w(x) = C_1 \cosh(kx) + C_2 - \frac{q_0}{2k^2} x^2. \tag{16}$$

where C_1 and C_2 are constants, which must be determined using BCs at either edges of the plate. Assume that the origin of the coordinate system is located at the plate mid-span; accordingly, the simply supported BCs yield

$$w(a) = w(-a) = 0 \tag{17a}$$

$$M_x(a) = M_x(-a) = 0 \tag{17b}$$

$$u(a) = u(-a) = 0 \tag{17c}$$

Since N_x is an unknown constant along the x-axis, the in-plane displacement u can be derived by integrating (6a) over the span of the plate, using the general solution shown in (16). The boundary conditions can be expressed as

$$w(a) = 0 \tag{18a}$$

$$M_x(a) = \frac{B_{11}}{A_{11}} N_x^0 - \left(D_{11} - \frac{B_{11}^2}{A_{11}}\right) w_{,xx}\bigg|_{x=a} = 0 \tag{18b}$$

$$u(a) = \int_0^a \left(\frac{N_x^0}{A_{11}} + \frac{B_{11}}{A_{11}} w_{,xx} - \frac{1}{2} w_{,x}^2\right) dx = 0 \tag{18c}$$

Substitution of (16) into (18) and evaluating the integral in (18c) yield

$$C_1 = \left(\frac{B_{11}}{A_{11}} + \frac{q_0}{k^4}\right) \frac{1}{\cosh(ka)} \tag{19a}$$

$$C_2 = \frac{q_0 a^2}{2k^2} - C_1 \cosh(ka) \tag{19b}$$

$$u(a) = \frac{N_x^0}{A_{11}}a + \frac{B_{11}}{A_{11}}kC_1\sinh(ka) - \frac{B_{11}q_0}{A_{11}k^2}a - \frac{1}{2}k^2C_1^2\left(\frac{1}{4k}\sinh(2ka) - \frac{a}{2}\right)$$
$$-\frac{q_0^2a^3}{6k^4} + C_1\frac{q_0}{k^3}[ka\cosh(ka) - \sinh(ka)] = 0 \quad (19c)$$

These three equations contain three unknown quantities: C_1, C_2 and N_x^0 and a numerical technique must be used to obtain a solution.

3.2 Linear Analysis

Considering the small deformation theory, the infinitesimal strains are applicable and the nonlinear terms of the von Karman strains are neglected. By canceling the nonlinear terms of the equilibrium equation, i.e. the second term of (14), the following solution for the linear analysis is obtained:

$$w(x) = \frac{1}{24}q_0\left(x^4 - a^4\right) + \frac{1}{12}\left(\frac{2B_{11}^2 - 3D_{11}A_{11}}{D_{11}A_{11}}\right)q_0a^2\left(x^2 - a^2\right) \quad (20)$$

$$u(x) == \frac{1}{6}q_0x^3\frac{B_{11}}{A_{11}} - \frac{1}{6}q_0a^2x\frac{B_{11}}{A_{11}} \quad (21)$$

4 Numerical Results and Discussion

The results obtained from the analysis are presented in dimensionless parametric terms of deflections and stresses as follows:

- side coordinate $\bar{x} = x/a$,
- thickness coordinate $\bar{z} = z/h$,
- deflection $\bar{w} = w/h$,
- axial stress $\bar{\sigma}_x = \sigma_x/\left[Q_{11m}(h/a)^2\right]$, ($Q_{11m}$) stiffness coefficient of metal plate),
- load parameter $q'' = q_0a^4/(E_mh^4)$.

It is assumed that the Young's modulus at the bottom surface of the E-FGM plate, E_1, at the top surface of the E-FGM plate, E_2, is 70 GPa, while that varies with the ratio of E_1/E_2. Note that Poisson's ratio is selected constant and equal to 0.3 for both of the constituents. The plate geometry is chosen so the $h = 5$ mm and $a = 0.5$ m.

Figure 1 show the non-dimensional deflections due to transverse load $q =$ 1 KN/m^2 versus non-dimensional length for different E_1/E_2 in non-linear analysis. It can be seen that the deflection w decrease with increasing of E_1/E_2 and the linear solution overestimates the deflections of the E-FG plate.

Figures 2 and 3 present the variation of the non-dimensional axial stress of the E-FG plate subjected to for different in linear and non-linear analyses, through the thickness distributions, respectively. In the case of linear analysis, it can be observed

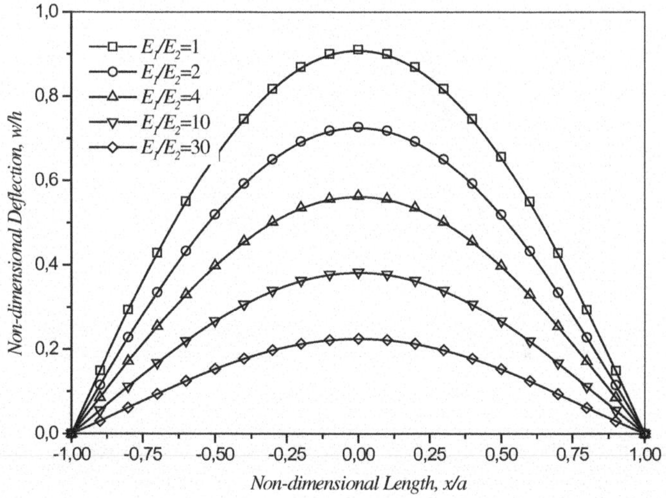

Fig. 1. Non-dimensional deflections due to transverse load $q = 1$ KN/m^2 versus non-dimensional length for different E_1/E_2 in non-linear analysis.

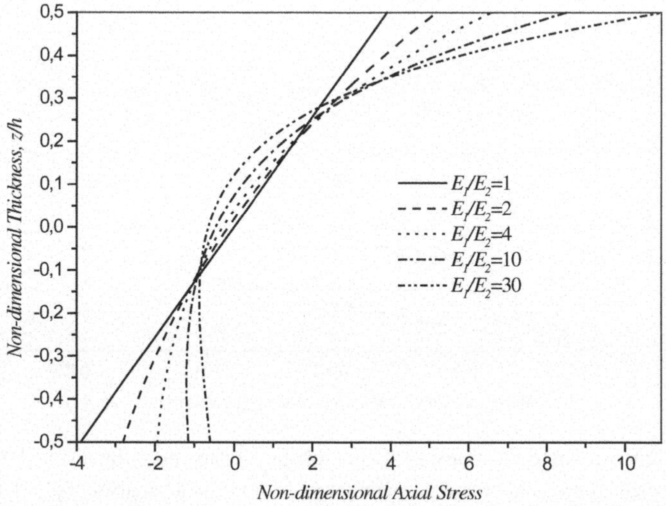

Fig. 2. Through the thickness distribution of non-dimensional axial stress $\overline{\sigma}_x$ of the E-FG plate subjected to $q = 1$ KN/m^2 for different E_1/E_2 in linear analysis.

that for a homogeneous plate in which, the magnitude of the tensile and compressive stresses are equal and located at the top and bottom surfaces. However, in the case of non-linear analysis, this remark is not checked. The stress profile for homogeneous plate changes linearly across the thickness for both linear and non-linear solutions.

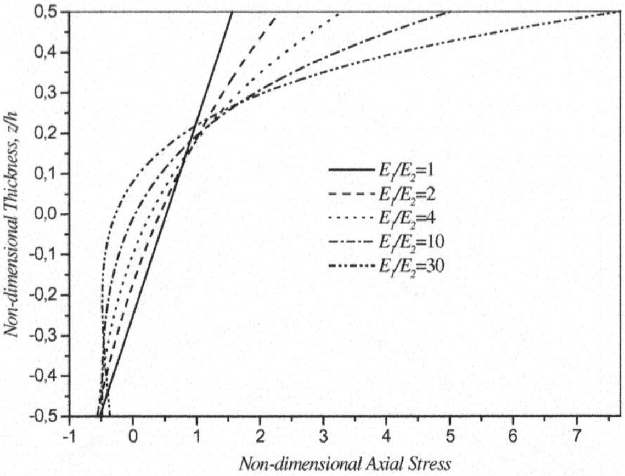

Fig. 3. Through the thickness distribution of non-dimensional axial stress $\overline{\sigma}_x$ of the E-FG plate subjected to $q = 1$ KN/m^2 for different E_1/E_2 in non-linear analysis.

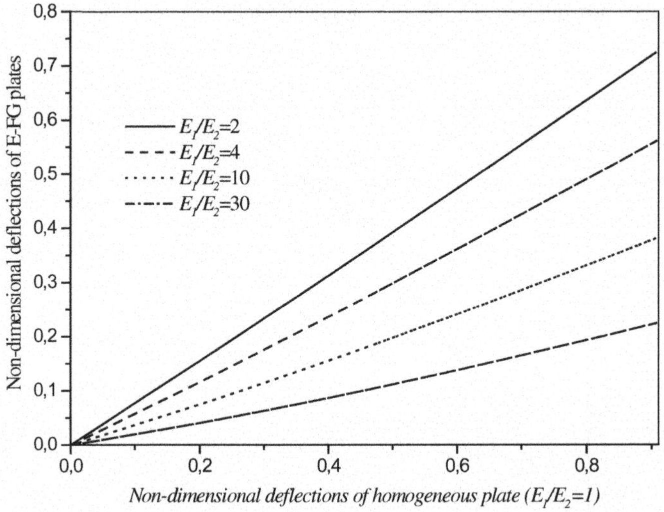

Fig. 4. Non-dimensional deflections of E-FG plates versus non-dimensional deflections of homogeneous plate for various material parameters in non-linear analysis ($q = 1$ KN/m^2).

The non-dimensional deflections through the length of E-FG plates are shown in Fig. 4. It can be observed that the non-dimensional deflections are directly proportional to those of the homogeneous plate. Therefore, one can predict the deflections of E-FG plates knowing that of similar homogeneous plates.

5 Conclusion

In this paper, the nonlinear cylindrical bending of the functionally graded (E-FG) plates under pressure loading was studied. The Von-Karman theory for large deflections used to obtain the fundamental equations for thin E-FG plates. The material properties of E-FG plates were assumed to vary discretely across the thickness of plates according to two power-law distributions. For the cylindrical bending problem, it is found that the plate Navier equations according to the large deflection theory can be expressed as linear equations for the deflection, leaving nonlinear boundary conditions. This linearity of the differential equations greatly simplifies the large deflection analysis. The non-dimensional stresses and deflection are computed for plates with two different ceramic-metal mixtures. Several numerical results shows that the non-linearity effect on the plate responses is significant. On the other hand, the results indicate that the effect of non-linearity is to lower the magnitude of the transverse deflection.

References

Niino, M., Maeda, S.: Recent development status of functionally gradient materials. ISIJ Int. **30**, 699–703 (1990)

Hirano, T., Yamada, T., Teraki, J., Niino, M., Kumakawa, A.: A study on functionally gradient material design system for a thrust chamber. In: Proceedings of the 16th International Symposium on Space Technology and Science, Sapporo, Japan (1988)

Yamanoushi, M., Koizumi, M., Hiraii, T., Shoita, I. (eds.): Proceedings of the First International Symposium on Functionally Gradient Materials, Japan (1990)

Koizumi, M.: The concept of FGM. Ceram. Trans. Funct. Grad. Mater., 3–10 (1993)

Tanigawa, Y., Akai, T., Kawamura, R., Oka, N.: Transient heat conduction and thermal stress problems of a nonhomogeneous plate with temperature dependent material properties. J. Therm. Stresses **19**, 77–102 (1996)

Reddy, J.N., Wang, C.M., Kitipornchai, S.: Axisymmetric bending of functionally grade circular and annular plates. Eur. J. Mech. A/Solids **18**, 185–199 (1999)

Praveen, G.N., Chin, C.D., Reddy, J.N.: Thermoelastic analysis of functionally graded ceramic–metal cylinder. J. Eng. Mech. ASCE **125**, 1259–1267 (1999)

Mizuguchi, F., Ohnabe, H.: Large deflections of heated functionally graded simply supported rectangular plates with varying rigidity in thickness direction. In: Proceedings of the 11th Technical Conference of the American Society for Composites, USA, pp. 957–66 (1996)

Praveen, G.N., Reddy, J.N.: Nonlinear transient thermoelastic analysis of functionally graded ceramic-metal plates. Int. J. Solids Struct. **35**, 4457–4476 (1998)

Reddy, J.N.: Analysis of functionally graded plates. Int. J. Numer. Methods Eng. **47**, 663–684 (2000)

GhannadPour, S.A.M., Alinia, M.M.: Large deflection behavior of functionally graded plates under pressure loads. Compos. Struct. **75**(1–4), 67–71 (2006)

Sun, C.T., Chin, H.: Analysis of asymmetric composite laminates. AIAA J. **26**(6), 714–718 (1988)

Sun, C.T., Chin, H.: Cylindrical bending of unsymmetric composite laminates. AIAA J. **30**(5), 1438–1440 (1991)

Navazi, H.M., Haddadpour, H., Rasekh, M.: An analytical solution for nonlinear cylindrical bending of functionally graded plates. Thin-Walled Struct. **44**, 1129–1137 (2006)

Markworth, A.J., Ramesh, K.S., Parks, W.P.: Modeling studies applied to functionally graded materials. J. Mater. Sci. **30**, 2183–2193 (1995)

Sallai, B.-O., Tounsi, A., Mechab, I., Bachir Bouiadjra, M., Meradjah, M., Adda Bedia, E.A.: A theoretical analysis of flexional bending of Al/Al_2O_3 S-FGM thick beams. Comput. Mater. Sci. **44**, 1344–1350 (2009)

Template Optimization of Block Copolymer Thin Films by Self-Assembly Process

M. Loucif Seiad[1,2,3,4(✉)], V. K. MKuppuswamy[2], M. Ferhat[3,5], and R. Gronheid[2]

[1] KACST-Intel Consortium Center of Excellence in Nano-manufacturing Applications (CENA), Riyadh, Saudi Arabia
mloucif@cdta.dz
[2] imec, Kapeldreef 75, 3001 Leuven, Belgium
[3] Semiconductor and Functional Materials Laboratory, University Amar Telidji of Laghouat, BP 37G, Ghardaïa Road, 03000 Laghouat, Algeria
[4] Centre de Développement des Technologies Avancées (CDTA), 20 Août 1956, Baba Hassen, 16303 Algiers, Algeria
[5] Department of Physics, The University of the West Indies, Mona Campus, Kingston 7, Jamaica

Abstract. We study in this work the possibility to fabricate a template using block copolymers (BCPs) thin films. The self-assembled (S-A) Poly (styrene-block-methyl-methacrylate) (PS-*b*-PMMA) films are a useful template used after removing the PMMA domains with well-defined pore dimensions. Indeed, after thermal annealing the BCP spontaneously separates into a hexagonal array of micro-domains oriented perpendicularly to the surface. Two BCP formulations with natural periodicities of 29 and 37 nm are compared. The optimization of S-A process parameters as film thickness, annealing conditions on quality of the template morphologies was the aim of this work. Large process window was obtained for the low molecular weight BCP (film thickness from 20.3 to 43.5 nm) compared to the BCP with higher molecular weight. Increasing the film thickness affect the micro-domain orientation. Mixed orientation was observed on the final template. When the annealing temperature was varied from 200 to 260 °C, the periodicity was slightly changed. Both BCP formulations present parallel domain orientation above 260 °C, probably due to polymer decomposition.

Keywords: Block copolymers · Self-assembly · Micro-phase separation Cylindrical phase · Template

1 Introduction

Block copolymers (BCPs) have received considerable attention during the last decade due to their potential applications in nanofabrication and nanolithography (Han et al. 2008). Under the influence of solvent vapor annealing, BCPs self-organize to minimize their free energy and form periodic nano-patterns. Three variables influence the morphology of BCPs: the degree of polymerization (N), the Flory-Huggins interaction parameter (χ) where χN is the segregation strength, and the relative volume fraction of the blocks (f).The self-assembly of BCPs can lead to various nanoscale structures, such

B. Abdelbaki et al. (Eds.): SMSD 2017, *Proceedings of the Third International Symposium on Materials and Sustainable Development*, pp. 157–165, 2018.
https://doi.org/10.1007/978-3-319-89707-3_19

as spheres, cylinders, bi-continuous gyroids, and lamellaes (Jeong et al. 2011). The resulting sizes of the microdomains range from 5 to 50 nm, and are tunable by varying the molecular weights of the BCPs. In thin films, where the interfaces strongly affect the final nanoscale morphology, commensurability between the film thickness (*Ft*) and the natural period (L_0) along with polymer-surface interactions can influence both the morphology's symmetry and orientation (Segalman 2005). The simplicity and cost-effectiveness of the spontaneous BCP self-assembly process is attractive and holds promise for applications in nanolithography (Li et al. 2005), with demonstrated capabilities in the fabrication of nanoscale devices (Bates et al. 1990; Park et al. 1997). Specifically, the use of cylindrical phase BCPs has been demonstrated for printing the capacitor layer of dynamic access memory devices (Bang et al. 2009; Guarini et al. 2001; Black et al. 2001). Poly (styrene-*b*-methyl methacrylate) (PS-*b*-PMMA) is a promising choice for nano-lithographic applications since the surface energies at the PS/air and PMMA/air interface are nearly equal under typical thermal annealing conditions. This allows for easy and therefore low-cost processing. Moreover, the nanodomains of PMMA can be degraded by radiation and then removed with an organic solvent creating a positive mask (Guarini et al. 2002a, b). Such wet development option for selective removal of one of the blocks furthermore has a cost-advantage over other BCP approaches which require dry etch to selectively remove one of the blocks (Xu et al. 2001). Such a wet development is especially amenable to cylindrical phase BCP materials, since the intrinsic geometric stability of the remaining structure prevents collapse during the drying process after wet development. The quality of the ordering in a BCP thin film depends on many factors, including substrate material, polymer film thickness, annealing conditions, and BCP molecular weight. In this study the impact of process parameters on two BCP formulations with different molecular weight has been investigated.

2 Experimental

2.1 Materials

As underlayer material a mat copolymer from TOK was used as received. Two commercial P(S-*b*-MMA) BCPs with different periodicity from AZ Electronic Materials and TOK Material were used as received. Organic solvent RER600 was purchased from Fujifilm and used as received.

2.2 Microphase Separation Process

All processes used in this study, were performed at imec on a TEL CLEAN TRACK ACT™12 system. A 300 mm silicon blanket wafer was coated with underlayer material in order to neutralize the substrate surface. It was spin coated at 1500 rpm (giving 10 nm film thickness) and annealed under nitrogen atmosphere at 250 °C for 60 s. Subsequently, the BCP was spin-coated on these neutralized substrates. By varying the spinning speed different BCP film thicknesses were obtained. The films were annealed at 240 °C for 600 s under nitrogen atmosphere to promote the self-assembly of the BCP.

Next, the films were exposed to deep ultraviolet radiation (DUV) in order to cross-link the PS and chain-scission the PMMA blocks. The degraded PMMA was selectively removed by solvent rinse. The samples were characterized in each step of processing by film thickness measurement and template quality evaluation using a scanning electron microscope.

2.3 Characterizations

An ellipsometer (ALERIS from KLA Tencor) was used to measure the polymeric film thickness. The morphology of the polymeric films after annealing and PMMA removal was studied using top-down two scanning electron microscopy (Hitachi CG4000). Domain periodicities of the BCP were determined via Frequency Analysis of top-down SEM images using in-house developed software.

3 Results and Discussion

3.1 Effect of Film Thickness

BCP A, was coated at film thicknesses ranging from 14 nm to 45 nm and annealed 240 °C. The morphologies of those films were evaluated qualitatively (Fig. 1a). Some

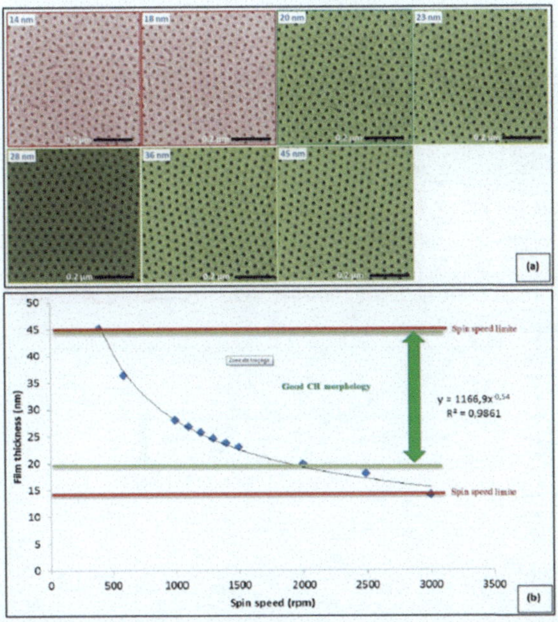

Fig. 1. (**a**) Top-down SEM images obtained after 600 s annealing at 240 °C under nitrogen, for PS-b-PMMA films having different thicknesses (red is microphase separation with defects, green is microphase separation without defects). BCPA film thickness and SEM magnification is shown in each image. (**b**) Process windows of BCP A.

defects are seen in some images (Fig.1 in red contour).The appearance of these defects delimits the required processing conditions to produce uniform perpendicular orientation of the block copolymer domains. The main morphology for these defects is bridging between adjacent cylinders. A process window is plotted to correlate film thickness and quality of self-assembly (Fig. 1b). Defect free perpendicular assembly was obtained for film thickness ranging from 20 to 45 nm, which corresponds to $0.7L_0$ to $1.5L_0$ (green contour). The main observation is that the template, exhibits highly ordered arrays of cylindrical domains that are oriented perpendicular to the substrate with nearly defect-free arrays of microdomains within this process window.

A similar study was performed for BCP B (Figs. 2a and b). Well-ordered perpendicular orientations were observed when the film thickness over the range of $0.8L_0$ to $1.0L_0$, corresponding to approximately 29 to 39 nm thick films (green contour). Beyond this, a small increment in film thickness destroyed the perfect arrangement of perpendicular domains and some parallel oriented domains appeared. The density of these domains further increases with increasing film thickness. A process window analysis (Fig. 2b) correlates the quality of assembly with film thickness. The main observation is that a good ordered film was obtained at lower film thickness ($<L_0$). In the comparison between both BCP formulations it is observed that at smaller periodicity the process window is at higher film thickness ($>L_0$). In contrast, with larger periodicity the process window is at lower film thickness ($<L_0$). Guarini et al. (2002a, b) found that one repeat period of the microstructure is an optimal thickness for PS-*b*-PMMA thin films to form

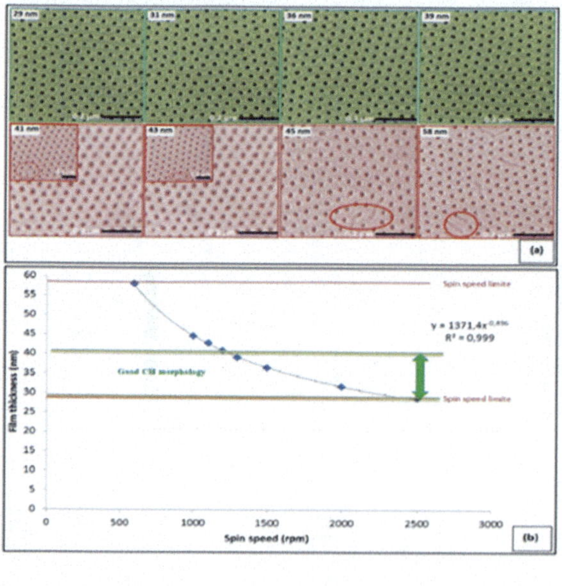

Fig. 2. (**a**) Top-down SEM images obtained after 600 s annealing at 240 °C under nitrogen, for PS-b-PMMA films having different thicknesses (red is microphase separation with defects, green is microphase separation without defects). BCP B film thickness and SEM magnification is shown in each image. (**b**) Process windows of BCP B.

a well ordered template. In our case, it is unclear how formulation of the commercial BCP samples contributes to the observed differences in the optimal film thickness range.

3.2 Effect of Annealing Temperature

a - Effect of annealing on morphology

In this part of work the effect of annealing was investigated. The samples coated at 1500 rpm (23, 36 nm for BCP A and B respectively) were annealed under nitrogen for a fixed time (600 s) at varying temperatures (240, 260, 280 and 300 °C). Fig. 3 shows the morphologies after annealing of BCP A and B (morphologies for annealing at 240 and 260 °C are indistinguishable). Both BCPs exhibit an apparent morphology change at 280 °C, with line structures being formed rather than the holes that are observed at lower temperatures. This is attributed to the appearance of parallel orientation of cylinders. At lower magnification (×100 k) mixed morphologies are observed. When the annealing temperature was further increased to 300 °C the parallel orientations largely covered the 300 mm wafer and the perpendicular cylinders mostly disappeared.

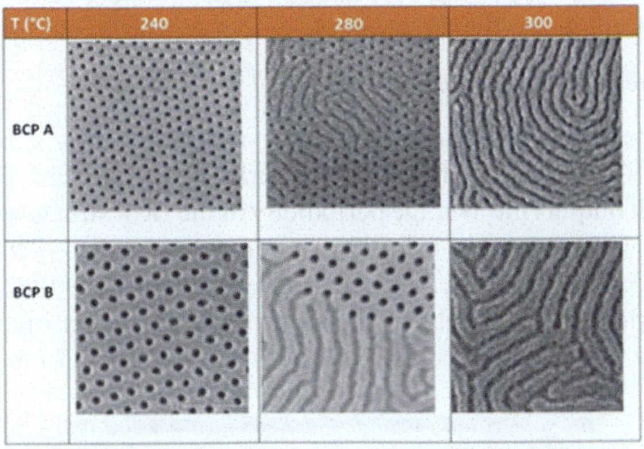

Fig. 3. Effect of temperature of annealing on pattern quality for BCP A (left) and BCP B (right). The annealing temperature and SEM magnification is shown in each image.

There are two potential explanations for these observations. At first, it could be a thermodynamic effect. An increase in annealing temperature has a direct effect on polymer chains diffusion D_0 ($D_0 \alpha T$). This is amplified by the fact that monomeric friction coefficient ξ(T), decreases by increasing temperature according to Vogel-Fulcher law (Byutner et al. 2001). This leads to better ordering of microdomains in the self-assembly process. The energy barrier ΔG, corresponding to unfavorable interactions between PS and PMMA for diffusion of perpendicular chains on the interface, increases with temperature according to the following Eq. (1) (Byutner et al. 2001):

$$\Delta G = \chi_{AB} N_0 K_0 T \quad (1)$$

where N_0, is the number of segments chemically different in the block. This energy barrier increase is particularly important for the PS-*b*-PMMA system, since its Flory-Huggins parameter evolves little with temperature ($\chi_{PS/PMMA} = 0.028 + 3.9/T$) (Russel et al. 1990). The increased energy barrier at elevated temperatures may significantly reduce chain mobility at 280 °C. SEM images analysis reveals a range of temperature to which the organization of the cylinders is optimal. This phenomenon was also observed by Guarini et al. (2002a, b). Beyond 250 °C these authors claim to reach the temperature of order-disorder transition for this BCP system.

As a second explanation, also thermal decomposition of BCPs may cause a change in domain orientation. Thermal decomposition of the BCP will change the χ parameter of the BCP and could result in a change in microphase structure and/or domain orientation. In order to verify this hypothesis an experiment was conducted by annealing the wafer after BCP coating at 300 °C in the same condition as before. After that, the same wafer was annealed at 240 °C. The results shows that the parallel orientation persist after the second annealing. This is interpreted as strong evidence for the thermal degradation hypothesis. If the orientation change is induced by the annealing thermodynamics, perpendicular domain orientation would be expected after the second 240 °C bake. We conclude that there is an upper limit for annealing temperature beyond which the BCP degrades and its domains no longer assemble perpendicular to the substrate.

b - Effect of annealing on periodicity

An in-house developed Frequency Analysis software at imec was used in order to characterize the SEM images of self-assembled BCP using fast Fourier transform methods (Fig. 4). It outputs the average periodicity of the BCP structures. The periodicity of BCP A and B were measured using this methodology at L_0 ~ 29 nm and ~37 nm, respectively. Fig. 5 shows the effect of annealing temperature over the range from 200 °C to 260 °C on periodicity. A slight decrease on L_0 with increasing temperature was observed, from 29.9 to 29.1 nm for BCP A and from 37.7 to 36.7 nm for BCP B.

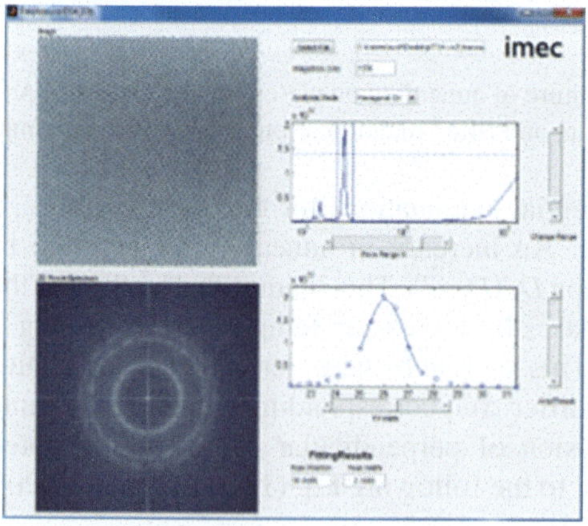

Fig. 4. Snapshot of the Frequency Analysis Metrology on DSA developed at imec.

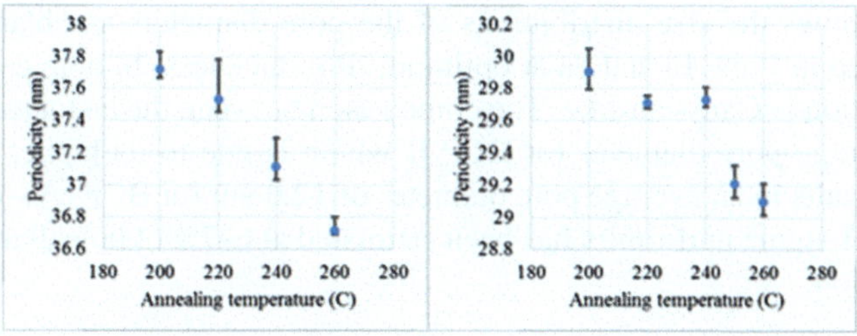

Fig. 5. Effect of annealing temperature on periodicity BCP A (left) and BCP B (right)

This change in periodicity contrasts with previous observations (Black et al. 2001); Guarini et al. 2002a, b), which claim that pore diameter and domain spacing is solely determined by the polymer properties. This periodicity L_0 is described by Eq. (2) (Men'shikov et al. 2009):

$$L_0 = aN^{2/2}\chi^{1/6} \tag{2}$$

where a is the characteristic segment length; N is the degree of polymerization. Based on this formula and the χ dependence on temperature, the variation of periodicity from 200 to 260 °C annealing temperature is not expected to exceed 0.2 nm for BCP A and B. It is unclear how surface effects or additives in the commercial BCP samples may impact the current observations.

3.3 Metrology

After selective removal of the PMMA cylinders by UV degradation and development in organic solvent, a nanoporous PS film is obtained. A software analysis of the SEM images was used (Loucif Seiad et al. 2017). It allowed us to determine the size and the spatial distribution of the pores in the PS matrix. The schematic diagram shown in Fig. 6, demonstrate the steps followed in this methodology. Initially the wafers morphologies were captured with required top-down SEM images, which contain several numbers (~130) of pores with respect to the applied magnification. In those images, noise smoothing filter is applied. After that, using the Otsu's method [24–25] (Otsu 1979; Sezgin and Sankur 2004), image threshold binarization is obtained and then the inner edges of the pores are detected. The best fitted circle to these points is defined and the coordinates of the center and the radius are obtained, the diameter and the circularity of pores are then calculated.

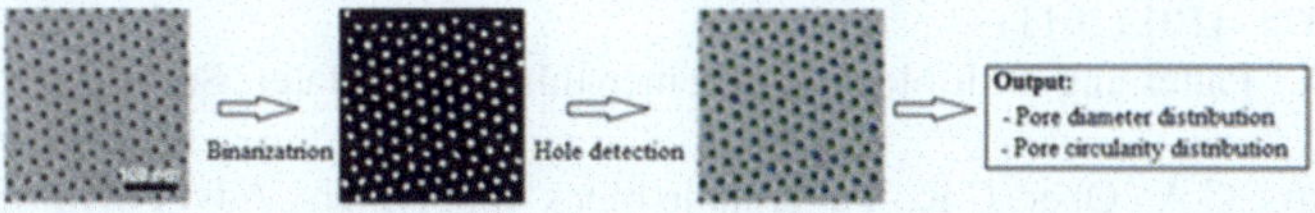

Fig. 6. Schematic flow diagram of pore size evaluation methodology as implemented in software which has been developed by our group. From top-down SEM images and pore dimensions are computed

Figure 7 shows the size distributions of the pore diameters as obtained from the images reported in Figs. 1a and 2a at optimum film thickness. In both cases the shape of the distribution is Gaussian-like. The comparison between the histograms highlights the larger average pore diameter for BCP B which is due to its higher L_0. The pore distribution results in an average pore diameter of 12.0 nm for BCP (A) of 14.4 nm for BCP (B). The 1 sigma uniformity has been estimated at 6–7% of hole diameter for both BCP formulation.

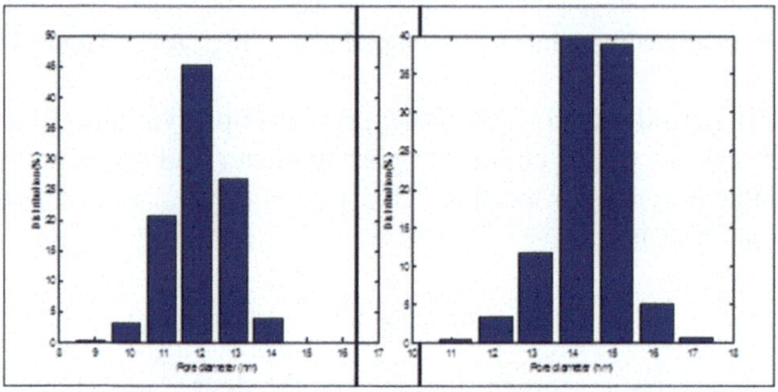

Fig. 7. Histogram of pore size distribution for BCP A (left) and BCP B (right).

4 Conclusion

We investigated in this work the effect of process parameters on self-assembly morphology for two commercial BCP formulation based on PS-*b*-PMMA. We conclude that high-quality structures of the self-assembly process depend strongly on film thickness and annealing temperature. The process window depends also on the BCP molecular weight. According to this study, BCP S-A was found to be a good template to generate well-ordered nanoscale feature perfectly suited for nanotechnological applications.

References

Han, E., Stuen, K.O., La, Y., Nealey, P.F., Gopalan, P.: Effect of composition of substrate-modifying random copolymers on the orientation of symmetric and asymmetric diblock copolymer domains. Macromolecules **41**, 9090–9097 (2008)

Jeong, J.W., Park, W.I., Kim, M., Ross, C.A., Jung, Y.S.: Highly tunable self-assembled nanostructures from a poly (2-vinylpyridine-b-dimethylsiloxane) block copolymer. J. Nano Lett. **11**, 4095–4101 (2011)

Segalman, R.A.: Patterning with block copolymer thin films. Mater. Sci. Eng. R **48**, 191–226 (2005)

Li, M., Coenjarts, C.A., Ober, C.K.: Patternable block copolymers. Adv. Polym. Sci. **190**, 183–226 (2005)

Bates, F.S., Fredrickson, G.H.: Block copolymer thermodynamics: theory and experiment. Ann. Rev. Phys. Chem **41**, 525–57 (1990)

Park, M., Harrison, C., Chaikin, P.M., Register, R.A., Adamson, D.H.: Block copolymer lithography: periodic arrays of 1011 holes in 1 square centimeter. Science **276**, 1401–1404 (1997)

Bang, J., Jeong, U., Ryu, D.Y., Russell, T.P., Hawker, C.J.: Block copolymer nanolithography: translation of molecular level control to nanoscale patterns. Adv. Mater. **21**, 4769–4792 (2009)

Guarini, K.W., Black, C.T., Milkove K.R., Sandstrom, R.L.: Nanoscale patterning using self-assembled polymers for semiconductor applications. J. Vac. Sci. Technol. B **19**, 2784 (2001)

Black, C.T., Guarini, K.W., Milkove, K.R., Baker, S.M., Russell, T.P., Tuominen, M.T.: Integration of self-assembled diblock copolymers for semiconductor capacitor fabrication. Appl. Phys. Lett. **79**, 409 (2001)

Guarini, K.W., Black, C.T., Zhang, Y., Kim, H., Sikorski, E.M., Babich, I.V.: J. Vac. Sci. Technol. B **20**, 2788–2792 (2002a)

Xu, T., Kim, H.C., DeRouchey, J., Seney, C., Levesque, C., Martin, P., Stafford, C.M., Russell, T.P.: The influence of molecular weight on nanoporous polymer films. Polymer **42**, 9091–9095 (2001)

Guarini, K.W., Black, C.T., Yeung, S.H.I.: Adv. Mater. **14**, 1290 (2002b)

Byutner, O., Smith, G.D.: Temperature and molecular weight dependence of the zero shear-rate viscosity of an entangled polymer melt from simulation and theory. J. Polym. Sci. Pt. B-Polym. Phys. **39**, 3067–3071 (2001)

Russell, T.P., Hjelm, R.P., Seeger, P.A.: Temperature dependence of the interaction parameter of polystyrene and poly (methyl methacrylate). Macromolecules **23**, 890 (1990)

Men'shikov, E.A., Bol'shakovaI, A.V., Yaminskii, V.: Determination of the Flory-Huggins parameter for a pair of polymer units from AFM data for thin films of block copolymers. Prot. Met. Phys. Chem. Surf. **45**, 295–299 (2009)

Loucif Seiad, M., MKuppuswamy, V.K., Cao, Y., Gronheid, R., Ferhat, M.: Dynamic behavior in self-assembly process of cylindrical phase PS-b-PMMA block copolymer. Mater. Sci. Semicond. Process. **66**, 26–32 (2017)

Otsu, N.: A threshold selection method from Gray-level Histograms. IEEE Trans. Syst. Man Cybern. **9**, 62–66 (1979). https://ieeexplore.ieee.org/search/searchresult.jsp?searchWithin=%22Authors%22:.QT.Nobuyuki%20Otsu.QT.&newsearch=true

Sezgin, M., Sankur, B.: A survey over image thresholding techniques and quantitative performance evaluation. J. Elec. Imag. **13**, 146–165 (2004)

Hydroxy Ethyl Cellulose-Graft-Acrylamide Superabsorbant Hydrogel for Controlled Drug Delivery of Cetirizine

N. Bouslah Mokhnachi(✉), L. Bounabi, N. Baa, F. Ouazib, and N. Haddadine

Faculty of Chemistry, Laboratory of Synthesis Macromolecular and Thio-Organic Macromolecular, University of Sciences and Technology HouariBoumediene, 16111 Algiers, Algeria
bouslah_naima@yahoo.fr

Abstract. There has been considerable growing in recent years in the use of hydrogels as drug delivery systems because they can extend the duration of drug release and thus can reduce dose to be administrated improving patient compliance. Hydrogels are high-water content materials prepared from cross-linked polymers that are able of holding large amounts of water in their three-dimensional networks. Due their biocompatibility, low toxicity, and biodegradability, natural polymers are excellent candidates to develop new hydrogels.

In the present work a crosslinked graft copolymer of acrylamide onto hydroxy ethyl cellulose (HEC-g-PAM) has been synthesized in order to examine its potential use in biomedical applications as drug carrier. The synthesis was performed using N,N′-methylene-bis-acrylamide (MBA) as a crosslinking agent and potassium persulfate (KPS) as initiator via free radical initiation. The structural properties of the hydrogel were investigated by Fourier transform infrared spectroscopy (FTIR), differential scanning calorimetry (DSC) and thermogravimetric analysis (TGA). Swelling behavior in two aqueous media, simulated gastric fluid (pH 1.2) and simulated intestinal fluid (pH 7.4) was also studied. The pseudo Fickian diffusion model was adopted to fit the results and a good linear relationship could be observed. Cetirizine, a highly water soluble drug, was chosen as drug molecule for the drug delivery study. For drug loading, the hydrogels discs were steeped in a saturated cetirizine (Ctz) solution for 48 h. Drug incorporation efficiency in the (HEC-g-PAM) hydrogels was determined by UV-vis spectroscopy. FTIR spectroscopy was used to prospect the compatibility of the drug with the hydrogel used as excipient. Differential scanning calorimetry DSC study revealed that the introduction of cetirizine (Ctz) in the hydrogel network induced the amorphization of the drug. The thermal stability of the (HEC-g-PAM)/Ctz formulation has been studied. The in vitro drug release study indicated that 100% of the drug is released after 10 h.

Keywords: Hydrogels · Grafting · Drug delivery · Acrylamide
Hydroxy ethyl cellulose

B. Abdelbaki et al. (Eds.): SMSD 2017, *Proceedings of the Third International Symposium on Materials and Sustainable Development*, pp. 166–174, 2018.
https://doi.org/10.1007/978-3-319-89707-3_20

1 Introduction

Hydrogels are hydrophilic, cross-linked, polymers capable of absorbing large amounts of water without dissolving due to the presence of cross-links between network chains (Ahmed 2015). The ability of hydrogels to absorb water arises from hydrophilic functional groups attached to the polymeric backbone. Hydrogels can be classified in several ways. As example they can be classified into two categories based on the chemical or physical nature of the cross-link junctions. Chemically cross-linked networks have permanent junctions while physical crosslinks are not permanent, but they are sufficient to make them insoluble in an aqueous media (Shetye et al. 2015). They have gained considerable attention in recent years in bioengineering (El-Sherbiny and Yacoub 2013), biomedical applications, cosmetic (Parente et al. 2015), water treatment (Jing et al. 2013), food industry (Roy et al. 2012) etc. Due to their unique properties, such as biocompatibility, compiliance with sterilization protocols and biodegradability, they have also emerged as effective drug delivery systems in the pharmaceutical field (AL-Kahtania and Sherigara 2014, McKenzie et al. 2015). Both natural polymers, such as alginate; chitosan, cellulose etc., and synthetic polymers such as polylactic acid (PLA), poly(ethylene glycol) (PEG) have been used to produce hydrogels (Siepmann and Peppas 2001). However these last decades several studies have been reported on hybrid hydrogels based on both natural and synthetic polymers which are developed to combine the biodegradable property of natural polymers with several functionalities of synthetic polymers (Sadeghi 2010; Kwon et al. 2015). Hydroxyethyl cellulose (HEC) is a polysaccharide that finds several applications in drug delivery (Vasquez and Terence 2015). PAM hydrogels have already been used in several applications and have proven capability of water absorption and biocompatibility with physiologic body fluids (Mahdavinia et al. 2009). Cetirizine dihydrochloride (Ctz) is an effective drug for the treatment of a wide range of allergic diseases (Ciurlizza et al. 2014).

In this study, a novel hydrogel prepared by grafting cross-linked polyacrylamide chains onto hydroxyethyl cellulose) via a free radical polymerization was developed to study its potential use as drug carrier of cetirizine (Ctz).

2 Materials and Methods

2.1 Materials

Hydroxy ethyl cellulose, the monomer acrylamide, the crosslinking agent N, N-methylenebis acrylamide (MBA), and the initiator potassium persulfate (KPS) were purchased from Sigma Aldrich. Cetirizine Ctz (Sigma–Aldrich Co.) was chosen as a model drug to characterize the release behavior from the hydrogel. HCl, KCl, K_2HPO_4 and NaOH used for the preparation of simulated gastric fluid (SGF, pH 1.2) and simulated intestinal fluid, phosphate buffer solutions, PBS, (SIF pH 7.4) were analytical grade and used as received.

2.2 Synthesis

Grafting of monomer onto polymer backbone is one of the effective and accessible methods for the chemical modification of polysaccharides. The hydrogel HEC-g-PAM has been synthesized by grafting acrylamide (2.81 mol/L) onto HEC polysacharide by free radical polymerization method in aqueous system using MBA (1.36×10^{-3} mol/L), and KPS (7.77×10^{-3} mol/L). Firstly the HEC solution was purged with nitrogen for 20 min. The KPS solution and AM monomer were added. Afterwards MBA solution was introduced into the flask dropwise. The reactional mixture was heated to 70 °C and kept for 20 min to complete polymerization. The resulting hydrogel was washed with distilled water several times and then dried at 50 °C to a constant weight. For comparative purpose the homopolymer PAM was synthesized by free radical polymerization.

2.3 Characterization

Fourier-transform infrared (FTIR) spectra were recorded at room temperature with a Perkin Elmer FTIR Spectrum Two, using 32 scans and a resolution of 2 cm^{-1}, from 4000 to 400 cm^{-1}. FTIR spectra were collected on samples in the form of KBr pellets.

A Perkin Elmer, DSC Diamond differential scanning calorimeter equipped with a cooling device was used to characterize the thermal behaviour of the samples. All measurements were carried out under nitrogen atmosphere at a heating/cooling rate of 10 °C/min.

Thermo gravimetric analysis (TGA) was carried out within 30–800 °C at the heating rate of 10 °C/min under nitrogen flow (20 ml/min) using a TA Q-50.

Swelling studies of the hydrogel discs were carried out in both simulated gastric fluid SGF (HCl solution pH 1.2) and in simulated intestinal fluid SIF (phosphate buffer solution pH 7.4) by gravimetric method. Each disc was immersed in excess of the adequate solution. The weights of swollen hydrogels were measured at consecutive time intervals after removing the surface water with filter paper. This procedure was repeated until the weight of the swollen samples remained constant.

The in vitro drug release behavior was carried out by placing the hydrogel pellets loaded with cetirizine in both simulated gastric fluid SGF (HCl solution pH 1.2) and in simulated intestinal fluid SIF (phosphate buffer solution pH 7.4) in a water bath with stirring at 37 °C. Aliquots (5 mL) were withdrawn at periodic intervals and were immediately replaced by the same volume of fresh medium. The aliquots, after suitable dilution, were analyzed using the UV–visible Perkin Elmer lambda 20 spectrophotometer. The UV–visible absorbance of cetirizine solutions were measured at $\lambda_{max} = 225$ nm. The release amount was calculated by previously established calibration curve.

3 Results and Discussion

3.1 Fourier Transform Infrared FTIR Study

Grafting efficiency was prospected by FTIR spectroscopy. The HEC spectrum (Fig. 1) showed a broad band at 3386 cm^{-1} corresponding to the hydroxyl groups of the HEC. The(C-H) elongation vibrations appeared at 2937 and 2875 cm^{-1}. The bands of the deformation vibrations appeared at 1355 cm^{-1} and 1453 cm^{-1} for the -O-H and C-H groups respectively. Other bands with relatively high intensities appeared at 1054 and 1020 and were attributed to the C-O-C elongation vibration while the band at 1145 cm^{-1} was assigned to the elongation vibration C-O. In the HEC-g-PAM hydrogel spectrum all the bands relating to the elongation and deformation vibrations of the O-H groups and the C-H groups of the HEC were observed. However, the bands at 1054 and 1020 cm^{-1} disappeared completely, reflecting the fact that the grafting reaction of acrylamide has indeed taken place. In addition, a shoulder at 3194 cm^{-1} relative to the N-H groups of PAM as well as an intense band at 1665 cm^{-1} relative to the carbonyl groups of the amide were clearly observed and confirmed the presence of the acrylamide groups in the hydrogel (Fig. 1).

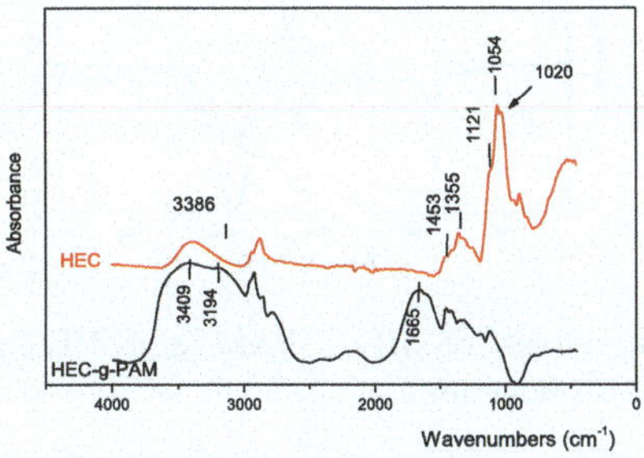

Fig. 1. FTIR spectra of HEC and HEC-g-PAM hydrogel

3.2 Thermogravimetric TGA Study

The TG-DTG curves of HEC, PAM and HEC-g-AM are depicted in Fig. 2. It can be noticed that HEC presented a first mass loss of about 10% from ambient to 100 °C corresponding to the elimination of sorbed water. The second weight loss between 252 and 383 °C is about 69% corresponding to polymer chains degradation. The homopolymer PAM has three zones of mass loss: the first one from ambient to 210 ° C corresponds to a loss of about 9% of absorbed and bonded water. The second stage is between 210 and 350 °C, with 20% mass loss corresponding to the formation of inter and intra-chains imides (Caulfield et al. 2002). The third mass loss above 350 °C of about 50% corresponds to the decomposition of the imides to form nitriles and the release of volatile substances such as CO_2 and H_2O.

The examination of the derivative curves revealed that the HEC decomposition peak shifts to higher temperatures in the grafted hydrogel thermogram indicating that the PAM chains slowed the HEC degradation.

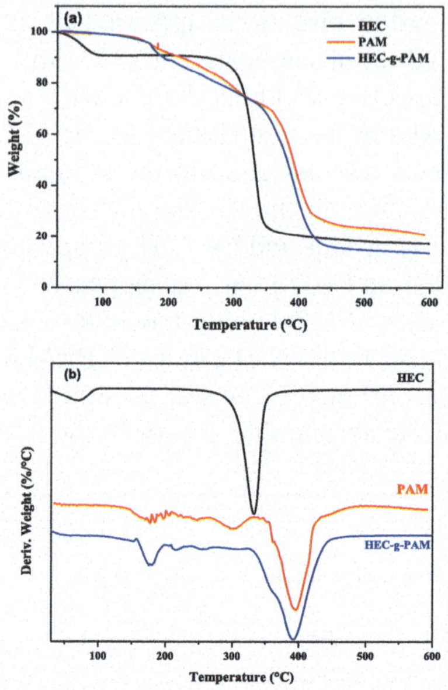

Fig. 2. Thermogravimetric curves of HEC, PAM and HEC-g-PAM hydrogel (**a**) TG thermograms, (**b**) derivative thermograms.

3.3 Differential Scanning Calorimetry DSC Study

As shown in Fig. 3, the glass transition temperature of the HEC appeared at 130 °C in accordance with the literature (Russo et al. 2010).

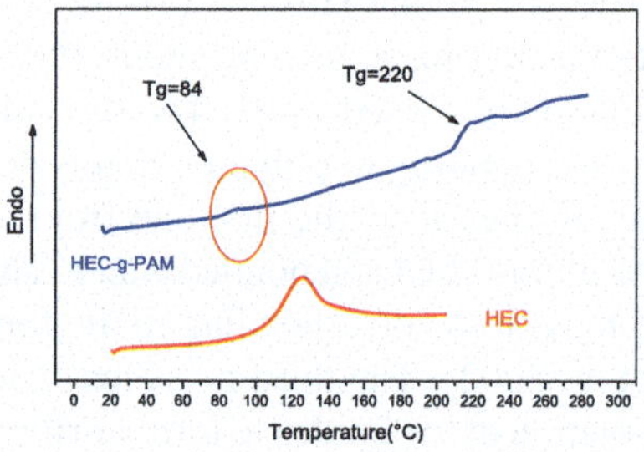

Fig. 3. DSC thermograms of HEC and HEC-g-PAM hydrogel

The thermogram of the HEC-g-PAM hydrogel, illustrated in the same Figure, showed an endothermic peak at 220 °C corresponding to the Tg of the HEC-g-PAM hydrogel. It should be noted that this temperature occurs at a higher temperature than that of pure HEC due to the chemical crosslinks which are present in the hydrogel and which have the effect of reducing the mobility of the chains. In order to allow this chain to move, a greater energy is therefore necessary and consequently the glass transition temperature will increase. A second, lower glass transition appeared at 84 °C. It corresponds to the chains of acrylamide grafted on the HEC.

Figure 4 shows DSC thermogram of Ctz drug ad drug loaded hydrogel (HEC-g-PAM)/Ctz. The Ctz thermogram exhibits a large endothermic peak at 223 °C corresponding to the melting of the drug. In the drug loaded hydrogel this peak is absent, indicating the uniform dispersion of drug in an amorphous form in the hydrogel. Moreover, the examination of the thermogram of the drug loaded hydrogel, shows a slight glass transition at 84 °C corresponding to the acrylamide chains grafted on the HEC main chain. The second transition relative to the HEC chains, initially situated at 220° undergoes a significant shift to lower temperatures up to 166 °C in the HEC-g-PAM thermogram (Fig. 3). This lowering of the Tg reflects the development of strong interactions between the hydrogel and the drug molecules.

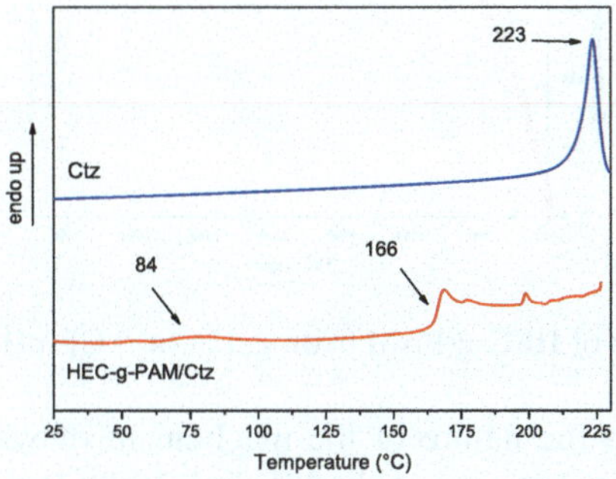

Fig. 4. DSC thermograms of the drug Ctz and drug loaded hydrogel (HEC-g-PAM)/Ctz

3.4 Swelling Behaviour

The capacity of swelling is one of the most important parameters to evaluate the appropriateness of hydrogels for drug delivery. In tis study, the swelling ratio was calculated as follows:

$$W\% = \frac{w_s - w_d}{w_d}.100 \quad (1)$$

Where w_s and w_d are the weights of swollen and dry gel respectively. As shown in Fig. 5., in both media the swelling rate for HEC-g-PAM hydrogel was high at the beginning of the swelling process and then levels off. The percentage of swelling reached up 670% within 9 h in both SGF (pH 1.2) and SIF (pH 7.4).

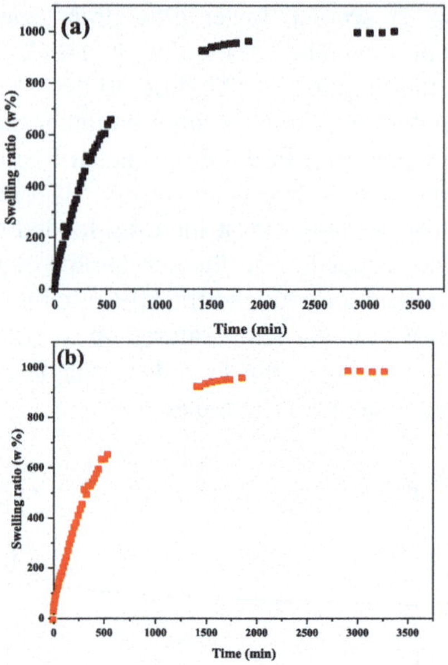

Fig. 5. Swelling behavior of HEC-g-PAM hydrogel in (**a**) SGF(pH 1.2) and (**b**) SIF(pH 7.4)

In order to determine the nature of the mechanism of water molecules transport within the hydrogel, power law was applied (Frisch 1980; Peppas 1985).

$$\frac{w_t}{w_\infty} = kt^n \tag{2}$$

w_t and w_∞ are the water intake at time t and at equilibrium respectively. k is a constant indicating the structural and geometric properties of the matrix, n is a number indicative of *the* type of diffusion. If $n \leq 0.5$, penetration of the water molecules follows the Fickian law and the rate of diffusion is much lower than the rate of relaxation. In cases, where n = 1, the diffusion is very fast. If $0.5 < n < 1$, the process corresponds to non-Fickian anomalous diffusion where structural relaxation is comparable to diffusion. The diffusion process was evaluated for the HEC-g-PAM hydrogel showing a pseudo-Fickian type diffusion (see Table 1).

Table 1. Diffusion parameters of HEC-g-PAM in SGF and SIF media

	Correlation factor	n	K
pH 1.2	0.99	0.39	0.05
pH 7.4	0.98	0.47	0.47

3.5 In Vitro Drug Release Study

The % cumulative cetirizine release from HEC-g-PAM hydrogel as a function of time is shown in Fig. 6. During the first phase of release, the formulation releases 77% Pa in the AMS and 68% Pa in the MIS in 4 h.

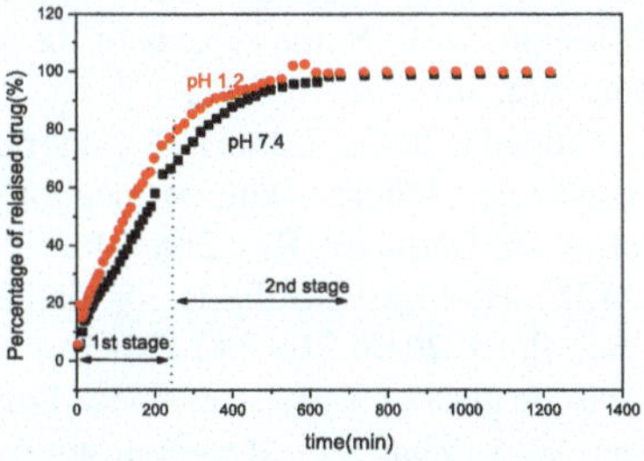

Fig. 6. Drug release profile of HEC-g-PAM hydrogel in (**a**) SGF(pH 1.2) and (**b**) SIF(pH 7.4)

During the second phase, the formulation allows to release much more slowly since 99% of the AP are released only after 10 h.

The system (HEC-g-PAM/CTR) developed in the context of this study is therefore suitable for a prolonged release of the active principle. The progressive swelling of the hydrogel delays the release of the drug. Once the polymeric matrix is in contact with the dissolution medium, it hydrates and swells and generates a gelling barrier through which the active principle diffuses towards the dissolution medium. The greater the swelling, the more the active principle is released. Thus, the rate of release of the active principle is dependent on the hydration rate of the matrix system.

We also observed that the drug levels released in the SGF are slightly higher than those obtained in the SIF. This can be attributed to the slightly higher CTR solubility in acid medium.

4 Conclusion

HEC-g-PAM hydrogel was prepared by grafting PAM homopolymer on HEC macromolecules. The success of grafting was confirmed by FT-IR, DSC and TGA analyses.

In vitro release experiments revealed Sustained release of Ctz as model drug suggesting that HEC-g-PAM hydrogel was suitable for controlled drug delivery. The study also showed that drug release from a polymer matrix is closely related to many factors such as swelling behavior of the hydrogel matrix, drug affinity to polymer chains and solubility of drug in water.

References

Ahmed, E.M.: Hydrogel: preparation, characterization, and applications: a review. J. Adv. Res. **6**, 105–121 (2015)

AL-Kahtania, A.A., Sherigara, B.S.: Controlled release of diclofenac sodium through acrylamide graftedhydroxyethyl cellulose and sodium alginate. Carbohyd. Polym. **104**, 151–157 (2014)

Caulfield, M.J., Qiao, G.G., Solomon, D.H.: Some aspects of the properties and degradation of polyacrylamides. Chem. Rev. **102**, 3067–3083 (2002)

Ciurlizza, C., Fernández, F., Calpena, A.C., Lázaro, R., Parra, A., Clares, B.: Semisolid formulations containing cetirizine: human skin permeation and topical antihistaminic evaluation in a rabbit model. Arch. Dermatol. Res. **306**, 711–717 (2014)

El-Sherbiny, I.M., Yacoub, M.H.: Hydrogel scaffolds for tissue engineering: progress and challenges. Glob. Cardiol. Sci. Pract. **2013**, 316–342 (2013)

Frisch, H.L.: Sorption and transport in glassy polymers-a review. Polym. Eng. Sci. **20**, 2–13 (1980)

Jing, G., Wang, L., Yu, H., Amer, W.A., Zhang, L.: Recent progress on study of hybrid hydrogels for water treatment. Colloids Surf. A **416**, 86–94 (2013)

Kwon, S.S., Kong, B.J., Park, S.N.: Physicochemical properties of pH-sensitive hydrogels based on hydroxyethyl cellulose–hyaluronic acid and for applications as transdermal delivery systems for skin lesions. Eur. J. Pharm. Biopharm. **92**, 146–154 (2015)

Mahdavinia, G.R., Mousavi, S.B., Karimi, F., Marandi, G.B., Garabaghi, H., Shahabvand, S.: Synthesis of porous poly(acrylamide) hydrogels using calcium carbonate and its application for slow release of potassium nitrate, eXPRESS. Polym. Lett. **3**, 279–285 (2009)

McKenzie, M., Betts, D., Suh, A., Bui, K., Kim, L.D., Cho, H.: Hydrogel-based drug delivery systems for poorly water-soluble drugs. Molecules **20**, 20397–20408 (2015)

Parente, M.E., Ochoa, Andrade A., Ares, G., Russo, F., Jiménez-Kairuz, Á.: Bioadhesive hydrogels for cosmetic applications. Int. J. Cosmet. Sci. **37**, 511–518 (2015)

Peppas, N.A.: Analysis of Fickian and non-Fickian drug release from polymers. Pharm. Acta Helv. **60**, 110–111 (1985)

Roy, N., Saha, N., Kitano, T., Saha, P.: Biodegradation of PVP–CMC hydrogel film: A useful food packaging material. Carbohyd. Polym. **89**, 346–353 (2012)

Russo, R., Malinconico, M., Santagata, G.: Effect of polyglycerol and the crosslinking on the physical properties of a blend alginate-hydroxyethylcellulose. Carbohyd. Polym. **82**, 1061–1067 (2010)

Sadeghi, M.: Synthesis and swelling behaviors of graft copolymer based on chitosan-g-poly(AA-co-HEMA). Int. J. Chem. Eng. Appl. **1**, 354–358 (2010)

Shetye, S.P., Godbole, A., Bhilegaokar, S., Gajare, P.: Hydrogels: introduction, preparation, characterization and applications. Hum. J. **1**, 51–55 (2015)

Siepmann, J., Peppas, N.A.: Modeling of drug release from delivery systems based on hydroxypropyl methylcellulose (HPMC). Adv. Drug Deliv. Rev. **48**, 139–157 (2001)

Vasquez, J.M.G., Tumolva, T.P.: Synthesis and characterization of a self-assembling hydrogel from water-soluble cellulose derivatives and sodium hydroxide/thiourea solution. Am. J. Chem. **5**, 60–65 (2015)

Improving the Power Efficiency of Organic Photovoltaic Cells with P3HT:ICBA Used as Active Layer

O. Ourahmoun(✉)

Laboratory of Advanced Technologies of Genie Electrics (LATAGE), Faculty of Electrical and Computer Engineering, Mouloud Mammeri University (UMMTO), BP 17 RP, 15000 Tizi-Ouzou, Algeria
ourahmounourida@yahoo.fr

Abstract. In this article the optical properties of poly (3-hexylthiophene): Indene-C60 bisadduct (P3HT:ICBA) layers were examined. The good absorption of the blend in the visible spectra of light allows using ICBA as acceptor material for the fabrication of organic solar cells. The structure of the realized device is glass/ITO/ZnO(np)/P3HT:ICBA/PEDOT:PSS/Ag.

Indene-C60 bisadduct (ICBA) has been used as acceptor molecule to achieve open-circuit voltage Voc = 0.8 V when poly (3-hexylthiophene) (P3HT) is used as donor polymer. Compared to [6,6]-phenyl-C61-butyric acid methyl ester (PCBM) a shift in open-circuit voltage $\Delta Voc \approx 0.2$ V is observed. This study shows that the parameters of the cells with ICBA are improved compared to PCBM. An efficiency of 5.85% is obtained with ICBA and 3.89% is obtained with PCBM.

Keywords: Polymers · ICBA · Renewable energy · Solar cells Encapsulation

Abbreviations

NP	Nanoparticle
ZnO	Zinc oxide
P3HT	Poly (3-hexylthiophene)
PCBM	1-(3-methoxycarbonyl)-propyl-1-phenyl- (6,6)C61
ICBA	Indene-C60 bisadduct
ITO	Indium Tin Oxide
Ag	Silver
OPVs	Organic photovoltaics
PEDOT:PSS	Poly(3,4-ethylenedioxythiophene):poly(styrene sulfonate)

B. Abdelbaki et al. (Eds.): SMSD 2017, *Proceedings of the Third International Symposium on Materials and Sustainable Development*, pp. 175–180, 2018.
https://doi.org/10.1007/978-3-319-89707-3_21

1 Introduction

Conjugated polymers are of increasing interest as new materials for electronic applications. Bulk heterojunction (BHJ) polymer solar cells (PSCs) based on composites of conjugated polymers (electron donor) and fullerene derivatives (electron acceptor) have attracted attention due to their potential as renewable energy sources. However, their efficiency is usually less than 5%. By using thin-film optical simulation modelling, Boland and Gon [1] demonstrated that optimizing device efficiency is strongly dependent on the simultaneous control of active layer thickness and the charge carrier mobility. One of the most important factors for high efficiency OPVs, is the fabrication of the appropriate morphology of this active blend. Ideally, the polymer and the fullerenes must be phase separated at 10–20 nm scale (typical exciton diffusion length, providing thus large interfacial area for exciton dissociation [2]. The different device geometries and interface morphologies need to be evaluated for the purpose of trapping more light, dissociating excitons more efficiently, increasing the exciton diffusion length and transporting charges with fewer barriers [3]. The internal quantum efficiency of a solar cell depends on its intrinsic material properties, such as its crystallinity, energy band gap, carrier transport behavior, and the number of defects and impurities. The external quantum efficiency is associated with the solar cell's interfacial reflection, photon absorption, and architecture, which all affect the number of collected electron/hole pairs [4]. Indium Tin Oxide (ITO), or Fluorine-doped Tin Oxide, is a transparent electrode present many advantages such as high transmission (85% in the visible spectrum), low resistivity due to a low sheet resistance (approximately 7 Ω/sq) and a high work function.

Inverting the polarity of the device and processing the electron collecting electrode onto the ITO and the hole collection electrode on top of the active layer has also been successfully used to increase the efficiency of devices made with the same active materials.

Among the n-type metal oxides used in inverted cells, ZnO is a promising candidate due to its relatively high electron mobility, environmental stability, and high transparency. Carrier transport effects by means of a simple equivalent circuit model directly derived from the inspection of impedance spectra corresponding to P3HT: ICBA bulk heterojunction solar cells is studied by Guerrero [5].

In this article, Indene-C60 bisadduct (ICBA) has been used as acceptor molecule to achieve open-circuit voltage Voc = 0.8 V when poly(3-hexylthiophene (P3HT) is used as donor polymer.

2 Realization of the Cells

2.1 Chemical Structure of the Materials Used in the Elaboration of the Cells

The chemical structure of the donor (P3HT), the acceptors (ICBA and PCBM) and the hole transport material PEDOT (F010) are shown in Fig. 1(a) and (b) respectively. $LUMO_{ICBA} = -3.67$ eV and $LUMO_{PCBM} = -3.85$ eV.

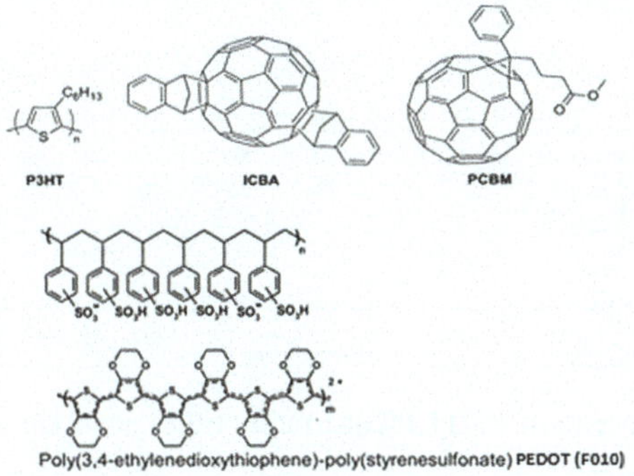

Fig. 1. Chemical structure of the materials used in the fabrication of the device: (a) active layer materials and (b) hole transport material PEDOT (F010).

2.2 Preparation of Solution of P3HT:PCBM and P3HT:ICBA

P3HT:PCBM and P3HT:ICBA are dissolved in 1–2 dichlorobenzene. Both solutions are let shake in box gloves at 70 °C for 30 min. The two solutions are left in the glove box for a period longer than 24 h before to be used.

2.3 Optical Properties of P3HT:ICBA and P3HT:PCBM

The absorption of P3HT:ICBA as function of the thickness of layer is given in Fig. 2. The absorption is measured by the SAFAS 200 instrument. The thickness of the layers is measured by the Dektat profilometer.

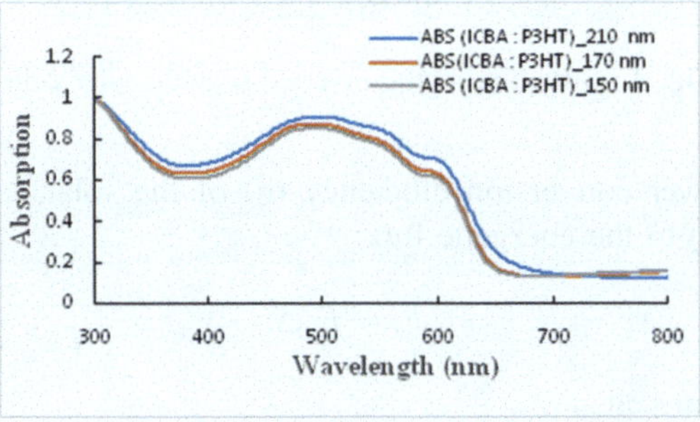

Fig. 2. Absorption of P3HT:ICBA depending in the thickness

Figure 3 shows the absorption spectra of P3HT:PCBM layer annealed at 110 °C for 30 min.

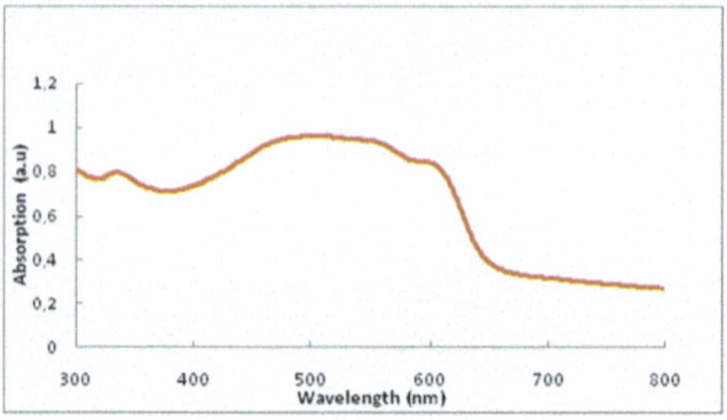

Fig. 3. Absorption spectra of P3HT:PCBM ratio (1:0.8) annealed at 110 °C for 30 min.

3 Structure of the Cells

A schematic of device architecture is shown in Fig. 4. The active layer is P3HT:PCBM or P3HT:ICBA. Indium Tin Oxide (ITO) is used as cathode, nano-particles of zinc oxide (ZnO) are used as electron transport layer, PEDOT F010) is used as hole transport layer and silver (Ag) is used as anode.

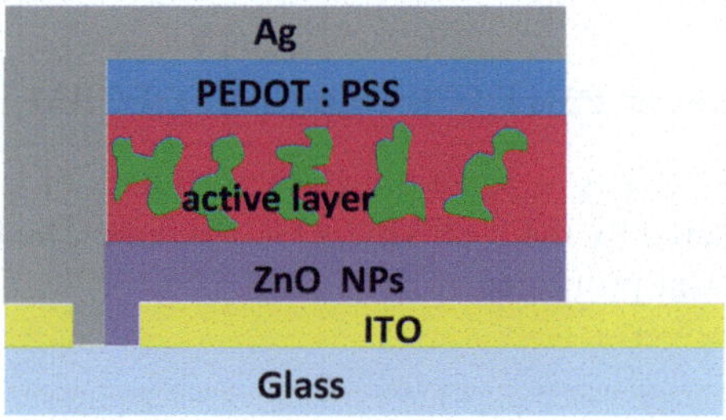

Fig. 4. Structure of the inverted organic solar cells.

The external power conversion efficiency (η) of the solar cell is defined as the electric power (P_{max}) to the energetic flux:

$$\phi_e = E_0.S \tag{1}$$

S: surface of the solar cell

$$\eta = \frac{P_{max}}{\phi_e} = \frac{FF.I_{sc}.V_{oc}}{\phi_e} \tag{2}$$

Where FF is the fill factor, I_{SC} the short-circuit current and V_{OC} is the open circuit voltage. E_0 and λ are the intensity (W/m^2) and the wavelength (nm) respectively.

4 Results and Discussion

Fabricated cells have a surface of 0.18 cm^2. The electrical characteristics are made in the glove box using a Solar Simulator under light intensity of 100 mW/cm^2. The electrical parameters of the produced cells are summarized in Table 1. Performed cell current densities are represented in Fig. 5.

Table 1. Comparison between performances of P3HT:PCBM (Cell1) and P3HT:ICBA (Cell 2).

Active layer	Voc (V)	Jsc (mA/cm^2)	η (%)	FF (%)
Cell 1	0.57	10.09	3.89	68
Cell 2	0.79	11.83	5.85	62

Indene-C60 bisadduct (ICBA) has been used as acceptor molecule to achieve open-circuit voltage Voc = 0.8 V when poly (3-hexylthiophene (P3HT) is used as donor polymer. Compared to [6,6]-phenyl-C61-butyric acid methyl ester (PCBM) a shift in open-circuit voltage ΔVoc ≈ 0.2 Vis observed (Table 1).

Cells with the active layer of P3HT:ICBA have better electrical performance than the active layer of P3HT:PCBM, the efficiency and the open circuit voltage of P3HT: ICBA are 5.85% and 0.78 V respectively.

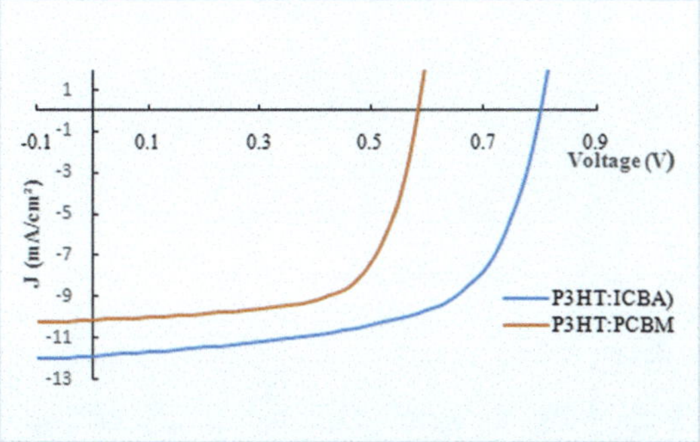

Fig. 5. Electrical characteristics of the cells depending on the type of the active layer.

5 Conclusions

Organic photovoltaic cells, with P3HT:ICBA have the best electrical and optical performances compared to cells with the P3HT:PCBM. This result allows replacing PCBM by ICBA in the realization of this type of cells. Different parameters affect the performances of the OPV such as solvents, concentration, annealing temperature, interface layer and the light intensity. More studies will be realized in order to understand the influence of each parameter in the performances and the stability of the organic solar cells based P3HT:ICBA. In order to improve the power conversion of the cells and to protect cells to degradation it is necessary to use the functional encapsulation.

References

1. Boland, P., Namkoong, G.: Optimization of active layer thickness in planar organic solar cells via optical simulation methods. Jpn. J. Appl. Phys. **49**(3), 030205 (2010)
2. Karagiannidis, P.G., Kassavetis, S., Pitsalidis, C., Logothetidis, S.: Thermal annealing effect on the nanomechanical properties and structure of P3HT:PCBM thin films. Thin Solid Films **519**, 4105–4109 (2011)
3. Keawaprajak, A., Piyakulawat, P., Klamchuen, A., Klaraska, P., Asawapiron, U.: Influence of crystallizable solvent on the morphology and performance of P3HT:PCBM bulk-heterojunction solar cells. Sol. Energy Mater. Sol. Cells **94**, 531–536 (2010)
4. Chuang, S.-Y., Yu, C.-C., Chen, H.-L., Su, W.-F., Chen, C.-W.: Exploiting optical anisotropy to increase the external quantum efficiency of flexible P3HT:PCBM blend solar cells at large incident angles. Sol. Energy Mater. Sol. Cells **95**, 2141–2150 (2011)
5. Guerrero, A., et al.: Series resistance in organic bulk-heterojunction solar devices: modulating carrier transport with fullerene electron traps. Org. Electron. **13**, 2326–2332 (2012)

Solar Cell Based on p-Si/n-Zno:Al Thin Films: Simulation and Parameters Extraction

Safia Kerour[1,2(✉)], Meryem Lasladj[1,2], and Abdeslam Bouloufa[1,2]

[1] Electrochemical and Materials Laboratory, Ferhat Abbas University of Setif-1, Setif, Algeria
kersafia@yahoo.fr

[2] Electronics Department, Faculty of Technology, Ferhat Abbas University of Setif-1, Setif, Algeria

Abstract. In this work, we propose to develop a method based on the single diode model circuit, which able to extract the five electric parameters: the photocurrent I_{ph}, the saturation current I_s, the ideality factor n, the series and shunt resistance (R_s and R_{sh}) of the solar cell based on silicon as an absorber and ZnO:Al as a transparent conductive oxide. In this context, an approach based on the genetic algorithms (GAs) was developed in order to model and optimize the simulated data. Optimal values of I_{ph}, I_s, n and R_s, R_{sh} between the experimental and calculated values were estimated by minimizing the quadratic error (6.7×10^{-5}).

Keywords: Solar cell · Extract · Genetic algorithms · Silicon ZnO:Al

1 Introduction

Transparent conductive Aluminium-doped Zinc Oxide ZnO:Al (AZO) layers are widely used as the front electrode with relatively low cost and to improve the conversion efficiency for Si thin-film solar cell [1–3].

The possibility to determine the important parameters of the solar cells plays a major role in their evaluating, in controlling their quality and in the fabrication and power performance of reliable solar panels [4–7]. The determining of the solar cell electrical parameters, from *J(V)* characteristic describing its operation, plays an important role in the design and manufacture of the cells, to better understand the physical mechanisms acting in the interior of the solar cell, not only to be able to simulate its behavior and optimize its different characteristics but also for the increase of efficiency [7]. However, the precise determination of these parameters is still a challenge for researchers, which led to a diversification in models and numerical methods dedicated to their characteristics [8]. Several algorithms extract the device parameters: the saturation current I_s, the series resistance R_s, the ideality factor n, the shunt conductance G_{sh} and the photocurrent I_{ph}, have been proposed. Those algorithms are of two types [9]: Those that use selected parts of the current-voltage curve [10]. This kind of algorithms involves selected points of the J-V characteristic, which is a disadvantage for this approach [9]. The second group of algorithms is those exploiting

B. Abdelbaki et al. (Eds.): SMSD 2017, *Proceedings of the Third International Symposium on Materials and Sustainable Development*, pp. 181–188, 2018.
https://doi.org/10.1007/978-3-319-89707-3_22

the whole characteristic [11]. This kind of algorithms is based on curve fitting and takes all the data in consideration. The fitting technique principle is to minimize a non-convex criterion; and that's why this technique used several optimization algorithms which guaranteed to obtain the best minimum value of the cost function [9].

The main objective of our work is to extract the five electrical parameters of a p-Si/n-ZnO:Al solar cell structure under illumination (AM1.5G) using Genetic Algorithms (GAs) strategy. The solar cell was firstly simulated using the AMPS-1D dispositive and taking out the current-voltage characteristic.

2 Solar Cell Structure and Simulation

The solar cell structure consisted of the following layers: n-AZO used as a transparent conductive oxide and p-Si used as an absorber (Fig. 1). The solar cell is completed by a back contact and a front grid.

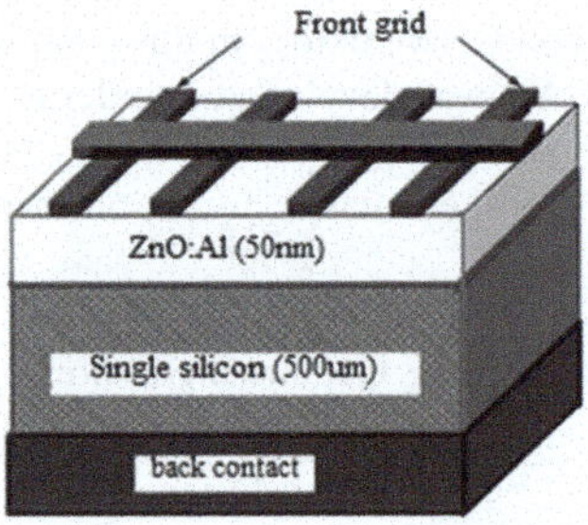

Fig. 1. Solar cell structure.

This solar cell is simulated by means of AMPS-1D (Analysis of Microelectronic and Photonic Structures) software which solves the three governing semiconductor device equations (the Poisson equation and the electron and hole continuity equations) and provides a convenient way to take out the *J(V)* characteristic. Figure 2 shows the software principal window (Table 1).

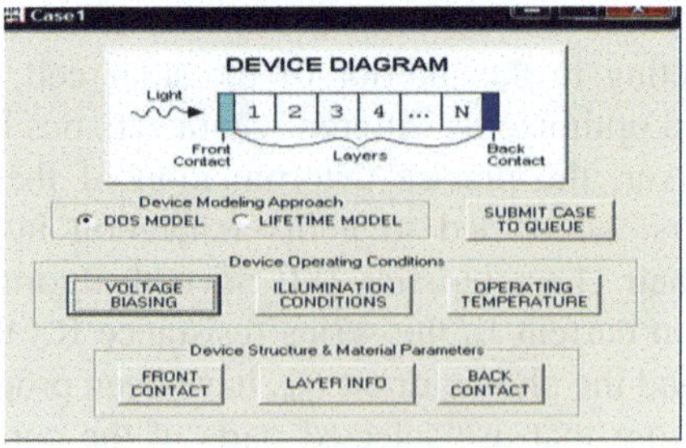

Fig. 2. AMPS-1D principal window.

Table 1. Parameter set for the simulation.

Parameters	n-ZnO:Al	p-Si
Dielectric constant	13.6	11.9
N_C (cm^{-3})	$2{,}22.10^{18}$	2.10^{19}
N_V (cm^{-3})	$1{,}78.10^{19}$	$1{,}04.10^{19}$
Band gap (eV)	3.30	1.12
Electron affinity (eV)	4.5	4.05
Electron mobility (cm^2/V.s)	100	1200
Hole mobility (cm^2/V.s)	25	400

After simulation, AMPS-1D achieved the *J(V)* characteristic of the simulated solar cell (Fig. 3).

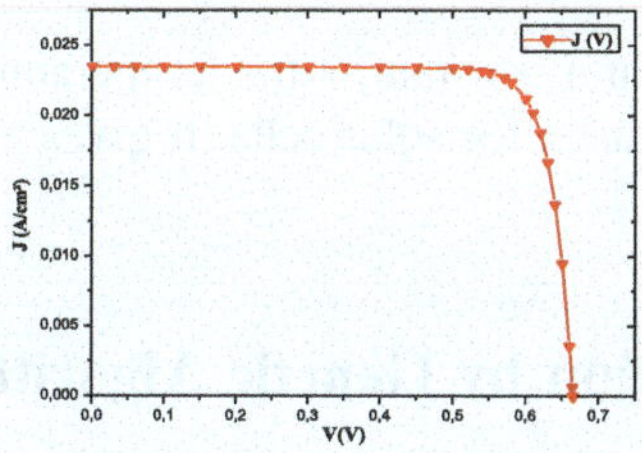

Fig. 3. *J(V)* characteristic of p-Si/n-ZnO:Al solar cell.

The solar cell performances obtained are summarized in Table 2.

Table 2. Solar cell photovoltaic parameters.

Performances	Values
Open-circuit voltage V_{oc} (V)	0.665
Short-circuit current I_{sc} (mA)	23.523
Fill factor *FF*	0.831
Efficiency η (%)	13.002

3 Solar Cell Modeling

The mostly model utilized to characterize the solar cell is the one diode model said single exponential model (SEM) where the cell behaves like a generator producing a current *I* and a voltage *V*. The solar cell has the combined advantage of a current and voltage generator. The equivalent circuit is shown as in Fig. 4 [12].

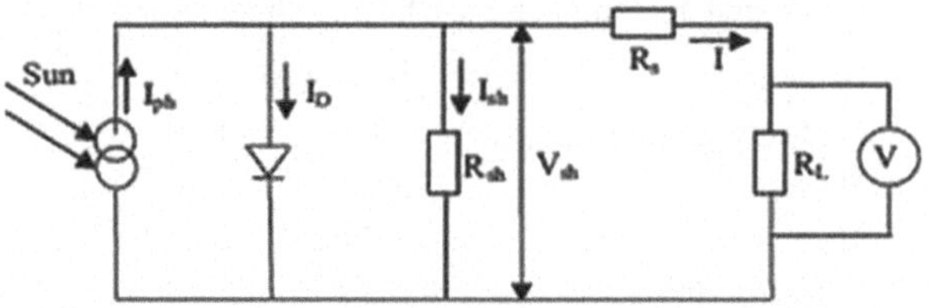

Fig. 4. *J(V)* characteristic of *p*-Si/*n*-AZO.

The mathematical model which describes the one diode model is characterized by the following equation:

$$I = I_{ph} - I_s \left[\exp\left(\frac{q(V + R_s I)}{nKT} \right) - 1 \right] - \frac{V + R_s I}{R_{sh}} \tag{1}$$

where, K denotes the Boltzmann constant, T the temperature and q the electron charge. The SEM was the first to be used for solar cells. It gives a good description of next to ideal devices [13].

4 Parameters Extraction by Genetic Algorithms

Two ways are there to use the GAs under MATLAB environment. The first one consists of using the interface window for continuous mathematical problems, and the second one is to formulate a program by exploiting the command on line for continuous and discrete problems which is our choice.

4.1 Problem Formulation

The solar cell parameter extraction is an optimization process which minimizes the difference between simulated and calculated values. The parameter calculation using optimization technique is implemented in the following approach:

- a set of data of *J(V)* for p-Si/n-AZO solar cell is carrying out by means of AMPS-1D device simulator.
- an objective function is defined for minimizing the difference between the simulated data and calculated values.
- the parameters are tuned by applying the optimization algorithm until the best objective function is obtained.
- after completing the optimization algorithm, the optimal value is extracted from the solution obtained by the optimization algorithm.

4.2 Cost Function

Before carrying out the GAs under MATLAB, it is imperative to create the problem's cost function. The problem is to find the optimal values of I_{ph}, I_s, R_s, G_{sh}, and n which minimize the quadratic error between the simulated and calculated values, Consequently, the cost function to be minimized is given by [4]:

$$\chi = \sum_{i=1}^{M} [I_i^{exp} - I(V_i, \theta)]^2 \tag{2}$$

where I_i^{exp} is the measured current at the V_i bias, $\theta = (I_{ph}, I_s, R_s, G_{sh}, n)$ is the set of parameters to carry out, M the number of considered data points and $I(V_i, \theta)$ is the predicted current.

4.3 Flow Chart of the Genetic Algorithm

Firstly, we define the chromosome by the vector given by $[I_{ph}, I_s, R_s, G_{sh}, n]$. In the same way, we have to define the initial population of chromosomes (IPOP) defined by the matrix given by [14]:

$$IPOP = (h_i - l_o).\ random\left[N_{ipop}, N_{par}\right] + l_o \tag{3}$$

Where:

N_{ipop}: is the initial number of chromosomes in IPOP;
N_{par}: number of parameters in the chromosome;
random $[N_{ipop}, N_{par}]$: the function which produces a $N_{ipop} \times N_{par}$ matrix from uniform numbers between zero (0) and one (1);
h_i: highest values of each parameter;
l_o: lowest values of each parameter

The GAs starts with an initial population seeking the optimal combination of the parameters corresponding to the best solution. With each generation, a new population is created with the same number of chromosomes. Progressively from the generations, the chromosomes will tend in general towards the cost function optimum. The algorithm cycle consists of a whole of stages which were used as a basis for the implementation of the genetic algorithm. This requires the definition of several parameters of a major importance to lead to the perfect performances. It is about the iteration count (generation max) and the dimension of the population, the probabilities of crossing and mutation.

After evaluating the cost function for each chromosome, we apply a selection in IPOP which means that only chromosomes corresponds to good values of the cost function was kept for the crossing (to reproduce). To make sure that the parameters space is explored, a mutation of 0.05 in the chromosomes was carried out. At the end, convergence was examined. If the result (the last value of χ) is not satisfied in comparison with a predefined cost minimum ($\chi = 10^{-5}$), All the preceding steps are repeated in the second generation and so on.

4.4 Test of the Genetic Algorithm

The developed algorithm must be sufficiently accurate and efficient, that's why a test is required. In the following, we will use the GAs to estimate the electrical parameters from the J(V) curve of a 57 mm diameter silicon solar cell. Then, we will compare these results to those in [4]. Table 3 summarizes results obtained and published.

Table 3. Summarized estimated parameters

Parameters	GAs	Reference [4]
G_{sh} (Ω^{-1})	0.0176	0.0186
R_s (Ω)	0.0364	0.0364
n	1.4823	1.4837
I_s (μA)	0.3227	0.3223
I_{ph} (A)	0.7603	0.7608

According to this relatively good value, we can assume that the GAs is very suitable for the estimation of the electrical parameters.

4.5 p-Si/n-ZnO:Al Electrical Parameters

We applied the GAs to extract the electrical parameters of p-Si/n-ZnO:Al solar cell. Then, we plot the both *J(V)* curve simulated and calculated one obtained using the GAs method. The race of the algorithm stopped once the cost function is equal to 6.7×10^{-5}. Table 4 summarizes obtained results.

Table 4. Summarized extracted parameters of *p*-Si/*n*-ZnO:Al solar cell.

Parameters	Values
R_{sh} (Ω)	2000
R_s (Ω)	0.195
n	1.43
I_s (A)	9.3376×10^{-10}
I_{ph} (A)	0.0235
χ	6.7×10^{-5}

The adjustment of the solar cell's simulated J–V curve to the calculated one using GAs method is plotted in Fig. 5.

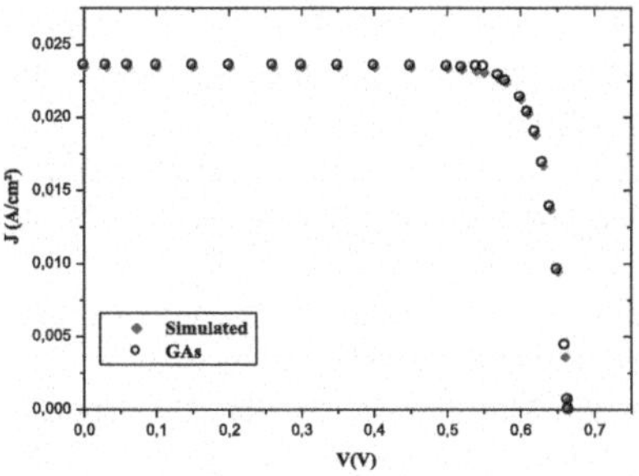

Fig. 5. Adjustment of the solar cell simulated *J–V* curve to the calculated one.

5 Conclusion

In this study, we simulate the p-Si/n-AZO solar cell by means AMPS-1D simulator and we propose a numerical technique based on GAs to estimate the electrical parameters (I_s, I_{ph}, R_s, R_{sh} and n). The used algorithm was firstly tested and validated then applied on the solar cell with minimized quadratic error at a value of 6.7×10^{-5}. According to the obtained values and the quality of the simulated-calculated adjustment curve; we assume that the GAs is a very efficient technique to estimate the electrical parameters of solar cells.

References

1. Yan, X., Li, W., Aberle, A.G., Venkataraj, S.: Investigation of the thickness effect on material and surface texturing properties of sputtered ZnO:Al films for thin-film Si solar cell applications. Vacuum **123**, 151–159 (2016)
2. Manea, E., Budianu, E., Purica, M., Cristea, D., Cernica I., Muller, R., Moagar Poladian, V.: Optimization of front surface texturing processes for high-efficiency silicon solar cells. Sol. Energy Mater. Sol. Cells **87**, 423–431 (2005)
3. Khaled, F., Bouloufa, A., Djessas, K., Mahamdi, R., Bouchama, I.: Aluminum doped zinc oxide wide band-gap n-type optical window for mc-Si superstrate solar cell. Vacuum **120**, 14–18 (2015)
4. Easwarakhanthan, T., Bottin, J., Bouhouch, I., Boutrit, C.: Nonlinear minimization algorithm for determining the solar cell parameters with microcomputers. Sol. Energy **4**, 1–12 (1986)
5. Ortiz-Conde, A., Garcia Sanchez, F.J., Muci, J.: New method to extract the model parameters of solar cells from the explicit analytic solutions of their illuminated I-V characteristics. Sol. Energy Mater. Sol. Cells **90**, 352–361 (2006)
6. AlRachidi, M.R., AlHajri, M.F., El-Naggar, K.M., Al-Othman, A.K.: A new estimation approach for determining the I-V characteristics of solar cells. Sol. Energy **85**, 1543–1550 (2011)
7. Chellaswamy, C., Ramesh, R.: Parameter extraction of solar cell models based on adaptive differential evolution algorithm. Renew. Energy **97**, 823–837 (2016)
8. Khezzar, R., Zereg, M., Khezzar, A.: Comparaison entre les différents modèles électriques et détermination des paramètres de la caractéristique I-V d'un module photovoltaïque. revues des Energies Renouvelables **13**, 379–388 (2010)

9. Zagrouba, M., Sellami, A., Bouaïcha, M., Ksouri, M.: Identification of PV solar cells and modules parameters using the genetic algorithms: application to maximum power extraction. Sol. Energy **84**, 860–866 (2010)
10. Charles, J.P., Ismail, M.A., Bordure, G.: A critical study of the effectiveness of the single and double exponential models for I-V characterization of solar cells. Solid-State Electron **28**, 807–820 (1985)
11. Chegaar, M., Ouennoughi, Z., Guechi, F.: Extracting dc parameters of solar cells under illumination. Vacuum **75**, 367–372 (2004)
12. Charles, J.P., Abdelkrim, M., Muoy, Y.H., Mialhe, P.: A practical method of analysis of the current–voltage characteristics of solar cells. Sol. Cells **4**, 169–178 (1981)
13. Shockley, W.: Bell Syst. Tech. J. **28**, 435–489 (1949)
14. Ketter, R.L., Sherwood Jr., P.: Modern Methods of Engineering Computation. McGraw-Hill, Inc., New York (1969). ISBN: 07-034423-x

Material Fracture of Dynamic VAWT Blade

S. Lecheb[1(✉)], A. Chellil[1], H. Mechakra[1], B. Safi[2], Y. Beggah[1], and Y. Lounici[1]

[1] Laboratory of Engines Dynamics and Vibroacoustics, University of Boumerdes, Boumerdes, Algeria
samir_lecheb@univ-boumerdes.dz

[2] Research Unit: Materials, Processes and Environment, Boumerdes, Algeria

Abstract. In this paper we studied fracture and dynamic behavior of vertical axis wind turbine blade, the VAWT is a historical machine, it has many properties, structure, advantage, component to be able to produce the electricity. we modeled the blade design then imported to Abaqus software for analysis the modes shapes, frequencies, stress, strain, displacement and stress intensity factor SIF, after comparison we chose the idol material. Finally the CTS test of glass epoxy reinforced polymer plates to obtained the material fracture toughness Kc.

Keywords: Material · Blade · Crack · Frequency · SIF

1 Introduction

The renewable energy is necessary energy in our life modern, the wind turbine produce the electricity by wind force applied on the blade. In this study we define damage and fracture of Vertical Axis Wind Turbine blade. The air in motion relative to the surface of the earth, the differences in atmospheric temperature, pressure and density, which in turn initiate forces that move air from one place to another. Wind Turbine is rotating machine which converts the wind potential energy to electricity, we have two types, HAWT and VAWT, Horizontal Axis Wind Turbine (HAWT) is a turbine with a rotor is parallel to the ground, Vertical Axis Wind Turbine (VAWT) is a turbine with a rotor is normally to the ground, this last don't require yaw processes, lower start up velocities, localization on the rooftops, mesas, shafts located close to the grounds where wind velocities are lower, lower efficiency as compared to HAWT. The principals components of VAWT are shaft, Blade, Generator, Tower, Foundation. We have many manufacturing of VAWT. Blades is the surface through it by the wind to produce aerodynamic forces for turn the rotor, we have many types of blade: straight, helical and circular. Profile of Blade has many shapes like: Rectangular, Circular and NACA profile. The materials used in production of blade wind turbine has been resistive, cheeped, low weight, like this materials for example Wood, Aluminium, Composite. Aluminium material have light weight, low tensile strength, good reliability. PVC is very light weight. Composite material have good thermo-mechanical and chemical properties.

B. Abdelbaki et al. (Eds.): SMSD 2017, *Proceedings of the Third International Symposium on Materials and Sustainable Development*, pp. 189–196, 2018.
https://doi.org/10.1007/978-3-319-89707-3_23

2 Design of Blade

General information of blade helical wind turbine geometry Lecheb et al. (2015) (Table 1):

Table 1. Power as function a wind speed

V (m/s)	2	3	4	5	10
P (W)	3.279	11.067	26.233	51.237	409.9

Number of blade: 3
Power: 1–6 kW
Diameter of wind turbine: D = 1600 mm.
Higher of rotor: H = 2300 mm.
S_{swept} area = 1.9125 m^2.
Profile: NACA0015
C = 200 mm
ρ_{air} = 1225 kg/m^3
C_p = 0.35.

There are four various materials used for Wind Turbine Blades in this analysis:

1. Aluminium alloy blades.
2. GFRP blade.
3. CFRP blades.
4. PVC Polyvinyl Chloride blades.

Material properties of composite lamina shown in Table 2:

Table 2. Material properties of lamina composite

	Carbon/epoxy	Glass/epoxy
E_1 (GPa)	135	36.8
E_2 (Gpa)	9	8.27
V_{12}	0.3	0.21
G_{12} (Gpa)	5	4.14
G_{23} (Gpa)	5	4.14
G_{31} (Gpa)	8	4.14

3 Loads and Boundary Conditions

The blade is loaded by aerodynamic force (Fig. 1):

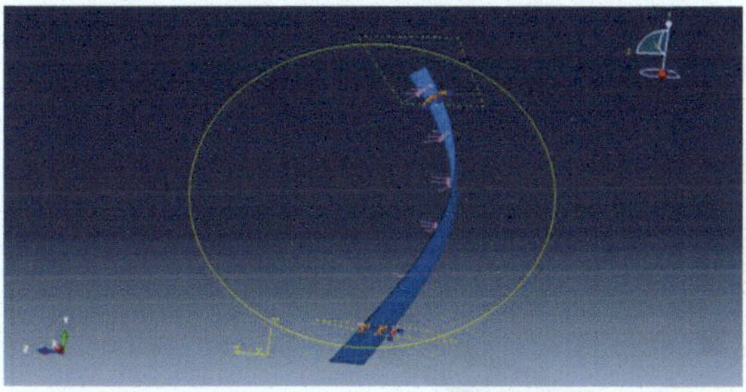

Fig. 1. Loads and boundary conditions picture

The mesh is obtained avec 11138 elements (Fig. 2):

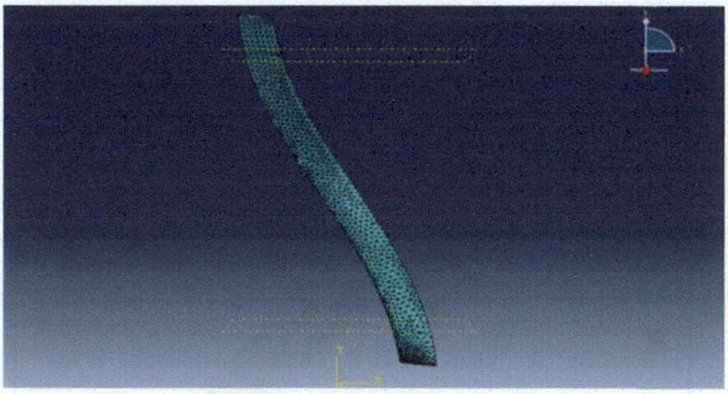

Fig. 2. Mesh of blade

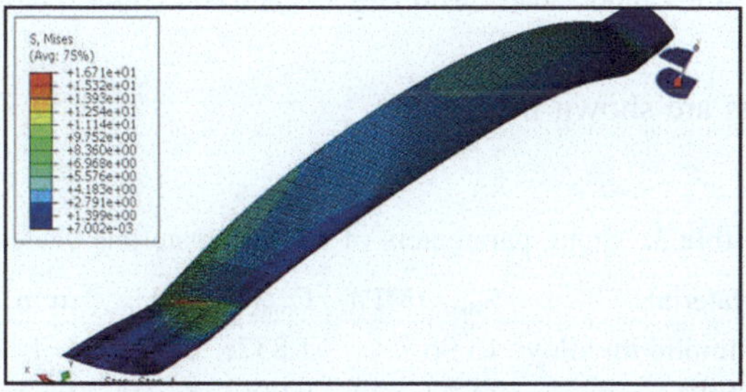

Fig. 3. Von Mises stress maximum S_{max} = 16.71 MPa

4 Results of Free GFRP

The Von Mises stress is obtained Lecheb et al. (2014) (Fig. 3):

Displacement (Fig. 4):

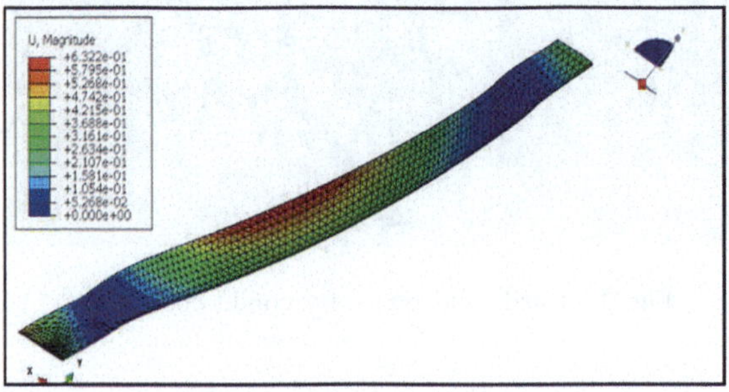

Fig. 4. Displacement maximum U_{max} = 6.322e−1 mm

We have shown first mode shape of blade and naturals frequencies Chellil et al. (2013) (Fig. 5):

Fig. 5. First mode shape with f_1 = 80.917 Hz, U_{1max} = 1.15 mm

The static results are shown in Table 3:

Table 3. Static parameters of various materials blades

Materials	S_{max} (MPa)	E_{max}	U_{max} (mm)
Aluminium alloy	15.96	1.832e−4	2.373e−1
CFRP Laminate	16.23	8.906e−5	1.1e−1
GFRP Laminate	16.71	5.255e−4	6.286e−1
PVC	16.08	8.921e−3	11.16

5 Crack Propagation

After crack initiation, we increase the crack length between 10 to 50 mm, The results obtained shown in Fig. 6.

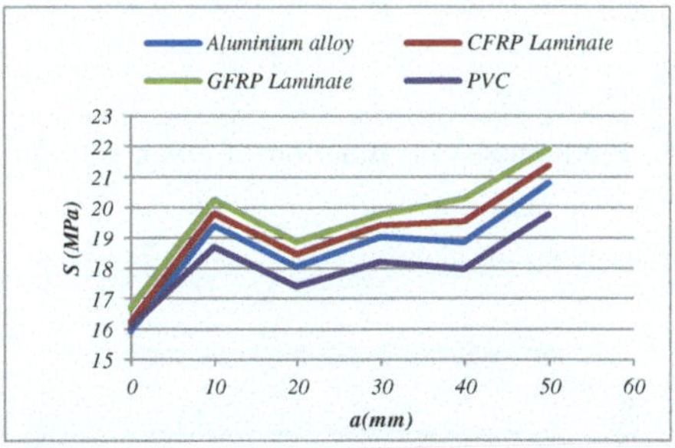

Fig. 6. Stress a function as crack size

6 Stress Intensity Factor

The equation of SIF is done by IRWIN:

$K1 = \sigma\ y\ \sqrt{\pi a}$

In our case we have edge crack with factor of geometry correction is y = 1.12 and the curves are shown in Fig. 7.

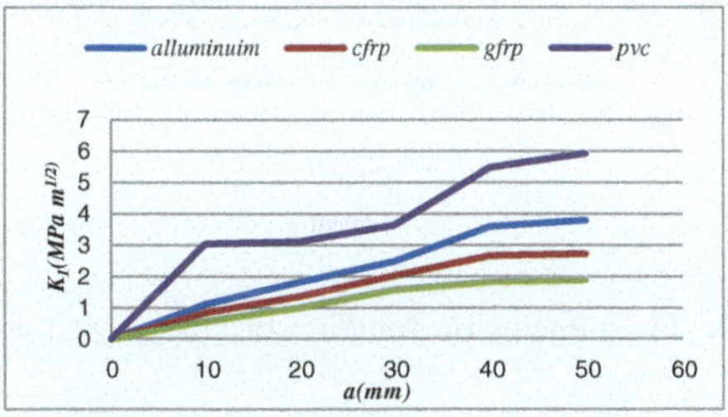

Fig. 7. SIF a function as crack size

7 Naturals Frequencies

The followings graphs present first six natural frequencies evolution as function a crack initiation of various materials (Figs. 8, 9, 10 and 11):

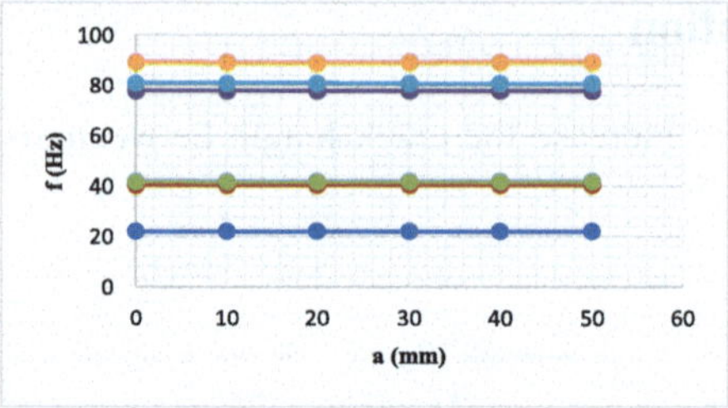

Fig. 8. Frequencies on function of crack size for PVC

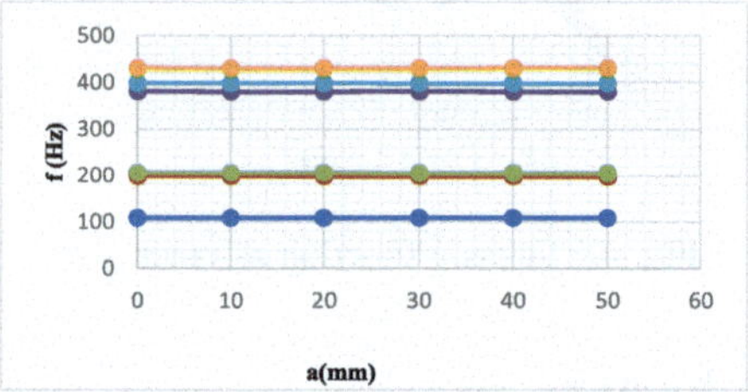

Fig. 9. Frequencies a function as crack size for aluminum alloy material

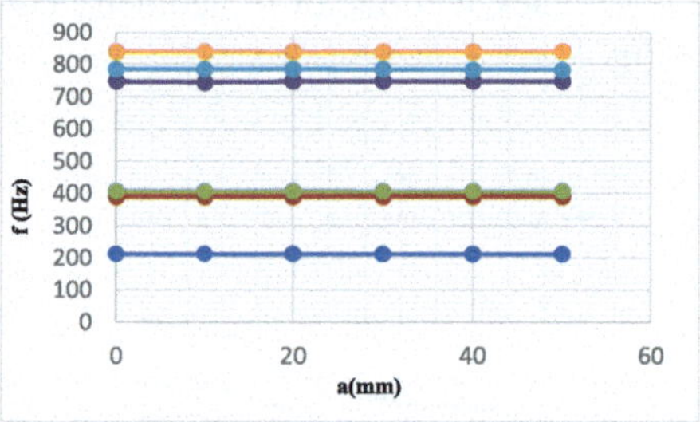

Fig. 10. Frequencies on function of crack size for CFRP

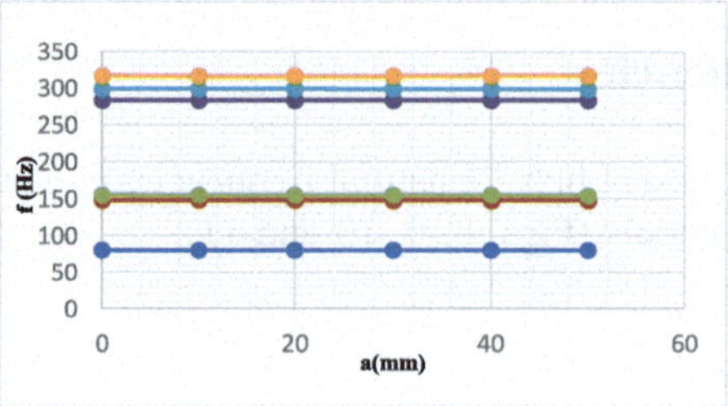

Fig. 11. Frequencies a function as crack size for GFRP

8 Experimental Study

In this experimental study we tested, calculation and compares experimentally the behavior results of the specimen of glass/epoxy cracked and un-crack then compared between the results obtained in the simulation. This result obtained by tensile machine (ZWICK ROELL 250 KN). So in our experiences we have cases to calculate (Fig. 12):

- Without crack in (glass/epoxy).
- With crack a = 5 mm and a = 10 mm.

Fig. 12. The specimen used in this test is plates of glass/epoxy (Mechakra et al. 2015)

Graph of the specimen with crack a = 5 mm (Fig. 13):

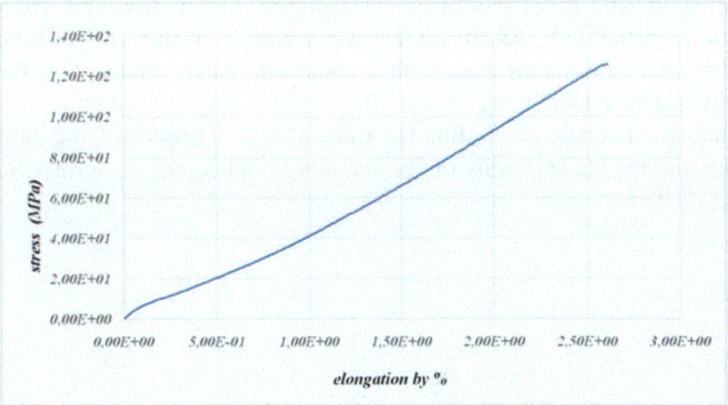

Fig. 13. Stress strain curve

This point of rupture has a pick value of stress which is S_{max} = 126 N/mm^2. In the experience of specimen with crack size a = 5 mm:

we have σy = 126 MPA and w = 25 mm:
Y(CTS) = 5.26 + α(5.22−5.88)/(1−(1.07 × α))

$\alpha = a/w = 5/25 = 0.2$
$Y = 5.26 + 0.2 \times (5.16-5.88)/((1-(1.07 \times 0.2)) = 5.092$
$K_{IC} = 126 \times 5.092 \times \sqrt{(}0.005 \times 3.14) = 80.391\ MPa{*}m^{1/2}$
$K_{IC} = 80.391\ MPa{*}m^{1/2}$

9 Conclusion

In this study we were treated an example of application of the method FEM in 3D with several crack size and we compared the blade without and with crack. The vertical axis wind turbine is oldest machine, the vertical axis wind turbine have special component and many types. The composite materials are playing an important role in industry, it's very useful now. The frequencies increase on function of displacement in cases with crack and without it. The stress, strain and displacement increase proportionally with crack size in different materials. From this study the CFRP and GFRP are best materials use to produce blade of wind turbine. The experimental test the glass/epoxy with increase crack size. The result of Fracture toughness in numerical simulation is equal then in experimental test.

References

Lecheb, S., Nour, A., Chellil, A., Mechakra, H., Ghanem, H., Kebir, H.: Dynamic prediction fatigue life of composite wind turbine blade. Steel Compos. Struct. Int. J. **18**(3), 673–691 (2015)

Mechakra, H., Nour, A., Lecheb, S., Chellil, A.: Mechanical characterizations of composite material with short Alfa fibers reinforcement. Compos. Struct. **124**, 152–162 (2015)

Lecheb, S., Nour, A., Chellil, A., Mechakra, H., Amarache, A., Kebir, H.: An advanced dynamic repair of edge crack aluminum plate with a composite patch. Mater. Sci. Forum **794–796**, 716–721 (2014). ISSN 1662-9752

Chellil, A., Nour, A., Lecheb, S., Kebir, H., Chevalier, Y.: Impact of the fuselage damping characteristics and the blade rigidity on the stability of helicopter. J. Aerosp. Sci. Technol. **29** (1), 235–252 (2013)

Effect of Dopant Material on the Performance UV Photodetector Based SnO_2 Thin Films Deposited by Sol-Gel Dip-Coating Method

Kaour Selma[1(✉)] and Rechem Djamil[1,2]

[1] Laboratory of Active Component and Material, University Larbi Ben M'hidi Oum El Bouagh, Oum El Bouaghi, Algeria
selmakaour185@gmail.com

[2] Department of Electrical Engineering, Faculty of Sciences and Applied Sciences, University Larbi Ben M'hidi Oum El Bouagh, Oum El Bouaghi, Algeria

Abstract. Metal oxide nanomaterials have been attracting growing interest for large domains applications such as gas sensors, photocatalysts, solar cell and UV photodetectors. In this works, the undoped and F, Al, Zn (3 at%) doped SnO_2 thin films were successfully deposited by sol-gel dip coating technique and characterized by X-ray diffraction (XRD), UV–Visible spectroscopy and photoconductivity study. Structural analysis showed that all films are polycrystalline with tetragonal rutile structure where the crystallite size is calculated by the debye Scherer's formula and it was obtained in the range of 6.49 to 9.33. Optical transmittance spectra of the films showed high transparency (>80%) in the visible range and gap energy values were obtained in the range 3.93 to 3.99. Finally, the variation of photocurrent with voltage and with time has been studied under UV illumination (λ = 365 nm). The high photocurrent is observed using F-doped SnO_2 sample while the film doped with Al had the highest photosensitivity. Current-Voltage characteristics of dark and photo current exhibited linear behavior. The phenomena of photoconductivity in SnO_2 thin films were interpreted by chemisorptions of oxygen molecules on the surface.

Keywords: Metal oxide · Thin films · Sol-gel · Photoconductivity Photodetectors

1 Introduction

In recent years, the transparent conducting oxide (TCO) such as zinc oxide, indium oxide, cadmium oxide, titanium oxide and tin oxide have attracted much attention because of their unique combination of low electrical resistivity and high optical transparency.

Tin oxide is IV VI_2 compound semiconductors with rutile structure. Thin films of SnO_2 are being studied for several applications such as gas sensor [1], solar cells [2], thin films transistors [3] and UV photodetectors [4]. SnO_2 thin film are attractive for UV photodetectors application because of their unique properties including large band

B. Abdelbaki et al. (Eds.): SMSD 2017, *Proceedings of the Third International Symposium on Materials and Sustainable Development*, pp. 197–205, 2018.
https://doi.org/10.1007/978-3-319-89707-3_24

gap (3.6 a 4 eV) [5, 6] the character n-type conduction [7], large excitonic binding energy of 130 meV [8], stoichiometric structure [9], and high UV absorption rate [10].

UV photodetectors have several applications in the field of biological analysis, environmental, optical communication, flam detection, astronomy lithography and detection of missile [11, 12]. A number of earlier reports showed the performance of UV photodetectors in SnO_2 can be dependent by time and temperature of annealing [4], method deposition and element dopant [13–15].

For enhancing the optoelectronic properties of SnO_2 some elements of group II_B, III_B, and VII_B of the periodic table are being used. Among these elements, Zinc, aluminium and fluorine. Numbers of research groups elaborated and studied structural, optical, and morphological proprieties of Zn, Al, and F doped SnO_2 thin films [16–18]. However, to the best of our knowledge, there are no reported works on SnO_2 thin films based photo-sensor and photoconductor doped with Zn, Al, and F.

In this works, our objective is to determine the influence of Zn, Al, and F atoms, in the constant doping ration, on the structural, morphological, optical properties and the performance of UV photodetectors. Sol-gel dip-coating technique was used to deposit the SnO_2 thin films on cleaned glass.

2 Experiment

2.1 Materials

Tin (II) chloride dihydrate 98% ($SnCl_2$:$2H_2O$), Aluminium trichloride hexahydrate ($AlCl_3$:$6H_2O$), zinc (II) chloride ($ZnCl_2$), fluorine ammonium (NH_4F), absolute ethanol 99.8% (C_2H_6O) and hydrochloride acid (HCl) were used to preparing undoped, Zn, Al and F doped thin films.

2.2 Preparation and Characterization

The films were prepared by sol-gel dip coating technique on ultrasonically cleaned substrates. The glass substrate was cleaned by acetone, ethanol and deionized water for 10 min in ultrasonic bath and then dried at room temperature. In another hand, the coating solution for undoped SnO_2 thin film was prepared by dissolving Tin (II) chloride dihydrate 98% ($SnCl_2$:$2H_2O$) (0.2 M) in 30 ml of ethanol absolute (C_2H_6 O), then a few drops of HCl was added to solution for accelerating the hydrolysis reaction between the precursor and solvent. The homogenous solution was obtained after stirring at 70 °C for 120 min and was kept in a covered for 24 h at room temperature. The film was deposited by the dip-coating technique with speed 60 mm/min. After dip coating process the film was dried 200 °C for 10 min. In order to increase the thickness of the film, the procedure was repeated 10 times and this film was annealed at 550 °C for 2 h. 3% wt Al, Zn and F doped SnO_2 thin films were elaborated under similar experimental conditions with adding ($AlCl_3$:$6H_2O$), ($ZnCl_2$) and (NH_4F) as a source of Al, Zn and F respectively.

X-ray diffraction (XRD, Bruker AXS-8D) with Cu Kα radiation (Cu Kα = 0.1541 nm) was used to study the structural properties, the surface morphology of the films was analysed with atomic force microscopy (A 100-AFM). The optical band gap of samples was performed by UV-Visible spectrophotometer (jasko V-630) in the range (200–1100).

2.3 Measurement of Photoconductivity

SnO_2-based UV photodetectors were fabricated with planar interdigital Ag electrodes. The Keithley source meter was used to measure UV photoresponse under illumination with 4 W, UV lamp available in tow wavelengths: 254 nm, 365 nm (VL- 4LC, vilber lourmat) and equipped. The Keithley was used to measure current-voltage (I-V) characteristics in dark and under UV source of 365 nm wavelength and intensity 350 $\mu W/cm^2$.

3 Results

3.1 Structural Analysis

Figure 1 shows the XRD patterns of the undoped, Zn, Al and F doped SnO_2 thin films. As can be seen five diffraction peaks corresponding to (110), (101), (200), (210), (220) direction without any impurity and secondary phase. These spectra confirm that all films have polycrystalline nature with tetragonal rutile structure with a preferred orientation along (110) plane. All diffraction peaks intensity varies with a dopant. The crystallite size (D) can be calculated from the Scherer's equation [19].

$$D = \frac{0.9\lambda}{\beta \cos\theta} \tag{1}$$

Where λ is the X-ray wavelength, β is the full width at half maximum of the XRD peak, θ is the Bragg diffraction angle.

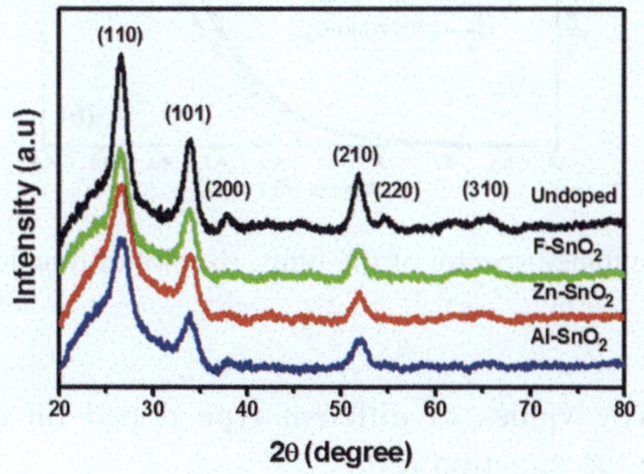

Fig. 1. X-ray Diffractions of the SnO_2 thin films.

The (110) peak was utilized to estimate the crystallite sizes (Table 1). It can be noticed from Table 1 that the crystallite size of the films decreases from 9.33 to 6.49 nm.

Table 1. Data obtained from the XRD and UV–Visible measurement for thin films.

Samples	2θ	FWHM (rad)	Crystalline size (nm)	Gap (eV)
Undoped	26.61	0.016	9.33	3.92
F-SnO_2	26.58	0.018	7.78	3.99
Zn-SnO_2	26.58	0.021	6.79	3.97
Al-SnO_2	26.70	0.023	6.49	3.95

3.2 Optical Properties

The optical transmission spectra of un-doped, fluorine, aluminium and zinc doped tin oxide thin films prepared and deposited by sol-gel dip coating technique on a lime glass substrate and annealed at 550 °C are plotted in Fig. 2a. All films showed high transparency (>80%) in the visible range.

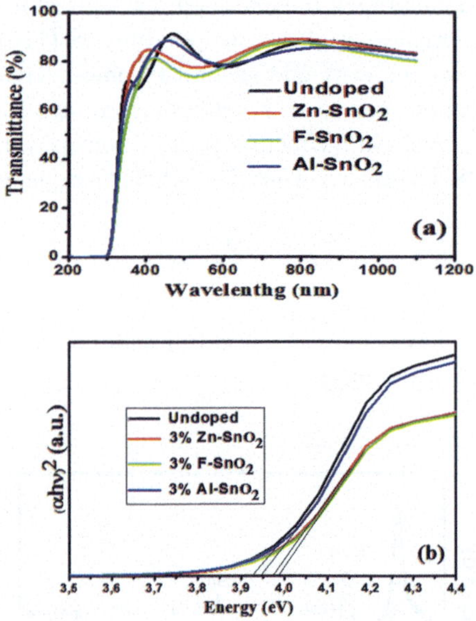

Fig. 2. (a) Optical transmittance spectra of the films, (b) plot of function of photo energy $h\upsilon$ of SnO_2 thin films.

The band gap energy values of different type doped tin oxide thin films were calculated by using the tuc equation [20].

$$\alpha h\upsilon = A(h\upsilon - E_g)^{1/2} \tag{2}$$

Where α The absorption coefficients, A is constant, $h\upsilon$ is photon energy and E_g is the band gap energy.

Figure 2b shows $(\alpha h\upsilon)^2$ variation as a function of photon energy $(h\upsilon)$. The values of optical energy band gap, which was, be estimated from the extrapolating the linear part of $(\alpha h\upsilon)^2$ versus $(h\upsilon)$ plot to $(\alpha h\upsilon)^2$ shown in Table 1. It can be seen that the band-gap energy increased for the all thin films doped with Al, Zn and F. These results are very similar to some reports [21, 22].

3.3 Photoconductivity Study

3.3.1 Voltage Dependence of Dark and Photocurrent

Figure 3 shows a variation of photocurrent and dark current with an applied voltage obtained from undoped as well as Al, Zn and F doped SnO_2 on a log-log scale. From the I-V measurements, it can be seen that the photocurrent increase of the applied bias voltage because of the augmentation of the carrier drift velocity [23]. The ln (I) vs. ln (V) curves are straight lines having different slopes with respect to varying voltage according to the power law relation I α V_r where r is the slope of a line. In the dark current and photocurrent, the linear behavior of curves is indicated the ohmic conduction. The ohmic behavior is very important to the photo sensing because the sensitivity of photodetector can be maximized when the metal-semiconductor junction is ohmic [24].

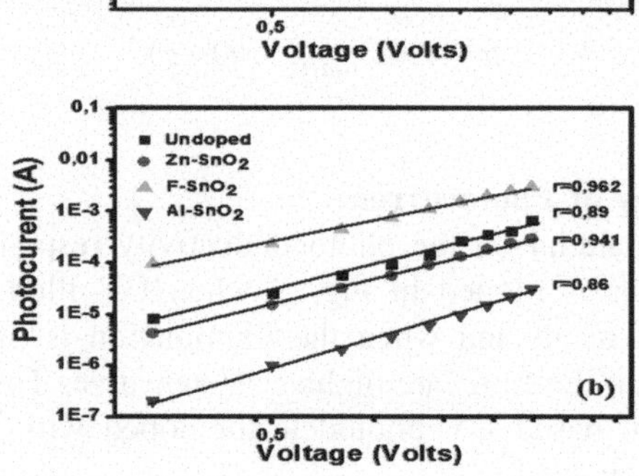

Fig. 3. Variation of (a) dark current and (b) photocurrent as a function of applied voltage of SnO_2 thin films on a logarithmic scale.

3.3.2 UV Detection Mechanism of SnO_2 Thin Films

The photoresponse mechanism of SnO_2 thin films to UV light can be explained in Fig. 4 with according to the following equations:

$$O_2(g) + e^- \rightarrow O_2^- \tag{3}$$

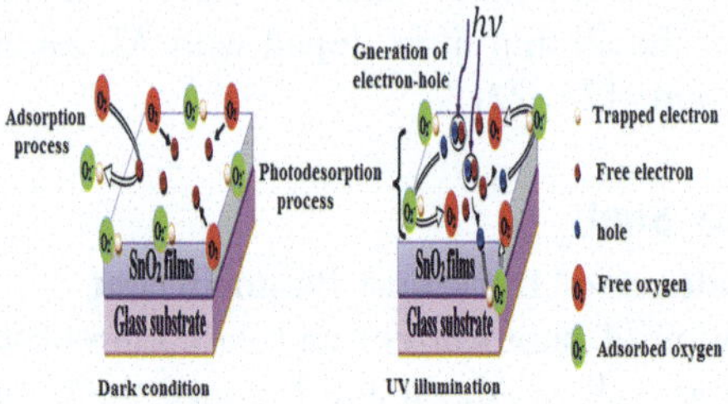

Fig. 4. A schematic of photoresponse mechanism of SnO_2 thin films.

The process of oxygen chemisorptions plays a vital role in UV photoconduction mechanism. Under dark condition, the oxygen molecule O_2 (g) is absorbed on the surface of as negatively charged ions by capturing free electron.

This process leads to the formation of depletion layer near the surface resulting in the bending of the conduction band and valence band, as well as trap center-related band or trap depth. Formation of a large number of ionized oxygen on the thin films surface enhanced the band bending, resulting in a very low conductivity. When the sample is illuminated with UV light, the electron-hole pair is generated by light absorption [25].

The photo-generated hole oxidizes negatively, charged oxygen ions adsorbed on the surface O_2^- (ads). While the remaining unpaired electron in the conduction band increases the conductivity.

$$h\upsilon \rightarrow e^- + h^+ \tag{4}$$

$$h^+ + O^-_{2\,(ads)} \rightarrow O_2(g) \tag{5}$$

3.3.3 Rise and Decay of Photocurrent

The effect of dopant material on the photoconductivity response of SnO_2 thin films under 2 V bias voltages is plotted in Fig. 5, when UV illumination is on, a large increase in the current result and when the illumination is off the current increase gradually, indicating that the films are highly UV sensitive. This figure clearly shows that the SnO_2 thin film based UV photodetector doped with fluorine displayed the maximum photocurrent value.

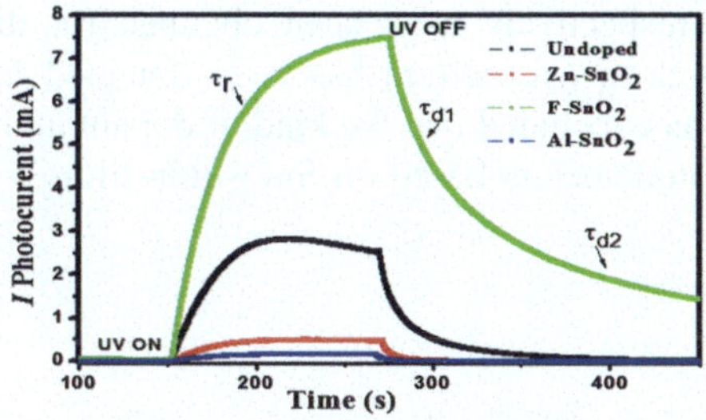

Fig. 5. The rise and recovery-current characteristics of SnO_2 thin films.

The highest on/off ratio was recorded by Al-doped SnO_2 about 5000.

Rise and decay time are estimated by the fitting curve using the exponential functions.

$$I_{ph} = I_0 A \exp^{-t/t_r} \quad (6)$$

$$I_{ph} = I_{ph(\infty)} + A_1 \exp^{-t/t_{d1}} + A_2 \exp^{-t/t_{d2}} \quad (7)$$

The times constants values are noted in Table 2, the minimum values of the rise time and the decay time constants were obtained for the sample doped with aluminum. On other hand, samples doped with fluorine which present a high photocurrent compared to other samples show high rise time and decay time constants.

Table 2. The values of photocurrent, dark current, I_{ON}/I_{OFF} ratio, rise time constant, decay time constants for the thin films.

Samples	I_{light}	I_{dark}	I_{ON}/I_{OFF}	τ_r (s)	τ_{d1} (s)	τ_{d2} (s)
Undoped	2.90×10^{-3}	5.50×10^{-5}	52	27.73	11.18	51.96
F-SnO_2	1.15×10^{-2}	4.90×10^{-4}	23.46	35.71	14.27	83.70
Zn-SnO_2	5.12×10^{-4}	1.77×10^{-6}	289.26	25.60	5.78	30.42
Al-SnO_2	1.78×10^{-4}	3.56×10^{-8}	5000	24.39	4.51	27.63

4 Conclusions

In summary, we have successfully fabricated undoped as well as Zn, Al and F doped SnO_2 UV photodetectors on glass substrates by sol-gel method using low cost dip coating technique. XRD and UV-visible transmission were used to characterize the elaborated films. The results indicated that our thin films show a polycrystalline nature with rutile structure and optical transmittance spectra of the films showed high transparency (>80%) in the visible range and gap energy values were obtained in the range 3.93 to 3.99. In photoconductivity study, the low dark current and high I_{ON}/I_{OFF} current ration with rise time constant τ_r and decay time constants τ_{d1} and τ_{d2} of about

24.39, 4.51 and 27.63 respectively have been obtained for the sample doped with aluminium while the highest I_{ON} current has been obtained for F doped SnO_2 thin films. In this study, it was concluded that the kind of dopant atoms change strongly the performance of UV photodetectors based on SnO_2 thin films.

References

1. Selma, M.H.: Influence of multilayer deposition on characteristics of nanocrystalline SnO_2 thin films produce by Sol-Gel Technique for gas sensor application. Optic **146**, 17–26 (2017)
2. Bu, I.Y.Y.: Sol-gel deposition of fluorine-doped tin oxide glasses for dye sensitized solar cell. Ceram. Int. **40**, 417–422 (2014)
3. Priyadarshini, D.M., Ramanjaneyulu, M., Ramachandra, M.S., Nandita, D.G.: Effect of annealing ambient on SnO_2 thin film transistors. Appl. Surf. Sci. **418**, 414–417 (2016)
4. Rechem, D., Khaial, A., Souifi, A., Djeffal, F.: Effect of annealing time on the performance of tin oxide thin films ultraviolet photodetectors. Thin Solid Films **623**, 1–7 (2017)
5. Khan, A.F., Mazhar, M., Aslam, M., Ashraf, M.: Characterisatics of electron beam evaporated nanocrystalline SnO_2 thin films annealed in air. Appl. Surf. Sci. **256**, 2252–2258 (2010)
6. Kumaril, N., Ghosh, A., Tewari, S., Bhattacharjeel, A.: Synthesis, structural and optical properties of Al doped SnO_2 nanoparticles. Indian J. Phys. **88**, 65–70 (2014)
7. Kucheyev, S.O., Baumann, T.F., Sterne, P.A., Wang, Y.M., Buuren, T., Hamza, A.V., Terminello, L.J., Willey, T.M.: Surface electronic state in three-dimensional SnO_2 nanostructures. Phys. Rev. B **72**(035404), 1–5 (2005)
8. Reddy, A.S., Figueiredo, N.M., Cavaleiro, A.: Nanocrystalline Au: Ag SnO_2 films prepared by pulsed magnetron sputtering. J. Phys. Chem. Solids **74**, 825–829 (2013)
9. Kilic, C., Zunger, A.: Origine of coexistence of conductivity and transparency in SnO_2. Phys. Rev. Lett. **88**(095501), 1–4 (2002)
10. Venkateswara Reddy, P., Venkatramana Reddy, S., Sankra Reddy, B.: Synthesis and properties of (Ni, Al) co-doped nanoparticles. J. Mater. Sci.: Mater. Electron. **27**, 10712–10719 (2016)
11. Liao, M.Y., Sang, L.W., Teraji, T., Imura, M., Alvarez, J., Koide, Y.: Comprehensive investigation of single crystal diamond deep-ultraviolet detectors. J. Appl. Phys. **51**, 090115:1–090115:7 (2012)
12. Sandvik, P., Mi, K., Shahedipour, F., McClintock, R., Yasan, A., Kung, P., Razeghi, M.: $Al_xGa_{1-x}N$ for solar-blind UV detectors. J. Cryst. Growth **231**, 366–370 (2001)
13. Wu, J.M., Kuo, C.H.: Ultaviolet photodetectors made from SnO_2 nanowires. Thin Solid Films **517**, 3870–3873 (2009)
14. Huang, S., Matsubara, K., Cheng, J., Li, H., Pani, W.: Highly enhanced ultraviolet photosensitivity and recovery speed in electrspun Ni-doped SnO_2 nanobelts. Appl. Phys. Lett. **103**(141108), 1–5 (2013)
15. Deng, K., Lu, H., Shi, Z., Liu, Q., Li, L.: Flexible three-dimensional SnO_2 nanowire arrays: atomic layer deposition-assisted synthesis, excellent photodetectors, and field emitters. Appl. Mater. Interfaces **5**, 7845–7851 (2013)
16. Xu, B., Ren, X.R., Gu, G.R., Lan, L.L., Wu, B.J.: Structural and optical properties of Zn-doped SnO_2 films prepared by DC and RF magnetron co-sputtering. Superlattices Microstruct. **89**, 34–42 (2016)

17. Benouis, C.E., Benhaliba, M., Mouffak, Z., Avila-Garcia, A., Tiburcio-Silver, A., Ortega Lopez, M., Romano Trujillo, R., Ocak, Y.S.: The low resistive and transparent Al-doped SnO_2 films: P-type conductivity, nanostructures and photoluminescence. J. Alloys Compd. **603**, 213–223 (2014)
18. Tran, Q.P., Fang, J.S., Chin, T.S.: Properties of fluorine-doped SnO2 thin films by agreen sol–gel method. Mater. Sci. Semicond. Process. **40**, 664–669 (2015)
19. Anandhi, R., Mohan, R., Swaminathan, K., Ravichandran, K.: Influence of aging time of the starting solution on the physical properties of fluorine doped zinc oxide films deposited by a simplified spray pyrolysis technique. Superlattices Microstruct. **51**, 680–689 (2012)
20. Tauc, J., Grigorovici, R., Vancu, A.: Optical properties and electronic structure of amorphous germanium. Phys. Status Solidi B **15**, 627–637 (1966)
21. Ahmed, S.K.F., Khan, S., Ghosh, P.K., Mitra, M.K., Chattopadhyay, K.K.: Effect of Al doping on the conductivity type inversion and electro-optical properties of SnO_2 thin films synthesized by sol-gel technique. Sol-gel Science Technology **39**, 241–247 (2006)
22. Pacheco, A.P., Acosta, D.R., Magana, C.: Effect of the amount of the starting solution on physical properties of SnO_2: F thin films. Surf. Interfaces **6**, 85–90 (2017)
23. Soci, C., Zhang, A., Xiang, B., Dayeh, S.A., Aplin, D.P.R., Park, J., Bao, X.Y., Lo, Y.H., Wang, D.: ZnO nanowire UV photodetectors with high internal gain. Nano Lett. **7**, 1003–1009 (2007)
24. Yoon, Y.J., Park, K.S., Heo, J.H., Park, J.G., Nahm, S., choi, K.J.: Synthesis of $Zn_XCd_{1-X}Se$ (06_X61) alloyed nanowires for variable wave length photodetectors. Mater. Chem. **20**, 2386–2390 (2010)
25. Kim, J.M., Lim, S.J., Nam, T., Kim, D., Kim, H.: The effects of exposure on the device characteristics of atomic layer deposited-ZnO: N thin films transistors. Electrochem. Soc. **158**, 150–154 (2011)

Non Local Vibration of Damped Viscoelastic—FGM-Beam

Souad Hamzi[1,3](✉), Ismail Mechab[2,3], and Yassine Senhadji[4]

[1] LSTE, Laboratory of Sciences and Water Techniques, University of Mustapha Stambouli, 29000 Mascara, Algeria
ing.hamzi@yahoo.fr

[2] LMPM, Laboratory of Mechanical and Physics of Materials, University of Sidi Bel Abbes, 22000 Sidi Bel Abbes, Algeria

[3] Department of Civil Engineering, University of Mustapha Stambouli, Mascara 29000, Algeria

[4] LAB-MAT, Laboratory of Materials, Department of Civil Engineering, ENPO, 31000 Oran, Algeria

Abstract. This study presents a mechanical model to predict the effective damped viscoelastic behavior of a functional grading material (FGM). A detailed parametric study is carried out to elucidate the dynamic response of an Euler-Bernoulli beam with finite length on the basis of the non-local Eringen theory which reveal the influence of the material proprieties, nonlocal parameter and the viscoelastic parameter on the vibratory frequency of the system, we obtain the analytical solutions for the vibratory frequency of the system with boundary conditions. The applicability and the effectiveness of the model presented are demonstrated in canonical numerical examples.

Keywords: Euler Bernoulli beam · FGM · Viscoelastic behavior
Nonlocal · Vibration

1 Introduction

Functionally graded materials (FGMs) are a class of composites which both the composition and the structure gradually change over the volume, resulting in a corresponding variation of the material properties. Typically, these materials consist of a mixture of ceramic and metal as presented in this paper, or a combination of different materials. As a result, FGM is considered of having an enormous application in modern technology and engineering. Consequently, studies devoted to understand the static and dynamic behaviors of FGM structure. Some of these investigations are mentioned by Zhou and Yang (1996), Alshorbagy et al. (2011), Kahrobaiyan et al. (2012), Asghari et al. (2010) Reddy (2000) and Kumar et al. (2016).

However, with the nanotechnology developed; Nano-structures have already drawn a great deal of attention and stimulated extensive studies (Zhu and Li 2017). As known, the concept of nonlocal theory of linear elasticity was initially introduced by Kröner (1967), and further developed by Eringen (1972). The basic feature of the nonlocal theories of elasticity is that the stress a teach point is related to the strain at all points in

B. Abdelbaki et al. (Eds.): SMSD 2017, *Proceedings of the Third International Symposium on Materials and Sustainable Development*, pp. 206–215, 2018.
https://doi.org/10.1007/978-3-319-89707-3_25

the domain. This influence decreases as the distance between the point of interest and the neighboring points increases. The Eringen nonlocal integral constitutive equation describes the dependence of the stress at a point on the strain in the rest of the domain through a positive-decaying kernel function. A differential constitutive theory, introduced by Eringen (1983), showed that for a specific class of kernel functions the nonlocal integral constitutive equation can be simplified into a differential form. Many researchers have applied the nonlocal elasticity concept for bending, buckling and vibration analyses of nano-sized structures such as beams (Aranda-Ruiz et al. 2012; Aydogdu 2009; de Sciarra 2014; Ebrahimi and Barati 2016; Wang et al. 2007), rods (Adhikari et al. 2014; Murmu and Adhikari 2010) and plates (Ansari et al. 2016; Daneshmehr et al. 2015; Zenkour and Sobhy 2013; Zenkour 2014; Lu et al. 2007).

Recently, the nonlocal viscoelastic models were introduced in order to consider the mechanical energy dissipation occurring in vibrating nanostructures. In those works, the authors mostly applied the nonlocal Kelvin–Voigt constitutive equation, but some of them applied other viscoelastic models as well. Lei et al. (2013) derived motion equations for bending vibration of viscoelastic Euler–Bernoulli beam using nonlocal Kelvin–Voigt and three parameter standard viscoelastic constitutive equations by using transfer function method, they obtained closed form expressions to analyze the dynamic behavior of beam model for different boundary conditions. Further, Karličić et al. (2014) studied the vibration of nonlocal multi-nanobeam system embedded in a viscoelastic medium, which represents the nanocomposite composed of multiple carbon nanotubes embedded in certain type of polymer matrix.

The analyzed system is subjected to the influence of longitudinal magnetic field. The authors found damped and magnetically affected natural frequencies and investigated the influence of different parameters on the dynamic behavior of the system. In addition, Ansari et al. (2015) analyzed the free vibration problem of fractional viscoelastic Timoshenko nano-beam using the nonlocal theory; they used single-walled carbon nanotube by adopting corresponding values of parameters.

Two solution procedures were proposed to solve the governing equations of system on the time.

In the present study, the free vibration analysis of FG beams is investigated using the viscoelastic models and Eringen's nonlocal elasticity theory. Governing equations and boundary conditions for this case have been derived via Hamilton's principle. These equations are solved using numerical solutions.

Mathematical analysis is presented whilst the emphasis is placed on investigating the impact of several parameters such as constituent volume fractions, mode number and boundary conditions on the dynamic response of FG nano-beams. Comparisons with analytical solutions and the results from the existing literature are provided for two-constituent metal–ceramic of nano-beams and the good agreement between the results of this article and those available in literature validated the presented approach. Numerical results are presented to serve as reference for the application of nano-electronic and nano-oscillators, and nano-sensors, in which nano-beams act as basic elements. They can also be useful as treasured sources for validating different approaches and approximate methods.

2 Problem Formulation

2.1 FG Nano-beam Equations

A functionally graded nano-beam of length L, width b, and thickness h is shown in Fig. 1.

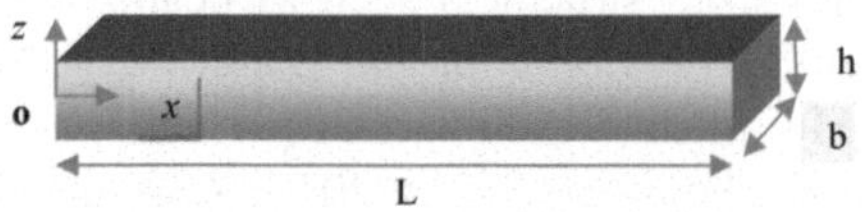

Fig. 1. Gradation of material properties through the thickness of the beam.

It is assumed that the material properties of the beam, such as Young's modulus (E) and mass density (ρ), vary continuously through the beam thickness according to power-law form, which can be described by

$$P(z) = (P_C - P_M)\left(\frac{z}{h} + \frac{1}{2}\right)^k + P_M \tag{1}$$

Where P_M and P_C are the corresponding material property at the upper and lower surfaces of the beam, the subscripts c and m are respectively denote ceramic and metal, and k is a material index, which indicates the material variation profile along the thickness of FGM-nano-beam, with k = 0 corresponding to an isotropic homogeneous ceramic and k = ∞ corresponding to an isotropic homogeneous metal; k is a non-negative number that dictates the material variation profile through the thickness of the beam.

2.2 Nonlocal Viscoelastic Constitutive Relation

According to the nonlocal elasticity theory, the stress at a point x is a function of the strains at all other points of the elastic body. Eringen (1972) defined the integral form of nonlocal constitutive relation for a three-dimensional structure as:

$$\sigma_{ij} = \int \chi(|x - x'|, \in) t_{ij}(x')\, dV(x'),\ \forall x \in V \tag{2}$$

Where σ_{ij} is the nonlocal stress tensor, t_{ij} is the local or classical stress tensor at a point x′, χ (|x − x′|, ϵ) is an attenuation function that incorporates nonlocal effects into the constitutive equation at a reference point x produced by local strain at a source x′. In above equation |x − x′| denotes a distance in Euclidean norm where $\epsilon = (e_0 a)/l$ and l denotes an external characteristic length (crack length or wave length), a is an internal characteristic length (lattice parameter, granular size, etc.) and e_0 is a material constant that can be determined from molecular dynamics simulations or by using the dispersive curve of the Born-Karman model of lattice dynamics. Later, Eringen (1983) proposed a differential form of the constitutive relation in the form:

$$(1-\in^2 l^2\nabla^2)\sigma_{ij} = t_{ij}(x'). \tag{3}$$

$$E\varepsilon_{xx} = t_{ij}(x'). \tag{4}$$

Where E denotes elastic modulus. Based on Eq. (3) we can write nonlocal constitutive equation for one dimensional elastic body as

$$\sigma_{xx} - \mu\frac{d^2\sigma_{xx}}{dx^2} = E(z)\varepsilon_{xx} \tag{5}$$

Where $\mu = (e_0 a)^2$ is the nonlocal parameter and σ_{xx} is the nonlocal stress. The constitutive relation for the nonlocal viscoelastic body can be obtained by combining nonlocal elasticity and viscoelasticity theory (Lei et al. 2013). Constitutive relations for the Kelvin Voigt viscoelastic model are given by

$$\sigma_{xx} - \mu\frac{d^2\sigma_{xx}}{dx^2} = E(z)(\varepsilon_{xx} + \tau_d\acute{\varepsilon}_{xx}). \tag{6}$$

Where $\tau_d = \eta/E$, is the rate of relaxation with η is the viscosity.

2.3 Governing Equations of Nano-beam

According to Euler hypothesis, the only nonzero strain is

$$\varepsilon_{xx} = -z\frac{\partial^2 w(x,t)}{\partial x^2} \tag{7}$$

The equation of motion for a damped nonlocal viscoelastic Euler-Bernoulli beam including external velocity dependent damping can be expressed by the following equation:

$$\begin{gathered} D\frac{\partial^4 w}{\partial x^4} + D\tau_d\frac{\partial^5 w}{\partial x^4\partial t} + C_1[1-\mu\nabla]\frac{\partial w}{\partial t} + \\ I_0[1-\mu\nabla]\frac{\partial^2 w}{\partial t^2} + I_1[1-\mu\nabla](\frac{\partial^2 w}{\partial x\partial t})^2 = 0 \end{gathered} \tag{8}$$

$$\text{Avec} \begin{cases} D = \int\limits_A E(z)z^2 \mathrm{dA} \\ (I_0, I_1) = \int\limits_A \rho(z)(1, z^2)\mathrm{dA} \end{cases} \tag{9}$$

Where ρ(z) denotes the mass density, a stands for the section area, C1 the damping coefficient and nonlocal shear force and bending moment can be described by:

$$M = D\frac{\partial^2 w}{\partial x^2} + D\tau_d\frac{\partial^3 w}{\partial x^2 \partial t} + \mu\left[I_0\frac{\partial^2 w}{\partial t^2} + C_1\frac{\partial w}{\partial t} + I_1\left(\frac{\partial^2 w}{\partial x \partial t}\right)^2\right] \tag{10}$$

$$Q = D\frac{\partial^3 w}{\partial x^3} + D\tau_d\frac{\partial^4 w}{\partial x^3 \partial t} + \mu\left[I_0\frac{\partial^3 w}{\partial x \partial t^2} + C_1\frac{\partial^2 w}{\partial x \partial t} + I_1\frac{\partial^5 w}{\partial x^3 \partial t^2}\right] \tag{11}$$

2.4 Analysis of Damped Natural Frequencies

The equations of motion derived in the last section can now be used to obtain the dynamic characteristics of the system. To obtain the natural frequencies (ω) assume that $w(x, t) = W(x)e^{i\omega t}$. Then, Eqs. (8), (10), (11) becomes:

$$[(1+\tau_d i\omega)D + I_1\omega^2\mu]\frac{d^4W}{dx^4} + [(I_0\omega^2 - i\omega C_1)\mu - I_1\omega^2]\frac{d^2W}{dx^2} + \left[i\omega C_1 - I_0\omega^2\right]W = 0 \tag{12}$$

$$M = [D(1+\tau_d i\omega) + I_1\omega^2\mu]\frac{dW}{dx^2} + \mu\left[C_1 i\omega - I_0\omega^2\right]W \tag{13}$$

$$Q = [D(1+\tau_d i\omega) + I_1\omega^2\mu]\frac{d^3W}{dx^3} + \mu\left[C_1 i\omega - I_0\omega^2\right]\frac{dW}{dx} \tag{14}$$

The general form of function W(x) can be expressed by:

$$W(x) = B_1\cosh\frac{\beta x}{L} + B_2\sinh\frac{\beta x}{L} + B_3\cos\frac{\beta x}{L} + B_4\sin\frac{\beta x}{L} \tag{15}$$

where Bi, i = 1, 2,…, 4, are unknown constants.

To investigate the variations of natural frequencies of the nano-beam it is necessary to introduce boundary conditions properly into the solution.

Finally, the natural frequency can be expressed as follows:

$$\omega_n = \pm \frac{(C_1(1-\mu\beta^2)+\tau_d D\beta^4)i}{2I_1(\mu\beta^4-\beta^2)+2I_0(\mu\beta^2-1)} + \frac{\sqrt{(C_1(1-\mu\beta^2)+\tau_d D\beta^4)^2 - 4D\beta^4(I_1(\mu\beta^4-\beta^2)+I_0(\mu\beta^2-1))}}{2I_1(\mu\beta^4-\beta^2)+2I_0(\mu\beta^2-1)} \tag{15}$$

For an undamped, local elastic isotropic beam, $C_1 = \mu = \tau_d = 0$. Then the natural frequencies are given by

$$\omega_n = \beta^2\sqrt{\frac{EI}{\rho A}} \quad \text{(Euler-Bernoulli beam)} \tag{19}$$

2.5 Result and Discussion

In this study the FG nano-beams are made of a ceramic and metal mixture that's the properties varies through the thickness according to power-law as shown in Fig. 2. The upper side of the nano-beam (z = +h/2) is pure ceramic (Alumina), while the lower side of the nano-beam (z = −h/2) is pure metal (Aluminum). The material properties which used in the present study are given in Table 1.

For this case a simply supported FG nano beams are presented with a varying nonlocal parameter (μ), viscoelastic parameter (τ_d) and material properties with $L/h = 10$, $C_1 = 0$.

Table 1. Material properties of the FGM constituents.

Properties	Unit	Aluminum (Al)	Alumina (Al_2O_3)
E	GPA	70	380
P	kg/m^3	2700	3800

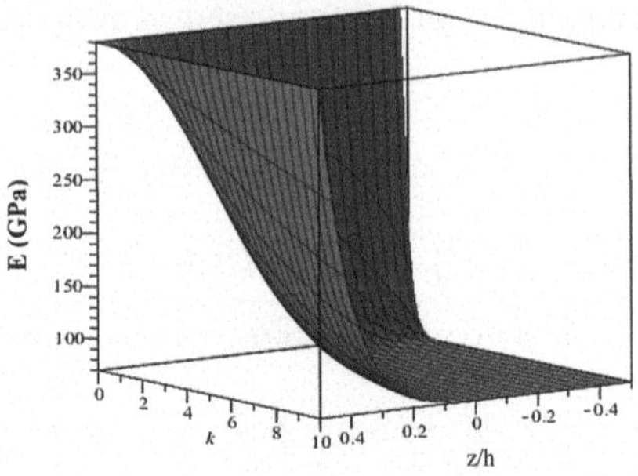

(a)

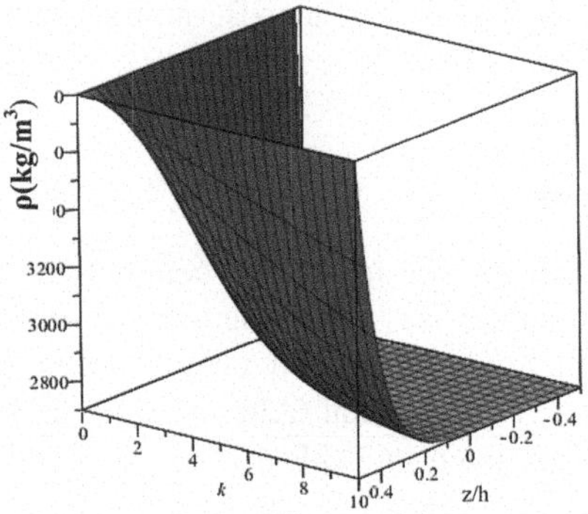

(b)

Fig. 2. The variation of Young's modulus (a) and the mass density (b) through the thickness direction of the beam.

2.5.1 The Effect of Young's Modulus Variation

It is observed in Fig. 3, that the effects of both of Yong modulus and mode shape 'n' on the imaginary part (relating to damping ratio) of the frequency are smaller value than the real part. The frequency increased significantly with increasing of young modulus, especially for higher value of the young modulus. This decreasing due to the transition of the beam from alumina which has a higher elastic modulus (k = 0 corresponding to an isotropic homogeneous ceramic) to steel which has the low elastic modulus especially for the first mode in the real part of the fundamental frequency. On the other hand, no significant changes occur in fundamental frequency for the second and the third mode shape and even the fourth one unlike the imaginary part.

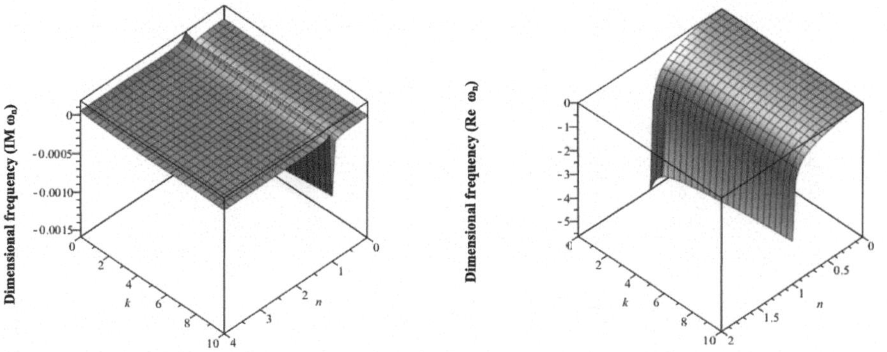

Fig. 3. Variation of the dimensional frequency with k the material index. With (L/h = 10, C1 = 0, τd = 10^{-4}, μ = *0.1*)

2.5.2 The Effect of the Nonlocal Parameter (μ)

Figure 4 present the dimensionless frequency of the nano-beam against the non-local parameter with different proprieties. It shows that on the real part of the dimensionless frequency increases, with the non-local parameter and the young's modulus. Even as on the imaginary part of the frequency. The non-local parameter as show in the figure has no signified effect after some value even with the increasing of the young modulus. The effect of the non locality parameter is more significant with the value 0.1.

2.5.3 The Effect of the Viscoelastic Parameter (τ_d)

The influence of viscoelastic parameter (τ_d) on the imaginary frequencies of S–S viscoelastic FG nano-beams is demonstrated in Fig. 5; which is increasing almost linearly with the viscoelastic parameter and young modulus. While on the real of the complex natural frequencies, all the values of the τ_d are much converged with the same values of rigidity The results presented show very good agreement with the results in (Lei et al. 2013).

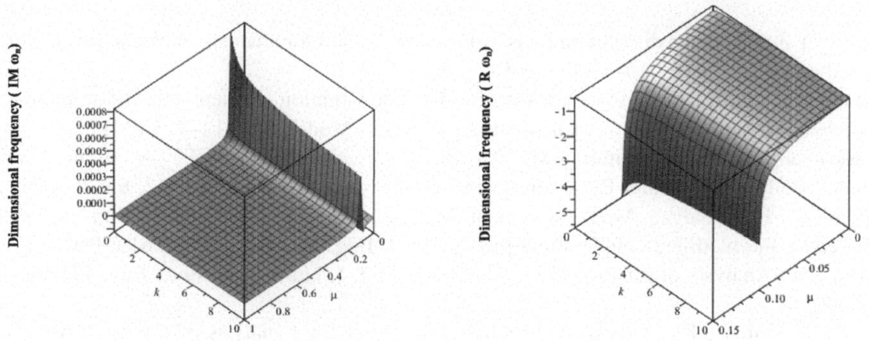

Fig. 4. Variation of the dimensional frequency with nonlocal parameter (μ)

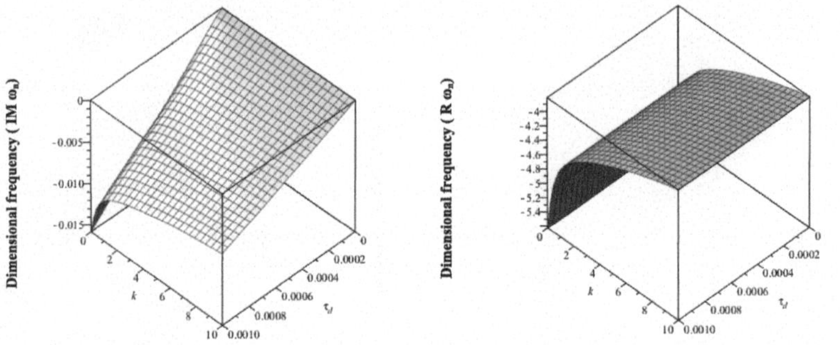

Fig. 5. Variation of the dimensional frequency with viscoelastic parameter (τ_d).

3 Conclusions

By considering the frequency and parameter-dependent properties of a damped viscoelastic nonlocal FG beams, a new approach is constructed and researched in presenting analysis, which could provide a relatively large effect not only in medium and high frequency, but also in low frequency, according to the research, it could conclude that:

The natural frequencies increase with the increasing of Yong modulus, and decrease with an increase in density.

The proprieties of the FG beam have no effect at the most important factor the viscoelastic parameter; which that affected the imaginary part of the complex natural frequency.

The nonlocal parameter has a signified effect on the fundamental frequencies with Specific values depends on the materials used.

We can predict the behavior of the FGM beams knowing that the behavior is similar to the homogeneous beams.

References

Zhou, J., Yang, B.: Strip distributed transfer function method for analysis of plates. Int. J. Numer. Methods Eng. **39**(191), 5–1932 (1996)

Alshorbagy, A.E., Eltaher, M.A., Mahmoud, F.F.: Free vibration characteristics of a functionally graded beam by finite element method. Appl. Math. Model. **35**, 412–425 (2011)

Kahrobaiyan, M.H., Rahaeifard, M., Tajalli, S.A., Ahmadian, M.T.: A strain gradient functionally graded Euler-Bernoulli beam formulation. Int. J. Eng. Sci. **52**, 65–76 (2012)

Asghari, M., Kahrobaiyan, M.H., Ahmadian, M.T.: A nonlinear Timoshenko beam formulation based on the modified couple stress theory. Int. J. Eng. Sci. **48**, 1749–1761 (2010)

Reddy, J.N.: Analysis of functionally graded plates. Int. J. Numer. Methods Eng. **47**, 663–684 (2000)

Kumar, V., Sinha, S.K., Darbari, A.S., Singh, S.K.: Study and analysis of sandwich beam with graded material for dynamic stability and free vibration. Int. J. Inf. Res. Rev. **01**, 1762–1768 (2016)

Zhu, X., Li, L.: Twisting statics of functionally graded nanotubes using Eringen's nonlocal integral model. Compos. Struct. (2017) (S0263-8223(17)31393-4)
Kroner, E.: Elasticity theory of materials with long range cohesive forces. Int. J. Solids Struct. **3**, 731–742 (1967)
Eringen, A.C.: Linear theory of nonlocal elasticity and dispersion of plane waves. Int. J. Eng. Sci. **10**, 425–435 (1972)
Eringen, A.C.: On differential equations of nonlocal elasticity and solutions of screw dislocations and surface waves. J. Appl. Phys. **54**, 4703–4710 (1983)
Aranda-Ruiz, J., Loya, J., Fernández-Sáez, J.: Bending vibrations of rotating nonuniform nanocantilevers using the Eringen nonlocal elasticity theory. Compos. Struct. **94**(9), 2990–3001 (2012)
Aydogdu, M.: A general nonlocal beam theory: its application to nanobeam bending, buckling and vibration. Phys. E. **41**(9), 1651–1655 (2009)
de Sciarra, F.M.: Finite element modelling of nonlocal beams. Phys. E. **59**, 144–149 (2014)
Ebrahimi, F., Barati, M.R.: Vibration analysis of nonlocal beams made of functionally graded material in thermal environment. Eur. Phys. J. Plus **131**, 279 (2016)
Wang, C.M., Zhang, Y.Y., He X.Q.: Vibration of nonlocal Timoshenko beams. Nanotechnology **18**, 105401, 9 p. (2007)
Adhikari, S., Murmu, T., McCarthy, M.: Frequency domain analysis of nonlocal rods embedded in an elastic medium. Phys. E. **59**, 33–40 (2014)
Murmu, T., Adhikari, S.: Nonlocal effects in the longitudinal vibration of double-nanorod systems. Phys. E. **43**(1), 415–422 (2010)
Ansari, R., Shahabodini, A., Shojaei, M.F.: Nonlocal three-dimensional theory of elasticity with application to free vibration of functionally graded nanoplates on elastic foundations. Phys. E. **76**, 70–81 (2016)
Daneshmehr, A., Rajabpoor, A., Hadi, A.: Size dependent free vibration analysis of nanoplates made of functionally graded materials based on nonlocal elasticity theory with high order theories. Int. J. Eng. Sci. **95**, 23–35 (2015)
Zenkour, A., Sobhy, M.: Nonlocal elasticity theory for thermal buckling of nanoplates lying on Winkler-Pasternak elastic substrate medium. Phys. E. **53**, 251–259 (2013)
Zenkour, A.M.: On the magneto-thermo-elastic responses of FG annular sandwich disks. Int. J. Eng. Sci. **75**, 54–66 (2014)
Lu, P., Zhang, P.Q., Lee, H.P., Wang, C.M., Reddy, J.N.: Non-local elastic plate theories. Proc. R. Soc. A **463**, 3225–3240 (2007)
Lei, Y., Murmub, T., Adhikari, S., Friswell, M.I.: Dynamic characteristics of damped viscoelastic nonlocal Euler-Bernoulli beams. Eur. J. Mech. A/Solids **42**, 125–136 (2013)
Karličić, D., Murmu, T., Cajić, M., Kozić, P., Adhikari, S.: Dynamics of multiple viscoelastic carbon nanotube based nanocomposites with axial magnetic field. J. Appl. Phys. **115**(23), 234303 (2014)
Ansari, R., Oskouie, M.F., Sadeghi, F., Bazdid-Vahdati, M.: Free vibration of fractional viscoelastic Timoshenko nanobeams using the nonlocal elasticity theory. Phys. E. **74**, 318–327 (2015)

Chemistry of Building Materials and Materials

Characterization of "Float Glass" Eroded by Sandblasting

Djamila Aboutaleb[1(✉)], Brahim Safi[1], Ahmed Hammouda[2], and Mohammed Hadj Ammar[2]

[1] Research Unit: Materials, Processes and Environment (UR-MPE), University M'hamed Bougara of Boumerdes, Boumerdes, Algeria
aboutalebdjamila@univ-boumerdes.dz

[2] University M'hamed Bougara of Boumerdes, Boumerdes, Algeria

Abstract. In this work, soda-lime glasses of the "Float Glass" type have been characterized before and after sanding. The latter process consists in producing erosion by sandy particles on the surface of the flat glasses. The impact of this erosion on the properties of these glasses was studied. Two types of sand blasting were carried out, one sandblasting machine was used in the first one, while in the second one sandblasting was carried out at the laboratory to assimilate the effect of the sand wind on the vehicle windscreens. Sandblasting has been found to have an enormous influence on the properties of glasses; the density increased, the chemical attack of the glasses in the acid solutions was greater than by bases, the mechanical resistances and the optical transmissions dropped.

Keywords: Soda-lime · Erosion · Density · Chemical etching
Optical transmission

1 Introduction

Flat glass is widely used in the construction and automotive sector; its transparency, its hardness and its shape make major parameters for this choice. Among the problems encountered in the use of glass in the automobile are often problems related to the nuisance of the surface state and therefore to the view of the windshields of the vehicles which have been exposed to the various external aggressions according to the climatic conditions (winds of sand and dust…). In the Saharan regions, the phenomenon of sand wind is often encountered and the maximum wind speed can reach 120 km/h [1], this natural phenomenon can unfortunately erode the glass and can therefore attack the outer surface of the glass and consequently the transparency and the visibility of this surface diminish. We thought of a characterization of "Float glass" manufactured at the Larabaa (Blida, Algeria) MFG (Mediterranean Float Glass) unit after its erosive wear by the sand to see its influence on the surface state and the variation of the properties of the glass which have flowed, following this process assimilated to the natural phenomenon of the sand wind.

B. Abdelbaki et al. (Eds.): SMSD 2017, *Proceedings of the Third International Symposium on Materials and Sustainable Development*, pp. 219–227, 2018.
https://doi.org/10.1007/978-3-319-89707-3_26

2 Experimental Study

2.1 Glass Used and Sample Preparation

The purpose of this work is to characterize a flat glass "Float glass" eroded by sanding. We used a flat silico-soda-lime type glass, manufactured by the MFG unit. The average chemical composition is given in Table 1.

Table 1. Average chemical composition of the glass used

Oxides	SiO_2	Na_2O	CaO	MgO	Al_2O_3	K_2O	TiO_2	Other
El. %	72.85	12.43	08.15	04.10	01.28	0.48	0.47	0.25

From the same glass plate, we took square samples of dimensions 30 × 30 mm with two different thicknesses. The glasses named type A have a thickness of 3 mm and those of thickness 5 mm are called glasses type B. The test pieces have been eroded by sandblasting on only one face.

2.2 Sand Used

A sand of the Algerian Sahara (Ghardaïa) was used. Its chemical composition mainly contains silica (74%) and sodium oxide (24%). The average grain size is 1000 µm.

2.3 Materials Used

(a) Industrial sandblasting device: An industrial sandblasting machine was used (Society GTP in Régaia Algeria), working with a pressure of 7 bar.
(b) Sandblasting device carried out at the laboratory [2]: A sandblasting device was produced consisting of the following elements:

- Variable speed compressor.
- A nozzle with a diameter of 4 mm and a length of 1 m.
- A sand feeding hopper that allows to release the sand particles with a variable flow rate.
- The sample holder is adjustable to 90°. It is fixed in an enclosure that allows to recover the sand used and to prevent the sand from scattering.
- Sand blasting test with a pressure of 7 Bar; two distances between the nozzle of the sandblaster and the eroded glass were chosen:
  - A distance of 30 cm, glasses called A30 (e = 3 mm) and B30 (e = 3 mm)
  - A distance of 50 cm, glasses called A50 (e = 5 mm) and B50 (e = 5 mm)
- Sand blasting test performed at a research laboratory with a pressure of 5 bar; two distances between the blasting nozzle and the eroded glass were chosen:
  - A distance of 30 cm, glasses called A′30 (e = 3 mm) and B′30 (e = 3 mm)
  - A distance of 50 cm, glasses called A′50 (e = 5 mm) and B′50 (e = 5 mm).

3 Results and Discussions

3.1 Samples Appearance

The aspects of the glasses studied before and after sandblasting are shown in Fig. 1.

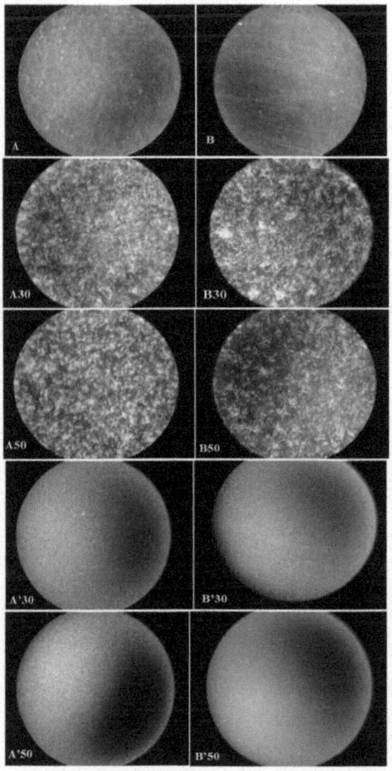

Fig. 1. The aspects of the glasses studied before and after sandblasting.

3.2 The Density

Glasses of small thickness are denser than thick glasses. Sandblasting increases the density of glass (densification by filling voids probably with fine sand particles or broken glass debris) and sandblasted glass at close distances (30 cm) are denser than sandblasted glasses at long distances (50 cm) due to the high impact of the projected particles (Table 2).

3.3 Chemical Resistance

(a) *Before Sandblasting:*

Acid (HCl) action: Acids are characterized by the presence of H^+ or H_3O ions and soda-lime glass is characterized by the presence of modifiers (Na_2O, K_2O...) [3–5]

Table 2. Density of the glass used

Sample	Before sanding	Industrial sandblasting	Sandblasting in the laboratory
A	2.4779	M_V(A30) = 2.4938	M_V(A′30) = 2.500
		M_V(A50) = 2.4934	M_V(A′50) = 2.4841
B	2.3809	M_V(B30) = 2.4828	M_V(B′30) = 2.4454
		M_V(B50) = 2.4774	M_V(B′50) = 2.3879

which can migrate from one vacuum to another and pass into the solution that surrounds them; A weak attack is observed in the acid (ion exchange glass modifier and H^+ ion solution)

Action of the base ($NaOH + Na_2CO_3$): In the basic solutions, the OH^- ion reacts with the lattice, thus there appear network cuts (dissolution of the glass)

Water action: The action of water on the glass begins first with an alkaline ion exchange with the H^+ ion which leads to a decrease in the number of H^+ of the water which becomes alkaline and which causes network cuts and consequently dissolution of the glass (Tables 3 and 4).

Table 3. Acidic attack of samples

Samples	M_0 (g)	M_1(g)	$\Delta M = (M_0 - M_1)/M_0.100\%$
A	1.9063	1.9055	0.0419
B	1.2795	1.2790	0.0390
A30	3.3168	3.3124	0.1326
A50	2.8634	2.8565	0.2409
B30	6.8625	6.8130	0.7210
B50	3.7359	3.7355	0.0107
A′30	1.4650	1.4515	0.9215
A′50	1.2460	1.2431	0.2327
B′30	1.8177	1.8172	0.0275
B′50	2.1900	2.1890	0.0456

For glasses A and B, it was found that the attack of the basic solution was stronger (network cuts), then the acid attack was lower and less by neutral solution (water). The A glasses are more attacked with respect to the glasses B because of their low thicknesses, resulting in a faster diffusion of the solutions of attacks in the glasses A (Table 5).

(b) *After Sandblasting:*

Usually base attack is stronger for glasses than acid attack because the former is carried out with dissolution of the vitreous network following the reaction of the OH^- ions of the base with the vitreous network [6, 7] (bonds of the bonds), whereas the second attack is effected by an ion exchange mechanism between H^+ or H_3O^+ ions of the acid

Table 4. Attack of samples in basic solutions

Samples	M_0 (g)	M_1(g)	$\Delta M = (M_0 - M_1)/M_0.100\%$
A	1.1967	1.1954	0.1086
B	2.0883	2.0872	0.0526
A30	3.2416	3.2404	0.0370
A50	2.4605	2.4602	0.0121
B30	4.1706	4.1694	0.0280
B50	3.7161	3.7159	0.0053
A′30	1.4650	1.4515	0.9215
A′50	1.2460	1.2431	0.2327
B′30	1.8177	1.8172	0.0275
B′50	2.1900	2.1890	0.0456

Table 5. Neutral Attack of samples

Samples	M_0 (g)	M_1(g)	$\Delta M = (M_0 - M_1)/M_0.100\%$
A	1.1252	1.1250	0.0177
B	1.6383	1.6383	0.000
A30	3.4911	3.4906	0.0143
A50	3.0622	3.0622	0.0000
B30	7.1553	7.1528	0.0340
B50	3.8684	3.8673	0.0280
A′30	1.2872	1.2870	0.0155
A′50	2.5148	2.5140	0.0318
B′30	3.8026	3.8021	0.0131
B′50	1.0894	1.0893	0.0077

and the network modifying ions, but in our case, for the sandblasted samples, has observed that the opposite is true, that is to say that the attack by acid is stronger than that by base, we can explain this by the fact that the grains of sand projected by the blasting mechanism have been inserted within the vitreous structure and produced a protective layer on the surface of the glass which consolidated the structure and the base attack was weaker (by comparing it with samples before sanding), whereas the H + ions of the acid have easily (small size) and have easily established ionic changes with modifiers (Na_2O, K_2O …).

In the case of base attack, the sand grains projected by sandblasting were inserted into the structure and protected the surface of the glass against OH^- ions; thereby consolidating the structure and attack per base was lower (comparing with samples before sanding).

Figures 2 and 3 show the glass float glass of 3 mm, sanded by the device of the laboratory at 30 cm and 50 cm and attacked in the three types of solutions.

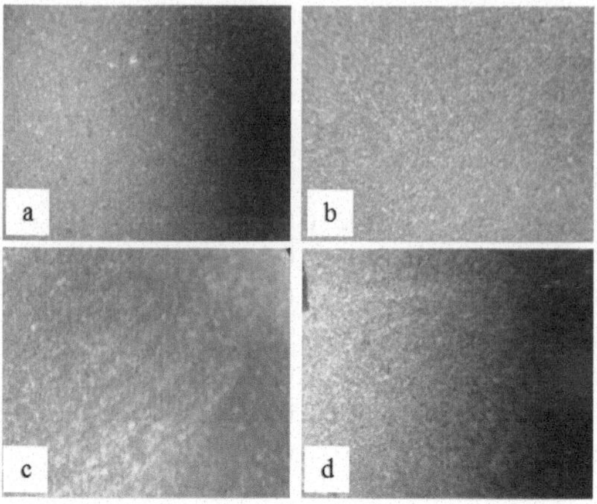

Fig. 2. The attack of the A30 glass in the three media; (a) Before attack (b) acid solution (c) basic solution (d) solution neuter

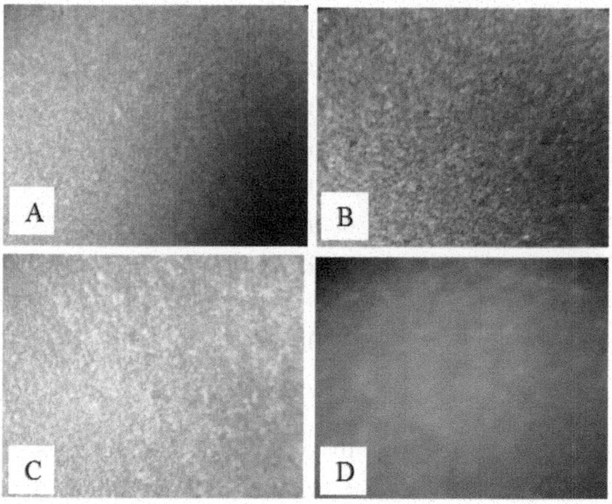

Fig. 3. The attack of the A50 glass in the three media: (a) before attack (b) acid solution (c) basic solution (d) solution neuter

3.4 Thermal Expansion

Generally, glasses with low thicknesses dilate less than glasses with large thicknesses. After sandblasting, the expansion of the glasses increases due to sanding defects.

According to the sandblasting distance, sandblasted glasses at 30 cm have coefficients of thermal expansion larger than sandblasted glasses at 50 cm; intensive sandblasting densifies the glass and increases its expansion.

According to the sandblasting distance, sandblasted glasses at 30 cm have coefficients of thermal expansion larger than sandblasted glasses at 50 cm; intensive sandblasting densifies the glass and increases its expansion.

3.5 Mechanical Strength

After sandblasting, the mechanical strength drops due to sandblasting defects that are generally 30 μm in size compared to crude glass (pre-existing microcracks). The greater thickness of the glass is the greater the mechanical resistance to bending which is due to the increase in the stressed section. Based on sandblasting distance; the

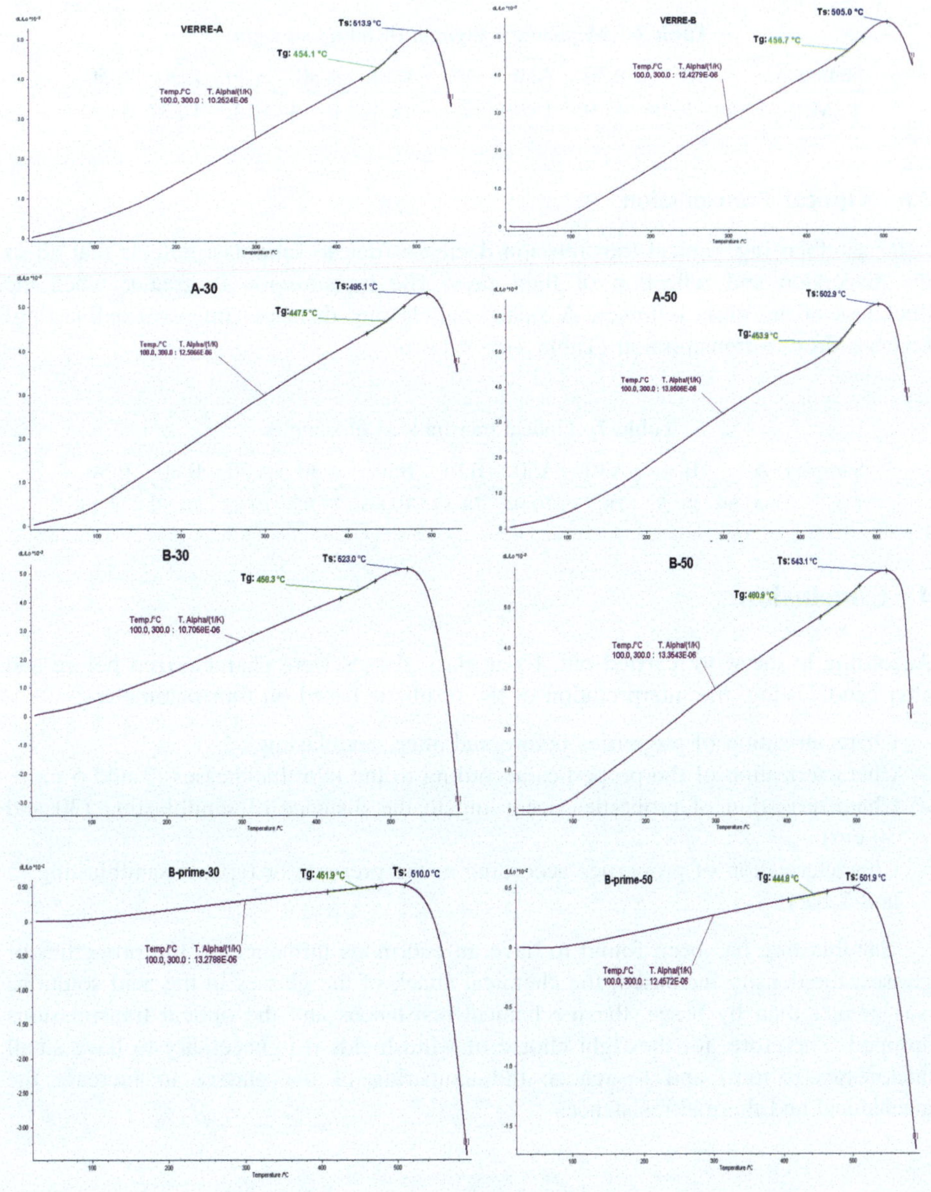

Fig. 4. Expansion curves of samples before and after sandblasting

sandblasted glasses at a greater distance (50 cm) are more resistant than sandblasted glasses at (30 cm), because a high intensity of blasting reduces the flexural strength. The industrially sanded glasses with 7 bar pressure have a lower flexural strength than the sandblasted glasses with the laboratory device (5 bar pressure), since high pressure increases the erosion of the surface and thus weakens the resistance mechanical (Fig. 4 and Table 6).

Table 6. Mechanical strength (bending strength)

Samples	A	B	A30	A50	B30	B50	A′30	A′50	B′30	B′50
σ [MP]	18.68	20.06	11.95	15.99	17.71	18.42	10.13	10.15	10.32	11.41

3.6 Optical Transmission

After sandblasting, optical transmission decreases due to sandblast defects that affect the dispersion and reflection of light rays. The transmission is greater when the thickness of the glass is lower. A small sandblasting distance (intense sandblasting) favors a drop in transmission (Table 7).

Table 7. Optical transmission of samples

Samples	A	B	A30	A50	B30	B50	A′30	A′50	B′30	B′50
T%	90,56	89,57	78,77	79,94	78,57	79,66	75,92	79,92	70,59	73,13

4 Conclusion

According to the work carried out, Float glass glasses were characterized before and after sandblasting, the interpretation of the results is based on four parameters;

- Characterization of properties before and after sandblasting
- Characterization of the properties according to the two thicknesses (3 and 5 mm)
- Characterization of properties according to the distance of sandblasting (30 and 50 cm)
- Characterization of properties according to the pressure or type of sandblasting (5 and 7 bar)

Sandblasting has been found to have an enormous influence on the properties of glasses; the density increased, the chemical attack of the glasses in the acid solutions was greater than by bases, the mechanical resistances and the optical transmissions dropped. Therefore, for the right choice of windshields it is necessary to have small thicknesses (3 mm) and to realize the tempering of the glasses to increase the mechanical and thermal resistances.

References

1. Benterki, S.: Etude de la diffusion par réflexion de la lumière par un pare-brise d’automobile érodé par sablage. Thèse doctorat (2015)
2. Kermiche, A.: Etude de l’effet de HF sur l’état de surface d’un verre érodé par sablage. Thèse de Magister (2010)
3. Zarzicki, R.: Le verre et l’état vitreux. Masson, Edition (1982)
4. Shelby, J.E.: Introduction to Glass Science and Technology. Royal Society of Chemistry, Cambridge (1997)
5. Scholze, H.: Glass: Nature, Structure and Properties. Springer Verlag, New York (1990)
6. Varshneya, A.K.: Fundamentals of Inorganic Glasses. Academic Press, Harcourt Brace & Co. Pub., New York (1994)
7. M EEDDM: Guide méthodologique pour évaluation des incidences sur l’environnement. Fabrication et façonnage du verre plat, du verre creux et des fibres de verre, Mai 2010 85383F_guide_v5/MP

Effect of Chemical Activation on the Surface Properties of Apricot Stones Based Activated Carbons and Its Adsorptive Properties Toward Aniline

Zoubida Kecira(✉), Asma Benturki, Mounir Daoud, and Oumessaâd Benturki

Laboratory of Physico-Chemical Study of Materials and Application to the Environment, Faculty of Chemistry, University of Sciences and Technology Houari Boumediène (USTHB), BP 32 16111, EL-Alia, Algeria
z_kecira@yahoo.com

Abstract. The present study reports the preparation of activated carbons from apricot stones (an agricultural waste) by chemical activation with phosphoric acid (H_3PO_4) and Zinc Chloride ($ZnCl_2$) in N_2 atmosphere. The potential use of the prepared activated carbons as adsorbents and the effects of chemical activation on the surface properties and adsorptive mechanism of Aniline were investigated. The physico-chemical properties of the obtained materials were characterized by Nitrogen adsorption–desorption isotherms, Scanning Electron Microscope coupled to energy dispersive X-ray (SEM-EDX), Fourier Transform Infrared Spectroscope (FTIR), Boehm titration and pH of point of zero charge (pH_{pzc}). The results show that the prepared samples present high apparent surface areas and pore volume with surface functional groups. Adsorption studies were carried out under varying conditions of contact time, initial concentration and temperature. Pseudo-first order and pseudo-second order kinetic models were applied, finding that the pseudo-second order model was the most suitable for the fitting of the experimental kinetic data. The equilibrium adsorption data were analyzed using Langmuir and Freundlich isotherm models. The equilibrium data were well fitted to the Freundlich isotherm. Based on the results, it is, therefore, suggested that the adsorption is mainly brought about by dispersive interactions between Aniline and activated carbon surface. The prepared activated carbons could be used as a low-cost alternative to commercial activated carbon for the removal of aniline from wastewater.

Keywords: Activated carbon · Chemical activation · Adsorption Aniline · Surface properties

1 Introduction

Aniline is an important compound in the petrochemical industry, used primarily for the synthesis of isocyanate, an intermediate in the polyurethane manufacture. Aniline derivatives are also key precursors for the production of accelerators and antioxidants in the rubber industry, herbicides and pharmaceuticals (Pardo et al. 2016; Chen et al. 2015).

B. Abdelbaki et al. (Eds.): SMSD 2017, *Proceedings of the Third International Symposium on Materials and Sustainable Development*, pp. 228–240, 2018.
https://doi.org/10.1007/978-3-319-89707-3_27

It is estimated that the world emission of aniline into the environment is 30,000 tons per year caused by accidental spills, illegal release of industrial and municipal wastewater, and excessive use of pesticides (Pardo et al. 2016; Xiao et al. 2015). Due to the high toxicity of aniline for humans and aquatic life it is necessary to remove it from wastewaters.

Previous works have studied the removal of aniline from aqueous solution using different treatments such as oxidation, extraction, biodegradation and adsorption. Among all these methods, adsorption has been widely used in industry due to its easy operation and wide adaptability (Pardo et al. 2016). Many investigations have analyzed different adsorbents for the removal of aniline such as polymers, activated carbon, inorganic layered materials, nanotubes and minerals. Activated carbon (AC) has been the most studied and used adsorbent in industrial wastewater treatments, especially for organic pollutants such as aniline.

AC is a porous carbonaceous material characterized by its well-developed porous structure and surface chemistry. It can be produced by either physical or chemical activation processes. Various categories of raw materials, rich in carbon, such as different agricultural wastes, coal, lignite, and wood can be used as precursors for the preparation of activated carbons. In recent years, agricultural residues have received an increasing interest for activated carbon production due to their natural abundance, renewability, cost effectiveness and eco-friendliness.

The aims of the present study are: (1) preparation and characterization of activated carbons from an agricultural waste (apricot stones) by chemical activation using H_3PO_4 and $ZnCl_2$ as activating agents; (2) investigation on the effect of chemical activation on the adsorption properties toward aniline.

2 Materials and Methods

2.1 Preparation of Activated Carbons

The apricot stones (AS) were used as raw materials to prepare activated carbons. First the raw AS material was carefully washed with distilled water (d-H_2O) and dried at 60 °C for 24 h. The dried stones were ground and sieved to obtain a particle size varying from 0.5 to 1.2 mm. The resulting product was then maintained in hermetically closed bottles.

For phosphoric acid activation, a granulated apricot stones were impregnated with 85% (vol) H_3PO_4 at a ratio of 1:1 (1 g of H_3PO_4:1 g of raw material) for 24 h. Then the impregnated sample was carbonized in a horizontal furnace under a nitrogen flow with a heating rate of 5 °C/min from room temperature to the final carbonization temperature of 400 °C for two hours. After cooling under a nitrogen flow, the obtained carbon was washed with distilled water until a neutral pH was reached. This is followed by drying the activated carbon overnight at 110 °C and storing it in the hermetically sealed package. The obtained activated carbon is labeled ASPAC.

For the zinc chloride activation, the raw material was impregnated with $ZnCl_2$ at a ratio of 1:1 (1 g of $ZnCl_2$:1 g of raw material) under manual agitation for ten minutes. Then the mixture was carbonized at 600 °C for one hour under nitrogen flow with a heating rate of 5 °C/min. The obtained carbon was treated with a hydrochloric acid solution (0.1 N) by refluxing for three hours then washed with distilled water until silver nitrate test becomes negative. Finally, the activated carbon oven dried at 110 °C for 24 h and marked as ASZAC.

2.2 Characterization

The specific surface area (BET) and the porosity of the activated carbons have been determined by N_2 adsorption at 77 K using the automatics adsorption volumetric apparatus (Micrometrics ASAP 2010). The Brunauere- Emmette-Teller (BET) is the most used method to characterize an activated carbon equation (Brunauer et al. 1938). The presence of porosity and the surface morphology of the activated carbons prepared in this study were examined using scanning electron microscopy (SEM) analysis. The surface functional groups of the activated carbons were examined by Fourier Transform Infrared Spectroscopy method. The samples were mixed with potassium bromide, pressed and analysed by using the FTIR spectroscopy (FTIR-4200, Jasco) at wave-length range of 400 and 4000 cm^{-1}. The Boehm titration method was used to measure the number of acid and basic functions in the samples produced (Boehm 1994). This method is based on the fact that the material superficial groups can react with $NaHCO_3$, Na_2CO_3, NaOH and HCl. The $NaHCO_3$ only neutralizes the carboxylic groups, Na_2CO_3 neutralizes the lactonic and carboxylic groups, while NaOH neutralizes the phenolic groups, with the acid groups mentioned before and HCl neutralizes all the basic groups. The pH_{PZC} of adsorbents was determined by the so-called pH drift method (Franz et al. 2000). Mixtures of 0.15 g of adsorbent sample and 50 ml of NaCl (0.01 mol L^{-1}) solutions of initial pH values varying from 2 to 12 were stirred for 48 h. The final pH values were plotted against the initial pH values and the pH at which the curve crosses the line $pH_{final} = pH_{initial}$ is taken as the pH_{PZC} of the given carbon.

2.3 Batch Adsorption Experiments

The adsorption of aniline from aqueous solutions onto prepared activated carbons (ASACP and ACACZ) was performed by using batch equilibrium technique. About 25 mg of activated carbon samples were introduced into conical flasks containing 25 ml of aniline solutions with initial concentrations ranging from 20 to 300 mg L^{-1}. The mixture was shaken at fixed temperature (25, 35 or 45 °C), using a temperature controlled water bath shaker for 24 h in order to reach the adsorption equilibrium. Then, samples where filtered and the concentrations of aniline in the filtrate were analyzed by using a UV-Visible spectrophotometer (PerkinElmer UV/VIS, Lambda 20) at maximum wave length of 280 nm.

The effect of solution pH was measured by carrying out the adsorption process at different pH values ranged between 2 and 12, using aniline initial concentration of 100 mg L^{-1} and shaking time of 24 h at 25 °C. The pH of solutions was adjusted by addition drops of 0.1 M HCl or NaOH.

The kinetic studies were carried out at 25 °C, using three initial concentrations of aniline (20, 100 and 200 mg L^{-1}) for times ranging from 5 to 480 min. For such studies, a series of 25 mg of adsorbent was added to 25 ml of adsorbate solutions.

The uptake at equilibrium, q_e (mg g^{-1}), was calculated according to the following equation:

$$q_e = \frac{(C_0 - C_e) * V}{W} \quad (1)$$

where C_0 (mg L^{-1}) is the initial concentration of solutions, C_e (mg L^{-1}) is the equilibrium concentration of solutions, V(L) is the volume of solution, and W(g) is the weight of activated carbon.

The adsorption capacity at time t, q_t (mg g^{-1}) was calculated using the following equation:

$$q_t = \frac{(C_0 - C_t) * V}{W} \quad (2)$$

where C_t (mg L^{-1}) is the equilibrium concentration of solutions at time t (min).

3 Results and Discussion

3.1 Characterization of Prepared Simples

The nitrogen adsorption–desorption isotherms of samples are shown in Fig. 1. The adsorption–desorption isotherms obtained are of type I according to the International Union of Pure and Applied Chemistry (IUPAC) classification (Bouras et al. 2015). This shape is characteristic of the presence of a microporous structure.

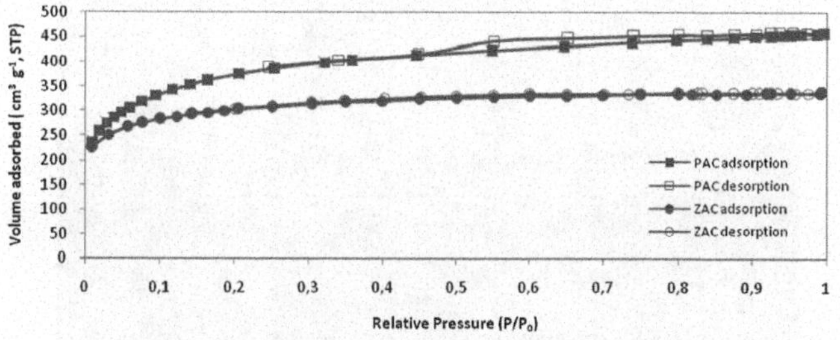

Fig. 1. Nitrogen adsorption and desorption isotherms of the ASPAC and ASZAC samples

The textural parameters of AC samples including specific surface area, pore volume and pore size are given in Table 1. It can be seen that the ASPAC is a highly-developed microporous carbon-based material in comparison with ASZAC.

The surface morphology of prepared carbons was examined using the scanning electron microscopy (SEM) and the micrographs are presented in Fig. 2. The visual appearance indicates that the external surface of the activated carbons ASZAC and ASPAC presents a series of cavities with different sizes and shapes distributed over the

Table 1. Textural characteristics of the activated carbon samples

Samples	BET surface area (m^2 g^{-1})	Total pore volume (cm^3 g^{-1})	Micro-pore volume (cm^3 g^{-1})	Meso-pore volume (cm^3 g^{-1})	External surface area (m^2 g^{-1})	Average pore diameter (nm)
PAC	1382	0.7103	0.6714	0.0389	20.6	2.06
ZAC	1111	0.5225	0.5121	0.0104	4.6	1.88

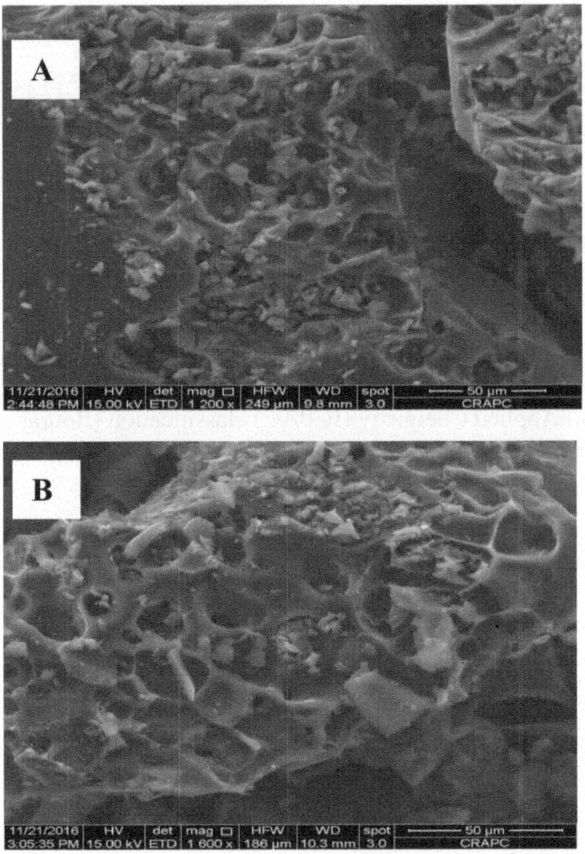

Fig. 2. SEM images of the (A) ASPAC and (B) ASZAC

surface. This texture is resulted from depolymerization and subsequent release of volatile organic substances during the carbonization process (Wang et al. 2008).

The surface functional groups are very important characteristics of the activated carbons because they determine the surface properties of the carbons and hence their feature as ion exchangers, adsorbents, catalysts, and catalyst supports (Budinova et al. 2006). Carboxyls, carbonyls, phenols, lactones, aldehydes, ketones, anhydrides, quinines and hydro quinones are some of the various forms of functional groups in the structure of which oxygen takes part. Delocalized electron of the graphitic structure and such functional groups determine the acidic or basic character of activated carbon structure (Boehm 1994).

Figure 3 shows the FTIR spectra of the produced activated carbons ASPAC and ASZAC. It was obvious that the samples exhibited similar FTIR spectra characteristic of carbonaceous materials and show that the same functional groups are present in the four oxidized activated carbons.

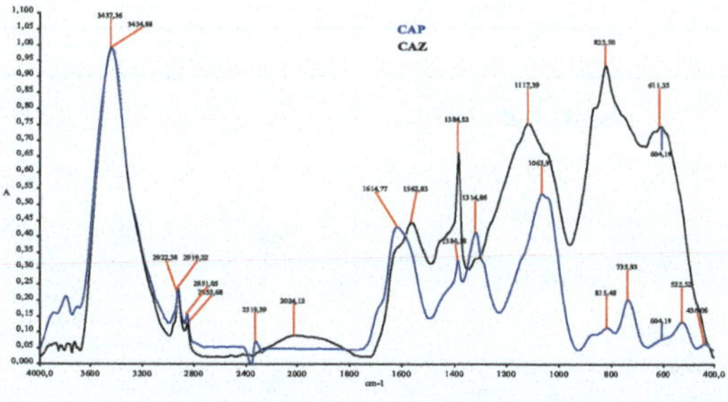

Fig. 3. FTIR spectra of PAC and ZAC

The Boehm's titration method was investigated to determine the amounts of surface functional existing in prepared activated carbons. The results obtained through the Boehm titration; show that ASPAC and ASZAC samples have approximately similar amount of basic groups, and exhibit relatively larger amount of acid groups than basic groups.

3.2 Adsorption Investigation

3.2.1 Effect of Contact Time and Initial Concentration

The adsorption of aniline onto ASPAC and ASZAC was studied in various contact times at initial concentrations of 20, 100 and 200 mg/l. The evolution of the concentration of aniline as a function of time, plotted as qt along time, is shown in Fig. 4. It can be observed that the adsorption was found to be extremely rapid at the initial stage and then gradually increases with rising the time until equilibrium was reached. This result was attributed to the large amount of adsorption sites available on adsorbent at the beginning of adsorption. Adsorption became less efficient, resulting in the second slower phase, with the gradual saturation of adsorption sites on adsorbent surface (Kayranli 2011; Chu et al. 2014; Daoud et al. 2017).

3.2.2 Kinetic Adsorption of Aniline

In order to assess the adsorption kinetics of aniline by ACs, the two mostly applied kinetic models, namely, pseudo-first order and pseudo-second order kinetic models were tested.

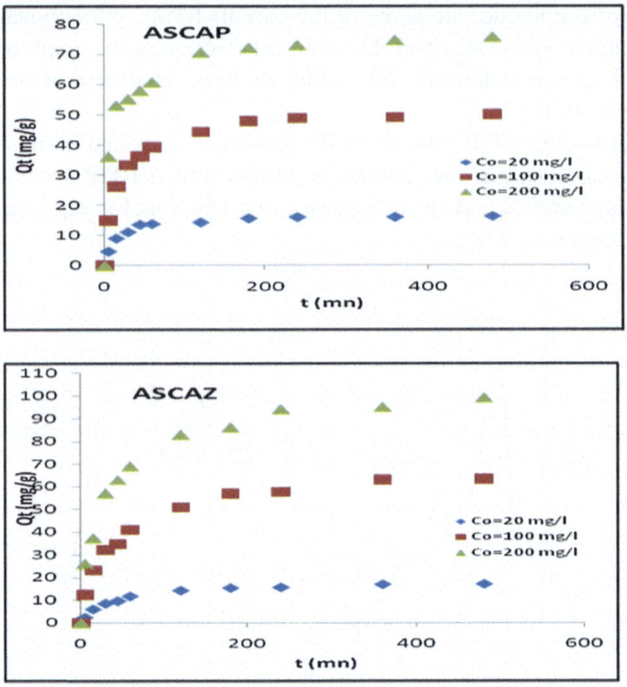

Fig. 4. Kinetic experimental data for the adsorption of aniline onto ASPAC and ASZAC (conditions: adsorbent dose = 1 g L^{-1}, temperature = 25 °C, agitation speed = 150 rpm and natural pH)

The pseudo-first order model proposed by Lagergren and Svanska (1898) is expressed as follows:

$$Ln(q_e - q_t) = lnq_e + K_1 t \quad (3)$$

where q_e and q_t are the amount of solute adsorbed at equilibrium and at time t, respectively; and k_1 (min^{-1}) is the pseudo-first order rate constant which could be obtained from lineal regression analysis of the $\log(q_e - q_t) = F(t)$ function.

The linear form of the pseudo-second order kinetic model (Tütem et al. 1998; Vinod and Anirudhan 2003) is expressed as follows:

$$\frac{t}{qt} = \frac{1}{k_2 q_e^2} + \frac{t}{q_e} \quad (4)$$

where k_2 is the rate constant of pseudo-second order model. Note that K_2 and qe could be obtained from lineal regression analysis of the $t/q_t = F(t)$ function.

The R_2 and constants values for those adsorption kinetic models were calculated and summarized in Table 2, where $q_{e,cal}$ and $q_{e,exp}$ are the calculated and measured adsorption values. It was found that the experimental data were well fitted by the

Table 2. Pseudo-first order and pseudo-second order kinetic model parameters for the adsorption of aniline on ASCAP and ASCAZ

		PFO model			
Acs	Co (mg/l)	Qe exp	Qe th	K1	R2
ASCAP	20	16.3	32.7	0.010	0.91
	100	50.3	26.6	0.010	0.906
	200	75.8	7.9	0.010	0.88
ASCAZ	20	20.0	62.6	0.008	0.926
	100	63.6	54.7	0.012	0.954
	200	99.2	13.6	0.011	0.969
		PSO model			
		Qe exp	Qe th	K2	R2
ASCAP	20	16.3	16.8	0.072	0.999
	100	50.3	51.8	0.061	0.999
	200	75.8	76.9	0.090	0.999
ASCAZ	20	20.0	20.9	0.048	0.999
	100	63.6	71.9	0.025	0.996
	200	99.2	102.0	0.050	0.998

pseudo-second order model with high correlation coefficient values ($R_2 > 0.999$), and the $q_{e,cal}$ value was also very close to that of experimental value. This result indicates that the adsorption of aniline onto the prepared activated carbons follows the pseudo-second order kinetic model. It is also seen that the pseudo-first order model cannot fit the experimental data accurately. In many cases, the first-order model of Lagergren does not fit well with the whole range of contact time and is generally applicable over the initial stage of the adsorption processes (Ma et al. 2015). Consequently, it can be concluded that the adsorption of aniline onto ASPAC and ASZAC obeys the pseudo second order kinetic model.

3.2.3 Intraparticle Diffusion Analysis

The intra-particle diffusion model proposed by Weber and Morris (1963), was used to identify the diffusion mechanism of aniline into the prepared activated carbons during the adsorption process. The rate of the adsorption process may be assumed by four steps. These steps include (1) molecular diffusion of aniline from the bulk solution to a film layer surrounding the adsorbent particle; (2) diffusion of aniline through the boundary layer to the surface of the solid phase (film diffusion); (3) migration of aniline inside the adsorbent particle by surface diffusion or diffusion within liquid filled pores (pore diffusion); (4) adsorption of aniline at an active site on the solid phase surface (adsorption reaction). The slowest step affects the rate of the adsorption process and furthermore, it is necessary to determine the rate-limiting step (Daoud et al. 2017).

The equation of the Intra-particle-diffusion model is shown below:

$$q_t = k_{id}\sqrt{t} + C \tag{5}$$

where q_t (mg g^{-1}) is the amount of solute adsorb at time t, K_{id} (mg/g $min^{1/2}$) is the initial rate of intra-particle diffusion and C (mg/g) is the intercept which is related to the thickness of the boundary layer. Plotting q_t versus $t^{1/2}$, provides an indication of the dependency of adsorption on intraparticle diffusion. If the plot produces a straight line, thus the adsorption process is controlled only by intraparticle diffusion. In contrast, if it exhibits multi-linear plots, then there are two or more steps affecting the adsorption process (Sun et al. 2012).

The plot of the Morris-Weber relationship for the sorption of aniline by ASPAC and ASZAC, at initial concentrations of 20, 100 and 200 mg g^{-1} (figure not shown). Based on this figure, it may be concluded that the intraparticle diffusion of aniline within the ASPAC and ASZAC occurred in three stages for all the range of aniline concentrations. The first linear portion is due to the instantaneous adsorption or external mass transfer or film diffusion (boundary layer). The second linear portion is due to the intraparticle diffusion and binding of aniline molecules into the internal active sites of the CAs. Finally, the third linear portion indicated a saturation of the adsorption process (the equilibrium stage).

3.2.4 Adsorption Isotherms

The adsorption isotherm is the equilibrium relationship between the amount of solute adsorbed onto the solid qe and the equilibrium concentration of the solute in solution Ce at a given temperature. The experimental equilibrium isotherms for the adsorption of aniline on prepared activated carbons are given in Fig. 5.

The experimental equilibriums data were described by optimum isotherm model. The linear forms of isotherm model of Langmuir and Freundlich equations were used to correlation coefficients R^2 to compare the equations applicability.

The linear expression of the Langmuir isotherm equation is given as:

$$\frac{C_e}{q_e} = \frac{1}{K_L q_{max}} + \frac{C_e}{q_{max}} \tag{6}$$

The linear expression of the Freundlich isotherm equation is given as:

$$lnq_e = lnK_F + \frac{1}{n} lnC_e \tag{7}$$

where q_e (mg g^{-1}) is the adsorption capacity at equilibrium; C_e (mg L^{-1}) is the concentration of aniline solution at equilibrium; q_{max} (mg g^{-1}) is the maximum amount of aniline that activated carbon can adsorb; K_L (L mg^{-1}) is the Langmuir isotherm constant that reflects the adsorption energy; K_F ($mg^{1-1/n}$ $L^{1/n}$ g^{-1}) is the Freundlich isotherm constant and 1/n is the value that is used to indicate the heterogeneity of the adsorbent's surface. As 1/n is closer to zero, the surface becomes more heterogeneous.

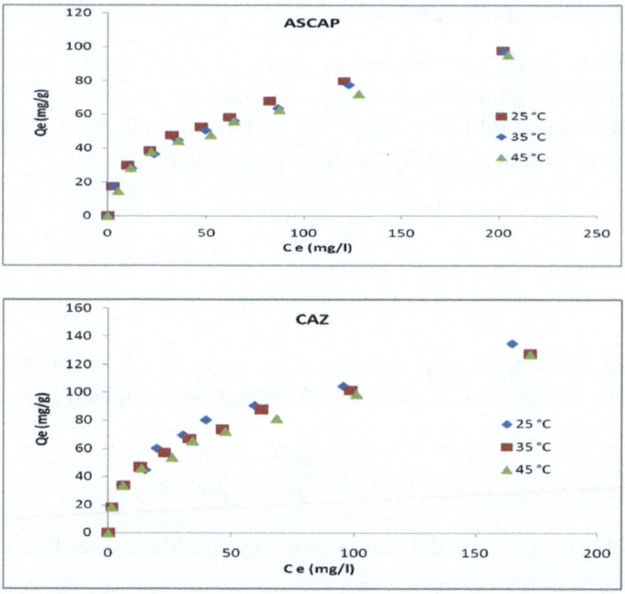

Fig. 5. Experimental adsorption isotherms of aniline onto prepared activated carbons (conditions: adsorbent dose = 1 g L^{-1}, contact time = 24 h, agitation speed = 150 rpm and natural pH)

The parameters of these models were calculated and summarized in Table 3. According to the results, the Freundlich model was best fitted equilibrium data since the correlation coefficient, R^2 values obtained Freundlich isotherm yielded the best fit (>0.99) at all temperatures in comparison to Langmuir isotherm. This means that the adsorbents (ASPAC and ASZAC) provide a heterogeneous surface. Similar trend was observed for adsorption of nitrobenzene onto commercial activated carbon (Jadhav and Srivastava 2013).

Table 3. Parameters obtained for the isotherm models used in the studies

Adsorbent	Langmuir parameters		
	Qm	KL	R2
ASCAP	115	0.02087833	0.9541
ASCAZ	147	0.03549061	0.9635
	Freundlich parameters		
	n	Kf	R2
ASCAP	2.54000508	11.8484849	0.998
ASCAZ	2.32666356	15.483888	0.996

3.3 Adsorption Thermodynamics

The thermodynamic study was realized by using the experimental data obtained from the effect of solution temperature studies carried out earlier. The thermodynamic parameters, that is, standard free energy (ΔG^0), standard enthalpy (ΔH^0), and standard entropy (ΔS^0), can be calculated by the following equations:

$$k_d = \frac{q_e}{C_e} \tag{8}$$

$$\Delta G^0 = -RTlnk_d \tag{9}$$

where R (8.314 J/mol K) is the universal gas constant, T is absolute temperature (K) and K_d is the distribution coefficient which can be calculated as:

$$Ln(K_d) = \frac{\Delta H^0}{R} + \frac{\Delta S^0}{RT} \tag{10}$$

where C_0 and C_e (mg/L) are the initial and the equilibrium aniline concentrations, respectively.

The values of ΔH^0 and ΔS^0 were calculated from the slope and intercept of the linear regression of Ln K_d versus 1/T.

The thermodynamic parameters are listed in Table 4. The negative values of ΔG^0 at different temperatures indicate the spontaneous nature of the adsorption process. These values became more negative as the temperature increased, indicating the exothermic nature of the adsorption of aniline by ASPAC and ASZAC. The values of ΔH^0 of adsorption were found to be negative, indicating exothermic process. In addition, the negative values of ΔS^0 showed the decreased randomness at the solid/liquid interface during the adsorption of aniline on the examined activated carbons.

Table 4. Thermodynamic parameters of adsorption of aniline onto ASPAC and ASZAC at different initial concentrations

	ΔH0 (kJ/mol)	ΔS0 (J/mol)	ΔG0 (kJ/mol)		
			25 °C	35 °C	45 °C
CAP	−3.4528	−10.7334	−0.2471	−0.1517	−0.0317
CAZ	−6.9613	−16.5532	−2.0462	−1.8155	−1.7180

4 Conclusions

The present work shows that local apricot stone based granular activated carbons with highly developed porosity can be prepared using chemical activation with H_3PO_4 or $ZnCl_2$. It was found that the chemical activation influences the textural characteristics and surface chemistry of prepared carbons. Aniline was selected as simple aromatic compound model to study the effect of surface properties on the adsorptive properties.

Isotherm studies favor Freundlich model suggesting multilayer adsorption of aniline over the heterogeneous surface of ASPAC and ASZAC. The maximum monolayer adsorption capacities Qm (mg/g) are 115 and 147 mg/g for ASCAP and ASZAC respectively. The kinetics modeling studies revealed the applicability of pseudo-second order model. Multi-linearity and deviation of intercept from the origin in Weber and Morris plots suggested that intra-particle diffusion was not the only rate-controlling step. The thermodynamic parameters suggested that $\Delta G^0 < 0$, $\Delta H^0 < 0$, and $\Delta S^0 < 0$, which indicated that the adsorption of aniline onto tested activated carbons were a spontaneous and exothermic adsorption process.

Therefore, the apricot stones based activated carbons was considered a high-efficiency adsorbent for aniline, which presents great potential in applications aiming the removal of aromatic pollutants from aqueous solutions.

References

Pardo, B., Ferrer, N., Sempere, J., Gonzalez-Olmos, R.: A key parameter on the adsorption of diluted aniline solutions with activated carbons: the surface oxygen content. Chemosphere **162**(Suppl. C), 181–188 (2016)

Chen, W., Zhang, L., Li, X., Ye, R., Li, Q., Zhu, J., Fang, N., Wang, L., Wu, Z., Horwath, W.R.: Elevated ozone increases nitrifying and denitrifying enzyme activities in the rhizosphere of wheat after 5 years of fumigation. Plant Soil **392**(1), 279–288 (2015)

Xiao, W., Zhou, P., Mao, X., Wang, D.: Ultrahigh aniline-removal capacity of hierarchically structured layered manganese oxides: trapping aniline between interlayers. J. Mater. Chem. A **3**(16), 8676–8682 (2015)

Brunauer, S., Emmett, P.H., Teller, E.: Adsorption of gases in multimolecular layers. J. Am. Chem. Soc. **60**(2), 309–319 (1938)

Boehm, H.P.: Some aspects of the surface chemistry of carbon blacks and other carbons. Carbon **32**(5), 759–769 (1994)

Franz, M., Arafat, H.A., Pinto, N.G.: Effect of chemical surface heterogeneity on the adsorption mechanism of dissolved aromatics on activated carbon. Carbon **38**(13), 1807–1819 (2000)

Bouras, H.D., Benturki, O., Bouras, N., Attou, M., Donnot, A., Merlin, A., Addoun, F., Holtz, M.D.: The use of an agricultural waste material from Ziziphus jujuba as a novel adsorbent for humic acid removal from aqueous solutions. Journal of Molecular Liquids **211**(Suppl. C), 1039–1046 (2015)

Wang, X., Zhu, N., Yin, B.: Preparation of sludge-based activated carbon and its application in dye wastewater treatment. J. Hazard. Mater. **153**(1–2), 22–27 (2008)

Budinova, T., Ekinci, E., Yardim, F., Grimm, A., Björnbom, E., Minkova, V., Goranova, M.: Characterization and application of activated carbon produced by H_3PO_4 and water vapor activation. Fuel Process. Technol. **87**(10), 899–905 (2006)

Kayranli, B.: Adsorption of textile dyes onto iron based waterworks sludge from aqueous solution; isotherm, kinetic and thermodynamic study. Chem. Eng. J. **173**(3), 782–791 (2011)

Chu, S.Y., Xiao, J.B., Tian, G.M., Wong, M.H.: Preparation and characterization of activated carbon from aquatic macrophyte debris and its ability to adsorb anthraquinone dyes. J. Ind. Eng. Chem. **20**(5), 3461–3466 (2014)

Daoud, M., Benturki, O., Kecira, Z., Girods, P., Donnot, A.: Removal of reactive dye (BEZAKTIV Red S-MAX) from aqueous solution by adsorption onto activated carbons prepared from date palm rachis and jujube stones. Journal of Molecular Liquids **243**(Suppl. C), 799–809 (2017)

Lagergren, S.Y.: Zur Theorie der sogenannten Adsorption gelöster Stoffe (1898)

Tütem, E., Apak, R., Ünal, Ç.F.: Adsorptive removal of chlorophenols from water by bituminous shale. Water Res. **32**(8), 2315–2324 (1998)

Vinod, V.P., Anirudhan, T.S.: Adsorption behaviour of basic dyes on the humic acid immobilized pillared clay. Water Air Soil Pollut. **150**(1), 193–217 (2003)

Ma, J., Huang, D., Zou, J., Li, L., Kong, Y., Komarneni, S.: Adsorption of methylene blue and Orange II pollutants on activated carbon prepared from banana peel. J. Porous Mater. **22**(2), 301–311 (2015)

Weber Jr., W.J., Morris, J.C.: Kinetics of adsorption on carbon from solution. J. Sanitary Eng. Div. Proc. Am. So. Civil Eng. **89**, 31–60 (1963)

Sun, Y., Yue, Q., Gao, B., Li, Q., Huang, L., Yao, F., Xu, X.: Preparation of activated carbon derived from cotton linter fibers by fused NaOH activation and its application for oxytetracycline (OTC) adsorption. J. Colloid Interface Sci. **368**(1), 521–527 (2012)

Jadhav, A.J., Srivastava, V.C.: Adsorbed solution theory based modeling of binary adsorption of nitrobenzene, aniline and phenol onto granulated activated carbon. Chem. Eng. J. **229**, 450–459 (2013)

Porosity Effect of a FG Porous Plate on the Behavior of Bonded Beam and Strengthened Structure Instability

A. Chedad[1(✉)] and N. Elmeiche[1,2]

[1] Université Mustapha Stambouli, Mascara, Algeria
abdelbassetl593@gmail.com
[2] Laboratoire de génie civil et d'environnement, LGCE, Sidi Bel Abbes, Algeria

Abstract. In this paper an improved theoretical interfacial stress analysis is presented for simply supported beam bonded with a FG plate. This solution take into account the dislocation of the neutral axes and this is helpful for introducing the inhomogeneity through the thickness of the plate and thus this solution is applicable for wide range of composite material. The plate is assumed to be isotropic with a constant Poisson's ratio and variation according to a power law for the elastic modulus through the plate thickness. Stresses distribution are presented in form of graphs, it is shown that both normal and shear stresses at the interface are influenced by the inhomogeneities due to the variation in mechanical characteristics or porosity distribution, also the geometric of the plate play an important role in the distribution of interfacial stresses. This research is helpful in understanding the mechanical behavior of the interface and strengthened structure instability. The results presented in the paper can serve as a benchmark for future analyses of bonded beams with composite plate.

Keywords: FG plate · Interfacial stresses · Porosity

1 Introduction

Retrofitting damage structure with bonding composite material has known a tremendous development in the last decade, this technique has replaced the ancient one which is the hermitian technique that uses steal plate because of the high characteristic of composite plate compared to the steel plate, and the possibility of tailoring the material depending on a desirable characteristics, even though the debonding of the plate due to the high concentration of stresses at the plate end still remarkable instability problem. This topic has been the subject of many experimental end theoretical researches, experimental studies Jones et al. (1988) and Etman and beeby (2000) have shown the difficulty of studying the interfacial stresses because of the complexity of local stresses. Important contributions on debonding models have been made, Ascione and Feo (2000) developed a finite element model to predict the shear and the normal stresses in the adhesive layer of plated reinforced concrete beams and the numerical results agreed with the experimental results obtained by Jones et al. (1988), and the theoretical results presented by Roberts (1989), Roberts and Haji-Kazemi (1989) proposed an analytical

B. Abdelbaki et al. (Eds.): SMSD 2017, *Proceedings of the Third International Symposium on Materials and Sustainable Development*, pp. 241–253, 2018.
https://doi.org/10.1007/978-3-319-89707-3_28

solution based on partial interaction theory, for predicting the shear and normal stress concentrations in adhesive joints. This solution is for a uniformly distributed load only. Shen et al. (2001) have developed a complementary energy method to study the interfacial stresses for simply supported RC beams and slabs bonded with a thin composite plate or steel plate. Rabinovich and Frostig (2000) and Yang et al. (2002) presented closed-form high-order analyses of reinforced concrete beams strengthened with FRP plates, which satisfy the free surface condition at the ends of the adhesive layer. Taljsten (1997) proposed a solution based on direct deformation compatibility to determine the interfacial stresses. Smith and Teng (2001), having carried out a review of the approximate closed-form solutions for interfacial stresses in literature, presented a new solution which was intended for application to beams with a bonded thin plate, it seems that it is the more accurate widely applicable solution, particularly when the flexural stiffness of the bonded plate becomes significant. Recently Tounsi (2006) proposed an improved solution for simply supported beam bonded with a FRP plate including the adherend shear deformation by assuming a linear shear stress through the thickness of the adherends. Tounsi et al. (2009) proposed an improved solution by assuming a parabolic shear stress through the thickness of both the concrete beam and the bonded plate. Meradjah et al. (2010) presented a method for determining the elastic shear and peel stresses in an adhesive joint between a strengthening plate and a functionally graded beam (FGB).

The objective of this study is to develop a procedure that predicts the shear and normal stresses in the interface between a beam and a FG plate with reasonable accuracy. The material inhomogeneity through the thickness was included by assuming the dislocation of the neutral axes and that helps to represent a wide range of composite material, Finally, a parametric study was carried out to show the effects of some design variables.

2 Method of Solution

2.1 Basic Assumption

Consider a simply supported RC beam with a span of L (Fig. 1). A FG plate with a length of L_p is symmetrically bonded to its soffit. The beam is subjected to a uniformly distributed load. The thickness of the plate is t_p and its width is the same as that of the beam. The thickness of the adhesive layer is t_a. The beam section area is b × h.

The main assumptions used in this research are as follows:

1. All materials considered are linear elastic.
2. No slip is allowed at the interface of the bond (i.e., there is a perfect bond at the adhesive FG plate interface).
3. Stresses in the adhesive layer do not change with the thickness.
4. The shear stress analysis assumes that the curvatures in the beam and plate are equal (since this allows the shear stress and peel stress equations to be uncoupled). However, this assumption is not made in the peel stress solution.
5. Bending deformations of the adhesive are neglected.

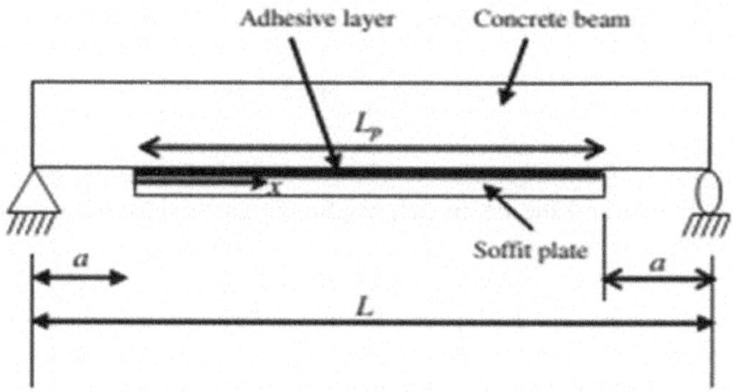

Fig. 1. Simply supported beam strengthened with bonded plate.

2.2 Love-Kirchhoff Plate Theory

We will follow the Love-Kirchhoff plate theory assumption that plane sections normal to the plate axis remain plane and normal after deformation. w displacements are independent of z. Then the displacements can be written as:

$$u(x,z) = u_0(x) - z\frac{dw}{dx} \text{ and } \quad w(x,z) = w_0(x) \tag{1}$$

It may be noted that u_0 and w_0 denote the middle surface displacement components along the x and z directions. Within the framework of small perturbations, the linear. Green–Lagrange strain tensor writes:

$$\varepsilon_x = u^0_{,x}(x) - zw^0_{,xx}(x) = \varepsilon_0 - zk \tag{2}$$

Then the strain component (Eq. (2)) is related to its corresponding stress component through the basic constitutive law:

$$\sigma_x = E(z)\varepsilon_x \tag{3}$$

The elastic modulus is assumed to vary continuously across the plate thickness and is given by:

$$E(z) = E_L + (E_U - E_L)\left(\frac{z}{h} + \frac{1}{2}\right)^P - (E_U - E_L)\alpha/2 \tag{4}$$

Where E_L and E_U are the elastic modulus of the lower and upper surface of the plate, P is the homogeneity index and α is the porosity index.

The axial force and bending moment resultants, N and M, are defined as:

$$(N, M) = \int_{-h/2}^{h/2} \sigma_x(1, z)dz \tag{5}$$

Substituting for σ_x from Eqs. (2) and (3) into Eq. (5), a relation between the force and moment resultants and the beam deformations can be derived as follows:

$$\begin{Bmatrix} N \\ M \end{Bmatrix} = \begin{bmatrix} A & B \\ B & D \end{bmatrix} \begin{Bmatrix} \varepsilon_0 \\ k \end{Bmatrix} \tag{6}$$

The definition of plate stiffness coefficients A, B and D are:

$$(A, B, D) = \int_{-h/2}^{h/2} E(z)\left(1, z, z^2\right)dz \tag{07}$$

The inverse relations corresponding to that in Eq. (6) are:

$$\begin{Bmatrix} \varepsilon_0 \\ k \end{Bmatrix} = \begin{bmatrix} A & B \\ B & D \end{bmatrix}^{-1} \begin{Bmatrix} N \\ M \end{Bmatrix} = \begin{bmatrix} A^* & B^* \\ B^* & D^* \end{bmatrix} \begin{Bmatrix} N \\ M \end{Bmatrix} \tag{8}$$

Since the axial force resultant N $\equiv$ 0, the expressions for the deformations take the form:

$$\varepsilon_0 = B^* M, \quad k = D^* M \tag{9}$$

Substituting the Eq. (2) into Eq. (3) the axial stresses in a FG plate take the form:

$$\sigma_x = E(z)M[B^* - zD^*] \tag{10}$$

From Eq. (10) one can recognize that the neutral axis is at:

$$z_{NA} = \frac{B^*}{D^*} \tag{11}$$

2.3 Adhesive Shear Stress

The shear stress in the adhesive can be expressed as follows:

$$\tau_a(x) = \frac{G_a}{t_a}\left[u_p(x) - u_b(x)\right] \tag{12}$$

where G_a, t_a, u_b and u_p denote the shear modulus, the thickness of the adhesive material, the longitudinal displacement at the base of the beam, and the longitudinal displacement at the top of FG plate, respectively. Equation (12) can be expressed in terms of the mechanical strain of the beam, $\varepsilon_b(x)$, and the FG plate $\varepsilon_p(x)$ after differentiating the equation with respect to x:

$$\frac{d\tau_a(x)}{dx} = \frac{G_a}{t_a}\left[\varepsilon_p(x) - \varepsilon_b(x)\right] \quad (13)$$

Where

$$\frac{du_b(x)}{dx} = \varepsilon_b(x) \quad \text{and} \quad \frac{du_p(x)}{dx} = \varepsilon_p(x) \quad (14)$$

The strain at the bottom of the beam is given as:

$$\varepsilon_p(x) = \frac{t_b}{2E_bI_b}M_b(x) - \frac{N_b(x)}{E_bA_b} \quad (15)$$

The strain at the top of the FG plate, using Eqs. (2) and (8) is given as:

$$\varepsilon_b(x) = \frac{1}{b}\left[A^* - \left(\frac{h}{2} + z_{NA}\right)B^*\right]N_p(x) - \frac{1}{b}\left[B^* - \left(\frac{h}{2} + z_{NA}\right)D^*\right]M_p(x) \quad (16)$$

Where E_b, A_b, I_b, t_b and b denote the elastic modulus, the cross-sectional area, the second moment of area, the thickness of the beam, and width of the FG plate, respectively.

Consideration of horizontal equilibrium gives:

$$\frac{dN_b(x)}{dx} = \frac{dN_p(x)}{dx} = b\tau(x) \quad (17)$$

Where

$$N(x) = N_b(x) = N_p(x) = b\int_0^x \tau(x)dx \quad (18)$$

Assuming equal curvature in the beam and the soffit plate, the relationship between the moments in the two adherends can be expressed as:

$$M_p(x) = \frac{b}{D^*E_bI_b}M_b(x) - \frac{B^*}{D^*}N_p(x) \quad (19)$$

Moment equilibrium of the differential segment of the plated beam in Fig. 2 gives:

$$M_T(x) = M_b(x) + M_p(x) + N(x)\left[\frac{t_p}{2} + t_a + \left(\frac{h}{2} + z_{NA}\right)\right] \quad (20)$$

The bending moment in each adherend, expressed as a function of the total applied moment and the interfacial shear stress, is given as:

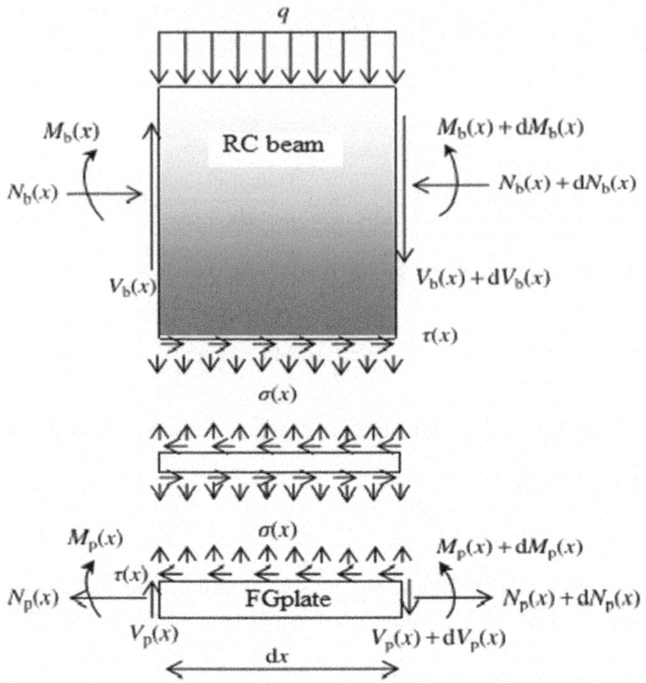

Fig. 2. Forces in infinitesimal element of a soffit-plated beam.

$$M_b(x) = \frac{D^* E_b I_b}{b + D^* E_b I_b}\left[M_T(x) - N(x)\left(\frac{t_b}{2} + t_a + \left(\frac{h}{2} + z_{NA}\right) - B^*/D^*\right)\right] \tag{21}$$

$$M_p(x) = \frac{b}{b + D^* E_b I_b}\left[M_T(x) - N(x)\left(\frac{t_b}{2} + t_a + \left(\frac{h}{2} + z_{NA}\right) + \frac{E_b I_b}{b} B^*\right)\right] \tag{22}$$

The first derivation of the bending moment in each adherend gives:

$$\frac{dM_b(x)}{dx} = \frac{D^* E_b I_b}{b + D^* E_b I_b}\left[V_T(x) - b\tau(x)\left(\frac{t_b}{2} + t_a + \left(\frac{h}{2} + z_{NA}\right) - B^*/D^*\right)\right] \tag{23}$$

$$\frac{dM_p(x)}{dx} = \frac{b}{b + D^* E_b I_b}\left[V_T(x) - b\tau(x)\left(\frac{t_b}{2} + t_a + \left(\frac{h}{2} + z_{NA}\right) + \frac{E_b I_b}{b} B^*\right)\right] \tag{24}$$

Substituting Eqs. (15) and (16) into Eq. (13) and differentiating the resulting equation once, yields:

$$\frac{d^2\tau(x)}{dx^2} = \frac{G_a}{t_a}\left[\frac{1}{b}\left(A^* - \left[\frac{h}{2} + z_{NA}\right]B^*\right)\frac{dN_p(x)}{dx} - \frac{1}{b}\left(B^* - \left[\frac{h}{2} + z_{NA}\right]D^*\right)\frac{dM_p(x)}{dx} + \frac{1}{E_bA_b}\frac{dN_b(x)}{dx} - \frac{t_b}{2E_bI_b}\frac{dM_b(x)}{dx}\right] \tag{25}$$

Substitution of the shear forces Eqs. (23) and (24) and axial forces Eq. (18) in both adherends into Eq. (25) gives the following governing differential equation for the interfacial shear stress:

$$\begin{aligned}&\frac{d^2\tau(x)}{dx^2} + \frac{\frac{G_a}{t_a}}{D^*E_bI_b + b}\left[\frac{t_b}{2}D^* + \left(B^* - \left[\frac{h}{2} + z_{NA}\right] - D^*\right)\right]V_T(x)\\ &- \frac{G_a}{t_a}\left[\frac{b}{E_bA_b} + \frac{t_bD^*b}{2(D^*E_bI_b + b)}\left(\frac{t_b}{2} + t_a + \left(\frac{h}{2} + z_{NA}\right) - \frac{B^*}{D^*}\right) + A^*\right.\\ &- \left[\frac{h}{2} + z_{NA}\right]B^* + \frac{b}{b + D^*E_bI_b}\left(\frac{t_b}{2} + t_a + \left[\frac{h}{2} + z_{NA}\right]\right) + \frac{E_bI_b}{bB^*}\\ &\times\left(B^* - \left[\frac{h}{2} + z_{NA}\right]D^*\right)\tau(x) = 0\end{aligned} \tag{26}$$

The general solution to Eq. (26) is given by:

$$\tau(x) = B_1\cosh(\lambda x) + B_2\sinh(\lambda x) + m_1V_T(x) \tag{27}$$

Where

$$\begin{aligned}\lambda^2 = &\frac{G_a}{t_a}\left[\frac{b}{E_bA_b} + \frac{t_bD^*b}{2(D^*E_bI_b + b)}\left(\frac{t_b}{2} + t_a + \left(\frac{h}{2} + z_{NA}\right) - \frac{B^*}{D^*}\right)\right.\\ &+ A^* - \left[\frac{h}{2} + z_{NA}\right]B^* + \frac{b}{b + D^*E_bI_b}\left(\frac{t_b}{2} + t_a + \left[\frac{h}{2} + z_{NA}\right] + \frac{E_bI_b}{b}B^*\right)\\ &\left.\times\left(B^* - \left[\frac{h}{2} + z_{NA}\right]D^*\right)\right]\end{aligned} \tag{28}$$

And

$$m_1 = \frac{G_a}{t_a}\frac{1}{\lambda^2}\frac{1}{D^*E_bI_b + b}\left[\frac{t_b}{2}D^* + \left(B^* - \left[\frac{h}{2} + z_{NA}\right]D^*\right)\right] \tag{29}$$

B_1 and B_2 are constant coefficients determined from the boundary conditions as:

$$B_1 = \frac{m_2}{\lambda}\frac{qa}{2}(L - a) - \frac{m_1}{\lambda}q = -B_2 \tag{30}$$

Where

$$m_2 = \frac{G_a}{t_a} \frac{t_b}{2E_b I_b} \tag{31}$$

Substitution of B_1 and B_2 into Eq. (27) gives an expression for the interfacial shear stress at any point:

$$\tau(x) = \left[\frac{m_2 a}{2}(L-a) - m_1\right] \frac{q e^{-\lambda x}}{\lambda} + m_1 q\left(\frac{L}{2} - a - x\right) \tag{32}$$

2.4 Adhesive Normal Stress

The governing differential equation for the interfacial normal stress is derived in this section. When the beam is loaded, vertical separation occurs between the beam and the bonded plate. This separation creates an interfacial normal stress in the adhesive layer. The normal stress, $\sigma(x)$, is given as:

$$\sigma(x) = \frac{E_a}{t_a}\left[w_p(x) - w_b(x)\right] \tag{33}$$

where $w_b(x)$ and $w_p(x)$ are vertical displacements of the beam and the plate, respectively. The equilibrium of the beam and the plate, neglecting second-order terms, leads to the following relationships.

Adherend 1 (beam):

$$\frac{d^2 w_b(x)}{dx^2} = -\frac{1}{E_b I_b} M_b(x) \tag{34}$$

$$\frac{dM_b(x)}{dx} = V_p(x) - \tau(x) b \frac{t_b}{2} \tag{35}$$

$$\frac{dV_b(x)}{dx} = -\sigma(x) b - q \tag{36}$$

Adherend 2 (plate):

$$\frac{d^2 w_p(x)}{dx^2} = -\frac{1}{b} B^* N_p(x) - \frac{1}{b} D^* M_p(x) \tag{37}$$

$$\frac{dM_p(x)}{dx} = V_p(x) - \tau(x) b \left[\frac{h}{2} + z_{NA}\right] \tag{38}$$

$$\frac{dV_p(x)}{dx} = \sigma(x) b \tag{39}$$

Based on the above equilibrium equations, the governing differential equations for the deflections of adherends 1 and 2, expressed in terms of the interfacial shear and normal stresses, are given as follows:

Adherend 1 (beam):

$$\frac{d^4 w_b(x)}{dx^2} = -\frac{1}{E_b I_b}\left[-\sigma(x)b - q - \frac{t_b}{2}b\frac{d\tau(x)}{dx}\right] \tag{40}$$

Adherend 2 (plate):

$$\frac{d^4 w_p(x)}{dx^2} = -B^*\frac{d\tau(x)}{dx} - \frac{1}{b}D^*\left[\sigma(x) - \left[\frac{h}{2} + z_{NA}\right]\frac{d\tau(x)}{dx}\right] \tag{41}$$

Substitution of Eqs. (40) and (41) into fourth derivative of the interfacial normal stress obtained from Eq. (33) gives the following governing differential equation for the interfacial normal stress:

$$\begin{aligned}&\frac{d^4\sigma(x)}{dx^4} + \frac{E_a}{t_a}\left(\frac{b}{E_b I_b} + D^*\right)\sigma(x) + \frac{E_a}{t_a}\left(\frac{b}{E_b I_b}\frac{t_b}{2} + B^* - D^*\left[\frac{h}{2} + z_{NA}\right]\right)\frac{d\tau(x)}{dx}\\&+\frac{E_a}{t_a}\frac{1}{E_b I_b}q = 0\end{aligned} \tag{42}$$

The general solution to this fourth-order differential equation is:

$$\begin{aligned}&\sigma(x)e^{-\delta x}[C_1\cos(\delta x) + C_2\sin(\delta x)]\\&+e^{\delta x}[C_3\cos(\delta x) + C_4\sin(\delta x)] - n_1\frac{d\tau(x)}{dx} - n_2 q\end{aligned} \tag{43}$$

For a large value of x it is assumed that the normal stress approaches zero, and as a result $C_3 = C_4 = 0$. The general solution therefore becomes:

$$\sigma(x) = e^{-\delta x}[C_1\cos(\delta x) + C_2\sin(\delta x)] - n_1\frac{d\tau(x)}{dx} - n_2 q \tag{44}$$

Where

$$\delta = \sqrt{4}\frac{E_a}{4t_a}\left(\frac{b}{E_b I_b} + D^*\right) \tag{45}$$

$$n_1 = \frac{b\frac{t_b}{2} + B^* E_b I_b - D^* E_p I_p\left[\frac{h}{2} + z_{NA}\right]}{D^* E_b I_b + b} \tag{46}$$

$$n_2 = \frac{1}{(D^* E_b I_b + b)} \tag{47}$$

C_1 and C_2 are constant coefficients determined from the boundary conditions as:

$$C_1 = \frac{E_a}{2\delta^3 t_a} \frac{1}{E_b I_b} [V_T(0) - \delta M_T(0)] - \frac{n_3}{2\delta^3} \tau(0) + \frac{n_1}{2\delta^3} \left[\frac{d^4 \tau(0)}{dx^4} + \delta \frac{d^3 \tau(0)}{dx^3} \right] \tag{48}$$

$$C_2 = -\frac{E_a}{t_a} \frac{1}{2\delta^2} \frac{1}{E_b I_b} M_T(0) - \frac{n_1}{2\delta^2} \frac{d^3 \tau(0)}{dx^3} \tag{49}$$

$$n_3 = \frac{E_a}{t_a} \left(-D^* \left[\frac{h}{2} + z_{NA} \right] + \frac{b}{E_b I_b} \frac{t_b}{2} + B^* \right) \tag{50}$$

3 Results and Discussions

3.1 Numerical Comparisons

The present model was examined by comparing the results obtained with the present theory and those determined from the famous method developed by Smith and Teng (2001). For this, an RC beam strengthened with CFRP soffit plate is analyzed. The beams are simply supported and subjected to a uniformly distributed load. The span of RC beam is L = 3000 mm, the distance from the support to the end of the plate is a = 300 mm, the UDL is 50 kN/m. A summary of the geometric and material properties is given in Table 1.

Table 1. Geometric and material properties

Component	Width (mm)	Depth (mm)	Young's modulus (MPa)	Poisson's ratio
RC beam	200	300	30000	0.18
Adhesive layer	200	2.0	3000	0.35
CFRP plate	200	4.0	140000	0.28
FG porous plate	200	4.0	E_U = 200000 E_L = 70000	0.3

The present model determines the mechanical characteristic based on the laminated plate theory or classical plate theory. Overall, the interfacial stresses predicted by the present method are in good agreement with those of Smith and Teng (2001) (Fig. 3).

3.2 Parametric Study

In this section, numerical results of the present model are presented to study the effect of various parameters on the edge interfacial stresses in an RC beam bonded with a non-homogenous FG plate, with E_U = 200000 MPa and E_L = 70000 MPa, the geometric and material properties is given in Table 1.

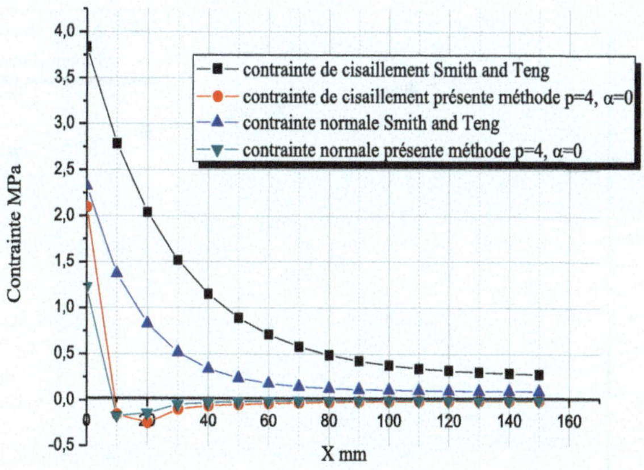

Fig. 3. Comparison of interfacial shear and normal stresses for an RC beam bonded and subjected to a UDL.

Figure 4 shows the effect of homogeneity index on interfacial stresses, for p = 0 the plate is considered stiff plate and the shear and normal stresses are remarkably high, as the value of “p” increases the plate becomes softer and the interfacial stresses become smaller.

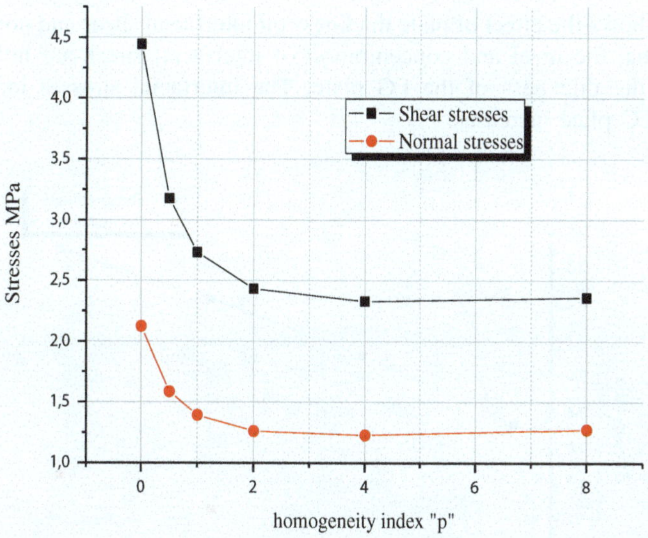

Fig. 4. The effect of homogeneity index on edge shear and normal stresses

Porosity effect on edge shear and normal stresses is shown in Fig. 5, porosity is considered harmful to composite material and it appears during fabrication, for this it is necessary to take it into account to design reinforced structure, Fig. 5 shows that the interfacial shear and normal stresses decrease as the porosity volume fraction increases.

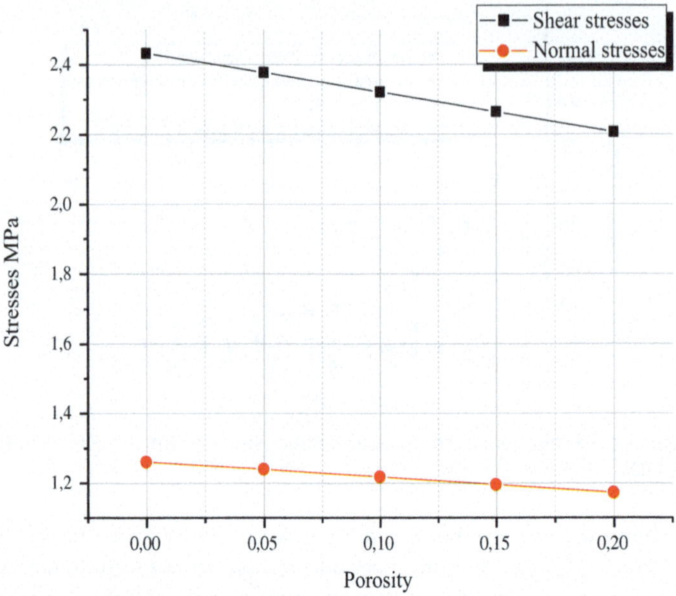

Fig. 5. Porosity effect on edge shear and normal stresses.

Figure 6 shows the effect of plate thickness on interfacial shear and normal stresses. It is shown that the level and concentration of interfacial stress are influenced considerably by the thickness of the FG plate. The interfacial stresses increase as the thickness of FG plate increases.

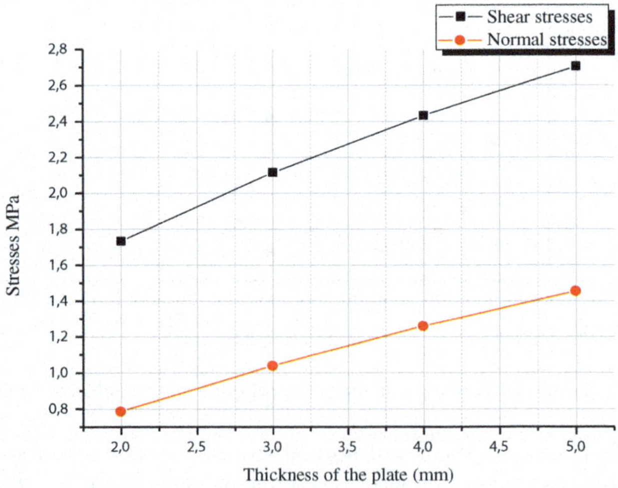

Fig. 6. Plate thickness effect on edge shear and normal stresses

4 Conclusions

An improved theoretical interfacial stress analysis has been presented for simply supported RC beams bonded with a thin composite FG plate and subjected to a uniformly distributed bending load. The salient features of this analysis include the consideration of the inhomogeneity of the bonded plate by assuming the dislocation of the neural axes of the plate. The solution methodology is general in nature and may be applicable to the analysis of other types of composite material.

In the final part of this paper, a parametric study was undertaken by using this solution for strengthened beams with various ratios of design parameters. The results show that geometry, inhomogeneity and porosity play an important role on the edge shear and normal stresses.

References

Ascione, L., Feo, L.: Modeling of composite/concrete interface of RC beams strengthened with composite laminates. Compos. Eng. Part B **31**, 535–540 (2000)

Etman, E.E., Beeby, A.W.: Experimental program and analytical study of bond stress distributions on a composite plate bonded to a reinforced RC beam. Cem. Concr. Compos. **22**(4), 281–291 (2000)

Jones, K.R., Swamy, R.N., Charif, A.: Plate separation and anchorage of reinforced concrete beams strengthened by epoxy bonded steel plates. Struct. Eng. **66**(5/1), 85–94 (1988)

Meradjah, M., Benyoucef, S., Tounsi, A., Adda Bedia, E.A., Merdaci, S.: Interfacial stresses in plated beams with exponentially-varying properties. J. Adhes. Sci. Technol. **24**, 1063–1081 (2010)

Rabinovich, O., Frostig, Y.: Closed-form high order analysis of RC beams strengthened with FRP strips. ASCE J. Compos. Constr. **4**(2), 65–74 (2000)

Roberts, T.M.: Approximate analysis of shear and normal stress concentrations in adhesive layer of plated RC beams. Struct. Eng. **67**(12), 229–233 (1989)

Roberts, T.M., Haji-Kazemi, H.: A theoretical study of the behaviour of reinforced concrete beams strengthened by externally bonded steel plates. Proc. Ins. Civil Eng. Part 2 **87**, 39–55 (1989)

Shen, H.S., Teng, J.G., Yang, J.: Interfacial stresses in beams and slabs bonded with a thin plate. ASCE J. Eng. Mech. **127**(4), 399–406 (2001)

Smith, S.T., Teng, J.G.: Interfacial stresses in plated beams. Eng. Struct. **23**(7), 857–871 (2001)

Taljsten, B.: Strengthening of beams by plate bonding. ASCE J. Mater. Civil Eng. **9**(4), 206–212 (1997)

Tounsi, A.: Improved theoretical solution for interfacial stresses in concrete beams strengthened with FRP plate. Int. J. Solids Struct. **43**, 4154–4174 (2006)

Tounsi, A., Hassaine Daouadji, T., Benyoucefa, S., Adda bedia, E.A.: Interfacial stresses in FRP-plated RC beams: effect of adherend shear deformations. Int. J. Adhes. Adhes. **29**, 343–351 (2009)

Yang, J., Chen, J.F., Teng, J.G.: Interfacial stresses in plated RC beams under arbitrary symmetric load: a higher-order closed form solution. In: Proceedings of the First International Conference on the Use of Advanced Composites in Construction, Southampton University, UK, pp. 153–163 (2002)

Evaluation of the Impact Sound Transmission Through Concrete Floor Using Numerical Simulation and Laboratory Experiment

Abdelouahab Bouttout(✉) and Mohamed Amara

National Center of Studies and Integrated Research of Building Engineering, CNERIB, Cité Nouvelle El-Mokrani, 16097 Souidania, Algeria
Bouttout@gmail.com, mamaral@yahoo.com

Abstract. The present paper examines the impact noise transmission through some floor building assemblies using different approaches. Firstly, the Acoubat software has been used to simulate the impact noise transmission through different floor configurations used in Algerian construction mode. The results are compared with the available measurements based on Brüel and Kjær 2260 sound level meter analyzer equipment. The results of the first method show that the different cases of floor configurations need some improvement to ensure the acoustic comfort in the receiving apartment. The addition of resilient underneath tile reduces the impact noise level from 67 dB to 57 dB. When the polyethylene and mineral wool are used, the results showed that the total level of the impact noise can be achieved $L'_{nT,w}$ = 56 dB and $L'_{nT,w}$ = 54 dB, respectively. In conclusion, the recommended value of the impact sound level in the receiving room should not exceed 58 db. Secondly, we have developed two experimental methods, (i) field method in multi-storey residential building, and (ii) laboratory method using the following equipment: sound level meter analyzer Brüel and Kjær 2270 with accelerometer Brüel and Kjær 5408, BZ 5503 software, and impact-tapping machine Brüel and Kjær 3204. The laboratory method is good way to estimate the weighted sound reduction index of resilient materials mixed with mortar or layer placed under tile. The results show that the weighted sound reduction index of resilient layer can reach 18 dB, 6.6 dB and 11.6 dB for the rubber, PVC and Bitumen layer, respectively. The important results obtained in this paper can be used as platform to improve the acoustic rehabilitation of the multi-storey residential building.

1 Introduction

Generally, the main interest in building regulation codes are focused on buildings structure stability against earthquakes and other natural disasters. The sustainable development which was introduced in the building sector, after petroleum crisis, leads the construction of multi-storey building with minimum energy consumption and sufficient thermal comfort for the occupants.

Pereira et al. [1] studied experimentally the noise reduction provided by floor coverings; using reduced sized slab following the procedure described in the ISO/CD 16251-1 document [2]. Many resilient coverings, floating floors and floating slabs are

B. Abdelbaki et al. (Eds.): SMSD 2017, *Proceedings of the Third International Symposium on Materials and Sustainable Development*, pp. 254–266, 2018.
https://doi.org/10.1007/978-3-319-89707-3_29

evaluated and the results are compared with those obtained using the standards EN ISO 140-8 [3] and ISO 717-2 [4]. The results show that the impact sound reduction provided by resilient coverings are close to those obtained using the methodology described in ISO 140-8 [3].

Rasmussen et al. [5] have investigated a comparison study of the legal requirements for sound insulation between dwellings in 24 European countries. They obtained significant differences in descriptors and the levels used by the European countries.

Schiavi et al. [6] investigated an experimental measurement of impact sound insulation of floor coverings with resilient layers according to the ISO 10140 standards series. They found that the accurate characterization of acoustical performances of resilient layers depends on several conditions, such as the state (clean or dirty) of the surfaces, and the fixation of the layer on the reference floor. They reported that these conditions are not stated by the standard ISO 10140.

Kim et al. [7] studied the manufacture of mineral damping materials using mineral binders and applied to reduce the effects of floor impact noises. Different configurations have been used depending on physical properties such as density, dynamic stiffness, and remnant strain. Their results show that the installation of damping materials between a concrete slab and a form concrete layer reduced very significantly the impact noises. Nevertheless, the damping materials currently used are mainly manufactured from organic materials with lower density and lower dynamic stiffness such as polystyrene (EPS). These products generate a floating structure due to heterogeneous material layers and cause problems such as resonance and amplification of some frequencies.

The low frequency vibro-acoustic characteristics of a massive floating floor with different resilient layer are investigated with experimental measurements and numerical simulations method by Cho et al. [8].

In Algeria, the main construction type is a multi-storey building, lacking sufficient acoustic insulation to dampen the noise transmission through the floors. The Algerian building acoustic regulations specifies the sound insulation calculation method for airborne noises in the "DTR 3.1.1" code.

The executive decree N° 93-184 of July 27th, 1993 [9] regulating noise emission states the following in its 2nd and third articles:

- **Article 2.** The maximum allowed noise levels in residential areas, roads and private or public places are 70 dB during the daytime (from 6 am to 10 pm) and 45 dB at night (from 10 pm to 6 am).
- **Article 3.** The maximum allowed noise levels in the immediate vicinity of hospitals, teaching, rest and relaxation areas as well as within their walls are 45 dB during the daytime (from 6 am to 10 pm) and 40 dB at night time (from 10 pm to 6 am).

In this paper, we examine both numerical simulation and experiments method to investigate the impact noise through floor assemblies. The main task is to develop a detailed method for acoustic evaluation behavior of the floor in multi-storey residential buildings. To our knowledge, there is no published work about the acoustical performance of the floor in residential multi-storey buildings in Algeria.

The paper is divided into six sections. After the Introduction (Sect. 1), mathematical formulations and numerical solution of the impact sound level are given in Sects. 2 and 3, respectively. The material and measurement methodology are presented in Sect. 4. Section 5 deals with the acoustic results and discussions. Finally, a conclusion is given in Sect. 6.

2 Mathematical Formulation

In multi-storey constructions, heavy-weight impact sounds caused by walking of the occupant, running or jumping are important sources of noise transmission through the floors junctions.

The impact sound insulation is calculated according to ISO 140-7 [5]. The evaluation of the sound transmission is divided in two quantities:

Normalized impact sound pressure level:

$$L'_n = L_i + 10\lg\frac{A}{A_0} \tag{1}$$

Standardized impact sound pressure level

$$L'_{nT} = L_i + 10\lg\frac{T}{T_0} \tag{2}$$

where; Li is the impact sound pressure level measured in the receiving room when the floor under test is excited by the standardized impact source (tapping machine or impact rubber balls), A is the equivalent sound absorption area in the receiving room, A_0 is the reference absorption area, $A_0 = 10\ m^2$. T is the reverberation time measured in the receiving room. T_0 is the reference reverberation time, $T_0 = 0.5$ s. A = 0.16 V/T, where V is the room volume in m^3.

3 Numerical Solution

The Acoubat software is used to calculate the acoustic behavior of the floor. This software is developed by CSTB center and is based on modeling the impact noise through a floor according to the method described in EN 12543-2 [10]. This standard describes the models used to calculate the impact sound insulation between apartments in multi-storey building based on the measured data characterizing the transmission of sound in the structural elements.

Acoubat software has a products database, where each product is disposed according to its function within the building such as masonry, covering layer, windows, door, ceiling, floor and others different construction materials. Figures 1 and 2 display the architectural sketch of the apartments under study.

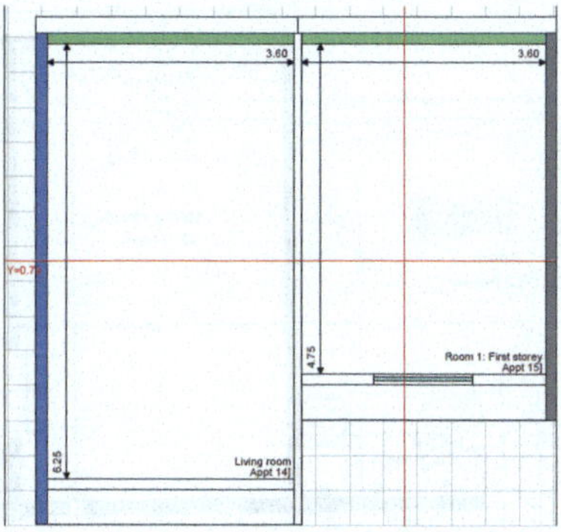

Fig. 1. Architectural sketch of the apartment under experiment.

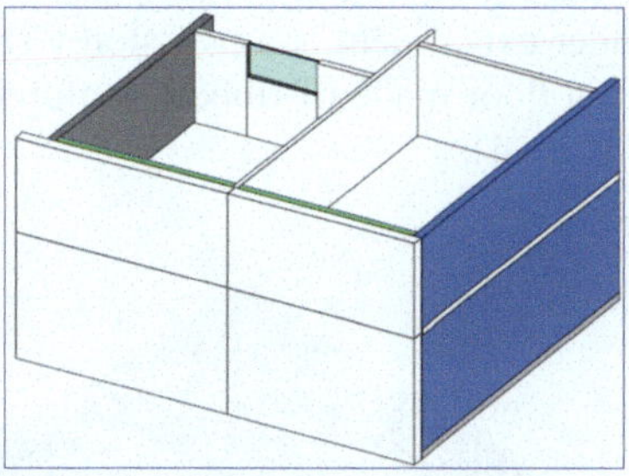

Fig. 2. Acoubat 3D model of the apartment under experiment.

Acoubat is a good tool for estimating different building acoustic descriptors such as airborne noise insulation, impact noise of the floor, noise equipment, and reverberation time of classrooms.

4 Experimental Work

4.1 Field Measurements

The experimental works are examined by national center of building researchers in different cases: projects research and in situ expertise [11]. The measurements are investigated in second storey apartment of residential building situated in Algiers using Brüel and Kjær sound level meter.

The impact sound is the sound level in the receiving room from standardized tapping machine containing five steel hammer of 0.5 kg weight, which is placed in the source room to simulate footsteps of the occupant.

The first measurements are made in five-storey residential building situated in the Algiers region. The floor is composed from different layers: 15 cm of concrete, 2 cm of sand, and 2 cm of tile.

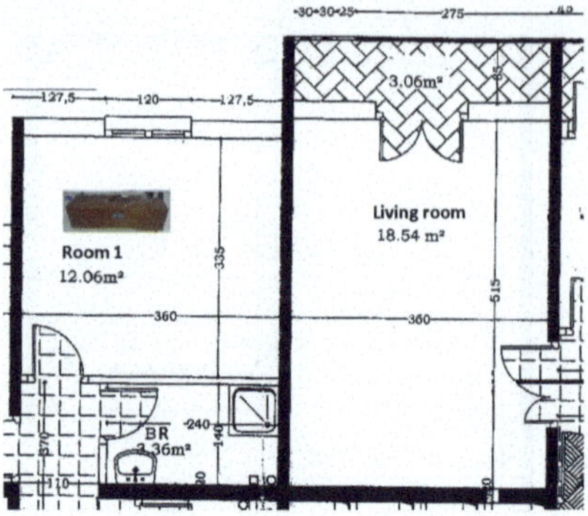

Fig. 3. Detailed plan configuration of the apartments under experiment [11].

Figure 3 illustrates the plan configuration of the apartment under investigation [11]. The second building under experiment is a two-storey residential building situated in the north of Algiers with a floor made of ribbed reinforced concrete and expanded polystyrene hollow blocks (Fig. 4).

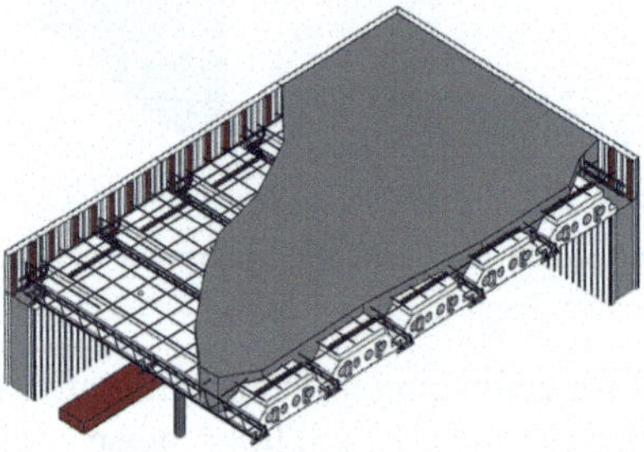

Fig. 4. Detailed of the floor of the second apartments under experiment.

Sound pressure levels obtained at different microphone positions must be averaged for all positions of the tapping machine using:

$$L = 10\lg\left(\frac{1}{n}\sum_{j=1}^{n} 10^{Lj/10}\right) \quad (3)$$

where; Lj is the sound pressure levels L_1 to L_n, n is the number of the different positions of the microphone in the receiving room.

4.2 Laboratory Measurements

The laboratory measurements are made on building elements such as wall panels, windows, and floor cover mounted in special test assemblies. Figure 4 illustrates the first step in the construction of a reinforced concrete floor to be used as an acoustic performance cover specimen. The measurements are done under a uniform and strict procedure [2]. The results are used by manufacturers to document the performances of their products.

A laboratory experimental procedure is developed in the national center for building research (CNERIB), it aims at measuring the impact sound level in residential building with the following equipment: the sound level meter analyzer Brüel and Kjær 2270 with BZ 5503 software and impact-tapping machine Brüel and Kjær 3204. The equipment is illustrated in Fig. 5.

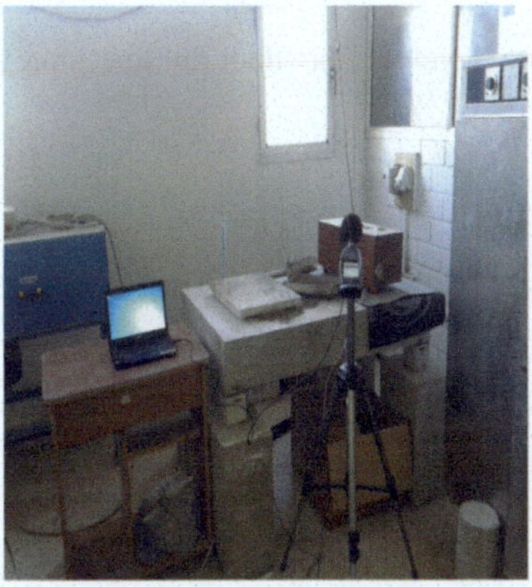

Fig. 5. Detailed reinforced floor (1.2 m × 0.8 m × 0.2 m) and acoustic equipment used for the measurements. Sound level meter Brüel and Kjær 2270 with accelerometer and tapping machine.

5 Results and Discussions

Measurements in the field are performed in situ on partition of buildings. The results are used to document conformance to building regulation. In the field measurements, sound and vibration propagate not just via the partition under investigation, but also via other partitions, structures, and leaks. This propagation is called flanking transmission.

The impact sound level for the vertical transmission direction according to the model of the Acoubat is $L'_{nT,w}$ = 67 dB (see Fig. 6). However, the value in the diagonal transmission is less than the one in the vertical direction, i.e. $L'_{nT,w}$ = 59 dB. The measurements reported in [11] showed that the global sound level is 66 dB (see Fig. 7). The difference between measurements in situ and numerical simulation with Acoubat software is 1 dB. There exist a good agreement between numerical simulation and experimental measurements.

The noise radiated by the structural element can be evaluated as the sum of the acoustic transmission noises in various ways. Each way can be identified by the element ***i*** for which the noise is incidental in the emission room and the element ***j*** radiating the noise in the received room [4].

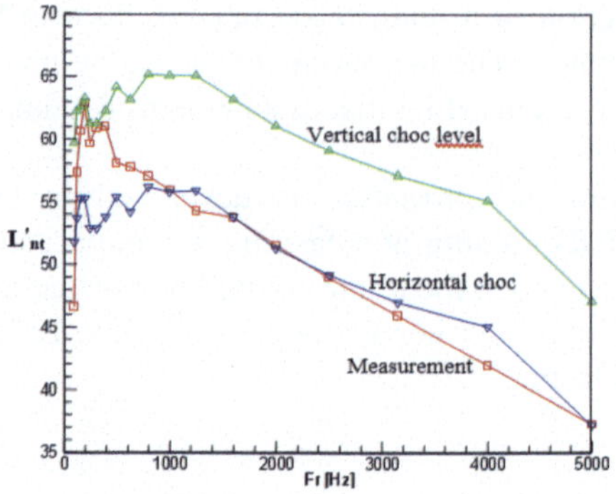

Fig. 6. Impact sound pressure level of the floor in five storey residential building in the horizontal and vertical transmission ways.

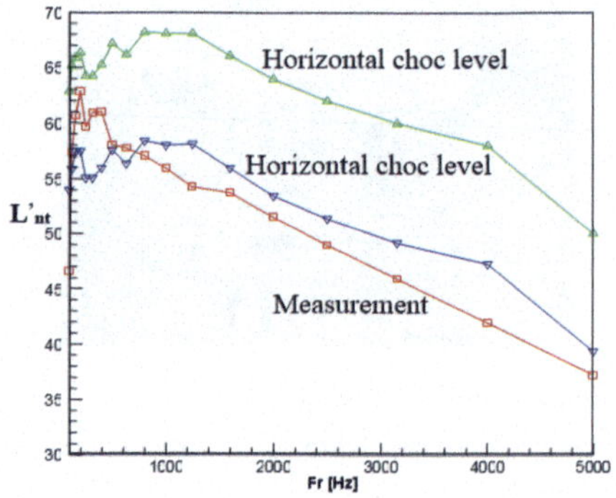

Fig. 7. Impact sound pressure level of the floor in five-storey building in the horizontal and vertical transmission ways when the dimensions of the façade are changed.

The dimensions of the facades play a significant role in the values of the sound insulation. Figure 6 shows the impact sound levels when the dimensions of the facade are changed. Generally, in the bottom room the noise levels are expressed relative to the equivalent absorption area measured in the receiving room, expressed in m^2 or the reverberation time (see Eqs. (1) and (2)).

In the second apartment which is located in a two-storey building, the measurements showed that the total level of the impact noise obtained is 79 dB(A), which is not in agreement with the recommended value, 58 dB(A) (Fig. 8). The impact sound level in vertical transmission direction simulated by Acoubat software is 76 dB(A). However, the value in the diagonal direction is 67 dB(A).

Numerical simulations were carried out in floor made of (bottom to top) 10 cm concrete, then 22 cm hollow block and a tile covering of 5 dB(A) acoustic reduction.

Figures 9, 10, 11 and 12 illustrate the numerical simulation of the impact sound pressure level through some floor with different resilient materials. In the first case: (from bottom to top) 17 cm of concrete, underneath layer of 16 dB acoustic reduction

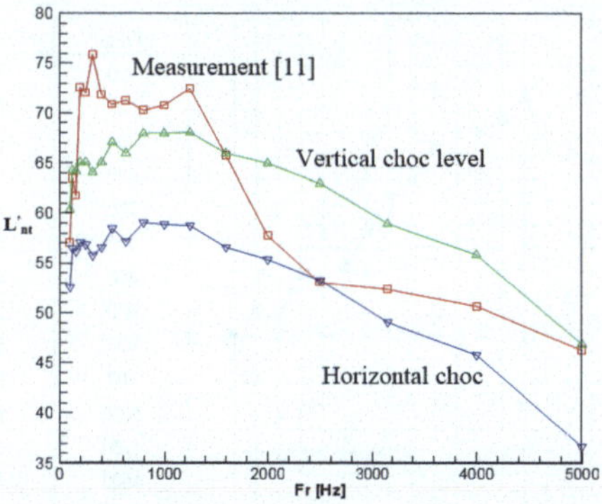

Fig. 8. Impact sound pressure level of the floor in two storey residential building in the horizontal and vertical transmission ways.

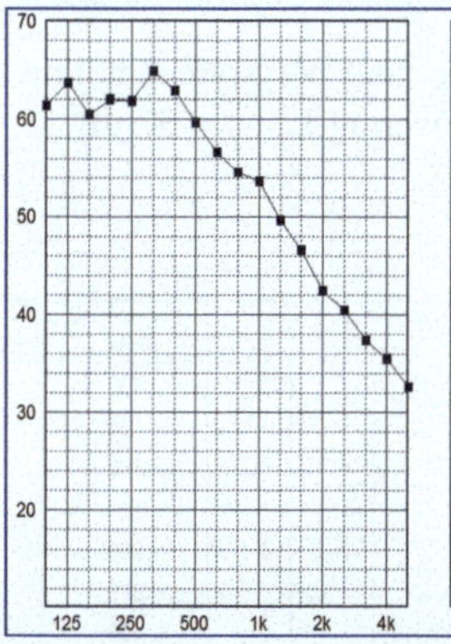

Fréq.	L'nT
100	61.4
125	63.7
160	60.4
200	62.0
250	61.8
315	64.9
400	62.9
500	59.7
630	56.7
800	54.6
1000	53.6
1250	49.6
1600	46.6
2000	42.5
2500	40.4
3150	37.4
4000	35.5
5000	32.6
Hz	dB

Fig. 9. Impact sound pressure level of the floor: first case (from bottom to top): 17 cm of concrete, underneath layer of 16 dB acoustic reduction and then a tile covering. $L'_{nT,w} = 57$ dB.

and then a tile covering. In the second case: (from bottom to top) 17 cm of concrete, polystyrene hollow blocks, and then a tile covering.

Figures 10 and 11 show the impact sound noise level in the case of the hollow block with polyethylene and mineral wool resilient materials respectively.

The global value obtained is 57 dB calculated according to ISO 717-2 [4] for two floors. The allowable sound level was found to be 58 dB according to the European regulation. In addition, the maximum of light-weight impact sound and heavy-weight impact sound must be lower than 58 dB and 50 dB, respectively, in other international building codes [12].

Resilience is defined as the capacity of a material to absorb energy when it is deformed elastically and then, upon unloading to have this energy recovered. It is the maximum energy per volume that can be elastically stored. It is represented by the area under the curve in the elastic region in the stress-strain diagram.

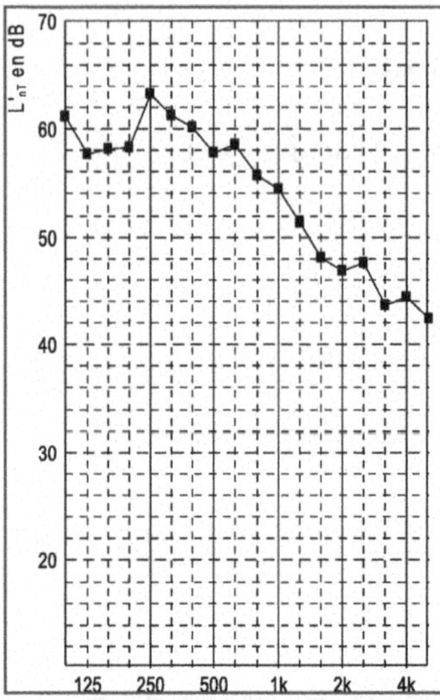

Fréq.	L'nT
100	61.2
125	57.7
160	58.2
200	58.2
250	63.2
315	61.2
400	60.1
500	57.8
630	58.6
800	55.7
1000	54.5
1250	51.4
1600	48.1
2000	46.9
2500	47.6
3150	43.6
4000	44.4
5000	42.3
Hz	dB

Fig. 10. Impact sound pressure level of the floor polystyrene hollow block, underneath layer of 16 dB acoustic reduction and then a tile covering $L'_{nT,w}$ = 57 dB.

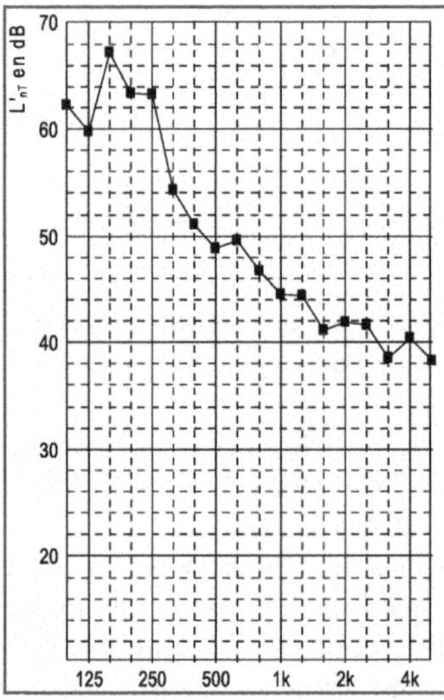

Fréq.	L'nT
100	62.3
125	59.7
160	67.3
200	63.3
250	63.3
315	54.3
400	51.1
500	48.8
630	49.6
800	46.7
1000	44.5
1250	44.4
1600	41.2
2000	41.9
2500	41.6
3150	38.6
4000	40.4
5000	38.3
Hz	dB

Fig. 11. Impact sound pressure level of the floor hollow block, mortar underneath layer of polyethylene of 16 dB acoustic reduction. $L'_{nT,w}$ = 56 dB

In case of floating floors, resilient materials are placed between concrete or wooden slab and finishing materials.

The resilient materials reduce impact sound through a floor by reducing the vibration generated whenever a body hits the floor [12]. The floor impact sound reduction increases with the decrease of dynamic stiffness of resilient material [12].

Results obtained in the literature indicate that dynamic stiffness, as a physical property of resilient materials, decrease as the thickness of the resilient materials increase.

Widely used resilient materials are made of EPS (Expanded Polystyrene), waste urethane series, EVA (Ethylene Vinyl Acetate), EPE (Expanded Polyethylene), EPP (Expanded Polypropylene), glass fiber and rock wool, waste tires, compressed polyester [12].

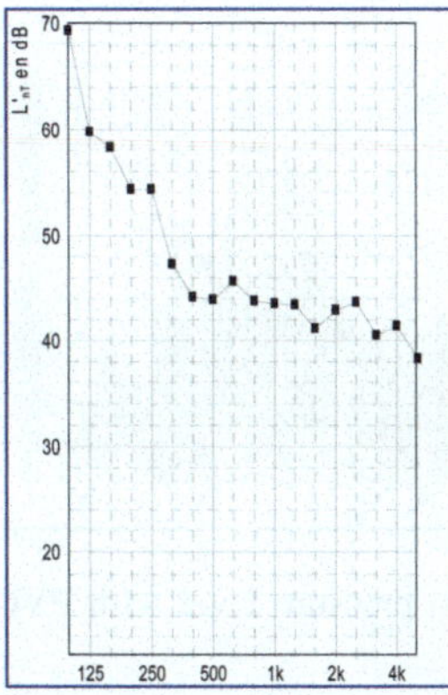

Fréq.	■ L'_{nT}
100	69.3
125	59.7
160	58.3
200	54.3
250	54.3
315	47.3
400	44.1
500	43.8
630	45.6
800	43.7
1000	43.5
1250	43.4
1600	41.2
2000	42.9
2500	43.6
3150	40.6
4000	41.4
5000	38.3
Hz	dB

Fig. 12. Impact sound pressure level of the floor hollow block, mortar underneath layer of mineral wool of 20 dB acoustic reduction. $L'_{nT,w} = 54$ dB.

Figures 13 and 14 show the impact signal of the tapping machine measured with sound level meter analyzer Brüel and Kjær 2270 and accelerometer Brüel and Kjær 5408 placed on the upper and lower surface of the concrete floor. The value of the impact sound level obtained is 103.0 dB on the upper surface without covering and it is 96.4 dB measured in the lower surface with PVC layer. The value of ΔL is **6.6 dB.**

The same results for the rubber layer are illustrated in Figs. 15 and 16. The value of ΔL is **18 dB**.

In the reinforced concrete apartment buildings, the floor impact sound from the floor above can be easily transferred to the floor below [12]. Due thin floors, neighboring occupants can hear conversations from the neighboring flats.

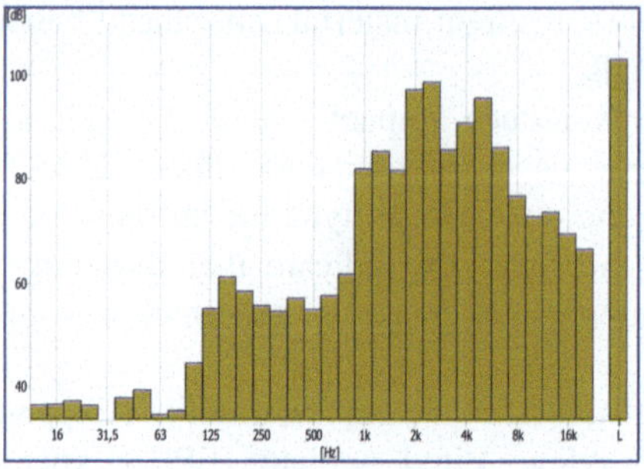

Fig. 13. Impact sound pressure level without covering.

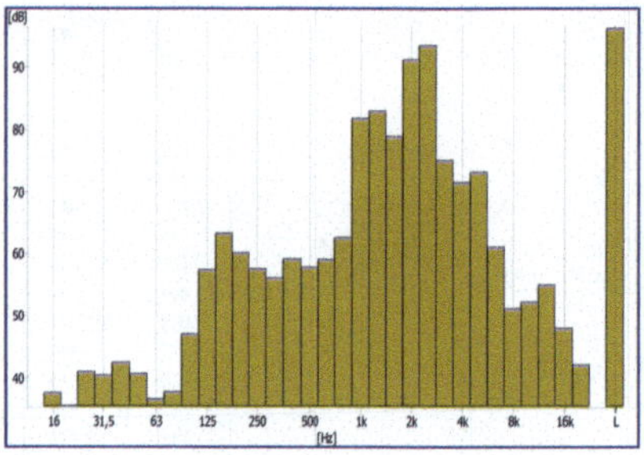

Fig. 14. Impact sound pressure level with PVC layer.

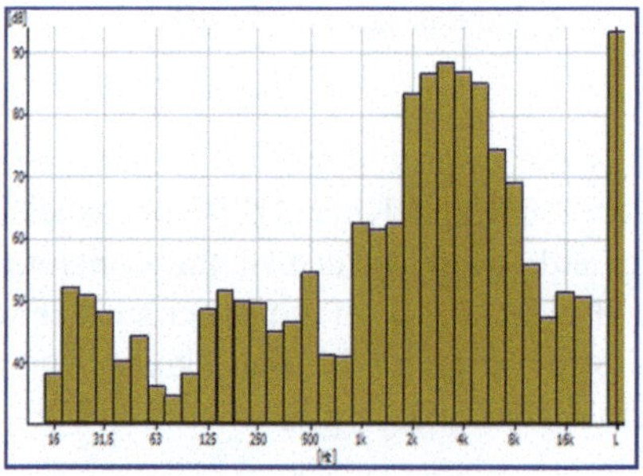

Fig. 15. Impact sound pressure level without covering near support.

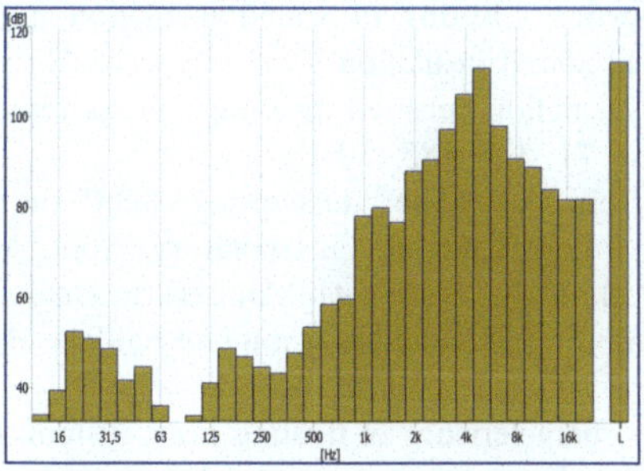

Fig. 16. Impact sound pressure level without covering near support with rubber layer

6 Conclusion

- The simulations display that the values of the simulated model with ACOUBAT software are close to the experimental results for frequencies below than 1000 Hz, but there exist a discrepancy for frequencies higher than 1000 Hz.
- The impact sound level of the first apartment for the vertical transmission direction according to the model of the Acoubat is $L'_{nT,w}$ = 67 dB. However, the value in the diagonal transmission is less than the one which is in the vertical direction, i.e. $L'_{nT,w}$ = 59 dB. The measurements showed that $L'_{nT,w}$ = 66 dB.
- For the second apartment which is located in two-storey building, the total level of the impact noise is $L'_{nT,w}$ = 79 dB, which is not with agreement with recommended value, i.e. $L'_{nT,w}$ = 58 dB.
- The addition of resilient underneath tile reduces the impact noise level from 67 dB to 57 dB.
- The weighted sound reduction index of resilient layer in laboratory method can reach **18 dB**, **6.6 dB** and **11.6 dB** for the **rubber**, **PVC** and **Bitumen** layer, respectively.

Acknowledgment. The authors gratefully acknowledge the financial support of this work (VIBROACOU project) by Algerian Ministry of Housing, Urbanism and the City.

References

1. Pereira, A., Godinho, L., Mateus, D., Ramis, J., Branco, F.G.: Assessment of a simplified experimental procedure to evaluate the impact sound reduction of floor coverings. Appl. Acoust. **79**, 92–103 (2014)
2. ISO 16251-2014: Acoustics-Laboratory measurement of the reduction of transmitted impact noise by floor coverings on a small floor mock-up – Part 1: Heavyweight compact floor
3. ISO 140-8:1997 Acoustics-Measurement of sound insulation in buildings and of building elements – Part 8: Laboratory measurements of the reduction of transmitted impact noise by floor coverings on a heavyweight standard floor

4. ISO 717-2: 2013 Acoustics - Rating of sound insulation in buildings and of building elements - Part 2: Impact sound insulation
5. Rasmussen, B.: Sound insulation between dwellings—requirements in building regulations in Europe. Appl. Acoust. **71**, 373–385 (2010)
6. Schiavi, A., Prato, A., Belli, A.P.: The "dust spring effect" on the impact sound reduction measurement accuracy of floor coverings in laboratory. Appl. Acoust. **97**, 115–120 (2015)
7. Kim, H.-S., Kim, B., Park, S.-K., Lee, S.-H.: An experimental study on the effects of floor impact sounds from damping materials using mineral binders. Build. Environ. **89**, 244–252 (2015)
8. Cho, T.: Vibro-acoustic characteristics of floating floor system: the influence of frequency-matched resonance on low frequency impact sound. J. Sound Vib. **332**, 33–42 (2013)
9. Algerian regulation Law: The executive decree N° 93-184 of 27 July 1993 of the regulation of the noise emission
10. EN 12543-2: Building acoustics – Estimation of acoustic performance of buildings from the performance of elements – Part 2: Impact sound insulation between rooms
11. Abdelghani, G., Mohamed, A.: Mesures in-situ et identification des défauts pour des bâtiments typiques. Phase 2 document interne CNERIB (2004)
12. Lee, J.Y., Kim, J., Chang, H.J., Kim, J.M.: Long-term structural behavior of resilient materials for reduction of floor impact sound. Int. J. Civ. Environ. Struct. Constr. Archit. Eng. **8**(5), 509–512 (2014)

An Overview of Functionally Graded Materials «FGM»

Adda Hadj Mostefa[1(✉)], Slimane Merdaci[2(✉)], and Noureddine Mahmoudi[1,2]

[1] Laboratoire des Matériaux Procédés et environnement (URMPE) université de Boumerdes, Département de Génie Civil, Institut des Sciences and Technologie, Centre universitaire de Rélizane, Rélizane, Algeria
addahadjmostefa@yahoo.fr

[2] Université Djillali Liabès, Laboratoire des Structures et Matériaux Avancés dans le Génie Civil et Travaux Publics, 89 Cité Ben M'hidi22000, Sidi Bel Abbés, Algeria
slimanem2016@gmail.com

Abstract. In this paper, we have presented an overview on functionally graded materials FGM. An overview has been exposed on the development of FGM, their concepts, their properties; their main manufacturing steps. Significant advanced techniques in manufacturing and processing over the last decade have produced FGM using processes that offer great latitude in making microscopic and material composition. In the future, the FGM with complex shapes and properties, including two and three-dimensional gradient will be produced using the production, a technique that requires the computer. This potential means that the designer is no longer limited to a range of existing homogeneous materials, although many studies have focused on the analysis of this material, engineers and other professionals engaged in the design process with FGM. Use of the FGM seems to be one of the most effective materials in the realization of the sustainable development in the industries. The paper will focus light on above issues and each plays an important role within the FGMs literature and ultimately influences on planning and development practices. It is expected that this comprehensive contribution will be very beneficial to everyone involved or interested in FGMs.

Keywords: Graded materials · Concept · Preparation
Application areas of FGM

1 Introduction

The Functionally Graded Materials **FGM** represent the latest developments; the revolution and the design of these materials have been defined in the 21st century. They are developed for their excellent mechanical properties. These materials have high performance and capable of resisted to ultra high temperatures.

The Functionally Graded Materials **FGM**, are composite materials produced by changing continuously the volume fractions in the thickness direction for a definite profile Fig. 1. These types of materials, have attracted much attention recently because of their ability to reduce the disparity in material properties and reduce thermal stresses

B. Abdelbaki et al. (Eds.): SMSD 2017, *Proceedings of the Third International Symposium on Materials and Sustainable Development*, pp. 267–278, 2018.
https://doi.org/10.1007/978-3-319-89707-3_30

[21]. The continuously variable mechanical property gives the material an optimized behavior. **FGM** are particularly used in high technology applications: aeronautics, aerospace, nuclear, semiconductor, and Civil Engineering and also include biomedical applications [6].

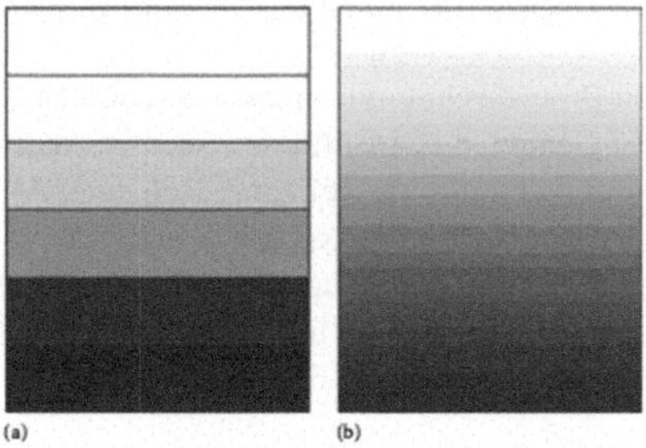

Fig. 1. Model of FGM.

2 Concept of Functionally Graded Materials FGM

The walls spacecraft constituent materials (space shuttle or hypersonic aircraft) are subjected to high temperatures. The most exposed parts are the inlet cone, the leading edges of the wings as well as certain lower surfaces. For this reason the wall materials, submitted on one side to about 1800 °C in the atmosphere, have to bear in their thickness of ten millimetres, a thermal gradient of about 1300 °C. There are no monolithic materials able to withstand such thermal stress [12].

The proposed solution is the implementation of composite materials, particularly the use of assessed gradient materials. One can imagine a material having a face exposed to high temperature would possess resistance properties to high temperatures and oxidation resistance, such as ceramic, and whose inner surface is very good conductor of heat and would possess a good mechanical strength and better toughness, such as metal.

However, if we consider a simple assembly of these two materials, they immediately have a break due to thermal stresses at the interface between the two types of materials having different thermal properties. Ideally remove this interface creating a seamless transition between the two sides.

Thus was born the concept of estimated gradient material in the 1980 by a group of researchers at the National Aerospace Laboratory (National Aerospace Laboratory, STA) in Japan. **FGM** is the combination of two materials with different structural and functional properties with an ideally continuous transition of the composition of the structure and distribution of pores between these materials.

Generally, **FGMs** are materials made up of several materials to mechanical characteristics containing different components such as ceramics and metals. The voids are also considered as a component of **FGM** materials [8]. So they are composite with inhomogeneous macroscopic characteristics.

Most **FGM** are composed of ceramics and metals, the mechanical properties are compared in Table 1.

Table 1. Comparison of properties of ceramics and metals.

The high temperature side	Ceramic	- *Good heat resistance* - *Good resistance to oxidation* - *Low thermal conductivity*
Continuity of material from one point to another "intermediate layers"	Ceramic-metal	- Knock of interface problems - Relaxes Thermal stresses
The low temperature side	Metal	- Good mechanical strength - High thermal conductivity - Very good toughness

The continuous change in the composition and therefore the microstructure of a **FGM** material is illustrated in Fig. 2, this results in a gradient which determine the properties of the **FGM**. In some cases, there may be a **FGM** comprised of a same material but of different microstructure [3].

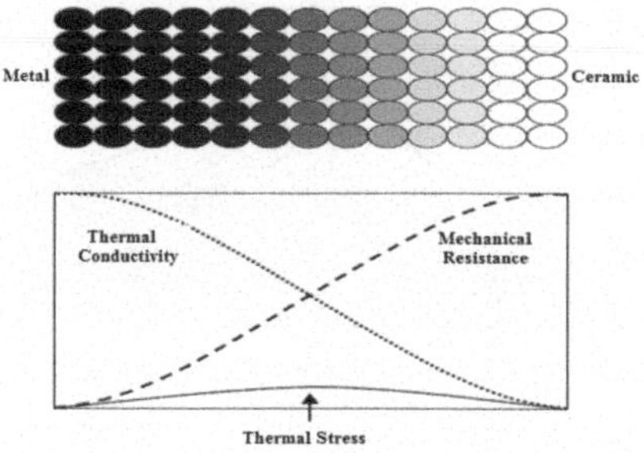

Fig. 2. The change in the composition of FGM.

Figure 3 shows the stress concentrations in the thermal barriers at the interfaces (sudden change in composition). This also shows how a **FGM** can lean these stress concentrations by gradually changing the material properties and always ensures thermal protection.

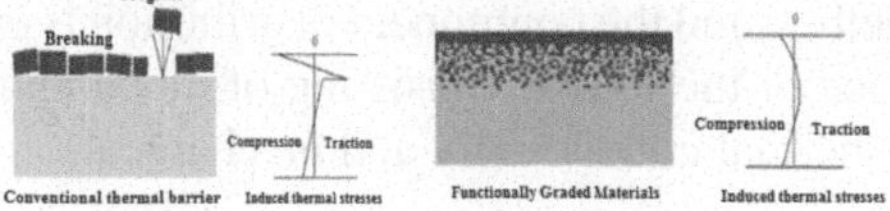

Fig. 3. Thermal protection.

3 Comparison of FGM Type Traditional Materials and Composite Materials

Generally, **FGMs** are materials made of different materials containing different components such as ceramics and metals. So they are composite with inhomogeneous macroscopic characteristics. The continuous change in the composition and therefore the microstructure of the material in the **FGM** distinguishes conventional composite materials as shown in Fig. 4 [9]. The result is a gradient determined by the material properties of the **FGM**. In some cases, there may be a **FGM** comprised of a same material but of different microstructure [3] (Fig. 5).

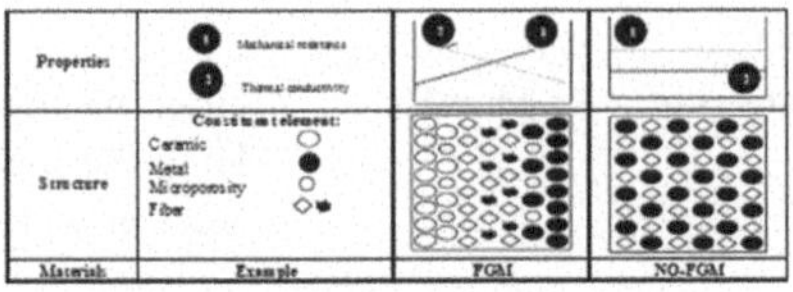

Fig. 4. Characteristics of composite materials FGM.

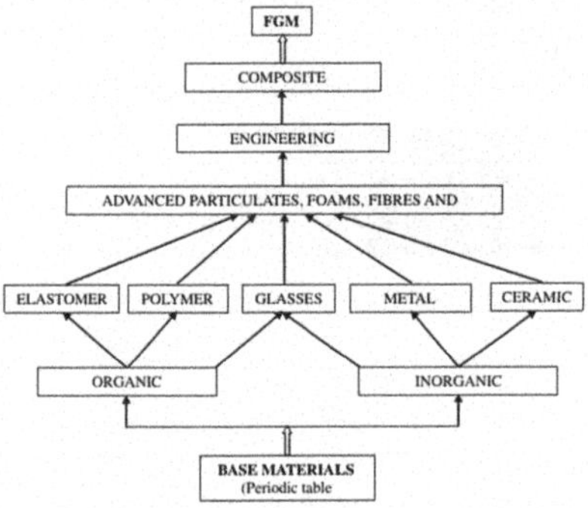

Fig. 5. Modern Hierarchy of FGM material.

Yoshihisa [20] to establish a simple model illustrating the differences between the estimated gradient materials **FGM** and conventional materials (Fig. 6). The planar material compound to a flat characteristic and the bound material has a border on the interface of two materials. **FGM** have excellent characteristics that differ from the compounds and related materials shots. Therefore, **FGM** attract attention in terms of their application in industrial areas. Since **FGM** have dual ownership of the two raw materials are mixed together, and the component distribution is graduated without interruption. For example, one of the **FGM**, consisting of metal and ceramics together the characteristics of good thermal conductivity and mechanical.

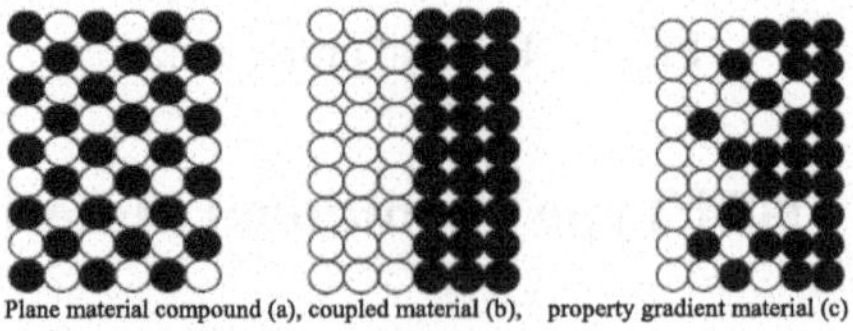

Fig. 6. The distribution component materials.

4 Review of the Methods of Preparation of FGM

The methods of manufacturing a rated gradient material can usually be divided into the building structure in a heterogeneous space (graded mixture) and transformation of this structure from a single material (solidification). The gradual mixing process can be classified as constituents, homogenization and segregation. Basic methods are based on the manufacturing step by structure precursor's gradual materials or powders. The advanced automation technologies in the last decades have rendered elementary process of technologically and economically sustainable growth. In the homogenization procedure that treats a sharp interface between two materials is converted into a gradient by transporting equipment. The methods of homogenization and segregation produce a continuous gradient, but have limitations regarding the types of gradients which can be produced.

Typically the drying and solidification of the following steps graded mixture. The need for these consolidation processes must be adapted to **FGM**. Particular attention should also be paid to the uneven shrinkage of **FGM** during consolidation [10].

In recent years, the work carried out by the researchers it possible to develop a novel method to prepare composites of continuous composition gradient. This method is based on co-sedimentation technique medium powders proportional to the material density and the square of the particle diameter. By controlling and adjusting the particle size distributions of each powder, it is possible to achieve different gradients of concentration in the deposit formed after sedimentation.

There are many development assessed gradients material methods, the most widely used techniques are explained briefly below,

4.1 Pouring in Band (Tape Casting or Doctor-Blade)

The tape casting a slurry of fine powders in aqueous or non-aqueous suspension (most commercial work using non-aqueous process) on a flat support in thin, even layers in Fig. 7. The products are sheets with controlled thicknesses (25–1000 μm). After strengthening of the dough, the sheets are removed from the mold and then cut.

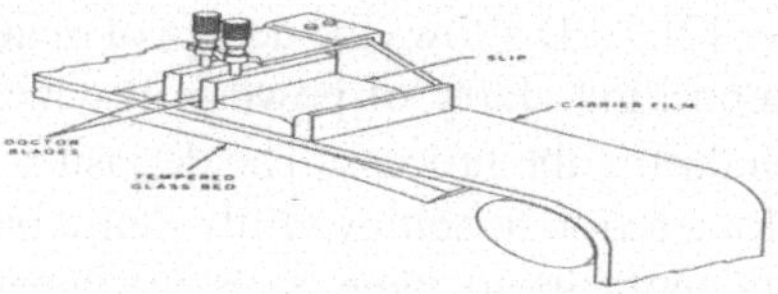

Fig. 7. Principle of the method in strip casting [17].

The solvent should have a low viscosity and a low boiling point. It should be soluble with the binder, the plasticizer and other additions but must be neither soluble nor reactive with the ceramic powder. The binder gives a high strength to the raw product by allowing its handling. Usually a plasticizer is added to the binder to decrease its viscosity. The binder, plasticizer and deflocculant must be fully released during fellmongery.

The strip casting process is widely used for producing composite materials laminar following two methods: either by direct realization of multilayer bands thanks to a system of multiple blades, this is the case of three-layer prepared by Mistler [11]; either by stacking layers developed separately, the cohesion of which is then provided by a thermo-compression step [3, 4].

4.2 Sequential Coupling in Barbotine (Slip Casting)

Slip casting consists in pouring slurry into a porous mold which will drain the liquid through capillary forces, leaving a shard (compact powder layer) on the mold surface. After drying, believed in the body is obtained.

So casting is divided into two essential steps:

- Training shard or "take";
- Consolidation shard or "stronger".

Filtering, that is to say the formation of the ceramic material during the casting can be considered a process of elimination of a portion of the water from the slurry; this water migrates through the shard layer already formed due to:

* The suction power of the platform (classical casting [12]);

* Or pressure applied to the slip (diecast).

In the case of the manufacture of multilayer, after formation of the first ceramic material deposition of the second layer is carried out such that the slurry does not penetrate the ceramic material formed. This process is successively repeated for the other layers.

4.3 Plating Electrophoresis

Another process is added to different techniques of manufacture of **FGM**, that deposition by electrophoresis call (EPD) is a fairly quick process at low cost, capable of producing the properties of gradient materials changing continuously with complex geometry. EPD is composed of two processes, that is to say the movement of charged particles suspended in an electric field between two electrodes (electrophoresis) and the deposition of particles on one of the electrodes [2]. EPD allows the design of materials with binary gradient plate-like properties by depositing slurry of powder in which a second suspension is added without interruption during the process. The deposit is a narrow packed powder contract that requires agglomeration to achieve fully dense hardware components.

Therefore, many IPAD applications have been developed for the manufacture of ceramics, including the application of gradients evaluated materials can be obtained such that the ceramic-ceramic ZrO_2/Al_2O_3, and metal ceramic, WC/Co. A general system of this principle is illustrated in Fig. 8.

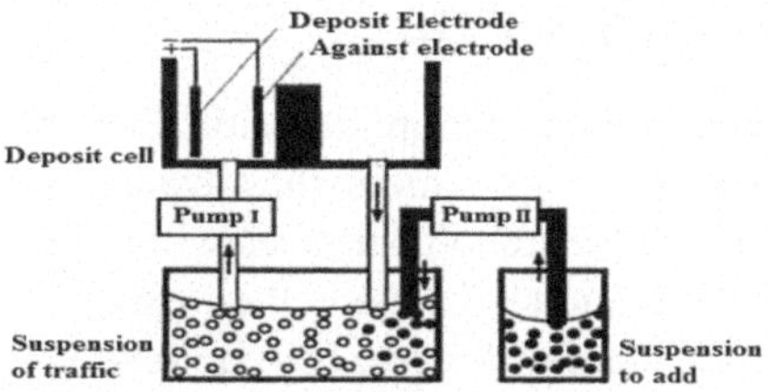

Fig. 8. The installation of EPD for the production of materials.

4.4 Dry Powders the Compaction

In this technique the powders are successively poured into a steel mold. Whenever a powder is poured, low compression is applied. Follows the compaction of all the layers will be performed. This process is followed generally by isostatic pressure and debinding. Densification is finally the final stage [5].

4.5 Plasma Projection

Plasma spraying involves introducing solid particles (of size less than 100 μm, approximately) in a jet of gas, the temperature usually reaches 11,000 °C and the speed from 1000 to 1500 m/s. The deposit is formed by the stacking of solidified shavings from the impact of particles on the previously prepared substrate [18]. This method is used in particular for depositing high melting temperature materials such as heat resistant alloys or ceramics. The characteristics of plasma spray deposits are achieved by the resistance to wear and corrosion as well as thermal and electrical insulation.

A detailed search was performed on the microstructure and property control of fine particles sprayed by plasma. Plasma spraying is a rapid process and provides film formation over a large area, but usually has a disadvantage with its rough microstructure [15]. Ichikawa [7] used the mechanically alloyed powders of thermoelectric materials, and managed to properly control the deposit content in the silicon (SiC) to form a thermoelectric material carrier concentration graduate with a homogeneous microstructure. Its spray pattern is shown schematically in Fig. 9.

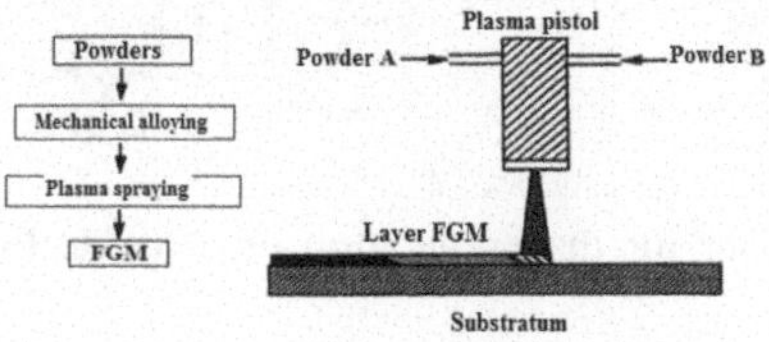

Fig. 9. Plasma projection to form a graded thermoelectric material.

4.6 Physical Vapour Deposition (PVD)

The technique of physical vapor deposition is a critical technology (promising) for the production of thermal barrier coatings. Among the advantageous characteristics of PVD is that the coatings produced have smooth surfaces which do not require the closing of cooling holes. However, the main advantage is their exceptional resistance to thermal shock, and the lifetime is considerably longer. The equipment used to deposit the thermal insulating coatings via PVD, consists of a plurality of containers for loading, preheating, and depositing a further rotating device and handling of samples… etc. [13].

Another approach to obtain a gradient structure is to use a mixture of aluminum, alumina, and zirconia, chemically controlled can be manufactured easily by spraying from multiple sources using one or more electronic guns. By adjusting the electron beam parameters, different deposition rates can be obtained for each source [13]. If the evaporation rate of the two components, for example alumina and zirconia, are changed continuously, a defined composition gradient is obtained, the **FGM**. For thermal barrier systems more biased to the oxidizing environment, such as moving blades of aeronautical or land turbines are, however, not routinely prepared by thermal spraying; in the case of yttria-stabilized zirconia ZrO_2 the another different method is also used. In this case, it is the technique of electron beam evaporation (EB-PVD) which leads to microstructures and thermo mechanical properties quite different from those obtained by the technique of thermal spraying [14].

4.7 Chemical Vapor Deposition (CVD)

In the art of chemical vapor deposition, shown schematically in Fig. 10, a deposit is formed on a substrate by subjecting the source of gas (e.g., hydrides, bromides or chlorides) occupy the room reaction to different types of energy such as heat, light, and plasma [13].

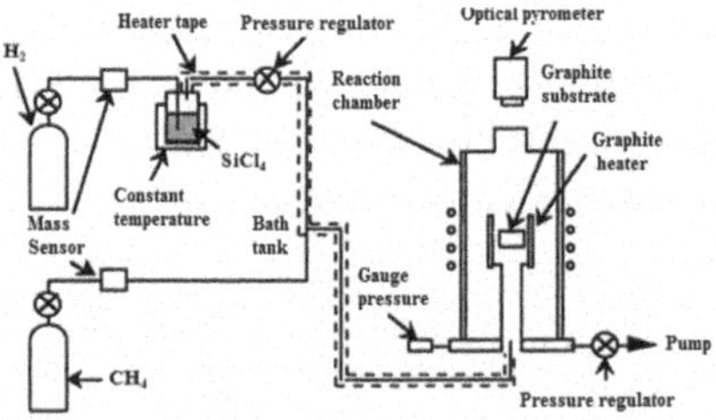

Fig. 10. Diagram of the plant for the preparation of a SiC/C FGM by the chemical vapor deposition (CVD).

FGMs can be synthesized slowly to moderate deposit rates by changing the mixing ratio of the source gases, or by controlling the deposition temperature, the gas pressure or gas flow. Because of its low deposition rate, CVD is widely used for infiltration of cells or for the manufacture of thin layers.

A typical example of a CVD **FGM** application is the formation of a graded coating of silicon carbide SiC on graphite wherein there is a continuous transition SiC of C. Compositions controlled source of mixtures of gases (CH_4–$SiCl_4$–H_2) are used for deposition temperatures between 1400 and 1500 °C and pressures between 1.3 and 6.5 kPa.

4.8 In Infiltration Sintering

This technique consists of two steps and is suitable for the manufactures of a function gradient composite compound of two materials having melting temperatures are very different. The first step is making a sintered body of high melting temperature material with a porosity gradient. The second is to fill these pores with infiltration by the second molten material. The result is excellent for reducing the thermal stress [16]. This technique can be generally applied for several combinations of materials which are chemically inert and which have very different melting points from each other reports.

4.9 Differential Laser Sintering

The laser thermal radiation allows the temperature control and the focusing point to be heated. The difference in the intensity of the radiation at different points of the material causes a differential sintering along the work piece, which results in different microstructures, depending on the position of the irradiated spot. YUKI, used this technique to develop an **FGM** PSZ/Mo. Figure 11 shows schematically the arrangement of the process used [19].

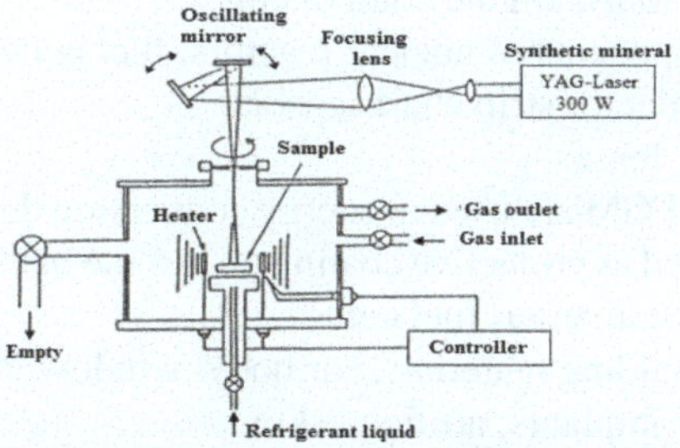

Fig. 11. Layout Process Laser Sintering Differential.

4.10 Ion Implantations (Sputtering)

This is an advanced technique for producing **FGMs** to only produce thin thicknesses (< 1 µm) on various substrates (plastic, ceramic, and metals). The treatment is carried out by energy ion beams or via reactive gas. The fabricated layers present a composition gradient can be very finely controlled [1].

5 Application Areas of the FGM

The concept of assessed graded materials **FGMs** is applicable in numerous and broad ranges of areas such as building materials, energy conversion materials, aeronautics, aerospace, nuclear and semiconductors. Thus, these materials have two conflicting properties: thermal conductivity and thermal insulation. Currently, they allow the production of lightweight, tough and durable (Fig. 12).

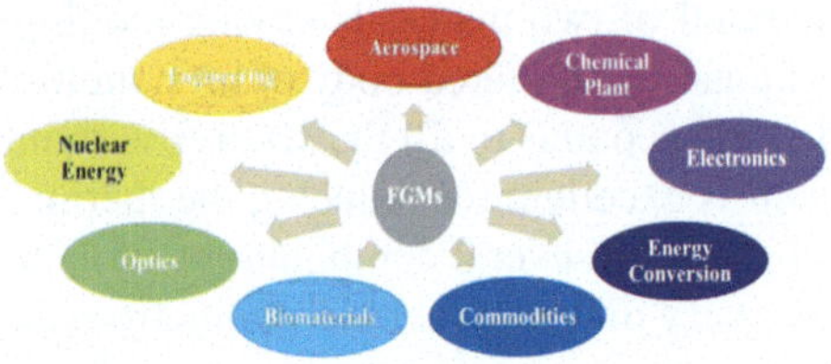

Fig. 12. Various fields of Application of FGMs.

The main application areas of **FGMs**:

- **Civil Engineering**: box girder bridges, chimneys, elastic foundations (such plate structures can be found in various types of industrial applications such as riffles), storage tanks, pools.
- **Aeronautics**: rocket engine components, body space planes.
- **Chemical**: the heat exchanger, heat pipe, reaction vessel.
- **Engineering**: cutting tools, engine components.
- **Nuclear energy**: components of nuclear reactors, fuel pellets.
- **Electronics**: semiconductor strip chart sensors.
- **Optics**: optical fibers, lenses.
- **Energy Conversion**: FGM are used in energy conversion devices. They also provide thermal barrier and used as protective coating on turbine blades in gas, thermoelectric generator, thermionic converter, fuel cell.
- **General products**: building materials, car body, window glasses.
- **Biological materials**: implants, artificial skin,
- **Biomaterials:** artificial bones joist, teeth, cancer prevention.
- **Power plant**: thermal barrier coatings, heat exchanger tubes.
- **Manufacturing**: machine tools, forming and cutting tools, metal casting and forging processes.
- **Applications of FGM as Thermal barrier**: One of the salient applications of FGM is in thermal Barrier Coatings. Thermal barrier coatings find their widespread applications in automotive and aircraft industries. They are specifically designed to reduce heat loss from engine exhaust system components including exhaust manifolds, turbocharger casings, exhaust headers, down pipes and tail pipes.
- **Defence**: One of the most important characteristics of functionally graded material is the ability to inhibit crack propagation. This property makes it useful in defines

application as a penetration resistant materials used for armour plates and bullet-proof vests.

- **Others**: baseballs cleats, razor blades, Titanium watches.

6 Conclusions

Functionally graded material is an excellent advanced material that will revolutionize the manufacturing world. There are a number of hurdles for realizing this objective. Cost is one of them, with substantial part of the cost expended on powder processing and fabrication method. Lots of studies have been conducted on behaviour of functionally graded materials and the literature is very rich on this because of the wide areas of application of this novel material. Functionally graded materials are very important in engineering and other applications but the cost of producing these materials makes it prohibitive in some applications.

The aim of the research is to develop a robust methodology for the adaptation of the spatial composition of a gradient material evaluated during the application of a high temperature or high heat flux.

References

1. Abdizadeh, H., Elaboration Et Caractérisation De Composites Duplex: « Composites Laminaires Tri-Couches à Base D'Alumine », Thèse de Doctorat d'Etat, Institut National des Sciences Appliquées de Lyon et l'Université Claude Bernard Lyon ŔFrance, 212p
2. Boccaccini, A.R., Van der Biest, O., Talbot, J.B.: Electrophoretic Deposition, Fundamentals and Applications: Proceedings of the International Symposium, Electrochemical Society, Electrodeposition Division, United Engineering Foundation (U.S.) (2002)
3. Boch, P., Chartier, T., Huttepain, M.: Tapa casting of AL_2O_3/ZrO_2 laminated composites (1986)
4. Boch, P., Chartier, T., Huttepain, M.: Tape casting of Al_2O_3/ZrO_2 laminated Composites, J. Am. Ceram. Soc. **69**(8):191–192 (1987)
5. Bishop, A., Lin, C.Y., Navaratnam, M., Rawlings, R.D., McShane, H.B.: A functionally gradient material produced by a powder metallurgical process. J. Materials Sci. Lett. **12**(19): 1516–18 (1993)
6. Baron, C., Naili, S.: Propagation d'ondes élastiques au sein d'un guide d'ondes élastiques anisotrope à gradient unidirectionnel sous chargement fluide, Compte Rendue Mécanique **336**(9):722–730 (2008)
7. Ichikawa K. (Ed), Functionally Graded Materials in the 21ST Century, A Workshop on Trends and Forecasts, Hardcover, 264p (2000)
8. Kokini, K., Takeuchi, Y.: Multilayer ceramic thermal barrier coatings under transient thermal loads. In Proceeding of the First International Symposium on Functionally Gradient Materials-FGM'90-Sendai- Japan; pp. 31–36 (1990)
9. Koizumi, M.: The concept of FGM. Ceram. Trans. Funct. Graded Materials **34**, 3–10 (1993)
10. Kieback, B., Neubrand, A., Riedel, H.: Processing techniques for functionally graded materials. Mater. Sci. Eng., A **362**, 81–105 (2003)
11. Mistler, R.E.: High strength alumina substrates produced by a multiple-layer casting technique. Am. Ceram. Soc. Bull. **52**(11), 850–854 (1973)

12. Moya, J.S., Sanchez-Herencia, A.J., Requena, J., Moreno, R.: Functionally gradient ceramics by sequential slip casting. Materials Lett. **14**(5–6), 333–35 (1992)
13. Miyamoto, Y., Kaysser, W.A., Rabin, B.H., Kawasaki, A., Ford, R.G.: Functionally graded materials: design, processing and applications, Materials Technology Series, vol. 5 (1999)
14. Mévrel, R.: Les systèmes barrières thermiques pour aubes de turbine. Département Matériaux Métalliques et Procédés, PlastOx **2009**, 255–264 (2007)
15. Omori, M., Kakita, T., Okubo, A., Hirai, T.: J. Jpn. Inst. Met. **62**, 986–991 (1998)
16. Takahashi, M., Itoh, Y., Kashiwaya, H.: Fabrication and evaluation of W/Cu gradient material by sintering and infiltration technique. In: Proceeding of the First International Symposium on Functionally Gradient Materials FGM, Sendai – Japan, pp. 129–134 (1990)
17. Williams, J.C.: « Doctor-Blade Process, in Treatise on Materials Science and Technology » , pp. 173–198. Academic Press, New York (1976)
18. Watremetz, B.: Modèle thermomécanique 3D d'un matériau à gradient de propriétés à l'aide de techniques multigrilles. Application aux moules d'injection de polymères, Thèse de doctorat, Ecole doctorale des sciences pour l'ingénieur de Lyon (2006)
19. Yuki, M., Murayama, T., Irisawa, T.: Temperature gradient sintering of PSZ/Mo functionally gradient material by laser beam heating. In: Proceeding of the First International Symposium on Functionally Gradient Materials FGM, Sendai – Japan, pp. 203–208 (1991)
20. Uchida, Y.: Properties of functionally graded materials, Manufactured by progressive lamination method for applications, 39–51-B, vol. 39-B, Mar. (2004)
21. Zhong, Z., Yu, T.: Analytical solution of cantilever functionally graded beam. Composites Science and Technology; 67 (3–4): 481–488 (2007)

Experimental and Numerical Damaged Characteristics in Composite Plate with Crack Growth by Vibration Analysis

A. Chellil[1(✉)], S. Lecheb[1], H. Mechakra[1], S. Chellil[1], and B. Safi[2]

[1] Laboratory of Engines Dynamics and Vibroacoustics, University of Boumerdes, Boumerdès, Algeria
a.chellil@univ-boumerdes.dz

[2] Research Unit: Materials, Processes and Environment, Boumerdès, Algeria

Abstract. The aim of this paper was to presents a identification of dynamic behavior and response analysis of a glass-epoxy plate. The numerical simulation were carried out on model plate made of glass-epoxy, and the corresponding frequency response functions have been calculate. An initial study into the dynamic loads of this method has been considered, the use of the finite element method makes it possible to develop the model of the plate, the model is used to evaluate the maximal (displacement, strain and stress) and the natural frequencies, mode shapes of plate in composite glass-epoxy under damage condition.

The stress should be increased to improve the strength of the plate. The experimental results can provide a reference for analysts and designers of composite material in aeronautical systems.

Keywords: Material · Crack · Frequency · Stress intensity factor

1 Introduction

The constant search for increasingly high performance in the aerospace field leads to the development of materials with specific rigidities and resistances increasingly high. In this new generation of materials, composites are trying to achieve a breakthrough.

Thus in aeronautics major global manufacturers have been able to make significant progress in developing cells and parts made of composite materials.

The use of new materials advantages are numerous, for example: they have an unlimited lifespan due to good fatigue strength, they are not susceptible to corrosion, they can be repaired easily, it is easier to level of production, etc.

All these benefits are responsive to the concern of aircraft manufacturers: performance and economy. In fact, the composite plates have offered compelling improvements over conventional weight saving solutions by Huang (2015), reducing the cost of production and maintenance, extension of operational possibilities, improved performance while providing a security enhancement.

B. Abdelbaki et al. (Eds.): SMSD 2017, *Proceedings of the Third International Symposium on Materials and Sustainable Development*, pp. 279–287, 2018.
https://doi.org/10.1007/978-3-319-89707-3_31

These new materials, all exhibit anisotropic behavior. In addition to achieve an efficient structure is necessary to achieve it with several materials by choosing the optimal directions for each axis orthotropic.

Synthesized research is conducted in order to take into account the crack propagation phenomenon fatigue in composite blades to assess the stability of the structure (Chellil et al. 2014).

The detection of the emergence and spread of structural damage is difficult. Nevertheless, a clear definition of the damage is not an easy task, for it Prashant and Ganguli (2007) we studied the feasibility of online fault detection for helicopter composite plates, damage modeled in the composite and the cracking of the matrix with stiffness and geometric properties similar to a composite plate.

The fracture processes induced depend upon the nature of the constituents, the architecture of the laminates and the type of mechanical loading imposed to the laminate.

Roy (2005) studied the effect of the growth of damage on modal frequencies of the system and evaluated their feasibility for monitoring damage and fatigue life of the blade.

The effects of matrix cracking, peeling/delamination and fiber breakage on the structural properties of the response of the helicopter rotor are studied by Prashant (Prashant 2005) in the case of forward flight. Firstly, the three major damage modes for a cracked matrix composite are determined based on the duality peeling/delamination, Mechakra et al. (2015).

The aim is therefore, from a simulation of a model representative of the composite plate, develop predictive numerical criteria of the propagation of a crack in the plate.

2 Analytical Formulation

The description of the mechanical behavior of a composite plate is based on the stress-strain relationships designated as the Hooke's law. For the model of the stress-strain response of a plate, we use the generalized version of the anisotropic body.

Generally these relationships have allowed two expressions; flexible and rigid as coefficients of stress-strain relations. Jones (1975) has provided a description of these relations. The law of the generalized Hooke constraints related to deformation can be written in the index notation contracted also:

$$\sigma_i = C_{ij}\varepsilon_j \qquad i,j = 1,\ldots,6$$

Where σ_i strain components, C_{ij} is the stiffness matrix, and ε_j is the deformation component. Components of rigidity and flexibilities often designated as the elastic constants. The two sets of components may be represented by constants E_i , υ_{ij} , G_{ij} which are the Young's modulus, Poisson's coefficient and the shear modulus, respectively.

These constants are especially useful in describing the behavior of the laminar layer. New relationships reduced to 21 independent constants can be represented by Eq. (1). These relationships represent an anisotropic material because there is no plane of symmetry for the material properties. This is the case where none of the blade axis is not aligned with the principal axes of symmetry of the material.

$$\begin{Bmatrix} \sigma_1 \\ \sigma_2 \\ \sigma_3 \\ \tau_{23} \\ \tau_{31} \\ \tau_{12} \end{Bmatrix} = \begin{bmatrix} C_{11} & C_{12} & C_{13} & C_{14} & C_{15} & C_{16} \\ C_{12} & C_{22} & C_{23} & C_{24} & C_{25} & C_{26} \\ C_{13} & C_{23} & C_{33} & C_{34} & C_{35} & C_{36} \\ C_{14} & C_{24} & C_{34} & C_{44} & C_{45} & C_{46} \\ C_{15} & C_{25} & C_{35} & C_{45} & C_{55} & C_{56} \\ C_{16} & C_{26} & C_{36} & C_{46} & C_{56} & C_{66} \end{bmatrix} \begin{Bmatrix} \varepsilon_1 \\ \varepsilon_2 \\ \varepsilon_3 \\ \gamma_{23} \\ \gamma_{31} \\ \gamma_{12} \end{Bmatrix} \tag{1}$$

The relationships stress-strain reduce to Eq. (2) in a plane of symmetry. These relationships are a plate in a coordinate system which rotates in its plane,

$$\begin{Bmatrix} \sigma_1 \\ \sigma_2 \\ \sigma_3 \\ \tau_{23} \\ \tau_{31} \\ \tau_{12} \end{Bmatrix} = \begin{bmatrix} C_{11} & C_{12} & C_{13} & 0 & 0 & C_{16} \\ C_{12} & C_{22} & C_{23} & 0 & 0 & C_{26} \\ C_{13} & C_{23} & C_{33} & 0 & 0 & C_{36} \\ 0 & 0 & 0 & C_{44} & C_{45} & 0 \\ 0 & 0 & 0 & C_{45} & C_{55} & 0 \\ C_{16} & C_{26} & C_{36} & 0 & 0 & C_{66} \end{bmatrix} \begin{Bmatrix} \varepsilon_1 \\ \varepsilon_2 \\ \varepsilon_3 \\ \gamma_{23} \\ \gamma_{31} \\ \gamma_{12} \end{Bmatrix} \tag{2}$$

For plane stress in the plane of the laminar layer, the Eq. (2) may be rewritten with respect to the condensed stiffness matrix $[Q]$.

Placing σ_3, τ_{23} and τ_{31} equal zero, which implies that ε_3, γ_{23} and γ_{31} are equal to zero, leading to the following ratio between the components of stress and strain components in the plan:

$$\begin{Bmatrix} \sigma_1 \\ \sigma_2 \\ \sigma_3 \end{Bmatrix} = \begin{bmatrix} Q_{11} & Q_{12} & 0 \\ Q_{12} & Q_{22} & 0 \\ 0 & 0 & Q_{66} \end{bmatrix} \begin{Bmatrix} \varepsilon_1 \\ \varepsilon_2 \\ \gamma_{12} \end{Bmatrix} \tag{3}$$

Where the elements of the reduced-stiffness matrix [Q] are given in terms of elements of the stiffness matrix in three dimensions [C].

Equation (3) defines the stresses and strains in the plane in the system of coordinates of a major hardware orthotropic or transversely isotropic material. Using the above relationships in Eq. (3) allows us to link the stresses in the plane deformation in the rotated coordinate system as follows:

$$\tilde{\sigma}' = [\bar{Q}]\, \tilde{\varepsilon}' \text{ ou } \tilde{\varepsilon}' = [\bar{S}]\, \tilde{\sigma}' \tag{4}$$

Or

$$[\bar{Q}] = [T_1]^{-1} [Q]\, [T_2] \tag{5}$$

And

$$[\bar{S}] = [T_2]^{-1}[S]\,[T_1] \tag{6}$$

3 Numerical Study for the Composite Plate

We use step module to create and configure analysis steps and associated output requests. The step sequence provides a convenient way to capture changes in a model.

In this work we study a helicopter fuselage made of glass/epoxy which have for properties:

$$E_1 = 36800\,\text{MPa}$$

$$E_2 = 8270\,\text{MPa}$$

$$\rho = 2600\,\text{Kg/m}^3$$

$$G_{1.2} = 4140\,\text{MPa}$$

$$\mu = 0.26$$

Where: E_1, E_2: two value of yong's module, ρ: Density, $G_{1.2} = G_{1.3} = G_{2.3}$: shears modules, μ: Poisson ration (Fig. 1).

The load module allows you to specify load, boundary conditions, and predefined

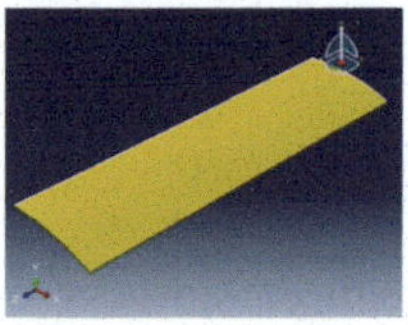

Fig. 1. The property of martial with 2ply

fields. Loads and boundary conditions are step-dependent objects, which mean that you must specify the analysis steps in which they are active; some predefined fields are step-dependent, while others are applied only at the beginning of the analysis. And in our case we apply a force equal to F = 100 KN.

For the displacement we have:

$$U1 = U2 = U3 = UR1 = UR2 = UR3 = 0$$

The mesh module contains tools that allow us to generate a finite element mesh on an assembly created within abaqus/CAE. The approximate global size equal to 10 (Fig. 2).

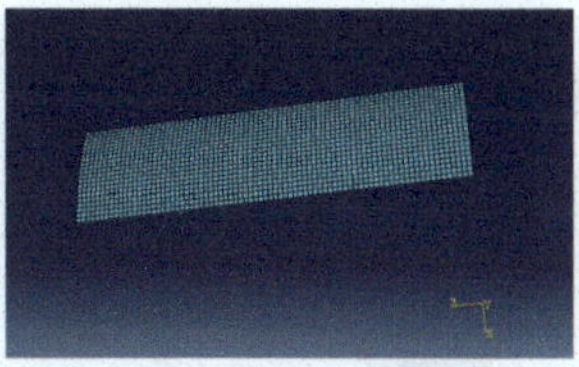

Fig. 2. The composite plate

3.1 Simulation without Crack

Modal analysis:

Mode01: See Fig. 3.

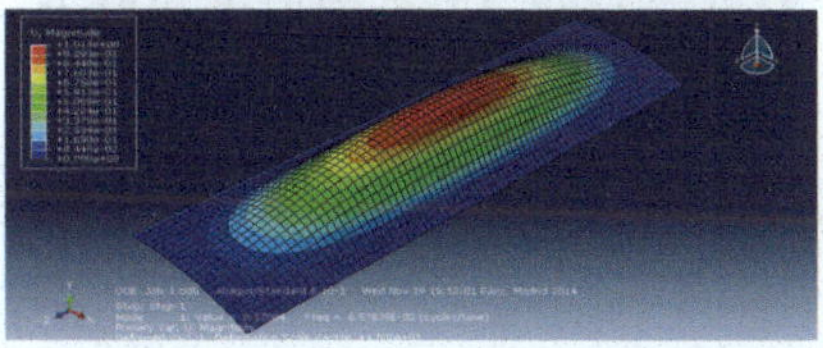

Fig. 3. The first Mode shape (U max = 1.014 mm, f = 0.657 Hz)

Mode02: See Fig. 4.

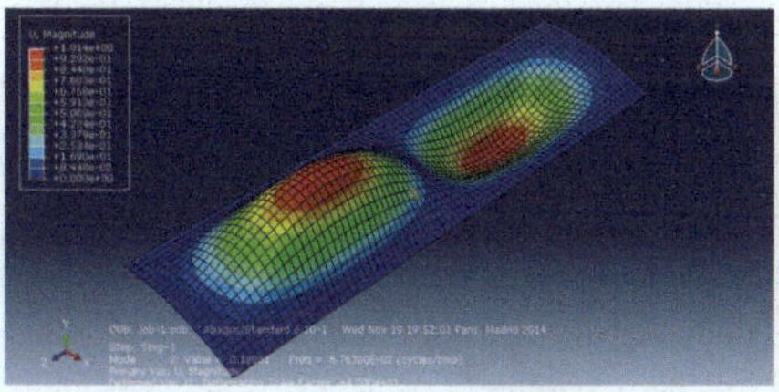

Fig. 4. The second Mode shape (U max = 1.014 mm, f = 0.676 Hz)

Von Misses Stress:

Figure 5 is observed that the critical load of the plate where the stress concentration is in the middle of the plate, and which is caused due to the critical loading. The value of maximum stress of 135 MPa.

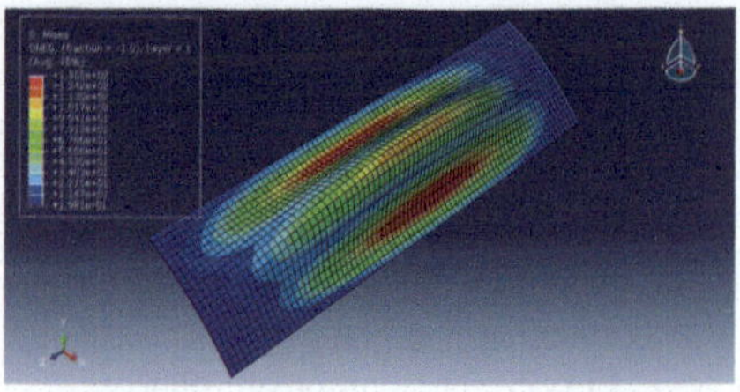

Fig. 5. Maximum stress of Von Misses (Smax = 1.35 e + 02 MPa)

Strain: See Fig. 6.

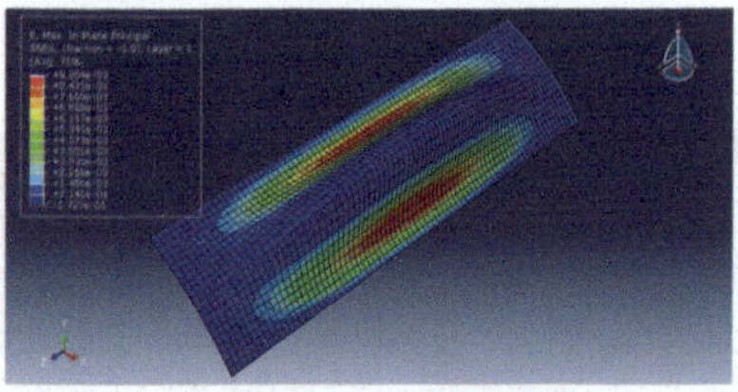

Fig. 6. Maximum strain (ε max = 0.920e − 02)

3.2 Simulation With Crack Take a = 10 Cm

Modal analysis:

Mode01: See Fig. 7.

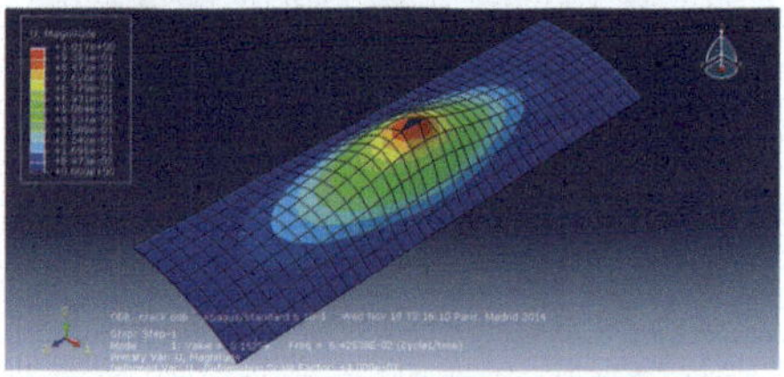

Fig. 7. The first Mode shape (U max = 1.017 mm, f = 0.642 Hz)

Mode02: See Fig. 8.

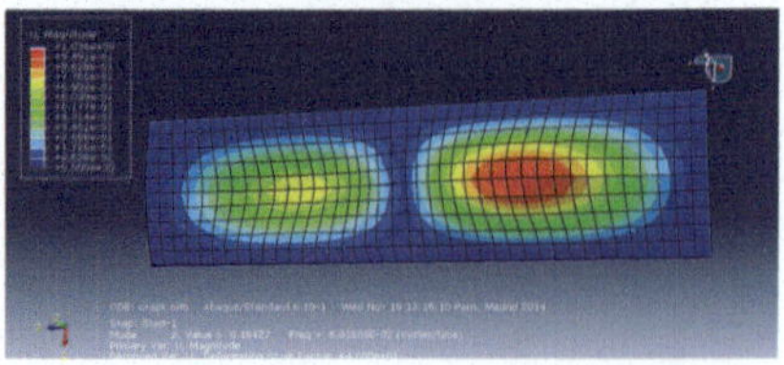

Fig. 8. The second Mode shape (U max = 1.035 mm, f = 0.683 Hz)

Von Misses Stress: Fig. 9.

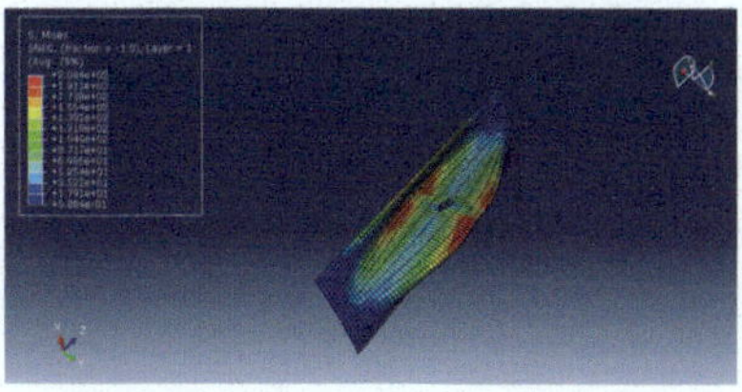

Fig. 9. Maximum stress of Von Misses (Smax = 2.08 e + 02 MPa)

There is a stress concentration at the edge of the crack. The strain value is σmax = 208 MPa; which can be interpreted by the appearance of a more sensitive area for the occurrence of damage.

Strain: See Fig. 10.

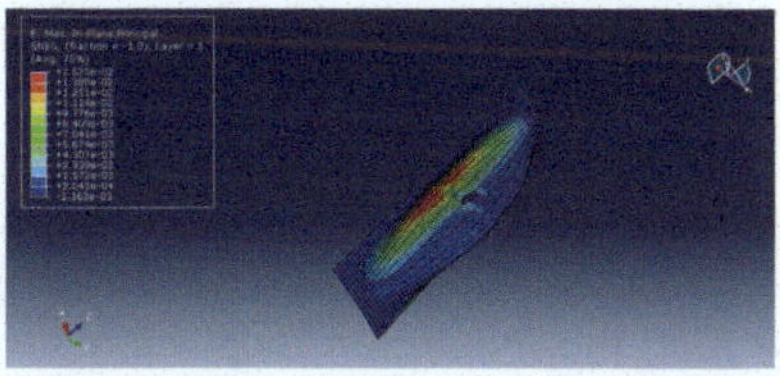

Fig. 10. Maximum strain (ε max == 1.52e – 20)

4 Experimental Study

Measurement hardware and instruments (Fig. 11) is a 8801 Instron machine. this 8801 servohydraulic fatigue machine meets the needs of a wide variety of fatigue testing. provides a complete test solution for testing of advanced materials. It is particularly suitable for testing in low and high number of cycles, with a capacity of up to 100 kN, high stiffness, very accurate alignment and with characteristics.

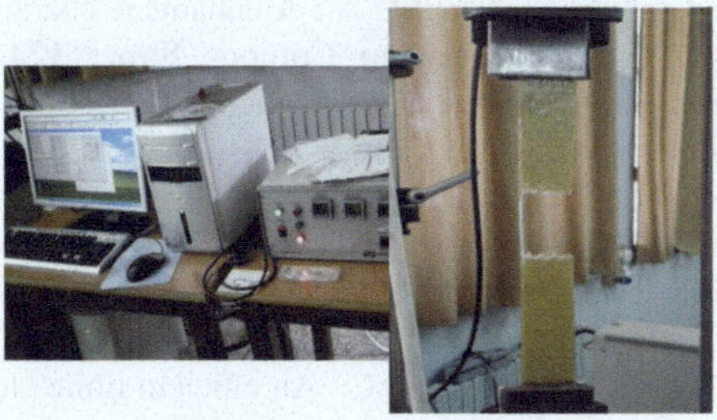

Fig. 11. Materials testing

- Axial Capacity ± 100 kN (22,500 lbf)
- inertia compensation force sensor
- Standard or Extra High Optional Built
- Wide range of grips, fixtures and test accessories

From experience the fracture is occurring after 3010 life cycles, and so tenuously rupture occur (fast fracture). So the fracture in composites occurs rapidly (Fig. 12).

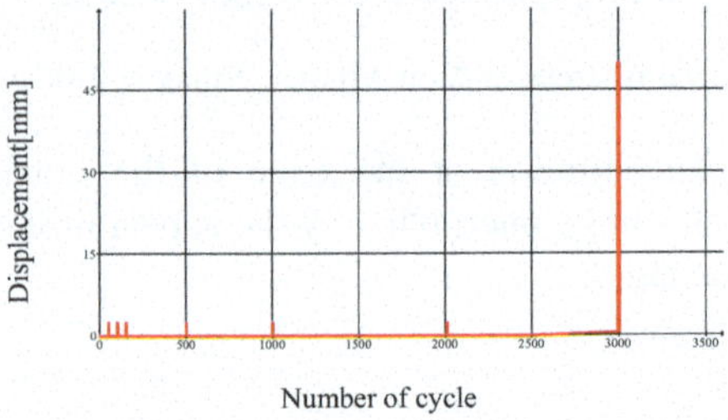

Fig. 12. The displacement according the cycles number

5 Conclusion

The results obtained show that: the maximum displacements are in the middle of the plate. Von Mises stresses it possible to observe the critical areas, where it is necessary to determine the maximum load point of the plate and stress concentrations that are in the middle of the plate, due to high stress. Experimentally the fracture in composites occurs rapidly.

Also, modal analysis in terms of frequencies and modes may be employed as an indicator of the health of composite plates for predictive maintenance of the variable of the damage.

References

Mechakra, H., Nour, A., Lecheb, S., Chellil, A.: Mechanical characterizations of composite material with short Alfa fibers reinforcement. Compos. Struct. **124**, 152–162 (2015)

Lecheb, S., Nour, A., Chellil, A., Mechakra, H., Amarache, A., Kebir, H.: An advanced dynamic repair of edge crack aluminum plate with a composite patch. Mater. Sci. Forum **794–796**, 716–721 (2014). ISSN: 1662-9752

Chellil, A., Nour, A., Lecheb, S., Kebir, H., Chevalier, Y.: Impact of the fuselage damping characteristics and the blade rigidity on the stability of helicopter. J. Aerosp. Sci. Technol. **29** (1), 235–252 (2013)

Huang, L., Sheikh, A.H., Ng, C.-T., Griffith, M.C.: An efficient finite element model for buckling analysis of grid stiffened laminated composite plates. Compos. Struct. **122**, 41–50 (2015)

Chellil, A., Nour, A., Lecheb, S., Chibani, M., Kebir, H.: Parameter identification of crack growth in helicopter blade. Key Eng. Mater. **577–578**, 673–676 (2014). ISSN: 1013-9826

Pawar, P.M., Ganguli, R.: On the effect of progressive damage on composite helicopter rotor system behavior. Composites Science and Technology **65**, 581–594 (2005)

Roy, N., Ganguli, R.: Helicopter rotor blade frequency evolution with damage growth and signal processing. J. Sound Vib. **283**, 821–851 (2005)

Pawar, P.M., Ganguli, R.: On the effect of progressive damage on composite helicopter rotor system behavior. Compos. Struct. **78**, 410–423 (2007)

Jones, R.M.: Mechanics of Composite Materials. Scripta Book Company, Washington, D.C. (1975)

Tensile Behavior of Bio-Composites: Date Palm Rachi/Epoxy, Date Palm Pits/Epoxy

A. Dakhli[2], R. Benzidane[1(✉)], R. Zehaf[2], L. M. Bennegadi[1], and Z. Sereir[1]

[1] Laboratoire Structure de Composite et Matériaux Innovants – LSCMI, Faculté de génie mécanique, USTOMB, BP 1505, El M'Naouer, Oran, Algeria
r-benzidane-usto@hotmail.com
[2] Département de génie maritime, Faculté de génie mécanique, USTOMB, BP 1505, El M'Naouer, Oran, Algeria

Abstract. The present paper deal within mechanical characterization of a composite material reinforced by rachis short fibers of date-palm wood and loaded by the flour of date-pits. The palm and pits collected from region of Biskra were crushed in an electric shredder. After that, they were finely sieved and then mixed with an epoxy resin according to different volume fractions. To predict the mechanical properties of these composite, normalized specimens are made and tested under tensile loading according to respected volumetric fraction of fibers. Obtained results, showed that, rachis short fibers and flour of date-palm have a significant effect on the elastic modulus increase. But an important volumetric fraction of fibers decreased the tensile strength of this composite material.

Keywords: Epoxy resin · Date pits · Date palm wood rachi · Bio-composite Tensile tests

1 Introduction

Recently, Composites based on natural fiber and thermoplastic matrixes have received a huge interest as innovative material in a wide range of industrial applications [1]. Some of the reasons are the renewability and abundance of natural fibers [2]. Various natural fibers such as henequen, Alfa, Palm, sisal or hemp been tested as new reinforcement agents in polymer-based composite materials [3]. The drawback is the weak compatibility between the polymer matrix and natural fibers that yields low properties, thus limiting possible applications [3, 4]. The use of natural fibers in organic matrices is highly beneficial because the strength and toughness of the resulting composites are greater than those of the unreinforced materials. Moreover, lignocelluloses natural fibers are usually strong, light in weight (low density), cheap, abundant and renewable. In addition, these composites based on natural fiber include the uses of treated and untreated fiber, the effect of the rigidity of natural fibers in composites, the manufacturing process, and the state of growth of the fiber [5]. The positive results indicate that natural fibers are good candidates for replacing glass fibers as reinforcement in polymer composites [6].

Mechanical properties of polymeric composites based on natural fibers strongly depend on the interface adhesion between the fibers and the polymer matrix [7–9]. This

B. Abdelbaki et al. (Eds.): SMSD 2017, *Proceedings of the Third International Symposium on Materials and Sustainable Development*, pp. 288–295, 2018.
https://doi.org/10.1007/978-3-319-89707-3_32

is mainly because natural fibers are rich in cellulose, hemicelluloses, pectin and lignin, which are hydroxy1 groups, i.e. natural fibers tend to be strong polar and hydrophilic materials whilst polymers exhibit significant hydrophobicity. In other words, there are significant problems of compatibility between the fiber and the matrix, which weakens interface area between natural fibers and matrices. However, many investigators reported that chemical treatments such as bleaching, acetylation and alkali treatment may improve the matrix–fiber interfacial adhesion [10, 11]. Fiber size also has a significant effect on the performance of a natural fiber composite [12].

The present study is focused upon the utilization of rachis short fibers of date-palm wood and the flour of date-pits as reinforcement for preparation of epoxy composites. These fibers have been extracted and crushed to obtain shorts fibers and flour after having been sifted. The composite samples have been prepared with different fiber content ranging from 10 to 30% by weight and their mechanical properties have also been investigated.

2 Experiments

2.1 Materials

(a) *Date pits and short fiber rachi*

Firstly, the wood palm rachi and date pits collected from the Biskra region were ground in an electric shredder. Subsequently, they were finely sieved (Fig. 1-a, b). The date palm rachis fibers have a diameter varies between 180 to 350 µm. The diameter of the fiber was measured using an electron scanning microscope driven by an image processing program. The section of the fiber was considered as a circular section (Fig. 1-c).

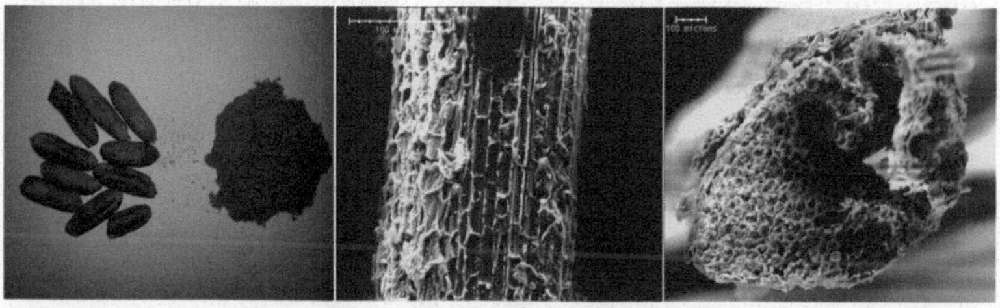

Fig. 1. (a) Date pits, (b) SEM view of a palm fiber rachi, (c) SEM view of a rachi palm fiber section.

(b) *Resin*

The synthetic matrix is an epoxy resin obtained by a combination of 2/3 MEDAPOXY STR EA and 1/3 of MEDAPOXY STR EB. The global behavior of this resin is viscoelastic (Fig. 2-a). The partially bio-resourced matrix is an epoxy resin with 56% of its molecular structure of vegetable origin (SR GREENPOXY 56) is also used to manufacturing the tensile specimens. The mechanical behavior of this resin is given in Fig. 2b.

It is noted that both resins have almost the same performances except that the MEDA-POXY STR resin has a more extensive relaxation zone.

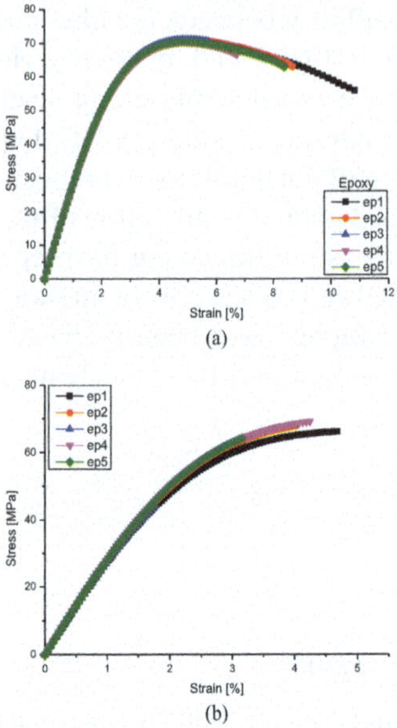

Fig. 2. Tensile stress-strain curves (a) Epoxy, (b) Greenpoxy.

2.2 Manufacture of Specimens

Composites were prepared using different volume fractions (30, 50 and 70%) of date pits and short rachi fibers using two types of resin (synthetic & partially bio-resourced) (Fig. 3-a). The dimensions of the test specimens of our study were taken according to the American standardization «ASTM D638-type1» (Fig. 3-b).

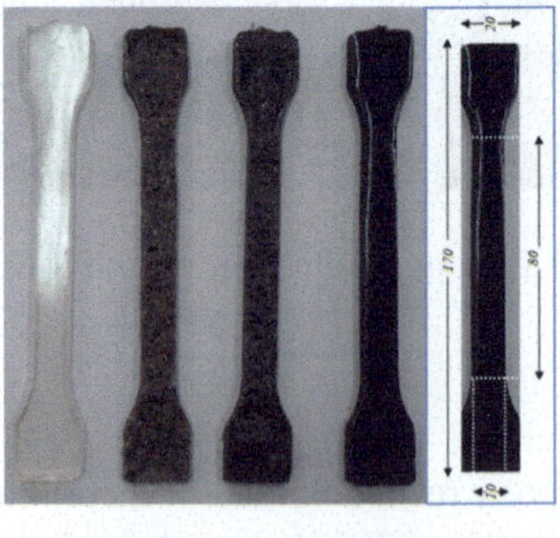

Fig. 3. (a) Tensile test specimens for different volume fractions, (b) dimensions of the standard (ASTM D638).

2.3 Tensile Test

The mechanical properties (stress, breaking strain and Young's modulus) of the composites were determined according to the standard ASTM D638-type1. Due to the variability, five specimens are tested for each case (i.e. a total of 50 test specimens). The tests were carried out on a universal traction machine equipped with a load cell with a capacity of 10 kN. The jaws clamps used on this work are manual clamping jaws. Figure 4 shows the mechanism used on the experiment.

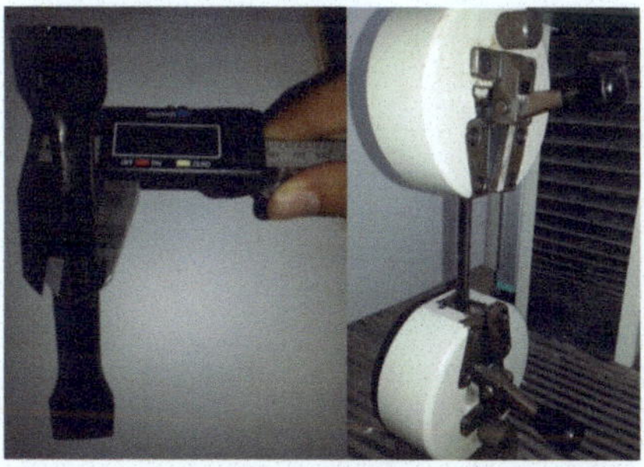

Fig. 4. Tensile testing of test specimens (date-pits/époxy, short fiber rachi/époxy)

Figure 5 illustrates the specimens after the monotonic tensile test. In this figure, it should be noted that the point of rupture is not uniform.

Fig. 5. Tensile specimen after tests.

3 Results and Discussion

The Figs. 6 a–b illustrates the theoretical stress-strain curves for two types of composite. These curves were recorded using a universal machine and indicated that the load used reinforced the plastic deformation. Consequently, these composites, in principle, can be considered as fragile materials.

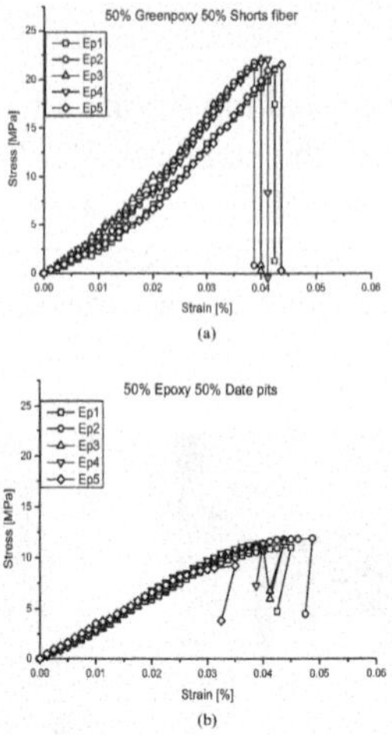

Fig. 6. Theoretical stress-strain tensile curves.

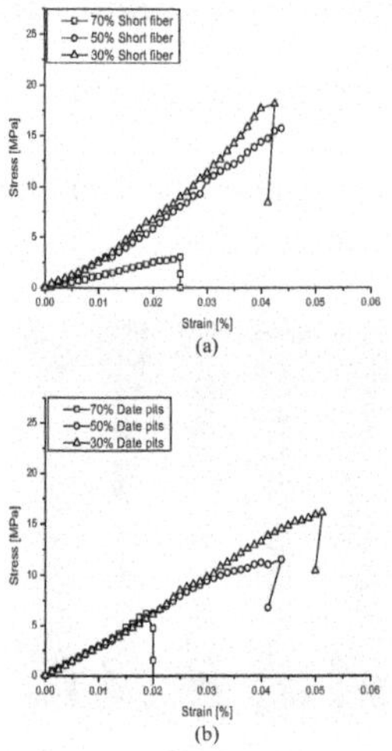

Fig. 7. Average theoretical stress-strain tensile curves synthetic epoxy resin.

From the mean theoretical stress-strain curves, tensile strength, deformation and Young's modulus were deduced. The histograms presented on the Figs. 8 a–c indicate the mean values for these tensile properties for the different volume fractions and the two types of fillers used (pits and short date palm fibers) studied. In these histograms, it should be noted that the use of flour (pits and short date palm fiber) does not reinforce the epoxy composites (Fig. 7).

The values of the tensile strength and the deformation decrease as the volume fraction increases. However, an insignificant increase in the Young's modulus corresponds to the increase in the volume fraction. Consequently, the optimum values of tensile strength and Young's modulus were reached with the composites loaded with short fibers. On

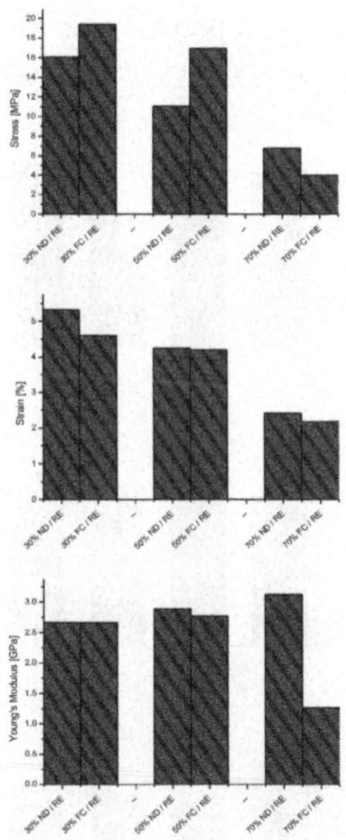

Fig. 8. Histogram of average mechanical properties of composites (synthetic epoxy resin)

the other hand, the optimal deformation values were reached with the composites loaded with date pits.

Figure 9 illustrates the theoretical stress-strain curves for composites based on two types of resin, a synthetic epoxy resin and a partially bio-resourced resin for two types of loading date pits and short fiber rachis.

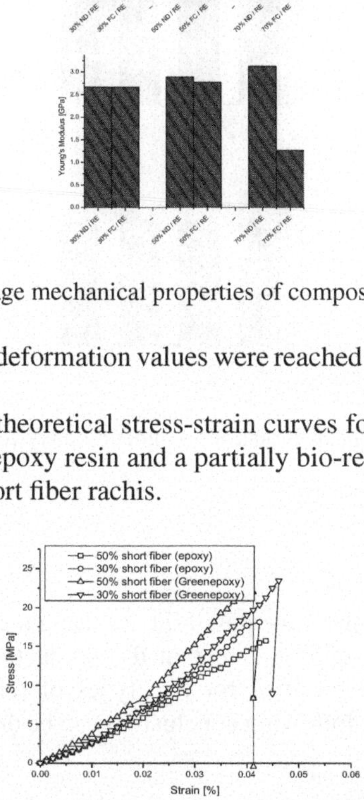

Fig. 9. Average theoretical stress-strain curve synthetic epoxy resin & partially bio-resourced resin

From the average theoretical curves of the two types of composite, the tensile strength, deformation and Young's modulus were deduced. The histograms in Fig. 10 a–c show the average values of the tensile properties for the two types of composites based on a synthetic epoxy resin and a partially bio-resourced resin with two volume fractions (30% and 50%) using the two types of flour (nuts and short palm tree fibers) studied. In these histograms, it is noted that the values of the tensile strength and the Young's modulus have been improved in composites based on a partially bio-resin, but the optimum deformation values have been reached with the composites at base of a synthetic epoxy resin.

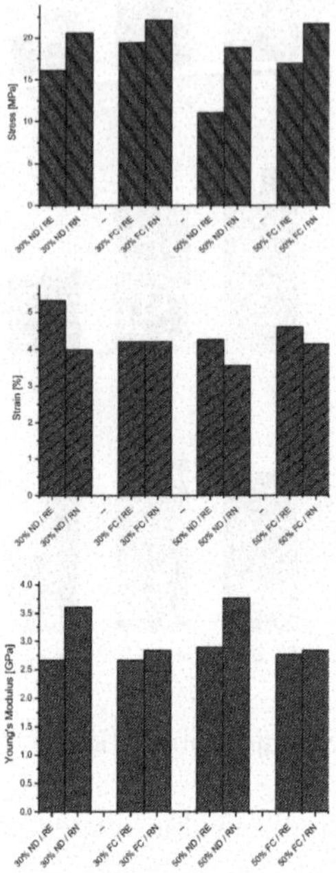

Fig. 10. Histogram of mechanical properties synthetic epoxy resin & partially bio-resourced resin.

4 Conclusions

The results of this research show the possibility of using the date palm waste (date-pits and palm wood rachis) as reinforcement in the production of bio-composites. The mechanical properties were verified for two types of reinforcements and different volume fractions. While the following conclusions can be drawn from the results of our research:

- Composites containing shorts fibers rachis have better mechanical properties compared to those composites containing date pits.
- The tensile strength and the deformation decrease as the volume fraction increases. However, a slight increase in Young's modulus is observed in parallel.
- The use of a partially bio-resin resin allows raising the mechanical properties compared to the use of a synthetic epoxy resin.

References

1. Pracella, M., Chionna, D., Anguillesi, I., Kulinski, Z., Piorkowska, E.: Functionalization, compatibilization and properties of polyolefin composites with natural fibers. Polymer **2**, 554–574 (2010)
2. Mulinari, D.R., Voorwald, J.C., Cioffi, O.H., da Silva, C.P., Luz, S.M.: Preparation and properties of HDPE/sugarcane bagasse cellulose composites obtained for thermokinetic mixer. Carbohyd. Polym. **75**, 17–21 (2009)
3. Herrera-Franco, P.J., Valadez-Gonzalez, A.: Mechanical properties of continuous natural fiber-reinforced polymer composites. Compos. Part A – Appl. Sci. Manuf. **35**, 339–345 (2004)
4. Nazrul, I., Rezaur, R., Mominul, H., Monimul, H.: Physicomechanical properties of chemically treated coir reinforced polypropylene composites. Compos. Part A – Appl. S **41**, 192–198 (2010)
5. Ochi, S.: Mechanical properties of kenaf fibers and kenaf/PLA composites. Mech. Mater. **40**, 446–452 (2008)
6. Nishino, T., Hirao, K., Kotera, M., Nalamae, K., Inagaki, H.: Kenaf reinforced biodegradable composite. Compos. Sci. Technol. **63**(9), 1281–1286 (2003)
7. Fu, S.Y., Feng, X.Q., Lauke, B., Mai, Y.W.: Effects of particle size, particle/matrix interface adhesion and particle loading on mechanical properties of particulate-polymer composites. Compos. B **39**, 933–961 (2008)
8. Nystrom, B., Joffe, R., Langstrom, R.: Microstructure and strength of injection molded natural fiber composites. Reinf. Plast. Compos. **26**, 579–599 (2007)
9. Rosa, M.F., Chiou, B.-S., Medeiros, E.S., Wood, D.F., Williams, T.G., Mattoso, L.H.C., et al.: Effect of fiber treatments on tensile and thermal properties of starch/ethylene vinyl alcohol copolymers/coir biocomposites. Bioresour. Technol. **100**, 5196–5202 (2009)
10. Chin, C.W., Yousif, B.F.: Potential of kenaf fibres as reinforcement for tribological applications. Wear **267**, 1550–1557 (2009)
11. Hepworth, D.G., Vincent, J.F.V., Jeronimidis, G., Bruce, D.M.: The penetration of epoxy resin into plant fibre cell walls increases the stiffness of plant fibre composites. Compos. Part: A Appl. Sci. Manuf. **31**, 599–601 (2000)
12. Alawar, A., Hamed, A.M., Al-Kaabi, K.: Characterization of treated date palm tree fiber as composite reinforcement. Compos. Part B Eng., 601–606 (2009)

Fabrication and Characterization of TiO_2 Nanoparticles

A. Bessi[1(✉)], B. Boudine[2], and C. Boudaren[3]

[1] Laboratory of Applied Chemistry, Mohamed Kheider University, PB 145, 07000 Biskra, Algeria
asieadz@yahoo.fr

[2] Laboratory of Crystallography, Physical Department, Mentouri University, Constantine, Algeria

[3] CHEMS Research Unit, Department of Chemistry, Mentouri University, Constantine, Algeria

Abstract. Titanium oxide (TiO_2) is an attractive semiconducting material due to its characteristics photochemical properties and high chemical stability. TiO_2 has been widely used for various applications such as a semiconductor in dye-sensitized solar cell, water treatment materials, catalysts, gas sensors, and so on. In this work, Titanium Dioxide (TiO_2) nanoparticles were prepared by sol-gel method using Titanium Isopropoxide (TiTP) as a source of Ti. Different solvents (isopropyl alcohol, ethanol and methanol) were used to prepare TiO_2 nanoparticles. The structural characterization of the samples is carried out by X-Ray Diffraction (XRD) And Fourier Transforms Infrared Spectroscopy (FTIR). (XRD) analysis reveals the anatase structure of the as prepared TiO_2 with nanometer size. The founded size of the particle is to be 15 nm when using isopropyl alcohol as a solvent. The increasing time of the calcination (at the same temperature) causes the size increasing. FTIR measurements confirm the anatase structure of TiO_2.

Keywords: TiO_2 · Sol-gel · Nanoparticles

1 Introduction

Sol gel process and colloid chemistry offer opportunities for synthesis of transparent materials embedding semiconductors nanocrystals Gacoin et al. (1997). One of the most important of these semiconductors is the titanium dioxide TiO_2 because of its various uses in wide applications: environment, purification, separation, photocatalytic and solar energy cells. Ishugoro et al. (2011) (O'regan and Grätzel 1991).

The proprieties of TiO_2 are connected with the crystalline structure, the size and morphology (Reyes-Coronado et al. 2008).

Several processes are used for the synthesis of titanium oxide. The sol-gel route (Akpan and Hameed 2010) the hydrothermal route (Li et al. 2008), or by chemical vapor deposition (CVD) Ding et al. (2001).

It was found that when synthesized TiO_2 nanoparticles under the same ambient conditions, the prepared powder via sol-gel route were highly crystalline and had

B. Abdelbaki et al. (Eds.): SMSD 2017, *Proceedings of the Third International Symposium on Materials and Sustainable Development*, pp. 296–300, 2018.
https://doi.org/10.1007/978-3-319-89707-3_33

smaller crystallite size as compared to the one prepared by hydrothermal method (Vijayalakshmi and Rajendran 2012). In other studies when using a mixture solvent containing methanol and ethanol, the existence of anatase and rutile is confirmed with the ratio 23% by weight of rutile Bessekhouad et al. (2003).

In the present study, we use sol gel method for preparation of TiO_2 nanoparticles with several solvents in order to investigate their effect on size, and morphology of synthesized samples.

2 Experimental Work

2.1 Materials and Synthesis

The Titanium Dioxide (TiO_2) nanoparticles were prepared by sol-gel process using Titanium Isopropoxide (TiTP) $Ti[OCH(CH_3)_2]_4$ as a source of Ti with a purity of 98% from SIGMA Aldrich, glacial acetic acid CH_3COOH with a purity of 99% and ethanol 99.8%, methanol 99.7% or 2-propanol 99.8% as solvents from SIGMA Aldrich.

The TiTp was added using a pipette (drop by drop) to a beaker which contains a mixture of 50 ml of ethanol (isopropyl alcohol or methanol), and 1/10 of the volume of glacial acetic acid. This solution was continually stirred by a magnetic stirrer at room temperature until the homogeneity. Then the obtained sol is put in an ultrasonic bath for 15 min, a yellow transparent gel was formed. This one is dried at 100 °C for a several hours. In the last step, a yellow block crystal is obtained. As showed in Fig. 1(a).

The powder was calcined at 500 °C for 5 h. As showing in Fig. 1(b).

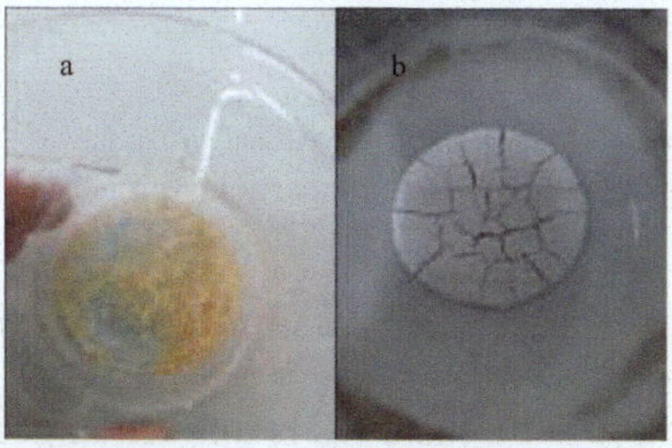

Fig. 1. (a) Powder TiO_2 before calcination, (b) powder of TiO_2 after calcination at 500 °C

2.2 Structural Characterization

The samples were analyzed by using a Burker D2 PHASER diffractometer equipped with an integrated PC and a flat screen monitor, with a high-speed transistor drive.

The average nanoparticles size was determinate using the Scheere's formula (1): (Patterson 1939)

$$D = \frac{0.9\lambda}{\beta cos(\theta)} \quad (1)$$

D : size of particles nm
λ : wave-lengths of X-ray 1.54 A°
β : Width at Half of Maximum FWHM
cos (θ) : Bragg's angles degree

The spectra of FTIR were recorded using a JASCO FTIR 4100 single-beam spectrophotometer equipped with a microcomputer. Infrared spectroscopy is used in transmission mode on pellets in KBr.

3 Results and Discussions

The X-ray analysis showed the anatase phase of the samples prepared using 2-propanol as showing in Fig. 2. Likewise in the ethanol and methanol, after calcination at 500 °C for 5 h. As showing in Figs. 3 and 4.

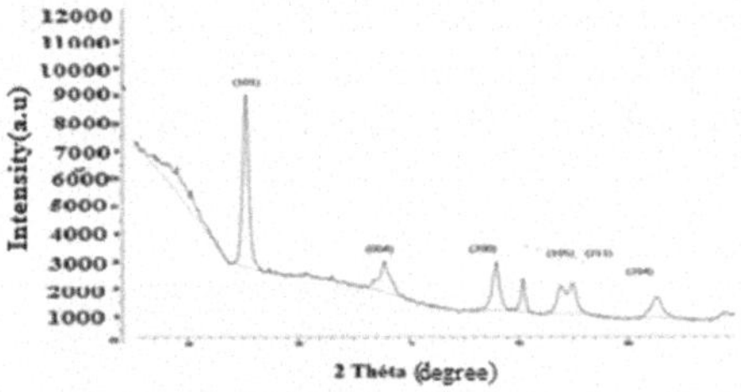

Fig. 2. The XRD patterns of TiO_2 nanoparticles prepared in isopropyl alcohol

The crystalline anatase phase was confirmed using the JCPDS (21-1272) as reference.

The plan (101) (004) (200) (105) and (211) diffraction peaks were presented in the DRX spectrum of samples as showing in Figs. 2, 3 and 4.

The other phases of TiO_2 rutile and brookite were not observed during the investigations which is in agreement with kheamrutai. Thamaphat et al. (2008).

Using Scheere's formula (1) the size grain was calculated and it was in the average of 15 nm, for the nanoparticles synthesized in isopropyl alcohol, but in the case of ethanol and methanol, this size was increased to 17 nm and 23 nm respectively. which is mentioned by Li et al. (2003).

Varying the time of calcination, to investigate its effect on the phase and size of nanoparticles synthesized. We found that after 10 h, at the same temperature of calcination 500 °C, the anatase phase is always present.

The TFIR spectrum of the synthesized particles is showed in Fig. 5.

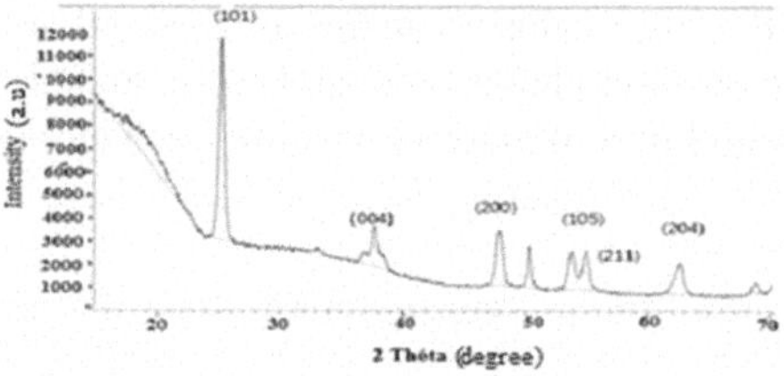

Fig. 3. The XRD patterns of TiO_2 nanoparticles prepared in ethanol

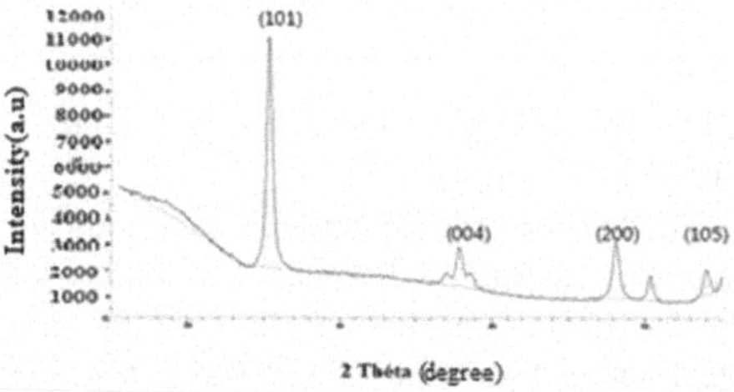

Fig. 4. The XRD patterns of TiO_2 nanoparticles prepared in methanol

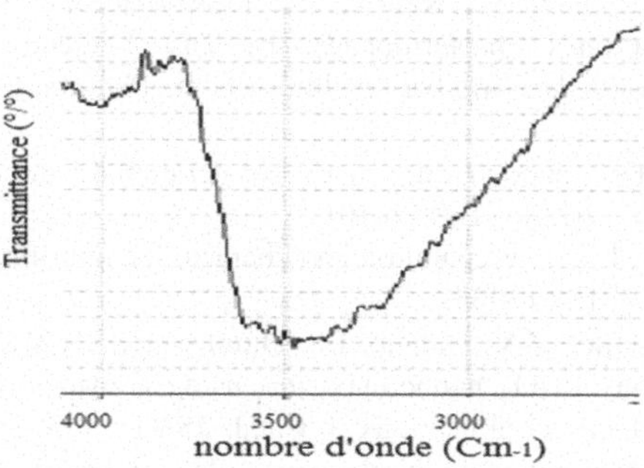

Fig. 5. Infrared transmission spectra in the region of the absorption band of hydroxyl of the TiO2 synthetized

The band of hydroxyls is situated at 3500 Cm^{-1} with strong intensity, which has an important role in photocatalytic process (Karami 2010).

4 Conclusion

Sol-gel processing is a very effective way to synthesize TiO_2 with nanometer size.

The nanoparticles of TiO_2 were synthesized using three kinds of solvents, ethanol, isopropyl alcohol and methanol, to examine their effect on the crystallinity, size of grain and yield of reaction.

The results showed that the anatase structure is obtained at calcination temperature of 500 °C when using isopropyl alcohol, ethanol or methanol.

It was found that the addition of isopropyl alcohol, ethanol or methanol to solutions of titanium isopropoxide significantly influences the size of the resultant TiO_2 nanoparticles.

An increase in the size of the as-prepared TiO_2, from 15 nm to 23 nm when using methanol.

The pure structure of TiO_2 particles was detected in all of the prepared samples.

References

Akpan, U.G., Hameed, B.H.: The advancements in sol–gel method of doped-TiO_2 photocatalysts. Appl. Catal. A **375**(1), 1–11 (2010)

Bessekhouad, Y., Robert, D., Weber, J.V.: Synthesis of photocatalytic TiO_2 nanoparticles: optimization of the preparation conditions. J. Photochem. Photobiol., A **157**(1), 47–53 (2003)

Ding, Z., Hu, X., Yue, P.L., Lu, G.Q., Greenfield, P.F.: Synthesis of anatase TiO_2 supported on porous solids by chemical vapor deposition. Catal. Today **68**(1–3), 173–182 (2001)

Gacoin, T., Malier, L., Counio, G., Boilot, J.-P.: CdS nanoparticles and sol-gel process. In: International Society for Optics and Photonics, Proceedings of Optical Science, Engineering and Instrumentation 1997, Sol-Gel Optics IV, vol. 3136, p. 8 (1997)

Ishiguro, H., Nakano, R., Yao, Y., Kajioka, J., Fujishima, A., Sunada, K., Kubota, Y.: Photocatalytic inactivation of bacteriophages by TiO 2-coated glass plates under low-intensity, long-wavelength UV irradiation. Photochem. Photobiol. Sci. **10**(11), 1825–1829 (2011)

Karami, A.: Synthesis of TiO_2 nano powder by the sol-gel method and its use as a photocatalyst. J. Iran. Chem. Soc. **7**(7), S154–S160 (2010)

Li, B., Wang, X., Yan, M., Li, L.: Preparation and characterization of nano-TiO_2 powder. Mater. Chem. Phys. **78**(1), 184–188 (2003)

Li, G., Ciston, S., Saponjic, Z.V., Chen, L., Dimitrijevic, N.M., Rajh, T., Gray, K.A.: Synthesizing mixed-phase TiO_2 nanocomposites using a hydrothermal method for photo-oxidation and photoreduction applications. J. Catal. **253**(1), 105–110 (2008)

O'regan, B., Grätzel, M.: A low-cost, high-efficiency solar cell based on dye-sensitized colloidal TiO_2 films. Nature **353**(6346), 737–740 (1991)

Patterson, A.L.: The scherrer formula for x-ray particle size determination. Phys. Rev. **56**(10), 978–982 (1939)

Reyes-Coronado, D., Rodríguez-Gattorno, G., Espinosa-Pesqueira, M., Cab, C., de Coss, R.D., Oskam, G.: Phase-pure TiO_2 nanoparticles: anatase, brookite and rutile. Nanotechnology **19** (14), 145605 (2008)

Thamaphat, K., Limsuwan, P., Ngotawornchai, B.: Phase characterization of TiO_2 powder by XRD and TEM. Kasetsart J. (Nat. Sci.) **42**(5), 357–361 (2008)

Vijayalakshmi, R., Rajendran, V.: Synthesis and characterization of nano-TiO_2 via different methods. Scholar Res. Libr. **4**(2), 1183–1190 (2012)

In Vitro Corrosion Study by EIS of Stainless Steel for Orthopaedic Applications

Faiza Kakaa(✉) and Mosbah Ferkhi

Laboratory of the Interactions Materials-Environment (LIME),
University of Jijel, Jijel, Algeria
kaka-faiza@hotmail.fr

Abstract. Stainless steel 316L is one of the most common biomaterials utilized for producing orthopedic implants. Since the device is in constant contact with the blood, which is a highly corrosive environment, it is necessary to know the effects of corrosion on the screw plates. Currently 316L stainless steel is the material most commonly used for osteosynthesis. To assess the sensitivity of the screw plates in physiological media of the corrosion test plate required. The 316L stainless steel plates were tested for corrosion characteristics. The corrosion behavior of orthopaedic implant stainless steel 316L has been studied in a simulated body fluid using potentiodynamic polarization and electrochemical impedance spectroscopy techniques. Polarization experiments were conducted after 4 h, 6 h, 24 h, 48 h, 168 h, 215 h, 10 days and 15 days, of immersion in Ringer's solution. Very low current densities were obtained, indicating the formation of a passive layer. Impedance spectra, represented in the Nyquist plan, exhibited a single constant system suggesting the formation of one layer.

Keywords: Stainless steel 316L · Ringer's solution · EIS · Impedance spectra Nyquist

1 Introduction

Stainless steel high resistance to corrosion derives from a dense film of chromium oxide (passive film) that naturally forms on the iron surface. Generally first requirement of any material serving in biological system is that it should be inert and not causing any undesirable reaction with its surrounding. When Stainless steel is placed inside a tissue, the interaction between the implant and tissue determines the degree of biocompatibility. It implies to the ability of the material to perform effectively with an appropriate host response for the desired applications [1]. Common medical devices include hip replacements, prosthetic heart valves, neurological prostheses and drug delivery systems [2].

Corrosion as an electrochemical process commences on the surface of Stainless steel implants and subsequently affects the body response. This is undesirably followed by release of ions such as chromium and nickel in surrounding tissue which can negatively affect the response of host body to biomaterial. The literature has proven that nature concentration of ions Cr + 3, Fe + 2, Ni + 1 inside the body may severely prevent the tissue functionality and can ultimately lead to severe health related

B. Abdelbaki et al. (Eds.): SMSD 2017, *Proceedings of the Third International Symposium on Materials and Sustainable Development*, pp. 301–305, 2018.
https://doi.org/10.1007/978-3-319-89707-3_34

consequences, therefore corrosion of material is the first issue to be considered when the material is designed as a bio implant. It is also necessary to establish an understanding of biological factors effecting corrosion. There are various in vivo and in vitro studies on corosion of metal implants which employed simulated body fluids [3].

2 Experimental

The corrosion behavior of orthopedic implant stainless steel 316L has been studied in a simulated body fluid (Ringer's solution), using potentiodynamic polarization and electrochemical impedance spectroscopy (EIS) techniques. Polarization experiments were conducted after 4 h, 6 h, 24 h, 48 h, 168 h, 215 h, 10 days and 15 days, of immersion in Ringer's solution.

A stainless steel 316L sample of the following chemical composition (as percentage) was served as working electrode (Table 1):

Table 1. Chemical composition of stainless steel 316L.

C	Mn	Si	Ni	Cr	Cu	Mo	V	Ti	Al
0.02	1.56	0.71	14.15	17.02	0.12	3.19	0.092	0.049	0.14

W	Sn	Co	B	Ta	Mg	Nb
0,25	0,023	0,037	0,009	0,041	0,021	0. 15

The working electrode was polished with different grades of SiC papers (500, 600, 800, 1200, 2000) and a finishing with diamond paste of 1 μm to obtain a mirror surface, degreased with acetone and rinsed with distilled water, before its immersion in a simulated blood medium (Ringer's solution). Potentiostatic polarization studies were carried out using Potentiostat/Galvanostat PGZ-301 (Table 2).

Table 2. Chemical composition of ringer's solution.

Composition	Concentrations (g/L)	Concentrations (mole/L)
NaCl	2.25	0.0384
KCl	0.105	0.0014
$CaCl_2$	0.12	0.0010
$NaHCO_3$	0.05	0.0005
H_2O	Graduated flasks 1000 ml	Graduated flasks 1000 ml

3 Results and Discussion

3.1 Absence of Agitation

See Fig. 1.

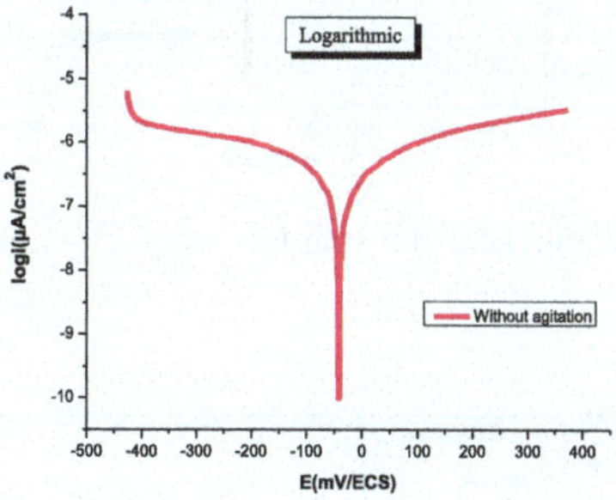

Fig. 1. Polarization curves obtained for the stainless steel 316 L immersed in Ringer's solution without agitation (T = 37 °C).

3.2 Presence of Agitation

In this work we followed the agitation action of orthopedic implant of stainless steel 316L in Ringer's solution (physiological solution blood simulated) at T = 37 °C (Table 3).

Table 3. Electrochemical parameters of stainless steel 316L in ringer's solution in the absence and presence of agitation.

	E_{corr} (mV/SCE)	i_{corr} ($\mu A.cm^{-2}$)	V_{corr} ($mm.an^{-1}$)	R_P ($K\Omega.cm^2$)
Without agitation	–43.9	0.5693	$6,658.10^{-3}$	170.79
With agitation	15.5	0.0974	$1,139.10^{-3}$	310.59

According to polarization curves (Figs. 1 and 2) we can be said that:

Stainless steel 316L in presence of agitation has a noble potential than unagitated solution and the polarization resistance of SS 316L in presence of agitation is twofold the resistance in solution of agitation (Fig. 3).

3.3 Impedance Characteristics Stainless Steel 316L

The existence of two capacitive arcs:

The first capacitive arc (high -frequencies) correspond to transfer of charge metal/solution interface.

The second capacitive arc (low- frequencies) corresponds to the formation and growth of oxide and hydroxide film (Table 4).

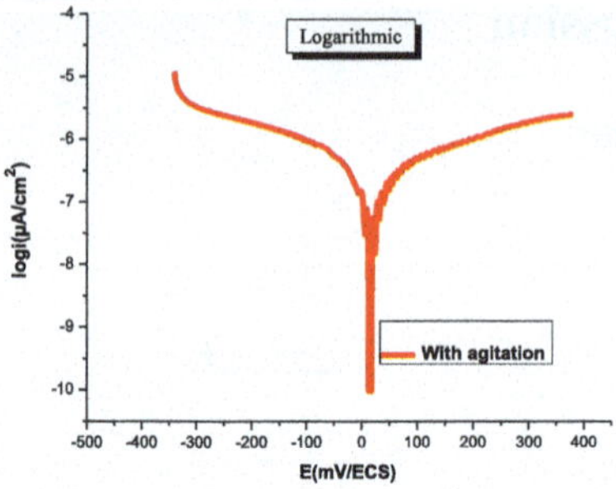

Fig. 2. Polarization curves obtained for the stainless steel 316 L immersed in Ringer's solution with agitation (T = 37 °C).

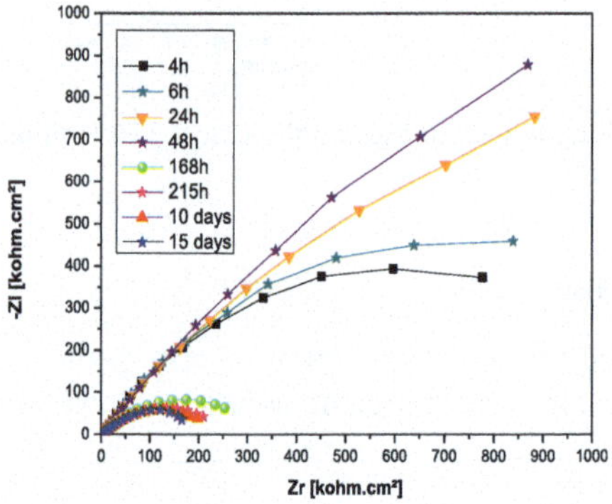

Fig. 3. Impedance spectroscopy analysis of 316L stainless steel in Ringer's solution for various immersion times at 37 °C.

Table 4. The equivalent circuit parameter obtained by adjusting electrochemical of impedance data for stainless steel 316L in Ringer's solution for various immersion times with a total error <4%.

		CPE_1			CPE_2		
Immesion time	R_e (Ωcm^2)	Q_d (Ω^1 Sn cm^{-2})	α_1	Rtc (Ωcm2)	Q_f (Ω^1 Sn cm^{-2})	α_2	R_f (Ωcm^2)
4 h	139,4	$1{,}32.10^{-5}$	0,785	$15{,}14.10^5$	$0{,}96.10^{-5}$	0,58	$0{,}827.10^4$
6 h	136,3	$1{,}12.10^{-5}$	0,787	19,37.105	$0{,}92.10^{-5}$	0,58	$0{,}939.10^4$
24 h	123,9	$0{,}53.10^{-5}$	0,806	$39{,}89.10^5$	$1{,}04.10^{-5}$	0,56	$1{,}828.10^4$
48 h	100,2	$0{,}78.10^{-5}$	0,722	$56{,}33.10^5$	$0{,}95.10^{-5}$	0,57	$1{,}586.10^4$
168 h	58,64	$1{,}35.10^{-5}$	0,655	$2{,}996.10^5$	$1{,}45.10^{-5}$	0,58	$2{,}864.10^4$
215 h	44,81	$1{,}56.10^{-5}$	0,685	1,984.105	$1{,}66.10^{-5}$	0,58	$5{,}842.10^4$
10 days	43,59	$2{,}02.10^{-5}$	0,658	$1{,}784.10^5$	$1{,}60.10^{-5}$	0,60	$5{,}979.10^4$
15 days	30,75	$3{,}62.10^{-5}$	0,674	$1{,}643.10^5$	$1{,}89.10^{-5}$	0,63	$47{,}7.10^4$

At the beginning of immersion the resistance transfer of charge is increased within 4 h, 6 h, 24 h, 48 h while the capacitance values of Qdc decreased slightly during the immersion time, due to construction of passive layer enriched with chrome; after 48 h of immersion, Rtc is decreased indicating deterioration of the passive layer.

4 Conclusion

Stainless steel surgical quality (316L stainless steel) is one of the materials that are widely used as implants in orthopedic surgery. The aim of this work is to study the biocompatibility of 316L stainless steel, for the manufacture of screw plates having high chromium and low carbon contents, in a simulated blood medium.

The electrochemical study reveals the best corrosion resistance of 316L stainless steel in Ringer's solution.

References

1. Leng, Y., Shi, D.: Introduction to Biomaterials, vol. 206, pp. 601–637. Tinsinghua University Press, Beijing
2. Kamachi Mudali, U., Sridhar, T.M., Raj, B.: Corrosion of bioplants, Sadhana, vol. 28, parts 3 and 4, pp 601–637, June/August 2003
3. Omanonic, S., Roscoe, S.G.: Electrochemical studies of the absorbtion behavior of bovine serum albumin on stainless steel. Langmuir **15**(23), pp. 8315–8321 (1999)

Application of the Rietveld Method in the Study of Phases in High Chromium Cast Irons

Fatma Zohra Boulassel[1(✉)], Benabbas Abderrahim[2(✉)], and Rekaik Mouna[1]

[1] Faculty of Science and Technology, University Mohamed Seddik Ben Yahia, Jijel, Algeria
boulassel.chimie@gmail.com
[2] Faculty of Science and Technology, University Akli Mohand Oulhadj, Bouira, Algeria
benabbas_am@yahoo.fr

Abstract. Grinding balls or crushing mills are elements used in cement industry. They require a high wear resistance under the action of abrasive products in the transformation of rock into fine particles smaller than a millimeter.

To satisfy the mechanical and technological requirements, these balls are made of high chromium cast irons (10 to 13% chromium).

The quantitative analysis of the phases containing these balls by the Rietveld method is an accurate and very important method to select their appropriate composition.

The characterization techniques used in this work are the DRX and the SEM.

Keywords: Grinding ball · High chromium cast irons · Rietveld
DRX · SEM

1 Introduction

The ball mills used in the cement industry allow the fragmentation of suspension by stirring a grinding zone. They are used to transform rock into fine caliber is less than one millimeter [1, 2]. Multiple impacts can damage balls and reduce grinding efficiency. The stresses they undergo are of various origins: breakage due to its contact with the wall of the mill, the balls between them, flaking caused by wear [3, 4].

To satisfy the mechanical and technological requirements, these balls are made of cast iron strongly alloyed with chromium (10 to 13% chromium) [5–8].

The significance of this study is the use of the Rietveld method to perform a quantitative calculation of the phases present in our samples by comparing the% Austenite/% Martensite ratios from the surface of the ball to the core.

B. Abdelbaki et al. (Eds.): SMSD 2017, *Proceedings of the Third International Symposium on Materials and Sustainable Development*, pp. 306–312, 2018.
https://doi.org/10.1007/978-3-319-89707-3_35

2 Experimental Process

2.1 Samples Studied

Two samples of two balls imported from India and premises supplied by the Tiaret Algeria smelter were cut and then well-polished (Fig. 1), and very similar chemical compositions (Table 1).

(a) imported ball

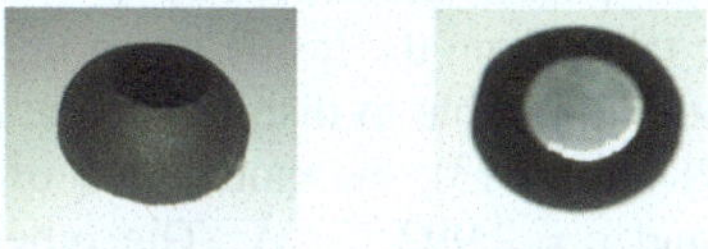

(b) local ball

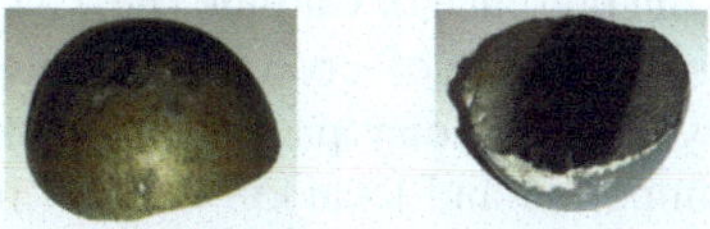

(a): imported ball with outer face (small) seen from the top and inner face below
(b): local ball by the internal face and the external part (bump)

Fig. 1. Representation of the samples studied.

Table 1. Chemical composition of the samples.

Element	Weight%
Fe	83.59–80.61
C	2.90–3.15
Si	0.69–0.89
Mn	0.37–0.38
P	0.067–0.11
S	0.032–0.052
Cr	9–11
Mo	0.2–0.4
Ni	3–3.2
Cu	0.15–0.20

2.2 Experimental Characterization Techniques

2.2.1 X-ray Diffraction

The X-ray diffractogram of JIJEL University of model D 8-ADVANCE of BRUKER-AXIS was used.

2.2.2 Scanning Electron Microscope (SEM)

The scanning electron microscope of Environmental Type XL from Philips was used at the University of Béjaia.

3 Results and Discussion

3.1 X-ray Diffraction Analysis (D.R.X)

To identify the nature of the phases present in these balls, we studied the X-ray diffraction pattern of a piece taken from the inside on an imported ball. Its comparison with the database PDF2 made it possible to distinguish the presence of the martensite phase (JCPDS No. = 44-1289 to 1293), the calculation of its crystalline parameters leads to a = 2,854 (2) Å and c = 2,992 (3) Å. The other phase is CFC structure austenite (JCPDS No. 31-0619) with a parameter equal to 3.659 (8) Å. These two phases are supplemented by the presence of Carbide (Fe, Cr) $_7C_3$ (JCPDS No. 05-0720) in a small proportion; the calculation of crystal parameters of this phase is made difficult by the low intensity and the noisy pace of its peaks. Then, according to the study of the spectra of the imported and local cut-out and polished balls, we note the presence of the same phases indicated above (Fig. 2a) However, there is a clear separation of the main peaks of residual Austenite and those of martensite, which

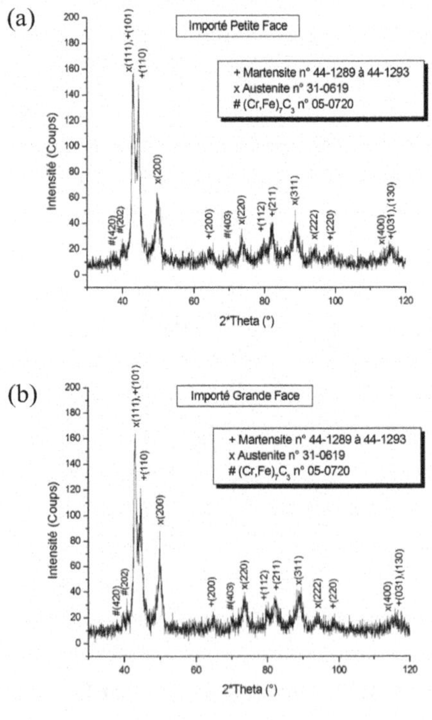

Fig. 2. (a) X-ray diffraction spectrum of the external face (small) of the imported ball, (b) X-ray diffraction spectrum of the inner (large) face of the imported ball.

highlights the quality of their formation with good crystallization and good homogeneity. The DRX Spectrum taken from the inner section (Fig. 2b); the same phases are also present here, with the same quality of formation.

These three phases are all present in the local ball (Fig. 3a). However, it is noted here that the peaks of the martensite are not completely solved. On the other hand, the fraction of the austenite phase is markedly smaller here than in the imported ball, and this is certainly related to the heat treatment undergone. Inside the ball, the quality of the formation of the two phases is better (Fig. 3b). This may be due to a temperature uniformity undergone by the inside of the ball.

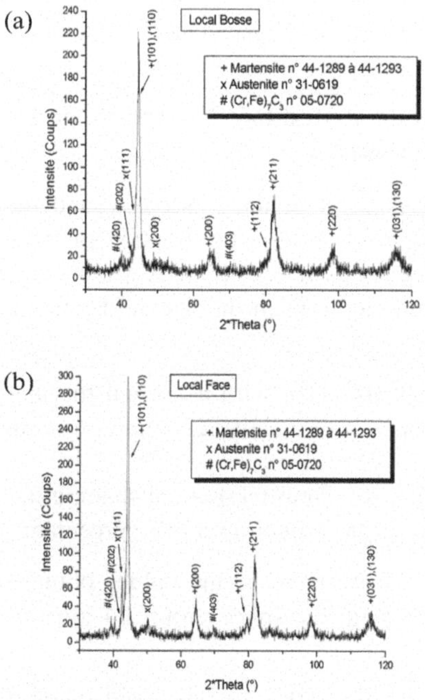

Fig. 3. (a) X-ray diffraction spectrum of the outermost part of the local ball, (b) Diffraction spectrum X of the inner face of the local ball

3.2 Microstructural Analyzes at M.E.B

The morphology of the imported ball (Fig. 4a) illustrates the presence of three different contrast phases from the EDX microanalysis. A homogeneous distribution of carbides in the form of slats (dark gray zone) in an austenite/martensite matrix; in the latter, the grains of the denser austenite appear lighter (white). On the contrary, a heterogeneous microstructure with a disorderly distribution of carbides (dark gray zone) in the austenite/martensite matrix is observed in the local ball (Fig. 4b) even the latter also presents a certain heterogeneity. This heterogeneity of the local bullet surely takes these sources in the distribution of the chemical elements during the production of the ball.

Probably, the austenitization temperature and the holding time before the quenching step are not optimized, which has prevented the best diffusion of the elements.

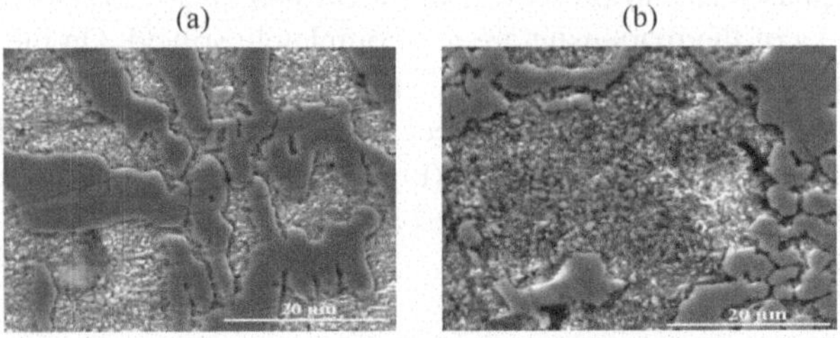

Fig. 4. (a) Micrography by SEM of the imported ball, (b) Micrography by SEM of the local ball

3.3 Quantitative Analysis of Residual Martensite and Austenite by the Rietveld Method

In our case, only the percentages of the Martensite and the residual Austenite were determined, without taking into account the fraction of the carbides M_7C_3 in view of their low peaks. The software used is RIETICA widely used by DRX specialists on powders. The structural models used in the Rietveld refinement calculation are:

- Martensite:
- Fe atom located at: 0.0, 0.0, 0.0, with a normal thermal agitation factor taken at B = 0.5 and an occupancy rate mainly reflecting chromium as the main secondary element.
- C atom at: 0.0, 0.0, 0.5, with a normal thermal shaking factor taken at B = 1 (lighter atom) and an occupancy rate reflecting a 4% percentage mentioned above.

The crystal parameters were fixed to the values obtained in a preliminary calculation and indicated above, with the space group I/4 mm.

- Residual austenite:
- Fe atom located at: 0.0, 0.0, 0.0, with a normal thermal agitation factor taken at B = 0.5 and an occupancy rate mainly reflecting chromium as the main secondary element.
- C atom at 0.5, 0.5, 0.5, with a normal thermal shaking factor taken at B = 1 (lighter atom) and an occupancy rate reflecting a 6% percentage.

The crystal parameters were fixed to the values obtained in a preliminary calculation and indicated above, with the space group Fm3 m.

Hence B is taken as an average value to avoid the divergence of the parameters sought.

The rest of the parameters are refined gradually to avoid the divergence of the calculation; the reliability factors obtained are between 12 and 14%, which is considered good in view of the strong background noise present in these diffractograms.

Since the number of form units in austenite is practically twice that of martensite, the following relationship is found:

(% Aust)/(% Mart) = (constant of the scale aust)/((mart scale constant)/2)

The complete results of the percentages of these two phases are summarized in the following Table 2:

Table 2. The percent ratios of austenite per martensite

Part of sample	%Aust/%Mart
The outer face of the imported ball	38/62
The inner face of the imported ball	44/56
The outer most part of the local ball	12/88
The inner face of the local ball	37/63

It is found that for both types of balls, the percentage of Martensite is greater outside than inside the sample; which was predictable as the temperature gradient front during quenching is large at the surface than in the mass. In comparison, the percentage of Martensite in the local ball is significantly higher than the imported ball. This is probably related to the non-homogeneity of the local ball mentioned above through the SEM images and the bad crystallinity as much of the Martensite as of the Austenite; this heterogeneity probably implies a high-temperature depletion of the austenite into added alloying elements. Thus, the local ball problem resides at the same time in the parameters of the heat treatment and in the heterogeneity of the samples.

4 Conclusions

The three phases martensite, residual austenite and carbides (Fe, Cr) $_7C_3$ are present with a good quality of formation, which improves the mechanical properties for the imported ball of the external (small) and internal (large) The martensite is predominant in the external phase whereas it is the austenite which is in excess in the internal face because of the quenching effect.

The same phases are present in the local ball. For the hump, the austenite phase is present in a reduced manner relative to the imported ball; this is due to the heat treatment conditions. For the internal surface, a good quality of formation is obtained.

Microcracks can be produced in the local ball because of a non-homogeneous distribution of the carbides in the austenite/martensite matrix, this heterogeneity is due to the austenitization temperature and the holding time.

The Rietveld calculation is very important for this quantitative study, which justifies that martensite is preponderant in the external surface as in the internal surface. The preponderance percentage is more pronounced in the local ball because of its non-homogeneity and the parameters of the heat treatment.

References

1. Romilliat, E.: Study of the modes of action of grinding agents on the grinding of clinker. Doctoral thesis. Ecole Nationale Supérieure des Mines of Saint-Etienne-France (2006)
2. Pintaude, G., Tschiptschin, A.P., Tanaka, D.K., Sinatora, A.: The particle size effect on abrasive wear of high-chromium white cast iron mill balls. Wear **250**, 66–70 (2001)
3. Chang, L., Lin, L., Liu, J.: Properties of cross-rolled low alloy white cast iron grinding ball. J. Iron. Steel Res. Int. **14**(5), 47–51 (2007)
4. Albertin, E., de Moraes, S.L.: Maximizing wear resistance of balls for grinding of coal. Wear **263**, 43–47 (2007)
5. Gates, J.D., Dargusch, M.S., Walsh, J.J., Field, S.L., Hermand, M.J.-P., Delaup, B.G., Saad, J.R.: Effect of abrasive mineral on alloy performance in the ball mill abrasion test. Wear **265**, 865–871 (2008)
6. Margerie, J.C.: Thermal treatment of cast iron engineering techniques metallurgical treaty, pp. M 1145-20; M 1145-24 (1994)
7. Asensio, J., Pero-Sanz, J.A., Verdeja, J.I.: Microstructure selection criteria for cast irons with more than 10 wt% chromium for wear applications. Appl. Mater. **49**, 83–93 (2003)
8. Abdelhamid, S.: Structural, microstructural and mechanical optimization of a highly alloyed chromium casting used in the manufacture of grinding balls. Doctoral thesis. University of Béjaia (2011)

New Materials Rigid Obtained by the Acid Hydrolysis of Starch Followed by a Grafting of Acrylamide with and Without Bis-acrylamide

Karima Boumerdassi(✉) and Aicha Serier

Laboratory of Coating, Materials and Environment, UMBB University, Boumerdes, Algeria
k.boumerdassi@yahoo.fr

Abstract. This work focuses on the development and characterization of new materials rigid based on non-plasticized starch. This material is obtained by the acid hydrolysis of starch followed by a grafting of acrylamide with and without bis-acrylamide. The results of infrared Fourier transform (FT-IR) showed the success of the destructuring of the starch and grafting. The rheological study by viscometer to confirm the shear thinning behavior of the two graft copolymers and rheology in dynamic mode showed that the graft copolymer in the presence of bis-acrylamide (D) is more elastic than the graft copolymer without the bis-acrylamide (E) and the stored energy is higher than the dissipated energy. The film analysis by XRD showed an improvement of the crystallinity of the starch after destruction and a change of the morphology by grafting. The analysis by scanning microscopy (SEM) confirmed the graft structure change. The gravimetric analysis (TGA) shows an improvement in thermal stability after the grafting of the starch acrylamide. However, the graft copolymer in the presence of bis-acrylamide (D) is more stable than the graft copolymer without the bis-acrylamide (E). Differential scanning calorimetry (DSC) confirmed that the graft allows the production of copolymers having specific characteristics.

Keywords: Starch · Hydrolyzed · Acrylamide · Films · Casting

1 Introduction

Simple or complex polysaccharides synthesized by living organisms form the composition of most cells (microbial, animal and plant). Among the best known are cellulose, starch, chitin, etc. These polysaccharides consist of 100 to 1000 D-glucopyranose units linked together by acetal bonds.

The chemical modification of polysaccharides such as cellulose or starch (Doi and Seinbuchel 2001), gives rise to the formation of charged polymers called polyelectolytes and thus they acquire remarkable physicochemical properties which result in numerous industrial applications in the food industry, medicine, cosmetics, the paper or oil industry… etc. Jammes (2007).

B. Abdelbaki et al. (Eds.): SMSD 2017, *Proceedings of the Third International Symposium on Materials and Sustainable Development*, pp. 313–329, 2018.
https://doi.org/10.1007/978-3-319-89707-3_36

Thus, in the context of the valorization of biomass and the development of new biodegradable materials, studies have been conducted on a very abundant polysaccharide on earth “starch”. It is a renewable polysaccharide, biodegradable and inexpensive; it is the reserve organ of energy of the higher plants and is in the form of a white powder constituted of micrometric grains. It is found in many botanical origins such as cereals, tubers and legumes. It is commonly used in the food, paper and textile industries. In recent years, it has also been used as a structural material after enzymatic or chemical modification (Rinoudo 2001), the introduction of new functional groups on its skeleton gives it different chemical and physicochemical properties.

Grafting is one of the chemical methods applied to improve the properties of starch. However, much attention has been given in the synthesis and study of reaction conditions for grafting starch with a variety of monomers such as acrylamide (Athawale and Lele 1998; Singh et al. 1984; Mostapha 1997), methacrylamide (Athawale et al. 1998), acrylonitrile and méthacrylnitrile (Athawale and Lele 2000).

Our work deals with the development and characterization of a new material based on hydrolyzed starch, grafted by acrylamide in the presence and without bis acrylamide. The elaborated samples are characterized by different analytical techniques in order to study their behavior, rheological, morphological and thermal.

2 Materials and Methods

2.1 Products and Reagents

Corn starch (soluble, Sigma-ldrich), HCl (C = 37%, d = 1.18 Prolabo) NaOH (M = 40 g/mol, C = 98%, Prolabo) Acrylamide (M = 71.08 g/mol, BIOCHEM chemopharma), N, N-methylene-bis (acrylamide) (M = 156 g/mol, Aldrich), potassium persulfate ($K_2S_2O_8$) M = 270.322 g/mol, BIOCHEM chemopharma), Ethanol (C = 96%, d = 0.803 g/ml, M = 84.01 g/mol Sigma-Aldrich), sodium carbonate (M = 105.988 g/mol, BIOCHEM Chemopharma).

2.2 Hydrolysis and Grafting Procedure

The processes of hydrolysis and grafting were performed according to the study conducted by Boumerdassi and Serier (2016).

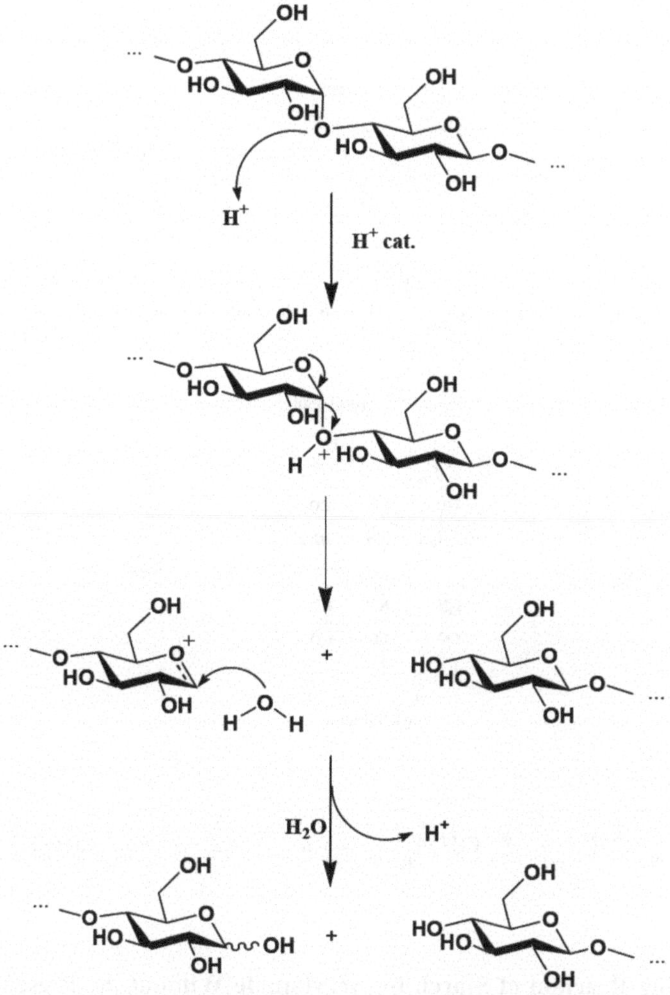

2.3 Grafting Reaction of Starch by Acrylamide in the Presence of Bis Acrylamide

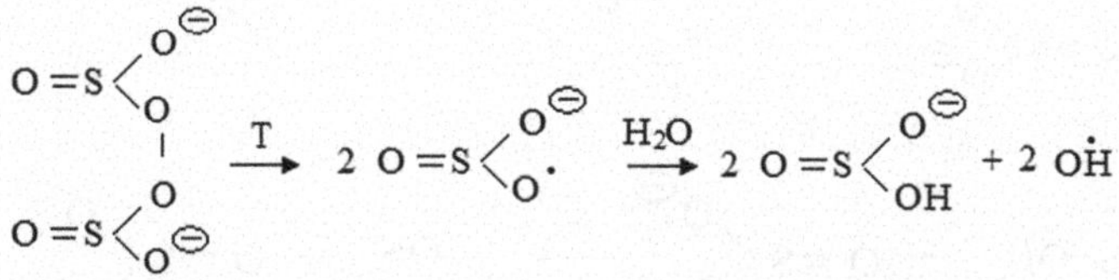

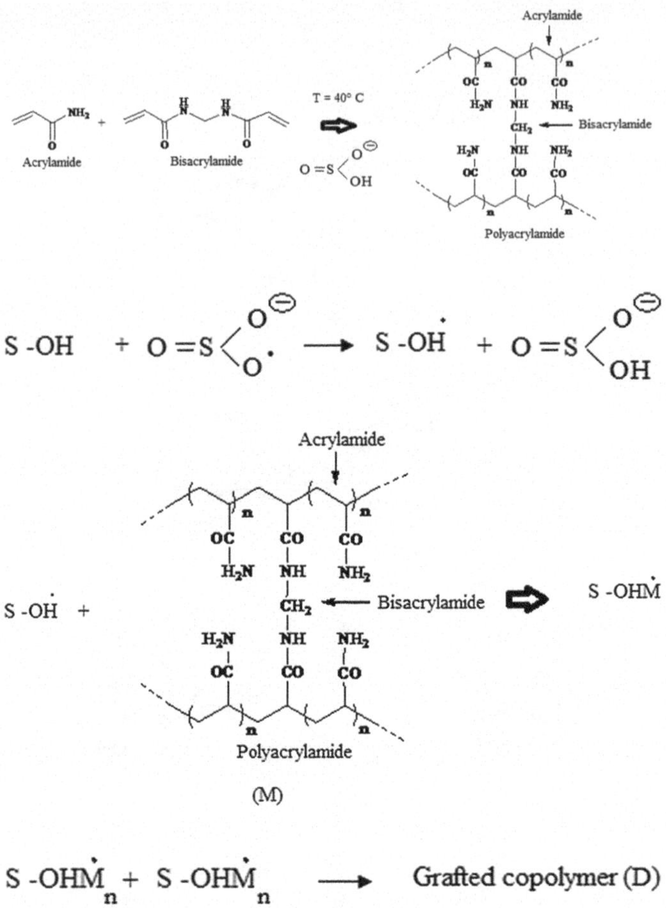

2.4 Grafting Reaction of Starch by Acrylamide Without the Presence of Bis-acrylamide

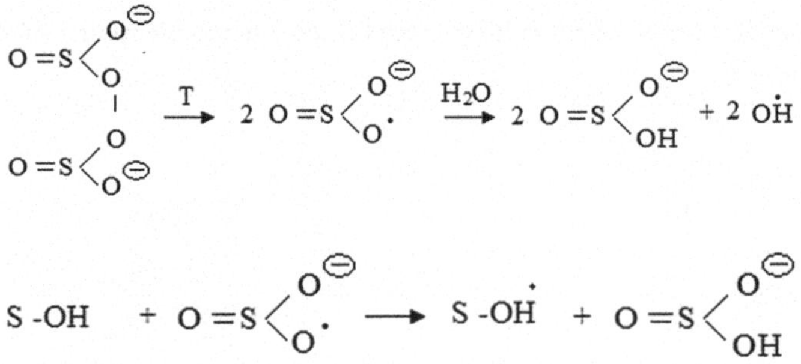

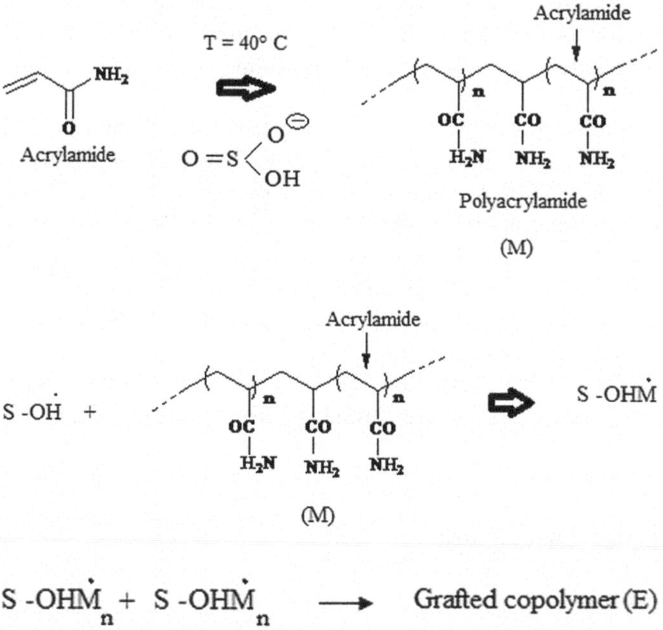

2.5 Characterization Methods

The prepared samples were characterized by various techniques.

Analysis of our samples was performed using a spectrometer of Fourier Transform Infrared brand "Perkin Elmer Spectrum One instruments" at room temperature with a resolution of 4 cm^{-1}.

To determine the rheological behavior of the individual suspensions, we use a viscometer "Thermo HAAKE VT 550" quilt system that monitors and characterizes fluid in rotation mode. A temperature controller type (HAAKE DC 30) is used.

We have chosen the quilt system using the NV geometry, the latter is chosen according to the HAAKE thermo standard, it is a cylinder of radius/length 20.1/60 mm and filling volume 9 cm, it is recommended for fluids of low viscosity (2-2000 mpa) and shear rates (27-27000 s^{-1}). All results are shown in the built-in LED digital display or in the optional HAAKE software rheowin.

As for test samples, they are obtained by setting a magnetic stirring volume of distilled water (20 ml) to which is added an amount of 0.2 g of material (powder) already prepared. To ensure a good homogenization, the suspension obtained is subjected to a continuous magnetic stirring for 30 min at 90 °C.

The microstructures of the various samples were analyzed with a scanning electron microscope type "JEOL JSM-6360LV scanning electron microscope".

Fragments of films were deposited on metal studs covered with adhesive. The composite material is insulating, the freshly fractured surfaces were rendered conductive by evaporating a layer of gold of 3 nm in thickness using a metallizer Cressington 108 Sputter coater.

The X-ray diffraction analysis (XRD studies) was carried out using a diffractometer PANALYTICALKV. The voltage, current and the wavelength of the X-ray source were 45 kV, 40 mA, and 1,5418 Å. Samples were analyzed on XRD between 2° and 70°

Thermogravimetric Analysis (TGA)

The analysis of our material was carried out in an apparatus for DSC/TGA coupled type NETZSCH STA409 PC/PG. A quantity of 10 mg of sample was analyzed. The scans are performed from 20 °C to 550 °C at a rate of 10 °C/min.

Differential scanning calorimetry (DSC)

We used a Q100 V8.1 Build 261 DSC device. A few mg (between 7.5 mg) of sample were analyzed. The scans are carried out from −80 °C. to 250 °C. at the rate of 10 °C/min.

3 Results and Discussion

3.1 FT-IR Study

3.1.1 Analysis of Native Starch and Hydrolyzed Starch

Infrared analysis was carried out on native starch (Fig. 1) and hydrolyzed starch (Fig. 2) by direct absorbance through KBr tablet.

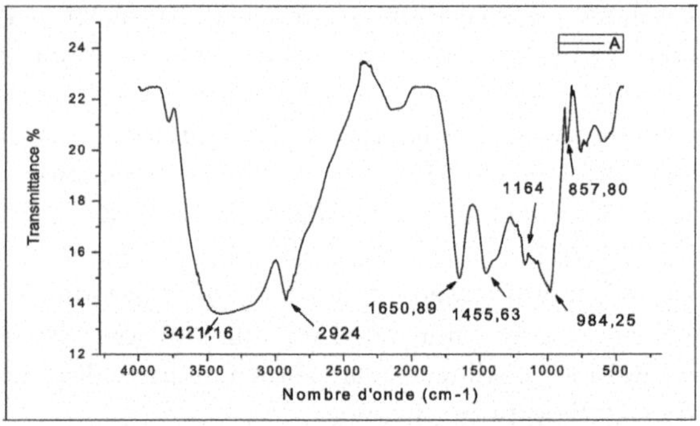

Fig. 1. IR spectrum of native starch

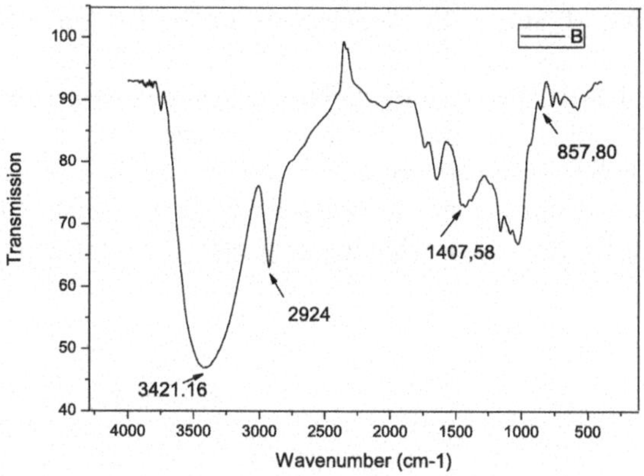

Fig. 2. FT-IR spectrum of hydrolyzed starch

In comparison between the native starch and hydrolyzed starch, we observed a shift of the stopper 3421,16 cm^{-1}. In the case of native starch to 3413,76 cm^{-1} in the case of the hydrolyzed starch, the appearance of a peak at 1407,58 cm^{-1} from deformation of the OH groups and the reduction of peak 857,80 cm^{-1} indicating the COC bond α (1-4) in the starch.

The acid hydrolysis cannot be performed at random, preferably it should be at the terminal connections of the chains (COC bond α (1–4)), especially those adjacent the non-reducing end units (Angellier 2005). This confirms the results.

The FT-IR spectrum of acrylamide (Fig. 3) shows two bands related to the unassociated valence vibrations of the NH2 bond of the primary amides at 3355,32 cm^{-1} and associated valence vibration to the (= NH) bond of the NH2 group of the amides at 3192,5 cm^{-1} cm^{-1}.

The presence of two peaks at 1617 cm^{-1} corresponds to the C= C bond and 1676,28 cm^{-1} assigned to CO elongation in CONH2.

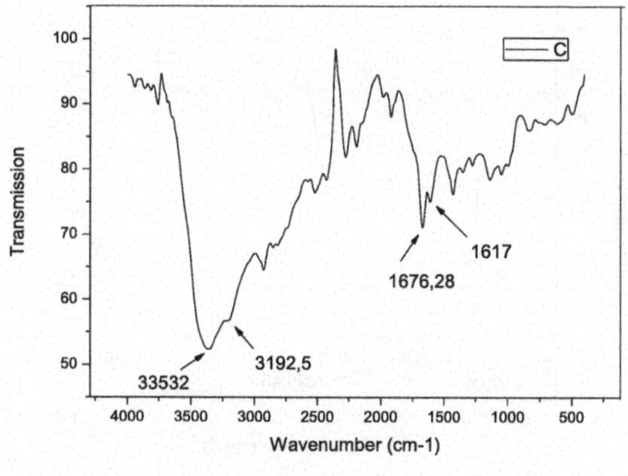

Fig. 3. FT-IR spectrum of acrylamide.

3.1.2 The Grafting of Starch by Acrylamide in the Presence of Bis acrylamide

Figure 4 shows the FT-IR spectrum of starch-g-acrylamide copolymer in the presence of bis-acrylamide (D), this spectrum shows the presence of a starch hydroxide broad OH group and amide NH group are presented overlapping the chains, leading to a peak 3435.39 cm^{-1} and the disappearance of peaks 1676.19 cm^{-1} and 1617 cm^{-1} attributed to CO elongation in $CONH_2$ of acrylamide suggesting the existence of grafting.

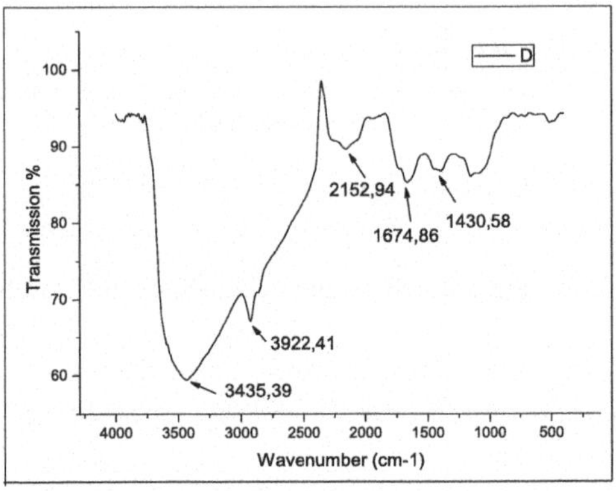

Fig. 4. FT-IR spectrum of acrylamide-grafted starch in the presence of bis-acrylamide

3.1.3 Analysis of Starch Grafted with Acrylamide Without the Presence of Bis-acrylamide

According to the FT-IR spectrum of the starch copolymer grafted with acrylamide without the presence of bis acrylamide (E) illustrated in Fig. 5, the presence of a very wide OH band of hydroxyl group of starch and NH of amide group and CH of the two materials, located at zone 3498–2932 cm^{-1}.

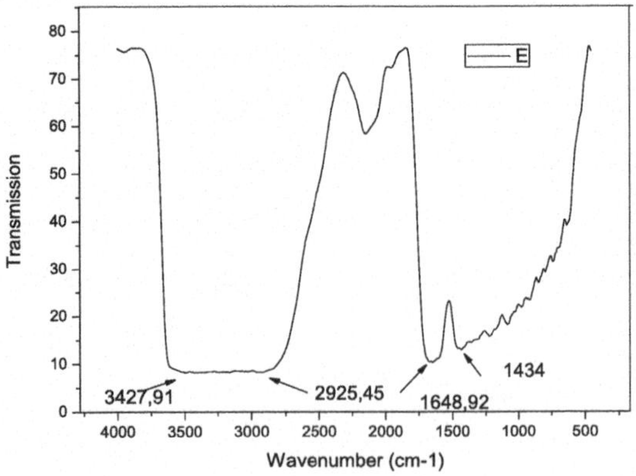

Fig. 5. FT-IR spectrum of starch grafted with acrylamide without the presence of bis-acrylamide

Moreover, an intense band observed at 1648.92 cm^{-1} at the elongation vibration of the C= O bond of the amide function and the disappearance of the peaks 1676.28 cm^{-1} and 1617 cm^{-1} attributed to CO elongation in $CONH_2$ of acrylamide suggesting the existence of grafting.

3.2 Rheological Properties

3.2.1 Viscosimetry Analysis

Figure 6 shows the flow curves of the suspensions: (B) the hydrolyzed starch, (D) the starch grafted with acrylamide in the presence of bis and (E) the starch grafted with acrylamide without bis acrylamide. The rheogram shows for the three suspensions a rheofluidifying behavior; the apparent viscosity decreases as a function of shear rate.

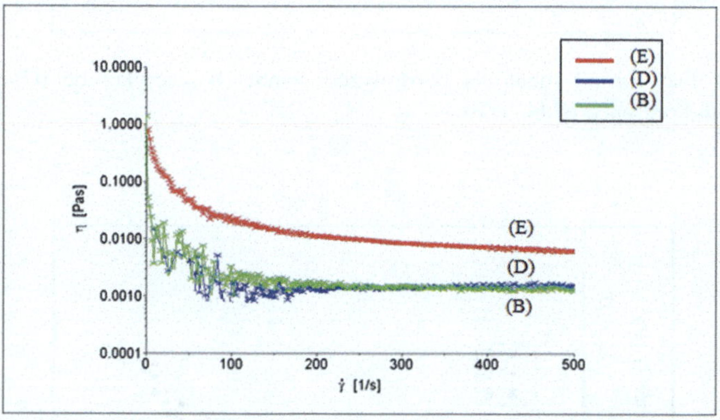

Fig. 6. Flow curves of (B) hydrolyzed starch, (D) starch- g- acrylamide in the presence of bis and (E) starch-g-acrylamide without bis

It is observed that the viscosity of the starch-g-acrylamide without bis-acrylamide (E) is higher than the viscosity of the hydrolyzed starch and starch-g-acrylamide in the presence of bis-acrylamide (B). The latter is more viscose than the grafted starch in the presence of bis-acrylamide (D).

3.2.2 Rheology in Dynamic Regime

3.2.2.1. Viscoelastic Properties

The results of the dynamic rheological study are illustrated in Figs. 7, 8, 9 and 10. These figures show, respectively, the storage modulus and the loss modulus at 60 °C.

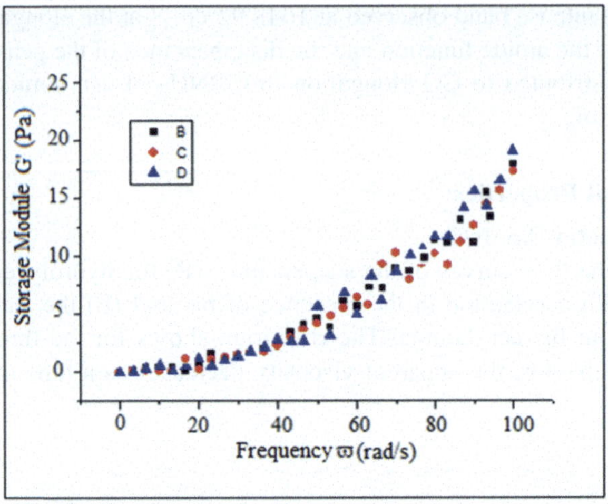

Fig. 7. Shows that for all materials (hydrolyzed starch (B), acrylamide (C) and starch-g-acrylamide in the presence of bis (D))

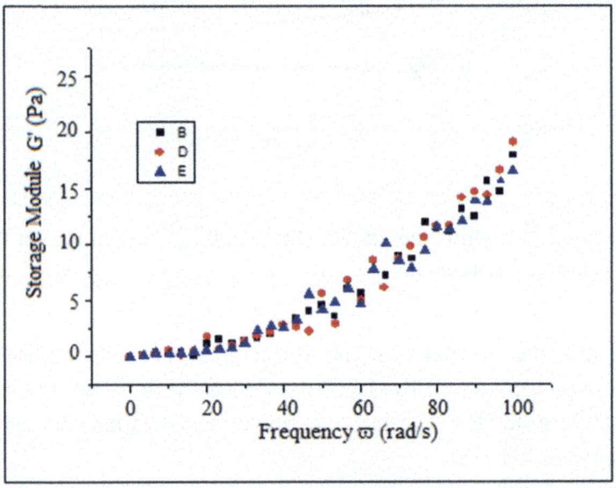

Fig. 8. Variation of storage modulus versus frequency for hydrolyzed starch (B), starch-g-acrylamide in the presence of bis (D) and starch-g-acrylamide without bis (E) (T = 60 °C).

The storage modulus increases with increasing frequency. This can be explained by the fact that at low frequencies macromolecular chains can reorganize, but as the frequency increases, they will have less time to relax (Shan et al. 2003).

It can also be noted that the curves are superimposed, especially at low frequency (≤ 50 rad/s), indicating that the three materials have the same storage modulus and therefore the same elasticity.

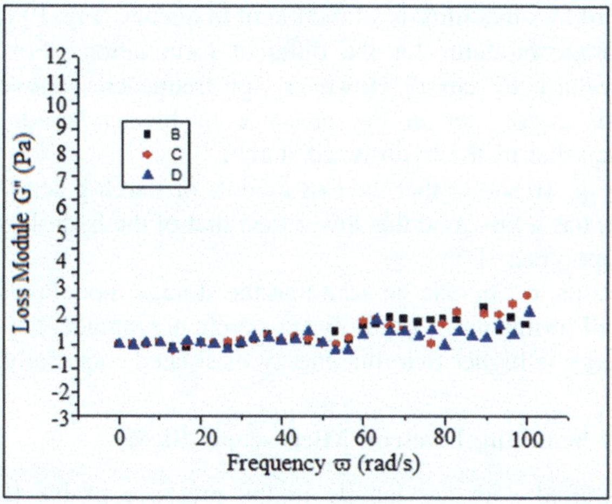

Fig. 9. Variation of loss modulus versus frequency for hydrolyzed starch (B), acrylamide (C) and starch-g-acrylamide in the presence of bis (D).

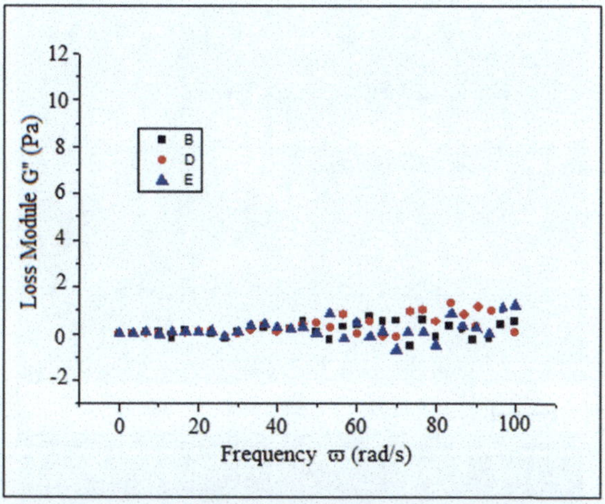

Fig. 10. Frequency-loss modulus variation for hydrolyzed starch (B), starch-g-acrylamide in the presence of bis (D) and starch-g-acrylamide without bis (E).

However, values tend to reverse at very high frequency (≥ 50 rad/s) where the starch-g-acrylamide copolymer in the presence of bis-acrylamide (D) has the highest value of the storage modulus.

For the starch-g-acrylamide copolymer without bis-acrylamide (E) (Fig. 8), the storage modulus values increase with increasing frequency but are lower than that of starch-g-acrylamide copolymer in the presence of bis (D) at high frequency. This result shows that the copolymer (D) is more elastic than the copolymer (E). (Fig 15).

The evolution of loss modulus as a function of frequency (Fig. 9) is small compared to that of the storage modulus for the different formulations. For low frequencies (<50 rad/s), the values are equal. However, for frequencies above 50 rad/sec, the starch-g-acrylamide copolymer in the presence of bis-acrylamide (D) has a loss modulus lower than that of the hydrolyzed starch.

Furthermore, Fig. 10 shows that the loss module of starch-g-acrylamide copolymer without the bis (E) has a loss modulus lower than that of the hydrolyzed starch and the high-frequency copolymer (D).

From these results, it can also be seen that the storage modulus is greater than the loss modulus for all formulations (hydrolyzed starch, acrylamide, copolymers), so that the conserved energy is higher than the energy dissipated, especially high frequency.

3.3 Analysis by Scanning Electron Microscope (SEM)

The copolymer grafted with acrylamide in the presence of bis (D) and the graft copolymer without the bis (E) showed a change in surface morphology due to the grafting of acrylamide on the starch (Fig. 11). The same results have been found in the literature (Singh et al. 2011; Mishra et al. 2011).

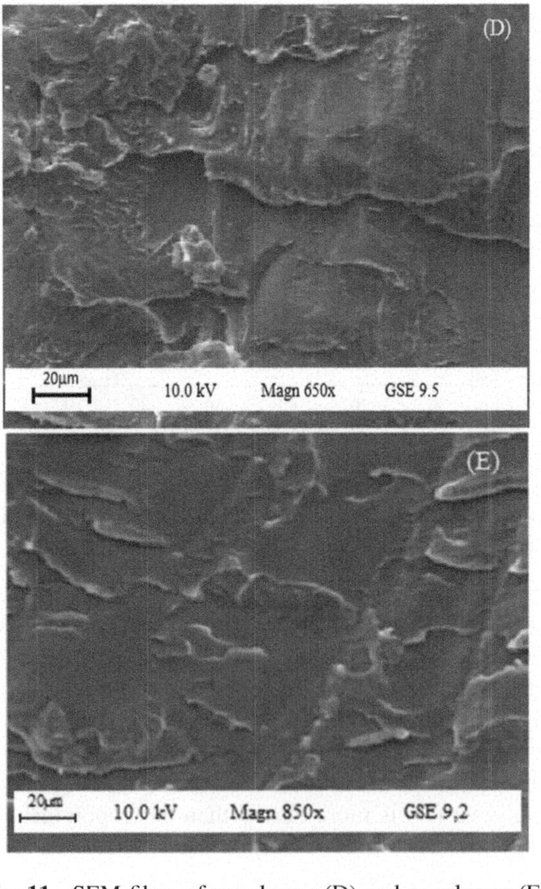

Fig. 11. SEM films of copolymer (D) and copolymer (E).

3.4 Structural Characterization by X-Ray Diffraction

The X-ray diffraction pattern of the hydrolyzed starch (A), the acrylamide (C) and the graft copolymer (D) is shown in Fig. 12. The diffraction pattern of the hydrolyzed starch showed the semi-crystalline structure, indicating a typical maize starch configuration. Pure acrylamide is crystalline in nature (Fig. 12 (C)).

During the grafting reaction, the semi-crystalline structure is transformed into an amorphous structure which is manifested by wide and dispersed regions for the two graft copolymers (D and E).

Microwave grafting by Singh et al. (2011) gave a crystalline copolymer, the same result was found by El-Hamshary et al. (2006). It should be noted that this is the reverse of what has been found, and is probably due to the grafting method.

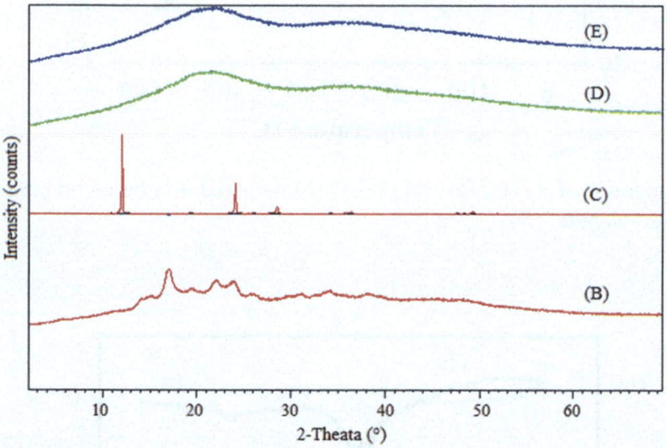

Fig. 12. X-ray diffraction spectra of hydrolyzed starch (B), acrylamide (C), graft copolymer in the presence of bis-acrylamide (D) and graft copolymer without bis-acrylamide (E).

In addition, we have noticed that copolymers (D) and (E) have the same type of (amorphous) structure with a large amorphous zone of copolymer (D) than of copolymer (E), which has been observed in scanning electron microscopy analysis (SEM).

3.5 Thermogravimetric Analysis

By examining the results of the TGA and DTG thermogrammes (Figs. 13, 14) it can be seen that the copolymer (D) is more stable than the copolymer (E) where the difference between the corresponding temperatures at the beginning of their degradation is about 12.30 °C (T1 (D) = 253.6 °C, T1 (E) = 265.90 °C).

On the other hand, decomposition of D takes place over a temperature range of 219.5 °C, which is greater than that of E (46.26 °C), indicating that the E decomposition reaction is faster than that of D.

Moreover, at the maximum decomposition temperatures of the two polymers, corresponding to the DTG peaks, the thermal degradation of E resulted in a greater mass loss (59.29%) than of the D (53.36%) (Fig. 15).

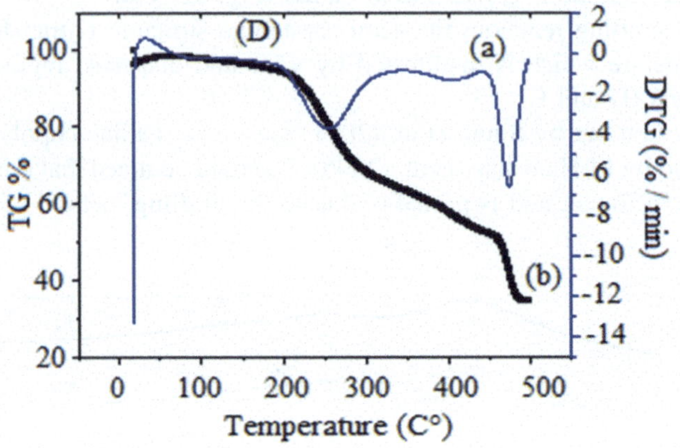

Fig. 13. Thermograms of (DTG) (a) and (TGA) (b) prepared film based on grafted starch in the presence of bis-acrylamide (D).

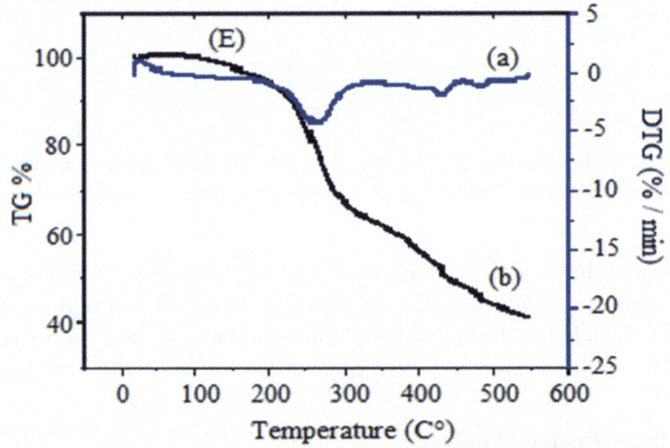

Fig. 14. Thermograms of (DTG) (a) and (TGA) (b) prepared film based on grafted starch without bis-acrylamide (E).

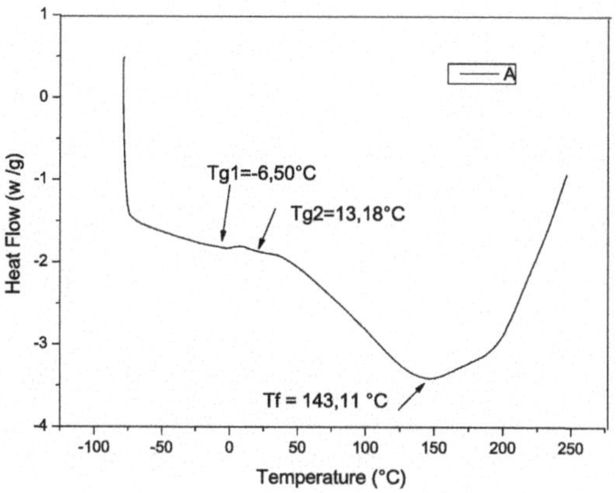

Fig. 15. DSC thermogram of native starch (A)

3.6 Differential Scanning Calorimetry (DSC)

After the acid hydrolysis of the starch (Fig. 16) we observed a single Tg = −6.40 °C and a melting temperature Tf = 236.81 °C. Compared with native starch, we observed an increasing of the melting temperature. This result is expected due to the increased crystallinity of the starch, confirmed by DRX. The acrylamide has a Tg = 179.57 °C and a Tf = 233.34 °C (Fig. 17).

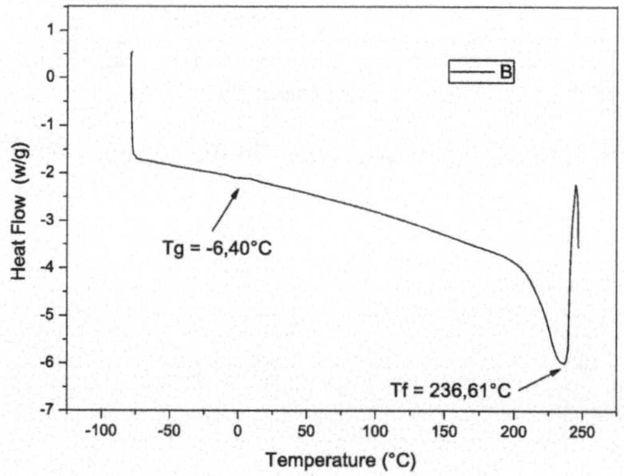

Fig. 16. DSC thermogram of hydrolyzed starch (B)

In the case of block or graft copolymers, phase separation often occurs. There are then two glass transitions. This is our case (Fig. 18), the starch-g-acrylamide copolymer in the presence of bis (D) has two glass transition temperatures Tg1 = −5.04 °C, Tg2 = 15.63 °C and Tf = 219.84 °C.

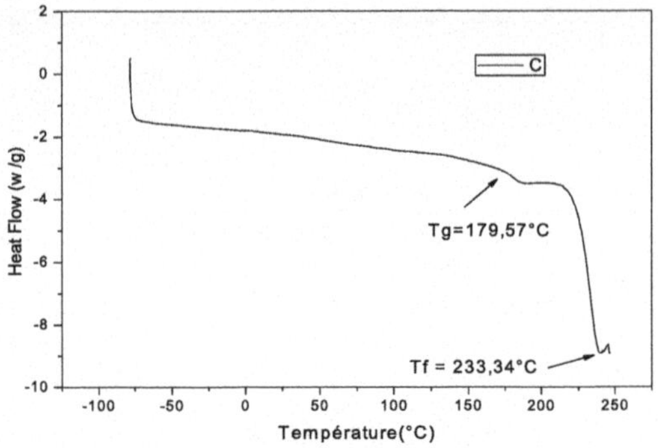

Fig. 17. DSC thermogram of acrylamide

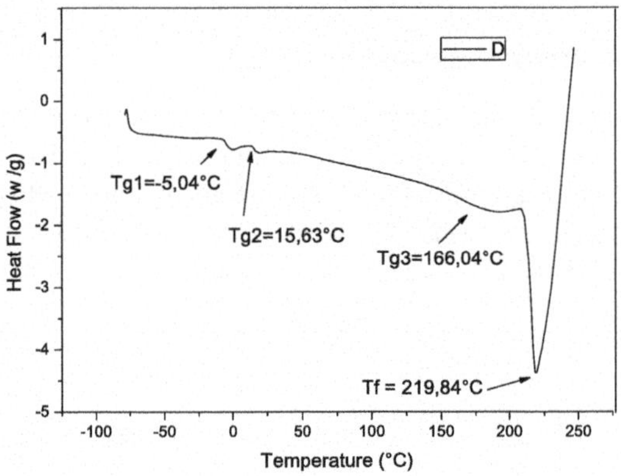

Fig. 18. DSC thermogram of graft copolymer in the presence of bis-acrylamide (D).

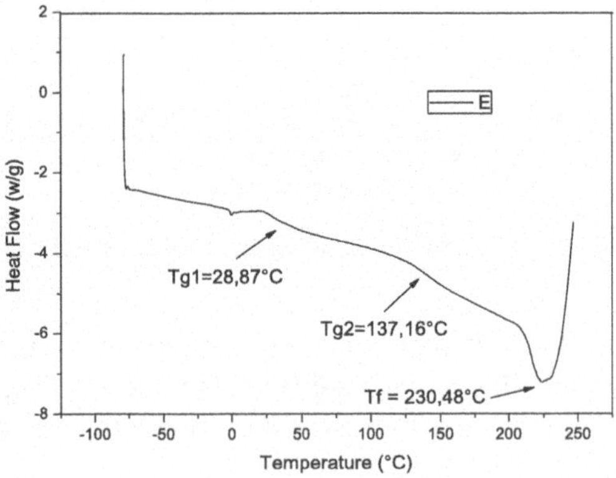

Fig. 19. DSC thermogram of graft copolymer without bis-acrylamide (E).

A third Tg3 = 166.04 °C can be observed, due to the chains of bis-acrylamide.

For starch-g-acrylamide copolymer without bis (E), Tg1 = 28.87 °C, Tg2 = 137.16 °C and Tf = 230.48 °C (Fig. 19).

4 Conclusion

From this work, it can be concluded that the grafting of the starch by acrylamide changes its semi-crystal structure to a totally amorphous and more stable structure. However, grafting in the presence of bis-acrylamide gives an elastic and more stable material relative to the material obtained by grafting without the bis-acrylamide.

References

Doi, Y., Steinbüchel, A. (eds.): Biological Systems and Bio-Technological Production, p. 75. Wiley-VCH, Weinheim (2001)

Jammes, C.: Thèse de doctorat. Université de Limoges, France, Chimie et microchimie de l'eau (2007)

Rinoudo, M.: Food Hydrocolloids **15**, 433–440 (2001)

Athawale, V.D., Lele, L.: Graft copolymerization onto starch. 3: grafting of acrylamide using ceric ion initiation and preparation of its hydrogels. Starch\Starke **50**, 426–431 (1998)

Singh, O.P., Sandle, N.K., Varma, I.K.: Graft-copolymerization of starch with acrylamide, II. Thermal behavior of graft copolymers. Angew. Makromol. Chem. **121**, 187 (1984)

Mostapha, K.H.M.: Synthesis of poly (acrylamide)-starch and hydrolyzed starch graft copolymers as a size base material for cotton textiles. Polym. Degrad. Stab. **55**, 125–130 (1997)

Athawale, V.D., Rathi, S.C., Lele, V.: Graft copolymerization on to maize starch I. Grafting of methacrylamide using ceric ammonium nitrate as an initiator. Eur. Polym. J. **34**, 159 (1998)

Athawale, V.D., Lele, V.: Thermal studies on granular maize starch and its graft copolymers with vinyl monomers. Starch/Starke **52**, 205 (2000)

Boumerdassi, K., Serier, A.: New biodegradable materials obtained by acid hydrolysis of starch flowed by grafting of acrylamide in presence of bis-acrylamide. Asian J. Chem. **28**(1), 91–94 (2016)

Angellier, H.: Nanocristaux d'amidon de maïs cireux pour applications composites. Thèse de Doctorat, docteur de l'universite joseph fourier Spécialité, Sciences et Génie des Matériaux, p. 245 (2005)

Shan, C.L.P., Soares, J.B.P., Penlidis, A.: HDPE/LLDPE reactor blends with bimodal microstructures—Part II: rheological properties. Polymer **44**(1), 177–185 (2003)

Singh, A.V., Nath, L.K., Guha, M.: Microwave assisted synthesis and characterization of phaseolus aconitifolius starch-g-acrylamide. Carbohydr. Polym. **86**, 872–876 (2011)

Mishra, S., Mukul, A., Sen, G., Jha, U.: Microwave synthesis of polyacryamide grafted starch (St-g-PAM) its applicability as flocculant for water treatment. Int. J. Biol. Macromol. **48**, 106–111 (2011)

El-Hamshary, H., Al-Sigeny, S., Ali, M.M.: Sunthesis and biological study of some amino acid functionalized starch-graft- polyacrylamide. Carbohydr. Polym. **64**, 282–286 (2006)

Copolymerization of B-Pinene with Limonene Using Heterogeneous Catalyst

K. Bekhemas(✉), A. Hachemaoui, and A. Yahiaoui

Laboratory of Organic Chemistry, Macromolecular and Materials, Faculty of Science and Technology, Road Mamounia, Mascara, Algeria
yayachimie@yahoo.fr, hachemaoui2002@yahoo.fr, yahmeddz@yahoo.fr

Abstract. The use of renewable resources (resources from biomass) is increasingly common these days. There is a strong market demand for products that are more environmentally friendly. In this work, we investigate a new method for the preparation of new material using the B-pinene and limonene belonging to terpene and a catalyst, more efficient, easy to implement it is Maghnite H^+, which is an Algerian montmorillonite treated by acid, is an effective catalyst for the cationic polymerization of many vinyl and heterocyclic monomers. We will try to determine the chemical structure of our product obtained by combining different spectroscopic methods (NMR and infrared spectroscopy (IR). In order to find the optimum conditions of the reaction and improve the productivity of polymer, we studied the influence of the amount of catalyst, time, temperature, solvent performance.

Keywords: B -pinène · Limonène · Heterogeneous catalyst · Terpene Maghnite H^+

1 Introduction

Polymers are large molecules, or macromolecules, consisting of many smaller chemically-linked molecules, termed monomer [1]. Polymers play an essential role in our lives owing to the variety of their properties. Up to now, fossil fuels have been the major source for the production of synthetic polymers. Of course, for a very long period, fossil fuels have provided an extraordinary variety of monomers at a relatively low price; however, things have changed. The depletion of fossil fuel reserves has led to a significant increase in the price of fossil fuel-based resources. In addition, growing environmental concerns have driven us to pursue a more sustainable way to produce polymer products.

Sustainable polymer reaction engineering is a field of application which seeks an effective and sustainable path towards the production of polymers, for instance, using renewable resources and developing solvent-free polymerization processes [2]. Actually, the rich diversity of renewable resources and their chemical structure indicate their strong potential for the production of more sustainable polymers in place of traditional fossil-based polymer products.

B. Abdelbaki et al. (Eds.): SMSD 2017, *Proceedings of the Third International Symposium on Materials and Sustainable Development*, pp. 330–336, 2018.
https://doi.org/10.1007/978-3-319-89707-3_37

In this project, we explored the use of B-pinene and limonene as monomers to make more durable polymers. Which belongs to the terpene family. B-pinene and limonene have been widely used in various applications, such as cosmetics, foods, drugs and solvents [3].

The purpose of this paper is to study the copolymerisation of B-pinene with limonene and examines the catalytic activity of an Algerian proton exchanged montmorillonite clay called "Maghnite-H^{+}" [4, 5]. The effect of some factors such as the amount of the Mag-H^{+} and effect of temperature are discussed.

2 Experimental Section

2.1 Materials

The preparation of the "Maghnite-H^{+}" was carried out by using a method similar to that described by Belbachir and al [6].

Dichloromethane, Chloroform, Carbon tetrachloride, Acetonitrile and toluene were purified following standard techniques and used after distillation over their respective drying agents. B-pinène and limonene were used as received.

2.2 Procedure and Polymer Characterization

In a flask containing a mass of beta-pinene and limonene, we pay an amount of Mag-H^{+} previously well dried for 30 min in the oven at 105°C , allowed all the stirring for 24 h at room temperature (scheme 1).

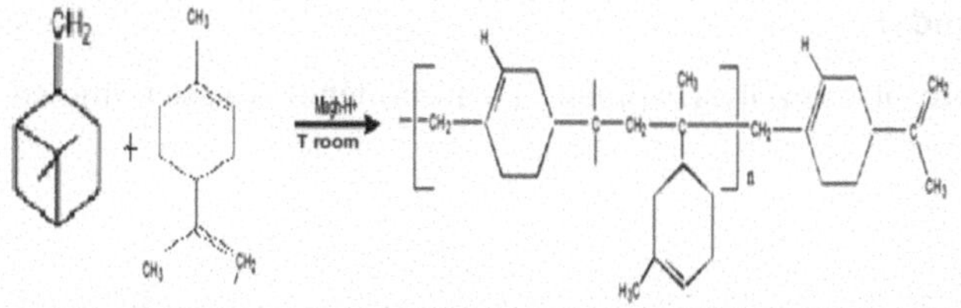

Scheme 1. Copolymerization of B-Pinene with Limonene by Maghnite-H.

The polymer obtained was a viscous honey liquid.

The experimental conditions and results are reported in Table 1.

Table 1. The experimental conditions and results of copolymerization.

Experiences	Mag- H^{+} (% by weight)	Température (°C)	Solvent	Yield%
1	25	Ambiante	–	32
2	25	100	Chloroforme	72
3	25	100	Dichlorometane	82

3 Results Section

3.1 Characterization of Products

- **FT-IR spectroscopy:**

The main IR bands with their attributions of the polymer prepared poly (B-pinene co Limonene) are shown in Fig. 1.

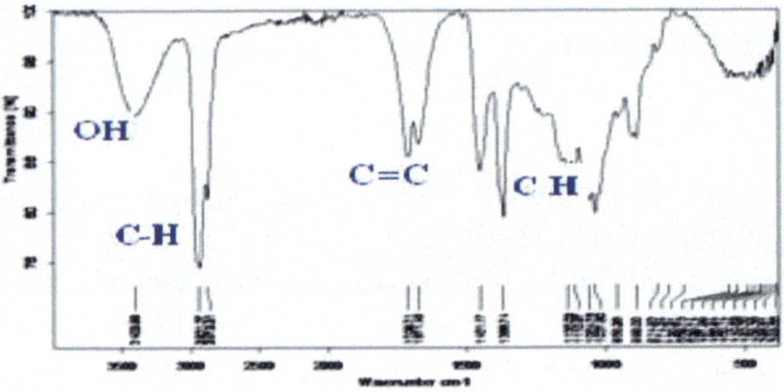

Fig. 1. IR spectrum of Poly (B-pinene-co-Limonene).

The copolymer is essentially characterized by the main bands situated at (3400 cm^{-1}, 2931 cm^{-1}, 1709 cm^{-1}, 1366 cm^{-1}, 1027 cm^{-1}) corresponds to the stretching vibration of bonds (O-H, C-H, C-H group of méthylènes, C = C, C-H groups of CH_2 and CH_3), respectively. Similar results are reported elsewhere [7, 8].

- **H-NMR Study:**

The structure of the poly (B-pinène co Limonène) was determined by 1H-NMR. Figure 2.

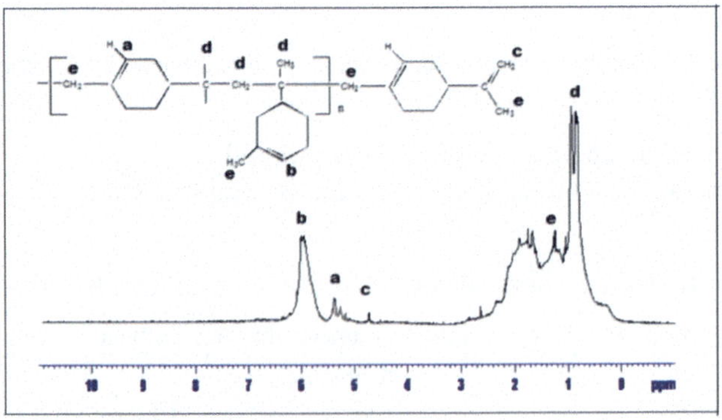

Fig. 2. ^{1}H-NMR spectrum of Poly (B-pinene-co-Limonene).

Figure 2 shows Signals (c) and (b) at 4.7–5.9 m which are attributed to the protons associated with the double bonds in limonene [9–11], which at 5.4 ppm corresponds to the endo-olefin in B-pinene isomers (a) [12].

^{13}C-NMR Study

The 13C-NMR spectrum Fig. 3 proves also the presence of both components in the copolymer shows numerous resonances in the regions of 133.93 125,00-128.81 32,00 46,92 ppm which characterize the carbons C_1, C_2, C_3, C_4 of poly B pinene and others at 11, 38, 36, 49, 44, 30, 129, 23, et 134 ppm assignable to C_5, C_6, C_7, C_8, C_9, C_{10}, C_{11}, C_{12}, C_{13} attributed to Poly limonene.

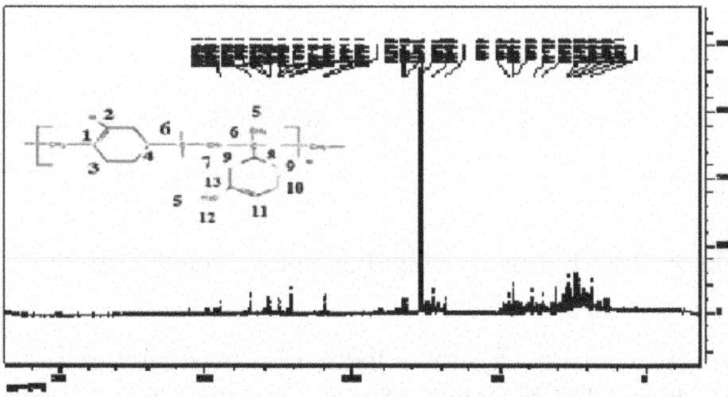

Fig. 3. ^{13}C-NMR spectrum of Poly(B-pinene-co-Limonene).

3.2 Effect of the Amount of Mag–H$^+$ on the Copolymerization

The effect of the amount of Maghnite-H$^+$ on the copolymerization was examined (Fig. 4). It can be noted that the yield increases with increasing "Maghnite-H$^+$" proportion. Which has a value of 24% for an amount of 25% maghnite-H$^+$. This phenomenon is probably the result of number of "initiating active sites" responsible of inducing polymerization, this number is prorata to the catalyst amount used in reaction Similar results are obtained by NJOPWOUO and al. [13].

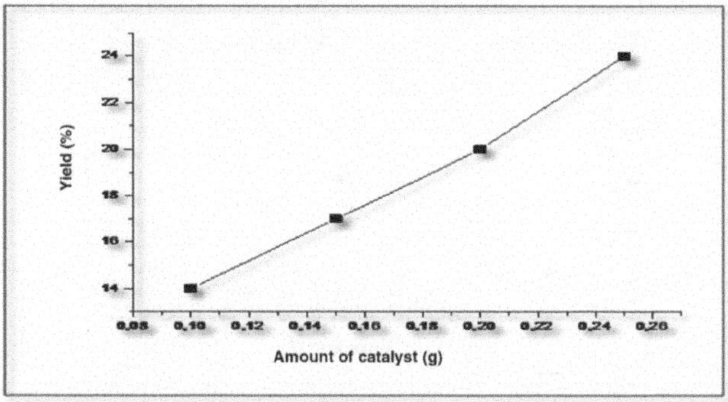

Fig. 4. Effect of the amount of Mag–H$^+$ on the yield of copolymerization.

3.3 Effect of the Reaction Time on the Copolymerization

Polymerization using 25% of Maghnite-H^+ was carried in bulk at room temperature and the reaction was monitored at various times.

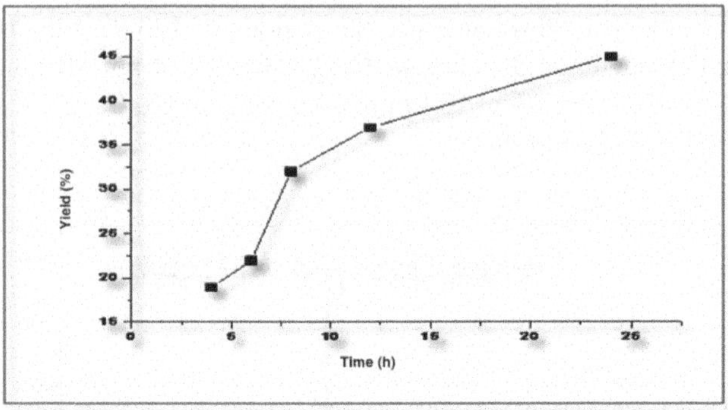

Fig. 5. Effect of the reaction time on the copolymerization.

The results are given in Fig. 5 shows the evolution of the yield of the copolymer with reaction time shows that the yield of the copolymerization reaction of B-pinène with Limonène evolves progressively to 32% after 8 h and then slows down until it becomes almost constant.

3.4 Effect of the Temperature on the Copolymerization

Figure 7 shows the experimental results for the copolymerization of B-pinene with Limonene induced by "Maghnite-H^+" proceed in bulk at different temperatures.

Figure 6 shows the influence of temperature on the yield of the copolymerization. It can be seen that a temperature increase leads to an increase in yield; the temperature is varied from 21 to 100 °C., the yield increases from 32 to 75%.

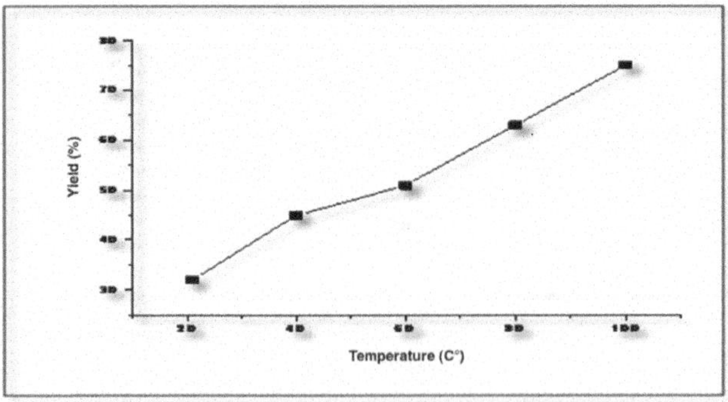

Fig. 6. Effect of the temperature on the copolymerization.

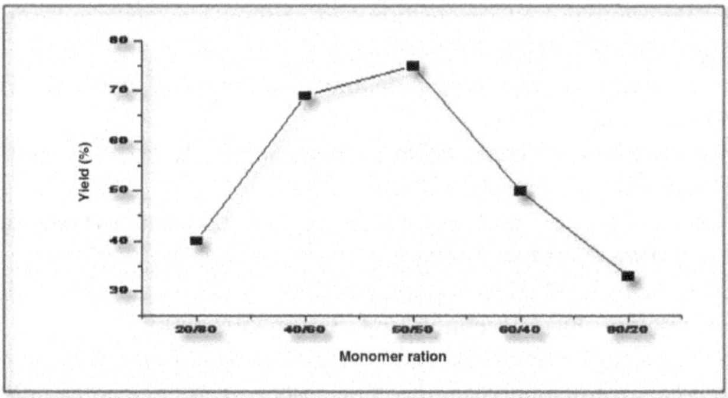

Fig. 7. Effect of Report (B-pinene/Limonene) on the copolymerization.

3.5 Effect of Report (B-Pinene/Limonene) on the Copolymerization

The results obtained in the figure showing the evolution of the yield as a function of the molar report.

It reaches a level of 75% for a molar report equal to 50/50, above this value, it can be seen that there is a drop in the values. This state of affairs can be interpreted by a polycondensation of type 1:1, that is to say one mole of beta-pinene with one mole of limonene.

4 Conclusion

In conclusion, it was proved that Mag-H^+ can be used to induce the copolymerization of B-pinene with Limonene. The copolymers were obtained by a very simple procedure via cationic transformation using Heterogeneous Catalyst which is a chemically and thermally stable catalyst having a proper mechanical resistance and non-toxic. The structure of copolymers was confirmed by ^{1}H-NMR, ^{13C}-NMR, and IR.

References

1. Odian, G.: Principles of Polymerization. John Wiley & Sons, Hoboken (2004)
2. Dubé, M.A., Salehpour, S.: Applying the principles of green chemistry to polymer production technology. Macromol. React. Eng. **8**(1), 7–28 (2014)
3. Silvestre, A.D.J., Gandini, A.: Terpenes: major sources, properties and applications. In: Belgacem, M.N., Gandini, A. (eds.) Monomers, Polymers and Composites from Renewable Resources, pp. 17–38. Newyork, ELSEVIER (2008)
4. Yahiaoui, A., Belbachir, M., Hachemaoui, A.: Cationic polymerization of 1,2-epoxypropane by an acid exchanged montmorillonite clay in the presence of ethylene glycol. Int. J. Mol. Sci. **4**, 572–585 (2003)
5. Ferrahi, M.I., Belbachir, M.: Preparation of poly (oxybutyleneoxymaleoyl) catalyzed by a proton exchanged montmorillonite clay. Molecules **9**, 968–977 (2004)

6. Belbachir, M ; Bensaoula, A.: Composition and method for catalysis using bentonites, US Patent No. 7, pp. 094–823 B2 (2006)
7. Norström, E.: Terpenes as renewable monomers for biobased materials Thèse, Master of Science Stockholm, Suède (2011)
8. Malika, A.: Contribution à l'étude de la Polymérisation de la β-pinène Catalysée Par la Maghnite-H+, mémoire de magister (2012)
9. Zhang, Y., Dubé, M.A.: Copolymerization of n-butyl methacrylate and d-limonene. Macromol. React. Eng. (2014, submitted)
10. Sharma, S., Srivastava, A.: Radical copolymerization of limonene with acrylonitrile: kinetics and mechanism. Polym. Plast. Technol. Eng. **3**, 485–502 (2003)
11. Sharma, S., Srivastava, A.: Synthesis and characterization of copolymers of limonene with styrene initiated by azobisisobutyronitrile. Eur. Polym. J. **9**, 2235–2240 (2004)
12. Li, A.-L., Zhang, W., Liang, H., Jiang, L.: Living cationic random copolymerization of b-pinene and isobutylene with 1-phenylethyl chloride/$TiCl_4$/$Ti(OiPr)_4$/nBu_4NCl. Polymer **45**, 6533–6537 (2004)
13. Njopwouo, D., Roques, G., Wandji, R.: Clay Miner. **22**(2), 145–156 (1987)

Synthesis, Characterization, Electrochemical of Ligational Behavior of Curcumin Drug Towards Some Transition Metal Ions

L. Abane Merzouk(✉) and A. Adkhis

Laboratoire de Physique et Chimie des Matériaux,
Université Mouloud Mammeri, Tizi-Ouzou, Algeria
abane_lamia@yahoo.fr

Abstract. Complexes of Cr(III), Mn(II), Fe(III) and Zn(II) with curcumin ligand as antitumor activity were synthesized and characterized by conductivity, infrared spectra and electronic spectra. The molar conductance data confirm that all the complexes are no electrolytic, The IR spectral data suggested that the ligand behaves as a monobasic bidentate ligand towards the central metal ion with an oxygen's donor atoms sequence of both –OH and C=O groups under keto–enol structure.

The electrochemical behavior of the M(II) and M(II) complexes was determined by cyclic voltammetry which shows that the chelate structure and electron donating effects of the ligands substituent are among the factors influencing the redox potentials of the complexes.

Keywords: Curcumin · Transition metal ion · IR spectroscopy UV-visible · Cyclic voltammetry

1 Introduction

Curcumin "cur" (1,7-bis-(4-hydroxy-3-methoxyphenyl)-1,6-heptadiene-3,5-dione) is a neutral yellow-orange dye. The medicinal activity of curcumin has been known since ancient times. It has also been used as a photodynamic agent useful for the destruction of bacteria and tumor cells. As such, curcumin was predominantly used to treat inflammatory conditions including bronchitis, colds, parasitic worms, leprosy, arthritis and inflammation of bladder, liver and to improve symptoms such as fever and diarrhea [1].

Curcumin exists in equilibrium with its enol tautomer (Fig. 1). The bis-keto form predominates in acidic and neutral aqueus solution. In contrast, above pH 8, the enolate form of the heptadienone chain predominates, and curcumin act mainly as an electron donor. The strong chelating ability of diketones has been widely invstigated towards some metal ions with differentiations in molar ratio between metal ions and curcumin ligand according to many items like solvent, metal salt, stoichiometric quantity and PH media (9–12). The goal of this paper is to get a wide understanding of the structural and spectral properties of curcumin and their Cr(III), Mn(II), Fe(III) and Zn(II) transition metal complexes [2].

B. Abdelbaki et al. (Eds.): SMSD 2017, *Proceedings of the Third International Symposium on Materials and Sustainable Development*, pp. 337–344, 2018.
https://doi.org/10.1007/978-3-319-89707-3_38

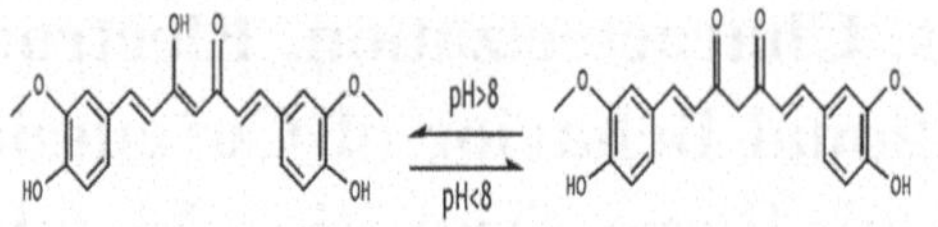

Fig. 1. Keto-enol equilibrium of curcumin ligand

2 Experimental

2.1 Synthesis of Cur Complexes

The complexes were synthesized with the same procedure. We have dissolved the Fe $(NO_3)_2$ $9H_2O$, $Fe(SO_4)7(H_2O)$, $Cr(NO_3)_29(H_2O)$, $Mn(NO_3)_24(H_2O)$ and $ZnCl_2$ salts respectively with 1 mol molar fraction in water distilled in a round-bottomed flask, then we have added curcumin with a 2 mol molar fraction previously dissolved in ethanol. The reaction mixture (salt + ligands) of each reaction is stirred and refluxed at 60 °C for 30–40 mn. The colored solid precipitated which isolated on cooling were filtered, and dried in the open air.

2.2 Physical Measurement

The molar Conductance values were obtained for 10^{-3} M in DMSO solution at 25 ± 1 °C determined using MPM-H2 melting point meter. The IR spectra were recorded with BRUKER TENSOR 27 spectrophotometer, between 4000 and 400 cm^{-1}. The electronic spectra were tested by thermo UV-Visible spectrophotometer (in DMSO solution). The All electrochemical experiments were carried out using a potentiostat-galvanostat EGG-273 a controlled with Power-suite software. The electrochemical cell was equipped with a modified carbon past disk as the working electrode, a platinum electrode as counter electrode and silver chloride (Ag/AgCl) electrode as the reference electrode. The electrolytic bath consisted of the complexes in DMSO solution (10^{-3} M) and sodium perchlorate ($NaClO_4$ 10^{-1} M) was used as a supporting electrolyte. The electrochemical characterization was performed using the cyclic voltammetry (CV) with a scan rate of 50 mV/s.

3 Results and Discussion

The analytical data and physical properties of the curcumin complexes are given in Table 1. The melting point of all complexes exceeds 360 °C indicating that all complexes are stable in air and can be preserved at room temperature. The molar conductance values of all complexes in DMSO at 10^{-3} M are in the range 2–12 $\Omega^{-1}.cm^{-1}.mol^{-1}$ and suggest that the complexes are non-electrolytes [3].

Table 1. Analytical data of the complexes

Complexes formula	Color	Molar mass (g/mol)	Yield (%)	Solubility	Melting point (°C)	Conductivities Λ (Ω cm−2 mol−1)
$[Fe(cur)_2(H_2O)_2]SO_4$	Marron	422,23	48	DMSO	300	5,54
$[Cr(cur)_2(NO_3)(H_2O)]$	Marron	418,38	33	DMSO	280	6,49
$[Mn(cur)_2(H_2O)_2]$	Orange	421,32	33	DMSO	300	5,62
$[Fe(cur)_2(H_2O)_2]SO_4$	Marron	422,23	48	DMSO	300	5,54

3.1 Infrared Spectra

The main absorption bands IR of the ligand and those of the complexes are represented in the Table 2. The IR spectra of the Fe (III), Cr (II), Mn (II) and Zn (II) complexes with curcumin are illustrated in Fig. 1. The latter with those of the corresponding ligands allowed us to note the following points:

Table 2. Main IR spectral data

Compound	Cur	$[Fe(cur)_2(H_2O)_n]$	$[Cr(cur)_2(H_2O)_n]$	$[Mn(cur)_2(H_2O)_n]$	$[Zn(cur)_2 (H2O)n]$
ν(OH)	3478	3452	3432	3740	3457
ν(MO)		466	463	466	473
ν(C=CH)	847	844	846	826	669
ν(CO) phenol	1281	1282	1262	1285	1125
ν(CO)	1431	1381	1415	1385	1378
ν(C=C) aliphatic	1511	1511	1514	1505	1508
ν(CO)	1627	1617	1625	1637	1637
ν(=C-H)	3085	3070	3085	3072	
ν(OH)	3478	3452	3432	3740	3457

In all the spectra analyzed we have observed the appearance of a broad band at a frequency [3475 cm^{-1}–3440 cm^{-1}] attributed to the vibrations ν (OH) of the phenol form of curcumin and to the hydration (OH) water molecules coordinated to the metal, the strip (C-O) is shifted slightly in the set of complexes towards 1284 cm^{-1}, suggesting that the oxygen in the carbonyl groups of the phynol form is not coordinated with the metal. Absence in the spectra of the complexes of the ν (CO) band at 1628 cm^{-1} indicated the coordination of the central metal with the oxygens of the carbonyl groups of the enol form. We also observe the appearance in each spectrum of a new band between [477 cm^{-1}–463 cm^{-1}] which does not exist in the spectrum of curcumin probably attributed to the (M-O) bond [4].

3.2 Electronic Spectra

The results observed on the electronic absorption spectra of the ligands and the complexes are shown in the Table 3. The comparison of the spectra of the curcuma alone with those of the complexes after complexation allows us to deduce the local symmetry of the metal ions. We observed in the spectra of the four complexes the appearance of intense peaks between [271 nm–275 nm] corresponding to the intraligand transition $\pi \rightarrow \pi^*$ in the ultraviolet range, confirming the presence of curcumin in the synthesized compounds. On the other hand, the spectra of the complexes exhibit bands in the [404 nm–431 nm] domain probably attributed to the transitions of $n \rightarrow \pi^*$ previously existed in the spectrum of curcuma indicating the involvement of the carbonyl group in the complexation of the metals. In the visible UV spectrum of the complex with Fe (III), we observe a shoulder at 584 nm due to the transition of the metal-ligand charge transfer (ML) and a broad band at 631 nm is observed for the transition d-d characteristic of octahedral geometry around the central ion Fe (III) [5] (Fig. 2).

Table 3. Electronic spectra data of cur complexes.

Compound	λ (nm)	Electronic transition
Cur	271	$\pi \rightarrow \pi$
	430	$n \rightarrow \pi^*$
$Fe(cur)_2$	271	$\pi \rightarrow \pi$
	404	$n \rightarrow \pi^*$
	584	Transfère de charge M-L
	631	Transition d-d
$Cr(cur)_2$	275	$\pi \rightarrow \pi$
	430	$n \rightarrow \pi^*$
$Mn(cur)_2$	277	$\pi \rightarrow \pi$
	431	$n \rightarrow \pi^*$
$Zn(cur)_2$	338	$\pi \rightarrow \pi$
	426	$n \rightarrow \pi^*$

3.3 Cyclic Voltammetry

The electrochemical study by cyclic voltammetry of curcumin and Fe (II) and Fe (III) iron complexes was carried out in solution in DMF (10^{-3} M). The supporting electrolyte is sodium perchlorate (NaClO 4) at 10^{-1} M, the measurements were carried out in a scanning range of –1.0 to +1.0 V. All the electrochemical results, in this case the redox potentials, ΔEp (potential difference between the anode peak and the cathodic peak) are illustrated in the Table 4.

Electrochemical Behavior of Curcumin

Before the electrochemical characterization of the complexes, we first studied the electrochemical behavior of the curcumin alone, the voltammogram of this ligand is represented in Fig. 4.

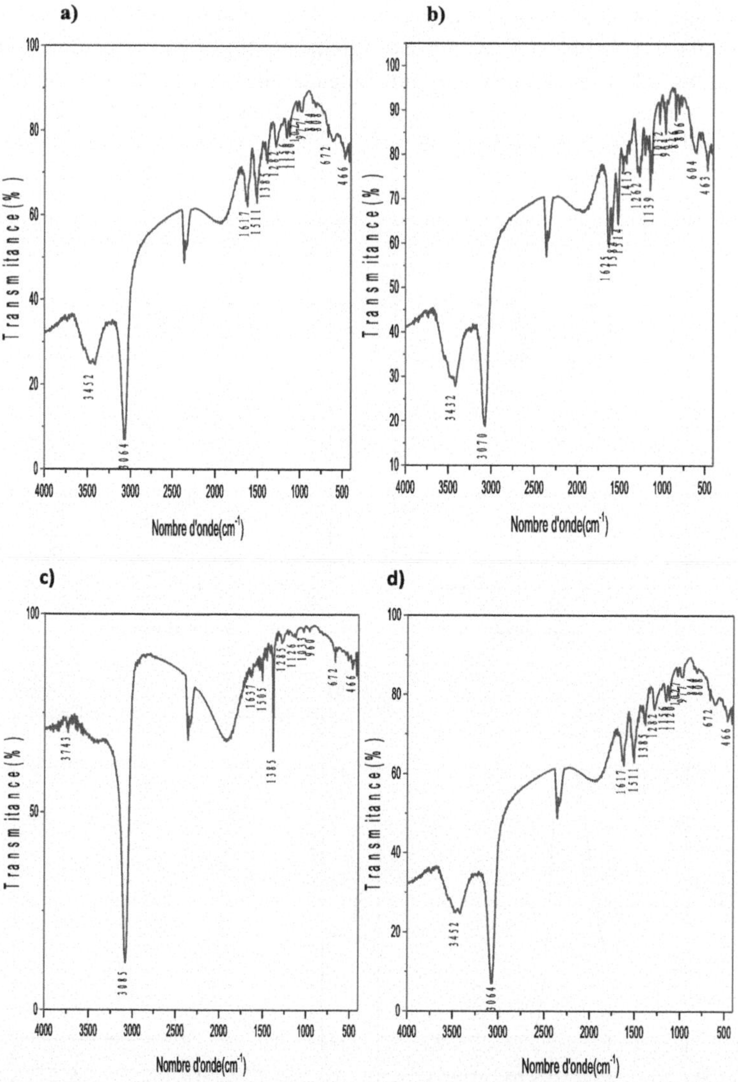

Fig. 2. IR spectrum complexes (a) $Fe(cur)_2$, (b) $Cr(cur)_2$, (c) $Mn(cur)_2$ and (d) $Zn(Cur)_2$.

Table 4. Potential values of the electrochemical processes of the Fe(III)

Compound	Couple red/ox	EPa(I) (mV vs. SCE)	EPa(I_) (mV vs. SCE)	ΔEp (mV)
Cur		–	426	426
		–	908	980
$[Fe(cur)_2]$	Fe^{2+}/Fe	−110	+840	730

The voltammograms show the appearance of two anodic oxidations, the first at an anode potential (I) of 0.420 V which is a characteristic of the enol function and the second at about 0.890 V (II) probably attributed to the phenol [6]. On the other hand, no peak is observed in a cathode potential, which makes it possible to deduce that there was no cathodic reduction process when the scanning potential was reversed (Fig. 3).

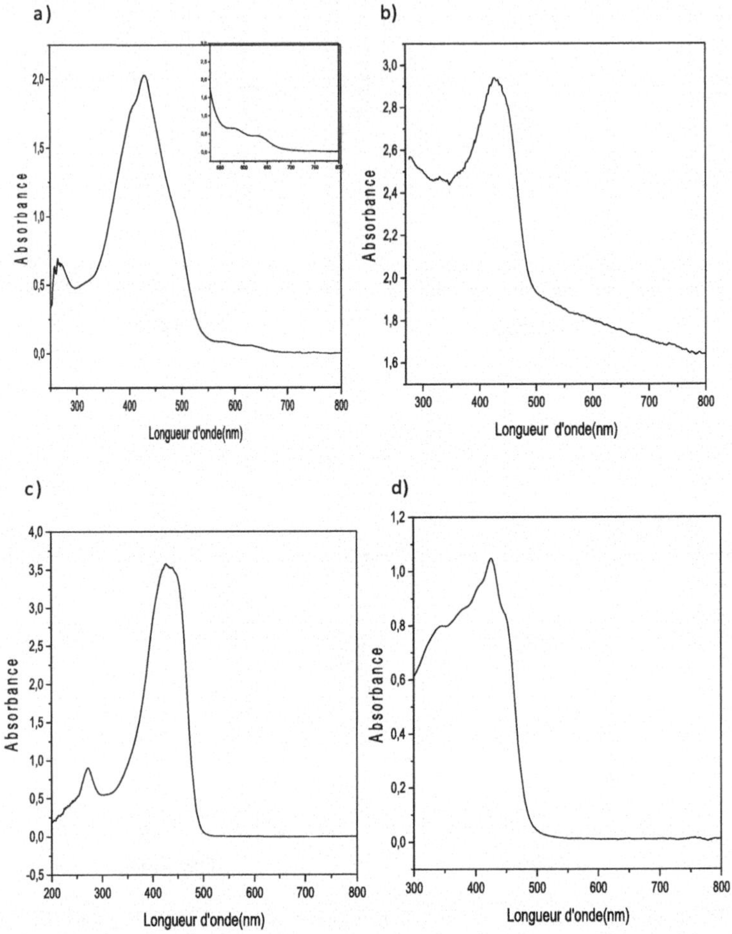

Fig. 3. UV-visible spectra of complexes (a) $Fe(cur)_2$, (b) $Cr(cur)_2$, (c) $Mn(cur)_2$ and (d) Zn $(Cur)_2$.

Electrochemical Behavior of Complexes

The cyclic voltammograms of the complexe are illustrated in Fig. 5. We found that in the case of Fe (III) when the analysis rate was initiated in the anodic direction, there was no oxidation process, however, when the analysis was reversed, a reduction process producing a tray-shaped trace has a potential –0.260 V which is attributed to the reduction of Fe (III) to Fe (II) [6].

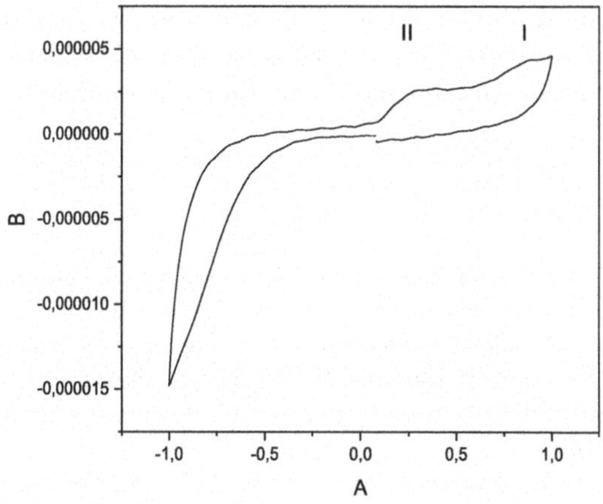

Fig. 4. Cyclic voltammograms of the curcumine [1.68×10^{-5} M] at pH 10 with a potential scan rate of 100 mVs^{-1}

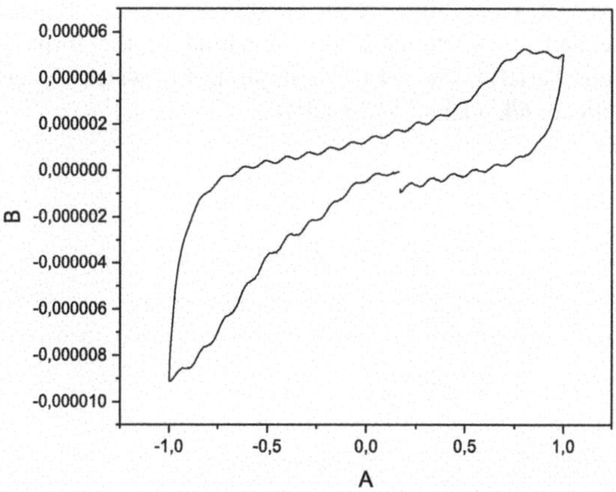

Fig. 5. Cyclic voltammograms of the [FeIII$(cur)_2$]

4 Conclusion

In the present study we have synthesized novel Fe (III), Mn(II), Cr(II) and Zn(II) complexes, which were characterized by several spectrometric methods. Spectral characterization show that the coordination number of all complexes is six and they have a octahedral geometry.

The IR spectral data suggested that the ligand behaves as a monobasic bidentate ligand towards the central metal ion with an oxygen's donor atoms sequence of both –OH and C=O groups under keto–enol structure.

The redox behavior was explored by cyclic voltammetry based on the metal-centred reduction for complexe Fe(III). The reduction/oxidation potential depends on the structure and conformation of the central atom in the coordination compound.

References

1. Moamen, S.R.: Synthesis and characterization of ligational behavior of curcumin drug towards some transition metal ion: chelation effect on the thermal stability and biological activity. Spectrochim. Acta Part A Mol. Biomol. Spectrosc. **105**, 326–337 (2013)
2. Sadaoui-Kacel, S., Zaater, S., Bensouilah, N., Djebbar, S.: Novel repaglinide complexes with Manganese(II), Iron(III), DFT characterization and electrochemical behavior. J. Struct. Chem. **57**, 1519–1530 (2016)
3. Subhan, M.A., Alam, K., Rahaman, M.S., Awal, M.R.: Synthese and characterization of metal complexes containing curcumin and study of their anti microbial activities and binding properties. J. Sci. Res. **6**, 97–109 (2014)
4. West, D., Swearing, J., Valdés-Martinez, J.: Spectral and structural studies of iron (III), cobalt and nickel complexs. Polyhedron **18**, 2919–2929 (1999)
5. Moore, M., Kinght, D., Zabetakis, D., Deschamps, J., Dressick, W.: Electronic effects on the reactivity of copper mono-bipyridine complexes. Inorg. Chim. Acta **388**, 168–174 (2012)
6. Bernabé-Pineda, M., Ramírez-Silva, M., González-Vergara, E., Rojas-Hernández, A.: Spectrophotometric and electrochemical determination of the formation constants of the complexes Curcumin–Fe(III)–water and Curcumin–Fe(II)–water. Spectrochim. Acta Part A Mol. Biomol. Spectrosc. **60**, 1105–1113 (2004)

Zinc Based Materials as Efficient Catalysts for a Clean Synthesis of Biginelli Product

L. Bennini[1], T. Mazari[1,2(✉)], C. Khiar[1], S. Menad[1], M. Halouane[1], and C. Rabia[2]

[1] Laboratoire de Chimie Appliquée et Génie Chimique LCAGC, Université M. Mammeri de Tizi-Ouzou UMMTO, 15000 Tizi-Ouzou, Algérie
tassa2002dz@yahoo.fr

[2] Laboratoire de Chimie du Gaz Naturel, Faculté de Chimie, USTHB, BP32 Bab-Ezzouar Alger, 16111 El-Alia, Algérie

Abstract. Ethyl6-methyl-2-oxo-4-phenyl-1,2,3,4 tetrahydropyrimidine-5-carb oxylate, Bigenilli product (noted DHPMs) was synthesized in the presence of chlorides and nitrates of metals as catalysts, in mild free solvent conditions. High DHPM yields were obtained with $Zn(NO_3)_2$ (97%), $ZnCl_2$ (90%) and $CuCl_2$ (76%). The reaction heptagram inspired by the principles of green chemistry show that in our reaction conditions, Biginelli process complies with the principles of sustainable development.

Keywords: Green chemistry · Biginelli reaction · Dihydropirimidinone Reaction heptagram

1 Introduction

Nowadays, environmental protection is the focus of the economic and industrial world concern. Green chemistry using catalytic processes represents a framework for the pollution prevention by minimizing waste and the toxic products use. Important research axis in organic chemistry focused on the development of new clean and efficient processes, particularly in the heterocyclic systems synthesis including the Biginelli reaction [1].

It is a multicomponent condensation reaction (MCR) of three reagents: a β-keto ester, an aldehyde and urea that leads to dihydropyrimidinones (DHPMs) formation in a single step, one-pot, (Fig. 1). DHPMs are an attractive class of organic compounds with important biological and pharmacological properties [2–6]. This reaction that dates back more than a century suffered from slowness (18 h) and low yields. In addition, the use of both corrosive mineral acids and toxic organic solvents in the original synthesis is far from the concepts of the ideal synthesis [7–10].

To remedy this, modern green practices give more attention towards the development of environmentally friendly, efficient and synthetic chemical procedures, in order to conquer the drawbacks of the above mentioned procedures and to accomplish most of the aspects of green and sustainable chemistry [11–17].

B. Abdelbaki et al. (Eds.): SMSD 2017, *Proceedings of the Third International Symposium on Materials and Sustainable Development*, pp. 345–352, 2018.
https://doi.org/10.1007/978-3-319-89707-3_39

The ideal synthesis in organic chemistry as described by Bataille [18] is based on the construction of a reaction heptagram inspired from twelve principles of green chemistry suggested by Anastas and Warner [17]. It is a combination of following parameters: carbon efficiency (CE), atoms economy (AE) [16], Yield (ρ), mass reaction efficiency (MRE), material recovery parameter (MRP) and hazard and toxicity coefficients, whose the values are close to 1.

In this study, the synthesis of DHPMs via Biginelli reaction was carried out in the presence of ethanol as solvent and in free solvent, using nitrates or chlorides of zinc, copper, nickel or iron as catalysts. Reaction heptagrams were realized to determine the better conditions of sustainable development.

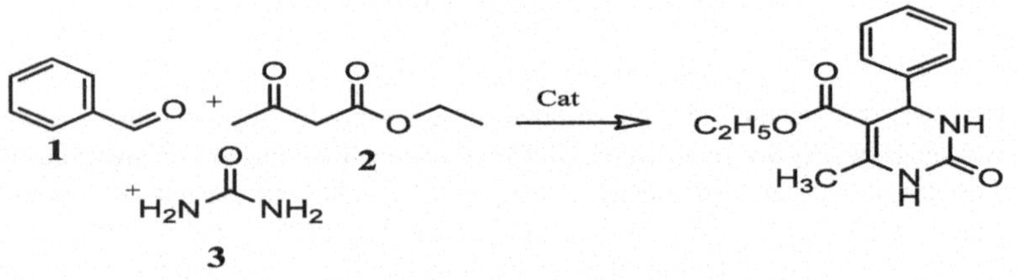

Fig. 1. Biginelli reaction

2 Materials and Methods

2.1 Biginelli's Reaction

In a 25 ml single-neck round bottom flask, a mixture of benzaldehyde 1 (2 mmol), ethyl acetoacetate 2 (2 mmol), urea3 (3 mmol) and 0.2 g of catalyst was kept under regular stirring at 100 °C for 1 h (Fig. 1). Completion of the reaction was monitored by thin layer chromatography. The reaction product was washed with boiled ethanol. DHPM product was purified and recrystallized in hot ethanol, then placed in an ice bath to obtain a pure product. Identification of DHPM is obtained via its melting temperature, infrared (IR) spectroscopic analysis and 1H and 13C nuclear magnetic resonance (NMR).

2.2 Characterization

The melting points of the products were taken using a Stuart SMP11 capillary apparatus. TLC was carried out on silica gel plates 60 F254 (thickness: 2 mm); the eluant is a mixture of chloroform-methanol.

IR spectra were recorded on a Shimadzu FTIR-8400 Fourier Transform Spectrometer using KBr pellets and the NMR ones on a Bruker FT-NMR DPX250 (250 MHz for the proton and 62.5 MHz for the 13C). Chemical shifts were expressed in parts per million (ppm) relative to TMS (δ = 0 ppm) and coupling constant J in Hertz (Hz). NMR multiplicities are reported using the following abbreviations: br = broad, s = singlet, d = doublet, t = triplet, q = quartet, m = multiplet.

2.3 Reaction Heptagram

For a reaction with this following equation:

$$aA + bB = pP + qQ$$

A: limiting reagent, B: co-reactant, P: main product, Q: identified byproduct a, b, p and q: stochrometric coefficients.

The used formulas for the reaction parameters calculations are as follows:

Carbon Efficiency:

$$CE = \frac{pn_C(P)}{an_C(A) + bn_C(B)}$$

Atom Economy:

$$AE = \frac{pM(P)}{aM(A) + bM(B)}$$

Yield:

$$\rho = \frac{a\,.\,n(P)}{p\,.\,n(A)}$$

Reaction Mass Efficiency:

$$RME = \frac{m(P)}{m(A) + m(B)}$$

Material Recovery Parameter:

$$MRP = \frac{\sum m_{recyclé}}{m_{tot} - \sum m_{réactifs}}$$

The danger and toxicity coefficients are also included in the heptagram.

3 Results and Discussion

3.1 Reaction Tests

After reaction conditions optimization (reaction time, temperature, reactants quantities and stoichiometric ratio of reagents), the parameters leading to the best yields were found to be: a mass of catalyst of 0.2 g, a stoichiometric ratio Benzaldehyde/ethyl acetoacetate/urea of 2/2/3 mmol under reaction temperature of 100 °C for one hour.

The results of the solvent free synthesis, summarized in Table 1 (entries 1–8) show that DHPMS yields are sensitive to both transition metal and anions natures. In the case of Zn and Ni, nitrate are more active than chloride parents (97 against 90 and 60 against 53% of DHPMS yields respectively). Contrarily, in the case of Fe and Cu, chloride are more active than nitrate parents (40 against 36 and 76 against 70% of DHPMS yields respectively). Among the tested salts, those based on zinc are the most efficient with 90–97 against 36–76% of DHPMS yields. Zinc is not a transition element; its effectiveness may be related to its metallic character similar to those of the alkaline and alkaline earth compared to Cu, Fe and Ni, transition metals, that have both acidic and oxidative character.

The same reaction has been carried out with addition of 5 ml of ethanol as solvent, in the presence of Zn based catalyst (Table 1, entries 9 and 10). The obtained results show a strong decrease from 90–97 to 10–50% of DHPMs yields, regardless of the anion nature. This is probably due to the polarity of the solvent molecules. Thus, the presence of hydrogen bonds, resulting from interaction between solvent hydrogen atoms and nucleophilic reagents, would lead to a slowdown of the condensation process.

Table 1. DHPMS yields

Entry	Catalysts	Yield (%)
1	$ZnCl_2$	90
2	$NiCl_2$	53
3	$FeCl_3$	40
4	$CuCl_2$	76
5	$Zn(NO_3)_2$	97
6	$Ni(NO_3)_2$	60
7	$Fe(NO_3)_3$	36
8	$Cu(NO_3)_2$	70
9[a]	$ZnCl_2$	10
10[a]	$Zn(NO_3)_2$	50

[a]With 5 ml of ethanol (solvent)

3.2 Heptagrams of Reactions

Obtained Figs. 2 and 3, from the parameter calculations (Table 2) show the heptagrams of the Biginelli's reactions, carried out in the presence of $Zn(NO_3)_2$ with and without solvent. The results show that the solvent-free mode is closer to the ideal synthesis as represented in Fig. 4.

Table 2. Obtained Biginelli reaction parameters

Catalysts	CE	AE	ρ(%)	MRE	MRP	Coef. tox.	Coef. danger
$Zn(NO_3)_2^a$	1	0.88	0.97	0.79	0	0.9	1
$Zn(NO_3)_2^b$/EtOH	1	0.88	0.50	0.41	0	0.8	0.8

[a]Without solvent
[b]With solvent

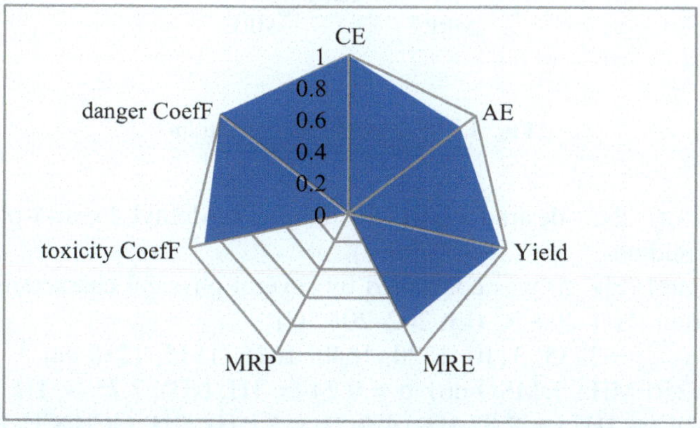

Fig. 2. Heptagram of synthesis reaction without solvent.

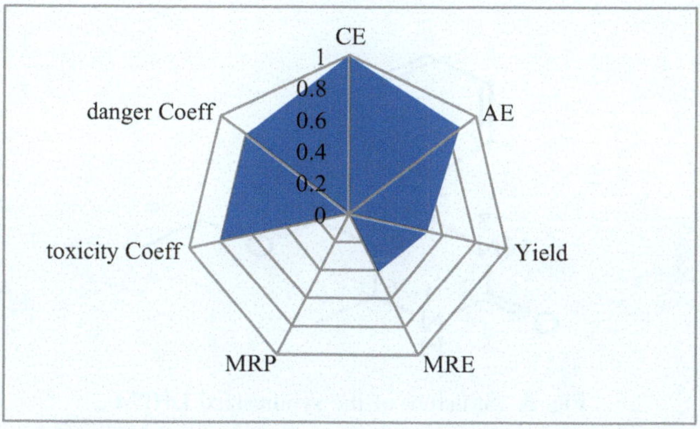

Fig. 3. Heptagram of synthesis reaction with solvent

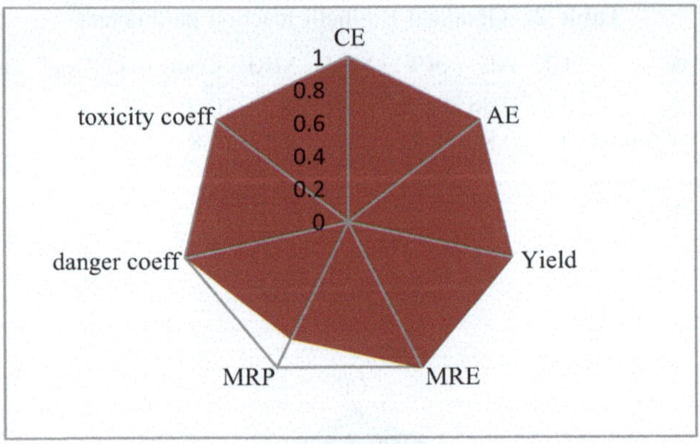

Fig. 4. Ideal synthesis heptagram

Structure of the desired DHPM (ethyl 6-Methyl-2-oxo-4-phenyl-1,2,3,4-tetrahydropyrimidine-

5-carboxylate) (Fig. 5) was confirmed by several physical characteristics:

Melting point: 200–202 °C (Lit 202–204 °C).

IR (KBr): υ_{max} = 3235, 3110, 1720, 1640, 1616, 1515, 1250 cm^{-1}.

^{1}H NMR (250 MHz, DMSO-d6): δ = 9.24 (s, 1H, NH), 7.75 (s, 1H, NH), 7.26 (s, 5H, C_6H_5), 5.14 (s, 1H, J = 2.87 Hz), 4 (q, J = 3.1 Hz, 2H, OCH_2CH_3), 2.24 (s, 3H, CH_3), 1.11 (t, J = 7 Hz, 3H, OCH_2 CH_3) ppm.

^{13}C NMR (62.5 MHz, DMSO-d6): δ = 165.4 (COOEt), 152.2 (C2), 148.4 (C6), 125.81–144.75 (C arom.), 99.83 (C5), 59.2 (OCH_2CH_3), 54 (C4), 17.8 (CH_3), 14.1 (OCH_2CH_3) ppm.

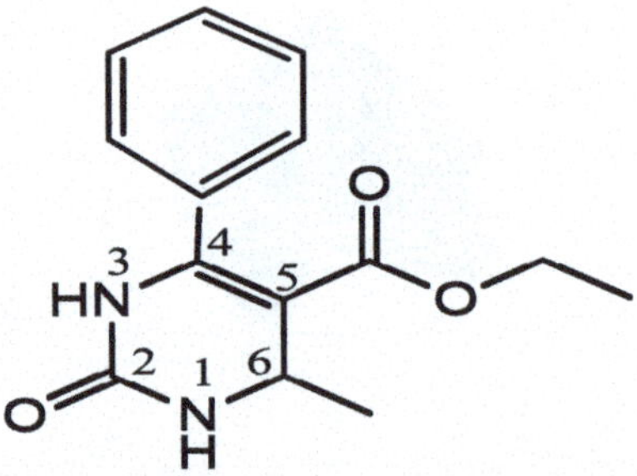

Fig. 5. Structure of the synthesized DHPM

4 Conclusions

In this study, the Biginelli reaction was performed in the presence of metal based nitrates and chlorides as catalysts. The zinc based system appear to be most efficient with high DHPMs yields, 97% in the presence of $Zn(NO_2)_3$ and 90% in the case of $ZnCl_2$. Reaction heptagrams are in favor of free solvent condition. The present methodology offers several advantages as high DHPMs yields, low reaction time (1 h), one pot synthesis and eco-friendly reaction conditions. It can be an alternative to the conventional processes which are very long, less efficient and more polluting.

References

1. Biginelli, P.: Aldehyde-urea derivatives of aceto- and oxaloacetic acids. Gazz. Chim. Ital. **23**, 360–413 (1893)
2. Kappe, C.O.: Biologically active dihydropyrimidones of the Biginelli-type – a literature survey. Eur. J. Med. Chem. **35**, 1043–1052 (2000)
3. De Fátima, A., et al.: A mini-review on Biginelli adducts with notable pharmacological properties. J. Adv. Res. **6**(3), 363–373 (2015)
4. Kappe, C.O.: Recent advances in the Biginelli dihydropyrimidine synthesis. Acc. Chem. Res. **33**, 879–888 (2000)
5. Kappe, C.O.: Synthesis of octahydroquinazolinone derivatives using silica sulfuric acid as an efficient catalyst. Eur. J. Chem. **35**, 1043–1052 (2000)
6. Rovnyak, G.C., et al.: Calcium entry blockers and activators: Conformational and structural determinants of dihydropyrimidine calcium channel modulators. J. Med. Chem. **38**, 119–129 (1995)
7. Sweet, F., Fissekis, J.D.: Synthesis of 3,4-dihydro-2(1H)-pyrimidinones and the mechanism of the Biginelli reaction. J. Am. Chem. Soc. **95**, 8741 (1973)
8. Sandhu, S., Sandhu, J.S.: Past, present and future of the Biginelli reaction: a critical perspective. ARKIVOC **66**, 66–133 (2012)
9. Heravi, M.M., et al.: A catalytic method for synthesis of Biginelli-type 3,4-dihydropyrimidin-2 (1H)-one using 12-tungstophosphoric acid. J. Mol. Catal. A: Chem. **242**, 173–175 (2005)
10. Debache, A., et al.: A one-pot Biginelli synthesis of 3,4-dihydropyrimidin-2-(1H)-ones/thiones catalyzed by triphenylphosphine as Lewis base. ChemInform **49**(42), 6119–6121 (2008)
11. Kouachi, K., et al.: Mo/γ-Al2O3 catalysts for the Biginelli reaction: effect of Mo loading. J. Mol. Catal. A: Chem. **395**, 210–216 (2014)
12. Khiar, C., et al.: Cobalt supported on alumina as green catalyst for Biginelli reaction in mild conditions: effect of catalyst preparation method, pp. 1–9 (2017)
13. Nasr-Esfahan, M., et al.: Efficient and green catalytic synthesis of dihydropyrimidinone (thione) derivatives using cobalt nitrate in solvent –free conditions. J. Chil. Chem. Soc. **59** (1), 2311–2314 (2014)
14. Barbero, M., et al.: A bønsted acid catalysed enantioselective Biginelli reaction. Green Chem. **19**, 1529–1535 (2017)

15. Shahid, A., et al.: Solvent-free Biginelli reactions catalyzed by hierarchical zeolite utilizing a ball mill technique. Catalysts **7**(3), 84 (2017)
16. Trost, B.M.: The atom economy—A search for synthetic efficiency, 254, 1471–1477 (1991)
17. Anastas, P.T., Warner, J.C.: In Green Chemistry: Theory and Practice. Oxford University Press, New York (1998)
18. Bataille, X.: Une initiation à la chimie verte, ENCPB (2007)

Study of the Electronic and Elastic Properties of ZrN and Zr_3N_4

M. Azibi(✉) and N. Saoula

CDTA, Cite 20 aout 1956 Baba Hassen, Algiers, Algeria
mazibi@cdta.dz

Abstract. The electron structure and elastic properties of c-ZrN and c-Zr_3N_4 compounds are studied using the functional density theory with the generalized gradient approximation (GGA) parameterized with the revised Perdew-Burke-Ernzerh as an exchange correlation function. The mesh parameters obtained, the dielectric and elastic constants, the isostatic modulus of elasticity, the shear modulus at ambient pressure are in agreement with the experimental and theoretical values.

Zr_3N_4 has a relatively high static dielectric constant. This high dielectric material can play an important role in the appropriate design of the next generation of microelectronic devices in which the reduced size requires that the gate insulators have high dielectric constants.

Keywords: ZrN · Zr_3N_4 · ab initio · DFT

1 Introduction

ZrN and Zr_3N_4 are known as refractory materials with other interesting properties. The electronic structure of ZrN in the rocksalt structure is well known (Neckel 1983; Schwarz et al. 1982; Stampfl et al. 2001). Its metallic band structure is manifested in the golden color of the crystal itself. In contrast, Zr_3N_4 is an insulator. It was first discovered by Juza et al. (1964). who showed it to be semitransparent with a brownish color. Johansson et al. (1984, 1986) have reported its insulating character with a band gap of approximately 2.2 eV.

First principle studies are optimistic approaches, which provide better experimental parameters leading to rapid material improvements economically.

In the present work we propose a comprehensive study about electronic and elastic properties of ZrN and Zr_3N_4 cubic phase by ab initio study.

2 Computational Details

CASTEP (Cambridge Serial Total Energy Package) software (Segall et al. 2002) was utilized in our calculations, based on density functional theory (DFT). CASTEP uses a plane wave basis set for the expansion of the single particle Kohn-Sham as implemented

B. Abdelbaki et al. (Eds.): SMSD 2017, *Proceedings of the Third International Symposium on Materials and Sustainable Development*, pp. 353–357, 2018.
https://doi.org/10.1007/978-3-319-89707-3_40

and ultra-soft pseudo potentials to describe ionic cores. The minimum total energy of the structure is achieved by relaxing automatically the internal coordinates using the Broyden-Fletcher-Goldfarb-Shanno (BFGS) algorithm.

The atomic configurations of Zr and N generated from the ultrasoft pseudopotential are $4s^2$ $4p^6$ $4d^2$ $5s^2$ and $2s^2$ $2p^3$, respectively. The exchange correlation energy is described by Generalized Gradient Approximation (GGA) of Perdew-Burke-Ernzerhof (Perdew et al. 1997). Brillouin zone sampling is performed using $9 \times 9 \times 9$ and $2 \times 2 \times 2$ Monkhorst-Pack k-points meshes for c-ZrN and c-Zr_3N_4, respectively. The plane wave cut off energy in our calculations is 380 eV, which assures a total energy convergence of 5.0×10^{-7} eV/atom.

3 Results and Discussion

3.1 Elastic Constant Calculations

Elastic constants of tetragonal crystal system can be calculated using a Taylor expansion of the total energy E (V,ε) (Fast et al. 1995; Tian 2004; Liu et al. 2010), is written as

$$E(V,\varepsilon) = E(V_0, 0) + V_0\left[\sum_i \tau_i\varepsilon_i\xi_i + \frac{1}{2}\sum_{ij} C_{ij}\,\varepsilon_i\xi_i\varepsilon_j\xi_j\right] \qquad (1)$$

Whereas E (V_0, 0) is the energy of the unstrained system with equilibrium volume V_0, τ_i is an element in the stress tensor, ξ_i is a factor to take care of Voigt index and C_{ij} represents elastic constants. The linear elastic constants form a 6×6 symmetric matrix, having 27 unlike components. These components will reduce to three elements C_{11}, C_{12}, and C44 according to the symmetry of cubic structure (Alton and Barlow 1967). The Born stability conditions (Mouhat and Coudert 2014) for the elastic constants of cubic structures are $(C_{11} - C_{12}) > 0$, $C_{11} + 2C_{12} > 0$, $C_{44} > 0$. Our c-ZrN and c-Zr_3N_4 satisfies these criteria showing its phase stability under zero pressure.

From the calculated elastic constants, we can find ZrN holds the largest C_{11}, C_{22} and C_{33} values among the c-Zr_3N_4, which indicate the very high compressibility along the axis directions. The bulk modulus B, shear modulus G and Young's modulus E were further obtained using the Voigt–Reuss–Hill averaging (Zhou et al. 2009), shown in Table 1. The calculated B values of zirconium nitrides shown that c-ZrN has the largest bulk and shear moduli than c-Zr_3N_4 (see Table 1).

Table 1. The single-crystal elastic stiffness constants C_{ij} of c-ZrN and c-Zr_3N_4. The poly-crystalline bulk modulus (B), shear modulus (G), and Young's modulus (E) obtained using the Voigt-Reuss-Hill averaging scheme compared to experiment and other theoretical results. All the elastic constants are in GPa

Compounds	C_{11}	C_{12}	C_{44}	*B*	*G*	*E*
ZrN	471	121	82	238	111	422
Theo.	591[a]	101[a]	105[a]	265[a]	147[b]	372[b]
Exp.	471[a]	88[a]	138[a]	249[c]	157[d]	
Zr_3N_4	217	126	86	228	66	125
Theo.	450[a]	134[a]	158[a]	239[a]	135[b]	341[b]
Exp.				224[e]		

[a]Oscar et al. (2015), [b] Yu et al. (2017), [c]Chen et al. (2013), [d]Sarioglu et al. (2006), [e]Gao et al. (2011)

3.2 Optical Properties

The optical properties of a system can be described by the complex dielectric function, which is given by:

$$\varepsilon(\omega) = \varepsilon_1(\omega) + i\varepsilon_2(\omega) \tag{2}$$

Where $\varepsilon_1(\omega)$ is the real part of the complex dielectric function following the Kramer–Kroning transformation and $\varepsilon_2(\omega)$ is the imaginary part.

Figure 1 shows the calculated dielectric function of the c-ZrN and c-Zr_3N_4 compounds for the energy range up to 25 eV.

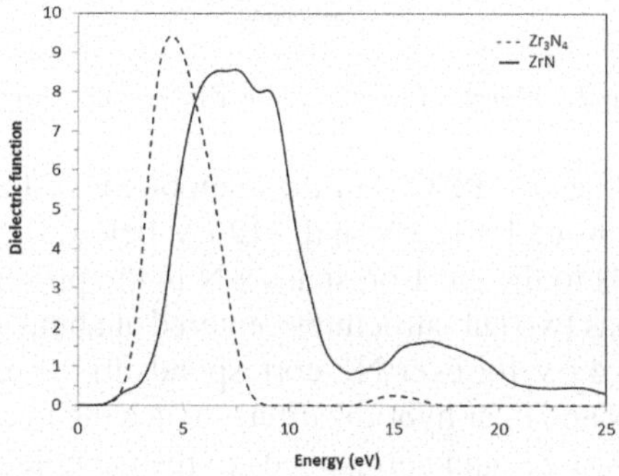

Fig. 1. The calculated imaginary part ε2(ω) of the dielectric function ε(ω) of ZrN and Zr_3N_4 compounds

The imaginary part of the dielectric function has an accentuated peak at 6.3 and 4.1 eV for the c-ZrN and c-Zr_3N_4 respectively with high intensity for c-Zr_3N_4. The origin of these peaks is due to the interband transition between the N-2p stats and the Zr-4d stats.

3.3 Density of States

Figure 2 show the partial density of states; s-N, p-N and d-Zr of c-ZrN and c-Zr_3N_4 compounds.

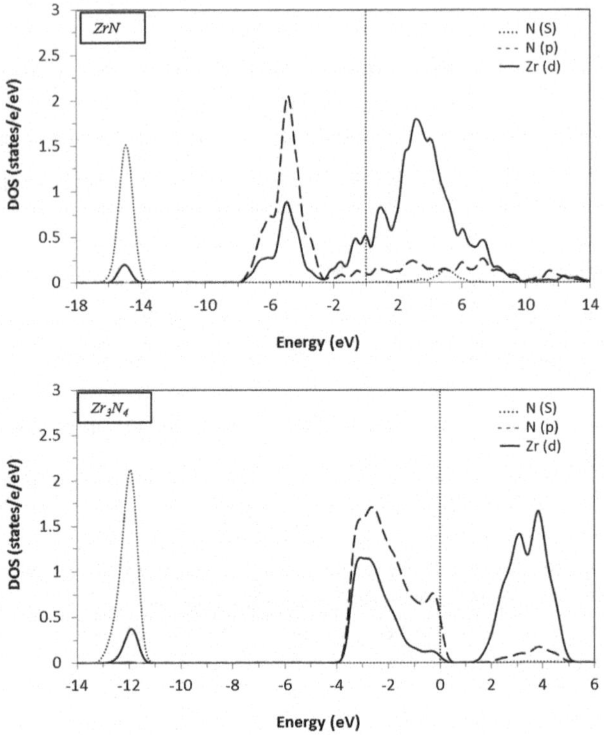

Fig. 2. PDOS of c-ZrN and c-Zr_3N_4 compounds

The partial density of states (PDOS) of the compounds is dominated by three structures: the first centered around −15 eV and −12 eV below E_F for c-ZrN and c-Zr_3N_4 respectively, corresponds to the electron states s-N in the presence of a low density of d-Zr states; the second has two sub-structures centered at about −6.5 eV and −5 eV, for c-ZrN, −2.8 eV and −0.4 eV for c-Zr_3N_4, corresponds to the electron states p-N; it is strongly hybridized (p-N and d-Zr hybridization). As for the third, it is overwhelmingly due to the electron states d-Zr; and corresponds to the occupied (<E_f) and unoccupied (>E_f): mostly due to the d-Zr and a low electron density of p-N. It is noted that above E_f there exists a non-negligible state density due to the electronic states p-N.

In c-ZrN, the E_f has a density of stats which explains its metallic behavior while for c-Zr_3N_4, bonding and anti-bonding states are separated by a gap, thus Zr_3N_4 is a semi-conductor and possesses a remarkable stability.

Its calculated band gap (0.6 eV) is consistent with previous theoretical results (Yablonskikh et al. 2014).

4 Conclusions

The electronic and mechanical properties c-ZrN and c-Zr_3N_4 are explored using the first principles total energy plane wave pseudo potential method based on density functional theory.

The elastic and mechanical properties for both compounds are in good agreement with available experimental and theoretical results.

c-Zr_3N_4 has a relatively high static dielectric constant. This high dielectric material can play an important role in the appropriate design of the next generation of micro-electronic devices.

References

Stampfl, C., Mannstadt, W., Asashi, R., Freeman, A.J.: Phys. Rev. B **63**, 155106 (2001)
Schwarz, K., Ripplinger, H., Neckel, A.: Z. Phys. B: Condens. Matter **48**, 79 (1982)
Neckel, A.: Int. J. Quant. Chem. **23**, 1317–1353 (1983)
Juza, R., Rabenau, H., Nitschke, I.: Z. Anorg. Allg. Chem. **332**, 1–4 (1964)
Johansson, B.O., Sundgren, J.E., Helmerson, U., Hibbs, M.K.: Appl. Phys. Lett. **44**, 670 (1984)
Johansson, B.O., Hentzell, H.T.G., Harper, J.M.E., Cuomo, J.J.: J. Mater. Res. **1**, 442–451 (1986)
Segall, M.D., Lindanp, J.D., Robert, M.J., Pickard, C.J., Hasnip, P.J., Clark, S.J., Payne, M.C.: J Phys.: Cond. Matt. **14**, 2717–2744 (2002)
Perdew, J.P., Burke, K., Ernzerhof, M.: Phys. Rev. Lett. **78**, 1396 (1997)
Tian, S.: Materials Physical Properties. University of Aeronautics and Astronautics Press, Beijing (2004)
Fast, L., Wills, J.M., Johansson, B., Eriksson, O.: Phys. Rev. B **51**, 17431–17438 (1995)
Liu, Q.J., Liu, Z.T., Feng, L.P., Tian, H.: Solid State Sci. **12**, 1748–1755 (2010)
Alton, W.J., Barlow, A.J.: J. Appl. Phys. **38**, 3817–3820 (1967)
Mouhat, F., Coudert, F.X.: Phys. Rev. B **90**, 224104 (2014)
Zhou, W., Liu, L.J., Li, B.L., Wu, P., Song, Q.G.: Comput. Mater. Sci. **46**, 921–931 (2009)
Oscar, U., Mota, O., Araujo, R.A., Wang, H., Çağın, T.: Phys. Procedia **66**, 576–585 (2015)
Yu, S., Zeng, Q., Oganov, A.R., Frapper, G., Huang, B., Niu, H., Zhang, L.: RSC Adv. **7**, 4697–4703 (2017)
Chen, Z.Q., Wang, J., Li, C.M.: J. Alloy. Compd. **575**, 137–144 (2013)
Sarioglu, C.: Surf. Coat. Technol. **201**, 707–717 (2006)
Gao, F.M.: Chin. Phys. Lett. **28**, 076102 (2011)
Yablonskikh, M., Dzivenko, D., Bourguille, J., Riedel, R., Magnano, E., Parmigiani, F., Zerr, A.: Phys. Status Solidi A **211**(4), 835–842 (2014)

Degradation Studies on PVC Plasticized Submitted to Gamma Radiation up to 50 KGy

N. Boutouchent-Guerfi[1,2(✉)], A. Benaboura[1], and A. Dib[3]

[1] Département Chimie, Laboratoire Synthèse Macromoléculaire et Thio-Organique Macromoléculaire, Université des Sciences et de la technologie Houari Boumédiène (USTHB), BP 32, El Alia, 16111 Bab Ezzouar, Algiers, Algeria
nabila.boutouchent@gmail.com, guerfinabila@yahoo.fr

[2] Centre de Recherche en Technologie des Semi-conducteurs pour l'Energétique (CRTSE), BP 140, 2 Bd Frantz Fanon, Alger-7-merveilles, 16038 Algiers, Algeria

[3] Centre de Recherche Nucléaire d'Alger, B.P. 399, 02 Bd, Frantz Fanon, Alger-Gare, Algiers, Algeria

Abstract. Many of the applications require sterilization by gamma radiation. When polymeric materials are submitted to gamma radiation, changes are observed on their molecular structure. Poly (vinyl chloride) (PVC) plasticized with DOP is one of the major bulk polymers which is widely consumed in various applications.

In this study, the effect of gamma irradiation up to 50 KGy on mechanical properties (tensile strength, elongation at break and Young's modulus) and the changes induced in the thermal analysis and morphology was investigated. The results revealed that the gamma irradiation affected considerably the PVC plasticized properties at 50 KGy. This degradation is due to the chain scission, which caused a reduction in the average molecular mass and cross-linking rate on one hand.

Keywords: ATR · ATG · Gamma radiation · PVC · Mechanical properties

1 Introduction

Poly (vinyl chloride) (PVC) is widely used material in various applications like consumer products as packaging, toys, furniture, windows frames, tubes, wire and cable… But it is the most dangerous plastic due to its high chlorine content and the highest amount of additives impairing its recyclability. PVC under recycling processes can generate dioxin, an extremely hazardous pollutant [1].

PVC is a glassy polymer well-known to be compatible with certain polymers to improve some properties for well-defined applications as an application in thermal aging conditions associated with gamma irradiations at nuclear centers. We know that most of the polymers subjected to gamma radiation undergo changes in their molecular structure crosslinking which results in degradation of the mechanical properties [2–4],

B. Abdelbaki et al. (Eds.): SMSD 2017, *Proceedings of the Third International Symposium on Materials and Sustainable Development*, pp. 358–364, 2018.
https://doi.org/10.1007/978-3-319-89707-3_41

degradation in physical and chemical properties [5] and modification of dielectric properties [6].

In this work, we study the mechanical, thermal, dielectric and morphological compaction of the PVC submitted to thermal aging and gamma radiation up to 50 KGy. In order to make PVC acceptable for the irradiation processing industry in the long time, non-toxic plasticizers should be used. PVC was added with dioctyl phtalate (DOP) [7] a plasticizer DOP contains an aromatic ring in its structure able to act as a quencher, was used, the aromatic groups absorb the excitation energy acting as traps and converting this energy into other non-reacting forms [8]. Zn and Ca are the recommended additives to react with the HCl from PVC decomposition and also approved by the FDA as additives for radiation sterilization [9].

We present an experimental analysis of the effect of thermal aging and gamma irradiation up to 50 KGy of PVC.

2 Experimental

2.1 Materials and Sample Preparation

PVC in powder type SHINTECH SE 1200 was procured from France, DOP were procured from SGP (Tunisia), chalk (OMAYACARB 2XT-KA) were procured from OMAYA (Turkish), stearate Zn/Ca (REAPAK B-CV/3037) were obtained from IACN (ITALY) and carbone black (BK 30) were procured from Wilson (USA).

The raw materials (PVC and additives) are mixed using a mixer with cylinders at 140 °C with 20 round/min for 5 min. After having obtained the homogeneous paste. It is then conveyed to the curing press at 170 °C at 300 kN for 2 min. Square plates with 25 cm of side and 2 mm of thickness are obtained. They are cut out into samples of different forms, according to the test to be carried out.

For the mechanical tests, the samples are cut out under a dumbbell form in accordance with IEC 60811.1.1 (1993).

Gamma radiation source of ^{60}Co is used. The pilot irradiator is composed with three COP_4 cylindrical ^{60}Co sources manufactured by ORIS (France), they are arranged vertically in a stainless steel source holder. Source dimensions are 384 mm height and 26.6 mm diameter thus arranged from an active height of 1152 mm.

Dumbbell samples were subjected to ageing at 80 °C for 7 days. The test is performed according to ISO 227 (2011).

2.2 Characterization

The density of samples was measured using a PRECIA MOLEN XB 220A, for this we proceeded to weigh the solid first in air and then immersed in water the density is calculated with the following formula:

$$\rho = A \times \rho_o/(A - B) \tag{1}$$

ρ: density of the solid;
ρ_o: density of water at the given temperature T;
A: weight of the solid in air;
B: weight of the solid immersed in water

The result of the density is average of 5 tests and according to NF T 51-561 (1990).

The measurement of water uptake allows the evaluation of possible variations in masses when polymer objects were used in contact with water or exposed to moisture. The absorption is determined by immersion in water at ambient temperature [10].

The tensile tests have been carried out using a ZWICK/ROELL type tensile machine of trademark, this machine is controlled by computer software PC/Test Xpert V10.11. The test consists in subjection the sample to tension at a constant controlled speed (300 mm/min in our case) up to rupture. The test was performed according to ISO 527-1/-2 and CEI 811.1 (1993). The result was an average obtained on five specimens.

The PVC was analyzed on FTIR instrument Spectrum two of Perkin ELMER, scanning between 400 and 4000 cm^{-1} for the IR spectrum.

The thermogravimetric analysis was performed on TGA 2 Star system from Mettler Toledo. Heating rate of 10 °C/mn from 25 °C up to 600 °C.

The morphology of samples was characterized by scanning electron microscopy (SEM), model Philis xl 30.

3 Reults and Discussion

3.1 Effect of Thermal Aging

The density of PVC is 1.0858 g/cm^3. The water uptake of PVC is 0.0024%, it is a very low rate, this means the presence of the microvoids which favors the penetration of water into the matrix of the polymer. According to the Fig. 1, the tensile strength of the PVC decreased at rate of 1.56% after aging, the elongation at break decreased at 8.32% and Young's modulus decrease also at 24.08%, this means that the PVC has lost its flexibility at 80 °C by the departure of plasticizer DOP after aging and DOP is sensitive to temperature

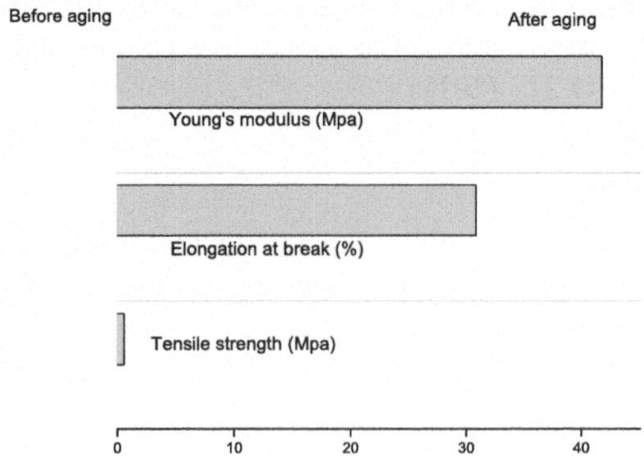

Fig. 1. Mechanicals Properties of PVC before and after aging.

These results show that the mechanicals properties after aging of PVC alone are poor. This means that, PVC is sensible to heat at 80 °C, despite additives added.

3.2 Effect of Irradiation Gamma on PVC Properties

The effect of irradiation gamma up to 50 KGy on the mechanicals properties of PVC has been studied.

The irradiated PVC darken, darkening is proportional to the radiation dose and is explained by the formation of double bonds along the chain in a process known as the zipper effect, illustrated in Fig. 2.

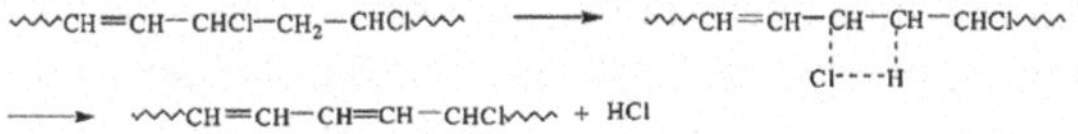

Fig. 2. Formation of insaturation at PVC chain by the "zipper" effect [11].

Among the irradiated poly (vinyl chloride), the PVC plasticized, have shown the highest degradation due to chain scission, particularly when exposed to the dose of 25 KGy which is the sterilization dosage, used for medical and pharmaceutical devices and packaging materials [11].

According to Fig. 3, the tensile of strength of PVC was decreasing as a function of the gamma radiation rate of 10 to 50 KGy. This means that the gamma radiation has an effect on this polymer which leads the latter to degradation; this degradation may be by the phenomenon of crosslinking and chain splitting or by oxidation.

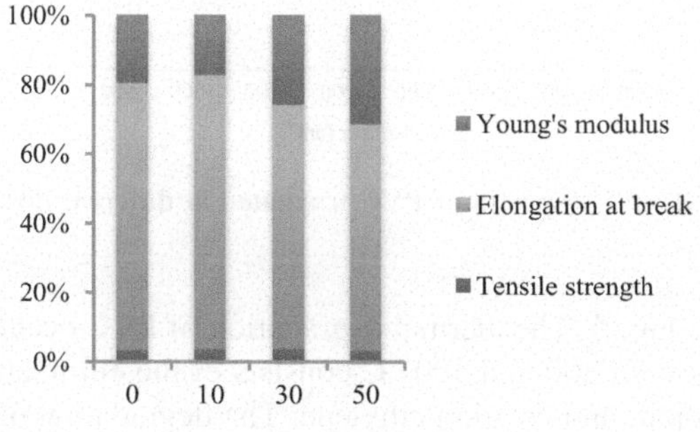

Fig. 3. Variation of tensile of strength, elongation at break and Young's modulus of the PVC exposed to radiation gamma.

According to Cota [12], he related the decrease in tensile of strength under the effect of gamma radiation at crosslinking, but Mckellop [13] explained that degradation of the polymer under the effect of gamma radiation is linked first to the oxidation of the air present in the sample with the formation of the radicals, as it has bound the oxidation rate to the thickness of the sample, and that the oxidation will be complete, thickness between 0.5 and 2 mm. These results were confirmed by Premnath [14] who adjusted the thickness of the sample so that the oxidation is complete under the effect of gamma radiation, and that the thickness should not exceed 1.6 mm and The thickness of our sample is 2 mm; it means that our degradation is justified by the oxidation which created free radicals followed by crosslinking.

The elongation at break is in increase under the effect of gamma radiation, this is linked to the PVC nature which leads to the increase its elasticity rate by inserting polar bonds between the macromolecular chains [15].

The Young's modulus is increasing for the PVC due to the degradation of this mixture under the effect of gamma radiation.

It remains to confirm these results on the macromolecular scale by the FTIR in ATR and ATG mode followed by morphological study.

Figure 4 have shown a decrease in the intensity of the bands attributed to the C-H methylene stretching, appearing at 2912 and 2970 cm^{-1}, an evidence of the loss of methylene protons which agree with the formation of double bonds. Also, we observe the new peaks appearing after irradiation this explains the PVC sensitivity and its oxidation under the effect of bombardment.

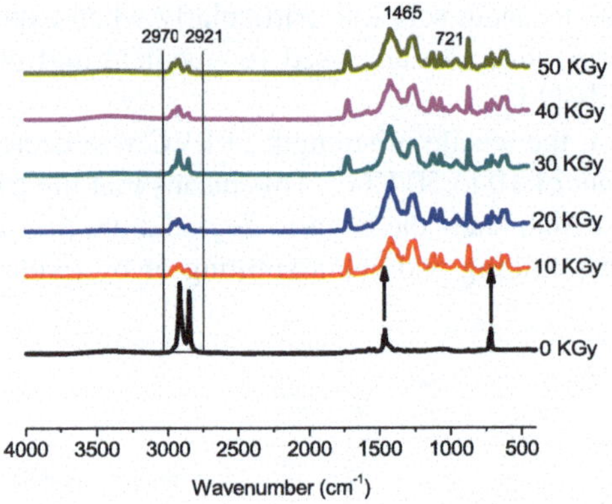

Fig. 4. ATR spectra of PVC irradiated at different dosages.

According to the Fig. 5. The thermal degradation of PVC occurs in two stages. The first stage comes between 200 and 300 °C consists of the elimination of a molecule of Cl (desydrochloruration) and creation ethylene. The degradation of PVC is initiated by thermal scission of the C-O bond of the chain. This break of the C-O bond is accompanied by the release of hydrogen atom of the adjacent carbon atom. A double bond is thus formed in the chain and the just adjoining C-O bond is breaked in the sequence that marks the origin of the propagation stage of the chain reaction.

Scanning electron microscopic analysis reveals the surface morphology of the PVC irradiated up to 50 KGy. According to Fig. 6, the irradiation effect is clearly observed on the PVC morphology, microvides appears at 50 KGy, it is the harmful effect irradiation on this polymer by his degradation.

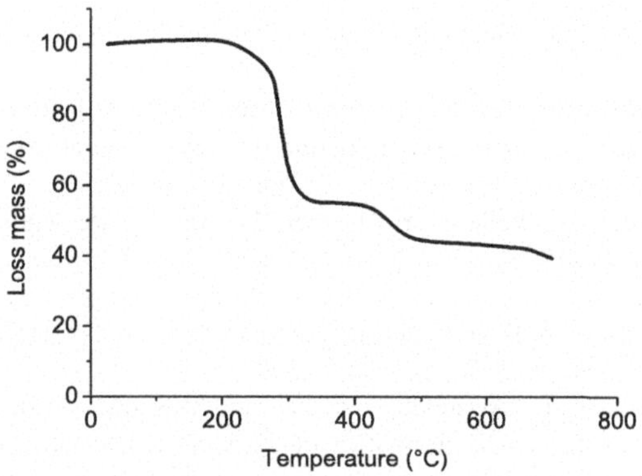

Fig. 5. ATG of PVC as a function of temperature.

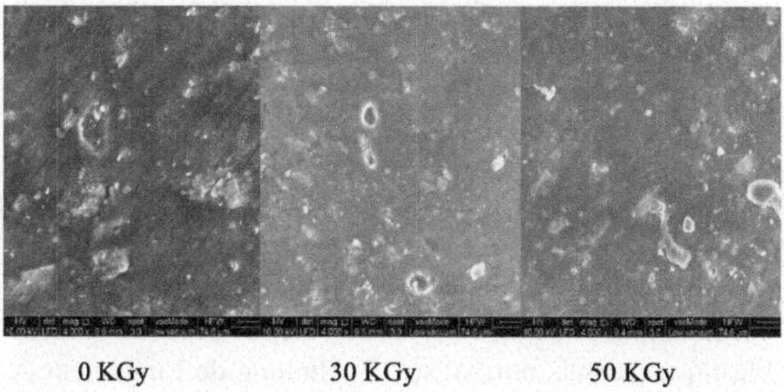

Fig. 6. Scanning electron microscopy of PVC as function of gamma irradiation.

4 Conclusion

The PVC plasticized with DOP was submitted to thermal aging and gamma irradiation up to 50 KGy. The mechanicals, thermals and morphological results show the mediocre properties of PVC and its high sensitivity to temperature and gamma irradiation, mainly chain scission which leads to reduction on molar mass, and reticulation which increases reduces solubility, these molecular alterations lead to changes on mechanical thermal and morphology material.

Acknowledgement. This study was funded by DGRSDT of the Ministry of Higher Education and Scientific Research (MESRS) of Algeria.

References

1. Hutzler, B.W., Machado, L.D.B., LugaÄo, A.B., Villavicencio, A.C.H.: Properties of irradiated PVC plasticized with non-endocrine disruptor. Radiat. Phy. Chem. **57**, 381–384 (2000)
2. Yagoubi, N., Baillet, A., Pellerin, F., Ferrier, D.: Beam Interactions with Materials and Atoms. **105**, 340–344 (1995)
3. Cote, S.S., Vasconcelos, V., Senne, Jr, M., Cravalho, L.L., Rezende, D.B., Côrrea, R.F.: Changes in mechanical properties due to gamma irradiation of high-density polyethylene (HDPE). Braz. J. Chem. Eng. **24**, 259–265 (2007)
4. Chai, C.K., Ratnam, C.T., Abdullah, L.C., Amin, M.S.S.M., Wan Zin, W.M.: Effect of gamma radiation on the tensile properties of epoxidized natural rubber latex. Int. J. Inst. Mater. Malaysia **1**, 223–232 (2013)
5. Byun, Y.J., Hong, S.I., Kim, K.B., Jeon, D.H., Kim, J.M., Whiteside, W.S., Park, H.J.: Physical and chemical properties of γ-irradiation EVOH film. Radiat. Phy. Chem. **76**, 974–981 (2007)
6. Ravi, H.R., Sreepad, H.R., Sajan, C.P.: Modification of dielectric and structural properties of PVC by gamma irradiation. Res. Rev. Polym. **3**, 89–92 (2012)
7. Hutzler, B.W., Machado, L.D.B., Lugao, A.B., Villavicencio, A.C.H.: Properties of irradiated PVC plasticized with non-endovrine discruptor. Radiat. Phy. Chem. **57**, 381–384 (2000)
8. Kojima, K., Kumafuji, H., Ueno, K.: Dicoloration of plasticized PVC by irradiation. Radiat. Phys. Chem. **56**, 859–863 (1981)
9. Saxena, A., Kalliyanakrishnan, V., Pal, V.N.: Studies on the effect of gamma irradiation on nontoxic PVC formulation. J. Appl. Polym. Sci. **34**, 1727–1738 (1987)
10. Labrosse, M.: Plastiques: essais normalisés. Technique de l'ingénieur A **3521**, 1–8 (1996)
11. Vinhasa, G.M., Souto-Maiorb, R.M., Lapac, C.M., Bastos de Almeidad, Y.M.: Degradation studies on plasticized pvc films submitted to gamma radiation. Mater. Res. **6**, 497–500 (2003)
12. Ikhuoria, E.U., Ogundele, K.T., Osinkolu, G.A.: Effect of beta irradiation on plasticized poly (vinyl chloride). Afr. J. Pure Appl. Chem. **5**, 333–338 (2011)
13. Mckellop, H.A.: Does gamma radiation speed or slow wear. Bull. Am. Acad. Orthop. Surg. **44**, 511–523 (2006)
14. Premnth, V., Bellar, A., Merrill, E.W., Jasty, M., Harris, W.H.: Molecular rearrangements in ultra-high molecular weight polyethylene after irradiation and long-term storage in air. Polymer **40**, 2215–2229 (1999)
15. Jimenez, A., Lopez, J., Iannoni, A., Kenny, J.M.: Formulation and mechanical characterization of PVC plastisols based on low- toxicit additives. J. Appl. Polym. Sci. **18**, 1881–1890 (2001)

Electronic and Thermoelectric Properties of $CdSnAs_2$

N. Si Ziani[1(✉)], H. Bouhani-Benziane[1], M. Baira[1], A. E. K. Belfedal[2], and M. Sahnoun[1]

[1] Laboratoire de Physique Quantique de la Matière et Modélisation Mathematique (LPQ3M), Faculté des Sciences Exactes, Université Mustapha Stambouli, Mascara, Algeria
sizianinacera@gmail.com

[2] Laboratoire de Chimie-Physique de Macromolécule et interface Biologique, Faculté des Sciences et de la vie, Université Mustapha Stambouli, Mascara, Algeria

Abstract. Electronic and thermoelectric properties of ternary chalcopyrite type CdSnAs2 were studied using the first principles density functional calculations performed in the full potential linear augmented plane wave (FP-LAPW) method as implemented in the WIEN2 k code. The thermoelectric properties are calculated by solving the Boltzmann transport equation within the constant relaxation time approximation. The calculated band gap using the Tran-Blaha modified Becke- Johnson potential (TB-mBJ) of CdSnAs2 compound is in good agreement with the available experimental data. Thermoelectric properties like thermopower, electricalconductivity scaled by relaxation time are calculated as a function of carrier concentration at différent temperatures.

Keywords: Electronic properties · Thermoelectric properties
FP-LAPW method · TB-mBJ

1 Introduction

Chalcopyrite-type materials are well-known semiconductors with energy band gap ranging from 1–3 eV [1, 2]. The semiconducting nature of these materials attracts the present researcher, as these materials find promising applications in solar cells, detectors, nonlinear optic devices and many more. The important application of the chalcopyrite-type compounds includes the energy conversion and serve as thermoelectric and photovoltaic energy converters. Thermoelectric (TE) materials with potential applications within power generation and refrigeration have been a thrust area of research for the past few decades [3, 4]. The efficiency of a TE material is quantified by dimensionless figure of merit ZT. The expression of figure of merit is defined as $ZT = S^2\sigma T/\kappa$, where S, σ, κ and T are thermopower, electrical conductivity, thermal conductivity.

y, and the absolute temperature, respectively. κ includes both the electronic κ_e and the lattice κ_l contributions. A good thermoelectric material should have high Seebeck coefficient, high electrical conductivity, and low thermal conductivity. To attain

B. Abdelbaki et al. (Eds.): SMSD 2017, *Proceedings of the Third International Symposium on Materials and Sustainable Development*, pp. 365–371, 2018.
https://doi.org/10.1007/978-3-319-89707-3_42

maximum Seebeck coefficient, material with high band mass is essential, and for achieving good electrical conductivity the mobility of the material should be high, leading to lower band mass carriers, implying the need for a mixture of heavy and light bands in the band structure [5].

Ternary chalcopyrite semiconductors and pnictide compounds of the form ABC_2 have attracted researchers because of their special physical properties like high melting point, high refractive index, high nonlinear optical susceptibility, and many more [6–8]. Because of the above mentioned diverse properties, both the chalcopyrites and pnictides are identified as promising materials for different device applications like electronic devices [9], nonlinear optical devices [10, 11], photovoltaic cell [12], thermoelectric applications [13] etc. These materials crystallize in the tetragonal structure with space group I_42d and are considered as the ternary analogue of zinc blende structure [14, 15]. In this work, we employ first-principles calculations to study thermoelectric properties of ternary chalcopyrite $CdSnAs_2$. After brief explanation of the method of calculations, the electronic band structure and then thermoelectric properties of the $CdSnAs_2$ compound will be presented. In Section ……, the thermoelectric properties will be discussed and the conclusions will be presented in the last section.

2 Computational Details

Present calculations are based on the first principles density functional theory. We have used full potential linearized augmented plane wave (FP-LAPW) method implemented in WIEN2 k package [16, 17]. The calculations using standard Generalized Gradient Approximation (GGA) schemes for the exchange-correlation potential underestimate the band gaps of semiconductors, and generally we have used the modified GGA known as the Tran-Blaha modified Becke-Johnson potential (TBmBJ) [18, 19] which is good in reproducing the experimental band gaps. TE properties like thermopower (S), electrical conductivity scaled by relaxation time (σ/τ) were calculated using the BOLTZTRAP code [20], within the Rigid Band Approximation (RBA) and the constant scattering time (s) approximation (CSTA), and these approximations have been successfully applied for several thermoelectric materials [21–25].

3 Results and Discussion

3.1 Structural and Electronic Properties

3.1.1 Structural Properties

The investigated compound $CdSnAs_2$ crystallize in tetragonal chalcopyrite type structure with space group $I4\bar{2}d$. Chalcopyrite crystal structure can be considered as the ternary analogue of zinc blende structure with slight tetragonal distortion.

The chalcopyrite structure is closely related to the zinc blende structure. Generally, the chalcopyrite lattice is characterized by three structural parameters: the lattice constants a and c, as well as the dimensionless anion displacement parameter u.

The calculated equilibrium lattice constant a, c/a ratio and distortion parameter u for CdSnAs2 chalcopyrite compound are presented in Table 1, together with some available theoretical [26, 27] data for comparison.

Table 1. Calculated equilibrium lattice constant (a, in Å) and c/a together with some available theoretical data for comparison.

	a	c/a
$CdSnAs_2$	6.1001	1.944
	6.094[a, b]	1.956[a, b]

[a] Ref. [26].
[b] Ref. [27].

3.1.2 Electronic Properties

It is a well known fact that LDA and GGA exchange-correlation potentials underestimate the band gap of semiconductors [28]. Therefore in order to obtain the band structure accurately, we have used mBJ potential [19]. In the absence of spin-orbit interactions our calculated energy band structure of CdSnAs2 along the high symmetry directions of Brillouin zone is presented in Fig. 1.

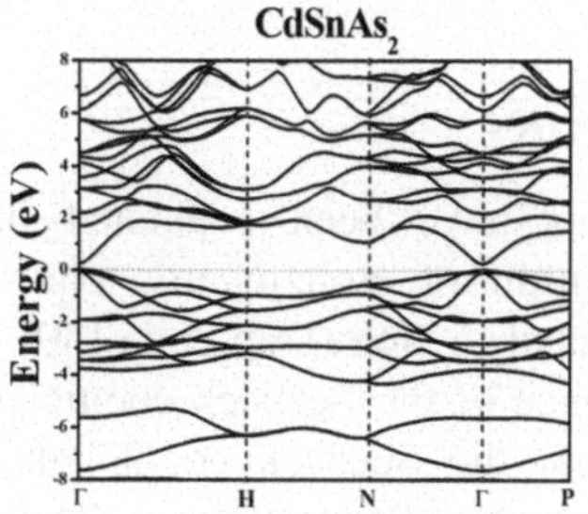

Fig. 1. Band structure of CdSnAs2.

Figure 1 shows the calculated band structure along the high symmetry directions at the equilibrium lattice parameter for $CdSnAs_2$ using the TB-mBJ. Both the top of the valence band and the bottom of the conduction band are found to be at Γ point in the Brillouin zone (BZ), indicating that the fundamental band gap is direct (Γ-Γ). The computed band gap value for the $CdSnAs_2$ is given in Table 2 along with experimental data and other theoretical results.

The total density of states (TDOS) and the partial density of state (PDOS) of CdSnAs2 chalcopyrite compound are plotted in Fig. 2.

In $CdSnAs_2$ chalcopyrite material, the Cd-s electrons contribute in the energy range from –5.0 to –3.5 eV. The Sn-p and As-P electrons have a major contribution for the states near the Fermi level (EF) in the energy range of –3.5 to 0 eV in the valance band. The conduction band consists predominantly of As-p and Sn-p states, but contains a small admixture of metal in Sn-s and Cd-s states.

Table 2. Values of the different bands energy compared with available theoretical and experimental data and previous works in eV for CdSnAs2.

Compound	Exchange	Band gap (eV)
$CdSnAs_2$	Tb.mbJ	0.28
	exp	0.28[a, b]

[a] Ref. [26].
[b] Ref. [27].

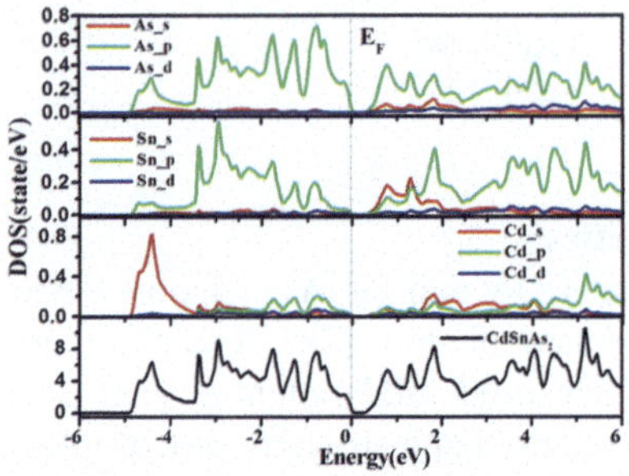

Fig. 2. The electronic total density of states (TDOS) and the electronic partial density of states (PDOS) for $CdSnAs_2$

3.2 Thermoelectric Properties

The thermoelectric properties have been evaluated based on the semi-classical approach given by the solution of Boltzmann's equation in the relaxation time approximation [29] implemented in BoltzTrap code [30].

The temperature dependence of the Seebeck coefficient for CdSnAs2 is shown in Fig. 3. It should be note that the Seebeck coefficient is affected by temperature, increased when temperature increased especially in the 200–450 K temperature range but after this value it starts to decreased when temperature increased. The Seebeck coefficient, S of CdSnAs2 compound is positive indicating p-type conduction. In addition as may be seen, a large S values (169 μVK − 1) observed at low temperature 450 K.

Figure 4 show the plots of the electrical conductivity over relaxation time, σ/τ as a function of temperature in range 200–1200 K, for CdSnAs2. The electrical conductivity increased with increasing temperature.

Figure 5 show the plots of the thermal conductivity over relaxation time, K/τ as a function of temperature in range 200–1200 K, for CdSnAs2. The thermal conductivity increased with increasing temperature.

From the Seebeck coefficient, electrical conductivity, and thermal conductivity values we have calculated the dimensionless figure of merit (ZT), of the $CdSnAs_2$, which evaluates the performance of thermo electric materials. Figure 6 shows the temperature dependencies of the figure of merit (ZT) for the CdSnAs2. From Fig. 6, we can see that the ZT increase when temperature increased, this due to increase of the

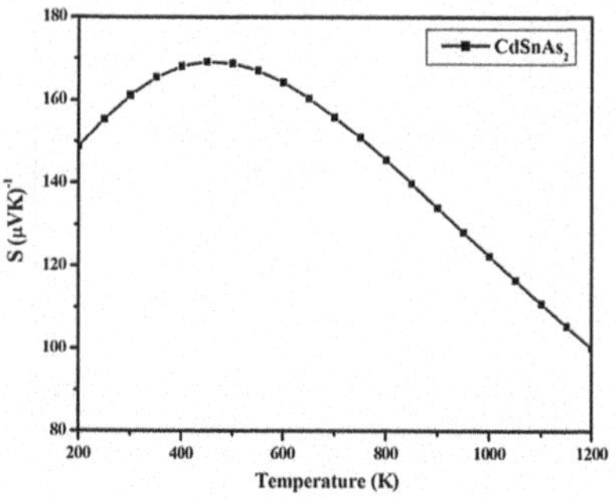

Fig. 3. Temperature dependence of the Seebeck coefficient

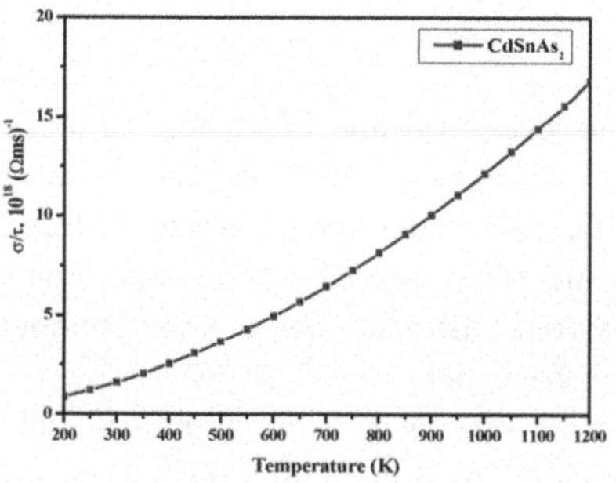

Fig. 4. Temperature dependence of the electrical conductivity

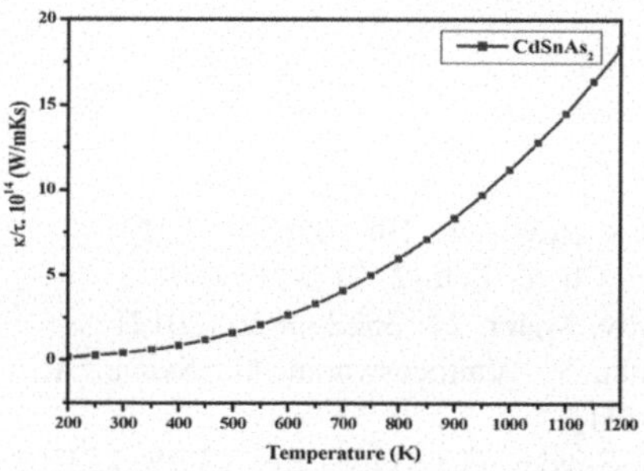

Fig. 5. Temperature dependence of the thermal conductivity

Seebeck coefficient and decrease of the thermal conductivity especially in the 200–450 K. After passing a temperature 450 K, the ZT value of CdSnAs2 decreased withe increasing of temperature the largest value of figure of merit ZTmax = 0.33, which is indicating this material is promising for low temperature applications.

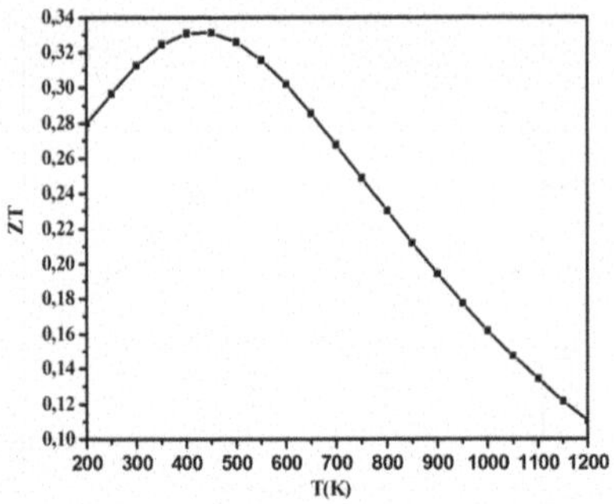

Fig. 6. Temperature dependence of the figure of merit.

4 Conclusion

In this work, we study the structural and electronic properties of pure CdSnAs2, by means of the density functional theory (DFT) and the linearized augmented plane wave method (LAPW). The calculated structural parameters obtained are in agreement with the experimental reports and other theoretical results. The calculation results of the electronic properties show that $CdSnAs_2$ has a semi conductor character, which is of interest for thermoelectric properties.

The electronic properties obtained from Wien2 K calculations and Boltzmann transport theory implemented in BoltzTrap code were employed to calculate the thermoelectric properties of CdSnAs2 as function of temperature raging from 200 to 1200 K. Our results reveal that CdSnAs2 has a largest figure of merit (ZTmax) with value of 0.33 at temperature.

References

1. Yu, L., Zunger, A.: Phys. Rev. Lett. **108**, 068701 (2012)
2. Xiao, H., et al.: J. Phys. Chem. Lett. **2**, 212–217 (2011)
3. Plirdpring, T., et al.: Adv. Mater. **24**, 3622–3626 (2012)
4. Gudelli, V.K., Kanchana, V., Vaitheeswaran, G., Svane, A., Christensen, N.E.: J. Appl. Phys. **114**, 223707 (2013)
5. May, A.F., Singh, D.J., Snyder, G.J.: Phys. Rev. B **79**, 153101 (2009)
6. Shay, J.L., Wernick, J.H.: Ternary Chalcopyrite Semiconductors Growth, Electronic Properties and Applications. Pergamon Press, Oxford (1975)
7. Lind, M.D., Grant, R.W.: J. Chem. Phys. **58**, 357–362 (1973)
8. Kumar, V., Tripathy, S.K.: J. Alloy. Compd. **582**, 101–107 (2014)
9. Turowski, M., Margaritondo, G., Kelly, M.K., Tomlinson, R.D.: Phys. Rev. B **31**, 1022–1027 (1985)
10. Medvedkin, G.A., Voevodin, V.G.: J. Opt. Soc. Am. B **22**, 1884–1898 (2005)
11. Das, S.: Infrared Phys. Technol. **69**, 13 (2015)
12. Xu, Y., Ao, Z.M., Zou, D.F., Nie, G.Z., Sheng, W., Yuan, D.W.: Phys. Lett. A **379**, 427–430 (2015)
13. Reshak, A.H.: J. Phys. Chem. Solid **78**, 46–52 (2015)

14. Yao, B., Yuan, J., Li, J., Dai, T., Duan, X., Shen, Y., Cui, Z., Pan, Y.: Opt. Lett. **40**, 348–351 (2015)
15. Zhang, Z., Reid, D.T., Kumar, S.C., Ebrahim-Zadeh, M., Schunemann, P.G., Zawilski, K.T., Howle, C.R.: Opt. Lett. **38**, 5110 (2013)
16. Blaha, P., Schwarz, K., Madsen, G.K.H., Kvasnicka, D., Luitz, J.: WIEN2 k an augmented plane wave þ local orbitals program for calculating crystal properties. In: Schwarz, K., Techn. Universitat Wien, Austria (2001)
17. Blaha, P., Schwarz, K., Sorantin, P.I., Tricky, S.B.: Comput. Phys. Commun. **59**, 399 (1990)
18. Becke, A.D., Johnson, E.R.: J. Chem. Phys. **124**, 221101 (2006)
19. Tran, F., Blaha, P.: Phys. Rev. Lett. **102**, 226401 (2009)
20. Madsen, G.K.H., Singh, D.J.: Comput. Phys. Commun. **175**, 67 (2006)
21. Ong, K.P., Singh, D.J., Wu, P.: Phys. Rev. B **83**, 115110 (2011)
22. Singh, D.J., Mazin, I.I.: Phys. Rev. B **56**, R1650 (1997)
23. Parker, D., Singh, D.J.: Phys. Rev. B **82**, 035204 (2010)
24. Madsen, G.K.H., Schwarz, K., Blaha, P., Singh, D.J.: Phys. Rev. B **68**, 125212 (2003)
25. Zhang, L., Du, M.-H., Singh, D.J.: Phys. Rev. B **81**, 075117 (2010)
26. Zunger, A.: Appl. Phys. Lett. **50**(3), 164 (1987)
27. Continenza, A., Massidda, S., Freeman, A.J.: Phys. Rev. B **46**(16), 10070 (1992)
28. Reshak, A.H., Khenata, R., Kityk, I.V., Plucinski, K.J., Auluck, S.: J. Phys. Chem. B **113,** 5803 (2009)
29. Allen, P.B., Pickett, W.E., Krakauer, H.: Phys. Rev. B **37**, 7482 (1988)
30. Madsen, G.K.H., Singh, D.J.: Comput. Phys. Commun. **75,** 67–71.30 (2006)

Milligels Synthesis and Characterization: Mebeverine Hydrochloride Uptake and Release

N. Baït[1,2(✉)], B. Grassl[3], and A. Benaboura[2]

[1] Centre de Recherche Scientifique et Technique en Analyses Physico-Chimiques (CRAPC), BP 384, Bou-Ismail, 42004 Tipaza, Algérie
wisali2003@yahoo.fr
[2] Laboratoire de Synthèse Macromoléculaire et Thio-Organique Macromoléculaire, Faculté de Chimie, USTHB, BP 32, El-Alia, Bab-Ezzouar, 16111 Alger, Algérie
[3] Institut Pluridisciplinaire Sur L'Environnement et Les Matériaux- LCABIE-UMR 5254, Hélioparc, 2 Avenue Président Angot, 64053 Pau Cedex 9, France

Abstract. This study describes the millifluidic-based synthesis of conventional Polyacrylamide (PAM), Polyacrylamide-Sodium Montmorillonite (PAM-NaMMT) and Polyacrylamide-Organo-modified Montmorillonite (PAM-OMMT) composite milligels. These hydrogel particles with millimeter size are intended to be used for the design of Drug Delivery Systems (DDS) combining hydrogels and Clay performances. The composite hydrogels structure, the interactions between the PAM chains and the Clay and the networks thermal behavior were evidenced by Fourier Transform Infrared Spectroscopy (FTIR) and Differential Scanning Calorimetry (DSC) respectively. PAM, PAM-NaMMT and PAM-OMMT milligels potential loading and release of Mebeverine Hydrochloride, taken as a model molecule, was investigated by UV Spectroscopy. This potential was examined as a function of the release medium, the networks chemical composition, the Montmorillonite (MMT) nature and its content.

Keywords: Millifluidic · Milligel · Montmorillonite · Drug

1 Introduction

For several years, there has been a growing interest in hydrogels prepared from naturally occuring polymers or synthetic polymers for various applications, ranging from controlled drug release to the tissue engineering developed for the reconstruction of diseased tissues and bone tissues (Alimardan et al. (2014); Tan and Marra (2010); Parida and Mishra (2012); Hoare and Kohane (2008); Thakur et al. (2011)). Hydrogels are macromolecular networks that swell, but not dissolve in water (Ferreira et al. (2001); Dafader et al. (2011); Saraydin et al. (2002)).

Multiparticle systems in controlling and sustaining of release rate of many active pharmaceutical ingredients are the greatest advance in the field of Drug Delivery Systems (DDS) design. As a result oral dosage forms such as microcapsules and microspheres have emerged as DDS for the treatment of various diseases (Patil et al. (2016)).

B. Abdelbaki et al. (Eds.): SMSD 2017, *Proceedings of the Third International Symposium on Materials and Sustainable Development*, pp. 372–377, 2018.
https://doi.org/10.1007/978-3-319-89707-3_43

Thus, hydrogels based on different polymers and prepared in the form of beads by different methods have emerged (Topuz and Okay (2009); Rajeswari et al. (2012); Ray et al. (2010); Wan (2012)). One of the innovative techniques for obtaining hydrogel beads is the *millifluidic* which is the « *science of fluids flow in tubes of small diameters* ». It is based on the generation of drops in millimeter-scale tubes from the flow of at least two immiscible fluids.

This study deals with the use of the millifluidic technique for the design of a drug delivery system combining hydrogels and Clay performances.

2 Experimental Part

2.1 Montmorillonite Sodium Exchange and Organo-Modification

- The Clay used in this study is the Montmorillonite of Maghnia (an Algerian Clay) (MMT). First, it was treated with an aqueous solution of NaCl in order to exchange its pristine interlayer cations by Na^+. The homoionized Montmorillonite (Sodium Montmorillonite) is noted NaMMT.
- The organo-modified Montmorillonite noted OMMT was obtained by exchanging the sodium ions of NaMMT by tetradecyltrimethyl ammonium ions $C_{17}H_{38}N^+$.

2.2 Milligels Synthesis

Polyacrylamide milligels PAM, and composite milligels PAM-NaMMT and PAM-OMMT were synthesized using millifluidic device, by in-situ free radical polymerization-cross-linking at 60°C, in aqueous solution and aqueous Clay suspensions respectively, using the initiator Potassium Persulfate (KPS) and the cross-linking agent N,N-methylene-Bis-Acrylamide (Bis). The weight ratios Clay/(Acrylamide monomer AM) selected were 0,5 and 1%. Composite milligels are noted PAM-NaMMT*i* and PAM-OMMT*i* (i = 0.5 and 1 = Clay/AM weight ratio).

Figure 1 shows the process of the milligels generation in capillary tubes.

Fig. 1. Milligels generation in a capillary tube.

For comparison, the macrogel PAM was synthesized in the same conditions by the conventionnal route.

2.3 Drug Loading and Release

The loading of Mebeverine Hydrochloride onto the three types of milligels was carried out by swelling equilibrium method. Each hydrogel was allowed to swell in the drug solution of known concentration. After the equilibrium loading, the drug release was investigated in acidic (pH = 1,2) and alkaline (pH = 7.4) media to simulate the stomach and the intestine environments. Milligels loading and release kinetics were monitored by UV Spectroscopy. The amount of the drug loaded or released at any given time and in a given medium was calculated from the absorbance of the sample and the calibration curves we previously constructed.

3 Results and Discussion

3.1 FTIR Analysis

3.1.1 PAM Analysis

Polyacrylamide IR spectrum is predominated by two large and intense bands. The first one extends between 3990 cm^{-1} and 2800 cm^{-1} and is assigned to asymetrical and symetrical stretching vibrations of N-H groups. The second one extends between 2000 cm^{-1} and 1480 cm^{-1} and is attributed to the stretching vibrations of the carbonyl groups ($\nu_{C=O}$) and deformation vibrations of the NH bonds ($\delta_{N\text{-}H\ scissoring}$). These bands were considered to highlight the effect of the Clay introduction within the PAM network.

The peaks at 2928 cm^{-1} and 2850 cm^{-1} are attributed, in the order, to asymetrical and symmetrical stretching vibrations of $-CH_2$ and $-CH$ groups (Fig 2).

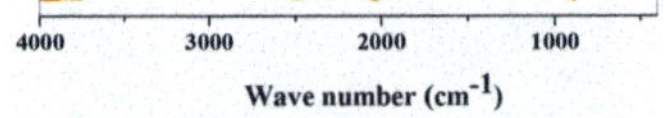

Fig. 2. FTIR spectrum of the PAM hydrogel.

3.1.2 The Clay Effect on PAM Structure

Interactions formation or destruction in the PAM network following the introduction of NaMMT or OMMT is demonstrated by the carbonyl regions of PAM-NAMMT*i* and PAM-OMMT*i* spectra. These regions showed an increase of C = O and N-H peak intensities ratio, as illustrated, for exemple, by PAM and PAM-NaMMT*i* regions shown in Fig. 3.

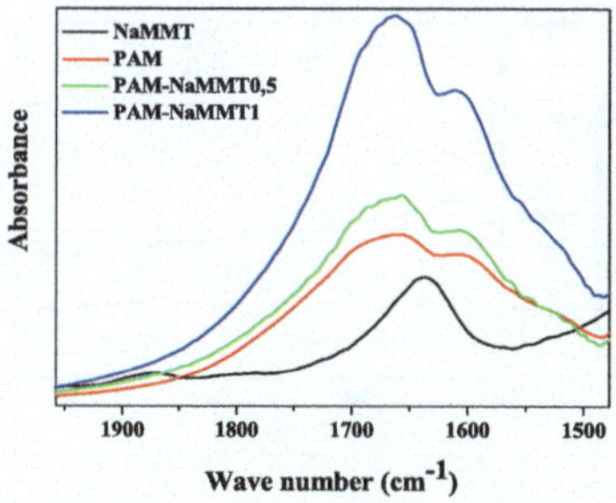

Fig. 3. Carbonyl groups region of NaMMT, PAM and PAM-NaMMT*i* milligels.

This result could be attributed to the destruction of the interactions between the functional groups of the PAM to the benefit of the establishment of interactions between the PAM amide groups and the negatively charged NaMMT or OMMT sheets.

3.2 Differential Scanning Calorimetry Analysis

Table 1 gathers the glass transition temperature (Tg) values of the PAM, and the composite networks PAM-NaMMT*i* and PAM-OMMT*i*. One can note that there is an increase in the PAM Tg value with the introduction of the MMT within its network, whatever is the homoionized or organo-modified nature of the MMT. In the presence of the Clay, polyacrylamide chains become less mobile. The composite networks are more rigid. This fact can be explained by an increase in the cross-links density, induced by the interactions between PAM chains and the MMT sheets.

Table 1. PAM, PAM-NaMMT*i* and PAM-OMMT*i* Tg values

Hydrogels	Tg (°C)
PAM	197
PAM-NaMMT0,5	203
PAM-NaMMT1	206
PAM-OMMT0,5	198
PAM-OMMT1	204

3.3 Mebeverine Hydrochloride Loading and Release

The loading kinetics obtained showed that the composite milligels load a more important quantity of the Mebeverine Hydrochloride than the PAM, as illustrated by Fig. 4.

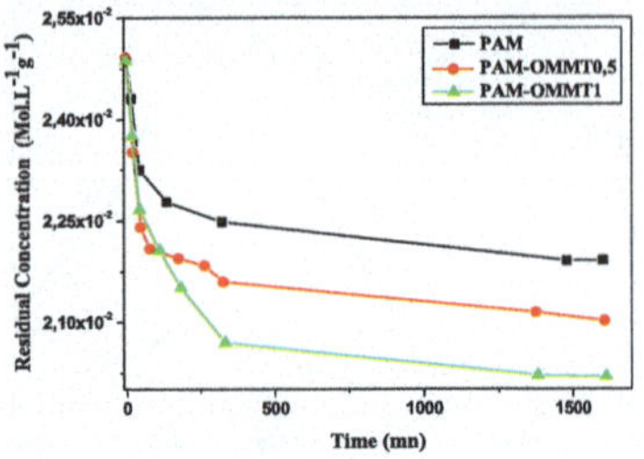

Fig. 4. Mebeverine Hydrochloride loading kinetics by PAM and PAM-OMMT*i* milligels.

The most important result of the Hydrochloride Mebeverine release study pointed out in this study is illustrated by Figs. 5(a), and (b) which show the release kinetics obtained for the composite milligels PAM-NaMMT1 and PAM-OMMT1 compared to the milligels PAM and the macrogel PAM, in the two different media (acid and alkaline). Our finding shows that:

- the behavior of the polyacrylamide network is insensitive to pH;
- the composite milligels are pH-sensitive.

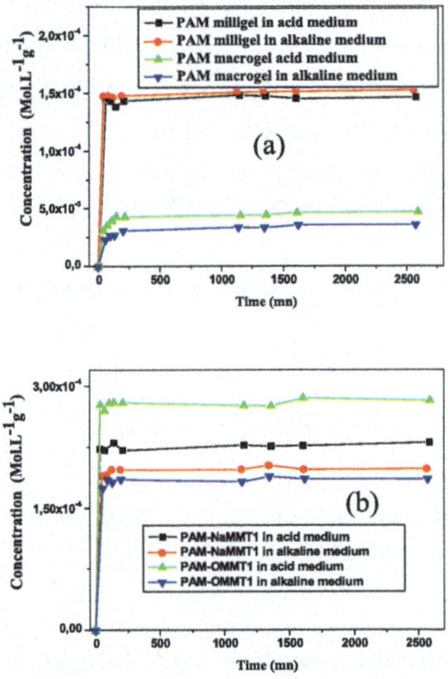

Fig. 5. Mebeverine Hydrochloride release kinetics obtained in acidic and alkaline media, for (a) Milligels PAM and Macrogel PAM and (b) PAM-NaMMT1 and PAM-OMMT1 composite milligels.

4 Conclusions

This work aimed the development of drug loading-release systems combining hydrogels and Clay performances. In this perspective we have synthesized the conventional hydrogel polyacrylamide PAM and the composite hydrogels Polyacrylamide-Sodium Montmorillonite PAM-NaMMT and Polyacrylamide -Organo-modified Montmorillonite PAM-OMMT. These hydrogels were synthesized in the form of millimetric sized beads on the basis of an adapted droplet-flow millifluidic process. The FTIR analysis highlighted the interactions developed between the functional amide groups of the polyacrylamide and the sheets of the Clay. This result was corroborated by the DSC analysis which showed the reduction of the PAM chains mobility in the presence of the Montmorillonite as a result of the establishment of new junctions. Mebeverine Hydrochloride loading-release kinetics showed that PAM, PAM-NaMMT*i* and PAM-OMMT*i* potential drug load and release is modulated by the networks chemical composition. The drug release kinetics showed that we succeeded in designing a pH-sensitive Drug Release System which can be used to control the Mebeverine Hydrochloride release.

References

Alimardan, Z., Darabi, M., Natori, R.: Drug delivery of indomethacin in use nature starch based nanocomposite hydrogel in vitro. Int. J. Biosci. **5**, 51–56 (2014)

Tan, H., Marra, K.G.: Injectable, biodegradable hydrogels for tissue engineering applications. Materials **3**, 1746–1767 (2010)

Parida, P., Mishra, S.C.: Synthesis, crosslinking and evaluation of graft polymer for biomedical use. Res. Rev.: J. Pharm. Sci. **3**, 37–45 (2012)

Hoare, T.R., Kohane, D.S.: Hydrogels in drug delivery: progress and challenges. Polymer **49**, 1993–2007 (2008)

Thakur, A., Wanchoo, R.K., Singh, P.: Hydrogels of poly (acrylamide -*co*-acrylic acid): in-vitro study on release of gentamicin sulfate. Chem. Biochem. Eng. Q. **25**, 471–482 (2011)

Ferreira, L., Vidal, M.M., Gil, M.H.: Design of a drug-delivery system based on polyacrylamide hydrogels. Evaluation of structural properties. Chem. Educ. **6**, 100–103 (2001)

Dafader, N.C., Adnan, M.N., Haque, M.E., Huq, D., Akhtar, F.: Study on the properties of copolymer hydrogel obtained from acrylamide/2-hydroxyethyl methacrylate by the application of gamma radiation. Afr. J. Pure Appl. Chem. **5**, 111–118 (2011)

Saraydin, D., Isikver, Y., Karadag, E., Sahiner, N., Guven, O.: In vitro dynamic swelling behaviors of radiation synthesized polyacrylamide with crosslinkers in the simulated physiological body fluids. Nucl. Insrum. Methods Phys. Res. B **187**, 340–344 (2002)

Patil, J.S., Kole, S.G., Gurav, P.B., Vilegave, K.V.: Natural polymer based mucoadhesive hydrogel beads of nizatidine preparation, characterization and evaluation. Indian J. Pharm. Educ. Res. **50**, 159–169 (2016)

Topuz, F., Okay, O.: Macroporous hydrogel beads of high toughness and superfast responsivity. React. Funct. Polym. **69**, 273–280 (2009)

Rajeswari, K.R., Abbulu, K., Sudhakar, M., Karki, R., Rajkumar, B.: Development and characterization of valsartan loaded hydrogel beads. Der Pharmacia Lettre **4**, 1044–1053 (2012)

Ray, R., Maity, S., Mandal, S., Chatterjee, T.K., Sa, B.: Development and Evaluation of a New Interpenetrating Network Bead of Sodium Carboxymethyl Xanthan and Sodium Alginate for Ibuprofen Release. Pharmacol. Pharm. **1**, 9–17 (2010)

Wan, J.: Microfluidic-based synthesis of hydrogel particles for cell microencapsulation and cell-based drug delivery. Polymers **4**, 1084–1108 (2012)

Dynamic Rheology Study of In-situ Gelation Process of Polyacrylamide-Montmorillonite Composite Hydrogels

N. Baït[1,2(✉)], B. Grassl[3], and A. Benaboura[2]

[1] Centre de Recherche Scientifique et Technique en Analyses Physico-Chimiques (CRAPC), BP 384 42004 Bou-Ismail, Tipaza, Algérie
wisali2003@yahoo.fr

[2] Laboratoire de Synthèse Macromoléculaire et Thio-Organique Macromoléculaire, Faculté de Chimie, USTHB, BP 32 El-Alia, 16111 Bab-Ezzouar, Alger, Algérie

[3] Institut Pluridisciplinaire sur l'Environnement et les Matériaux- LCABIE-UMR 5254, Hélioparc, 2 Avenue Président Angot, 64053 Pau Cedex 9, France

Abstract. This study describes the in-situ gelation kinetics of polyacrylamide PAM and PAM-Clay networks precursory reactional mixtures and the rheological properties of the resulting hydrogels. Elastic *G'* and loss *G"* moduli measurements via oscillatory shear in dynamic rheology and the structural parameters calculation of the networks using the rubber elasticity theory enabled us to apprehend the relationship between the hydrogels structure and the composition of their precursory mixtures: *(i)* the presence of the Clay and its absence, *(ii)* its content and its homoionized or organo-modified nature and *(iii)* the presence of the cross-linking agent and its absence.

Keywords: Polymerization-cross-linking · Rheology · Elasticity Montmorillonite

1 Introduction

Hydrogels are polymer networks swollen with a large amount of water. They can absorb water up to hundreds times their weight while maintaining their three-dimensional structure. The term "water" is used in a broad sense, in fact, the solvent stored in the network can be pure water, a biological fluid, aqueous solutions of salts or polymers (Peppas (1986); Brannon-Peppas (1999); Hamidi et al. (2008)). These materials are used in a wide range of applications including agriculture and soil conditioning, water treatment, sensors and wound dressing (Karadag et al. (2000); Kasgoz et al. (2003); Singh et al. (2006); Schneider et al. (2007); Zohuriaan-Mehr et al. (2010)). In the biomedical field, cross-linked polyacrylamide PAM hydrogels have been widely studied. The variable porosity of their networks is ideal in electrophoresis (Abd Alla et al. (2007)) and also in the drug release systems (Singh et al. (2008); Niamlang and Sirivat (2009)). Moreover, the ease of modulating its hydrogels rigidity, has shown that the PAM is an excellent substrate model for fundamental studies to the development of biomaterials

B. Abdelbaki et al. (Eds.): SMSD 2017, *Proceedings of the Third International Symposium on Materials and Sustainable Development*, pp. 378–384, 2018.
https://doi.org/10.1007/978-3-319-89707-3_44

for tissue engineering applications (Calvet et al. (2004)). Therefore, it is highly desirable to investigate the kinetics of in-situ polymerization reactions in order to better elucidate the effects of PAM polymerization- cross-linking conditions on structure- property relationships to optimize its functional performances. Among the experimental techniques employed to study the gelation kinetics of PAM-based hydrogels, dynamic rheology has emerged as an effective and powerful approach for gel characterization. Indeed, in-situ experiments are relatively easy to carry out and viscoelastic properties could directly reflect the hydrogels structure and morphology (Calvet. et al. (2004); Zhou et al. (2011)).

Having in mind the existence of the broad distribution of network chain lengths, the dangling chains, the cyclizations, the cross-link agglomerations in PAM hydrogels synthesized via free radical polymerization, (Gurtovenko and Gotlib (2001); Naghash and Okay (1996); Kizilay and Okay (2003)), we aimed, in this study, to describe the effects of the homoionized and organo-modified Montmorillonite on the PAM network and its inhomogeneities.

2 Experimental Part

2.1 Montmorillonite Organo-Modification

The Montmorillonite (MMT) is a smectite mineral clay which general chemical composition is $M_x(Al_{4-x}Mg_x)Si_8O_{20}(OH)_4$.

Once homo-iononized, the Montmonrillonite of Maghnia (an Algerian Clay) was organo-modified using the tetradecyltrimethyl ammonium bromide surfactant. In the following, the two kinds of the Clay will be denoted NaMMT and OMMT respectively.

2.2 Reactive Mixtures Formulation

The reactive mixtures were prepared by adding, under stirring, the monomer acrylamide AM, the cross-linker N,N-methylene-bis-acrylamide Bis and the initiator potassium persulfate KPS to salt-free Milli-Q water in the case of polyacrylamide hydrogel and to a suspension of NaMMT or OMMT for the composite hydrogels synthesis. The weight ratios Clay/(Acrylamide monomer AM) selected were 0,5 and 3% .

Other reactive solutions were prepared in the same manner without the cross-linking agent Bis.

2.3 Rheology Monitoring of Gelation Process and Spectromechanical Analysis

Rheological monitoring gelation was performed using two parallel plates of a Gemini Rheometer system (Bohlin instruments) and oscillatory time sweep testing mode.

During measurements, a sample of 1 mL of the reactive mixture was deposited on the lower plate and then the upper plate was set at a distance (typically 1 mm) at which a disc is formed between the two plates, without normal force.

To measure elastic *G'* and loss *G"* moduli as function of time, an angular frequency of 10 rad s^{-1} and a stress of 10 Pa were selected. These dynamic oscillatory measurements were within the linear domain previously detected by stress sweeps.

The hydrogels rheological behavior was evaluated from spectromechanical analyses carried out at 25 °C and performed using the gel samples prepared within the rheometer plates after the oscillatory time sweeps.

3 Results and Discussion

3.1 In-situ Gelation of (AM/KPS/Bis) and (AM/KPS/Bis/Clay) Mixtures

Figure 1 illustrates the in-situ polymerization-cross-linking kinetics of (AM/KPS/Bis) and (AM/KPS/Bis/Clay) mixtures precursors of PAM, PAM-NaMMT and PAM-OMMT conventional and composite hydrogels.

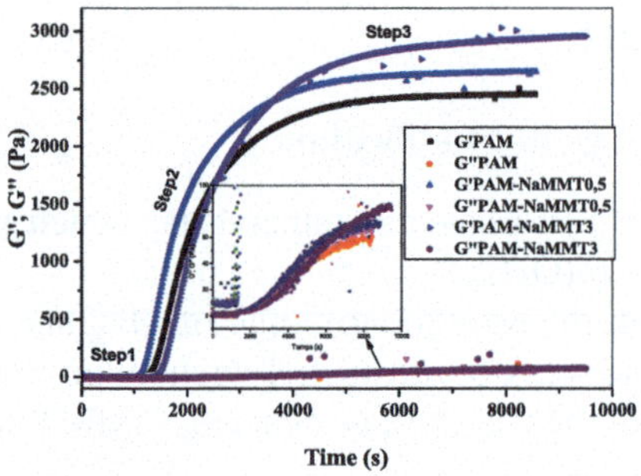

Fig. 1. Variation of elastic and loss moduli of PAM and composite hydrogels as a function of polymerization-cross-linking time.

The curves of elastic and loss moduli *G'(t)* and *G"(t)* show that :

- The gelation process proceeds in three steps: A first step where the elastic modulus is close to zero due to an inhibition period explained by the absence of the reactive solution degassing. The reaction mixture is in its *"sol"* form. A second step where *G'* value increases strongly compared to *G"* one, it corresponds to the *sol-gel* transition. The most important point in this region is the *gel point.* A third stage corresponding to the formed gel and where *G'* value increases weakly until reaching a plateau.
- The presence of the Clay in its two forms (curves with OMMT are not shown here) does not alter the PAM network formation. In addition, the elastic modulus of the PAM increases in the presence of the Clay. We suppose that its presence during the polymerization-cross-linking reaction contributes to the formation of additional cross-linking nodes which leads to the increase of *G'* value.

3.2 PAM and PAM-Clay Spectromechanical Analysis

Frequency sweeps carried out at 25 °C, allowed us, in addition to characterizing the hydrogels, to examine the effect of the mixtures composition on the elasticity of the obtained networks after the gelling process is completed.

Figure 2 shows, as an exemple, the frequency sweeps of the conventionnal hydrogel PAM and the composite hydrogels PAM-NaMMT and PAM-OMMT with 0,5 and 3% of each one of the two clays.

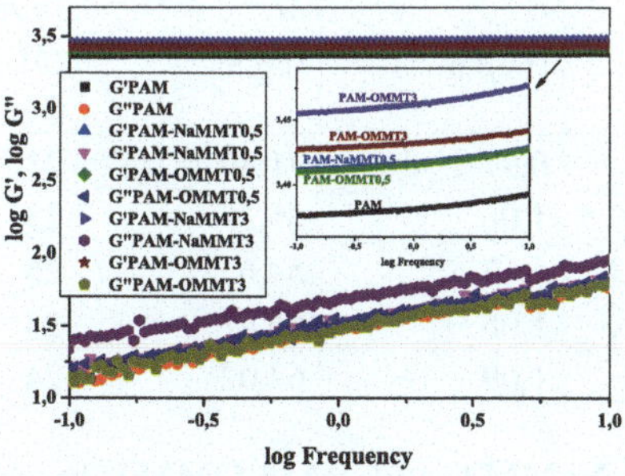

Fig. 2. Frequency sweeps of PAM and composite hydrogels.

Our samples are characterized by a storage (elastic) modulus $G'(\omega)$ which exhibit a pronounced plateau in the frequency range investigated and a loss modulus $G''(\omega)$ which is considerably smaller than the elastic modulus. This is an indication that our materials are cross-linked hydrogels (Baker et al. (2010); Bousmina (1999)).

The insert of Fig. 2 shows that:

- whatever is its nature (homoionized or organo-modified) and its content, the presence of the Clay in the reaction medium improves the elasticity of the PAM network;
- regardless of the Clay nature, *G'* value increases with increasing the Clay content;
- this increase is greater with the homoionized Montmorillonite (the NaMMT).

3.3 Structural Parameters and Inhomogeneities of the PAM, PAM-NaMMT and PAM-OMMT Networks

On the basis of the rubber elasticity theory developed by Flory (Flory (1953)), we determined the density of effective junctions (nodes density) n_e and the molar mass of a polymer chain binding two adjacent cross-links (sub-chain) *Me* for all the synthesized hydrogels. These two parameters are those which define a polymer network and determine its elasticity. Moreover, we calculated the theoretical values of n_e and *Me* (noted $n_{e\ theo}$ and $M_{e\ theo}$) predicted for an ideal network with no structural defects and where each molecule of the cross-linking agent Bis generates a single cross-linking node. The calculated values are 4,54 (mol.m^{-3}) and 15,42 (kg.m^{-3}) for $n_{e\ theo}$ and $M_{e\ theo}$ respectively.

The obtained results gathered in Table 1 show that in the case of PAM network n_e value counts for less than 21% of the theoretical cross-linking density predicted by the cross-linking agent (Bis) concentration. This is an indication that Bis molecules are not all effective in the generation of effective junctions. In fact the majority of these molecules are consumed in the multiple cross-linking and cyclization reactions and in the formation of dangling chains. This demonstrates the existence of inhomogeneities in the PAM network which are characteristic of hydrogels synthesized by free radical polymerization (Calvet et al. (2004); Naghash and Okay (1996)).

Table 1. Structural parameters of PAM and composites networks

Hydrogel	n_e (mol.m^{-3})	M_e (kg.m^{-3})	$(n_e/n_{e\ theo}) = (M_{e\ theo}/M_e)$ (%)
PAM	0,97	71,89	21
PAM-NaMMT-0.5	1,06	66,25	23
PAM-NaMMT-3	1,17	59,63	26
PAM-OMMT-0.5	1,05	66,41	23
PAM-OMMT-3	1,09	64,02	24

In the case of PAM-NaMMT and PAM-OMMT hydrogels, the ratio $n_e/n_{e\ theo}$ increases which means that the cross-linking density is greater in the composite hydrogels. This result could be interpreted in two manners : *(i)* the reactivity of the cross-linker Bis is modified with the presence of the Clay whatever is its nature and *(ii)* the Clay contributes to the formation of additional junctions, as we assumed in Sect. 3.1, by interacting with the acrylamide functional groups.

3.4 Spectromechanical Analysis of the Uncross-Linked Mixtures

To conclude on the Clay effect, we performed spectromechanical analyses on samples obtained after time sweeps carried out on AM-KPS and AM-KPS-Clay mixtures without the cross-linker Bis. Figure 3 shows, as an exemple, the frequency sweeps recorded for the linear polymers PAM and PAM-NaMMT3. The curves *G'(ω)* and *G''(ω)* of the two samples show a crossover point which delimits two regions in which one or the other of the viscous or the elastic behaviors predominates.

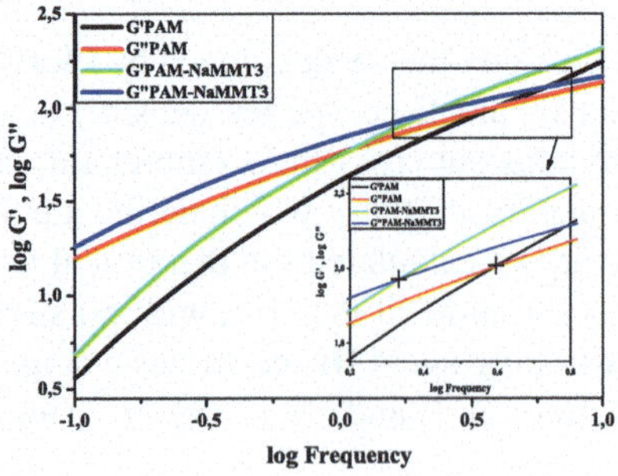

Fig. 3. Frequency sweeps of uncross-linked PAM and PAM-NaMMT3 polymers.

At lower frequencies, *G''* exceeds *G'* indicating a liquid-like behavior. Here polymer chains have sufficient time to free themselves from entanglements. At higher frequencies, a solid-like behavior is observed (*G'* > *G''*) indicating that polymer chains are unable to slip past one another and to free themselves from the entanglement constraints.

The insert of Fig. 3 shows that, when polymerized in the presence of the Montmorillonite, the crossover point of the PAM shifts to lower frequencies. This indicates that the PAM chains lenght and their distribution are affected by the presence of the Clay and that there are much more entanglements in the PAM-NaMMT compared to the PAM. This result could be explained by the contribution of the Clay in generating chain cross-links.

4 Conclusions

The influence of the Clay on the polymerization-cross-linking process of the acrylamide was studied using rheological monitoring. A serie of in-situ rheology experiments: time, stress and frequency sweeps, were performed to investigate the effect of the Montmorillonite, its sodium and organo-modified nature and its content on the elastic *G'* and loss *G''* moduli of the polyacrylamide hydrogel and its inhomogeneous structure. Spectromechanical analyses of the in-situ formed hydrogels PAM, PAM-NaMMT and PAM-OMMT in the presence of the cross-linking agent N,N-methylene-Bis-Acrylamide Bis showed that the elasticity of the PAM network is enhanced when the acrylamide is polymerized in aqueous NaMMT or OMMT suspensions. Moreover, the structural parameters calculated for the formed hydrogels on the basis of the rubber elasticity theory and the spectromechanical analyses carried out on Acrylamide aqueous solution and Acrylamide-Clay aqueous supensions in the absence of Bis showed that the Clay contributes to the formation of additional junctions in the PAM network by interacting with the acrylamide functional groups and could be at the origin of enhancing the Bis efficiency and thus lowering the PAM inhomogeneities.

References

Peppas, N.A.: Hydrogels in Medecine and Pharmacy. In: Mokos, A.G., Peppas, N.A. (eds.) vol. 1. CRC Press, Boca Raton (1986)

Brannon-Peppas, L.: Preparation and Characterization of cross-linked hydrophilic networks. In: Harland, R.S. (eds.) Polymer Technology, pp. 45–66. Elsevier, Amsterdam (1999)

Hamidi, M., Azadi, A., Rafei, P.: Hydrogel nanoparticles in drug delivery. Adv. Drug Deliv. Rev. **60**, 1638–1649 (2008)

Karadag, E., Saraydin, D., Caldiran, Y., Guven, O.: Swelling studies of copolymeric acrylamide/crotonic acid hydrogels as carriers for agricultural uses. Polym. Adv. Technol. **11**, 59–68 (2000)

Kasgoz, H., Ozgumus, S., Orbay, M.: Modified polyacrylamide hydrogels and their application in removal of heavy metal ions. Polymer **44**, 1785–1793 (2003)

Singh, B., Chauhan, G.S., Bhatt, S.S., Kumar, K.: Metal ion sorption and swelling studies of psyllium and acrylic acid based hydrogels. Carbohydr. Polym. **64**, 50–56 (2006)

Schneider, H.-J., Kato, K., Strongin, R.M.: Chemomechanical polymers as sensors and actuators for biological and medicinal applications. Sensors **7**, 1578–1611 (2007)
Zohuriaan-Mehr, M.J., Omidian, H., Doroudiani, S., Kabiri, K.: Advances in non-hygienic applications of superabsorbent hydrogel materials. J. Mater. Sci. **45**, 5711–5735 (2010)
Abd Alla, S.G., Kandeel, M.M., El-Naggar, A.W.M.: Characterization of hydrogels prepared by the gamma irradiation of aqueous solutions of acrylamide and N-Vinyl-2-pyrrolidone comonomers. J. Appl. Polym. Sci. **106**, 4202–4208 (2007)
Singh, B., Chauhan, N., Kuma, S., Bala, R.: Psyllium and copolymers of 2-hydroxyethylmathacrylate and acrylamide-based novel devices for the use in colon specific antibiotic drug delivery. Int. J. Pharm. **352**, 74–80 (2008)
Niamlang, S., Sirivat, A.: Electrically controlled release of salicylique acid from poly(p-phenylene vinylene)/polyacrylamide hydrogels. Int. J. Pharm. **371**, 126–133 (2009)
Calvet, D., Wong, Y.J., Giasson, S.: Rheological monitoring of polyacrylamide gelation: importance of cross-link density and temperature. Macromolecules **37**, 7762–7771 (2004)
Zhou, C., Wu, Q., Zhang, Q.: Dynamic rheology studies of in situ polymerization process of polyacrylamide–cellulose nanocrystal composite hydrogels. Colloid Polym. Sci. **289**, 247–255 (2011)
Gurtovenko, A.A., Gotlib, Y.Y.: Dynamics of inhomogeneous cross-linked polymers consisting of domains of different sizes. J. Chem. Phys. **115**, 6785–6793 (2001)
Naghash, H.J., Okay, O.: Formation and structure of polyacrylamide gels. J. Appl. Polym. Sci. **60**, 971–979 (1996)
Kizilay, M.Y., Okay, O.: Effect of hydrolysis on spatial inhomogeneity in poly (acrylamide) gels of various crosslink densities. Polymer **44**, 5239–5250 (2003)
Baker, B.A., Murff, R.L., Milam, V.T.: Tailoring the mechanical properties of polyacrylamide-based hydrogels. Polymer **51**, 2207–2214 (2010)
Bousmina, M.: Rheology of polymer blends: linear model for viscoelastic emulsions. Rheol. Acta **38**, 73–83 (1999)
Flory, P.J.: Principles of Polymer Chemistry. Cornell University Press, Ithaca (1953)

Temperature Effects on Corrosion Inhibition of Mild Steel in 3% NaCl Solution by EDTA and Sodium Benzoate

N. Brixi(✉), L. Sail, A. Bezzar, O. Sebouai, and S. Benmesmoudi

Laboratory EOLE, Department of Civil Engineering,
AbouBekrBelkaid University, Tlemcen, Algeria
Brixinezha@live.com

Abstract. Carboxylates are known as corrosion inhibition products of steels in medium contaminated with chlorides. Gravimetric method was used to study the temperature effects on corrosion inhibition of carbon steel in a medium of 3% NaCl simulating the marine environment of Mediterranean by two organic carboxylate products which are Ethylene diaminetetraacetic acid (EDTA) and sodium benzoate. The maximum inhibition efficiency for EDTA and sodium benzoate are about 86% at a concentration of 0.15 mol/l and 73% at 0.2 mol/l respectively at a temperature of 298°K. While for 308°K and 318°K of temperature, the results recorded for EDTA are 77% and 73%, and 69% and 46% for sodium benzoate at the same concentration 0.2 mol/l at the respective temperatures. Consequently, it is deduced that the inhibition efficiency decrease with increasing temperature. The Frumkin model correctly adjusts the fixation mechanism of the inhibition molecules on the surface steel for the two tested products. The free energy values indicate the spontaneity of the adsorption of the inhibitor on the surface of the metal indicating a physisorption mechanism.

Keywords: EDTA · Efficiency · Inhibition · Sodium benzoate
Temperature

1 Introduction

Corrosion of steel in concrete is one of the main causes of degradation of reinforced concrete structures. The durability of the structures depends on their opposition to the physical and chemical attacks of the environment, as well as on the ability of the building materials to protect themselves against these aggressions. Corrosion develops mainly as a result of two processes: carbonation of concrete and attack by chlorides, when they are present in sufficient quantity on the steel surface, generating corrosion pitting. Due to the importance of corrosion damage, new challenges arise today, such as the development of new technology methods of protection which are economic conserving so as the environment.

$C_{10}H_{16}N_2O_8$ ethylene diaminetetraacetic acid (EDTA) and Sodium Benzoate ($C_7H_5NaO_2$) are organic chemicals ascending carboxylates non-toxic and non-dangerous to the environmental and which had been found as a suitable agents in studies for prevent corrosion products from metals in multiple studies (Huda 2002),

B. Abdelbaki et al. (Eds.): SMSD 2017, *Proceedings of the Third International Symposium on Materials and Sustainable Development*, pp. 385–397, 2018.
https://doi.org/10.1007/978-3-319-89707-3_45

Keny et al. (2006), although researchers have used these products to inhibit corrosion of steel in hydrochloric acid mediums (HCl), while studies in alkaline medium remain very restricted (Okeniyi 2014).

This work is devoted to test EDTA and sodium benzoate as corrosion inhibitors of steel in 3% NaCl medium simulating the marine environment at different temperatures 298°K, 308°K and 318°K, using gravimetric tests based on weight loss measurements. These methods are the first approach to study the corrosion inhibition of a metal in an electrolytic solution in order to determine the inhibition efficiency of the tested products (Sail 2013).

2 Materials and Methods

2.1 Preparation of Specimens

The steel specimens are used in a circular section form 27 ± 2 mm of diameter and 2 ± 0.2 mm of thickness, subjected to polishing with a polisher at 500 rpm, with abrasive paper at grade from 120 to 1000, then, they will be rinsed in distilled water, degreased with acetone and dried using an electric dryer before being used for weight loss tests (Sail 2013).

Their chemical composition is given in Table 1.

Table 1. Chemical composition of steel specimens (Sail et al. 2011).

Chemical elements	C	Si	Mn	P	S	Cr	Mo
Compositions (w%)	0.259	0.271	1.25	0.0127	0.0402	0.05	0.01
Chemical elements	Ni	Al	Cu	V	Su	Fe	
Compositions (w%)	0.114	0.00363	0.29	0.0685	0.0177	97.65	

2.2 Test Solutions

The effect of the corrosion inhibition of steel by the two inhibitory products EDTA and sodium benzoate was tested in 3% NaCl medium simulating marine environment of the Mediterranean.

2.3 Tested Inhibitors

Ethylene diamine tetraacetic acid (EDTA) and Sodium benzoate are used as organic corrosion inhibitors. Their chemical structure is shown in Fig. 1.

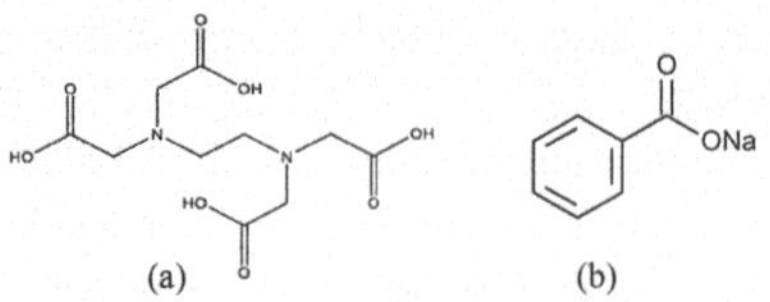

Fig. 1. Chemical structure (a) EDTA (Azooz 2016) and (b) Sodium benzoate (Wan Nik et al. 2010).

These two inhibitors were used in powder form and added to the study medium at different concentrations, Table 2 regroups together their molecular formulas and molar mass.

Table 2. Molecular formulas.

Compounds	Chemical formula	Physical forms	Molar mass (g/mol)
EDTA	$C_{10}H_{16}N_2O_8$	White, Odourless	372.24
Benzoate de Sodium	$C_7H_5NaO_2$	White, Odourless	144.11

2.4 Weight Loss Study

The weight loss method is of a simple use and does not require any important equipment, the corrosion rate is determined after 24 h of immersion at a temperature of 298°K provided by a thermostated bath, which is calculated by formula (1):

$$V_{corr} = \frac{\Delta \mathrm{m}}{\mathrm{S.t}} \tag{1}$$

Δm (m1 – m2) represents the difference between the initial mass m1 and the final mass m2 (mg), t is time of the test (hours), S is the surface area of the metal exposed to the study solution (cm^2).

This value of the corrosion rate is the average of three tests carried out under the same conditions for each concentration.

The value of the inhibitory efficiency is given by formula (2):

$$\mathrm{EI}(\%) = \frac{V_{cor} - V_{inh}}{V_{cor}} \times 100 \tag{2}$$

V_{cor} and V_{inh} respectively represent the corrosion rates in the absence and presence of inhibitor.

3 Results and Discussion

3.1 Effects of Inhibitor Concentration and Temperature

3.1.1 EDTA Inhibitor

The evolution of corrosion rates and inhibitory efficiency obtained from the weight loss measurements for steel specimens immersed in the test solution for 24 h as a function of inhibitor concentrations at different temperatures: 298°K, 308°K and 318°K are shown in Fig. 2.

We note that the corrosion rate decreases with increasing inhibitor concentration for different temperatures. The lowest corrosion rate is obtained for 0.15 Mol/l at a temperature of 298°K, for the other temperatures 0.2 Mol/l offer the lowest corrosion rates.

The maximum inhibitory efficiency was 86.65%, obtained at 0.15 mol/l at 298°K, 77.35% and 73.53% respectively at 308°K and 318°K of temperature at 0.2 mol/l.

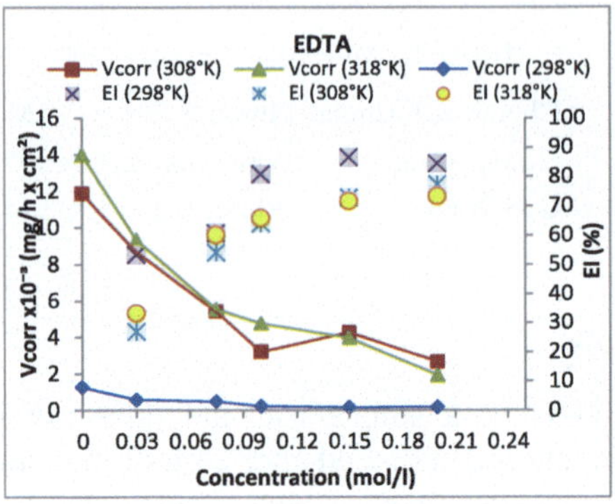

Fig. 2. Variation of corrosion rates and inhibitory efficiencies according to the concentration of EDTA at 298, 308 and 318°K.

3.2 Sodium Benzoate Inhibitor

Figure 3 shows the values of corrosion rate (Vcor), and inhibitory efficiency (EI %) in the absence and presence of the organic inhibitor Sodium benzoate at various temperatures studied previously.

From Fig. 3 we note that the corrosion rate is maximum without inhibitor and decreases in the presence of inhibitor while the inhibitory efficiency increases with the concentration of inhibitor in the solution at the different temperatures studied.

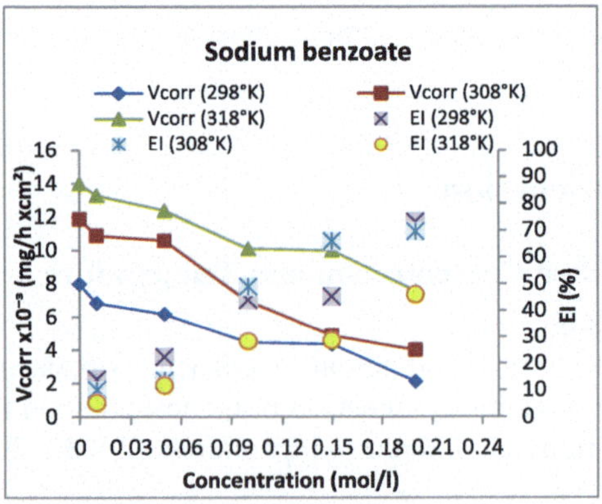

Fig. 3. Variation of corrosion rates and inhibitory efficiencies according to the concentration of Sodium benzoate at 298, 308 and 318°K

We obtain the maximum efficiency for the various temperatures at 0.2 mol/l but these efficiencies (EI %) depends on the temperature and are respectively of the order of 73.38, 69.68 and 46.22% for the temperatures of 298, 308 and 318°K.

3.3 Adsorption Isotherms

The influence of adsorption concentration by the inhibitors on the metal surface can be studied by laws of variation of the adsorbed amount as a function of the inhibitor concentration using the different isotherm models such as Langmuir, Frumkin, Temkin and Freundlich (Nwabanne and Okafor 2012).

3.3.1 Langmuir Isotherm

Langmuir isotherm describes the adsorption phenomenon of the adsorbate (A) on the surface of the adsorbent (S), it has proved that the adsorption is proportional to the concentration of the inhibitor, it's also as a function of the unoccupied adsorption sites (1 − θ).

The Langmuir equation is given as follows:

$$Log\left(\frac{C_{inh}}{\theta}\right) = \frac{1}{K} + Log(C_{inh}) \tag{3}$$

C_{inh} is the inhibitor concentration, θ is the recovery rate of the inhibitor at the corrosion sites, K is the equilibrium constant of the inhibitor, if the K values are high the inhibitory efficiency is better, which is due to a strong electrical interaction between the inhibitor and the steel (Pathak and Mishra 2015).

3.3.2 Temkin Isotherm

Temkin Eq. (4) is given as follows (Joseph and Vincent 2012):

$$\theta = \frac{-2,303 * \mathrm{LOG}(\mathrm{K})}{2\mathrm{a}} - \frac{2,303 * \mathrm{LOG}(\mathrm{C_{inh}})}{2\mathrm{a}} \tag{4}$$

θ is the inhibitor recovery rate, K: the equilibrium constant, a: coefficient of attraction.

3.3.3 Freundlich Isotherm

Freundlich presents an equation of a logarithmic form between the recovery rate and the inhibitor concentration: (Benali 2007a)

$$\mathrm{Ln}(\theta) = \mathrm{Ln}(\mathrm{K}) + \mathrm{nLn}(\mathrm{C_{inh}}) \tag{5}$$

C_{inh}: concentration of the inhibitor, θ: The recovery rate of the inhibitor, K: the adsorption equilibrium constant, n: adsorption intensity, reflecting the isothermal tendency: (Zarrouki 1990).

3.3.4 Frumkin Isotherm

Frumkin shows the change of recovery rate versus concentration as the equation given below (6) (Benali et al. 2007b):

$$Log\left[C\left(\frac{\theta}{1-\theta}\right)\right] = 2,303.Log(K) + 2\alpha\theta \tag{6}$$

Where K is the adsorption-desorption constant, and (α) is the lateral interaction term describing the interaction in adsorbed layer.

By these results related to variation in concentration C and the recovery rate θ, we have found that the adsorption isotherm follows the Frumkin model.

Figures 4 and 5 shows that the Frumkin model has correlation coefficients R^2 which is close to 1 for the two inhibitors tested.

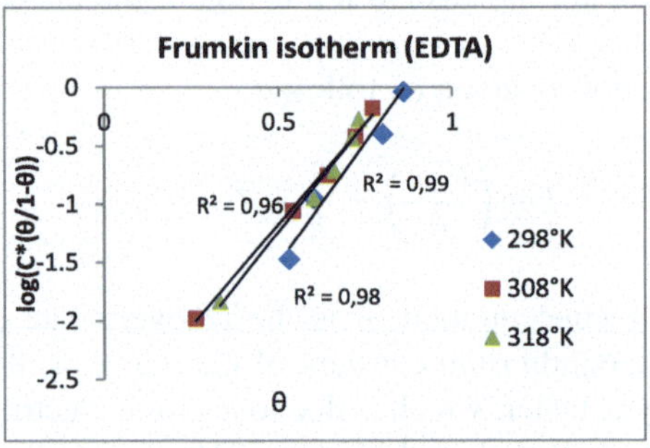

Fig. 4. Frumkin isotherm for EDTA

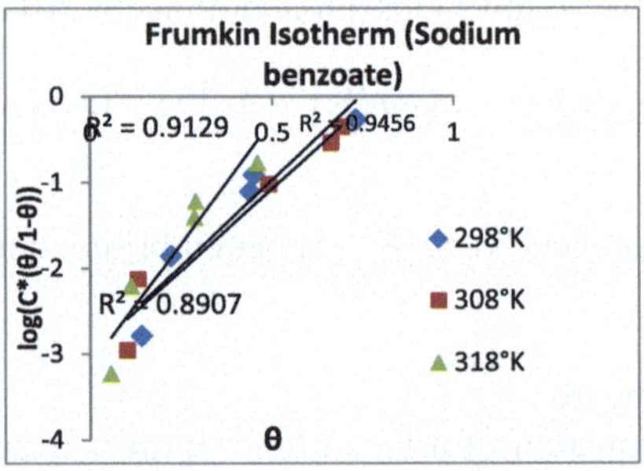

Fig. 5. Frumkin isotherm for Sodium benzoate

K value is related to the free enthalpy of adsorption ΔGads which is determined by Eq. (7) (Nwabanne and Okafor 2012):

$$\Delta G = -2,303\,RTLog(55,5K) \tag{7}$$

55, 5: Concentration of water in solution (mol/l).

The negative values of ΔGads indicate the spontaneity of the adsorption process and the stability of the adsorbed layer on the metal surface. Generally, the values of ΔGads, which are close to −20 $kJ.mol^{-1}$ or less negative, are linked to electrostatic interactions between charged molecules and the charged metal (physical adsorption), while those close to −40 $kJ.mol^{-1}$ or more negative ones involve a transfer of charge between the organic molecules and the metal surface (chemisorption) (Sail 2013).

Table 3 summarizes the Log K and ΔG values obtained with the Frumkin adsorption isotherm which gives us more suitable results for the two inhibitors "EDTA and Sodium benzoate" (Kamal and Sethuraman 2012; Li et al. 2012). It is found that the free energy values are negative (ΔG < 0) and less than −20 kJ/mol, hence the adsorption of EDTA and sodium benzoate on the steel surface is spontaneous suggests that it is a mechanism based on physical adsorption (physisorption), which is related to electrostatic interactions between charged molecules and metal (Faustin 2013; Okeniyi 2014).

It is exanimate that the values of the adsorption parameter (α) are positive, indicating that the behavior of the inhibitory molecules is attractive on the metal surface (Nwabanne and Okafor 2012).

Table 3. Adsorption parameters.

Inhibitor	T (°K)	R^2	Log k	α	ΔG (kJ/mol)
EDTA	298	0.96	−1.56	2.085	-6.07
	308	0.99	−1.26	1.73	−7.05
	318	0.98	−1.34	1.87	−7.05
Sodium benzoate	298	0.91	−1.29	2.00	−6.75
	308	0.94	−1.29	1.90	−6.98
	318	0.89	−1.36	2.85	−7.02

4 Thermodynamic Adsorption Parameters

The inhibition of corrosion is effected by adsorption of the inhibitors on the active sites of the metal in order to avoid corrosion. This adsorption which is only a set of chemical reactions is governed by thermodynamic parameters which are the enthalpy ΔH, the standard free enthalpy ΔG, the entropy variation ΔS and the activation energy Ea.

4.1 Enthalpy (ΔH)

In any chemical reaction, the heat is absorbed or released. This enthalpy ΔH is indirectly measured from weight loss tests carried out in different temperature ranges. It is calculated from Eq. (5) according to Van't Hoff, (Deng et al. 2009):

$$Ln\left(\frac{Vcor}{T}\right) = -\frac{\Delta H}{RT} + \frac{\Delta S}{RT} Ln\left(\frac{R}{NH}\right) \tag{5}$$

Where R is the constant of perfect gas, T is the temperature (°K), V_{cor} is the corrosion rate, H is the Planck constant, and N is the Avogadro number.

The enthalpy ΔH can be obtained by plotting the Ln (Vcorr/T) versus 1/T Figs. 6 and 7 for EDTA and sodium benzoate respectively, the slope of the line indicates (−ΔH/R), which makes it possible to calculate ΔH.

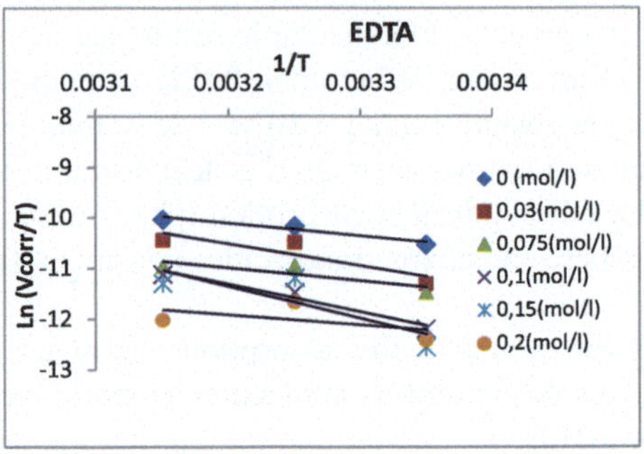

Fig. 6. Ln (Vcor/T) versus (1/T) in the absence and presence of EDTA.

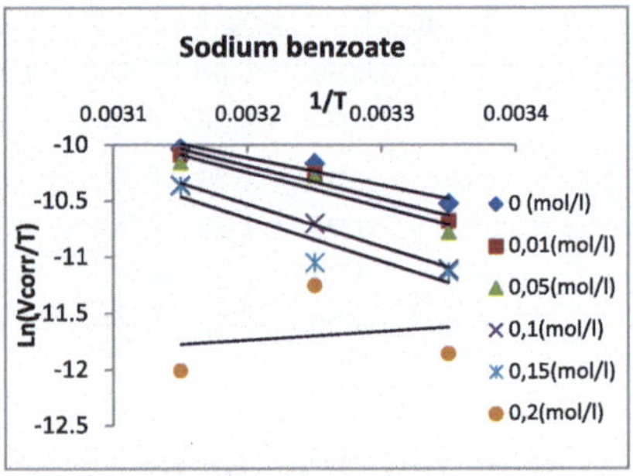

Fig. 7. Ln (Vcor/T) versus (1/T) in the absence and presence of the Sodium benzoate

4.2 Entropy (ΔS)

Entropy is a quantity that directly measures the disorder of molecules in a chemical reaction, while the disorder is larger, and the entropy is higher, it's calculated by the following Eq. (6) (Hamani 2015):

$$CR = \frac{RT}{NH} EXP\left(\frac{\Delta S}{RT}\right) EXP\left(-\frac{\Delta H}{RT}\right) \tag{6}$$

After transformation, we obtain Eq. (7):

$$Ln\left(\frac{CR}{T}\right) = -\frac{\Delta H}{RT} + \frac{\Delta S}{RT} Ln\left(\frac{R}{NH}\right) \quad (7)$$

Where CR is corrosion rate. The variation of Ln (CR/T) versus the inverse of temperature is a straight line with a slope of (−ΔH/R) and the intersection of this line with the ordinate axis Ln (CR/T) gives the value of ((ΔS/RT) Ln (R/NH)), and the value of ΔS in Figs. 6 and 7 can be concluded.

4.3 Activation Energy (Ea)

The activation energy value shows the nature of the reaction velocity, fast reactions have low activation energies those that are slow have high energies activation. The activation energy (Ea) is determined using the following Arrhenius law (Muthukrishnan et al. 2014):

$$Ln(CR) = \frac{-Ea}{R.T} + Ln(A) \quad (8)$$

Where CR is the corrosion rate, Ea is the apparent activation energy, R molar gas constant, T the absolute temperature, A is the frequency factor which is proportional to the frequency of the shocks between molecules; it not depends on the temperature, so it can be considered as constant for a given reaction (an ambient temperature range).

By plotting the graph Ln (CR) versus the inverse of the temperature in Figs. 8 and 9, the values of the activation energy can be obtained from the slope of the line multiplying to R (the gas constant perfect).

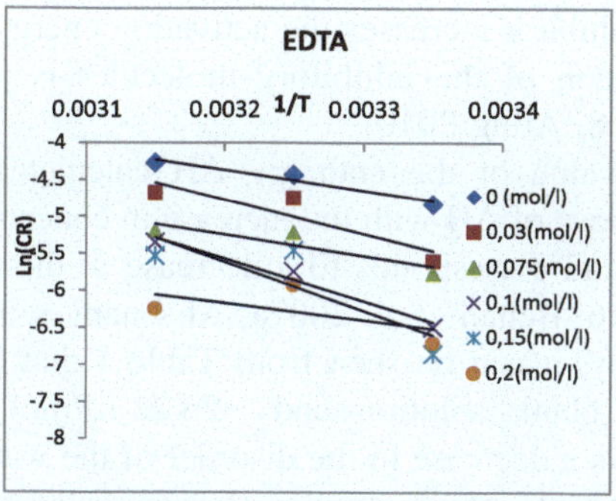

Fig. 8. Ln(CR) versus (1/T) in the absence and presence of the EDTA

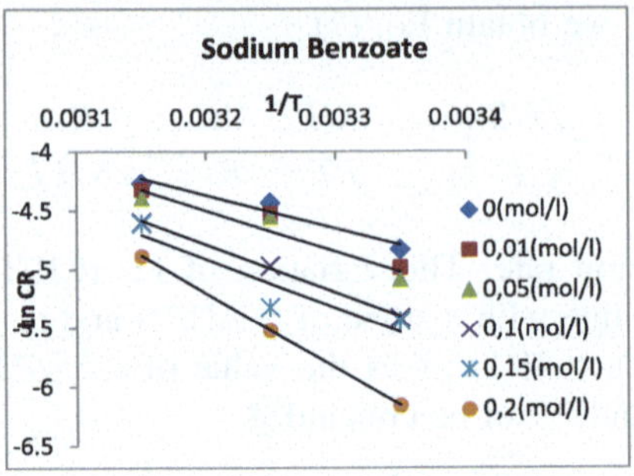

Fig. 9. Ln(CR) versus (1/T) in the absence and presence Sodium benzoate

Tables 4 and 5 combine the various thermodynamic parameters Ea, ΔH and ΔS for EDTA and Sodium benzoate, respectively.

Table 4. Thermodynamic parameters in the study medium "3% NaCl" in presence and absence of EDTA.

Concentration (Mol/l)	Ea (KJ/mol)	ΔH (KJ/mol)	ΔS (J/mol)
0	22.38	19.83	−20.64
0.03	36.70	34.15	21.77
0.075	22.92	20.36	−26.22
0.1	45.13	42.57	42.05
0.15	52.35	49.80	64.81
0.2	54.46	51.91	70.41

Analysis of results shown in Table 4 reveals that the apparent activation energy (Ea) for the dissolution reaction of the steel without inhibitor is 22.38 kJ/mol, the addition of the EDTA inhibitor increases the activation energy, this increase indicates that the type of adsorption of the inhibitory molecules is physical (Physisorption) (Badiea and Kikkeri 2008; Ating 2010).

It's noted that the value of the enthalpy ΔH calculated varies from 19.83 to 51.91 $kJ.mol^{-1}$; the increase of ΔH with the increase in concentration indicates that the process is endothermic and corresponds to a decrease in the dissolution of the metal (Singh 2016; Azooz 2016; Benali et al. 2007a; Al-senani et al. 2015).

While for the entropy, it can be seen from Table 4 that the two negative values −20.64 kJ/mol for the blank solution and −26.22 kJ/mol at a concentration of 0.075 mol/l, which shows a decrease in the disorder of the water molecules during the transformation of the reagents into a complex (Fe) activated molecule in the solution.

At the same time, it can be seen for the other concentrations that these values are positive and increase with the increase of the inhibitor concentration, indicating a disorder of the water molecules by replacing them with inhibitory molecules on the metal, this disorder results from the desorption of several molecules of water at the metal surface and the adsorption of a single inhibitor molecule (Hamani 2015; Ashassi-Sorkhabi and Shaabani 2005).

Table 5. Thermodynamic parameters in tested medium “3% NaCl” in presence and absence of Sodium benzoate.

Concentration (Mol/l)	Ea (KJ/mol)	ΔH (KJ/mol)	ΔS (J/mol)
0	22.39	19.83	−20.64
0.01	26.16	23.61	−9.18
0.05	27.49	24.94	−5.41
0.1	31.97	29.41	6.46
0.15	32.44	29.88	6.95
0.2	49.97	47.41	60.58

From Table 5, it is noted that:

The positive sign of the enthalpy reflects the endothermic nature ($\Delta H > 0$), and the increase of the enthalpy (ΔH) with the increase of the concentration corresponds to a decrease in the dissolution of the metal (Zohdy 2015; Pathak and Mishra 2015).

From the results of the entropy (ΔS) we note that there are two parts:

The first part [0 to 0.05 Mol/l]: the negative entropy values mean a decrease in the disorder which takes place starting from the reagents to the activated complex (Ben Hmamou et al. 2012).

The second part [0.1 to 0.2 Mol/l]: The values of ΔS is positive indicating an increase in disorder takes place from the reagent to the adsorbed species to the metal of the reaction complex (Zohdy 2015; Azooz 2016).

The increase in activation energy is proportional to the concentration of the inhibitor, indicating that the energy barrier for the corrosion process is also increased thus indicates a physical adsorption (Faustin 2013).

5 Conclusions

EDTA inhibitor presents greater inhibition efficiency than Sodium benzoate at different temperatures studied.

The results shows that the temperature increases the corrosion rate of the two inhibitors and it’s concluded than the inhibitory efficiency decreases, with increasing temperatures.

These two inhibitors are fixed in the same way on the metal surface by decreasing the dissolution of the metal; with, physical adsorption following Frumkin isotherm.

The inhibitory molecules for the two tested products react with the metal by an endothermic reaction.

References

Ashassi-Sorkhabi, H., Shaabani, B., Seifzadeh, D.: Effect of some pyrimidinic Shciff bases on the corrosion of mild steel in hydrochloric acid solution. Electrochim. Acta **50**, 16–17 (2005)

Ating, E.I., Umorena, S.A., Udousoroa, I.I., Ebensob, E.E., Udoh, A.P.: Leaves extract of Ananas Sativum as green corrosion inhibitor for aluminium in hydrochloric acid solutions **3** (2), 61–68 (2010)

Azooz, R.E.: EDTA as a corrosion inhibitor for Al in 0.5 M HCl: adsorption, thermodynamic and theoretical study. J. Electrochem. Sci. Eng. **6**, 235–251 (2016)

Badiea, A.M., Kikkeri, N.M.: The effect of sodium benzoate and sodium4-(phenylamino) benzenesulfonate on the corrosion behavior of low carbon steel **140**, 1–8 (2008)

Ben Hmamou, H., et al.: Prickly pear seed oil extract: a novel green inhibitor for mild steel corrosion in 1 M HCl solution. Int. J. Electrochem. Sci. **7**, 1303–1318 (2012)

Benali, O., Larabi, L., Traisnel, M., Gengembra, Y., Harek.: Theoretical and XPS studies of -mercapto-1-methylimidazole adsorption on carbon steel in 1 M HClO4. Appl. Surface Sci. **253**, 6130–6139 (2007a). (ISSN 0169-4332)

Benali, O.: Etude éléctrochimique et gravémitrique de l'inhibition de la corrosion du cuivre et deux acier au carbone par quelques composés organique soufrés en milieu acides. University Abou-bakrbelkaid Tlemcen, doctorat thesis (2007b)

Deng, S., Li, X., Li, T., Fu, H.: Adsorption and inhibition effect of 6-benzylaminopurine on cold rolled steel in 1.0 M HCl. Electrochim. Acta **54**, 4089–4098 (2009)

Faustin, M.: Etude de l'effet des alcaloïdes sur la corrosion de l'acier C38 en milieu acide chlorhydrique 1 M: Application à Aspidosperma album et Geissospermumlaeve. Université des Antilles et de la Guyane, Thèse, Cayenne (2013)

Al-senani, G.M., Al-Saeedi, S.I., Rasmiah, A.: Green corrosion inhibitors for carbon steel by green leafy vegetables extracts in 1 M HCl. Orient. J. Chem. **31**(4), 2077–2086 (2015)

Hamani, H.: Synthèse, caractérisation et étude du pouvoir inhibiteur de nouvelles molécules bases de Schiff. setif: Université Ferhat Abbas – Setif, These de doctorat (2015)

Huda, K.: A note on the efficacy of ethylenediaminetetra-acetic acid disodium salt as a stripping agent for corrosion products of copper. Stud. Conserv. **47**, 211–216 (2002)

Joseph, T.N., Vincent, N.O.: Adsorption and thermodynamics study of the inhibition of corrosion of mild steel in H_2SO_4 medium using vernonia amygdalina. J. Miner. Mater. Charact. Eng. **11**, 885–890 (2012)

Kamal, C., Sethuraman, M.G.: A novel green inhibitor for acid corrosion of mild steel. Arab. J. Chem. **5**, 155–161 (2012)

Keny, S., Manjanna, J., Venkateswaran, G., Kameswaran, R.: Dissolution behavior of synthetic Mg/Zn-ferrite corrosion products in EDTA and NTA based formulations. Corros. Sci. **48**, 2780–2798 (2006)

Muthukrishnan, P., Jeyaprabha, B., Prakash, P.: Mild steel corrosion inhibition by aqueous extract of Hyptis Suaveolens leaves. Int. J. Ind. Chem. **5**, 5 (2014)

Nwabanne, J., Okafor, V.: Adsorption and thermodynamics study of the inhibition of corrosion of mild steel in H_2SO_4 medium using Vernonia Amygdalina. Miner. Mater. Charact. Eng. **11**, 885–890 (2012)

Okeniyi, J.: $C_{10}H_{18}N_2Na_2O_{10}$ inhibition and adsorption mechanism on concrete steel-reinforcement corrosion in corrosive environments. J. Assoc. Arab Univ. Basic Appl. Sci. **20**, 39–48 (2014)

Pathak, R.K., Mishra, P.: Drugs as corrosion inhibitors. Int. J. Sci. Res. (IJSR) **5**(4), 671–677 (2015)

Sail, L., Ghomari, F., Bezzar, A., Benali, O.: Mass loss for assessment of the inhibitory efficiency of products to basis of phosphate. Publ. Can. J. Environ. Constr. Civ. Eng. **2**(05), 111–117 (2011)

Sail, L.: Etude de la performance d'inhibiteurs de corrosion a base de phosphate pour les constructions en béton armé. Doctorat thesis, University of Tlemcen. Algeria (2013)

Singh, A.K., Mohapatra, S., Pani, B.: Corrosion inhibition effect of Aloe Vera gel: gravimetric and electrochemical study. J. Ind. Eng. Chem. **33**, 288–297 (2016)

Nik, W.B.W., Sulaiman, O., EngGiap, S.G.: Evaluation of inhibitive action of sodium benzoate on corrosion behaviour of AA6063 in seawater. Int. J. Technol. **1**, 20–28 (2010)

Li, X., Shuduan, D., Fu, H.: Inhibition of the corrosion of steel in HCl, H_2SO_4 solutions by bamboo leaf extract. Corros. Sci. **62**, 163–175 (2012)

Zarrouki, M.: Etude de l'adsorption dans un système liquide-solide: solution d'ion dicyanoaurate-charbon actif. Ecole Nationale Supérieure des Mines de Saint-Etienne, Saint-Etienne (1990)

Zohdy, K.M.: Surface protection of carbon steel in acidic solution using ethylenediaminetetraacetic disodium salt. Int. J. Electrochem. Sci. **10**, 414–431 (2015)

Dynamic Behavior of the Composite Rotor Blade Using an Adaptive Damper

S. Chellil(✉), A. Nour, A. Chellil, S. Lecheb, H. Mechakra, and A. T. Settet

Laboratory of Engines Dynamics and Vibroacoustics, University of Boumerdes, Boumerdes, Algeria
sadek.chellil@yahoo.fr

Abstract. The reduction of helicopter vibration is becoming increasingly important to the helicopter structure, due to increasingly higher cruise speeds and improved comfort objectives. The adaptive damper technique has been shown to be capable of reducing fuselage vibration during steady flight conditions, and also maintaining reduced vibration levels during severe manoeuvres. The work is based on modeling by finite element method the blade with viscoelastic damper by applying an aerodynamic force. To realize this modeling we have used the numerical simulation by ABAQUS software to analyze the aerodynamic forces of the helicopter blade; it calculates the frequencies and Eigen valuer, stress, strain and the displacement, however the stress increased with crack propagation. Numerical calculations prove that the elastomeric damper of viscoelastic type produces better results compared to other ordinary systems.

Keywords: Stability · Composite · Damper · Crack · Blade

1 Introduction

The ground resonance instability is a result of a resonance of the blade regressive lag mode frequency with one of the fuselage mode frequencies. for solving this type problems, a viscoelastic damper is added at the links rotor-blade. As a means of active control of vibration and noise in rotary blade, two different approaches have been considered in the literature. The first is the blade pitch angle control introduced in the non-rotating frame through the swash plate, termed higher harmonic control (HHC). The second approach is an individual blade control (IBC) technique, which has been proposed by Kretz [1] and Ham [2]. In this technique, by mounting actuators either on the swash-plate or on the individual blade, each blade can be independently actuated on its flap, pitch, and lead-lag degrees of freedom. Incorporating smart structures technology into this technique can result in a highly effective means of reducing vibrations in helicopter blades.

To this effect, three main directions have been investigated in developing a "smart" rotor using adaptive structures. The first direction involves the incorporation of flaps actuated by piezo ceramic beam elements [3], resulting in a change of the airfoil affective camber. The second approach uses distributed actuators attached under the

B. Abdelbaki et al. (Eds.): SMSD 2017, *Proceedings of the Third International Symposium on Materials and Sustainable Development*, pp. 398–407, 2018.
https://doi.org/10.1007/978-3-319-89707-3_46

blade skin or piezoelectric fibers in a composite blade to control the blade twist [4], Rotor blade–lag damper system is of significant interest in the design and manufacture of helicopters previous modeling of this combined system has been purely numerical which has proved challenging [5].

Initially, these models were linear zed or simplified models of the damper's dominant characteristic behavior. More recently, Eyras [6] has developed a parametric model of the damper, based on an assumed piecewise linear force–velocity profile. Simulations carried out using this model when excited by recorded flight data have enabled designers to improve the modeling of the nonlinear behavior present in the blade–lag damper system, in above aeroelastic analysis of helicopter rotor blade [7], the influence of fuselage on vibration motion of rotor blade has not been taken into account. In reality, fuselage motion is known to have an influence on hub loads [8, 9]. Coupled rotor-fuselage models are therefore needed to investigate the rotorfuselage interaction and many researchers have proposed various methods [12]. Rutkowski [10] investigated the effect of rotor/fuselage coupling on vibration predictions using a simplified structural model of a helicopter in hover according to a two degree-of-freedom (dof) beam finite element. Hsu and Peter [11] developed a new impedance-matching method by the matching of rotor and fuselage impedances for coupled rotor/fuselage vibration analysis [13].

This study is to simulate helicopter blade with adaptive damper with crack by following parameters identifications; determine the distribution of the stress, strain field, appropriate criteria the frequencies and the modes shapes of a helicopter blade.

2 Modelling of the Rotor-Blade Interaction

Formulation of an orthotropic composite rotor blade structure is discussed in the following section.

2.1 Aerodynamic Force

Based on the two-dimensional aerodynamic model, Hudges and Ormiston [6] have determined the corresponding aerodynamic loads for hovering flight conditions to be:

$$L_v \cong \frac{\rho_\infty ac}{2}\{v_1^2 - \Omega^2 x^2 \frac{c_{d0}}{a} - \Omega x v_1(\theta+\phi) - [2\Omega x \frac{c_{d0}}{a} + (\theta+\phi)v_1]\dot{v} + [2v_1 - \Omega x(\theta+\phi)]\dot{w}\} \tag{1}$$

$$L_w \cong \frac{\rho_\infty ac}{2}\{-\Omega x v_1 + \Omega^2 x^2(\theta+\phi+\int_0^x v'w''dx) - \Omega^2 xv(\beta_{pc}+w') + \Omega^2\frac{xc}{2}(\beta_{pc}+w') + [2\Omega x(\theta+\phi) - v_1]\dot{v} - \Omega x\dot{w} + \frac{3}{4}c\Omega x\dot{\phi} - \frac{c}{4}\ddot{w}\} \tag{2}$$

$$M_\phi \cong \frac{\rho_\infty ac}{2}(\frac{c^2}{8}\Omega x\dot{\phi}) \tag{3}$$

Formulations
The kinetic energy of the rotating composite rotor blade is:

$$T = \frac{1}{2}\int_0^{\pi}\iint_A \rho\,\vec{V}.\vec{V}\,d\eta\,d\zeta\,dx \tag{4}$$

Where ρ is the mass density of the blade and Ω is the angular velocity of the rotor. And the variation of the kinetic energy for the blade is:

$$\delta T = \int_0^{\pi}\iint_A \rho\,\vec{V}.\delta\vec{V}\,d\eta\,d\zeta\,dx \tag{5}$$

The corresponding strain energy of the rotating blade is:

$$U = \int_0^{R}\iint_A (\sigma_{xx}\varepsilon_{xx} + \sigma_{x\eta}\varepsilon_{x\eta} + \sigma_{x\zeta}\varepsilon_{x\zeta})\,d\eta\,d\zeta\,dx \tag{6}$$

And the variation is:

$$\delta U = \int_0^{R}\iint_A (\sigma_{xx}\delta\varepsilon_{xx} + \sigma_{x\eta}\delta\varepsilon_{x\eta} + \sigma_{x\zeta}\delta\varepsilon_{x\zeta})\,d\eta\,d\zeta\,dx \tag{7}$$

Where

$$\sigma_{xx} = E\,\varepsilon_{xx}\,,\ \sigma_{x\eta} = G\,\varepsilon_{x\eta}\,,\ \sigma_{xx} = G\,\varepsilon_{x\zeta} \tag{8}$$

The virtual work for the aerodynamic loads may be expressed in the following form:

$$\delta W = \int_0^{R} \left(L_u\delta u + L_v\delta v + L_w\delta w + M_\phi\,\delta\phi\right) dx \tag{9}$$

Where L_u, L_v, L_w are distributed loads in the x, y, z directions respectively and M_ϕ distributed torque about the elastic axis.

Substituting Eqs. (5), (7) and (9) into Hamilton equation:

$$\int_{t_1}^{t_2} [\delta(U - T) - \delta W]dt = 0 \tag{10}$$

The differential equations governing the nonlinear lead-lag, flapping and torsion of composite rotor blade are:

Lead-lag equation:

$$-\frac{m\Omega^2}{2}\left[v'(R^2-x^2)\right]'-2m\Omega\left(v'\int_x^R \dot{v}dx\right)'+\left[EI_{z'}-(EI_{z'}-EI_{y'})\sin^2(\Re\theta)\right]v''''$$
$$+(EI_{z'}-EI_{y'})\frac{\sin(2\Re\theta)}{2}w''''+(EI_{z'}-EI_{y'})\left[-\sin(2\Re\theta)(\phi v'')''+\cos(2\Re\theta)(\phi w'')''\right]$$
$$-2m\Omega\beta_{pc}\dot{w}-2m\Omega\int_0^x(v'\dot{v}'+w'\dot{w}')dx+m(\ddot{v}-\Omega^2 v)=L_v \tag{11}$$

Flap equation:

$$-\frac{m\Omega^2}{2}\left[w'(R^2-x^2)\right]'-2m\Omega\left(w'\int_x^R \dot{v}dx\right)'+\left[EI_{y'}+(EI_{z'}-EI_{y'})\sin^2(\Re\theta)\right]w''''$$
$$+(EI_{z'}-EI_{y'})\frac{\sin(2\Re\theta)}{2}v''''+(EI_{z'}-EI_{y'})\left[\cos(2\Re\theta)(\phi v'')''+\sin(2\Re\theta)(\phi w'')''\right]$$
$$+2m\Omega\beta_{pc}\dot{v}+m\ddot{w}=L_w-m\Omega^2\beta_{pc}x \tag{12}$$

Torsion equation:

$$-\frac{m\Omega^2}{2}k_A^2\left[\phi'(R^2-x^2)\right]'-GJ\phi''+mk_m^2\ddot{\phi}+m\Omega^2(k_{m2}^2-k_{m1}^2)\phi\cos(2\theta)$$
$$+(EI_{z'}-EI_{y'})\left[(w''^2-v''^2)\frac{\sin 2\Re\theta}{2}+v''w''\cos 2\Re\theta\right]$$
$$=M_\phi-m\Omega^2(k_{m2}^2-k_{m1}^2)\frac{\sin 2\theta}{2} \tag{13}$$

2.2 Numerical Solutions

The equations of motion are solved by the Galerkin method. The solution in lead-lag flapping and twisting are expressed in dimensionless form in terms of series, based on generalized coordinates and mode shape functions are given by:

$$\bar{v}=\sum_{J=1}^{N}V_J(\psi)\,\Psi_J(\bar{x})$$
$$\bar{w}=\sum_{J=1}^{N}W_J(\psi)\,\Psi_J(\bar{x}) \tag{14}$$

$$\phi = \sum_{J=1}^{N} \phi_J(\psi)\,\Theta_J(\bar{x})$$

Where $\psi = \Omega t$, $\bar{x} = x/R$ and N is the number of modes; $\Psi_J(\bar{x})$ and $\Theta_J(\bar{x})$ are the mode shapes for the bending and torsion of the cantilever beam.

By employing the Eq. (14) into Eqs. (11) to (13). The equations of motion for the orthotropic composite blade can be rearranged in matrix form as:

$$[M]\{\ddot{X}\} + [C]\{\dot{X}\} + [K]\{X\} = \{0\} \quad (15)$$

The coefficients $[M]$, $[C]$ and $[K]$ represent the mass, damping and stiffness matrices.

2.3 Viscoelastic Damper Model

We now move onto another two-element model, the Kelvin model, which consists of a spring and dash pot in parallel. It is assumed there is no bending in this type of parallel arrangement, so that the strain experienced by the spring is the same as that experienced by the dash pot. However, here there is no reason why the stresses in both portions should be the same. We thus have the following three equations (Fig. 1):

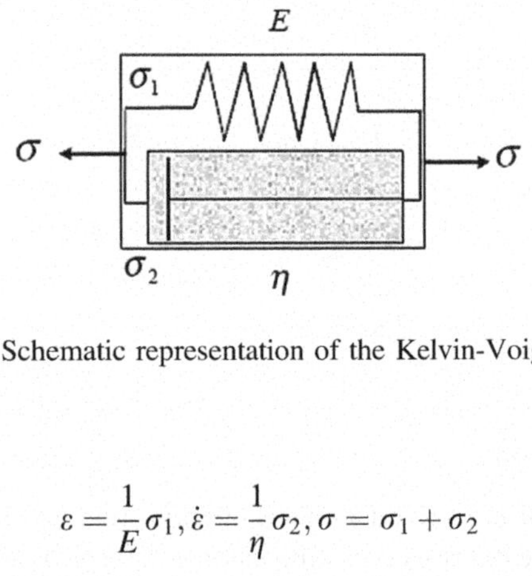

Fig. 1. Schematic representation of the Kelvin-Voigt model

$$\varepsilon = \frac{1}{E}\sigma_1, \dot{\varepsilon} = \frac{1}{\eta}\sigma_2, \sigma = \sigma_1 + \sigma_2 \quad (16)$$

Eliminating σ_1, σ_2 *then* leaves us with the constitutive law, which is in standard form.

$$\sigma = E\varepsilon + \eta\dot{\varepsilon} \quad (17)$$

3 Model of the Blade

The construction of the grid is made with ABAQUS software starting from the characteristics of the various elements of the blade (Fig. 2), is equipped with NACA23012 airfoil with active part radius 7.56 m and chord length 0.421 m. This rotor blade consists of D-spar made of unidirectional GFRP (Glas Fiber Reinforced Polymer), skin made of +450/–450 GFRP, foam core, MFC actuators and balance weight. The thickness of skin GFRP layer is 0.125 cm and thickness of MFC (Fig. 3) layer is 0.3 cm.

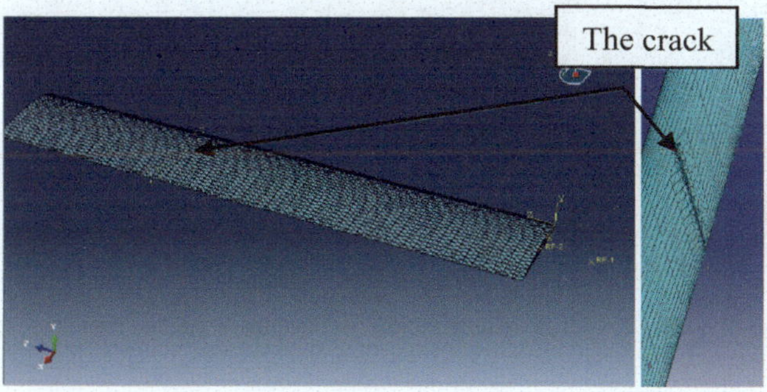

Fig. 2. Model of helicopter blade with crack

The following photo shows us formation the outer layer of the blade as well as the angles (3ply 45.90.45).

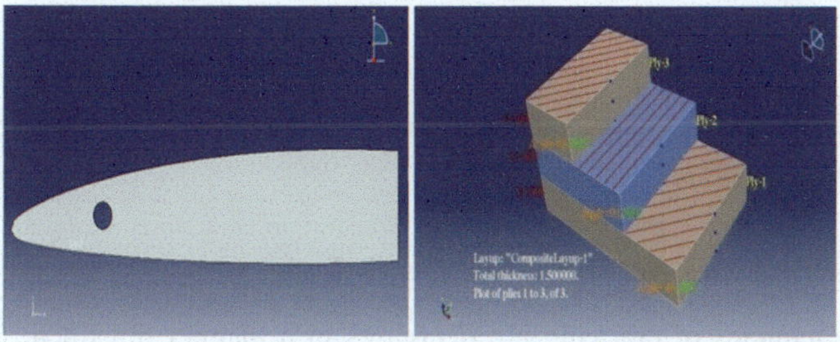

Fig. 3. Macro fiber composite (MFC).

In this case we take the crack into account, so we will mesh it with small size of the mesh element which equal to 35 element with tetrahedral element around the crack finally we get 124,668 element as illustrate in the following Fig. 2.

4 Parameter Identification of the Damage of the Blade

4.1 Modal Analysis

The objective of this study is to determine the eigen frequencies and eigen modes of the rotor blade (Fig. 4).

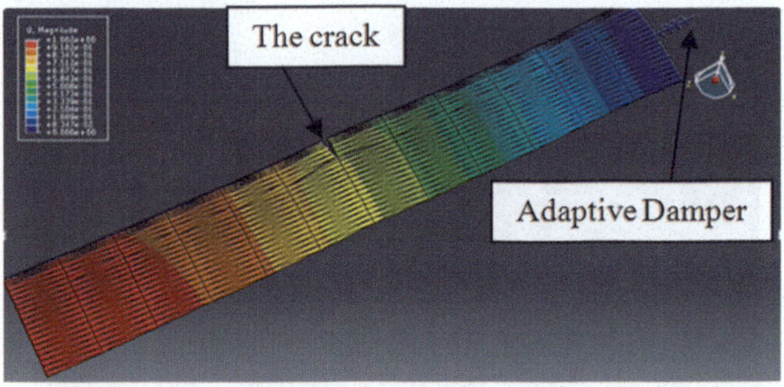

Fig. 4. Mode shape of the helicopter blade in flap

The eigen frequencies are represented on the Table 1 below:

Table 1. The eigen frequencies

Mode number	1	2	3	4	5	6
Frequency without crack [Hz]	0.135	0.322	0.543	1.736	2.674	4.129
Frequency with crack [Hz]	0.095	0.313	0.378	1.743	1.840	2.533

The frequencies in the blade Without crack is high than the case of the cracked blade, so monitored by reduction of frequency is a good indicator for monitoring the status of aircraft structures.

4.2 Modal Properties According the Crack Long

The Eigen frequencies of cracking are decreases in a manner not linear (Fig. 5).

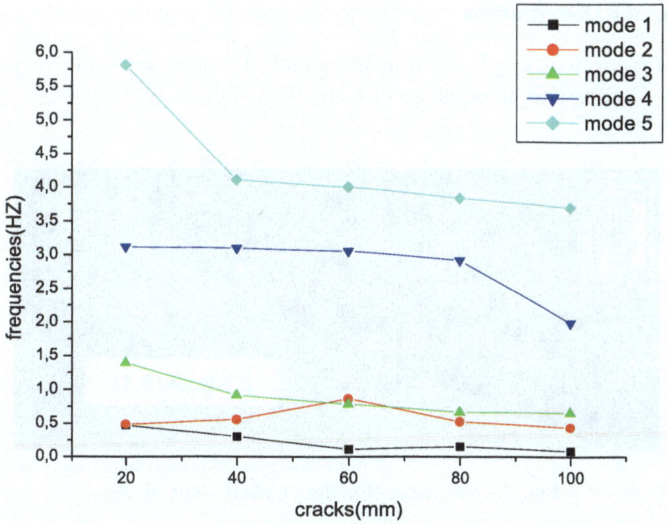

Fig. 5. Represents the difference of crack

4.3 Von Mises Stress

- The result for the cracked rotor blade (Fig. 6).

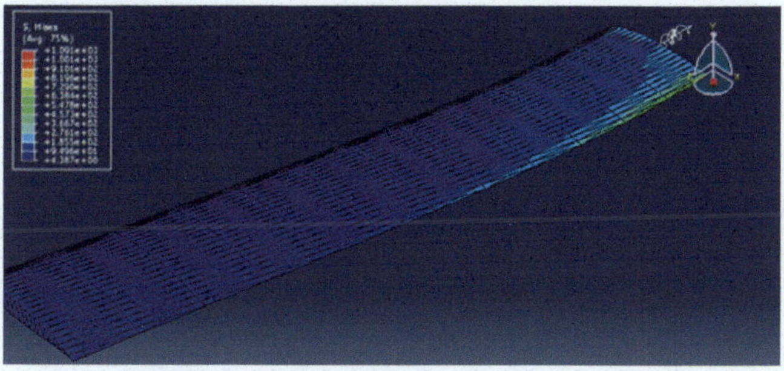

Fig. 6. Von Mises stress for cracked rotor blade

In this case we notice that the stresses in the crack. Which equal to 10.91 MPa which can make the crack propagate. So it's an increase this represents the concentration of stress.

4.4 The Strain Distribution

In this case of blade with crack, we note the strain of von Mises is the very higher at the end the party Crack. This is equal to (14.16) (Fig. 7).

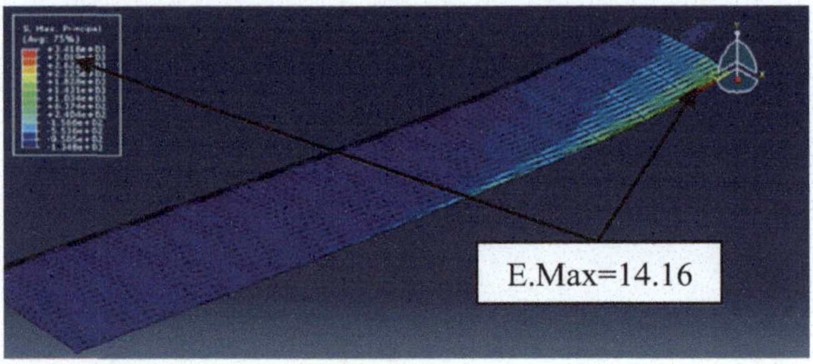

Fig. 7. The strain for the cracked rotor blade

5 Conclusion

The work is based on modeling by finite element method the helicopter blade. The ABAQUS software is used to analyze the crack monitoring of the helicopter blade; it calculates the frequencies and Eigen values, stress, strain and the displacement, as well as we noticed after the simulation of the blade without crack and with crack; the frequencies in the blade without crack are very high comparison with the blade with crack and stress and displacement of the blade with crack are very large comparison with the blade without crack. So monitored by reduction of frequency is a good indicator for monitoring the status of aircraft structures.

References

1. Kretz, M.: Research in multicyclic and active control of rotary wings. Vertica **11**(1/2), 95–105 (1987)
2. Ham, N.D.: Helicopter individual blade control at MIT 1977–1985. Vertica **11**(1/2), 109–122 (1987)
3. Spangler, Jr., R.L., Hall, S.R.: Piezoelectric actuators for helicopter rotor control. In: Proceedings: AIAA/ASME/ASCE/AHS/ASC 31. Structures, Structural Dynamics and Materials Conference, AIAA, Washington, DC, Part 3, pp. 1589–1599
4. Chopra, I.: Status of application of smart structures technology to rotorcraft systems. J. Am. Helicopter Soc. **45**(4), 228–252 (2000)
5. Eyres, R.D., Piiroinen, P.T., Champneys, A.R., Lieven, N.A.J.: Grazing bifurcations and chaos in the dynamics of a hydraulic damper with relief valves. SIAM J. Appl. Dyn. Syst. **4** (4), 1076–1106 (2005)
6. Eyres, R.E.: Vibration reduction in helicopters using lag dampers. Ph.D. Thesis, University of Bristol, UK (2005)

7. Viswamurthy, S.R., Ganguli, R.: An optimization approach to vibration reduction in helicopter rotors with multiple active trailing edge flaps. Aerosp. Sci. Technol. **8**(3), 185–194 (2004)
8. Papavassiliou, I., Friedmann, P.P., Venkatesan, C.: Coupled rotor-fuselage vibration reduction using open-loop blade pitch control. Math. Comput. Model. **18**(3–4), 131–156 (1993)
9. Yeo, H., Chopra, I.: Coupled rotor/fuselage vibration analysis for teetering rotor and test data comparison. J. Aircraft **38**(1), 111–121 (2001)
10. Rutkowski, M.J.: Assessment of rotor-fuselage coupling on vibration predictions using a simple finite element model. J. Am. Helicopter Soc. **28**(3), 20–25 (1983)
11. Hsu, T.K., Peters, D.A.: Coupled rotor/airframe vibration analysis by a combined harmonic-balance, impedance-matching method. J. Am. Helicopter Soc. **27**(1), 25–34 (1982)
12. Chellil, A., Nour, A., Lecheb, S., Kebir, H., Chevalier, Y.: Impact of the fuselage damping characteristics and the blade rigidity on the stability of helicopter. Aerosp. Sci. Technol. **29** (1), 235–252 (2013)
13. Chellil, A., Nour, A., Lecheb, S., Chibani, M., Kebir, H.: Parameter identification of crack growth in helicopter blade. Key Eng. Mater. **577**, 673–676 (2014)

Micro-Meso Mechanics Based Modeling of Damage Evolution in Cross Ply Laminates Composites

S. Benbelaid[1(✉)] and B. Bezzazi[2]

[1] Faculty of Engineering Mechanics, University of Sciences and Technology Houari Boumediene, Algers, Algeria
sam_ben0062@yahoo.fr

[2] Unité de Recherche Matériaux Procédés Environnement (UR-MPE), University M'Hamed Bougara, Boumerdès, Algeria

Abstract. This paper develops a numerical model to study the crack density evolution and delamination initiated from the transverse crack tips for microcracked ply in cross-ply laminates subjected to tensile loading. A representative unit cell using an energy based model is performed to study the growth of a new transverse crack and delamination process between two existing cracks in the 90° layer of a cross-ply laminate. However, an improved damage mesomodel for laminates allowing the calculation an energy standpoint, to the damaged laminate micromodel. A shear lag model was adopted to evaluate the stress distribution. The solution presented is obtained by using finite element analysis which implements micro-meso progressive failure analysis, while the effect of the stacking sequences has been done by varying the thickness of the 90° plies.

Keywords: Damage · Crack · Density · Delamination · Laminate

1 Introduction

The composite materials are one of the most widely used engineering materials in mechanical structures due to their high strength and relatively light weight. Such materials, despite the misunderstanding of certain aspects of their behavior. To exploit this design freedom, proper predictive tools are needed. However, the development of reliable numerical models is a challenging task, because the failure process is complex, involving different processes such as transverse ply cracking, fiber/matrix debonding, delamination, fiber breaking, and fiber pull out (see e.g. Green et al. 2007; Hallett and Wisnom 2006; Pierron et al. 2007). Often the damage evolution occurs over an important part of the loading, and a significant interaction is observed between the different mechanisms, leading to final fracture.

When a cross-ply laminate is loaded with the tensile axis aligned parallel to the 0° ply direction, it is generally believed that transverse matrix crack first appears in the weaker 90° plies. These small cracks quickly grow to extend laterally to span the entire 90° layer and then penetrate into neighbouring 0° plies. These cracks are more or less uniformly spaced. With applied loading, their number increases and then reaches a

B. Abdelbaki et al. (Eds.): SMSD 2017, *Proceedings of the Third International Symposium on Materials and Sustainable Development*, pp. 408–419, 2018.
https://doi.org/10.1007/978-3-319-89707-3_47

saturation level. This level has been termed as Characteristic Damage State (CDS). Transverse cracks induce local stress concentrations at crack tips and involve interlaminar delamination between 0° and 90° layers. Delamination develops from the crack tips and releases the local stress concentrations.

In order to predict the laminate failure which involves all these damage mechanisms evolution until final fracture, numerical tools are needed. However, this work intend to the development of such a predictive methodology to study the damage evolution in composite laminates. The existing approaches are based either on refined concepts on the microscale (which thus, provide precise information on specific mechanisms, but whose formalisms do not convene for structural analysis). The micromodel model is directly linked to the microscale and is able to take into account several mechanisms observed at microscopic level (the level of microcracking is quantified by a cracks density and delaminated process). However, The Damage Mesomodel based for Laminates developed since twenty years is one of those computational approaches. This model provides global measures of the degradation and no information on the microstate of degradation. The intact ply is modelled as a continuum, a relatively straightforward way to represent failure is with a continuum damage model, in which the stiffness of the material is reduced gradually after a certain failure criterion has been violated.

The present study develops both precise micromechanical models and mesocontinues models to evaluate the damage mechanisms and their evolution. The micromechanics is performed on a representative unit cell using an energy based model to study growth of a new transverse crack and delamination process between two existing cracks in the 90° layer of a cross-ply laminate. The objective is to build a continuum damage mechanics model which is quasi equivalent, from an energy standpoint, to the damaged laminate micromodel. Consequently, the potential energy stored in any part of a complete structure must be the same on the microscale and on the mesoscale.

A numerical application has been carried to cross-ply laminate of type $[0_n/90_{\ m}]_s$ under tensile loading. In addition, the solution algorithm using finite element analysis which implements progressive failure analysis is presented. Effect of stacking sequence is focused particularly on the 90° ply thickness which mostly affects the transverse cracks initiation, propagation, cracks density and delamination process.

2 Computational Damage Model

2.1 Mesocontinuum Damage Model

In the approach of "mesomodel damage of laminates" (see Ladeveze 1986; Ladeveze et al. 2000; Allix and Ladeveze 1992; Ladeveze and Lubineau 2000), one assumes that the behaviour of any composite laminate for any loading and any stacking sequence can be modelled using ply failure criterion. The ply takes into account diffuse damage, transverse microcracking, and fibre breakage. The level of each mesoconstituent is quantified by damage indicators, whose evolution is governed by damage forces through an assumed damage law. An important point of the model is that the state of damage is assumed to remain constant throughout the thickness of a single layer (of course, it can vary from one layer of the laminate to the next one). In other words, the main hypothesis of this computational model is that the damage evolution law is intrinsic to a ply and does not depend on the stacking sequence in which the ply is placed.

One of the most widely used approaches of continuum damage mechanics models based on energy potentials. The plies are taken to be homogeneous and orthotropic and the damage to be constant throughout the ply thickness. This mesoscale model is based on thermodynamic expressions, where the damage can be described by three internal variables d_{11}, d_{22} and d_{12} corresponding to the loss of stiffness in the fiber, transverse and shear directions, respectively:

$$\begin{aligned} E_{11} &= E_{11}^0(1 - d_{11}) \\ E_{22} &= E_{22}^0(1 - d_{22}) \\ E_{12} &= E_{12}^0(1 - d_{12}) \end{aligned} \tag{1}$$

where E_{11}^0, E_{22}^0 and E_{12}^0 are the initial stiffness of the material.

Assuming the existence of plane stresses and small perturbations, the associated elastic strain energy of a ply at a given stage of damage is:

$$E_D = \frac{1}{2}\left[\frac{\sigma_{11}^2}{E_{11}^0(1 - d_{11})} - \frac{2\nu_{12}^0\sigma_{11}\sigma_{22}}{E_{11}^0} + \frac{\langle\sigma_{22}\rangle_+^2}{E_{22}^0(1 - d_{22})} + \frac{\langle\sigma_{22}\rangle_-^2}{E_{22}^0} + \frac{\sigma_{12}^2}{2G_{12}^0(1 - d_{12})}\right] \tag{2}$$

The transverse tension energy and compression energy are split in order to describe the unilateral feature resulting from the opening and closing of the micro-cracks. In the case of a transverse compression load, micro-cracks close up and thus the transverse direction behaviour remains undamaged. From this potential, thermodynamic forces associated are defined in each unidirectional ply:

$$\begin{aligned} Yd_{11} &= \frac{\partial E_D}{\partial d_{11}} = \frac{\sigma_{11}^2}{2E_{11}^0(1 - d_{11})^2}, \\ Yd_{22} &= \frac{\partial E_D}{\partial d_{22}} = \frac{\langle\sigma_{22}\rangle_+^2}{2E_{22}^0(1 - d_{22})^2}, \\ Yd_{12} &= \frac{\partial E_D}{\partial d_{12}} = \frac{\sigma_{12}^2}{2E_{12}^0(1 - d_{12})^2} \end{aligned} \tag{3}$$

In order to account for the coupling between the transverse and the shear during the development of the damage, an equivalent associated force Y was defined in each time during the history of the loading in ply:

$$Y_{eq}(t) = Yd_{12} + bYd_{22} \tag{4}$$

Generally, the coupling factor b is determined from experiments.

The development of the internal variables depends on these equivalent thermodynamic forces. When tensile is applied, the d_{11} develops sharply because of the brittle behaviour of the fibres in the unidirectional plies:

$$d_{11} = 0 \quad if \quad Yd_{11} < Yd^{c}_{11} \quad else \quad d_{11} = 1 \tag{5}$$

Yd^{c}_{11} is the critical parameter, which defines the ultimate force corresponding to the tensile failure.

The evolutionary laws governing the internal variables d_{12} and d_{22} in each unidirectional ply, which depends on the associated equivalent forces (Eq. 4), still remain to be defined. In shear and transverse direction, a first approximation of the damage development, which corresponds to a linear law with respect to the square root of the equivalent associated force Y, shows a good fit with the experimental data:

$$d_{12} = \frac{\langle \sqrt{Y} - \sqrt{Y_0} \rangle}{\sqrt{Y_c} - \sqrt{Y_0}} \tag{6}$$

$$d_{22} = c d_{12} \quad \text{if} \quad Yd_{22} < Yd^{c}_{22} \quad \text{else} \quad d_{22} = 1 \tag{7}$$

where the constant parameters Y_0 and Y_c correspond to the threshold and the critical values of the development of d_{12} (which varies from 0 to 1) and the coefficients c stands for the ratio between the transverse damage and the shear damage, generally it equals 1. Yd^{c}_{22} is the critical parameter, which defines the ultimate force corresponding to transverse failure.

2.2 Micromechanics Damage Model

Up to now, there has been numerous theoretical and experimental works on the micromechanics of laminates (see Dvorak and Laws 1987; Hashin 1985a, b; Nairn and Hu 1994; Berthelot 2003); the micromechanics approach provides a relatively good understanding of damage mechanisms. A series of researchers attempted to quantify the stiffness reduction in a transversely cracked laminate. Aveston et al. (1971) proposed a so called "shear lag model" to modelling the stress transfer between fibers and matrix. This model is based on one dimensional analysis and was employed by Garrett and Bailey (1977) and others (Highsmith and Reifsnider 1982; Smith and Ogin 1999) to estimate the effective stiffness in a cracked laminate. Hashin (1985a, b) derived a variational solution to the stress field in a cross ply laminate under the assumption of uniformly distributed transverse cracks in 90° layer. His approach was a two dimensional analysis and adopted and modified by many authors to obtain the closed form solution of the normal and shear stresses as well as the axial stresses in the cracked 90° layers. More recently, the extended shear-lag analysis and variational approach studied the longitudinal displacements in the 0° and 90° layers, and considers the delamination initiation and growth from the tips of the transverse cracks (Nairn and Hu 1992; Takeda and Ogihara 1994; Berthelot and Le Corre, 2000).

2.2.1 Stress Analysis

Considering a baseline cross-ply laminates $[0_n/90_m]_s$ with the midplane 90° plies, the loading state considered here is only the simple axial tensile load (σ_a), as shown in Fig. 1.

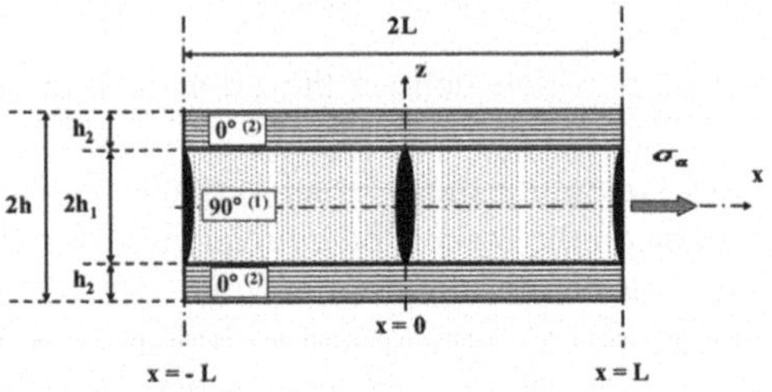

Fig. 1. Cross ply laminate under uniaxial tensile load.

A shear lag model is employed to estimate the average stress of the 0° and 90° plies. The average stress of the 0° and 90° plies is supposed constant along their thickness:

$$\sigma_{22}(x) = \sigma_a \frac{E_{22}^0}{E_{xx}^0} \left[(1 - \frac{ch\left(\beta \frac{x}{h_1}\right)}{ch\left(\beta \frac{L}{h_1}\right)} \right] \tag{8}$$

$$\sigma_{11}(x) = \frac{1}{h_2} [h\sigma_a - h_1 \sigma_{22}(x)] \tag{9}$$

With:

$$E_{xx}^0 = \frac{t_{90} E_{22}^0 + t_0 E_{11}^0}{t_{90} + t_0} \tag{10}$$

where β is the shear-lag parameter (Laws and Dvorak 1988):

$$\beta = \frac{(\alpha + 1) G_{23} (h_2 E_{11}^0 + h_1 E_{22}^0)}{E_{11}^0 E_{22}^0 h_2 h_1^2} \tag{11}$$

h_1, h_2 is the thickness of the 90° and 0° plies. α is the assumed shape index of the crack open displacement. For the cross-ply laminates $[0_n/90_m]_s$, it varies from 0 to 2 in different shear-lag models and reflects the assumed displacement function of the cracked 90° plies. It can be used as a global parameter to characterize the matrix crack effect.

The equilibrium of longitudinal applied force in an element of the 90° layer leads to the relation (Berthelot and Le Corre 2000):

$$\frac{d\sigma_{22}(x)}{dx} = -\frac{1}{h_1} \tau(x) \tag{12}$$

where $\tau(x)$ is the shear stress at the interface between 0° and 90° layers.

2.2.2 Computational Micromodel

For cross-ply laminates, 90° plies are susceptible to the transverse cracks and they result delamination at the interfaces of the transversely cracked 90° plies and the adjacent 0° plies. The micromodel is characterized by periodic micro patterns described by cracking density (*d*), at least locally, which is consistent with most practical situations. The level of microcracking is quantified by a cracking rate *a* defined by $a = h_1/L$. The local delamination is described at each transverse crack tip by a local delamination ratio r defined by $r = l_d/h_1$ ($r \in [0; 0.4]$) (Fig. 2).

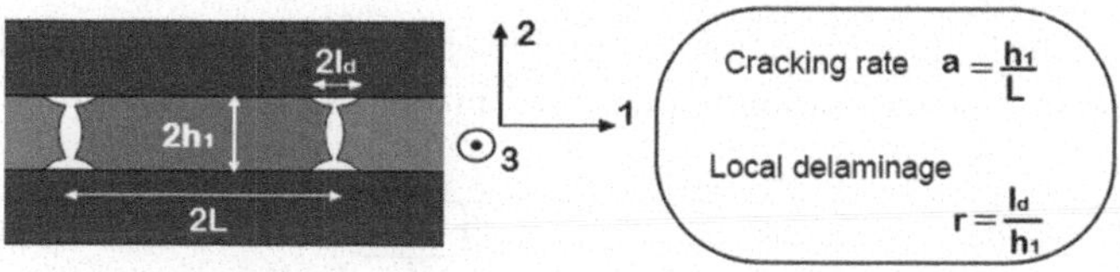

Fig. 2. Parameters of the micromodel

The evolution of these microvariables is governed by energy release rates. Considering the elementary cell of length 2L of cracked laminate limited by two pre-existing cracks, x = L and x = −L (Fig. 1). Expression of the strain energy release rate based on Nairn model (Nairn 1989; Nairn and Mendels 2001) which associated to nucleation of the new transverse crack between the existing cracks is a function of applied stress σ_a and of cracking density:

$$G_{\max}(\sigma_a, d) = G_{\max}(\sigma_a) \times f_d(d) \tag{13}$$

With:

$$G_{\max}(\sigma_a) = \frac{1}{\beta}\frac{E_{22}^0}{E_{xx}^0}\frac{1}{E_{11}^0}(1 + \frac{h_1}{h_2})(\sigma_a)^2 \tag{14}$$

$$f_d(d) = 2\tanh\left(\frac{\beta}{4d}\right) - \tanh\left(\frac{\beta}{2d}\right) \tag{15}$$

We suppose that the strain energy release rate reached the critical value, us a following forme:

$$G_{\max}(\sigma_a, d) = G_c \tag{16}$$

Replacing the Eqs. (13)–(15) in (16), we can find the final form for estimation the cracks density:

$$2\tanh\left(\frac{\beta}{4d}\right) - \tanh\left(\frac{\beta}{2d}\right) = \frac{G_c}{\left[\frac{1}{\beta}\frac{E_{22}^0}{E_{xx}^0}\frac{1}{E_{11}^0}(1 + \frac{h_1}{h_2})\right](\sigma_a)^2} \tag{17}$$

3 Application and Result

A numerical application has been carried out on the cross-ply laminates constituted from continuous carbon fibres and epoxy resin of type IM7/977-2 with long carbon fibres of intermediate reference module IM7 and with epoxy matrix of reference 977-2. The different stacking sequence $[0_n/90_m]$ is the quasi-isotropic laminate (dimensions 140 mm × 12.7 mm × 1.1 mm). The stacking sequence effect can be studied by varying the structural stiffness ratio of the two plies (0° and 90°). This can be done by varying the thickness of the 90° plies. In the present application, three configurations are analyzed: $[0_3/90_3]_S$, $[0_2/90_4]_s$ and $[0/90_5]_S$.

A computational flow chart for the progressive failure analysis of the cross ply laminates used in this application is presented in Fig. 3.

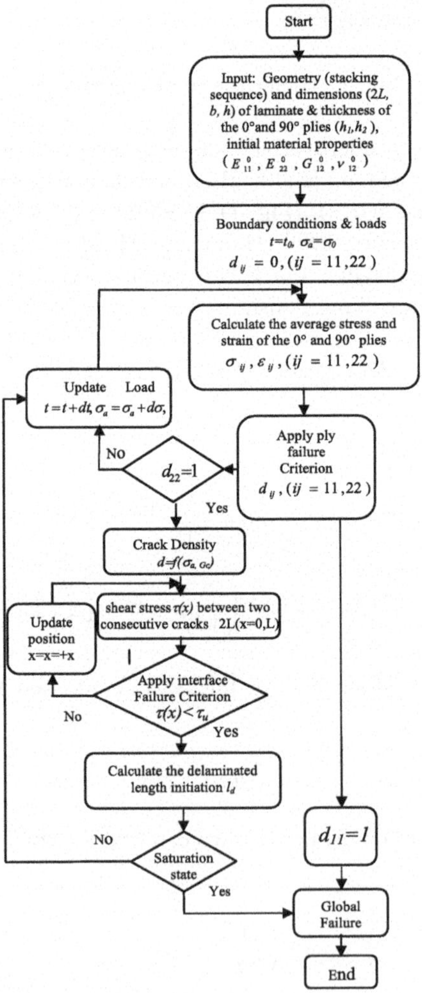

Fig. 3. Flow chart of progressive failure analysis used for modelling the damage in cross-ply laminates.

All these numerical steps can summarised by six points: (1) Finite element modelling is used by including the geometry (stacking sequence) and specimen dimensions (l, b, h), thickness of the 0° and 90° plies (h_1, h_2) and initial material properties (E_1^0, E_2^0, E_{12}^0, ν_{12}^0) (2) For each time (t) and load (F) step, the finite element analysis is performed and the on-axis stresses/strains at each element are obtained. (3) The stresses/strains at each element are compared with the material allowable values and are used to determine whether some elements have failed according to the failure criteria or not. (4) In mesoscopic scale if $d_{22} = 1$, the first transverse crack is created, which provide to calculate the density cracks. (5) For a given elementary cell between two consecutive cracks, the delaminated length was estimated if the shear stress at the interface with the position x reached the ultimate shear stress τ_u. The delamination process was continued until saturation state of microcracks, else the applied load is increased and the analysis continues. (6) finally, In macroscopic scale, the global failure is happen by fibre breakage in the 0° plies ($d_{11} = 1$).

3.1 Effect of Stacking Sequence on the First Ply Failure

Materials are initially degraded by micro-cracks which are associated mainly with matrix micro-cracking (diffuse damage). This zone is characterized by non linear evolution (Fig. 4). Further loading will cause micro-cracks to coalescence and form a macro-crack leading to brittle fracture which is assumed to propagate into the entire thickness of the off-axis and will extend until it is stopped by fibres crossing the crack paths. Initiation and propagation of the first transverse crack is significant if a 90° layer thickness becomes more than the thickness of the 0° plies.

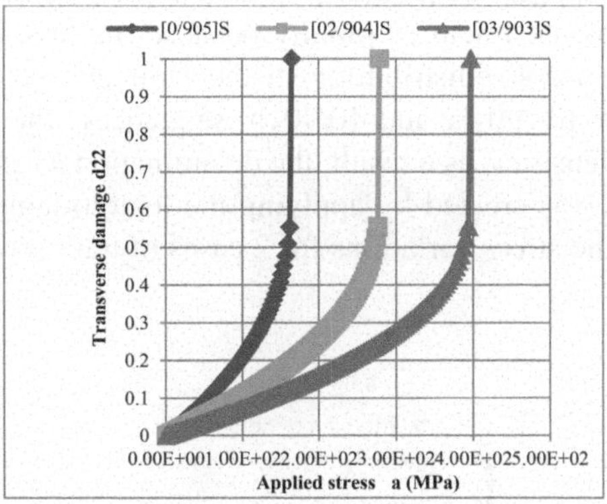

Fig. 4. Effect of stacking sequence on the first ply failure

3.2 Effect of Stacking Sequence on Crack Density Evolution

Figure 5 shows the crack density as a function of applied load. Continued loading will generate new transverse macro-cracks until a saturated level of matrix macro-cracks is reached an asymptotic state which called the characteristic damage state (CDS). Hence more stress needs to be applied to form a new crack. The crack density (cracks/length) increases if a 90° layer thickness becomes less than the thickness of the 0° plies (Fig. 5).

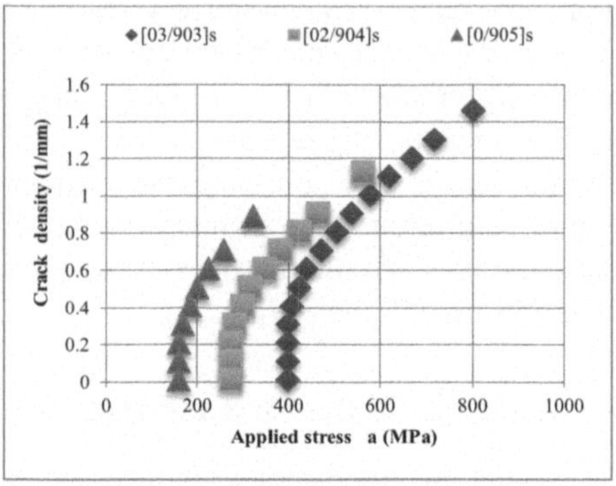

Fig. 5. Effect of stacking sequence on the crack density

3.3 Effect of Stacking Sequence on the Delamination Process

Figure 6 shows the delamination initiated length from the transverse crack tips versus cracks density. It is observed that When the thickness ratio $h_1 = h_2$ is equal to 1 (case of the $[0_3/90_3]_s$ laminate), the delamination process initiates only for high crack densities. when the tensile load applied to laminate is increased, delamination process do not progress more and laminate fractures rapidly because the higher crack density in the 90° plies causes local stress redistribution in the laminate which leads to additional loads on 0° plies. For $[0_2/90_4]$s and $[0/90_5]$s sequences, the delamination process initiates for low crack densities, as a result, the delamination length is considered before a new transverse crack was created by applying the load to laminate. So, Fig. 7 and 8 reports un example of the stress variations for 2 cases of transverse cracking density for $[0/90_5]$s laminate.

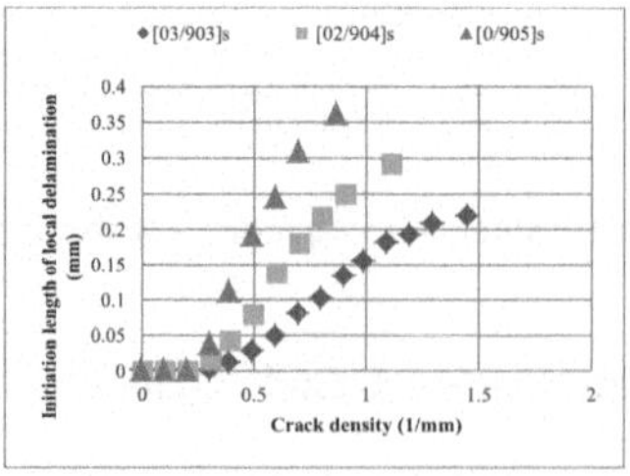

Fig. 6. Effect of stacking sequence on the delamination process

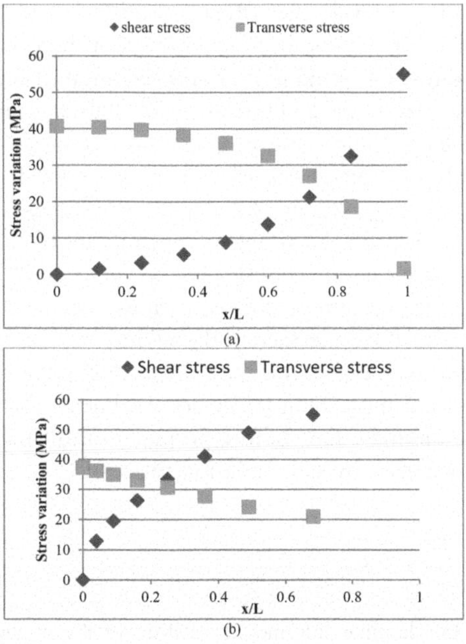

Fig. 7. Variation of shear and transverse stress in $[0/90_5]$s laminate: (a) d = 0.3, (b) d = 0.88

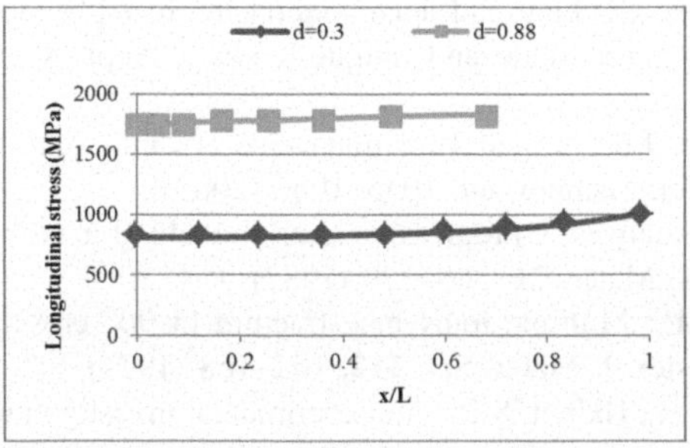

Fig. 8. Variation of longitudinal stress in $[0/90_5]$s laminate for two cases: d = 0.3 and d = 0.88

4 Conclusions

To study the damage mechanisms in composite laminate, it is absolutely essential to use a multiscale computational strategy. The mesomodel provides global measures of the degradation and no information on the microstate of degradation. However, the micromodel is able to take into account several mechanisms observed at microscopic level, such that the nucleation and growth of a transverse crack and delamination process from the transverse crack tips.

The modelling results showed the effects of different stacking sequences with thickness variation of the 90° layers. It is observed that there is a significant effect of 90° plies thickness on the stress initiation and propagation of transverses cracking. As the thickness of 90° plies decreases and becomes less than the thickness of the 0° plies, the stress to micro-cracks initiation and propagation decreases and the level of characteristic damage state (CDS) increases. However, when the 90° layers thickness are less than 0° layers thickness, the cracks density increases and affects the entire microcracking process and leads to instantaneous fracture events as the cracks in the 90° plies act to raise the stress in adjacent 0° plies to cause early fibre fracture. It is also found that the number of broken fibres did not increase because of the development of local delamination at the s0°/90° interface in the vicinity of the crack. At loads close to failure in the quasi-static tensile test, the same local delamination appeared. So, when the 90° layers thickness are more than 0° layers thickness, the micro-cracks are suppressed entirely and the laminate fails before propagation of micro-cracking, while the final failure is mainly occurred by delamination process.

References

Allix, O., Ladeveze, P.: Interlaminar interface modelling of the elementary ply for laminated composites. Compos. Struct. **22**, 235–242 (1992)

Aveston, J., Cooper, G.A., Kelly, A.: Single and multiple fracture. In: The Properties of Fiber Composites, Conference Proceedings of the National Physical Laboratory, Surrey, England, pp. 15–26 (1971)

Berthelot, J.M.: Transverse cracking and delamination in cross-ply glass fiber and carbon-fiber reinforced plastic laminates: static and fatigue loading. Appl. Mech. Rev. **56**(1), 111–147 (2003)

Berthelot, J.M., Le Corre, J.F.: A model for transverse cracking and delamination in cross-ply laminates. Compos. Sci. Technol. **60**, 1055–1066 (2000)

Dvorak, G.J., Laws, N.: Analysis of progressive matrix cracking in composite laminates, II. First ply failure. J. Compos. Mater. **21**, 309–329 (1987)

Garrett, K.W., Bailey, J.E.: Multiple transverse fracture in 90° cross-ply laminates of a glass fiber-reinforced polyester. J. Mater. Sci. **112**, 157–168 (1977)

Green, B.G., Wisnom, M.R., Hallett, S.R.: An experimental investigation into the tensile strength scaling of notched composites. Compos. A **38**(3), 867–878 (2007)

Hallett, S.R., Wisnom, M.R.: Experimental investigation of progressive damage and the effect of layup in notched tensile tests. J. Compos. Mater. **40**(2), 119–141 (2006)

Hashin, Z.: Analysis of cracked laminates: a variational approach. Mech. Mater. **4**, 121–126 (1985a)

Hashin, Z.: Analysis of cracked laminates: a variational approach. Mech. Mater. **4**, 121–136 (1985b)

Highsmith, A.L., Reifsnider, K.L.: Stiffness reduction mechanisms in composite laminates. ASTM Spec. Tech. Publ. **775**, 103–117 (1982)

Ladeveze, P.: On damage mechanics of composites. In: Bathias, C., Menkes, D. (eds.) Comptes-Rendus Des JNC5, pp. 667–683. Pluralis, Paris (1986)

Ladeveze, P., Allix, O., Deu, J.F., Leveque, D.: A mesomodel for localization and damage computation in laminates. Comput. Method. Appl. Mech. Eng. **183**, 105–122 (2000)

Ladeveze, P., Lubineau, G.: On a damage mesomodel for laminates: micro–meso relationships, possibilities and limits. Compos. Sci. Technol. **61**, 2149–2158 (2000)

Laws, N., Dvorak, G.J.: Progressive transverse cracking in composite laminates. J. Compos. Mater. **22**, 900–916 (1988)

Nairn, J.A., Hu, S.: The initiation and growth of delaminations induced by matrix microcracks in laminated composites. Int. J. Fract. **57**, 1–24 (1992)

Nairn, J.A.: The strain energy release rate of composite microcracking: a variational approach. J. Compos. Mater. **23**, 1106–1129 (1989)

Nairn, J., Hu, S.H.: Matrix Microcracking Damage Mechanics of Composites Materials. Elsevier, pp. 187–243 (1994). Taljera, R. (ed.)

Nairn, J.A., Mendels, D.A.: On the use of planar shear-lag methods for stress-transfer analysis of multilayered composites. Mech. Mater. **33**, 335–362 (2001)

Pierron, F., Green, B., Wisnom, M.R., Hallett, S.R.: Full field assessment of the damage process of laminated composite open-hole tensile specimens. Part II: experimental results. Compos. Part A **38**(11), 2321–2332 (2007)

Smith, P.A., Ogin, S.L.: On transverse matrix cracking in cross-ply laminates loaded in simple bending. Compos. A **30**, 1003–1008 (1999)

Takeda, N., Ogihara, S.: Initiation and growth of delamination from the tips of transverse cracks in CFRP cross-ply laminates. Compos. Sci. Techol. **52**, 309–318 (1994)

Thermoelectric Properties of Single Filled Skutterudites $Yb_xCo_4Sb_{12}$

Z. Benhalima[1(✉)], A. Bekhti-Siad[2], and M. Sahnoun[1]

[1] Laboratoire de Physique Quantique de la Matière et Modélisation Mathematique (LPQ3M), Faculté des Sciences Exactes, University of Mascara, Mascara, Algeria
benhalimazoulikha@gmail.com

[2] Laboratoire de Matériaux Applications et Environnement (LMAE), Faculté des Sciences Exactes, University of Mascara, Mascara, Algeria

Abstract. Filled Skutterudites are one of the most promising thermoelectric materials for power generation applications. The choice of the concentration of filler atoms is key aspect for achieving low thermal conductivity and high thermoelectric figure of merit value. We report an investigation of the thermoelectric properties of Yb filled skutterudites ($Yb_xCo_4Sb_{12}$). The insertion of Yb filler within the voids of the Skutterudite structure provides a wide range of resonant phonon scattering and therefore a strong reduction of the thermal conductivity is obtained.

Keywords: Thermoelectric materials (TE) · Filled Skutterudite
Thermal conductivity

1 Introduction

Under the challenges of the global energy crisis and climate change, thermoelectric (TE) materials which can be used to directly convert heat in solid state, have gained renewed interest in these two past decades. The selection of high-performance TE materials requires the research of materials which are able to operate in widest possible temperature range and with high value of dimensionless figure of merit defined as $ZT = \frac{S^2\sigma}{\kappa_t}T$; where S is the Seebeck coefficient, σ is the electrical conductivity, T is the absolute temperature and κ_t is the total thermal conductivity which is given by $\kappa_t = \kappa_e + \kappa_P$, with κ_e is the contribution due to the charge carriers and κ_P is the phonon contribution. Therefore the decrease in thermal conductivity is the most important issue in TE material development. For semiconductors (most TE materials), the phononic contribution generally exceeds the electronic one. So, reducing κ_P becomes more important. Various approaches have been taken to decrease κ_P such as the new concept of rattling ions in nano-cages structure.

Antimony based skutterudites have been attracting substantial interest because of their promising thermoelectric performance in the intermediate temperature range due to their excellent electrical properties including high mobility, high atomic masses, low electrical resistivity and large Seebeck coefficients. More specifically, $CoSb_3$ was found

B. Abdelbaki et al. (Eds.): SMSD 2017, *Proceedings of the Third International Symposium on Materials and Sustainable Development*, pp. 420–424, 2018.
https://doi.org/10.1007/978-3-319-89707-3_48

to have a high mobility 3445 $cm^2V^{-1}s^{-1}$, a charge concentration of 4 * 10^{17} cm^{-3} and the thermal conductivity 8.9 $Wm^{-3}K^{-1}$ at room temperature [1, 2]. Filling atoms into the voids of skutterudites results in a manifestation of the "phonon-glass-electron-crystal" (PGEC) [3–17] shows effective strategy for decreasing the thermal conductivity of $CoSb_3$.

In this work, we present theoretical study of the thermoelectric properties of Yb filled Skutterudites (Yb_xCoSb_{12}) with (x = 0.25, 0.5 and 1). A brief description of the computational method will be presented in next section. In Sect. 3, we present the main calculated results and discuss them as well as some experimental and theoretical validations. Finally, a short summary is given in Sect. 4.

2 Computational Details

All the calculations of structural and electronic properties have been performed in the framework of density functional theory (DFT). We have employed full potential Linearized augmented plane wave (FP-LAPW) method which is implemented in the WIEN2K code [19]. The exchange-correlation potential has been calculated using the Perdew-Burke-Ernzerhof generalized gradient approximation (PBE-GGA) and the convergence criterion for energy was set to 10^{-4} Ry and that for charge $10^{-3}e^{-}$. The self-consistent potentials were calculated on a 9 × 9 × 9 k-mesh in the first Brillion zone. All lower states were treated as core states. Non-spherical contributions to the charge density and potential within the muffin tin (MT) spheres were considered up to l_{max} = 10. The plane wave cut-off parameter was R_{MT} * K_{max} = 7 (Where RMT denotes the smallest muffin tin radius of atoms K_{max} is the maximum value of the wave vector). The different MT sphere radii (RMT) used were 2.09 a.u., 1.99 a.u. and 2.02 a.u. for Co, Sb and Yb atoms respectively. Transport properties are calculated by applying the Boltzmann transport equation within the constant relaxation time (τ) approximation as implemented in the BoltzTrap code [20].

3 Results and Discussion

The void radius of $CoSb_3$ is 1.892 A° and the atomic radius of Yb is 1.74 A°. The Yb filler position is determined to be located at the position of (0, 0, 0), whereas the Co position is (0.25, 0.25, 0.25) and antimony position is (0, 0.334, 0.157). These values are in agreement with the ones reported elsewhere [16, 21, 22].

3.1 Seebeck Coefficient

We have calculated the Seebeck coefficient or thermo power (S) of the Yb filled skutterudite (Yb_xCoSb_{12}) as a function of temperature at a certain value of chemical potential ($\mu = E_F$) as shown in Fig. 1. It is clear that the calculated S of the Yb filled skutterudite (Yb_xCoSb_{12}) represents negative values indicating that the n-type conduction is the dominant; because its electron mobility is much larger than the hole mobility, i.e. its electron effective mass is much smaller than the hole effective mass.

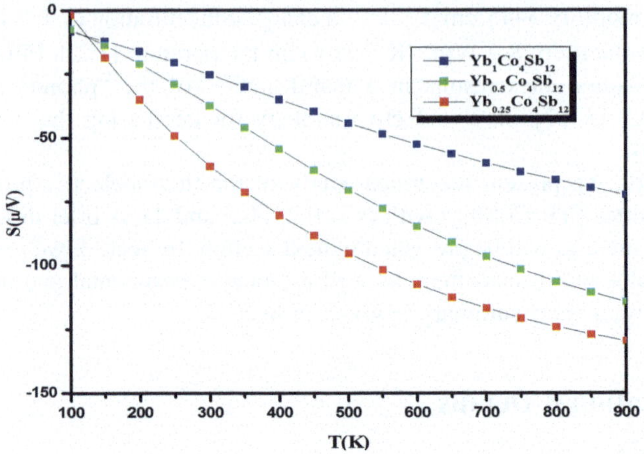

Fig. 1. The temperature dependence of seebeck coefficient of Yb filled skutterudite (Yb_xCoSb_{12})

3.2 Electrical Conductivity

Figure 2 shows the temperature dependence of electrical conductivity of Yb filled skutterudite (Yb_xCoSb_{12}). We notice that the electrical conductivity of Yb filled skutterudite (Yb_xCoSb_{12}) increased with increasing temperature. Which means that the Yb filler atoms have an influence on the reduction of electrical resistivity. The temperature dependences of electrical conductivity of the filled $CoSb_3$ skutterudites were proved to be nearly degenerate semiconductors.

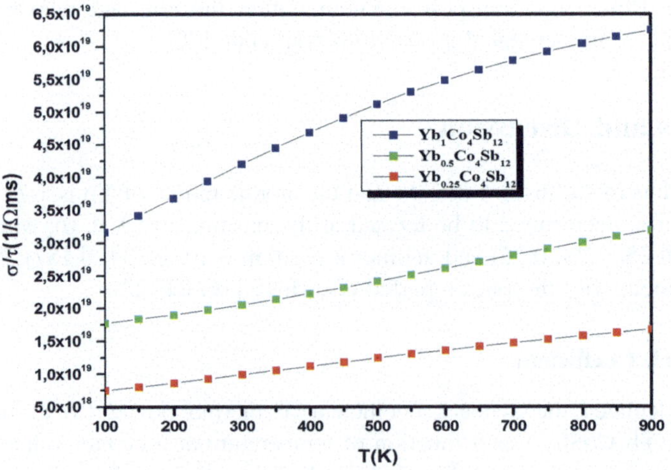

Fig. 2. The temperature dependence of electrical conductivity of Yb filled skutterudite (Yb_xCoSb_{12})

The Yb filled skutterudite (Yb_xCoSb_{12}) lies in the intrinsic semiconductor and semi metallic range which confirms that the material is a degenerate semiconductor. This is due to the decrease in the Fermi level and the increase in the electron concentration by adding the filler Yb atoms.

3.3 Thermal Conductivity

Figure 3 shows the temperature dependence of thermal conductivity of Yb filled skutterudite (Yb_xCoSb_{12}). We observe that the thermal conductivity was drastically reduced by Yb filling and the minimum value was obtained for the $Yb_{0.25}Co_4Sb_{12}$ because the Yb filler is completely inside the void and it can rattle well, which means that the rattling of Yb filler atoms inside the voids absorbs the heat (lattice vibration) and has a frequency that reduces the mean free path of phonons and the thermal conductivity. Where the voids are fully filled ($Yb_1Co_4Sb_{12}$), there is less effect on mean free path of phonons; and therefore the thermal conductivity increases.

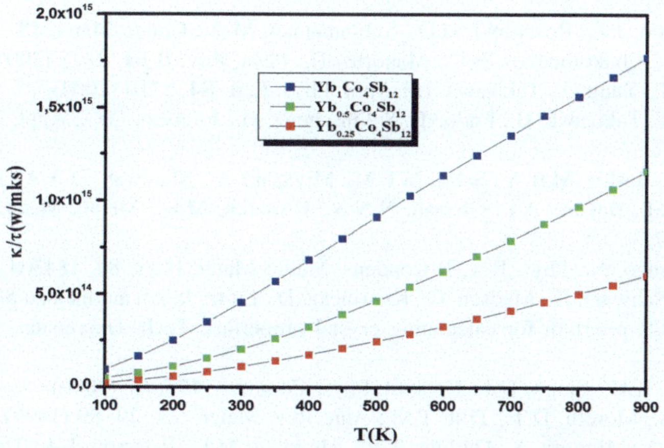

Fig. 3. The temperature dependence of thermal conductivity of Yb filled skutterudite (Yb_xCoSb_{12})

4 Conclusion

We investigated the thermoelectric properties of Yb filled skutterudite (Yb_xCoSb_{12}) based on the density functional theory. The calculations reveal some interesting findings. The first finding is that the Yb filled skutterudite (Yb_xCoSb_{12}) exhibits negative seebeck coefficient which confirmed that the Yb filler is n-type conducting. The second finding is that Yb filled skutterudite (Yb_xCoSb_{12}) exhibits a high electrical conductivity, low thermal conductivity and a large Seebeck coefficient. Based on these findings, the investigated crystal can be considered as an efficient thermoelectric material.

References

1. Caillat, T., Fleurial, J.P., Borshchevsky, A.: J. Cryst. Growth **166**, 722 (1996)
2. Morelli, D.T., Caillat, T., Fleurial, J.P., Borshchevsky, A., Vandersande, J., Chen, B., Uher, C.: Phys. Rev. B: Condens. Matter Mater. Phys. **51**, 9622 (1995)
3. Slack, G.A.: In: Powe, D.M. (ed.) Thermoelectric Handbook, p. 407. CRC, Boca Raton (1995)
4. Slack, G.A., Tsoukala, V.G.: J. Appl. Phys. **76**, 1665 (1994)
5. Nolas, G.S., Cohn, J.L., Slack, G.A.: Phys. Rev. B **58**, 164 (1998)
6. Morelli, D.T., Meisner, G.P., Chen, B.X., Hu, S.Q., Uher, C.: Phys. Rev. B **56**, 7376 (1997)
7. Kuznetsov, V.L., Kuznetsova, L.A., Rowe, D.M.: J. Phys.: Condens. Matter **15**, 5035 (2003)
8. Pei, Y.Z., Bai, S.Q., Zhao, X.Y., Zhang, W., Chen, L.D.: Solid State Sci. **10**, 1422 (2008)
9. Ballikaya, S., Uher, C.: J. Alloys Comp. **585**, 168 (2014)
10. Chen, L.D., Kawahara, T., Tang, X.F., Goto, T., Hirai, T., Dyck, J.S., Chen, W., Uher, C.: J. Appl. Phys. **2001**, 90 (1864)
11. Zhao, X.Y., Shi, X., Chen, L.D., Zhang, W., Pei, Y.Z.: J. Appl. Phys. **99**, 053711 (2006)
12. Puyet, M., Lenoir, B., Dauscher, A., Dehmas, M., Stiewe, C., Müller, E.: J. Appl. Phys. **95**, 4852 (2004)
13. He, T., Chen, J.Z., Rosenfeld, H.D., Subramanian, M.A.: Chem. Mater. **18**, 759 (2006)
14. Sales, B.C., Chakoumakos, B.C., Mandrus, D.: Phys. Rev. B **61**, 2475 (2000)
15. Nolas, G.S., Yang, J., Takizawa, H.: Appl. Phys. Lett. **84**, 5210 (2004)
16. Nolas, G.S., Takizawa, H., Endo, T., Sellinschegg, H., Johnson, D.C.: Appl. Phys. Lett. **77**, 52 (2000)
17. Siad, S.M., Bashir, M.B.A., Sabri, M.F.M., Miyazaki, Y., Shnawah, D.A.A., Hakeem, A.S., Shimada, M., Bakare, A.I., Ghazali, N.N.N., Elsheikh, M.H.: Metall. Mater. Trans. A **48**, 3073 (2017)
18. Li, W., Mingo, N.: Phys. Rev. B Condens. Matter Mater. Phys. **89**, 184304 (2014)
19. Blaha, P., Schwarz, K., Madsen, G., Kvasnicka, D., Luitz, J.: An augmented plane wave plus local orbitals program for calculating crystal properties. Tech. Universitat, Wien, Austria (2014)
20. Madsen, G.K.H., Singh, D.J.: Comput. Phys. Commun. **175**, 67 (2006)
21. Nolas, G.S., Morelli, D.T., Tritt, T.M.: Ann. Rev. Mater. Sci. **29**, 89 (1999)
22. Mandrus, D., Migliori, A., Darling, T.W., Hundley, M.F., Peterson, E.J., Thompson, J.D.: Phys. Rev. B **52**, 4926 (1995)

Waste Management, Recycling and Environment

Recycling of Plastic Waste and Valorization of Clay Material in the Road Area

Oum Essaad Ouaaz(✉) and Bachir Melbouci

Laboratory Geomaterials, Environment and Development (LGEA), The University M. Mammeri, BP 17 RP, DZ15000 Tizi-Ouzou, Algeria
ouaazmimi89@gmail.com, melbouciba@hotmail.fr

Abstract. The experimental study carried out has a dual objective: valorize the clay material in the road area and improving the bearing capacity of the shaped layers by reinforcement with plastic waste (granules plastic). To do this, six mixtures of clay and sand of different percentages were studied: 100% Clay, 95% Clay + 05% Sand, 90% Clay + 10% Sand, 85% Clay + 15% Sand, 80% Clay + 20% Sand, 75% Clay + 25% Sand. Proctor compaction tests were conducted on mixtures (sand + clay + plastic waste). The results obtained show a clear evolution of the characteristics of the Proctor test according to the different types and percentages of the recycled plastic.

Keywords: Valorization · Recycling · Mixing · Soil material
Mechanical testing

1 Introduction

During these last few years, the protection of the environment has become a fundamental factor to be taken into account in the process of economic and social development (Touiki 2013) in which studies have been launched in some countries with the aim of highlighting the danger that is passing through the world by the rate of waste that threatens our environment. Plastic materials have a very important role in our daily lives; their use becomes more and more important and the world demand for these materials increases from one year to another (Sadoun 2007).

For more than a century, the consumer society generates more and more waste diverse:

Domestic, hospital, industrial… 80% of hazardous waste, or 500 million tons per year, is produced by industrial countries. In poor countries, 80% of waste is released without treatment.

In rich countries, pollution is more insidious because garbage is often buried or incinerated.

Informal recovery circuits have often been set up to capture, under unsatisfactory conditions (Sorting directly into garbage or dumps), high value-added recyclable materials such as metals.

On the other hand, with the development of new design and realization technologies (Gaye et al. 2001), new recycling methods are being updated to save our environment and exploit our waste product waste.

B. Abdelbaki et al. (Eds.): SMSD 2017, *Proceedings of the Third International Symposium on Materials and Sustainable Development*, pp. 427–438, 2018.
https://doi.org/10.1007/978-3-319-89707-3_49

Research has shown that the use of recycled plastic is becoming more and more interesting in the field of civil engineering has known its exploitation in concrete in various forms (fiber, reinforcement, mortar, aggregates, etc.) (Samir et al. 2014; Safi et al. 2013; Rossignola and Angnesini 2014).

However, it is necessary to see the possibility of recycling plastic waste in the field of geotechnical, for this, our study will focus on the use and recycling of plastic waste in the formulation of soil mortars as an addition of aggregates in view of the availability of natural resources such as sand. An experimental study was conducted to evaluate the dry densities of clay samples with different proportions of substitution of sand and plastic waste.

1.1 Definition of Plastic

A material is called plastic, whether natural or synthetic, if after being deformed by an external action (temperature + stress, pressure…); it retains the form acquired at the end of this action.

The word plastic comes from the Greek word plastikos which means modelable.

Scientifically it is a polymer made of an organic material composed essentially of carbon atoms and hydrogen (Sadoun 2007).

1.2 Definition of Recycling

Recycling is the set of techniques aimed at recovering waste and reintroducing it into the production cycle from which it originates.

2 Materials and Test

2.1 Materials for Study

Three materials were used for this study (Fig. 2):

- A reworked clay extracted at a place called "Ait Chafaa" 30 m from the National Road;
- Fine sand with grains less than 2 mm;
- Plastic in the form of grains.

These two materials (clay and sand) were mixed with different percentages to obtain six (6) samples. Their main characteristics are summarized in Table 1 (Fig. 1 and Table 2).

Table 1. Optimal water content of the 6 study samples.

Type of mixture	100% Clay	95% Clay + 5% sand	90% Clay + 10% sand	85% Clay + 15% sand	80% Clay + 20% sand	95% Clay + 5% sand
WOP (%)	15	15	15	14	14	13

Fig. 1. Materials used are respectively: clay, granulated plastic and sand.

Table 2. Clay characteristic.

W (naturelle) %	w_P	I_P	I_C	W_L (%)
11.37	13.82	19.91	0.734	26

Fig. 2. Illustration for the tests proctor.

2.2 Test Carried Out and Principle

To possibly use these materials in the road domain, Proctor compaction tests were carried out.

2.2.1 Principle of Test

The Proctor compaction characteristics of a material are determined from the so-called "normal Proctor" or modified Proctor test. The two tests are identical in principle, only the parameter values that define the applied compaction energy differ.

The principle of these tests consists in humidifying a material with several water contents and in compacting it, for each of the water contents, according to a conventional method and energy. For each of the water content values considered, the dry density of the material is determined and the curve of the variations of this density is plotted as a function of the water content.

In this study, we varied three parameters:

- The material consisting of 6 mixtures;
- Compaction energy (25 strokes, 55 strokes, 75 strokes);
- The percentage of added plastic in the form of grains of (3%, 5% and 7%);
- The objective of this work is to study the influence of these three parameters on the Proctor characteristics (the dry density and optimum water content).

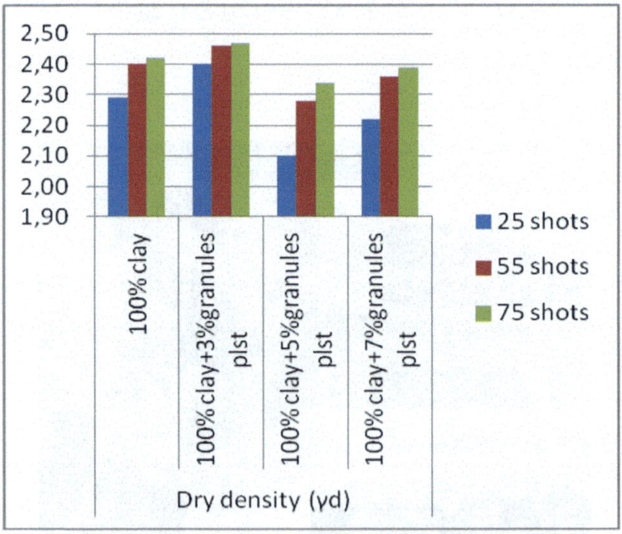

Fig. 3. Maximum dry density of clay with addition of plastic

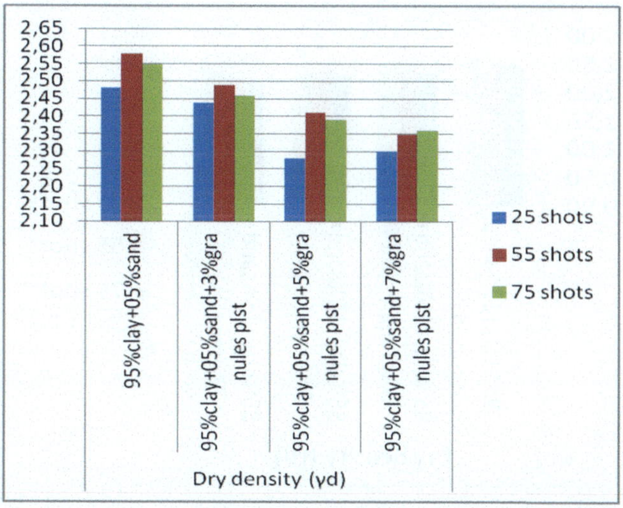

Fig. 4. Maximum dry density of 95% clay + 05% sand with addition of plastic.

3 Presentation and Interpretation of Results

The experimental study showed that the dry density depends on 3 essential parameters (Figs. 5, 6, 8 and Tables 3, 4, 5, 6):

- Compaction energy
- The percentage of sand
- Percentage of addition of granulated plastic

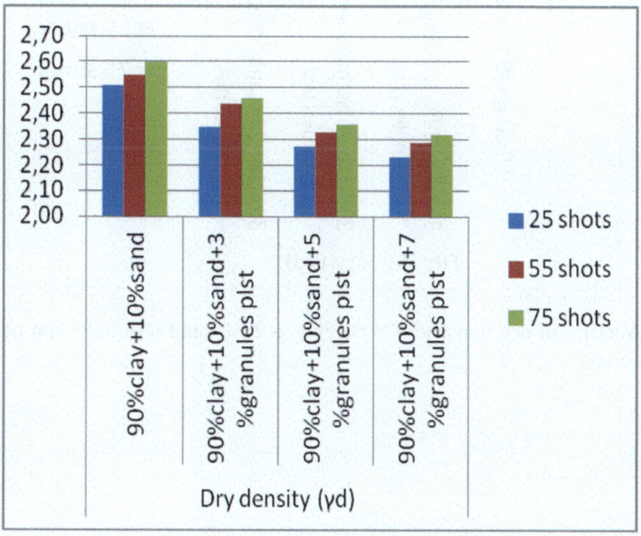

Fig. 5. Maximum dry density of 90% clay + 10% sand with addition of plastic.

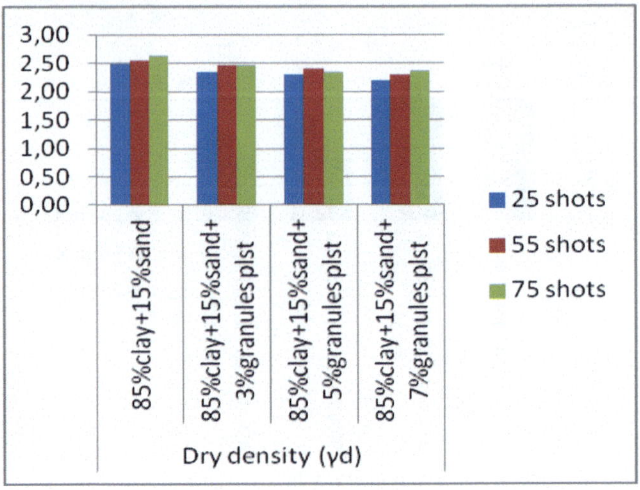

Fig. 6. Maximum dry density of 85% clay + 15% sand with addition of plastic.

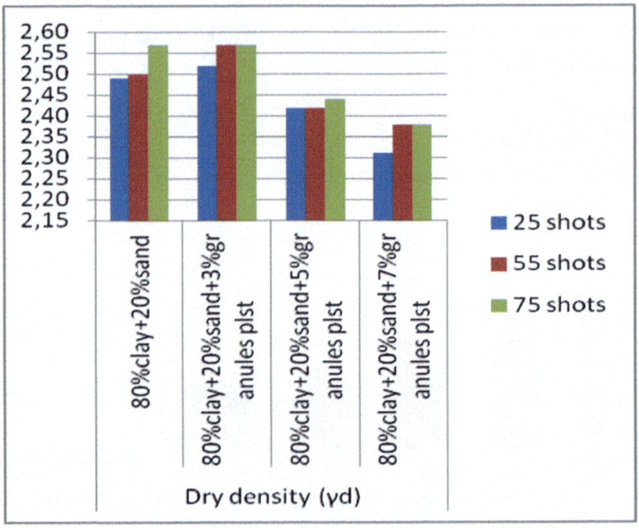

Fig. 7. Maximum dry density of 80% clay + 20% sand with addition of plastic.

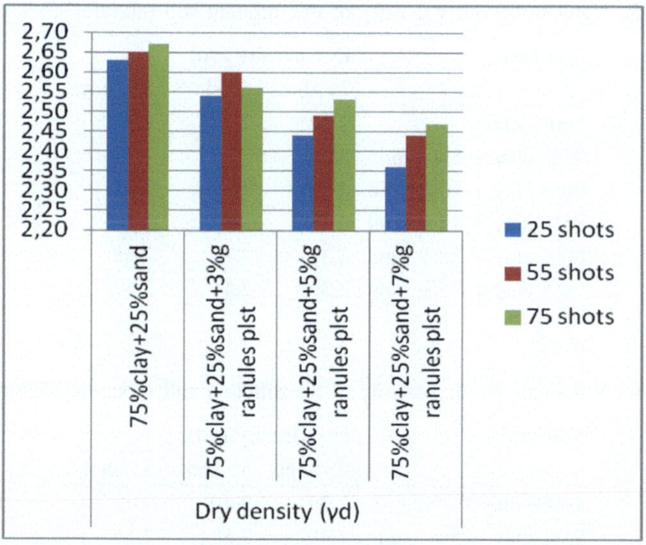

Fig. 8. Maximum dry density of 75% clay + 25% sand with addition of plastic.

Table 3. Values of the dry density of the different soil mixtures without the addition of plastic.

Soil type	Dry density (γd)		
	25 shots	55 shots	55 shots
100% clay	2.40	2.46	2.47
95% clay + 5% sand	2.44	2.49	2.46
90% clay + 10% sand	2.35	2.44	2.46
85% clay + 15% sand	2.34	2.46	2.47
80% clay + 20% sand	2.52	2.57	2.57
75% clay + 25% sand	2.54	2.60	2.56

Table 4. Values of the dry density of the different soil mixtures +3% plastic.

Soil type	Dry density (γd)		
	25 shots	55 shots	55 shots
100% clay	2.10	2.28	2.34
95% clay + 5% sand	2.28	2.41	2.39
90% clay + 10% sand	2.27	2.33	2.36
85% clay + 15% sand	2.30	2.40	2.35
80% clay + 20% sand	2.42	2.42	2.44
75% clay + 25% sand	2.44	2.49	2.53

Table 5. Values of the dry density of the different soil mixtures +5% plastic.

Soil type	Dry density (γd)		
	25 shots	55 shots	55 shots
100% clay	2.22	2.36	2.39
95% clay + 5% sand	2.30	2.35	2.36
90% clay + 10% sand	2.23	2.29	2.32
85% clay + 15% sand	2.19	2.30	2.36
80% clay + 20% sand	2.31	2.38	2.38
75% clay + 25% sand	2.36	2.44	2.47

Table 6. Values of the dry density of the different soil mixtures +7% plastic.

Soil type	Dry density (γd)		
	25 shots	55 shots	55 shots
100% clay	2.29	2.40	2.42
95% clay + 5% sand	2.48	2.58	2.55
90% clay + 10% sand	2.51	2.55	2.60
85% clay + 15% sand	2.50	2.56	2.65
80% clay + 20% sand	2.49	2.50	2.57
75% clay + 25% sand	2.63	2.65	2.67

Compaction energy is one of the known parameters that affect γd, the larger the latter, the greater the latter is the more the material is denser what was observed in the results above (Figs. 12, 13, 14 and 15) except in the case of sample 95% clay + 05% sand, or lower densities are observed for the compaction energy 75 strokes with respect to the compaction energy 55 strokes (Fig. 4).

Percentage of sand added proportionally influences the improvement of the density or the densities are higher in the case of samples which carries a higher percentage of last (Figs. 12, 13, 14 and 15).

The addition of 3% of the plastic showed an improvement in the dry density for the samples with 100% clay and (80% clay + 20% sand) (Figs. 3 and 7).

The other samples have γd fairly close and lower than the basic sample and this is in all 3 cases the compaction energy.

Samples with added percentages of 5% and 7% showed a sharp decrease in the dry density where it varies between 2.28 and 2.45.

3.1 Evolution of the Dry Density According to the Percentage of Plastic and Sand Added

The best densities were found for the materials 80% clay + 20% sand and 75% clay + 25% sand specially for the compaction energies 25 shots, 55 shots and 75 coups with a percentage of additions of 3% (Figs. 9, 10 and 11).

The lower the percentage of sand in the blends, the lower the densities decrease for plastic additions of 5% and 7%, the increase in the dry density of the samples is inversely proportional to the increase in the percentage of plastic additions.

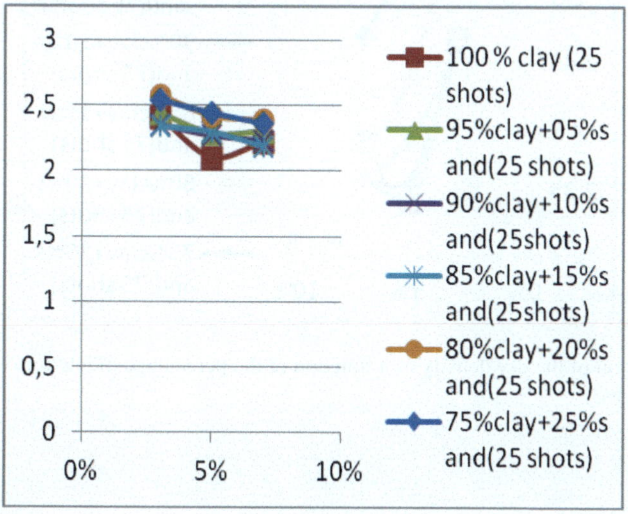

Fig. 9. Evolution of the dry density as a function of the percentage of addition of plastic for the energy of 25 shots.

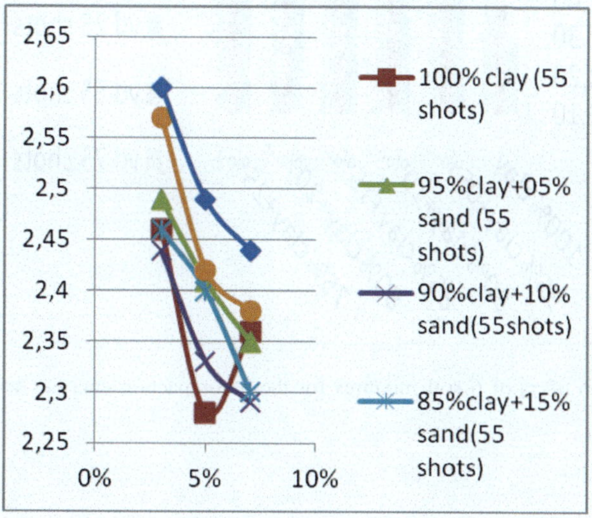

Fig. 10. Evolution of the dry density as a function of the percentage of addition of plastic for the energy of 55 shots.

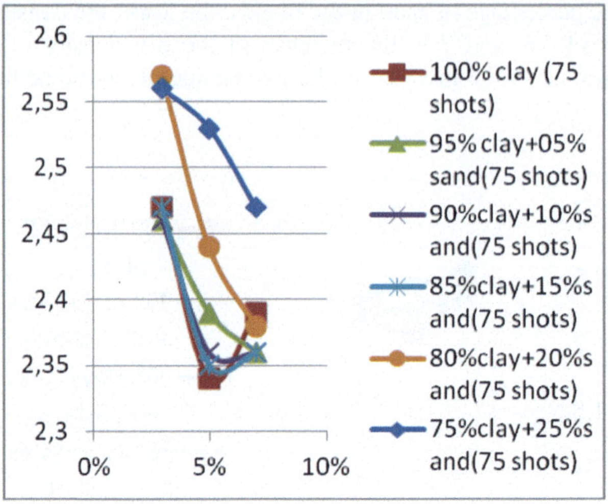

Fig. 11. Evolution of the dry density as a function of the percentage of addition of plastic for the energy of 75 shots.

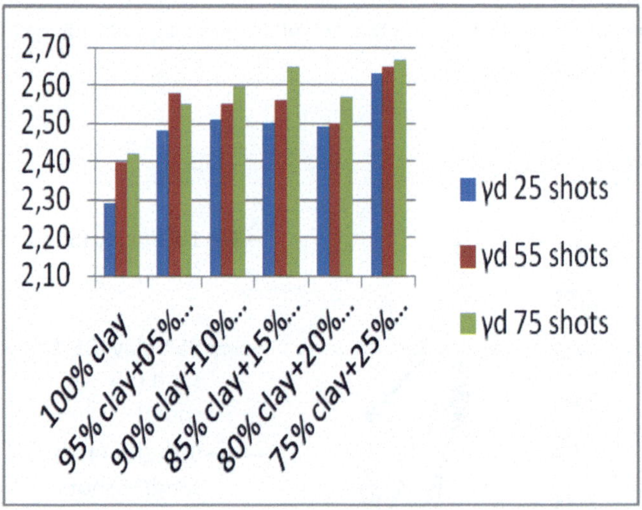

Fig. 12. Evolution of γ_d of 6 soil mixtures for the 3 compaction energies without addition of plastic.

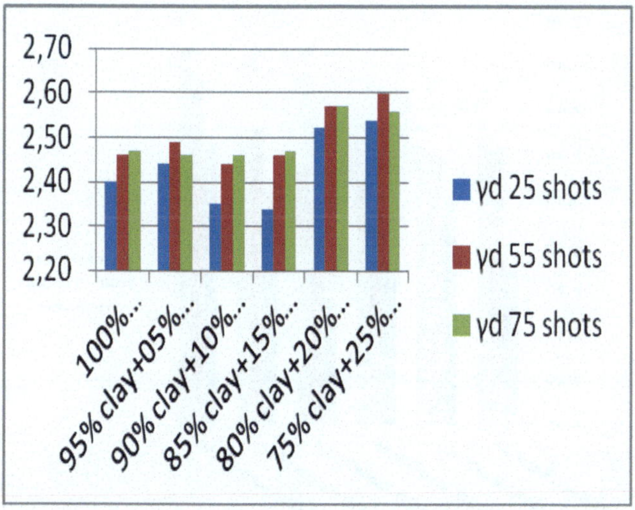

Fig. 13. Evolution of γ_d of 6 soil mixtures for the 3 compaction energies +3% plastic.

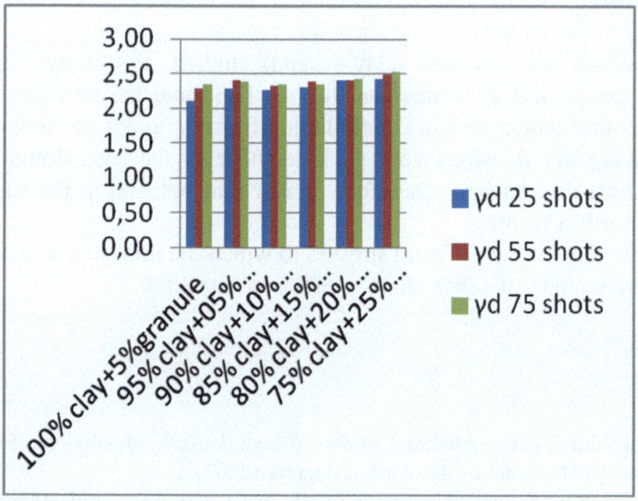

Fig. 14. Evolution of γ_d of 6 soil mixtures for the 3 compaction energies +5% plastic.

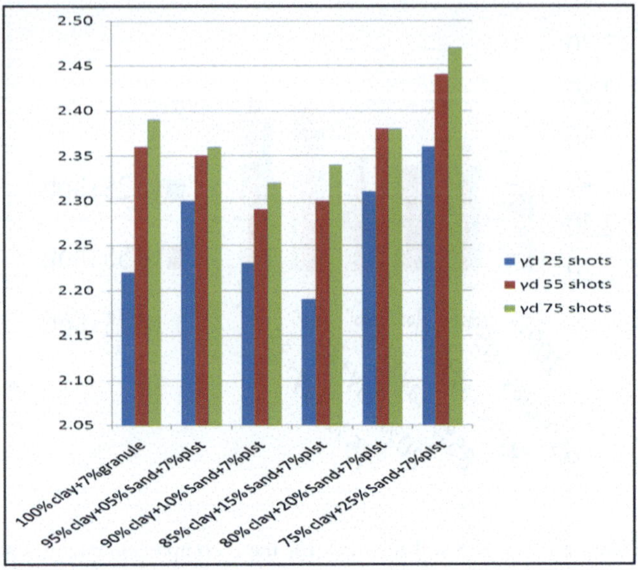

Fig. 15. Evolution of γ_d of 6 soil mixtures for the 3 compaction energies +7% plastic.

4 Conclusion

By varying, in the six mixtures (clay + sand) studied, the energy of compaction (25 coups, 55 coups and 75 coups) on the one hand and the addition of the plastic granule on the other hand, we can conclude that from the point of view of dry density our mixtures have dry densities very close to those of the road domain, 3% plastic blend has the best dry density. Therefore, plastic can be used in the road sector in a maximum proportion of 3%.

However, in view of the different stresses to which the mixtures will be subjected in practice, the knowledge of other characteristics is essential.

References

Sadoun, S.: Contribution to experimental studies of high density polyethylene (PEHD), effect of temperature and deformation rate, Anaba, Algeria (2007)

Touiki, H.: Waste flow management case plastic bags and bags, end studies Memory for obtaining a specialized masters degree. Université Mohamed V Souissi, MAROC (2013)

Gaye, S., et al.: Thermal and mechanical characterization of recycled polymer concrete. J. Sci. **1**, 53–66 (2001)

Safi, B., et al.: The use of plastics waste as fine aggregate in the self-compacting mortars: effect on physical and mechanical properties. Constr. Build. Mater. **43**, 436–442 (2013)

Rossignola, J.A., Angnesini, M.V.C.: Durability of polymer-modified hight weight aggregate concrete. Cem. Concr. Compos. **26**, 375–380 (2014)

Samir, B., et al.: Valorization and recycling of plastic waste in concrete. EDP Sci. (2014). https://doi.org/10.1051/matconf/20141101033

Experimental Study of an Inclined-Plane Electrostatic Separator

A. Nadjem(✉), M. Kachi, K. Rouagdia, and M. Remadnia

Electrical Engineering Laboratory – LGEG, University of Guelma, Guelma, Algeria
nadjem_kd@hotmail.com

Abstract. Free-fall electrostatic separator is the most basic and a standard solution to separate insulators mixtures. In this paper, an inclined configuration based on free-fall separator was studied. The performances of the studied separator were evaluated through recovery and purity of a separated mixture composed of Polyamide (PA) and Polycarbonate (PC) granular materials. Mainly, the applied voltage to the electrodes and the angle of inclination were the two studied factors. These two factors were varied respectively between 5 kV and 10 kV for the voltage and 25° to 90° for the inclination angle. The obtained results show that recovery rate is increased as the voltage increase so that the more is the voltage the more is the recovered quantity. Conversely, the best results are obtained for lowest inclination angle (25°). At 10 kV and 25°, the recovery rate was more than 70% with products purity roughly 97% for the PA and 98% for the PC. The inclined-plane allowed the increase in recovery rate and purity in comparison to the standard free-fall configuration.

Keywords: Triboelectrostatic separation · Granular plastics · Free-fall separator
Angle of inclination

1 Introduction

The need to the development of information technology leads to the progress of electrical and electronic equipments (Osibanjo and Nnorom 2007; Ongondo et al. 2011). In fact, this used equipments, whether broken or obsolete, such as computer, mobile phones, printers, televisions and other appliances have become an increasingly important source of waste composed of electrical and electronic equipments (WEEE) that can pose health risks to individuals and contribute to harmful impacts on the environment (Sepúlveda et al. 2010). Nowadays, the best way to reduce these risks and to enhance resources efficiency it consists in the recycling of WEEE properly in order to make new products (Ylä-Mella et al. 2014).

The electrostatic separation is considered as a multifactorial technique that is difficult to control, suitable for recycling (Wu et al. 2013; Remadnia et al. 2014; Haga et al. 1995). This non-pollutant technique is an important class for the selective sorting of plastic granular materials. In this class of applications, the final result is dependent on several factors, such as the parameters of the separator itself (Dwari et al. 2015;

B. Abdelbaki et al. (Eds.): SMSD 2017, *Proceedings of the Third International Symposium on Materials and Sustainable Development*, pp. 439–450, 2018.
https://doi.org/10.1007/978-3-319-89707-3_50

Miloudi et al. 2013), tribocharging device efficiency (Dascalescu et al. 2005; Buda et al. 2013; Li et al. 2015; Zelmat et al. 2013), surface proprieties of the granular materials (Nadjem et al. 2017), and the relative humidity which also have an important impact on separation results (Lungu 2004; Németh et al. 2003).

The free-fall electrostatic separator is one of most widely used devices for separating granular insulating mixtures by the combined action of mechanical and electric forces (Bendimerad et al. 2009). Therefore, the high-voltage applied to the electrodes should be as high as possible so that the electric Coulomb force could overcome the mechanical and gravitational forces and consequently better separate the constituents of the mixture. However, the main disadvantage of the free-fall separation system is the missing of controllability of the gravitational force, which related to the weight of the particles. In this context, previous carried studies have already tried to improve the separation devices and optimize their operation. The increasing of charging time, air speed and the high-voltage applied to the electrodes were the used solutions in most cases (Bendimerad et al. 2009; Li et al. 2013; Dascalescu et al. 2011; Rahou et al. 2013; Aksa et al. 2013). On the other hand, the adjustments of such parameters are always limited: long charging time would reduce the hourly quantities of treated material. Likewise, an intensive applied voltage and a high air speed would increase the energy consumption to expensive levels unacceptable for industry application.

The triboelectricity is a physical phenomenon defined as the exchange of surface electric charges between two materials during their contact (Greason et al. 2004; Bhushan 2000), it is the essential mechanism used for the charging of granular insulating materials in view to their separation in an intense electric field.

The aim of the present work is to contribute to the improvement of the free-fall separator by proposing a rather inclined configuration that has been designed for enhancing the recovery rate of the separated products. By comparing the results obtained with the two separators configurations, it would provide insight into the efficiency of the proposed installation.

2 Experimental Procedure

2.1 Materials

The experiments were performed on two granular insulators, Polyamide (PA) and Polycarbonate (PC) (Table 1). The mechanical separation of a mixture composed of these materials is difficult due to the similarity in their shapes and mass densities. The difference between the colors of the two materials facilitates the analysis of the purity of the separated products. The aspect of the used PA and PC granular materials is shown in Fig. 1.

Table 1. Characteristics of the PA and PC granules of plastics employed in the experiments

Granule	PA	PC
Color	Blue	Orange
Form	Cylindrical	Cylindrical
Height [mm]	Ø 2.5 × 3.5	Ø 3 × 3.6
Mass [mg]	20	24
Density [kg/m³]	1100	1200

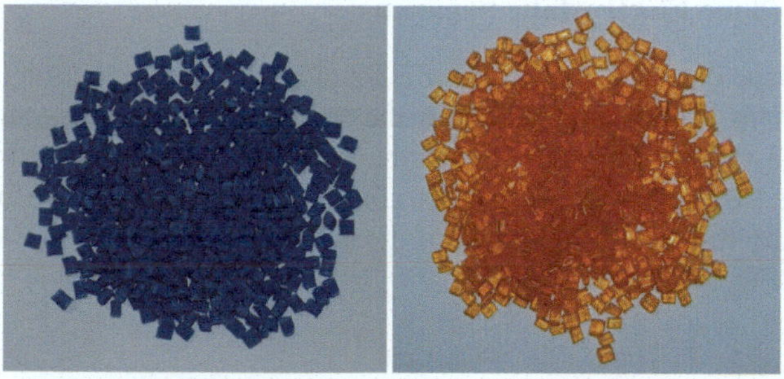

(a) (b)

Fig. 1. Aspect of the granular materials: (a) PA; (b) PC.

2.2 Fluidized Bed Tribocharging Device

The tribocharging of the granular mixtures was carried out in a fluidized bed device (Fig. 2). The fluidization air is obtained by a variable-speed blower. A custom-designed air diffuser ensures the uniformity of the fluidization bed of particles, which is a

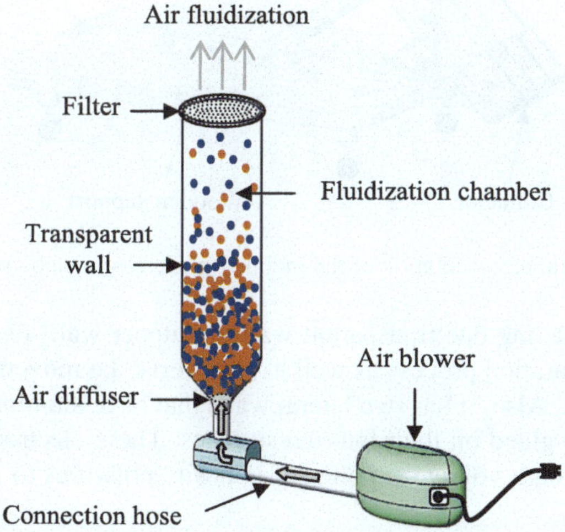

Fig. 2. Fluidized-bed device for the tribocharging of plastic granules.

perforated plastic plate situated at the bottom of the chamber. At the upper end, the chamber is equipped with a filter that prevents the granules to escaping out. The charge acquired by the particles is produced by the triboelectric effect in a cylindrical chamber (diameter: 100 mm; length: 450 mm), made of Polyethylene terephthalate (PET). The chamber has a transparent walls in order to permit the visualization of tribocharging process. In this device, the impacts of particle-to-particle and particle-to-wall result in the tribocharging of particles.

2.3 Inclined-Plane Electrostatic Separator

The inclined-plane electrostatic separator is intended for the separation of plastic granular mixtures. It is closely inspired from the standard free-fall separator. The device consists of a plastic parallelepiped enclosure (height: 775 mm; width: 160 mm; depth: 100 mm) placed over a wooden support combined by two movable bars that allow varying manually the inclination angle of the enclosure from 0° to 90° with respect to the horizontal (Fig. 3).

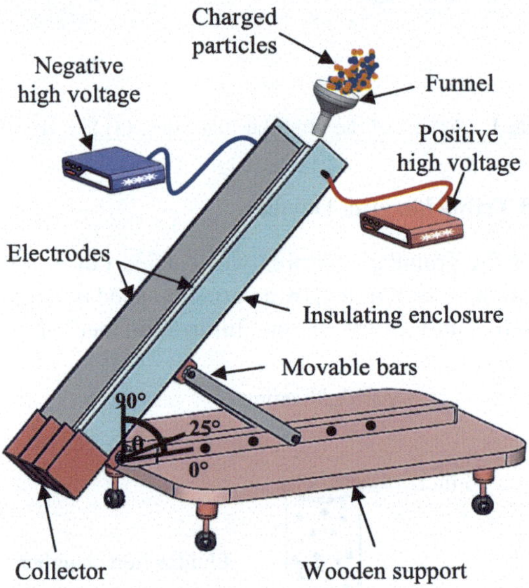

Fig. 3. Schematic representation of the inclined-plane electrostatic separator device.

The enclosure having one transparent wall (the upper wall) in order to permit the visualization of separation process as well as to observe the movement of the granules inside the enclosure. Also, it has two lateral walls that have aluminum plate electrodes (750 mm × 95 mm) glued on their internal surfaces. These electrodes are supplied by two adjustable DC high voltage supplies of opposite polarities to generate an intense electrical field.

The charged binary mixture may roll or slide on the bottom wall, and they separate under the combined action of electrical, normal, frictional and gravitational forces. The friction force is negligible due to the smooth surface of the plane. The two separated products are collected in three boxes placed at the lower end of the separator (right and left boxes for pure products and one in the center for middling).

2.4 Methods

The experiments were carried out on a binary mixture composed of 50 g of PA and 50 g of PC. Particles of this mixture were charged in the fluidized bed device. In all experiments, the variables of tribocharging system were maintained constant: the fluidization time: 2 min; the speed of the air fluidization supplied by the blower: V = 3.6 m/s.

The charged particles introduced in the separation enclosure by a funnel with a feed rate: D = 1.1 g/s, and were slid on the inclined bottom wall at various angle ranging between 25° and 90°. The minimum value (25°) was chosen after numerous preliminary experiments. Thus, it was observed that the particles cannot slide on the plane at angles of inclination lower than 25°, and the maximum value (90°) correspond the case of free-fall electrostatic separator. The high-voltage applied to the electrodes was adjusted at various values in the range 5 and 10 kV.

When the separation was over, the mass and the electric charge of the two products recovered in the PA and PC collecting boxes were measured. Finally, the recovery and the purity rates were measured for both products to evaluate the separation efficiency.

All experiments reported in this paper were performed under relatively stable ambient conditions: temperature T = 21 ± 1 °C, and relative humidity RH = 40 ± 1%. The electric charge of the samples was measured with a Faraday pail connected to an electrometer (Keithley Instruments, model 6514), while the mass of the recovered products in each compartment was measured with an electronic balance (resolution: 0.01 g).

The recovery rate of a product was calculated as:

$$\mathrm{Rec}\,(\%) = \frac{m_{ic}}{m_{it}} * 100 \tag{1}$$

Where m_{ic} is the quantity of product *i* that is collected in the compartment reserved for it, and m_{it} is the total mass of product *i* at the inlet of the separator.

The purity analysis was carried out by manual selective sorting; it was calculated as:

$$\mathrm{Pur}\,(\%) = \frac{m_{ic}}{m_{tc}} * 100 \tag{2}$$

Where m_{tc} is the total mass (both products) collected in the same compartment.

3 Results and Discussion

3.1 The Effect of High-Voltage

The results of the separation experiments as a function of the high-voltage applied to the electrodes are shown in Figs. 4 and 5.

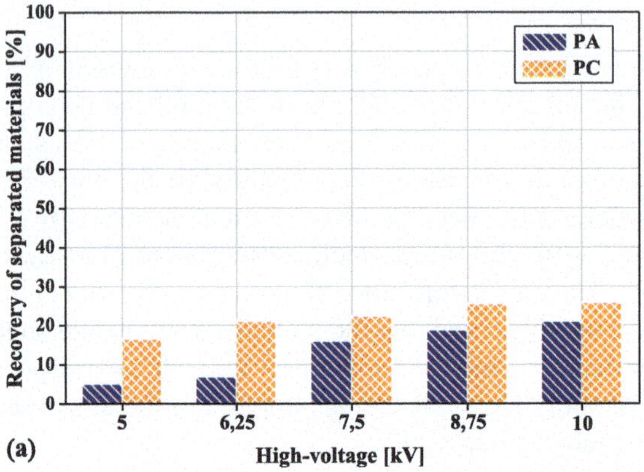

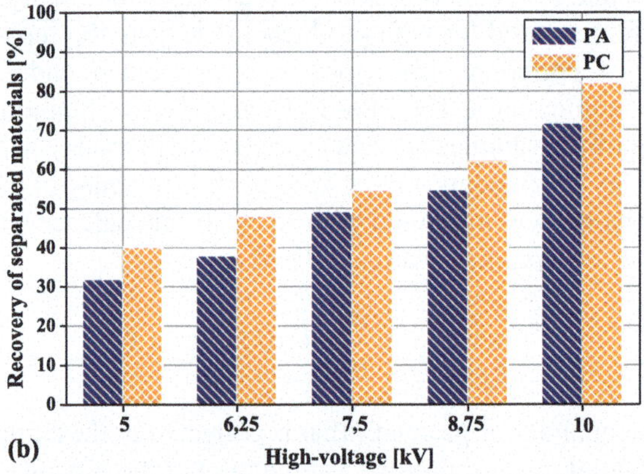

Fig. 4. Recovery rate of PA and PC products obtained after the separation as a function of the applied high voltage: (a) $\theta = 90°$; (b) $\theta = 25°$.

The recovery rate recorded for the two products are shown in Fig. 4a, b. In the preliminary experiments, it was observed that both PA and PC granules acquire a negative charge when contacting the PET chamber walls. But, when the mixture of the both materials charged in the chamber, PA granules acquired a positive charge while PC granules were charged negatively because of contact between them. Firstly, it can be seen that the recovery rates increase with the applied high-voltage, whatever the used separator. Indeed, in the case of the free-fall separator (Fig. 4a), the recovery rate of PA product increased from 5.26% with electrodes energized at ±5 kV, to 21.14% with

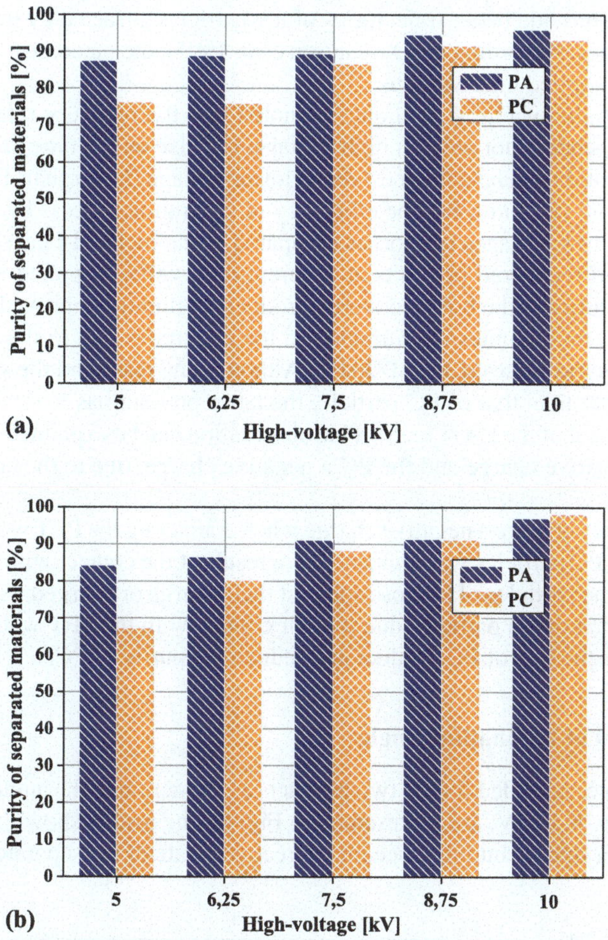

Fig. 5. Purity of PA and PC products obtained after electrostatic separation as a function of the applied high voltage: (a) $\theta = 90°$; (b) $\theta = 25°$.

electrodes energized at ±10 kV. As for the PC product, the recovery rate was increased from 16.8% for ±5 kV, to 26.18% for ±10 kV. In the case of the inclined-plane separator (Fig. 4b), the high-voltage applied to the electrodes has a high effect on the quantity of recovered products; the recovery rate of PA increases from 32.06% for ±5 kV to 71.94% for ±10 kV. For PC, the recovery rate increases from 40.43% for ±5 kV to 82.46% for ±10 kV.

The charged particles projected into the electrostatic separation zone between the two electrodes are subjected to the combined action of mechanical and electric forces. In addition, the intensity of the electric field established in the zone of separation is proportional to the high-voltage applied. Indeed, the increase of the recovery rates in the two lateral compartments can be explained by the fact that the electric field exerts a stronger action that can influence the movement of the charged particles and its good distribution can provide better trajectories of the particles attracted by the electrodes of opposite polarities. According to (Dascalescu et al. 1995), the outcome of the separation depends on the applied high-voltage.

From the results of Fig. 4, it should be noted that the recovery rates of separated products are dependent not only on high-voltage, but also on the type of the used separator. The separation can be obtained even at lower voltage (lower than 10 kV) by using the inclined-plane separator, but the efficiency of separation process by using the free-fall separator at low voltages is poor, as compared to the inclined-plane separator.

The purity of recovered products at different high-voltage applied is presented on Fig. 5a, b. The purity of the both recovered products is slightly improved with the high-voltage applied, the granules are attracted to the electrodes due to the strong electric field and the electric forces (Li et al. 2013). Although the mass and the size of each PA particle is smaller than that of PC particle, the latter product has low values of purity, compared with that of the PA (Fig. 5). This observation can be explained by the fact that the PA has a positive charge and the PC a negative charge, due to the contact between them.

However, they acquire a negative charge when contacting the PET walls. The charge acquired by the PC particles was enhanced as a result of the contact with the PET walls. Conversely, some particles of PA lose some of their charge or charged with the opposite polarity of the rest of the particles due to their contact with the PET walls; which leads to their move to the box of PC product and reduce the purity of PC.

3.2 The Effect of Inclination Angle

The recovery rates recorded for the two products as a function of the angle of inclination for high-voltage of 10 kV, are represented in Fig. 6. The results show that the angle of inclination has a significant influence on the recovery rate of the separated products. At

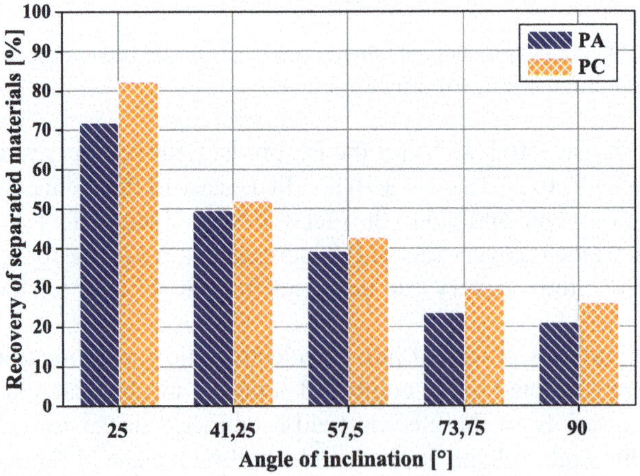

Fig. 6. Recovery of PA and PC products obtained after electrostatic separation as a function of the angle of inclination: V = 10 kV.

first glance, the recovery rate decreases with the angle of inclination (Fig. 6). The increasing of inclination angle increases the velocity as well as acceleration and displacement of particles, that leads to decrease in terminal time due to gravity (Jalaal and Ganji 2010). This indicates that gravitational forces could play a role in particles movement on the inclined-plane during the separation process, as a result, most particles of the mixture are oriented into the compartment in the middle of the collector.

The relative recovery rate due to the angle of inclination was computed as follows:

$$\Delta Rec_{(\theta i)}\% = \frac{(Rec_{(\theta i)} - Rec_{(90^\circ)})}{Rec_{(90^\circ)}} * 100 \tag{3}$$

Where $Rec_{(90^\circ)}$ is the recovery rate of separated products in free-fall electrostatic separator and $Rec_{(\theta i)}$ is the recovery rate of separated products in inclined-plane separator at an angle θ_i determined with respect to the horizontal. The Fig. 7 shows the variation of the recovery rate as a function of the angle of inclination for the two materials PA and PC. The angle parameter influences to a different degree the changes in recovery rate for the two materials. For both materials, the relative recovery rate tends to diminish with increasing the angle of inclination.

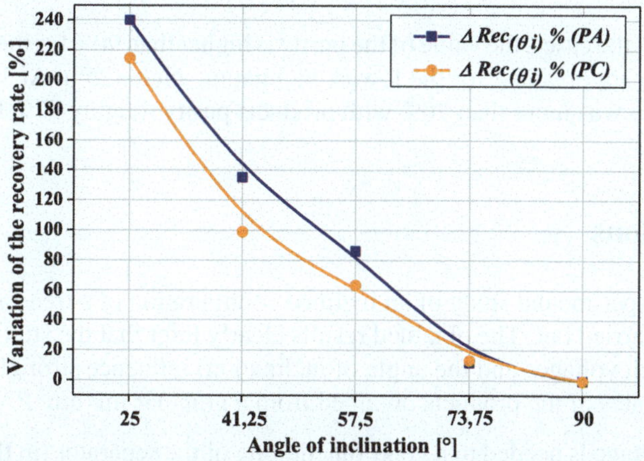

Fig. 7. Recovery variation rates of PA and PC separated products as a function of the angle of inclination: V = 10 kV.

Moreover, it should be noted that all the values of $\Delta Rec_{(\theta i)}$ are positive.

These values correspond to the situations in which the reduction in the angle of inclination can improve the recovery rate. The relative recovery rate of PA particles is higher than that of the PC, for most angles of inclination. Regardless of the angle of inclination selected by using the inclined-plane separator the recovery rates of the obtained products are the best comparing to the free-fall separator.

The separation results in terms of purity rate as a function of the angle of inclination are shown in Fig. 8. The results presented in Fig. 8 appear to be contradictory with the

Fig. 6. From the results of Fig. 8, the angle of inclination has no significant influence on product purity.

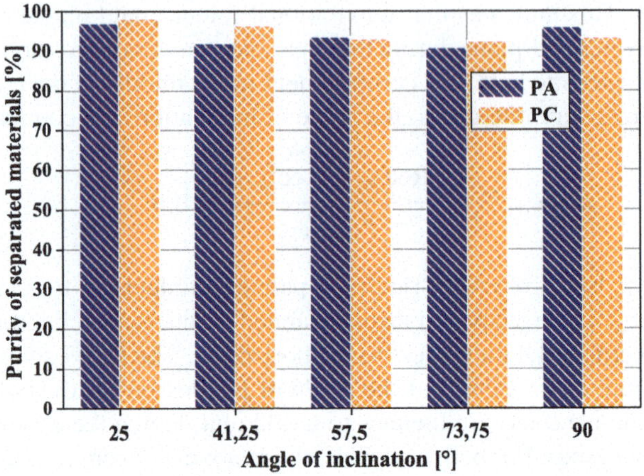

Fig. 8. Purity of products as function of the angle of inclination: V = 10 kV.

Indeed, in all the cases, the value of the purity is higher than 90% for the two products. The best results are obtained for the lowest inclination angle (25°). At 10 kV and 25°, the recovery rate was more than 70% with products purity roughly 97% for the PA and 98% for the PC.

4 Conclusions

In this paper, experimental study of an inclined configuration of a free-fall electrostatic separator was carried out. The obtained results clearly infer that the studied parameters (the applied high voltages and the angle of inclination) influence strongly the recovery and the purity rates of the products obtained from a granular mixture PA-PC. Mainly:

- The high voltage is needed to a good functioning of the separator (in the range of the studied voltages).
- The decrease of inclination angle can lead to a better separation of granular mixture. A small angle of inclination allows the increase of the recovery rate without affecting the purity of the recovered products. This means that, the vertical configuration of the free-fall separator can simply improved by tilting.

References

Aksa, W., Medles, K., Rezoug, M., Boukhoulda, M.F., Bilici, M., Dascalescu, L.: Two stage electrostatic separator for the recycling of plastics from waste electrical and electronic equipment. J. Electrostat. **71**(4), 681–688 (2013)

Bendimerad, S., Tilmatine, A., Ziane, M., Dascalescu, L.: Plastic wastes recovery using free-fall triboelectric separator. Int. J. Environ. Stud. **66**(5), 529–538 (2009)

Bhushan, B.: Modern Tribology Handbook, two volume set. CRC Press, Boca Raton (2000)

Buda, G., Samuila, A., Atroune, S., Bilici, M., Dascalescu, L.: Set point identification of a tribocharging process for mixed granular solids. J. Electrostat. **71**(3), 407–412 (2013)

Dascalescu, L., Mizuno, A., Tobazeon, R., Atten, P., Morar, R., Iuga, A., Samuila, A.: Charges and forces on conductive particles in roll-type corona-electrostatic separators. IEEE Trans. Ind. Appl. **31**(5), 947–956 (1995)

Dascalescu, L., Urs, A., Bente, S., Huzau, M., Samuila, A.: Charging of mm-size insulating particles in vibratory devices. J. Electrostat. **63**(6), 705–710 (2005)

Dascalescu, L., Fati, O., Bilici, M., Rahou, F., Dragan, C., Samuila, A., Iuga, A.: Factors that influence the efficiency of a fluidized-bed-type tribo-electrostatic separator for mixed granular plastics. J. Phys: Conf. Ser. **301**(1), 012066 (2011)

Dwari, R.K., Mohanta, S.K., Rout, B., Soni, R.K., Reddy, P.S.R., Mishra, B.K.: Studies on the effect of electrode plate position and feed temperature on the tribo-electrostatic separation of high ash Indian coking coal. Adv. Powder Technol. **26**(1), 31–41 (2015)

Greason, W.D., Oltean, I.M., Kucerovsky, Z., Ieta, A.C.: Triboelectric charging between polytetrafluoroethylene and metals. IEEE Trans. Ind. Appl. **40**(2), 442–450 (2004)

Haga, K., Chang, J.S., Kelly, A.J., Crowley, J.M.: Applications of the electrostatic separation technique, pp. 365–386. Marcel Dekker, New York (1995)

Jalaal, M., Ganji, D.D.: An analytical study on motion of a sphere rolling down an inclined plane submerged in a Newtonian fluid. Powder Technol. **198**(1), 82–92 (2010)

Li, J., Dascalescu, L., Bilici, M., Xu, Z.: Numerical modeling of the trajectories of plastic granules in a tribo-aero-electrostatic separator. J. Electrostat. **71**(3), 281–286 (2013)

Li, J., Wu, G., Xu, Z.: Tribo-charging properties of waste plastic granules in process of tribo-electrostatic separation. Waste Manag. **35**, 36–41 (2015)

Lungu, M.: Electrical separation of plastic materials using the triboelectric effect. Miner. Eng. **17**(1), 69–75 (2004)

Miloudi, M., Dascalescu, L., Li, J., El-Mossouess, S., Medles, K.: Tribo-aero-electrostatic separator for coarse granular insulating materials. IEEE Trans. Dielectr. Electr. Insul. **20**(5), 1510–1515 (2013)

Nadjem, A., Kachi, M., Bekkara, F., Medles, K., Zeghloul, T., Dascalescu, L.: Triboelectrification of granular insulating materials as affected by dielectric barrier discharge (DBD) treatment. J. Electrostat. **86**, 18–23 (2017)

Németh, E., Albrecht, V., Schubert, G., Simon, F.: Polymer tribo-electric charging: dependence on thermodynamic surface properties and relative humidity. J. Electrostat. **58**(1), 3–16 (2003)

Ongondo, F.O., Williams, I.D., Cherrett, T.J.: How are WEEE doing? A global review of the management of electrical and electronic wastes. Waste Manag. **31**(4), 714–730 (2011)

Osibanjo, O., Nnorom, I.C.: The challenge of electronic waste (e-waste) management in developing countries. Waste Manag. Res. **25**(6), 489–501 (2007)

Rahou, F., Tilmatine, A., Bilici, M., Dascalescu, L.: Numerical simulation of the continuous operation of a tribo-aero-electrostatic separator for mixed granular solids. J. Electrostat. **71**(5), 867–874 (2013)

Remadnia, M., Kachi, M., Messal, S., Oprean, A., Rouau, X., Dascalescu, L.: Electrostatic separation of peeling and gluten from finely ground wheat grains. Part. Sci. Technol. **32**(6), 608–615 (2014)

Sepúlveda, A., Schluep, M., Renaud, F.G., Streicher, M., Kuehr, R., Hagelüken, C., Gerecke, A.C.: A review of the environmental fate and effects of hazardous substances released from electrical and electronic equipments during recycling: examples from China and India. Environ. Impact Assess. Rev. **30**(1), 28–41 (2010)

Wu, G., Li, J., Xu, Z.: Triboelectrostatic separation for granular plastic waste recycling: a review. Waste Manag. **33**(3), 585–597 (2013)

Ylä-Mella, J., Poikela, K., Lehtinen, U., Keiski, R.L., Pongrácz, E.: Implementation of waste electrical and electronic equipment directive in Finland: evaluation of the collection network and challenges of the effective WEEE management. Resour. Conserv. Recycl. **86**, 38–46 (2014)

Zelmat, M.E.M., Rizouga, M., Tilmatine, A., Medles, K., Miloudi, M., Dascalescu, L.: Experimental comparative study of different tribocharging devices for triboelectric separation of insulating particles. IEEE Trans. Ind. Appl. **49**(3), 1113–1118 (2013)

Effect of Liquid Type on the Hydraulic Characteristic of Compacted Local Geomaterials for Use as Hydraulic Barriers

Abdellah Demdoum[1(✉)], M. K. Gueddouda[1], I. Goual[2], and H. Berkak[1]

[1] Department of Civil Engineering, Research Laboratory of Civil Engineering, Amar Telidji University, Laghouat, Algeria
a.demdoum@lagh-univ.dz

[2] Department of Civil Engineering, Structures Rehabilitation and Materials Laboratory, Amar Telidji University, Laghouat, Algeria

Abstract. In landfill applications, compacted sand-bentonite mixtures are often used for constructing hydraulic barriers. This paper studies the valorization of local materials containing calcareous sand, tuff obtained from Laghouat region, to associate with bentonite in order to improve their hydraulic characteristics for use as landfill liner material. Firstly, a geotechnical characterization of mixtures chooses from a fixed percentage to 10% bentonite and daren't percentages of calcareous sand and tuff so that they are complementary to 90% (by not 10%). Thereafter, the hydraulic characteristics study was conducted through free swelling test and the oedometric (K_{id}, indirect measure) test of all compacted mixtures have been carried out using both permeates by tap water and a landfill leachate in order to simulate long-term conditions. The results showed that the saturated hydraulic conductivity of landfill leachate is relatively lower than the one saturated by tap water. The $B_{10}CS_{20}T_{70}$ mixture has satisfied the hydraulic conductivity criterion of bottom barriers and relatively stable against chemical attack from leachate (i.e. with water permeated: $G(\%) = 12.21\%$ and k_{id} from 7.10^{-9} to $1.83.10^{-10} < 10^{-9}$m/s; with leachate permeated: $G(\%) = 10.42\%$ and k_{id} from 7.10^{-9} at $1.44.10^{-10} < 10^{-9}$ m/s). Finally, a comparison between direct measurements of the saturated hydraulic conductivity by triaxial (K_d) test and oedometer test (K_{id}) in the range of effective stress applied 100–800 kPa led to propose equations of correlations between these two methods. In conclusion, adopted formulation $B_{10}CS_{20}T_{70}$ perfectly meets the regulatory requirements in force and constitutes an economic product based on available local materials for engineers barriers.

Keywords: Landfill liners · Bentonite (B) · Calcareous sand (CS) · Tuff (T) Hydraulic conductivity

1 Introduction

A voluminous amount of literature in favor of using local materials available locally blended with bentonite for use as a landfill liner. On the other hand, it is to have a local

B. Abdelbaki et al. (Eds.): SMSD 2017, *Proceedings of the Third International Symposium on Materials and Sustainable Development*, pp. 451–464, 2018.
https://doi.org/10.1007/978-3-319-89707-3_51

availability of earthen material for covering the works or for covering the different steps of the landfill. Landfills need a very large volume of such material to cover the waste. For that, it is eventually partially possible to use inert construction materials but also local quarries are necessary to have the required material. These local quarries can be the next cells of the landfill. Accordingly, the different percentages of bentonite added to sand vary depending on the bentonite type: Ca or Na forms (Montañez 2002). Fan et al. (2014) have found this percentage range between 5 and 8% for Na-bentonite and from 9 to 15% for Ca-bentonite. The typical pH of Ca-bentonite suspension ranges from 7 to 8 in contrast to Na-bentonite with pH values ranging from 9 to 10 (Kaufhold et Dohrmann 2008).

Based on the recommendations of the Ministry of Planning, Environment and Tourism (MPET) of Algeria (2013), the saturated hydraulic conductivity must be less than 10^{-9} m/s from the household and hazardous waste landfill bottom.

The Laghouat area geology (south Algeria) is represented by very thick a marl- calcareous rock complex at the base (more than 50% of $CaCo_3$) and a crystalline rock and calcareous and/or gypsum tuff group at the top cover (from about 20 cm to around 2 m). The Tuffs have been exploited as road materials has become increasingly prevalent in the Maghreb countries (Algeria, Morocco, Tunisia and Libya). Consequently, tuff in Algeria covers an area of about 300 000 km^2. In 2009, tuff production reached an exceptional level of about 12 209 345 m^3 (Direction générale des mines, 2009). The exploitation of crushed calcareous sand was about 13 360 241 m^3 in the same year. It is mainly found in arid and semi-arid regions; tuff and calcareous sand (crushed rock waste) present a cheap alternative to sand as an additive to bentonite to be used as a bottom barrier satisfying the recommendations for landfill bottom liners design (Demdoum et al. 2016).

In the compaction of 20% calcareous sand –80% tuff mixture use modified Proctor is benefits a saturated hydraulic conductivity of 6.1×10^{-10} m/s (Goual, 2012). Imaizumi et al. (2009) showed that the hydraulic conductivity at permeameters with flexible walls of the tuff sand–bentonite mixture decreases with the increase in bentonite. For an addition of 10% bentonite, hydraulic conductivity is 5×10^{-10} m/s and it is reduced up to 1×10^{-10} m/s for an addition of 33% bentonite. One of these previous results came the idea of using percentages calcareous sand- tuff mixture. Consequently, swelling potential, hydraulic conductivity and compressibility (C_c, C_v and m_v) of light soil–bentonite mixtures are important parameters in mineral liners designing (Abeele, 1986; Imaizumi et al. 2009; Demdoum et al. 2017). Moreover, chemistry of the pore fluid has a great impact on the hydraulic conductivity of bentonite-based materials. An increase of the chemical concentration of the permeant produces an increase of its hydraulic conductivity (Studds et al. 1998). When salty water is added, the swelling and sealing capacity of the bentonite is compromised (Romero, 2013). Saline solutions reduce the swelling capacity and increased hydraulic conductivity of bentonite (Sun et al. 2013).

The work presented in this paper the refers to valorization of the local material base on the calcareous sand, tuff from Laghouat region and bentonite of careers in the design of the landfill bottom barrier. Bentonite percentage is fixed at 10% and the percentages of calcareous sand and tuff are varied. In order to determine the saturated hydraulic conductivity, oedometric method was used to the effect of two liquid types: tap water and landfill leachate collected from a domestic waste landfill site in Laghouat region.

The scientific approach integrates the physicochemical and hydraulic characterization of these mixtures (10% bentonite + X% calcareous sand + Y% + tuff). The results have led to propose a formulation that meets the regulatory requirements.

2 Materials and Methods

2.1 Materials and Characteristics

Three basic materials have been used in this work. The first material is the bentonite of Maghnia that is extracted from the Hammam Boughrara deposit (Maghnia–Tlemcen). Table 1 summarizes the results of the chemical analyzes. Its main mineral constituents are silica (55–65%) and the alumina (12–19%). The systematic control of the pH of bentonite revealed the basic nature of this material. According to the pH results, bentonite can be classified as Na-bentonite. The bentonite of Maghnia has content a large amount of montmorillonite on the structural formula: $(Na, Ca)_{0.3}(Al, Mg)_2 Si_4O_{10}(OH)_2.nH_2o$ (Fig. 1). The liquid limit is 141%, the plastic limit is 48%, the specific gravity 2.75, and 60% of particles are smaller than 2 μm (Fig. 2). The value obtained for the total specific surface area using the blue methylene technique is 462 m^2/g.

Table 1. Chemical analysis of bentonite Maghnia (Demdoum et al. 2016)

B	SiO_2	Al_2O_3	Na_2O	CaO	K_2O	MgO	Fe_2O_3	PF
%	55–65	12–19	1–3	1–2	0,5–1,18	2–3,4	1–3,4	8,2

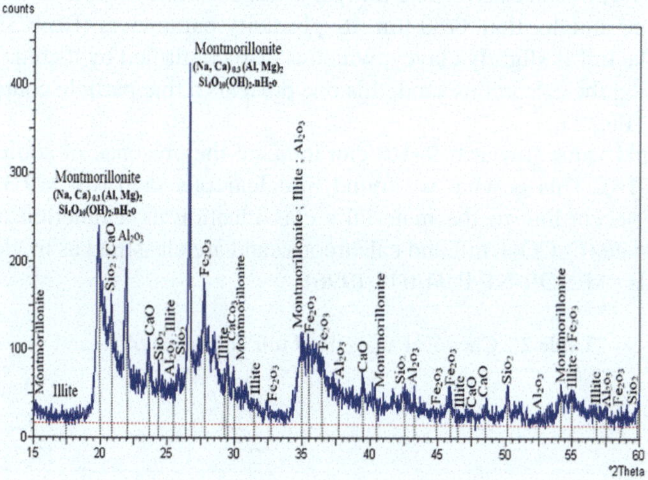

Fig. 1. XRD result for bentonite of Maghnia

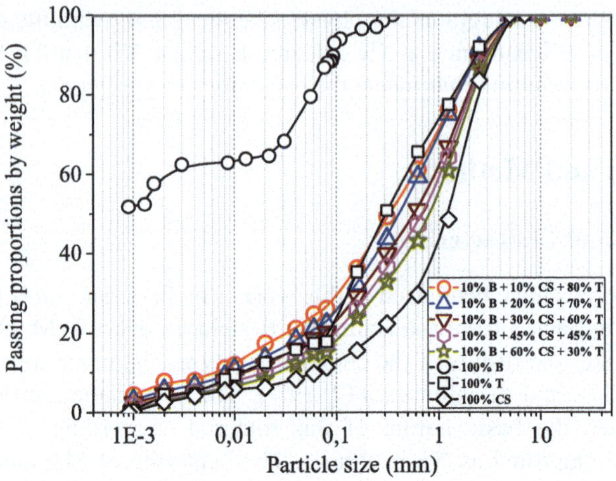

Fig. 2. Grain size distribution curves of all the mixtures, bentonite (B), tuff (T) and calcareous sand (CS)

Tuff and calcareous sand used in this study come from the Laghouat region, located 400 km South of Algiers, in Algeria. The second material is tuff, which is available within the Laghouat region. It is often used in road construction of low traffic. The three materials are the calcareous sand, which is a residue of the crushing stations of calcareous rocks located in the north of Laghouat City. Tuff particles present a maximum diameter (D_{max}) of 5 mm with sand fraction (<2 mm) of about 85%, and more than 18% of particles have a diameter smaller than 0/80 μm. Its plasticity parameters (LL = 33%, PI = 8%) indicate that the tuff is slightly clayey, which is also confirmed by its blue value (VB = 0.8). Concerning the calcareous sand, this one presents a fine particle content (<80 μm) of about 11% (Fig. 1).

A higher pH value (towards 9–10) can indicate the presence of sodium carbonate (Fan et al. 2014). This is what we found in calcareous sand (pH = 9.32) (Table 2). Consequently, according to the material's classification as a function to the weight percent carbonate ($CaCO_3$), tuff and calcareous sand are classified as moderately calcareous materials (AFNOR NF P 94-048 1996).

Table 2. Chemical analysis of tuff and calcareous sand

Elements	Tuff	Calcareous sand
PH	7.64	9.32
Insoluble (%)	27.7	17.2
$CaCo_3$ (%)	66.00	55.00
$CaSo_4.2H_2o$ (%)	0.46	0.52
CO_2 Aggressive (mg/kg)	56.00	Traces
Classification of aggressive environments according to the NF P 18-011 standard	Weakly aggressive	Weakly aggressive

The materials used are symbolized as follows: Bentonite by B, Calcareous Sand by CS and Tuff by T. Five mixtures were selected for the investigation. For each mixture, the percentage of bentonite is fixed at 10%. The mixtures are named $B_{10}CS_XT_Y$, where X and Y are the percentages of calcareous sand and tuff used complementary to 90%. In this sense, 10% Bentonite + X% Calcareous sand + Y% Tuff mixtures have been formulated with different percentages, and for each formula, the testing of mechanical and physical properties are performed (Figs. 2 and 3). Are used:

- *Mixture 1: 10% B + 10% CS + 80% T,* $B_{10}CS_{10}T_{80}$
- *Mixture 2: 10% B + 20% CS + 70% T,* $B_{10}CS_{20}T_{70}$
- *Mixture 3: 10% B + 30% CS + 60% T,* $B_{10}CS_{30}T_{60}$
- *Mixture 4: 10% B + 45% CS + 45% T,* $B_{10}CS_{45}T_{45}$
- *Mixture 5: 10% B + 60% CS + 30% T.* $B_{10}CS_{60}T_{30}$

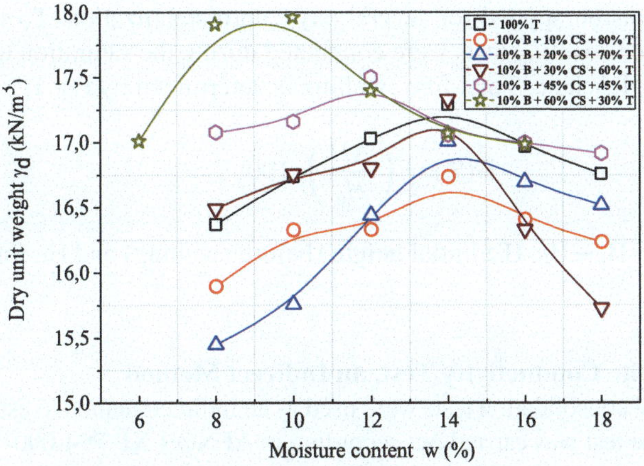

Fig. 3. Normal proctor curves of all the mixture

According to USCS classification, the Mixtures can be classified as silty sands (*Cu* > 6, $\Phi_{80\mu m} > 12\%$ *et* $\Phi_{2mm} < 50\%$) (Fig. 2). Figure 3 shows that the incorporation of calcareous sand improved the material by increasing its maximum dry density (γ_{dmax}) of 16.625 to 17,825 kN/m^3 and reducing its optimal content (w_{opt}) of 14.4% to 9%.

Two types of liquid were used in this study (1) tap water; (2) Leachate collected from a domestic waste landfill in Laghouat region that is in operation for 6 years. The samples were conserved in the icebox, taken to the university laboratory and stored at 4 °C. This landfill leachate is a saline liquid (ph = 8.04).

2.2 Experimental Procedures

The most important mechanisms for contaminants transport that typically need to be considered for modern facilities are permeability and diffusion. Permeability in landfill liners is a physical process whereby contaminants introduced into a groundwater flow system migrate in solution (as solutes) or in suspension (e.g. silt and fine sand particles, microbes) along with the movement of leachate or groundwater. It is governed by Darcy's Law, with the Darcy flux. In order to simulate the real site, we experimented with the following:

2.2.1 Free Swelling Tests

The swelling of clay may also narrow water-conducting pores in soil (Ishiguro 2010). In this sense, the free swelling tests are carried out using a classic oedometer on samples of the $B_{10}CS_XT_Y$ mixtures and each material separately, whose sizes are respectively: Height H = 20 mm, diameter d = 50 mm. This free swelling test presented by Cowland in 1990 shows a good correlation with the hydraulic conductivity tests. For the first time, an ad for each mixture, samples were preparatory by a static compaction at the Optimum Normal Proctor moisture content. The sample is placed in a cylindrical cell between two porous stones. Then, during the saturation, it is authorized to swell under vertical the pressure of the piston applied for several days, until stabilization. To see the effect of leachates, the free swelling tests were conducted during the saturation using two liquid types: water and leachate. Total free swelling is determined by Eq. 1:

$$G(\%) = \left(\frac{\Delta H}{H_0}\right)x100 \tag{1}$$

Where $\Delta H = H_f - H_0$; H_0: initial height (before swelling) and H_f: final height (after swelling).

2.2.2 Hydraulic Conductivity Test, an Indirect Method

One-dimensional consolidation tests were used as an indirect method to estimate hydraulic conductivity. The test was carried out according to AFNOR XP P94-090-1 standard. Pros of this experience compared to triaxial cells is that it offers the best means for quantitatively assessing the effect of chemical interaction on the permeability of clay soils (Madsen 1994). In order to prepare the mixture samples, the same procedures to measure the free swelling are considered to measure the saturated hydraulic conductivity, keeping the same characteristics (height, diameter, dry density and water content). Then entire assembly was placed in a consolidation cell and positioned in the loading frame and inundated with water or respective leachate under a nominal pressure of 5 kPa and allowed to swell. Once the swelling was completed, the sample is subject to increments of successive and increasing loads following by a geometric progression $\sigma_{i+1}/\sigma_i = 2$.

The hydraulic conductivity (k_{id}) is obtained from both the coefficient of consolidation C_V (m^2/s) evaluated by Taylor's approach and the coefficient of volume compressibility m_v (m^2/KN) (see Eq. (2)). In this method, the coefficients C_V and m_v are deduced from compressibility curves and consolidation curves, respectively, in order to obtain the hydraulic conductivity. In this study, the hydraulic conductivity k is written as:

$$k_{id} = C_v \times m_v \times \gamma_w \tag{2}$$

2.2.3 Hydraulic Conductivity Using Triaxial Test (Direct Method Test: Flexible Wall)

The directly saturated hydraulic conductivity with a flexible wall is carried out by a triaxial Plexiglas cell of the Controls Triax-50 type, controlled using two controllers pressure-volume GDS. The triaxial cell permits applying isotropic confining stress ($\sigma_3 \leq 1700$ kPa). The flow is done vertically (from bottom to top). The specimen is prepared by static compaction with a double piston at optimum Normal Proctor condition (w_{opt} and γ_{dmax}). Sample dimensions are: H = 78 mm; d = 38 mm. The displacement speed of the press is about 1.27 mm/s. After the establishment of the sample, a low pressure is applied to the base $u_e = 20$ kPa whose aim is to eliminate air bubbles between the membrane and in the sample (Fig. 4). Then, a confining stress total (σ_3) equal to 100 kPa is applied to the sample, and water pressures of $u_e = 40$ kPa and $u_s = 0$ kPa are applied respectively to the base (u_e), and the head (u_s) of the sample for 24 h. This corresponds to a pressure gradient i = 76.9. This method may be used to increase the degree of saturation from low values in soils that are relatively permeable (Percolation with water) (Lada 2016). A visual confirmation of the release of air bubbles being necessary. It is estimated that the sample is saturated as soon as it gets the slope is constant and the entering and outgoing flow are equal. Nevertheless, the vertical deformation of the samples was recorded versus time and the samples were considered as saturated when the displacement of the piston became constant (less than 0,01 mm in 24 h) (Souli et al. 2008). After the saturation phase, the samples were consolidated according to an experimental program in order to study the effect of effective stresses which vary from 100 to 800 kPa with a constant hydraulic gradient ($\Delta U = 100$ kPa, i = 128.2). Saturated hydraulic conductivity of the mixtures is based on Darcy's law. It is expressed by Eq. 8:

$$v = \frac{Q}{A} = k.i \tag{3}$$

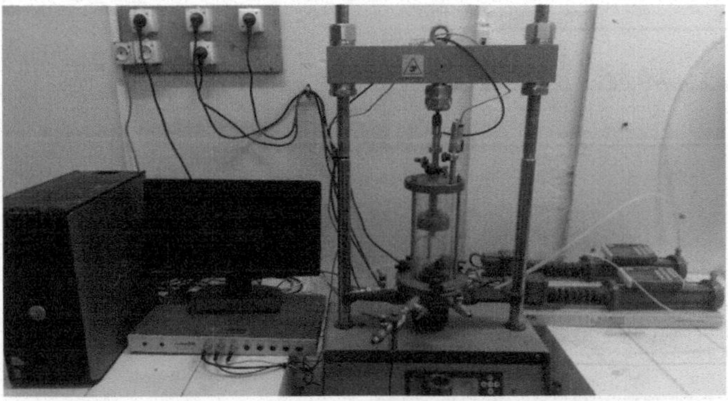

Fig. 4. Schematic diagram of the experimental set-up.

Where, v: is the Darcy flux (m/s) Q: is the volumetric flow rate [m^3/s]; A: is the flow area [m^2]; i: is the hydraulic gradient, which can be expressed as ($\Delta U/H.\gamma_w$), with Δu: is the difference of pressure ($u_e - u_s$) [kPa]; the weight of water [kN/m^3]; H: height of the sample [= 7.8.10^{-2} m].

3 Results and Discussions

3.1 Free Swelling Results

The free swelling was determined by both saturation cases for water (Fig. 5a) and leachate (Fig. 5b). According to these curves, there are two phases of swelling: primary swelling that takes place very early; its end is characterized by an inflection point, permitting the change of the curve allure towards a secondary swelling, which is characterized by the curve linear portion. The primary swelling occurs during the hydration of dry bentonite particles, water is absorbed in successive monolayers on the surface and pushes the particle or the unit layers of the montmorillonite apart (referred to as interlayers or intercrystalline swelling) (Dakshanamurthy 1978). The secondary swelling known as microscopic swelling is due to double-layer repulsion, which affected by the type of minerals present in the interstitial fluid, and it is influenced by the confining pressure (Sivapullaiah et al. 1996). Moreover, the variations of primary swelling with water and leachate solution for all the mixtures are generally less than log t = 20 min

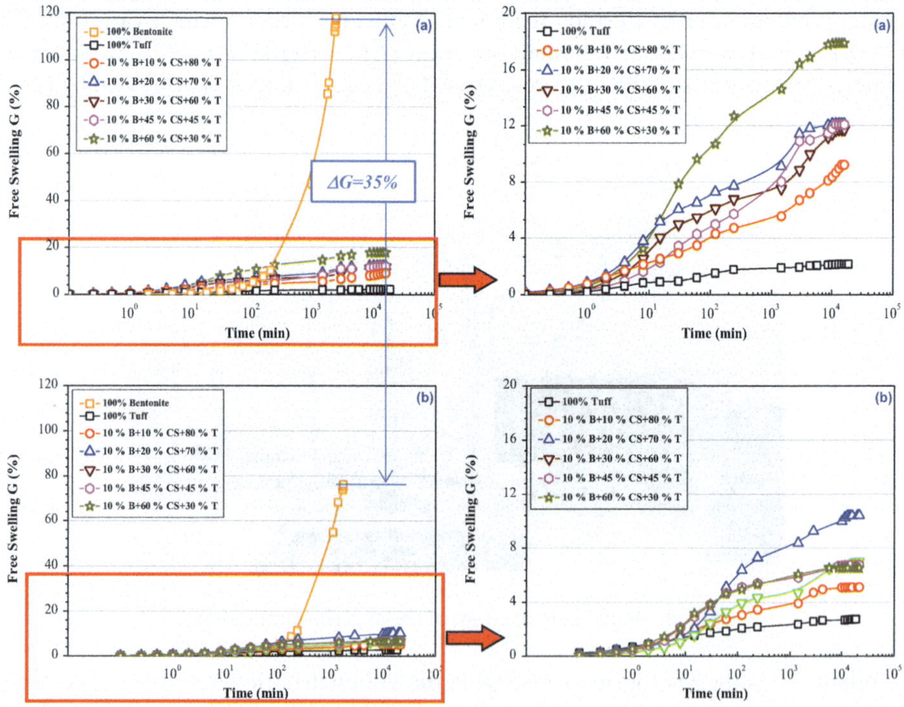

Fig. 5. Free swelling behavior versus time of the $B_{10}CS_XT_Y$ mixtures in (a) saturation case bytap water and (b) saturation case by landfill leachate

and log t = 40 min, respectively. The difference in the maximum swelling values of tap water for various mixtures this may be due to the fabric of calcareous sand (angular sand) and particle rearranging during swelling. For calcareous sand, whose shape is relatively angular, the effect of a rearrangement of particles is very significant (distribution and size of pores). Thus, the bentonite swelling is highly dependent on the fabric as the size and shape of the non-swelling fractions.

The total free swelling tendency varies with time is the same for all mixtures (approximately 12 days). Comparatively, free swell values with tap water for all mixtures were higher than the values obtained with the leachate. Maximum free swelling was, 0.1, 2.16 and 118.25, respectively, for calcareous sand, tuff and bentonite when the hydration liquid by tap water and 0.1, 2.7 and 76.5, respectively upon saturation with landfill leachate (Fig. 5). This allows concluding that calcareous is a non-swelling material (inert) (by tap water or leachate). For the tuff, the free swelling value obtained by leachate and by tap water is similar. It has a low swelling rate of the same values for the two cases, which makes it possible to conclude that the liquid types do not affect the tuff. Concerning bentonite, according to Fig. 5a, b, it can be seen that the ratio between the two swelling rates of water and landfill leachate is about 35%. The free swells redact of bentonite in, saturated by leachate is due to the chemical composition of landfill leachate. Since chemical compositions were determined in terms of cation concentrations, that landfill leachate contains a considerable amount of, Ca^{2+}, Mg^{2+}, Na^{+}, and NH_4^+ (Ruhl 1994).

The maximum values of water saturated swelling rates correspond to $B_{10}CS_{20}T_{70}$ and $B_{10}CS_{60}T_{30}$ mixtures with 12.20% and 18%, respectively, and for the maximum values of the leachate saturated swelling rate corresponds to $B_{10}CS_{20}T_{70}$ and $B_{10}CS_{30}T_{60}$ mixtures with 10.42 and 7.05%, respectively. For the same mixture, leachate leads to a decrease in the swelling rate varying from 40 to 60% except for $B_{10}CS_{20}T_{70}$ mixture, where the reduction is about 14%. Moreover, for pure bentonite by saline saturated solutions, several researchers show that free swelling decreases with the increase of the saline concentration (Sun et al. 2013; Dutta and Mishra 2016). Accordingly, for all the mixtures are relatively unstable against chemical attack from leachate except the $B_{10}CS_{20}T_{70}$ mixture. Mishar et al. (2011) indicated that the hydraulic conductivity is inversely proportional to the exchangeable sodium percentage (ESP) and the free swelling bentonite. Therefore, there is a possibility that the mixtures, which provides low permeability value, are mixtures continent 20%CS and 60%CS.

3.2 Hydraulic Conductivity Results, an Indirect Method

The results of the indirect saturated hydraulic conductivity (K_{id}) of the $B_{10}CS_XT_Y$ mixtures based on the vertical applied stresses in both saturation cases: by tap water (Fig. 6a) and by landfill leachate (Fig. 6b) are presented graphically in Fig. 6.

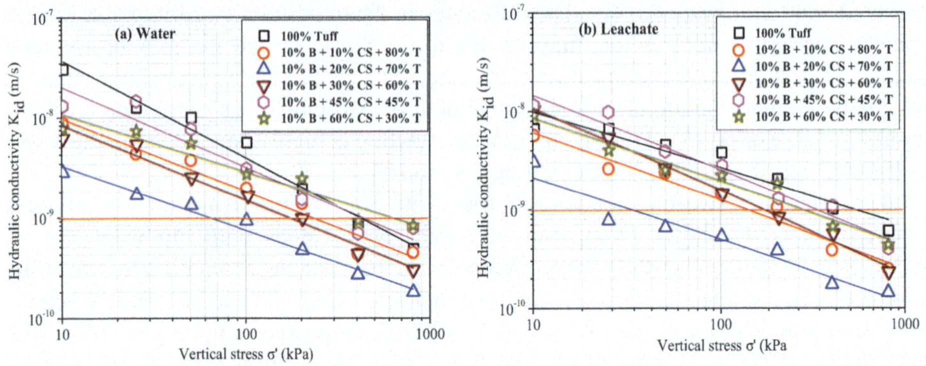

Fig. 6. Indirect hydraulic conductivity versus vertical stress of the $B_{10}CS_XT_Y$ mixtures using (a) tap water and (b) landfill leachate

The result of indirect hydraulic conductivity above shows that:

1. For all the mixtures and in the two saturation cases, indirect hydraulic conductivity (K_{id}) decreases with applied vertical stress increase (Demdoum et al. 2016). The hydraulic conductivity is generally expressed in two different orders of magnitude (10^{-10} and 10^{-9} m/s);
2. For vertical stress above 100 kPa and for all the mixtures, the indirect hydraulic conductivity values by landfill leachate are slightly lower than the values by tap water. These results are in agreement with those found by, Li et al. (2015);
3. Generally, the indirect hydraulic conductivity by leachate is more similar than that found in the tap water case. Ordinarily, pores clogging mechanisms could also restrict the leachate displacement inside mixture pores. Because the microorganisms are gradually adsorbed on the soil surface (Li et al. 2013). This similarity in indirect hydraulic conductivity values can be explained by the difficulty of transmission of the microorganisms in the void spaces, which decreases with each consolidation pressure increase (i.e. pore clogging) (VanGulck and Rowe 2004; Francisca and Glatstein 2010).
4. The mixtures which satisfied the regulatory requirements of the hydraulic conductivity of bottom barriers ($k < 10^{-9}$ m/s) are $B_{10}CS_{20}T_{70}$ and $B_{10}CS_{30}T_{60}$;
5. In both saturation cases, the $B_{10}CS_{20}T_{70}$ mixture presents the minimum indirect hydraulic conductivity values. In tap water case, indirect hydraulic conductivity decreases from by 7.10^{-9} to $1.83.10^{-10}$ m/s, for the vertical stresses from 10 to 800 kPa. For landfill leachate, and for the same vertical stress, hydraulic conductivity decreases from 7.10^{-9} to $1.44.10^{-10}$ m/s. The $B_{10}CS_{30}T_{60}$ mixture and in both cases presents saturated hydraulic conductivity about 2.10^{-10} m/s for vertically applied stress of 800 kPa.

3.3 Hydraulic Conductivity Results, an Direct Method

Figure 7 present the variation of direct hydraulic conductivity to a flexible wall (k_d) according to effective stress (σ_3' for the $B_{10}CS_XT_Y$ mixtures. For all the mixtures, direct hydraulic conductivity decreases with the increase of effective stress. These results are consistent with those found by Gueddouda et al. 2016. Generally, the direct hydraulic conductivity values for all the mixtures meet the basic criterion of landfill bottom liners ($k < 10^{-9}$ m/s) in the range of effective stress applied from 100 to 800 kPa. The $B_{10}CS_{10}T_{80}$ mixture presented the minimum value of the direct hydraulic conductivity (k_d). Its k_d value decreases by 4.10^{-10} m/s to 2.10^{-10} m/s. For all the mixtures, hydraulic conductivity decreases with the decrease in the tuff percent, except for the $B_{10}CS_{60}T_{30}$ mixture. In other words, the direct hydraulic conductivity values increase when the percentage of calcareous sand rises to 45% where $k_d = 3.3.10^{-10}$ m/s. Beyond this percentage, the hydraulic conductivity decreases in the case of 60% calcareous sand where $k = 2.10^{-10}$ m/s. These results are consistent with those found on sand-bentonite, tuff sand - bentonite and clay-sand mixtures (Imaizumi et al. 2009; Gueddouda et al. 2016).

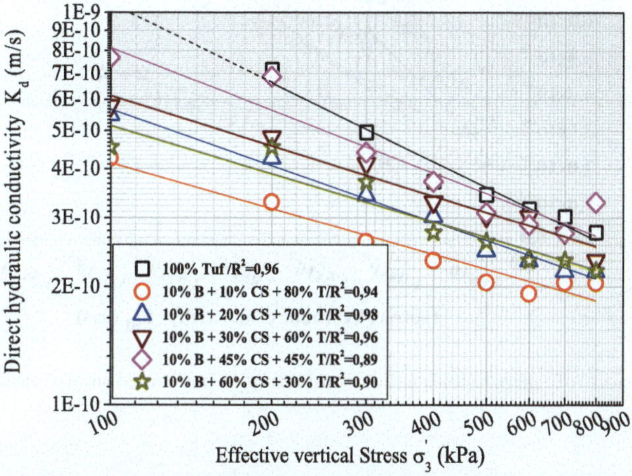

Fig. 7. Direct hydraulic conductivity versus effective stress of the $B_{10}CS_XT_Y$ mixtures

The indirect hydraulic conductivity test results were used to establish a correlation to the directly measured hydraulic conductivity at different effective stresses. The actual hydraulic conductivity is assumed to be similar to that obtained in triaxial tests set out to simulate actual field conditions. A comparison plot between direct and indirect hydraulic conductivity measurements is presented in Fig. 8. The ratio between them reached 5.10^{-1}. Moreover, triaxial tests are difficult compared with oedometer tests. Six linear equations have been established for effective stresses of 100, 200, 400 and 800 kPa. These equations will help to reduce the number of triaxial or flexible wall tests to be carried out. These equations can be used for 10% bentonite and different percentages of calcareous sand and tuff for different mixture proportions.

$$\text{For 100\% Tuff:}k_d = 1,488x10^{-11} + 0,411x(k_{id}) \tag{4}$$

$$\text{For } B_{10}CS_{10}T_{80}\text{mixture :}k_{d10\%} = 1,60x10^{-10} + 0,128x(k_{id}) \tag{5}$$

$$\text{For } B_{10}CS_{20}T_{70}\text{ mixture :}k_{d20\%} = 1,81x10^{-10} + 0,41x(k_{id}) \tag{6}$$

$$\text{For } B_{10}CS_{30}T_{60}\text{mixture:}k_{d30\%} = 3,11x10^{-10} + 0,163x(k_{id}) \tag{7}$$

$$\text{For } B_{10}CS_{45}T_{45}\text{mixture:}k_{d45\%} = 2,69x10^{-10} + 0,1749x(k_{id}) \tag{8}$$

$$\text{For } B_{10}CS_{60}T_{30}\text{mixture:}k_{d60\%} = 1,435x10^{-10} + 0,118x(k_{id}) \tag{9}$$

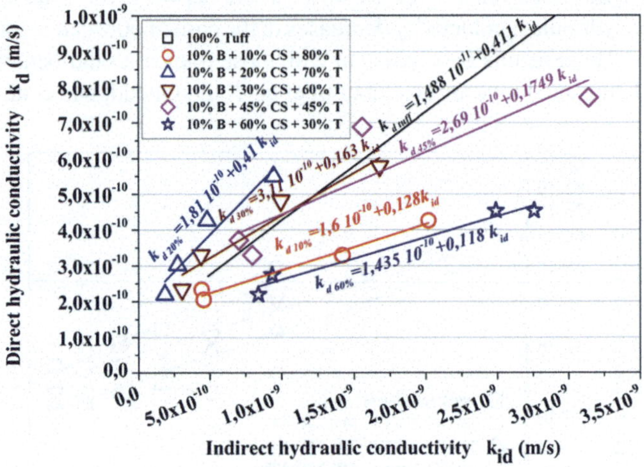

Fig. 8. Correlation of direct hydraulic conductivity from oedometer tests for different confinement stresses

4 Conclusions

This study considered a valorization of local materials for use as landfill bottom liners containing calcareous sand, tuff and bentonite. It has essentially environmental and economic benefits. Several conclusions can be made as a result of this study:

i. For some formulation, leachate cause a reduction in the free swell more than 40% to that of the water;
ii. The $B_{10}CS_{20}T_{70}$ mixture is relatively stable against chemical attack from leachate ($G_{water} = 12.20\%$ and $G_{Leach} = 10.425\%$ for 12 days of saturation);
iii. The leachate cause a slightly decreased in hydraulic conductivity obtained by the oedometer tests;

iv. The optimal mixture that satisfies hydraulic conductivity criteria (indirect method) is the $B_{10}CS_{20}T_{70}$ mixture in both cases of saturation with water ($k_{id} = 1.83.10^{-10}$ m/s) and with leachate ($k_{id} = 1.44.10^{-10}$ m/s);
v. In the case of triaxial tests with flexible walls, the saturated hydraulic conductivity values by water (direct method) show that all mixtures satisfy the hydraulic conductivity criterion ($k < 10^{-9}$ m/s) and the optimal mixture which presents the minimum hydraulic conductivity value is the $B_{10}CS_{10}T_{80}$ mixture with $k_d = 2.06.10^{-10}$ m/s;
vi. The saturated hydraulic conductivity obtained by the indirect tests are slightly higher than that obtained by direct testing. The correlation equations between these two methods permit to avoid the use of triaxial tests that are practically difficult compared to those of oedometer tests.
vii. The state of confinement and effective stresses were found to influence the hydraulic conductivity.

Finally, the adopted formulation of $B_{10}CS_{20}T_{70}$ perfectly meets the requirements of the regulations in force and is an economic product of local materials available in abundance for the design of bottom barriers especially in arid and semi-arid areas.

References

Abeele, W.V.: The influence of Bentonite on the permeability of sandy silt. Nucl. Chem. Waste Manag. **6**, 81–88 (1986)

AFNOR NF P 94-048: Sols: reconnaissance et essais - Détermination de la teneur en carbonate - Méthode du calcimètre. Paris (1996)

AFNOR XP P94-090-1: Sols: reconnaissance et essais - Essai oedométrique - Partie 1: essai de compressibilité sur matériaux fins quasi saturés avec chargement par paliers (1997)

Cowland, J.W., Leung, B.N.: Bentonite landfill liner. Waste Manage. Res. **9**, 277–291 (1990)

Dakshanamurthy, V.: A new method to predict swelling using a hyperbolic equation. Geotech. Eng. **14**, 79–87 (1978)

Demdoum, A., Gueddouda, M.K., Goual, I., Benabed, B.: Geotechnical characterization of geomaterial blends with bentonite of Maghnia for use as landfill liners. J. Materials Process. Environ. **4**, 48–54 (2016)

Demdoum, A., Gueddouda, M.K., Goual, I.: Effect of water and leachate on hydraulic behavior of compacted bentonite, calcareous sand and tuff mixtures for use as landfill liners. Geotech. Geol. Eng. **2017**(35), 2677–2696 (2017)

Direction generale des mines: Bilan des activites minieres, Edition 2010 (2009)

Dutta, J., Mishra, A.K.: Consolidation behaviour of bentonites in the presence of salt solutions. Appl. Clay Sci. **120**, 61–69 (2016)

Fan, R.D., Du, Y.J., Liu, S.Y., Chen, Z.B.: Compressibility and hydraulic conductivity of sand/clay–bentonite backfills. In: Reddy, K.R., Shen, S.J. (eds.) Geo-Shanghai 2014: Geoenvironmental Engineering ASCE GSP, vol. 241, pp. 21–30. ASCE, Reston (2014)

Francisca, F., Glatstein, D.A.: Long term hydraulic conductivity of compacted soils permeated with landfill leachate. Appl. Clay Sci. **2010**(49), 187–193 (2010)

Goual, I.: Comportement mécanique et hydrique d'un mélange de tuf et de sable calcaire de la région de Laghouat: Ph.D. Thesis. University of Telmcen, Algérie (2012)

Gueddouda, M.K., Goual, I., Benabed, B., et al.: Hydraulic properties of dune sand-bentonite mixtures of watertight barriers for hazardous waste facilities. J. Rock Mechan. Geotech. Eng. **8**(4), 541–550 (2016)
Imaizumi, S., Wataru, M., Takahashi, N., Doi, Y.: Estimation of critical ratio of bentonite addition to tuff sand. J. Mater. Cycles Waste Manage. **11**, 299–304 (2009)
Ishiguro, M.: Soil permeability as affected by repulsive potential energy. In: 19th World Congress of Soil Science, Soil Solutions for a Changing World, Brisbane, Australia, 1–6 August 2010
Kaufhold, S., Dohrmann, R.: The pH of aqueous bentonite suspensions. Clays Clay Miner. **56**, 338–343 (2008)
Lade, P.V.: Triaxial Testing of Soils. Wiley, West Sussex (2016)
Li, B., Li, L., Grace, J.R.: Adsorption and hydraulic conductivity of landfill-leachate perfluorinated compounds in bentonite barrier mixtures. J. Environ. Manage. **156**, 236–243 (2015)
Li, J.S., Xue, Q., Wang, P.: Influence of leachate pollution on mechanical properties of compacted clay: a case study on behaviors and mechanisms. Eng. Geol. **167**, 128–133 (2013)
Madsen, F.T.: Clays and synthetic liners-durability against pollutant attack. In: Proceedings of the 13th International Conference on Soil Mechanics and Foundation Engineering, New Delhi, vol. 5, pp. 287–288 (1994)
Ministry of Planning, Environment and Tourism (MPET) of Algeria: Technicians Guide for the management of municipal waste and assimilated (2013). Available: http://www.mate.gov.dz
Mishar, A.K., Masami, O., Loretta, Y.L., Takahiro, H., Junboum, P.: Controlling factors of the swelling of various bentonites and their correlations with the hydraulic conductivity of soil-bentonite mixtures. Appl. Clay Sci. **52**, 78–84 (2011)
Montanez, J.E.C.: Suction and volume changes of compacted sand-bentonite mixtures. Ph.D. Thesis, University of London, London (2002)
Romero, E.: A microstructural insight into compacted clayey soils and their hydraulic properties. Eng. Geol. **165**, 3–19 (2013)
Ruhl, J.L.: Effects of leachates on the hydraulic conductivity of GCLs, M.S. Thesis, Texas Univ., Austin, p 226 (1994)
Sivapullaiah, P.V., Sridharan, A., Stalin, V.K.: Swelling behaviour of soil bentonite mixtures. Can. Geotech. J. **33**(5), 808–814 (1996)
Souli, H., Fleureau, J.M., Trabelsi, A.M., Besnard, M.: Physicochemical analysis of permeability changes in the presence of zinc. Geoderma **145**(1–2), 1–7 (2008)
Studds, P., Stewart, D., Cousens, T.: The effects of salt solutions on the properties of bentonite-sand mixtures. Clay Miner. **33**(4), 651–660 (1998)
Sun, D., Zhang, J., Zhang, J., Zhang, L.: Swelling characteristics of GMZ bentonite and its mixtures with sand. Appl. Clay. Sci. **83–84**, 224–230 (2013)
VanGulck, J.F., Rowe, R.K.: Evolution of clog formation with time in columns permeated with synthetic landfill leachate. J. Contam. Hydrol. **75**(1), 115–139 (2004)

Application of Factorial Plan for the Recovery of Phosphate Rejects +15 mm by Flotation Method Case of the Mine of Djebel Onk, Tebessa - Algeria

M. Ali[1(✉)], I. Abdelaziz[1], O. H. Malek[2], and M. Mohamed[3]

[1] Mineral Processing and Environment Laboratory, Department of Mines, Faculty of Earth Sciences, The University of Badji Mokhtar, 23000 Annaba, Algeria
ali.messai@hotmail.com

[2] Mining Engineering Research Laboratory, Department of Mines, ENP, 16000 Algiers, Algeria

[3] Environment Laboratory, Department of Mines, Faculty of Science and Technology, The University of Larbi Tebessi, 12000 Tebessa, Algeria

Abstract. The mining complex of Djebel Onk specializes on phosphate mining. Between the faced problems put down in this establishment, the phosphate rejections (+15 mm) considered low-grade issues of the treatment cycle, which represent 20% of sieved product. Actually, it occupies large surfaces in the region, which provoke negatively on the environment under the effects of climate changes (wind, rain, snow). The main objective of this study is to do a chemical characterization to know their composition and to seek the possibility of upgrading them in one hand and to recover the maximum occupied surfaces to protect the environment of the region in the other hand. It was noted according to the chemical analysis that they are rich in P_2O_5 (19–25%), treating this later makes it a new source of phosphate. The proposed method of flotation required the application of a statistical method by using design of experiments, which proved the influence of some chosen factors on (tenor) as a response.

Keywords: Phosphate · Environment · Flotation · Design of experiments Statistical method

1 Introduction

Phosphate is one of much operated minerals in the world because it wide uses, especially in agriculture [1]. In Algeria, the phosphate reserves estimated billions of tons; mostly exist in the mine of Djebel Onk. The mining complex of Djebel Onk plays an important role in the national economic development and foreign trade with a production of 1.5 million of tons of concentrated phosphate per year [2].

During phosphate treatment, the +15 mm rejections constitutes more than 20% in the weight and 21% in P_2O_5 which shows that the current applied method have a great loss of phosphate.

According to [3, 4], the most efficient method to treat carbonated phosphate is opposite flotation, which is influenced by many factors, which are crushing times, flotation pulp density, reactive specific consumption, flotation period, flotation scheme…etc.

B. Abdelbaki et al. (Eds.): SMSD 2017, *Proceedings of the Third International Symposium on Materials and Sustainable Development*, pp. 465–480, 2018.
https://doi.org/10.1007/978-3-319-89707-3_52

Design of experiments (DOE) methods seems to be very useful exploratory tool at many points of material engineering. It is necessary to have a clear view on how the things are to have the processes under costs, used materials, production time and material properties can be optimized. In many cases, there are more than three known input factors in the process. For that reason, designing experiments and using statistical methods are very effective ways how to achieve the main aims in research and business [5–8] (Fig. 1).

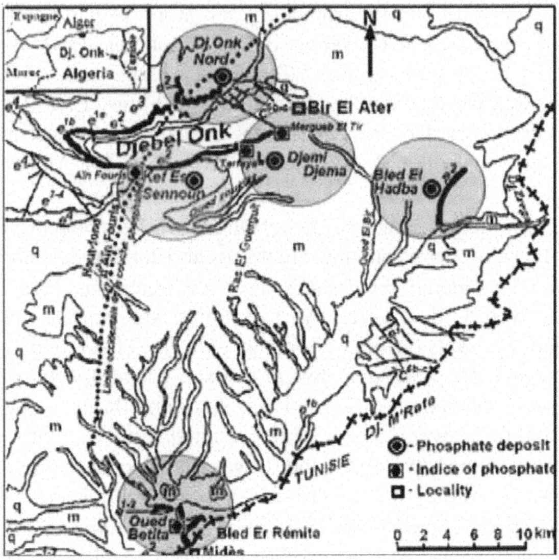

Fig. 1. The geographic situation of Djebel Onk region [9]

The main goal of the present study is therefore to study the influence of chose factors on flotation using design of experiments.

2 Materials and Methods

2.1 Geographic Situation

Djebel Onk region is located on the northwest of Algerian Sahara, specifically in the east of Nemamecha Mountain, on twenty miles of Algero-Tunisian borders; it is far about 100 km of Tebessa state. The Djebel Onk mining complex is remote 7 km from the city of Bir El Ater and 340 km of Annaba port facilities.

2.2 Collect and Preparation of Samples

It base of taking a certain representativeness quantity amount of substance in order to do different analysis (physical, chemical…etc.).

To ensure this representativeness, it is necessary to know the sample minimal mass according to the following formula: [10]

$$Q_{min} = K \times D \times a \tag{1}$$

Q_{min}: sample minimal mass (kg)
K: Proportionality coefficient (0.1 from the experiments)
D: Particles dimension in mm
a: Degree index (1.1–1.2).

The taken quantity is routes to samples preparation laboratory where dried and divided with quartering method until obtaining 1.5–2 kg for sieving and chemical analysis.

2.3 Sieving Analysis

It is carried out with following sieves apertures: +20, +10, +8, +6.3, +5, +4, +2, +1 et −1 mm.

2.4 Chemical Analysis

The purpose of chemical analysis is to determine the various elements which are contains on a the study component.

2.5 Determining of P_2O_5 Tenor

It based using automatic spectrophotometer method. It allows determining of P_2O_5 allows finding out Tricalcareous Phosphate tenor (TPL), whether:

$$\%P205 \times 2.185 = \%TPL \tag{2}$$

2.6 Determining of MgO Tenor

It based using Atomic Adsorption method with presence of lanthanum oxide or lanthanum chloride.

2.7 Determining of CO_2 Tenor

It based using BERNARD method, which consists in the measurement of CO_2 cleared by hydrochloric acid action (HCl) on calcium carbonate ($CaCO_3$) of a sample with taking account the temperature and pressure in which the operation placed, then calculates the CO_2 proportion according to the following formula:

$$\%CO_2 = V\,CO_2 \times K \quad (3)$$

V: Volume
K: Proportionality coefficient, it calculates with temperature T °C and pressure P mmHg at balance (in our analysis, it is equal to 0.65).

3 Results and Discussion

3.1 Sampling Quantity

$$Q_{min} = 0.1 \times 15^{1.15}; Q_{min} = 2.250\ \text{kg}$$
$$Q_{min} = 2250\ \text{g}$$

The sampling method consists of taking a test quantity of 1 kg each 4 h during 7 days at the +15 mm rejections band (3 test quantities in a single post) (Table 1)

Table 1. Rejections sieve analysis results

Classes (mm)	Weight (g)	Yields (%)		
		$\sum\gamma$	$\sum\gamma\nearrow$	$\sum\gamma\swarrow$
+20	446.05	31.82	100	31.82
+10	474.62	33.86	68.18	65.68
+8	271.99	19.4	34.32	85.08
+6.3	14.19	1.01	14.92	86.09
+5	4.66	0.33	13.91	86.42
+4	5.85	0.42	13.58	86.84
+2	7.46	0.53	13.16	87.37
+1	5.58	0.4	12.63	87.77
0	171.43	12.23	12.23	100
Total	1401.83	100	0	100

3.2 Chemical Analysis

See (Fig. 2).

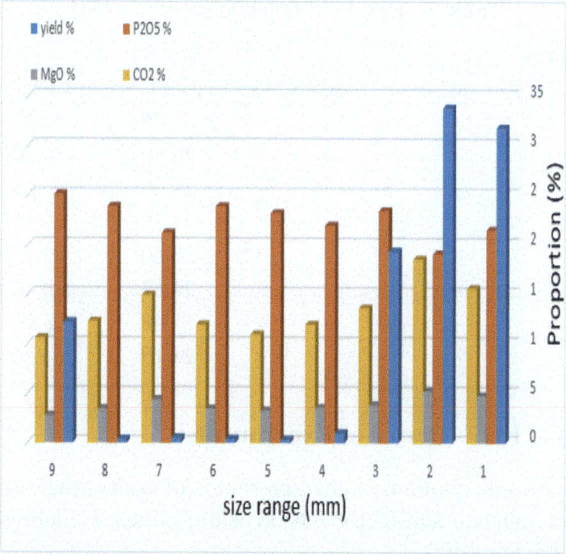

Fig. 2. Histogram of yield by weight, P_2O_5, MgO and CO_2 proportions

3.3 Flotation Procedure

Influential Factors Choosing

According to the bibliographical research [11, 12] and the flotation method application in the world, the factors most influential are:

1st: crushing times (8–14 min)
2nd: Depressor consumption (0.4–0.6 $kg.t^{-1}$)
3rd: collector consumption (1.0–1.8 $kg.t^{-1}$)

During the experiment, the three actors only changes while the other factors must be constant (Ghiani and Saba 1982).

Result of Flotation

The experiments number is in relation with factors number and determined by the following formula:

$$n = q \times k \quad (4)$$

n: Experiments number
q: Level number of each factor (equal to 2)
k: Factors number (equal to 3)
$n = 23 = 8$ experiments (Table 2).

Table 2. Results of flotation procedure [11]

Experiments	Factors			Results		
	A	B	C	β (%)	ɣ (%)	ε (%)
1	+	+	+	30.65	57.42	85.02
2	−	+	+	26.10	75.20	94.82
3	+	−	+	28.60	66.61	91.70
4	−	−	+	25.38	70.61	86.57
5	+	+	−	26.60	70.33	90.37
6	−	+	−	24.62	78.70	92.85
7	+	−	−	25.68	68.96	85.55
8	−	−	−	24.10	64.26	74.81

3.4 Application of DOE (Design of Experiments)

Under certain conditions, qualitative characteristics of concentrate issue of flotation is influenced by three independent factors (A: crushing times, B: depressor consumption and C: collector consumption)

The main design used in this work was the 2-level factorial design. At first, one DOE for the experiments with concentrate tenor, yield and extraction degree. It was concluded that in each of these qualitative characteristics, the operational behavior for a flotation is mainly influenced by three independent factors and their interactions.

In Figs. 3, 4 and 5, there are Pareto charts of these three designs. The same significant level alpha = 0.05 was used. The bar of the factor or interaction, which exceeds the red line in Pareto chart, is significant and the others are non-significant.

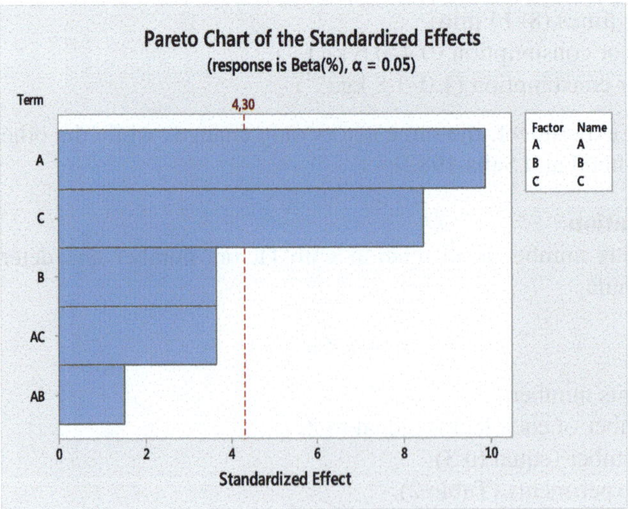

Fig. 3. Pareto chart of standardized effects (concentrate tenor)

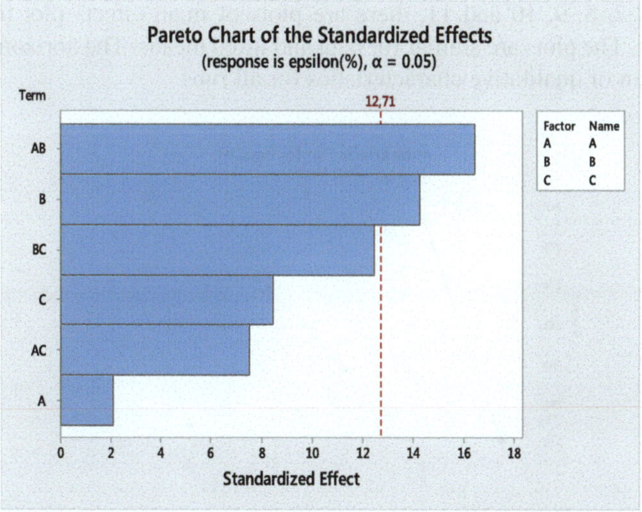

Fig. 4. Pareto chart of standardized effects (yield)

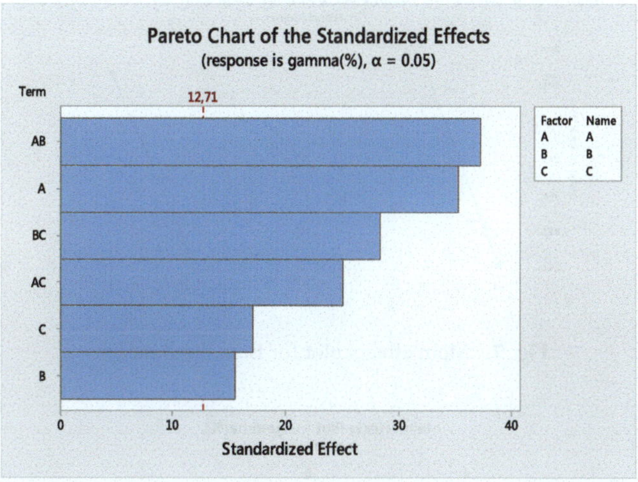

Fig. 5. Pareto chart of standardized effects (extraction degree)

For concentrate tenor which illustrated in Fig. 3, the most significant factors are crushing times and collector consumption. For yield, it is noted that depressor consumption and the interaction between depressor consumption and crushing times are the most significant factors. As for the extraction degree, all factors and Interactions are significant where crushing times and interactions between factors (AB, AC and BC) are the most considerable.

In Figs. 6, 7, 8, 9, 10 and 11, there are plots of main effects plot for qualitative characteristics. The plots are similar for data and fitted means. The horizontal centerline shows the mean of qualitative characteristics for all runs.

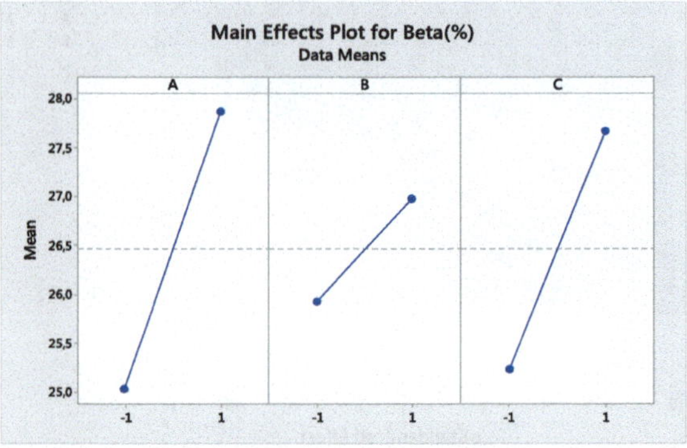

Fig. 6. Main effects plot for Beta data means

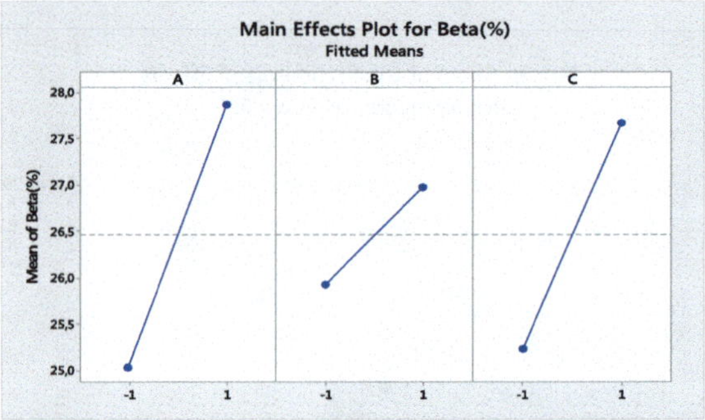

Fig. 7. Main effects plot for Beta fitted means

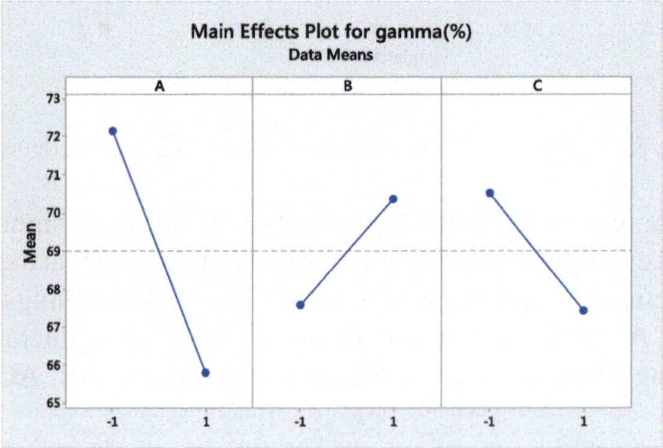

Fig. 8. Main effects plot for Gamma data means

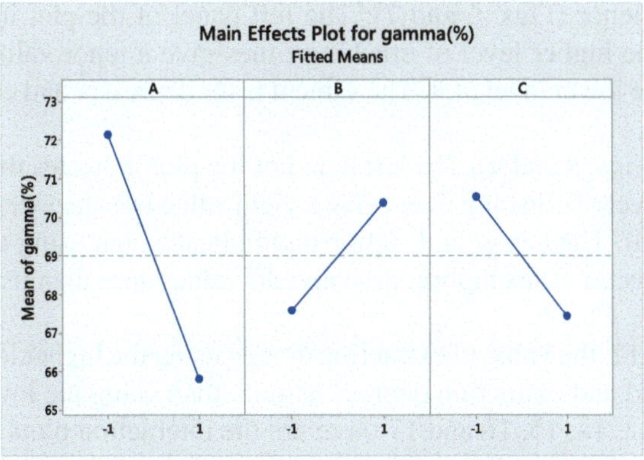

Fig. 9. Main effects plot for Gamma fitted means

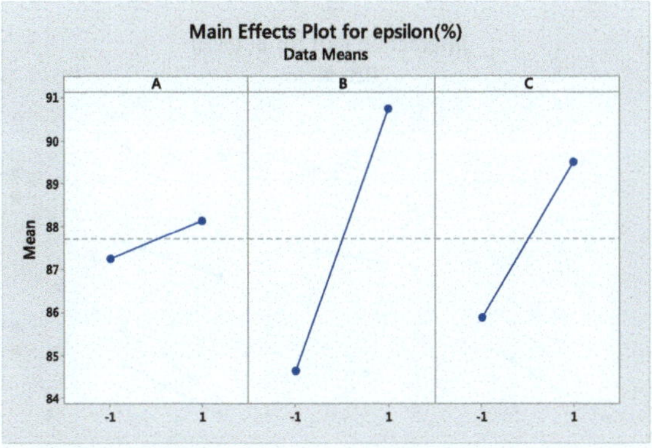

Fig. 10. Main effects plot for Epsilon data means

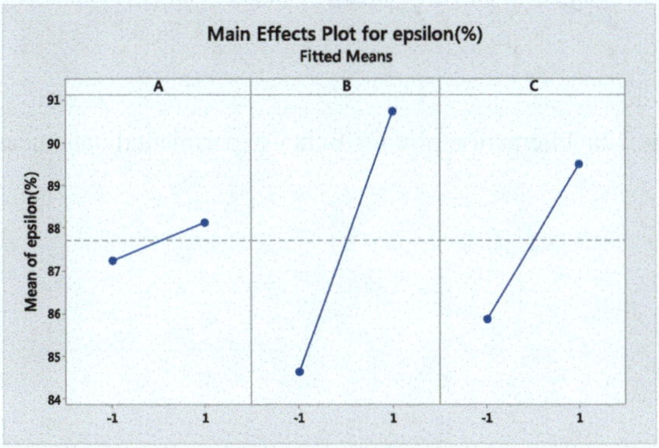

Fig. 11. Main effects plot for Epsilon fitted means

In the case of tenor (Figs. 6 and 7); The left panel of the plot indicates that were processed using the higher level of crushing times give a tenor value more than were processed using the lower level of it. The same it is for depressor and collector consumption.

For the yield (Figs. 8 and 9), The left panel of the plot indicates that were processed using the higher level of crushing times give a yield value less than were processed using the lower level of it. The center and right panel indicates that using the higher level of depressor and collector consumption give a yield value more than using lower level of them.

In Figs. 10 and 11, the value of extraction degree using the higher level of influencing factors (tenor, yield and extraction degree) is more than using the lower level of them.

The Figs. 12, 13, 14, 15, 16 and 17 there are the interaction plots show the mean of qualitative characteristics at different combinations of factor levels. If the lines are not parallel, the plot indicates that there is an interaction between the two factors.

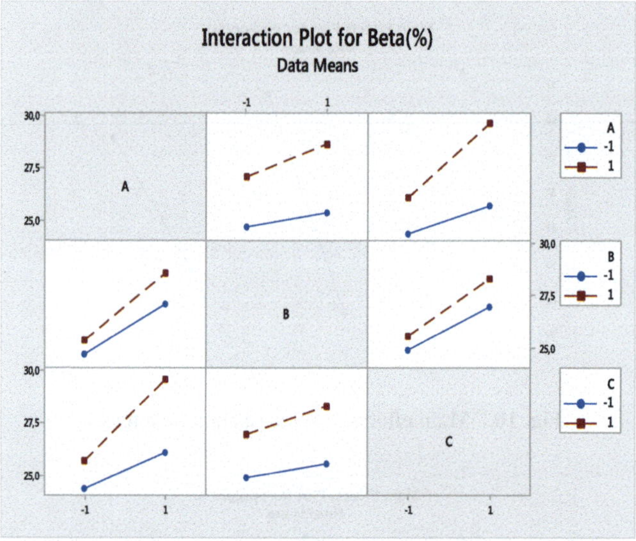

Fig. 12. Interaction plot for Beta - experimental data means

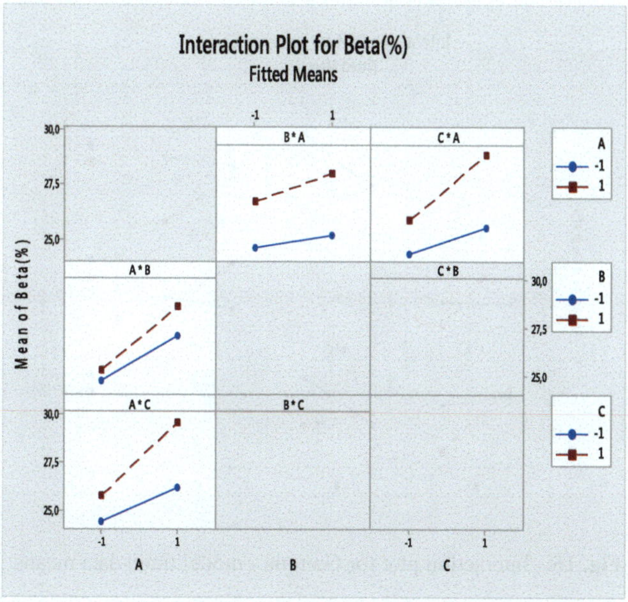

Fig. 13. Interaction plot for Beta - model fitted data means

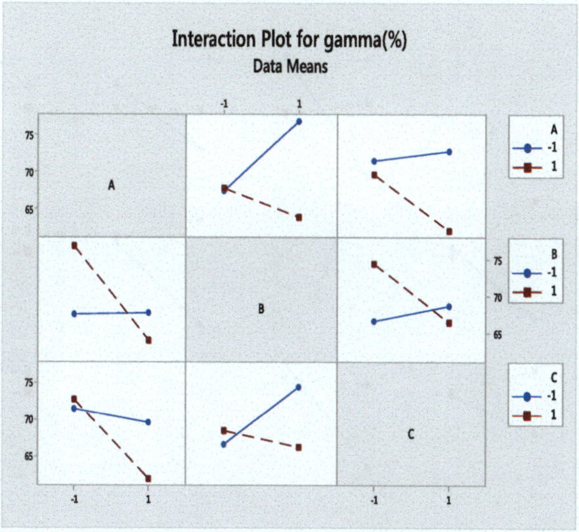

Fig. 14. Interaction plot for Gamma - experimental data means

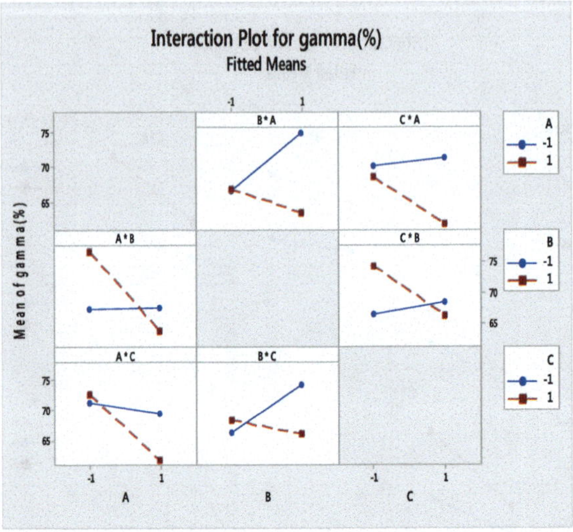

Fig. 15. Interaction plot for Gamma - model fitted data means

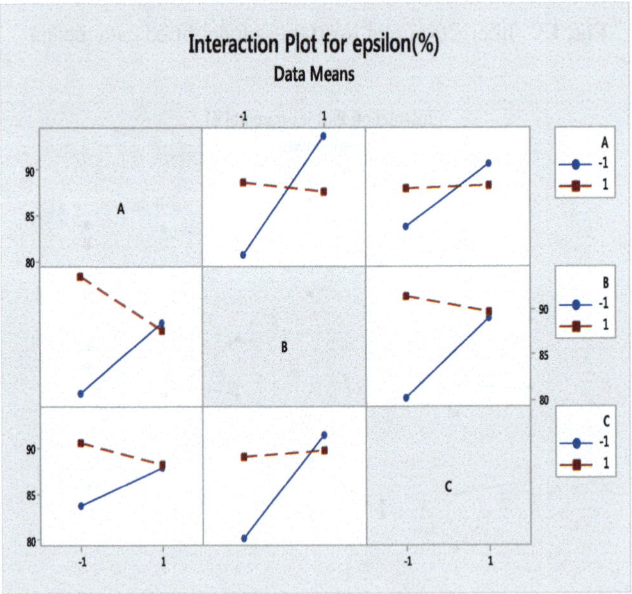

Fig. 16. Interaction plot for Epsilon - experimental data means

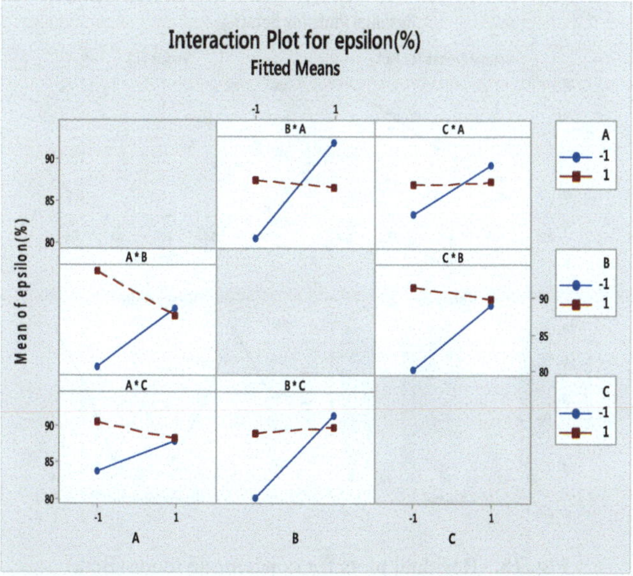

Fig. 17. Interaction plot for Epsilon - experimental fitted means

In the case of interaction plot for tenor (Figs. 12 and 13), there is an interaction between crushing times and collector consumption which mean that each of factors influencing the other factor while interaction plots for yield and extraction degree (Figs. 14, 15, 16 and 17), it seems a significant interactions between each factor with the other factors.

The residual plots (Figs. 18, 19 and 20) contain four different parts; normal probability plot to detect non-normality, the histogram to detect multiple peaks and non-normality, versus fits to detect non-constant variance and outliers, and versus order which detect the independence of the residuals.

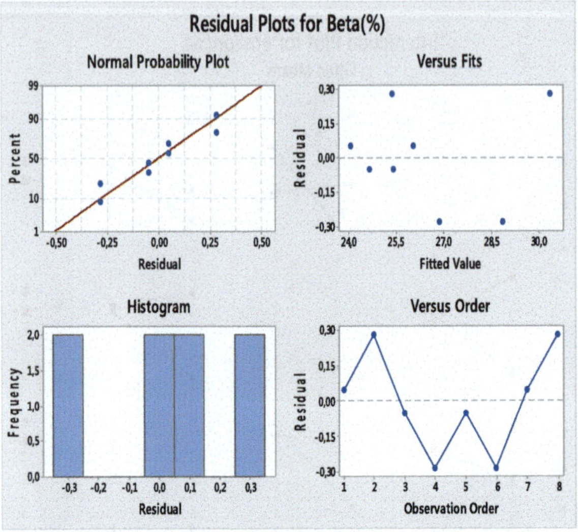

Fig. 18. Residual plots for concentrate tenor (Beta)

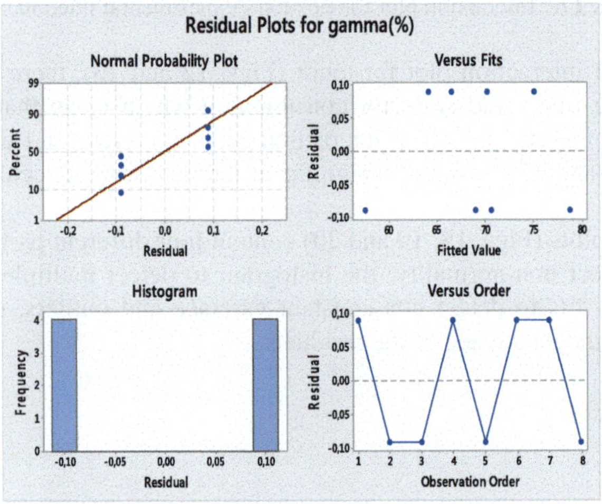

Fig. 19. Residual plots for yield (Gamma)

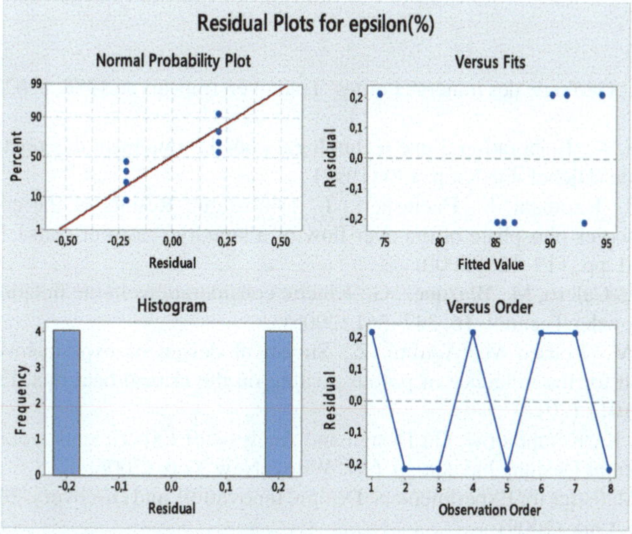

Fig. 20. Residual plots for extraction degree (Epsilon)

Normal probability plots show that the residuals are normally distributes because of points approximately follow a straight line.

The histograms are approximately symmetric and bell-shaped which mean that residuals are normally distributed.

We note in versus fits for the three plots that approximately random pattern of residuals on the both sides of zero, which indicate that our assumption of error terms having meant zero in valid.

Versus order of the three plots does not reveal any pattern, which means that the independence assumption is satisfied.

4 Conclusions

Design Of Experiments (DOE) using Minitab were performed to get better understanding of flotation conditions (choosing factors) in concentrate characteristics (tenor, yield and extraction degree).

This study reveals that, the most influencing factors in concentrate tenor value are crushing times and collector consumption in addition that the choosing of depressor and collector quality and their consumption allow the best recuperation of phosphate (yield and extraction).

This study is worthy of further investigation which allow up more procedure optimization and than going to the pilot phase for check the validity of obtained results.

References

1. Smirnov. V.: Géologie des minéraux utiles. Traduit en français en 1988, p. 623. Mir, Moscou (1982)
2. Bouzenzana, A.: Elaboration d'une technologie pour le traitement des rejets du phosphate dans le cas de (Djebel -ONK), p. 130 (1992)
3. Benamar. Z., Ezzahani. S., Heemsherk, J., Tshibahgu.: Results of the concentration by flotation reserves phosphate of the over-flow of a laundry (basin of Gafsa, Tunisia). Tech. Minerals Ind. pp. 119–128 (2000)
4. Hernainz, F., Calero, M., Blazquez, G.: Kinetic considerations in the flotation of phosphate ore. Adv. Powder Technol. **16**, 347–361 (2005)
5. Mebarkia, M., Louafi, M., Aoulmi, Z.: Statistical design of experiments as a tool for investigation for the influence of porous coating on the critical heat flux. Energy Procedia **119**, 1003–1011 (2017)
6. Hinkelman, K., Kempthorne, O.: Design and Analysis of Experiments, vol. 1. Introduction to Experimental Design, 2nd edn., p. 631. Wiley, New York (2008)
7. Box, G.E.: Statistics for Experimenters: Design, Innovation, and Discovery, 2nd edn., p. 1024. Wiley, New York (2009)
8. Cox, D.R., Reid, N.: The Theory of Design of Experiments, 1st edn., p. 336. Chapman and Hall/CRC, Boca Raton (2000)
9. Mezghache, H.: Détermination, localisation et estimation globales des différents types de minerais de phosphate dans le gisement de Kef Essennoun, Djebel Onk, Algérie. Rapport E.N.FERPHOS, p. 42 (1991)
10. Bouzenzana, A.: Recovery and valorization by flotation of treatment rejections of the phosphates case of Djebel Onk Algeria. J. Appl. Sci. **7**, 2554–2555 (2007)
11. Ghiani M,. Saba P.: The study of the flotation valorization in an acid milieu of the some minerals with a carbonates gangue. In: Round Table of the 11th mining World Congress, Belgrade (1982)
12. Viniti: Valorization of phosphate minerals. Result of sciences and techniques (paper). Tome **18**, 66–69 (1984)

Nanofiltration Performance for Synthetic and Natural Water Defluorination: Application to South-Algeria Groundwater

Amina Ramdani[1,3(✉)], Sana Gassara[2], André Deratani[2], and Safia Taleb[3]

[1] Department of Chemistry - Faculty of Sciences, University Dr. Moulay Tahar, 20000 Saida, Algeria
ramdaniamina010@gmail.com

[2] Institut Européen des Membranes, IEM – UMR5635, ENSCM, CNRS, Univ. Montpellier, Montpellier, France

[3] Laboratory of Materials & Catalysis, Faculty of Exact Sciences, Djillali Liabès University, BP 89, 22000 Sidi Bel Abbès, Algeria

Abstract. Currently, nanofiltration (NF) is gaining increasing importance in water treatment and can replace reverse osmosis especially in the case of brackish water. It is generally appreciated for its low energy consumption, good selectivity and high productivity. The purpose of this study aimed at evaluating the best performance of NF to remove fluoride ions using model solutions in order to treat a South-Algerian groundwater in the optimal conditions. WHO guidelines recommend a limit concentration of 1.5 $mg.L^{-1}$ in drinking water. A commercial NF membrane (NF90) was investigated to study its efficiency in treating synthetic and natural brackish waters in a cross-flow NF membrane process at a pressure range from 3 to 15 bar using a bench scale NF pilot. The membrane performance was determined by the permeation flux and the rejection capacity as a function of trans-membrane pressure. The results showed that productivity and rejection increase with pressure for NF90. The latter parameter leveled off at a trans-membrane a pressure of 11 bar. The defluorination efficiency was about 98% from a synthetic solution with a permeate flux of 64.8 $L/h.m^2$, an F^-concentration of 10^{-3} M (42 mg/L) and pH = 7.2. In the same conditions, NF90 can retain more than 88% of fluoride ions from natural groundwater water (Robbah) at its natural pH (7.85) in the presence of other ions. The F^-content was 0.35 mg/L well below the WHO threshold.

Keywords: Membrane technology · Water treatment · Nanofiltration
Defluorination · Brackish groundwater

1 Introduction

Many countries, more particularly those of the Great Maghreb, have some areas where waters are characterized by very high fluoride content and excessive mineralization (Bejaoui et al. 2014). Algeria is living at present this critical issue especially South-east

B. Abdelbaki et al. (Eds.): SMSD 2017, *Proceedings of the Third International Symposium on Materials and Sustainable Development*, pp. 481–491, 2018.
https://doi.org/10.1007/978-3-319-89707-3_53

Algerian that is affected by fluorosis locally known as 'Darmous' (Ramdani et al. 2015). For instance, this disease impacts nearly 75% of El Oued population (Mameri et al. 1998). Water tables in this area (Robbah, Kouinine, Tiksebt and Bayada) have high concentrations of fluoride ions, between 2 and 5 mg/L (Ahour 1990), largely exceeding that recommended by WHO guidelines and Algerian standards (1.5 mg/L). Faced with this problem, treatment of fluorinated waters is necessary in order to reduce the fluoride levels from groundwater and, consequently, to minimize the consequences of the fluoride impact on the health of consumers. Several studies and reviews have been made describing water defluorination technologies such as coagulation–precipitation, adsorption, ion exchange, electrochemical, and membrane process (Owusu-Agyeman et al. 2017; Mobeen and Kumar 2017). Among these techniques, membrane technology appears to be the most suitable for reducing the fluoride levels in drinking water because of its easy and compact implementation without heavy use of chemicals. This work deals with nanofiltration (NF) for the purpose of treatment of highly fluorinated groundwaters from South Algeria. NF technology uses membranes of nano-scale porosity with low molecular weight cut-offs (200–1000 Da) (Kumar et al. 2013) in contrast to reverse osmosis using dense membranes. NF is generally appreciated for its low energy consumption, good selectivity and high productivity (Mu et al. 2012). NF membranes exhibit selectivity mechanism based on sieving and Donnan exclusion related to the size ratio solute/pore and to electric interaction between solute and membrane charge, respectively (Elazhar et al. 2012). Since pore size in NF is small, trans-membrane pressure needs to be higher than in ultrafiltration (Chakrabortty et al. 2013). Two mass transfer mechanisms are involved in NF filtration: diffusion and convection, the latter being prevalent at high permeate flux (Diawara et al. 2003; Chakrabortty et al. 2013). Most of the commercially available membranes are organic thin film composite consisting of three layers with a non-woven support, an intermediate polymeric porous layer and an ultra-thin filtration top film (Fig. 1). The membrane permeability is related to the thickness of the selective layer and its surface roughness (Mänttäri et al. 2004). Owing to this structure, NF membranes have highwater permeability and salt rejection (Talaeipour et al. 2017).

Fig. 1. Scanning electron microscope observation of the cross-section of the NF90 membrane

Currently, NF is a process of choice for the treatment of surface and brackish water (Diawara 2008). Of great interest for this study, it allows to preferentially remove the fluoride ion while maintaining sufficient mineralized water (partial softening) to be directly drinkable (Diawara et al. 2003; Lhassani et al. 2001). This study focuses on the defluorination of synthetic fluoride solutions with controlled parameters (concentration and pH) and brackish groundwater with salinity of about 3 g/L using a bench scale pilot under identical conditions.

The selectivity of a commercial NF membrane NF90 to the transfer of fluoride ions was examined. This first part allowed us to determine the NF 90 performance parameters of flux and retention with respect to the fluoride concentration and pH. The second part was devoted to the defluorination of samples from Robbah well water by using NF90 membrane in order to achieve quality in accordance with drinking water standard. The comparison between the performances obtained and those expected from synthetic single NaF solute was analyzed in the last section to determine the influence of other solutes present in the natural water on the transfer mechanism and how they can affect the fluoride rejection.

2 Materials and Methods

2.1 Pilot and Membrane

The filtration experiments were carried out on a bench scale pilot (Fig. 2) in batch mode. It consists of a feed tank of 4 L, a positive displacement pump which offers a pressure of up to 20 bar and a temperature of 25 ± 0.5 °C, valves for regulating the flow meter pressure on permeate, retentate and feed circuits. The filtration module can accommodate 150 cm^2 flat organic membranes. The NF90 (gift from Dow Filmtech) is a thin-film composite (TFC) membrane comprising a macroporous polyester mechanical support, a polysulfone intermediate support and a filtration thin layer (Fig. 1). The molecular weight cut-off (MWCO) of the NF 90 is about 200 Da. It is negatively charged in the range of the pH studied.

2.2 Filtration Experiments

Synthetic single salt solutions of NaF at 10^{-3} and 10^{-2} M of fluoride ions (42 and 420 mg/L, respectively) were prepared at fixed pH and filtered with NF90 at trans-membrane pressure between 3 and 15 bar. The temperature of the feed was maintained at 25 °C. The sample of Robbah groundwater (2.9 mg/L of F^-) was treated under conditions identical to those of the synthetic solutions. Two analytical methods were used to determine the ion concentrations in permeate and feed samples: conductivity for model single solutions with a conductivity meter (Cond 315i, WTW) and ion chromatography (ICS-900, Dionex) for the complex matrix of the natural water.

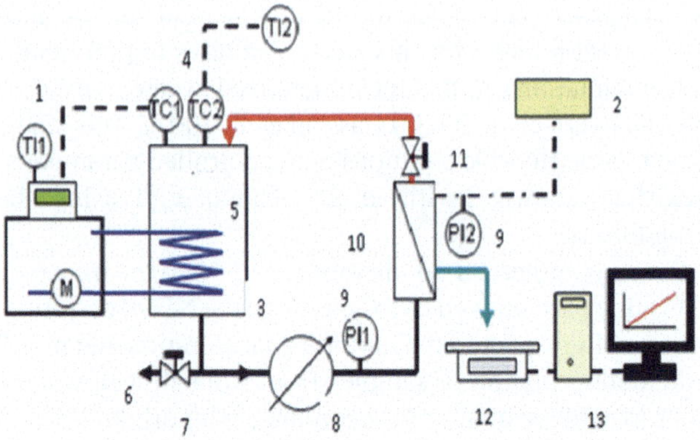

1-Circulating thermal bath, 2- preference thermometer, 3-Feed tank, 4-Temperature probe, 5-Coil, 6-Purge, 7-Three-way valve, 8-High-pressure pump, 9-Manometer, 10-Sepa CF II Cell, 11-Pressure control valve, 12-Balance, 13-Computer

Fig. 2. Synoptic of the nanofiltration pilot.

2.3 NF Performance

In order to evaluate the behavior and performance of NF membrane for application in defluorination of brackish water, Eqs. 1–3 were used to quantify the pure water and permeate flux J_p^i (L/h * m^2) (Eq. 1), permeability L_p^i(L/h * m^2 * bar) (Eq. 2) and the rejection rates of solute R_i (Eq. 3):

$$J_p^i = \frac{Q_p^i}{A} \tag{1}$$

$$L_p^i = \frac{J_p^i}{P} \tag{2}$$

$$R_i = 1 - \frac{C_p^i}{C_f^i} \tag{3}$$

where Q_p^i is the flow rate (L/h), A the active surface area of the membrane, P the trans-membrane pressure (bar) and C_p^i and C_f^i the permeate and feed concentration (M) of solute i, respectively.

3 Results and Discussion

3.1 Characteristics of the Studied Water

Table 1 lists the physico-chemical characteristics of the brackish groundwater from Robbah well. It is characterized by hardness and salinity that exceed the WHO and Algerian drinking water guidelines. The high contents in sulfate, bicarbonate, chloride, fluoride and sodium ions need to be reduced by an appropriate treatment.

Table 1. Physico-chemical characteristics of the raw water from Robbah

Parameters	Raw water	WHO guidelines	Algerian standards
Conductivity, μS/cm	3190	-	2800
pH	7.85	6.5 and ≤8.5	5 and ≤9
Ca^{2+}, ppm	310	-	250
Mg^{2+}, ppm	119	50	-
Na^{+}, ppm	514	150	200
K^{+}, ppm	21.5	12	12
NO_3^-, ppm	26.5	50	50
Cl^-, ppm	944.6	250	500
F^-, ppm	2.9	1.5	1.5
SO_4^{2-}, ppm	683.6	250	400
HCO_3^-, ppm	263.6	-	-

3.2 Pure Water Permeability

Pure water permeability (L_p^0) is an important parameter of membrane characterization. Its determination was done before and after filtration of solutes. L_p^0 is the slope of the linear regression in the plot of J_p *vs* the transmembrane pressure. Before measurement, the membranes were washed with deionized water. It was found that L_p^0 (initial value of 9.9 ± 1.0 L/h.m^2.bar) remained unchanged after filtration of synthetic NaF solutions.

On the other hand, the water permeability showed a value of 9.6 ± 1.0 L/h.m^2.bar after the raw water filtration and membrane rinsing. It was concluded that the NF90 fouling by the Robbah groundwater can be considered as limited and could be controlled by a suitable microfiltration or ultrafiltration pretreatment. This is due to the fact that the foulants are primarily comprised of inorganic solids in small amounts.

3.3 Effect of Operating Conditions on NaF Rejection

Figure 3 shows the variation of the permeate flux as a function of the pressure applied for both model solutions and raw Robbah water. The permeate flux increases linearly with the applied pressure indicating that no concentration polarization at the membrane surface in the range of pressure studied leading to a limitation of mass transfer. The pressure minimum to observe a permeate flux is a function of the osmotic pressure (Π) of feed and the membrane resistance (pore diameter and hydrophilicity). Π is related to the salt concentration contained in solution. The theoretical osmotic pressure values were 0.049, 0.49 and 1.56 bar for 10^{-3} M, 10^{-2} M NaF solution and for raw Robbah water, respectively. In these conditions, the intercept with the x-axis in Fig. 3 is other than zero and increases with an increasing of salinity. In addition, the slope of the linear regression decreases due to the interaction of ions with pore charges that affect the permeate viscosity (Chmiel et al. 2006). This electro viscous effect can be evidenced by the permeability values of 5.6, 5.5 and 5.3 L/h.m^2.bar for 10^{-3}, 10^{-2} M NaF solution, and for raw Robbah water, respectively.

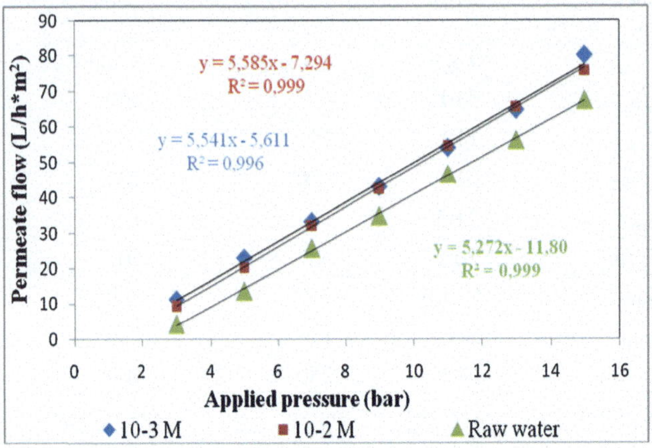

Fig. 3. Variation of permeate flux as a function of applied pressure

The rejection of fluoride ions by NF90 was plotted against the applied pressure as presented in Fig. 4. Rejection first increases with pressure and then reaches a plateau for transmembrane pressure higher than 11 bar. This corresponds to the limiting rejection of fluoride ion in the given operating conditions. Beyond this value, the permeate flux increases linearly with an increase of applied pressure (Fig. 3) while ion permeation is only a function of the feed composition (Lu et al. 2007). At low applied pressure, the permeate flux is low and the ion diffusion dominates the NF transfer resulting in a high passage of fluoride ion and low rejection value (Chmiel et al. 2006). On the other hand, transfer by convection becomes predominant for the higher permeate flux enhancing significantly the rejection.

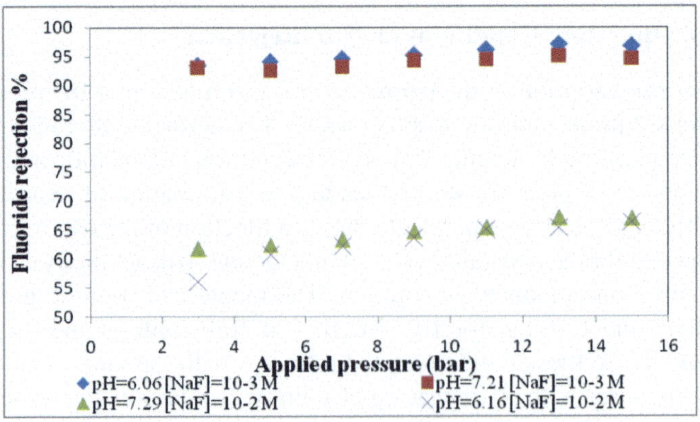

Fig. 4. Evolution of fluoride rejection as a function of applied pressure at different pH and concentration

As can be seen in Fig. 4, the fluoride rejection reaches values higher than about 95% and 65% for respective concentrations of 10^{-3} M (42 ppm) and of 10^{-2} M (420 ppm).

An increase of pH from 6 to 7 does not improve significantly the fluoride rejection even if data exhibited slightly better values at pH 7.21. NF90 is negatively charged in this pH range due to the ionization of carboxylic groups present on its surface explaining the high rejection of fluoride (Azaïs et al. 2014). The zeta potential values are reported to be from about −55 to −60 mV for pH between 6 and 7. This slightly downward sloping trend could explain the observed increase of rejection.

3.4 Membrane Performance in Robbah Natural Water Treatment

Figure 5 shows the desalination performance of NF90 for Robbah brackish water. The total salt rejection determined by the conductivity reduction of permeate is represented as a function of the applied pressure. Again, it can be seen that the optimum was observed at an applied pressure of about 11 bar with a retention of more than 90% of ions. It can be conclude that the NF90 membrane enables to reduce salinity to obtain partially demineralized water. The conductivity of the treated water was less than 400 μS/cm complying with the WHO guidelines and the Algerian standards.

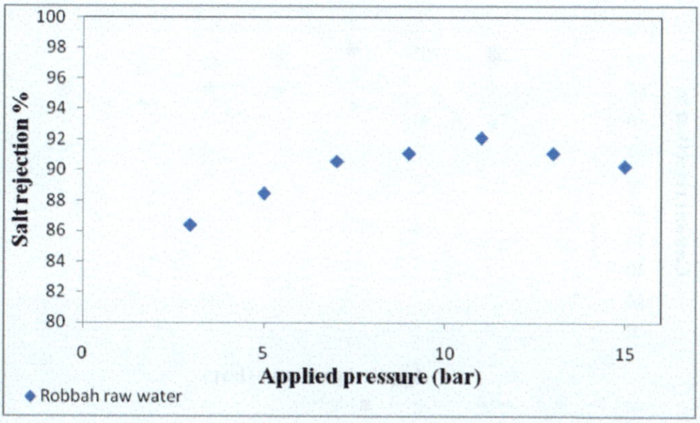

Fig. 5. Evolution of salt rejection as a function of applied pressure for Robbah brakish water

The NF selectivity with respect to ions depends on different parameters including the membrane characteristics (pore size and surface charge) and the operating conditions (Elazhar et al. 2012; Lhassani et al. 2001; Diawara et al. 2003).

Figures 6 and 7 show the anion and cation rejection *vs* applied pressure for the major monovalent and divalent ions present in Robbah water, i.e. F^-, Cl^-, SO_4^{2-}, Na^+, Ca^{2+} and Mg^{2+}. The higher rejection is generally observed for medium to high transmembrane pressures depending on the considered ions. On the other hand, the data show that the rejection decreased at high pressures. This trend was also seen in Fig. 5. This effect can be explained by an accumulation of ions at the membrane surface (concentration polarization) related to the combination of high permeate flux and high ion rejection. It leads to the decrease of convection transfer allowing the diffusion transfer to gain in importance. The conjunction of the two trends induces the observed decrease of rejection.

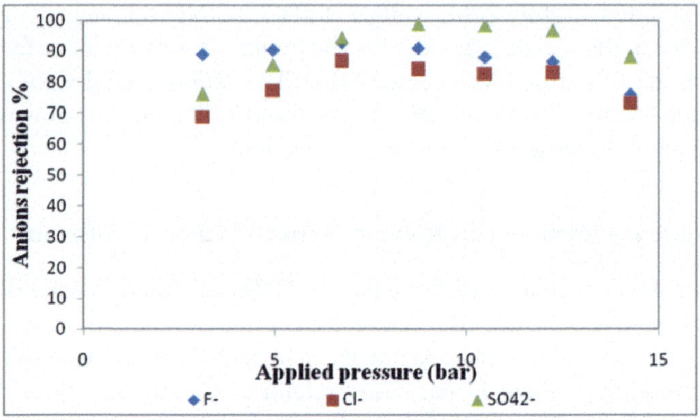

Fig. 6. Evolution of the rejection of major anions as a function of the applied pressure in the case of the Robbah brackish water

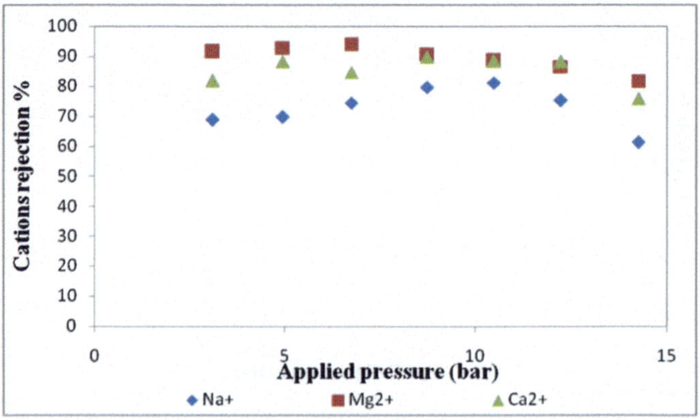

Fig. 7. Evolution of the rejection of major cations as a function of the applied pressure in the case of the Robbah brackish water

NF is known to generally retain more the divalent ions than the monovalent ones due to the Donnan exclusion. Negatively charged membranes (general case of NF membranes) tend to repel polyvalent anions to a larger extent than monovalent anions (Mnif et al. 2010). Similarly, with respect to cations, the separation order follows the order of the hydration energies: $R_{Mg^{2+}}$ (1921 kJ/mol) > $R_{Ca^{2+}}$ (1584 kJ/mol) > R_{Na+} (407 kJ/mol) (Dolvar et al. 2011). However, if these statements are true for single salt solutions, the case of complex matrix is non-trivial. As a matter of fact, electroneutrality of permeate must be respected involving that rejection of cations is dependent on that of anions, and vice versa. The presence of a high amount of divalent ions in the Robbah groundwater entails a good rejection of monovalent ions in the range of 80–90%

($R_{Mg^{2+}} \geq R_{Ca^{2+}} > R_{Na^{+}}$). In the same way, rejection of anions is in the expected order from their hydration energy ($R_{SO_4^{2+}}$(1138 kgJ/mol) > R_{F^-}(515 kgJ/mol) > R_{Cl^-}(376 kgJ/mol)) (Dolvar et al. 2011). SO_4^{2-} ions were retained almost quantitatively whereas the rejection of Cl^- ions is about 85%. The optimum rejection of F^- ions by the NF 90 membrane appears to be slightly lower in the case of Robbah groundwater (91%) than that expected from the study of synthetic solutions (about 97%). Indeed, the electroneutrality of permeate is maintained by the passage of the most mobile ions (Cl^- and F^-), which contributes to the slight decrease of F^- rejection.

Table 2 summarizes the major ion concentrations obtained in the treated water from Robbah well by using NF90 with transmembrane pressure of 8.75 and 10.5 bar. The data show that all these parameters are in line with the WHO Guide Values and the Algerian standards indicating that the NF with the NF90 membrane could be a valuable process to provide drinking water of good quality in the area of El Oued in South Algeria.

Table 2. Physico-chemical characteristics of the treated water from Robbah well

Ion	Pressure (bar)	Ca^{2+}	Mg^{2+}	Na^+	SO_4^{2-}	F^-	Cl^-
Content (ppm)	8.8	31	11	103	7	0.26	151
	10.5	34	14	98	14	0.35	161

4 Conclusion

Groundwaters in South-Algeria is often contaminated by fluoride ions having a predominantly geogenic origin.

The toxicity of fluoride ions has resulted in stricter regulations for drinking water.

In order to comply with these standards, the use of the NF technology was proposed. Successful defluorination tests were performed by treating the drilling groundwater of the commune Robbah with NF90 membrane. The data reported in this paper indicate that the defluorination efficiency was about 97%, for a synthetic solution with an F^- concentration of 42 ppm. The natural water contains 2.9 ppm of F^- in the presence of other ions at much higher concentrations. More than 91% of F^- can be removed from this water at natural pH. The F^- content of the filtered water at 8.8 bar was less than 0.3 mg/L. Removal of the other ions also took place during the same operation leading to desalted water with all the parameters within the regulatory limits. Increasing the transmembrane pressure to 11 bar enables to enhance the productivity by 28% without altering the quality of the water produced.

In summary, NF enables to provide access to water intended for human consumption, in the sense of the WHO guidelines owing to its ability to selectively remove fluoride ions and to demineralize brackish water. NF offers the great advantage to be operated at relatively pressure with high flux and retention without the need of remineralization, which makes the process cost effective.

References

Ahour, S.: La qualité des eaux du Sahara septentrional en Algérie: étude de l'excès en fluor. Tribune de l'Eau **42**(542), 53–57 (1990)

Azaïs, A., Mendret, J., Gassara, S., Petit, E., Deratani, A., Brosillon, S.: Nanofiltration for wastewater reuse: counteractive effects of fouling and matrice on the rejection of pharmaceutical active compounds. Sep. Purif. Technol. **133**, 313–327 (2014)

Bejaoui, I., Mnif, A., Hamrouni, A.: Performance of reverse osmosis and nanofiltration in the removal of fluoride from model water and metal packaging industrial effluent. Sep. Sci. Technol. **49**, 1–11 (2014)

Chakrabortty, S., Roy, M., Pal, P.: Removal of fluoride from contaminated groundwater by cross flow nanofiltration: transport modeling and economic evaluation. Desalination **313**, 115–124 (2013)

Chmiel, H., Lefebvre, X., Mavrov, V., Noronha, M., Palmeri, J.: Computer simulation of nanofiltration, membranes and processes. In: Rieth, M., Schommers, W. (eds.) Handbook of Theoretical and Computational Nanotechnology, vol. 5, pp. 93–214. American Scientific Publishers, Los Angeles (2006)

Diawara, C.K., Sidy, L., Rumeau, M., Pontié, M., Sarr, O.: A phenomenological mass transfer approach in nanofiltration of halide ions for a selective defluorination of brackish drinking water. J. Membr. Sci. **219**(1–2), 103–112 (2003)

Diawara, C.K.: Nanofiltration process efficiency in water desalination. Sep. Pur. Rev. **37**, 302–324 (2008)

Dolvar, D., Košutić, K., Vučić, B.: RO/NF treatment of wastewater from fertilizer factory-removal of fluoride and phosphate. Desalination **265**, 237–241 (2011)

Elazhar, F., Tahaikt, M., ElJalili, Z., Taky, M., Amor, Z., Boughriba, A., Hafsi, M., Elmidaoui, A.: Comparison of the performances of two commercial membranes in fluoride removal by nanofiltration from ground water. Cahiers de l'ASEES **17**, 27–34 (2012)

Kumar, P., Sharma, N., Ranjan, R., Kumar, S., Bhat, Z.F., Jeong, D.K.: Perspective of membrane technology in dairy industry: a review. Asian-Australas J. Anim. Sci. **26**, 1347–1358 (2013)

Lhassani, A., Rumeau, M., Benjelloun, D., Pontie, M.: Selective demineralization of water by nanofiltration application to the defluorination of brackish water. Water Res. **35**(13), 3260–3264 (2001)

Lu, G.Q., Diniz da Costa, J.C., Duke, M., Giessler, S., Scolow, R., Williams, R.H., Kreutz, T.: Inorganic membranes for hydrogen production and purification: a critical review and perspective. J. Colloid Interface Sci. **314**, 589–603 (2007)

Mameri, N., Yeddou, A.R., Lounici, H., Belhocine, D., Grib, H., Bariou, B.: Defluoridation of septentrional Sahara water of North Africa by electrocoagulation process using bipolar aluminium electrodes. Water Res. **32**, 1604–1612 (1998)

Mänttäri, M., Pekuri, T., Nystrom, M.: NF270 anew membrane having promising characteristics and being suitable for treatment of dilute effluents from the paper industry. J. Membr. Sci. **242**, 107–116 (2004)

Mnif, A., Ben Sik Ali, M., Hamrouni, B.: Effect of some physical and chemical parameters on fluoride removal by nanofiltration. Ionics **16**, 245–253 (2010)

Mobeen, N., Kumar, P.: Defluoridation techniques - a critical review. Asian J. Pharm. Clin. Res. **10**(6), 64–71 (2017)

Mu, T., Cong, Y., Wang, W., Zhang, B.: Preparation and characterization of novel chitosan composite nanofiltration membrane containing mesogenic units. Desalination **298**, 67–74 (2012)

Ramdani, A., Taleb, S., Benghalem, A., Deratani, A., Ghaffour, N.: Enhancement of Saharan groundwater quality by reducing its fluoride concentration using different materials. Desalin. Water Treat. **54**, 3444–3453 (2015)

Talaeipour, M., Nouri, J., Hassani, A.H., Mahvi, A.H.: An investigation of desalination by nanofiltration, reverse osmosis and integrated (hybrid NF/RO) membranes employed in brackish water treatment. J. Environ. Health Sci. Eng. **15**, 18–26 (2017)

Owusu-Agyeman, I., Jeihanipour, A., Luxbacher, T., Iris Schäfer, A.: Implications of humic acid, inorganic carbon and speciation on fluoride retention mechanisms in nanofiltration and reverse osmosis. J. Membr. Sci. **528**(15), 82–94 (2017)

Treatment of Colored Waters by Beads Chitosan, Extracted from Shrimp Waste

D. Aili[1(✉)], W. Arbia[2], and L. Adour[3]

[1] Faculty of Sciences, University Mouloud Mammeri, 15 000 Tizi-Ouzou, Algeria
aili.dihia@gmail.com
[2] Faculty of Sciences, University Dr. Yahia Fares, Medea, Algeria
[3] Faculty of Sciences, University of Algiers, 16 000 Algiers, Algeria

Abstract. The valorization of shrimp *Parapenaeus longirostris* waste, in the reduction of the dye: solophenyl blue (BS) in aqueous solution was studied. To date, the principal valorization of this flow of waste remains the isolation of chitin. Indeed, its transformation into chitosan, allowed its solubility in diluted acid medium. The extraction of chitin requires several conventional steps (demineralization, deproteinisation, bleaching and desacetylation), which explains the high cost of the product. In order to overcome this problem, chitin was isolated by biological way thanks to a bacterium *Lactobacillus helveticus*. In this work, the chitin purified by lactic fermentation (ChitiBio), was used for the desacetylation under new soft conditions. (NaOH 40%, T = 30 °C during 10 days). The physicochemical analyses, spectroscopy FTIR and micrography SEM have allowed characterizing the chitin deacetylated under our particular conditions. Compared to the commercial chitosan (Fluka), spectrum FTIR and external morphology present a strong similarity. Thanks to the solubility of the prepared chitosan in acid medium, beads of chitosan were formed then tested in order to prove their effectiveness in the reduction of dye BS. Comparisons were made with the commercial chitosan (ChitosCom). Experiments of adsorption were carried out, in batch mode, with free pH (6) and room temperature (22–25 °C). For each adsorbent, kinetics of adsorption to various initial concentrations while colouring (BS) (10–300 mg/l) were followed during several days. The beads of chitosan (ChitiBioDA10dT30 and ChitosCom) have allowed reducing, considerably, the residual concentration of the dye to values lower than 1 mg/l. Thus, rates of abatement of 99% were reached and this for all the studied concentrations. The soft desacetylation is thus recommended in the production of the chitosan, starting from the chitin purified by biological way.

Keywords: Chitosan beads · Shrimp shells · Textile dye · Deacetylation

1 Introduction

Faced with an unstable environment and technological advances, the necessary awareness leads to the development of products that reduce pollution and respect the environment.

B. Abdelbaki et al. (Eds.): SMSD 2017, *Proceedings of the Third International Symposium on Materials and Sustainable Development*, pp. 492–505, 2018.
https://doi.org/10.1007/978-3-319-89707-3_54

Water is essential to life on earth. It is, also, essential to the industrial and agricultural development of human societies. Therefore, this accelerated development has been accompanied by the pollution of atmosphere and water which poses a real problem for the environment. Water pollution occurs when materials are discharged into the water, which degrades the quality of the water, making its use dangerous and disrupting the aquatic environment, particularly fish life.

Activated carbons, the main adsorbents used in water treatment, have many advantages, they allow the elimination of a wide range of pollutants in different types of dyes, as well as other organic and inorganic pollutants such as phenols, metal ions, pesticides, detergents, etc. Despite their excellent adsorbent capacity, their use is usually limited because of their high cost. Recently, research has focused on the use of chitinous products such as chitin, chitosan and its derivatives. They are the most economically attractive biopolymers in water treatment because of their intrinsic properties, abundance and low cost. In addition, they are biodegradable and bioresorbable, both of which are paramount factors in our times when environmental protection plays an important role.

The main objective of this original study is the valorization of marine co-products, particularly the shrimp waste (shells), by the preparation of chitosan from biologically purified chitin, in the reduction of the dye: blue solophenyl (BS), in aqueous solution.

We have considered characterizing the chitosan prepared by the determination of two most important parameters: the degree of deacetylation (DDA) by FTIR spectroscopy and the viscosimetric molecular weight (PMv). A comparison of the results was carried out with commercial chitosan (Fluka) to demonstrate the efficiency of chitosan prepared in the laboratory.

2 Methods and Materials

2.1 Materials

- MD-2100uv UV-visible Medline Spectrophotometer.
- pH meter HANNA instruments pH 211.
- Caliper: brand: STAINLESS HARDENED (mm/inch).
- Acetic acid 99–100% purity, d = 1.5 (Analar NORMAPUR).
- Magnifier made by JEULIN WF1 10X/Q20.
- Sodium hydroxide (BIOCHEM Chemopharma).
- Commercial Chitosan (Fluka).
- Dye (BS) used in Algeria, in the textile industry.

2.2 Methods

2.2.1 Preparation of Chitin by Biological Means

In crustaceans, chitin is intimately associated with minerals, lipids, pigments and proteins by covalent bonds.

The extraction of chitin requires two main steps: demineralization and deproteinization. On the industrial scale chitin is produced by the chemical method using strong acid and base concentrations at high temperatures which explains the alteration of the intrinsic properties of the biopolymer. In order to overcome this problem, a chitin isolated from shrimp *Parapenaeus longirostris* shells by biotechnological method using lactic bacteria *Lactobacillus helveticus* carried out by Arbia (2014) works (named ChitiBio).

In this work, the chitin purified by lactic fermentation was used for the desacetylation under new “soft” conditions (NaOH 40%, T = 30 °C during 10 days), completely different from the commercial chitosane. Thus, the recovered product is named ChitiBioDA10dT30.

2.2.2 Preparation of Chitosan Beads

For the preparation of these beads, chitosan was dissolved in a 2N acetic acid solution. The mixture is then dropped into a 2N NaOH solution through a 20G syringe needle at a relatively slow rate and with stirring so as to obtain gel beads.

2.3 Characterization

2.3.1 Colorant

The solvent blue dye (BS) is a direct CI direct blue 71: BS dye and is intended for the printing of cellulosic fibers. Its structure is shown in Fig. 1.

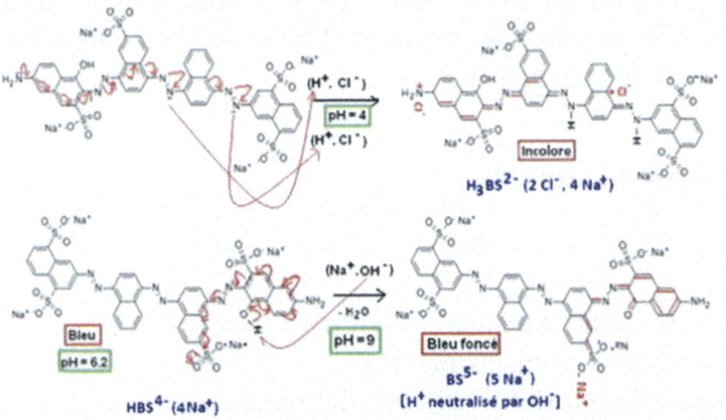

Fig. 1. Structure of solvent blue (BS) in neutral medium (HBS4-), acid (H3BS2-) and basic (BS5-)

Spectrophotometry is the technique used to assay the dye. The solophenyl blue assay was performed at the maximum adsorption wavelength of 587 nm. Dye solutions in concentrations of 10 to 300 mg/l were prepared from a stock solution with a concentration of 1 g/l.

2.3.2 Chitosanes (ChitiBioDA10dT30 and ChitosCom)

2.3.2.1. Solubility Test

One of the characteristics of chitosan in relation to chitin is its solubilization in a dilute acid medium, the most frequently used acid being acetic acid Crini et al. (2009). In this study, 50 mg of deacetylated chitins were solubilized in 50 ml of 1% acetic acid.

2.3.2.2. Determination of DDA of Deacetylated Chitins by Infrared Spectroscopy (FTIR)

Infrared spectroscopy techniques are used for the determination of the functional groups which make it possible to represent the structure. Each link exhibits characteristic vibrations of well-defined frequencies.

The analyzes were carried out using infrared spectroscopy of BRUKER type (TENSOR27). The spectrum obtained was recorded between 4000 and 500 cm^{-1}. Before the spectroscopic analysis, the samples were carried in the oven at 40 °C for 24 h.

The simplest and fastest method for calculating the DDA degree is that of IR spectroscopy proposed by Khan et al. (2000). The DDA of chitin DA was determined using the following formula (1) proposed by Brugnerotto et al. (2001a, b).

$$DDA = 100 - \left[\frac{\left(A_{1320}/A_{1420} \right) - 0.3822}{0.03233} \right] \quad (1)$$

A_{1320}: absorbance at 1320 cm^{-1} corresponding to amide III.

A_{1420}: absorbance at 1420 cm^{-1} corresponding to the symmetrical deformation of CH_3 and CH_2.

2.3.2.3. Scanning Electron Microscopy (SEM)

Scanning electron microscopy (SEM) is a technique based on the principle of electron-matter interactions, capable of producing high-resolution images of the surface of a sample.

The objective of the SEM observations carried out in this study was to observe the morphology of chitin prepared biologically.

2.3.2.4. Determination of the Molecular Weight of Deacetylated Chitin by Viscosimetry

The molecular weight of a chitosan sample is calculated from its intrinsic viscosity denoted [η] by applying the Mark-Houwink-Sakurada relation (2) Rinaudo (1993).

$$[\eta] = KM_{v^a} \quad (2)$$

[η]: intrinsic viscosity (dL/ g).

M_v: viscosimetric average molecular weight of the polymer (Da).

K et a: parameters which depend on the polymer-solvent system at a given temperature.

In our study, T = 30 °C, a = 0.72, K = 4.74 × 10^{-5} cm^3/dl) was used (Kasaai 2007).

[η] is determined by measuring the dynamic viscosity η of the solution and that of the pure solvent η_0 such that the specific viscosity is written (3):

$$\eta_{sp} = \frac{\eta - \eta_0}{\eta_0} \tag{3}$$

The graph plots of the inherent viscosity (5) or the reduced viscosity (4) as a function of the concentration of the chitosan solution (0.02–0.1 g/dL) makes it possible to deduce [η] (corresponding to the ordinate at the origin of the right) by:

$$\eta_{red} = \frac{\eta_{sp}}{c} \tag{4}$$

$$\eta_{\mathrm{inh}} = \frac{ln\eta_{\mathrm{rel}}}{c} \tag{5}$$

The relative viscosity can be written as (6):

$$\eta_{\mathrm{rel}} = \frac{\eta}{\eta_0} \tag{6}$$

C: polymer concentration (g/ dL)

- The intrinsic viscosity of ChitosCom (Fluka) is determined by two methods.
  - ball viscometer
  - capillary tube
- The intrinsic viscosity of chitin DA is determined experimentally using the ball viscometer.

(a) The method of calculating dynamic viscosity by ball viscosity is based on the measurement of the time required for a ball to fall by a defined distance. The rotational and sliding movements of the ball through the sample filling a slightly inclined cylindrical measuring tube are described by the fall time. The dynamic viscosity is expressed using the international standardized absolute unit of "milli Pascal · seconds" (mPa · s).

The dynamic viscosity η is calculated using the following Eq. (7):

$$\eta = \mathrm{k} \times (\rho_1 - \rho_2) \times t \tag{7}$$

Or:
K: bead constant (mPa · s cm^3/g · s)
ρ_1: density of the bead in (g/cm^3)
ρ_2: density of the liquid to be measured at the measuring temperature (g/cm^3)
t: fall time of the ball (seconds)

The densities of the solutions were determined using a pycnometer (10 ml) at a given temperature (30 °C).

With (8):

$$\rho_2 = \frac{M}{V} \tag{8}$$

(b) The method of calculating the dynamic viscosity by a capillary viscometer is carried out using a screw-on tube by measuring the flow time of the same volume of solvent (AA 1%) or ChitosCom solution (Fluka) contained between the two points (a) and (b) using a stopwatch. The equation is given by (9):

$$\eta_{sp} = \frac{t - t_0}{t_0} \tag{9}$$

2.4 Study of the Kinetics of Biosorption

In order to demonstrate the effectiveness of the prepared chitosan beads (ChitiBioDA10dT30), batch adsorption experiments were carried out at room temperature (22–25 °C) in a 50 ml reaction volume. Comparisons were made with commercial chitosan (ChitosCom). Adsorption kinetics at initial BS dye concentrations varying from 10 mg/l to 300 mg/l were plotted.

3 Results and Discussion

3.1 Characterization of Chitosans

3.1.1 Solubility Test

There was solubility of the deacetylated chitin, which is reflected in the effective presence of chitosan (Fig. 2).

This rapid and simple test showed that the basic and homogeneous deacetylation of the different chitins could be carried out under conditions totally different from those mentioned in the bibliography Mima et al. (1983). These conditions involved deacetylating chitin in a concentrated NaOH solution (up to 50% by mass) under an inert atmosphere (generally N_2). These new conditions are considered milder, they respect the environment (because the NaOH concentration is lower) and they consume less energy.

3.1.2 Determination, by FTIR and SEM, of the Structure of ChitiBioDA10dT30

The spectrum of biologically purified chitin (Fig. 3) is identical to that of commercial chitin (Fig. 4). The main characteristic peaks of chitin are: 1650 cm^{-1} corresponding to amide I, 1583 cm^{-1} to amide II, 1420 to the symmetrical deformation of $-CH_3$ and $-CH_2$, 1320, 3200 cm^{-1} and 3100 cm^{-1} to the intermolecular hydrogen bond $(C)_{(6)}$ OH..OH–$C_{(6)}$.

Chitin before DA Chitin after DA

Fig. 2. The solubility of chitin before and after deacetylation

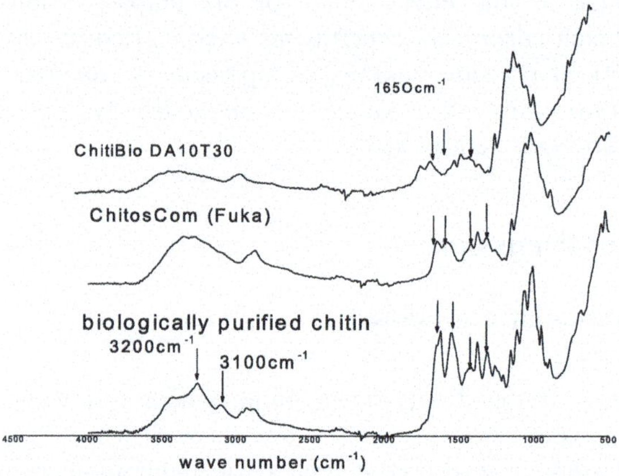

Fig. 3. FTIR spectra: of ChitiBioDA10dT30, Chitin purified biologically and ChitosCom (Fluka)

- The spectrum obtained from the ChitiBioDA10dT30 shows a decrease in the intensity of all the peaks and mainly those located at 1650 cm^{-1} and 1550 cm^{-1} and the disappearance of two peaks located at 3200 cm^{-1} and 3100 cm^{-1}. The spectrum of ChitiBioDA10dT30 is similar to that of ChitosCom (Fluka). These results confirm that ChitiBioDA10dT30 is indeed chitosan.

The deacetylation degree (DDA) of the deacetylated chitin was calculated. It has been found that at 10 days, the DDA of deacetylated chitin at 30 °C. is greater than 50% (it is 82%). It is also noted that chitin DA has a higher DDA than commercial chitosan (Fluka), which is 75%. This demonstrates that the "soft" conditions of deacetylation used in this study allowed to produce a large number of free –NH 2 groups. These results demonstrate once again that Chitin DA is chitosan.

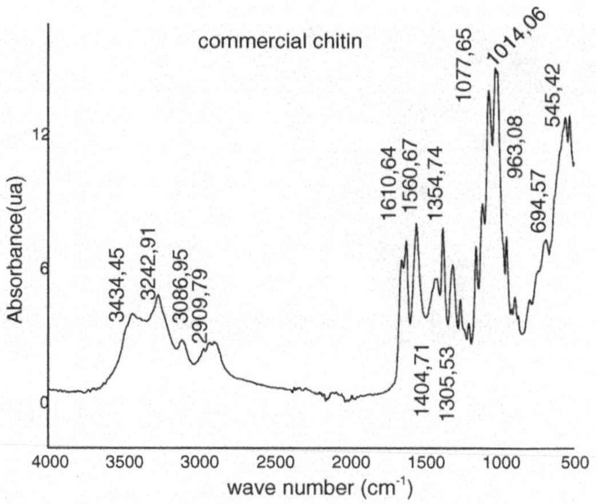

Fig. 4. Commercial chitin FTIR spectrum (Sigma Aldrich)

3.1.3 SEM Micrographs of the ChitiBioDA10dT30

Chitin purified by biological means is in the form of fibrils (Fig. 5) comparable to commercial chitin (Fluka) presented in the form of fine powder and taken as standard (Fig. 6). Treatment with the bacterium *Lactobacillus helveticus* eliminated minerals and proteins in the shrimp shell *Parapenaeus longirostris*. Therefore, chitin is free. It is noted that some bacterial cells still remain in the residue despite repeated washing with distilled water.

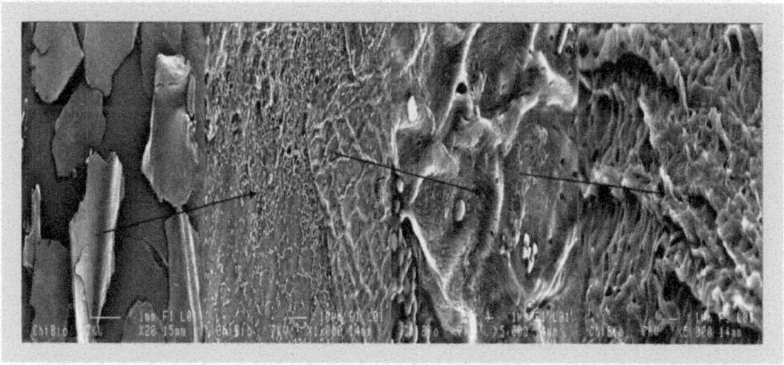

Fig. 5. SEM micrographs of biologically purified chitin

The photographs of Fig. 7 show the morphology of the ChitiBioDA10dT30. We observe a heterogeneity of the surface, some smooth and some porous places that resembles the ChitosCom (Fluka) (Fig. 8).

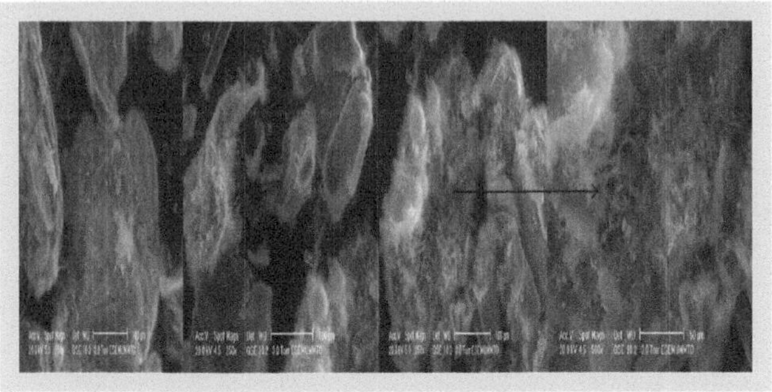

Fig. 6. SEM micrographs of commercial chitin (Fluka)

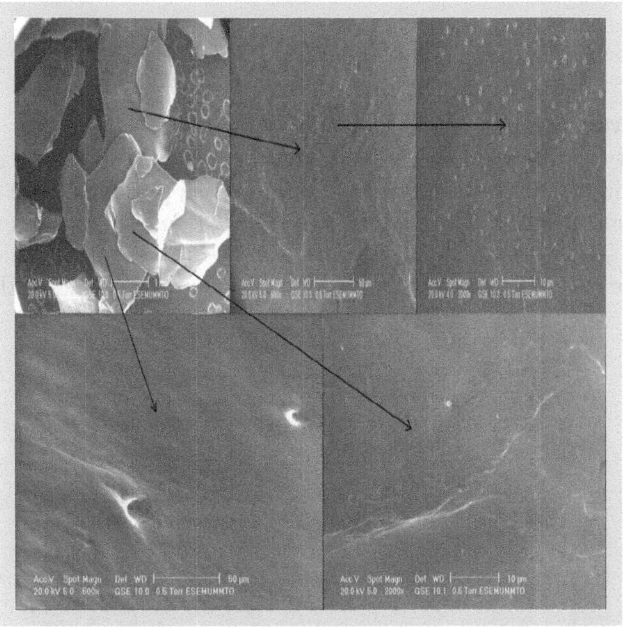

Fig. 7. Micrographs SEM of the ChitiBioDA10dT30

3.1.4 Determination of the Molecular Weight of Deacetylated Chitin by Viscosimetry

In the Mark-Houwink-Sakurada formula (Rinaudo 1993), the determination of PM is made by knowing the intrinsic viscosity of the sample calculated by the capillary tube. In this work, it has therefore been proposed to recalculate the intrinsic viscosity (value obtained if calculated by the capillary tube) of each sample by taking commercial (pure) chitosan as a standard. In other words, determine the intrinsic viscosity of the

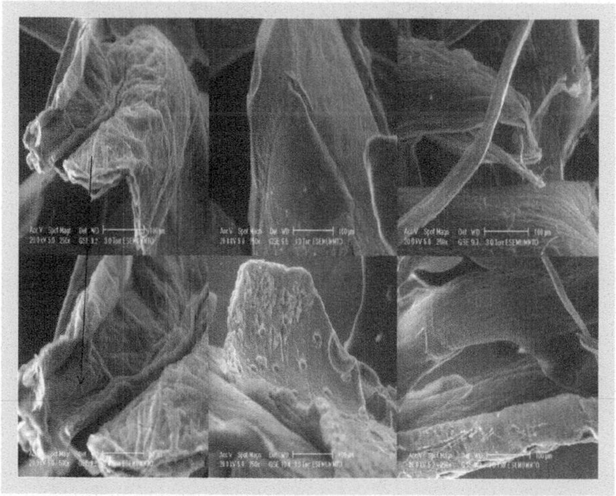

Fig. 8. Micrographs SEM of commercial chitosan (Fluka)

commercial chitosan (Fluka) by measuring the specific viscosity in the concentration range of 0.2 to 1 g/dL by the capillary tube and in the concentration range of 0.02 to 0, 1 g/dL by the ball viscometer. The results are summarized in Table 1 below:

Table 1. The intrinsic viscosities and molecular weight of the ChitiBioDA10dT30 and ChitosCom (Fluka)

ChitosCom (Fluka)			
[η] (dL/g) (capillary tube)	[η] (dL/g) (ball viscometer)	PM (KDa) (calculated)	PM (KDa) (commercial data)
0,558	37,922	450,401	400
ChitiBio DAT30			
t (days)	[η] (dL/g) (ball viscometer)	[η] (dL/g) (capillary tube)	PM(KDa)
10	37,198	0,547	434,857

3.1.5 Mass, Morphology and Size of Beads

The prepared wet beads are all spherical (Fig. 9). For each chitosan (ChitiBioDA10dT30 and ChitosCom), the total mass of the beads obtained was weighed, the diameter was determined with the aid of a caliper and the size (listed in Table 2). The beads observed by a magnifying glass have a smooth surface and a solid internal volume (filled balls).

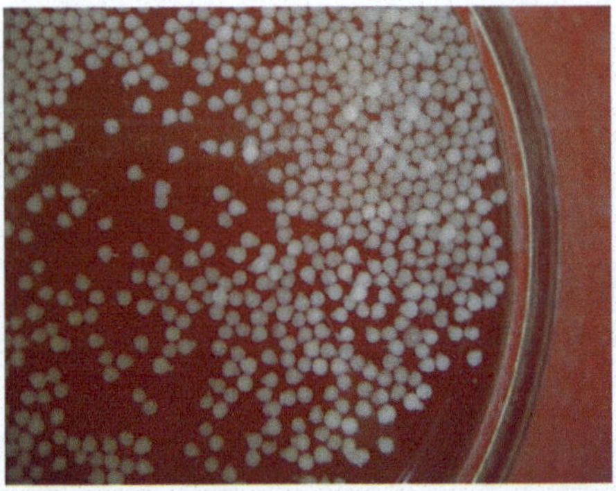

Fig. 9. Beads of ChitiBioDA10dT30

Table 2. Mass, morphology and sizes of the beads of each chitosan

Chitosan beads	Form	Total mass of the beads (g)	Size (mm)
ChitosCom	Spherical	11.64	1.76–2.29
ChitiBioDA10dT30	Spherical	11.6	1.85–2.27

3.2 Evolution of the Residual Concentration of BS Dye

Figure 10 shows the evolution of the residual dye concentration as a function of the contact time for the various Chitosan: ChitosCom and ChitiBioDA10dT30.

Both chitosan reduced all initial residual BS dye concentrations from 10 to 300 mg/l to very low values.

1. A rapid drop in the residual dye during the first 4 h of contact for all residual concentrations studied (10 to 300 mg/l).
2. After 24 h:

   – For the residual concentration of 10 mg/l: a reduction rate of 98.68% is achieved with a residual dye concentration of 0.099 mg/l for chitosan ChitosCom. With chitosan ChitiBioDA10dT30, it is 85.04% with a residual dye concentration of 1.50 mg/l.
   – For the residual concentration of 50 mg/l: with chitosan ChitosCom, a reduction rate of 87.47% is achieved with a residual dye concentration of 5.78 mg/l. With chitosan ChitiBioDA10dT30, it is 80.48% with a residual dye concentration of 8.75 mg/l.
   – For the residual concentration of 100 mg/l: with ChitosCom chitosan, a reduction rate of 90.23% with a residual dye concentration of 6.99 mg/l was achieved. With chitosan ChitiBioDA10dT30, it is 67.15% with a residual dye concentration of 25.85 mg/l.
   – For the residual concentration of 200 mg/l: with ChitosCom chitosan, the reduction rate is 86.50% with a residual dye concentration of 29.47 mg/l; with chitosan ChitiBioDA10dT30, it is 76.50% with a residual dye concentration of 52.83 mg/l.

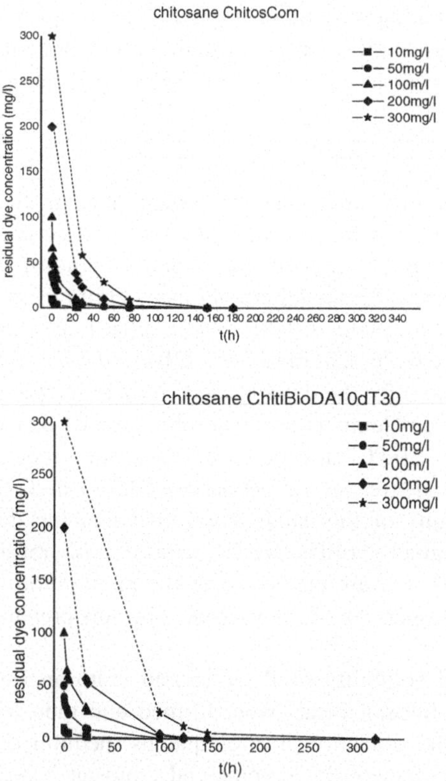

Fig. 10. Evolution of the residual concentration of BS dye as a function of the contact time for the different chitosan beads prepared (ChitiBioDA10dT30) and (ChitosCom) at different concentrations (10–300 mg/l)

3. The rate of abatement continues to increase to reach maximum rates at different times and according to the residual concentration used (10–300 mg/l):
   - With the concentration of 10 mg/l: with chitosan ChitiBioDA10dT30 (99.39%) at 4 days. With ChitosCom chitosan, increasing the contact time beyond one day does not improve the rate of adsorption of the BS dye.
   - With the concentration of 50 mg/l: For chitosan ChitosCom, it is 99.38% at 3 days chitosan ChitiBioDA10dT30 (99.38%), it is 4 days.
   - With the concentration of 100 mg/l: For chitosan ChitosCom, it is 99.00% at 3 days. For chitosan ChitiBioDA10dT30 (99.35%), it is 6 days.
   - With the concentration of 200 mg/l: the reduction rate is 99.93% for chitosan ChitosCom and ChitiBioDA10dT30.
   - With the concentration of 300 mg/l: For chitosan ChitosCom, the reduction rate is 99.32% at 7 days. With chitosan ChitiBioDA10dT30 (99.67%), it is 13 days.

Compared with all the results obtained, it was demonstrated in this work that all the chitosan beads prepared from the commercial chitosan or from the chitosan prepared in the ChitiBioDA10dT30 laboratory make it possible to completely eliminate the BS dye from the water regardless of the initial concentration (low or high).

4 Conclusions

Among the natural polymers that have been mentioned in the bibliography, we are interested in the chitosan which is considered as the first derivative of chitin extracted from crustaceans. Due to the high price of commercial chitosan and its non-availability, it was proposed to prepare it for the laboratory using chitin from shrimp shells fished on the Algerian coastline: *Parapenaeus longirostris*. After purification of the chitin carried out from the shrimps by biological means (ChitiBio), the homogenous deacetylation in a basic medium was tested for the deacetylated chitin at temperatures of 30 °C, chosen because it is near ambient temperature. The objective of this work is the characterization of chitin DA at 30 °C and 10 days. Two important intrinsic parameters in chitosan were studied: degree of deacetylation (DDA) and viscosimetric molecular weight (PM). The results of this study confirmed that the deacetylation conditions described as soft conditions (40% NaOH, chitin/ base ratio = 1 g/50 ml, without stirring, temperature 30 °C.) are envisaged in the preparation of soluble chitosans in 1% of acetic acid. A comparison of the various chitosans prepared was carried out with ChitosCom (Fluka).

Thanks to the acid solubility of the prepared chitosan (polymer symbolized by ChitiBioDA10dT30), chitosan beads were formed and then tested in order to prove their effectiveness in the reduction in the aqueous medium of the textile dye: solophenyl blue (BS). Simultaneously, commercial chitosan beads (symbolized by ChitosCom) were also tested.

The beads prepared from ChitBioDA10dT30 chitosan are as effective as those obtained with ChitosCom. The ChitosCom and ChitiBioDA10dT30 beads significantly reduced the residual dye concentration (BS) (<1 mg/l) and achieved very high abatement rates (>99%) for all concentrations initials studied (10 mg/l–300 mg/l).

References

Arbia, W.: Etude de l'optimisation de la recuperation de la chitine par Fermentation de dechets marins. Thèse de doctorat en sciences. Ecole Nationale Polytechnique d'Alger, Algérie (2014)

Brugnerotto, J., Desbrières, J., Roberts, G., Rinaudo, M.: Characterization of chitosan by steric exclusion chromatography. Polymer **42**(25), 9921–9927 (2001a)

Brugnerotto, J., Lizardi, J., Goycoolea, F.M., Argüelles-Monal, W., Desbières, J.: Characterisation of thermogelling chitosan/glycerol-phosphate solutions. Carbohydrate **42**, 3569–3580 (2001b)

Crini, G., Guibal, E., Morcellet, M., Torri, G., Badot, P.: Chitin and Chitosan. Preparation properties and main applications Chitin and Chitosan in the biopolymer application. Besançon University Press Franche-Comté, pp. 19–54 (2009)

Kasaai, M.R.: Calculation of Mark-Houwink–Sakurada (MHS) equation viscosemtrie constants for chitosane data. J. Carbohydr. Polym. **68**, 477–488 (2007)

Khan, T.A., Peh, K.K., et al.: Mechanical biomechanical bioadhesive strength (2000)
Mima, S., Miya, M., Iwamoto, R., Yoshikana, S.: Highly deacetylated chitosan and its properties. J. Appl. Polym. Sci. **28**, 1909–1917 (1983)
Rinaudo, C.: The Lino Ventura theater. Social assessment and impact assessment. Study Report. A.R.E.S: Nice (1993)

Use of the Potato Peels and the Bentonite as Additives of Flocculation in the Wastewater Treatment of the STEP of Saïda

Fatima Zohra Choumane[1,2(✉)], Fatma Kandouci[1], and Bouhana Maachou[1]

[1] Department of Process Engineering, Faculty of Technology, Dr Moulay Tahar University of Saida, Saida, Algeria
chimief@yahoo.fr

[2] Laboratory of Inorganic Chemistry and Environment, Department of Chemistry, Faculty of Sciences, Abou Bakr Belkaid, University of Tlemcen, Tlemcen, Algeria

Abstract. The objective of this study is to eliminate the pollutant load of wastewater from the STEP of Saida by the coagulation-flocculation tests using a jar-test bench. The analysis revealed a high COD and BOD_5 content, which can reach 529.08 mg O_2/l and 317.03 mg O_2/l respectively. Conventional treatment based on coagulation-flocculation made it possible to reduce considerably this loading in the presence of the coagulant aluminum sulfate by means of natural flocculation auxiliaries, potato peels and bentonite. The efficiency of the treatment depends strongly on the optimization and control of all physico-chemical parameters. Turbidity is the best indicator of the effectiveness of this treatment. The optimal dose of aluminum sulphate is equal to 0.6 g, it is estimated at 10 mn, with a turbidity abatement of the order of 99.7%. The flocculation tests show that, with a mass of 0.5 g of bentonite and for an optimal dose of aluminum sulphate, the percentage reduction in turbidity is 99.66%. The treatment carried out in combination of aluminum sulphate and potato peels leads to good flocculation and clarification with a turbidity abatement of 98.01%. Examination of the results suggests the use of bentonite and potato peels as flocculation adjuvants.

Keywords: Waste water · Coagulation-flocculation · Valuation · Mineral salts Turbidity

1 Introduction

The wastewater treatment means the ensemble of physical, chemical, biological and bacteriological processes, in order to reduce the load of organic and inorganic pollutants contained by water (Nutiu 2015). Coagulation and flocculation process is one of the best options to treat pulping wastewater using alum, ferric chloride, lime, ferrous sulphate and poly aluminium chloride (PAC) and can be considered as an efficient treatment method to reduce COD, TSS and colour of pulping effluent (Rahbar et al. 2006; Deegan et al. 2011; Wong et al. 2006; Garg et al. 2005). The major advantage of chemical treatment is that most of the COD and TSS are being reduced during this process

B. Abdelbaki et al. (Eds.): SMSD 2017, *Proceedings of the Third International Symposium on Materials and Sustainable Development*, pp. 506–515, 2018.
https://doi.org/10.1007/978-3-319-89707-3_55

therefore it can lead to make it more cost effective before secondary treatment as well as removal of colour for effluent. Flocculation is an essential phenomenon in industrial wastewater treatment. Inorganic coagulants (salts of multivalent metals) are being commonly used due to its low cost and ease of use. However, their application is constrained with low flocculating efficiency and the presence of residue metal concentration in the treated water. Organic polymeric flocculants are widely used nowadays due to its remarkable ability to flocculate efficiently with low dosage. However, its application is associated with lack of biodegradability and dispersion of monomers residue in water that may represent a health hazard. Therefore, biopolymers based flocculants have been attracting wide interest of researchers because they have the advantages of biodegradability and environmental friendly. But, natural flocculants are needed in large dosage due to its moderate flocculating efficiency and shorter shelf life (Lee et al. 2014).

2 Materials and Methods

2.1 Potato Peel

The food residue from the potato may represent the source of the starch of vegetable origin, this type of starch swells well, and gives a high viscosity. On the other hand, the adsorption of the blue dye of methylene in an aqueous solution, on a low cost adsorbent, potato peel, was studied by the author Boumchita et al. (2016), the results showed that the percentage of elimination goes from 61.4 to 87.6% during an increase in the adsorbent mass. All these results reveal that potato peelings could be used effectively as a adsorbent material for the removal of the methylene blue dye from a solution aqueous. In this context, this waste “potato peelings” can be valorised in the treatment of waste water by coagulation flocculation as an agent flocculant (source of starch) (Fig. 1).

Fig. 1. Potato peel

2.2 Bentonite

Bentonite is an argillaceous rock whose main mineralogical component is montmorillonite. Clay minerals have the ability to bind some cations and some anions exchangeable with other cations and other anions. The capacity to exchange is due to the fact that the existing links around the vertices of the structural unit silica-alumina give rise to charges which are neutralized by the absorption of other cations. This ion exchange property is of great importance when bentonite. Bentonite is an alumino-silicate of calcium, sodium, magnesium, and iron hydrate commonly used as a coagulant aid in the treatment of water. The crude bentonite used for experiments extracted from the Hammam Boughrara (Tlemcen) deposit, it supplied us in the finely ground state by ENOF of Tlemcen in the form of gray powder. It is used as a coagulant and as a focalizer. Bentonite has a low negative charge by which the flocs are pooled flocks to produce larger, heavier and more resistant flocs, decant more quickly.

In addition, before starting wastewater treatment by coagulation tests-flocculation by jar test, it is essential to characterize raw sewage and different natural materials proposed in this study (Gidas 1998).

3 Experimental

3.1 Wastewater Sampling

The wastewater to be studied was brought from the wastewater treatment plant of the town of Saida located in the north west part of town near the river Saida at the town Rebahia. The main characteristics of wastewater collected are presented in Table 1.

Table 1. The average values of the various analyzed parameters of wastewater

Parameter	Units	Average values	AFNOR standards based	Standards Algerian
T	°C	17.52	-	30
pH	-	8.12	NF T90-008	5.5–8.5
Turbidity	NTU	350	NF T90-033	-
Conductivity	mS/cm	2.410	NF T90-031	-
Suspended solids	mg/l	233.89	NF EN 872	35
BOD_5	mg/l	317.03	NF EN:1899-2	30
NO_3^-	mg/l	2.8	ISO 7890-3	-
COD	mg/l	529.08	ISO15705	125
NO_2^-	mg/l	0.036	NF EN 26777	-
PO_4^{3-}	mg/l	42.5	NF EN ISO 6878	10
Oxidation by $KMnO_4$	mg/l	29	NF EN ISO 8467	-

3.2 Characterization

In order to characterize the potato peel and bentonite, analysis the FT-IR spectra were taken to obtain information on the nature of functional groups at the surface of the natural flocculant, using potassium bromide pellets. Infrared analysis was carried-out with Perkin Elmer spectrum version 10.4.1.

3.3 Coagulation-Flocculation Experiments

The coagulation-flocculation process were carried out through a conventional Jar Test. They were filled with 300 ml of the sample to be treated. Next, the propellers of the flocculator were immersed into the beakers. Then, increasing doses of coagulant were then introduced in the samples, under fast stirring, with the speed 150 rpm for 10 min, 20 min, 30 min and 1 h. At the end of the experiment, the parameters that achieve the clearest water, the biggest and best decanted flocs were determined. Wastewater physico-chemical treatment is led by use of coagulants based on aluminum sulphate $Al_2(SO_4)_3.18H_2O$.

In order to determine the effect of coagulant dosage, the coagulant is introduced into the wastewater under study, with different amounts (0.1, 0.2, 0.4, 0.6, 0.8, 1, 1.2, 1.6 and 2 g). The samples are allowed to decant for a period extending from 30 min to 1 h, the flakes are examined during and after agitation. After decantation, a sample is taken from each beaker to measure its pH, turbidity, oxidation of organic molecules by $KMnO_4$ and BOD_5. The most effective coagulant was the one with the highest percentages of removal in each of the evaluated parameters.

4 Results and Discussion

4.1 FTIR Analysis

The spectrum of the bentonite (Fig. 2) shows that the band extending between 3200–3700 cm^{-1} shows an intense band with two shoulders at 34527 cm^{-1} and 3623 cm^{-1} corresponding to the stretching vibrations of the group hydroxyl -OH bonded water of constitution.

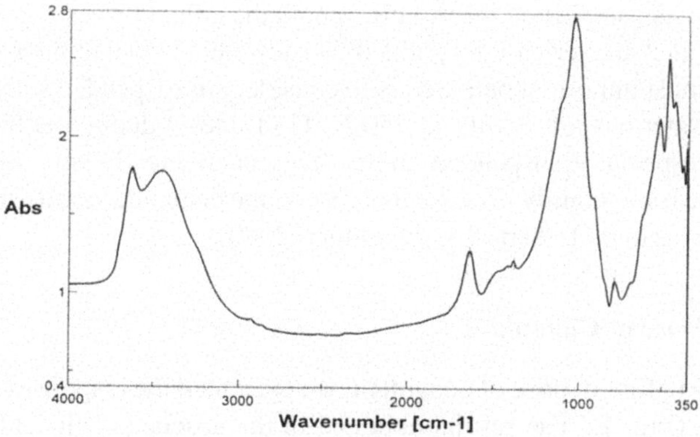

Fig. 2. FTIR spectrums of bentonite

The band between 1300–1700 cm^{-1} shows two peaks at 1639 cm^{-1} and 1378 cm^{-1} corresponds to the deformation vibrations of the O-H bond due to water molecules adsorbed between the layers.

The IR spectrum of potato peel (Fig. 3) shows a broad band near 3421.5 cm^{-1} indicating the presence of a hydroxyl group, another band recorded at 2925.8 cm^{-1} characterizing the stretching vibrations of CH_2 group, As for the band at 1647.1 cm^{-1}, it corresponds to the stretching vibrations of the aromatic group C = C. The spectrum revealed the presence of two bands at 1537.27 and 1548.7 cm^{-1} characteristic of the C-H deformation vibration bond. We also note the presence of two bands at 1461.97 cm^{-1} and 1388.7 cm^{-1} characterizing the vibrations nitro group (aromatic). On the other hand, several bands are observed which extend between 1157.27 and 1026.1 cm^{-1} corresponding to the C-OH bond of tertiary and primary alcohol respectively.

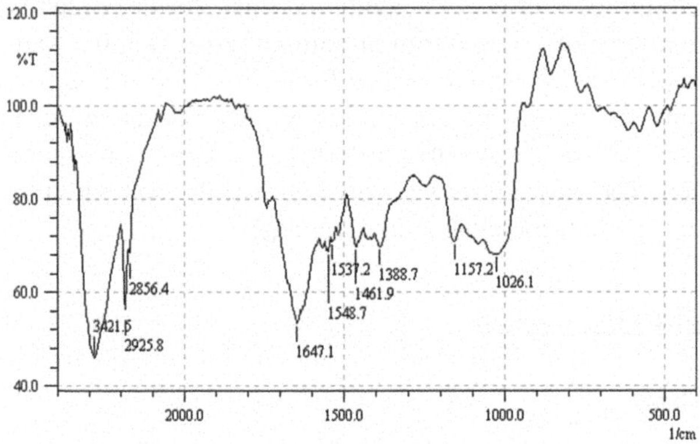

Fig. 3. FTIR spectrums of potato peel

4.2 Characteristics of Wastewater Sample

The Table 1 provides important information about the effluent before the treatment. The physico-chemical characterization results of wastewater show that the pollution level is fairly high, the most important parameters to consider are suspended solids, (BOD_5) and (COD). Wastewater has a turbidity of 350 NTU (Table 1) due to the high presence of solid material, especially suspended solids. This parameter is very important in the coagulation processes, mainly because it is easy to measure and sensitive to coagulation processes (Ministerio de Desarrollo Económico 2000).

4.3 Optimal Dose of Coagulant

To determine an optimum dose of coagulant, we presented the responses in terms of rate of turbidity (%) (Fig. 4). The results obtained in the absence of flocculants show that aluminum sulphate with a mass of 0.6 g, leads to good flocculation. For the stirring time of 10 min with a turbidity abatement of the order of 99.7% (Table 2). Determination of the coagulant dose is an essential parameter for the destabilization of colloids. An overdose of the polymer inevitably produces a restabilization of the colloidal particles which are susceptible to the surface charge. Therefore, the adsorption sites of colloidal particles would no longer be available for the formation of interparticle bridges (Gallard and Gunten 2002).

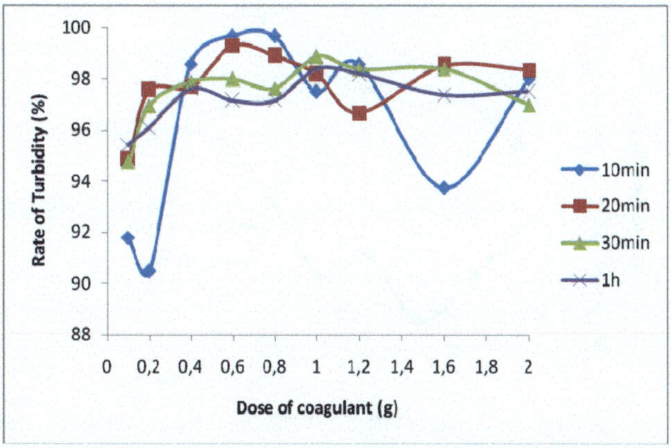

Fig. 4. Evolution of turbidity rate of wastewater according to the dose of coagulant aluminum sulphate

Table 2. Values of parameters of wastewater of the optimal dose of coagulant

Dose (g)	PH	BOD_5 (mg/l)	Rate of turbidity (%)	Permanganate oxidation (mg/l)	Stirring time (min)
0.6	6.1	86	99.7	0.9	10

It is therefore well established that there is an optimum coagulant dose which corresponds to the maximum abatement of turbidity. Thus, beyond a certain dose, a re-stabilization of the colloidal suspension is observed due to the possible inversion of the colloid charge, corresponding to a drop in the purification yields.

The results found can be explained that there is a pH range for which coagulation takes place rapidly. This range depends on the nature of the coagulant used, its concentration and the composition of the water to be treated. When coagulation occurs outside this optimum range, the amount of the coagulant should be increased (Bachouche 2013).

4.3.1 Optimal Dose of Flocculants

Two types of flocculants were used potato peel and bentonite. Various masses of flocculants were used (0.1, 0.2, 0.3, 0.4, 0.5, 0.7 and 1 g).

The coagulant was first added, 0.6 g of aluminum sulphate, The mixture obtained was stirred with a speed of 150 rpm during 10 min. Then the flocculants were added separately. The stirring speed was kept low, between 40 and 50 rpm, for a period of 15 min. The flocs were examined during and after agitation. After decantation (30 min to 1 h), the supernatant was recovered for analysis (measurement of turbidity, pH, the permangante oxidation, and BOD_5 was measured for the best turbidity levels. The results identified during the flocculation tests on wastewaters from the WWTP of Saida are reported in the following Figs. 5, 6 and 7.

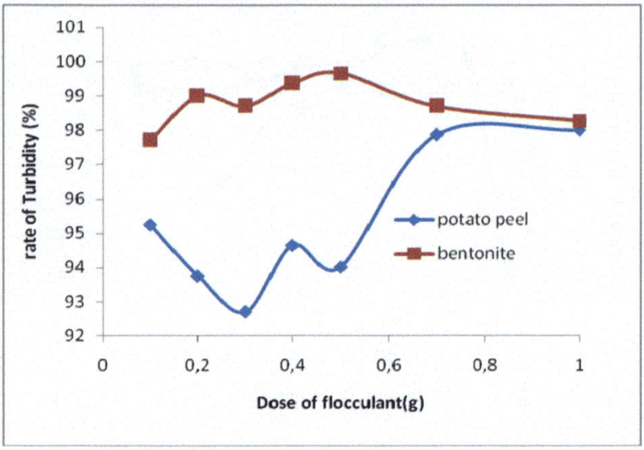

Fig. 5. Evolution of turbidity rate of wastewater according to the dose of flocculant in presence of coagulant.

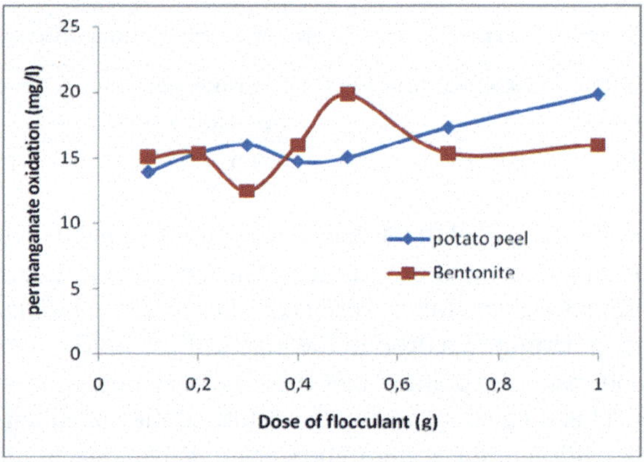

Fig. 6. Evolution of permanganate oxidation of wastewater according to the dose of flocculant in presence of coagulant

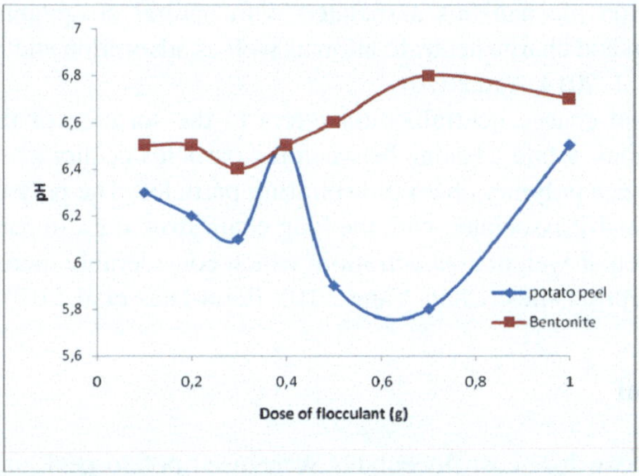

Fig. 7. Evolution of pH of wastewater according to the dose of flocculant in presence of coagulant

The presence of the flocculants gave better results of elimination of turbidity with aluminum sulphate (Fig. 5). The efficacy of treatment is greater than 95%. From the observations made during the manipulations and confirmed by our results, it should be emphasized that there is a disappearance of the harmful odor with a clarification of the waters in the presence of coagulant.

The coagulation-flocculation tests with the two adjuvants bentonite and potato peelers proved to be effective for the depletion of organic matter according to the dose of flocculant in presence of coagulant (Fig. 6).

The results obtained with the aluminum sulphate/potato peel combinations revealed a decrease in pH followed by an increase, this fluctuation being due to the hydrolysis of the salts used. It is also noted that the aluminum sulphate solutions have an acidic character in the presence of bentonite (Fig. 7).

The results showed that flocculation is good for a given mass of aluminum sulphate, and the level of pollution had decreased,

For a dose of aluminum sulphate of 0.6 g, the results obtained show that a dose of bentonite of 0.5 g makes it possible to obtain a good clarification, corresponds to a yield of 99.66%, they act by adsorption and bridging between the colloidal particles. However, an excess of flocculant isolates the surface of the micelles which could have constituted link bridges, which inhibits flocculation (Metchat et al. 2006). While a dose of 1 g of potato peel flocculant allows a good elimination of turbidity with a rate of 98.01% (Table 3).

Table 3. Optimization of the flocculant dose

Parameters/flocculants	Bentonite	Potato peel
Optimal dose of flocculant (g)	0.5	1
Rate of turbidity (%)	99.66	98.01
BOD_5 (mg/l)	94	94

The coagulation mechanisms associated with natural coagulants-flocculants are mainly adsorption and charge neutralization as well as adsorption and bonding between particles (Yen et al. 2014; Yin 2010).

Adsorption and charge neutralization refers to the sorption of the particles with opposite charge ions, while a bridge between the particles occurs when the coagulant-flocculant provides a polymer chain of aspirating particles. The polymeric coagulants/flocculants are usually associated with the long chain structures, in particular polymers with a high molecular weight concentration, which considerably increases the number of available adsorption sites (Chun-Yang 2010; Feria-Díaz et al. 2016).

5 Conclusion

In this research, two low cost flocculants were used potato peel and bentonite. The characterization results by FTIR have shown the functional groups of these flocculants. Flocculation- coagulation process using aluminum sulphate as coagulant was applied to investigate the effect of potato peel and bentonite on the removal of organic matter and turbidity from wastewater of the town of Saida. The effect dose of coagulant used was highly significant, the best results were achieved with aluminum sulphate amount of 0.6 g for the stirring time of 10 min.

For the best coagulation-flocculation treatment, turbidity is considered as a good efficiency indicator in wastewater treatment. The main effect of dose of flocculant showed a removal of turbidity of 99.66% and 98.01% for bentonite and potato peel respectively, the BOD_5 was reduced from 317.03 mg/l to 94 mg/l. Examination of the parameters, measured after the coagulation-flocculation tests on wastewaters showed that the level of pollution had decreased. This jar-test analysis leads to better treatment of water. These low cost flocculants are economically cheaper and easily available making this process very sustainable. This flocculants can be highly recommended for water purification.

References

Bachouche, A.: Traitement des eaux usées de la station de déshuilage de Haoud Berkaoui, mémoire master academique, université de Kasdi rbah, Ouargla (2013)

Boumchita, S., et al.: Removal of cationic dye from aqueous solution by a food waste: potato peel. J. Mater. Environ. Sci. **7**(1), 73–84 (2016)

Lee, C.S., et al.: A review on application of flocculants in wastewater treatment. Process Saf. Environ. Prot. **92**(6), 489–508 (2014)

Yin, C.-Y.: Emerging usage of plant-based coagulants for water and wastewater treatment. Process Biochem. **45**, 1437–1444 (2010)

Deegan, A.M., et al.: Treatment options for wastewater effluents from pharmaceutical companies. Int. J. Environ. Sci. Technol. **8**, 649–666 (2011)

Nutiu, E.: Waste water treatment using a new type of coagulant. In: 8th International Conference Interdisciplinarity in Engineering, INTER-ENG 2014, 9–10 October 2014. Procedia Technol. **19**, 479–482 (2015)

Garg, A., et al.: Thermochemical precipitation as a pretreatment step for the chemical oxygen demand and color removal from pulp and paper mill effluent. Ind. Eng. Chem. Res. **44**(7), 2016–2026 (2005)

Gallard, H., Gunten, U.V.: Chlorination of natural organic matter: kinetics of chlorination and of THM formation. Water Res. **36**, 65–74 (2002)

Feria-Díaz, J.J., Rodiño-Arguello, J.P., Gutiérrez-Ribon, G.E.: Behavior of turbidity, pH, alkalinity and color in Sinú River raw water treated by natural coagulants, Revista Facultad de Ingeniería, Universidad de Antioquia, No. 78, pp. 119–128 (2016)

Gidas, M.-B.: Etude de la performance du chitosane comme coagulant pour l'enlèvement du cuivre et de la turbidité des des eaux usées, université de Montréal (1998)

Metchat, M., et al.: Analyse des eaux résiduaires industrielles. Mesure de pollution de l'unité céramique de Ténès selon la réglementation en vigueur, mémoire d'ingéniorat en biologie, U.H.B.C (2006). p. 1

Ministerio de Desarrollo Económico: Reglamento Técnico Normativa del Sector de Agua Potable y Saneamiento Básico, Ministerio de Desarrollo Económico, Colombia (2000). p. 114

Rahbar, M.S., et al.: Color removal from industrial wastewater with a novel coagulant flocculant formulation. Int. J. Environ. Sci. Technol. **3**(1), 79–88 (2006)

Yan, S., Nagendra, K., Yeong, T., Eshwaraiah, M., Nagasundara, R.: Utilization of plant-based natural coagulants as future alternatives towards sustainable water clarification. J. Environ. Sci. **26**(11), 2178–2189 (2014)

Wong, S.S., et al.: Treatment of pulp and paper mill wastewater by polyacrylamide (PAM) in polymer induced flocculation. J. Hazard. Mater. **135**(1), 378–388 (2006)

Demineralization Effect on the Physicochemical and Interfacial Properties of Sweet Whey

K. Acem[1(✉)], A. Choukri[2], N. Saoudet[3], S. Tergou[3], and H. Trari[3]

[1] Vegetal Physiology Laboratory Applied in Hydroponics, Faculty of Nature Sciences and Life, Ibn Khaldoun University, PB 78, 14000 Tiaret, Algeria
kamel_acem@yahoo.fr

[2] Ziane Achour University, PB 3117, Ain Chih City, Djelfa, Algeria

[3] Faculty of Nature Sciences and Life, Ibn Khaldoun University, PB 78, 14000 Tiaret, Algeria

Abstract. In Algeria, the rejection of whey into the environment is a favorable factor in the biological pollution of freshwater ecosystems due to its native biochemical composition (lactic bacteria, lactose and proteins), the crude sweet whey is an effluent and a co-product of cheeses of pasta cooked, pressed and the casein; rejected in the environment without prior treatment (case of cheese industry of SIDI SAADA, YELLEL, RELIZANE.Algeria), the whey subsequently affects the quality of freshwater ecosystems (wadi of Mina.Relizane.Algeria). The objective of the work has focused on the characterization of the physicochemical and interfacial properties of demineralized sweet whey. The results showed that the demineralization of sweet whey modified its physicochemical properties on the one hand and improved its interfacial properties on the other hand. In perspective, the valorization of the whey in question remains one of the major preoccupations of the local cheese industries.

Keywords: Sweet whey · Ion exchange resin · Interfacial properties · Proteins Valorization

1 Introduction

Crude sweet whey is a co-product and effluent of cheese at paste cooked and pressed and casein; according to Mollea et al. (2013) global whey production was estimated at 180–190 tons/year; in Algeria and in other countries of the world, this rejection accounts for about 85% of milk processed into cheese (FAO 1995; Gana and Touzi 2001); rejected in the environment without prior treatment (case of dairy and cheese industry SIDI SAADA, YELLEL, RELIZANE.Algeria), the latter may be a favorable factor in the biological pollution of freshwater ecosystems following its native biochemical composition (Acem et al. 2016, 2017a, b), its valorization is an economic and ecological stake since one liter of the whey has a Biological Demand in Oxygen (BDO) is between 30 and 45 g/L of oxygen and therefore requires the oxygen of 4500 L of oxygen unpolluted water (Juillerat and Badoud 2010); making it a pollutant that can no longer be released into the environment.

B. Abdelbaki et al. (Eds.): SMSD 2017, *Proceedings of the Third International Symposium on Materials and Sustainable Development*, pp. 516–526, 2018.
https://doi.org/10.1007/978-3-319-89707-3_56

Whey proteins are widely used in the food industry because they possess, inter alia, good functional properties (Bottomley et al. 1990; Acem et al. 2017a, b), the whey proteins have in particular the properties to retain water, to act as foaming and/or emulsifying agent and to form aggregates (or gels) (Kinsella and Whitehead 1989; Acem and Choukri 2016, 2017).

According to Cheftel et al. (1985), Acem and Choukri (2012a, b) the whey proteins have functional properties which vary according to the preparation treatments they have undergone (ultrafiltration, fixation on ion exchangers, precipitation by chemical or thermal agents).

The whey is very mineralized (8 to 20% of the dry matter), which can impair its transformation and its properties (Daufin et al. 1998; Acem et al. 2008; Acem and Choukri 2012a), for some applications it is therefore important to demineralize it, for example in infant formulas and in dietary applications (Hoppe and Higgins 1992). In this context, the present study focuses on analyzing the impact of demineralization on the physicochemical and interfacial behavior of sweet whey.

2 Materiel and Methods

2.1 Crude Sweet Whey

The crude sweet whey is prepared at the laboratory level using a skim milk powder (0% MG) made from cow's milk by FONTERRA Ltd., 9 Princes Street, Auckland, New Zealand (Acem et al. 2008; Acem and Choukri 2012a), adding 2 volumes of rennet 1%, and heated to 35 °C/40 min (Baumy and Brule 1986); filtered by filter paper (German Folded Filters: 185 mm diameter) and stored at 4 °C (Acem et al. 2008; Acem and Choukri 2012a).

2.2 Demineralized Sweet Whey

According to Acem and Choukri (2012a) the demineralized sweet whey is prepared at the laboratory using the data sheet of the cation exchange resin (Amberlite IRC-86, weak acid, H^+ counter-ion, Fluka brand, product of France). The physico-chemical tests applied to both types of whey are: pH (pH meter CG 822 GHS), proteins (Lowry et al. 1951), viscosity by a ball drop viscometer (Viscometer: HOEPPLER BH2), the ash and dry matter content by the AFNOR method (1980).

2.3 Dispersed System

2.3.1 Emulsion

The emulsion is composed of a dispersed phase (virgin olive oil: it is a virgin olive oil from trade, production 2017 brand CHIALI of Sidi-Bel-Abbes, Algeria) and a dispersing phase (whey); the dispersions are prepared in the presence or absence of sodium caseinates (stabilizing agent) and Tween 80 as surfactant (Fluka product of France) in an

O/W ratio equal to 0.0526; each mixture is homogenized at 25 °C to 18000 rpm during 1 min by a homogenizer (Bomann, STABMIXER SM 354 CB-hand blender).

2.3.2 Foam

For the formation of the foam, 20 mL of the whey is whipped by a Blinder (Bomann, STABMIXER SM 354 CB-hand blender) at 18000 rpm during 1 min. The evaluation of the whey emulsifying power was carried out by measuring the mean diameter of the fat globules of virgin olive oil (Acem and Choukri 2012a) and the creaming index (Klinkesorn et al. 2004), and the estimation of its foaming power was made by controlling the mean diameter of the air bubbles and the foaming capacity (Martínez et al. 2011) whose average diameter of the fat globules and air bubbles is observed and determined under a photonic microscope (PHYWE, hund WETZLAR, GERMANY) using an ocular micrometer graduated from 0 to 10 whose graduations are distant from each other others of 0.1 μm.

2.4 Statistical Analysis

The experimental data are analyzed statistically by R software.

3 Results and Discussion

3.1 Demineralized Sweet Whey

Figure 1 gives the average physicochemical properties of the demineralized sweet whey.

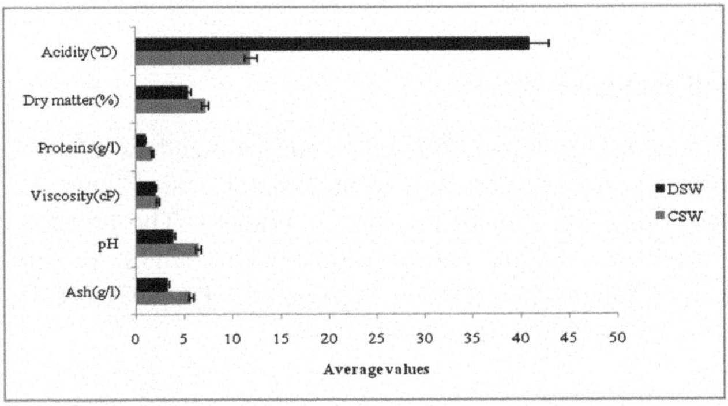

Fig. 1. Average values of the physicochemical parameters of crude sweet whey (CSW) and demineralized sweet whey (DSW).

Figure 1 shows that demineralization has modified the contents of controlled physicochemical parameters compared to those labeled for crude sweet whey, with the following values: acidity (DSW:41, CSW:12°D), dry matter (DSW:5.5, CSW:

7.25 g/L), proteins (DSW:1.06, CSW:1.81 g/L), viscosity (DSW:2.12, CSW:2.35cP), pH (DSW:3.93, CSW:6.48) et ash (DSW:3.31, CSW:5.72 g/L).

According to Adrian et al. (1981), the viscosity depends on the temperature, the nature of the solvent, the size, the shape, the concentration, the electrical charge of the dispersed particles and their affinity for the solvent. According to Lorient et al. (1985), the viscosity of most proteins increases in an alkaline environment. The pH and acidity values are comparable to those found by Acem and Choukri (2012a); the pH is a decreasing function with the acidity, the latter evolves with the composition and the high content of acid substances (Mathieu 1998). In addition, the ash content of the crude sweet whey studied is lower than that found by Croguennec et al. (2008) that they obtained the following results: 7.5 g/L for sweet whey, 12 g/L for acid whey and 9 g/L for acid whey of casein industry. According to Sottiez (1990) the variation in the dry matter content of whey depends mainly on the quality of the milk used and the process of separation of the whey.

3.2 Dispersed System

3.2.1 Emulsifying Power

Figure 2 shows the kinetics of the mean diameter of the fat globules of the emulsions studied.

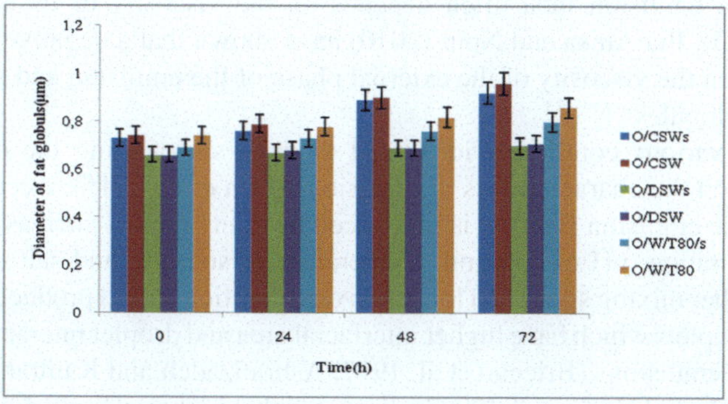

Fig. 2. Kinetics of the average diameter of the fat globules of the emulsions studied.

We note that the lower average diameter of the fat globules is recorded in the demineralized sweet whey emulsions in the presence of sodium caseinates (O/DSWs) and in the absence of sodium caseinates (O/DSW), which has been characterized its stability over time: emulsion (O/DSWs): 0.66 at time t0 h to 0.70 µm after 72 h; emulsion (O/DSW): 0.66 at time t0 h to 0.71 µm after 72 h.

Figure 3 illustrates the kinetics of the creaming index of the emulsions over time.

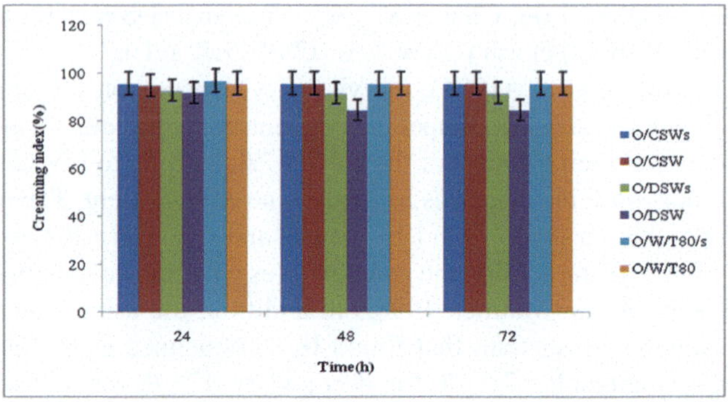

Fig. 3. Kinetics of the creaming index of the emulsions studied.

We note that the creaming index of the emulsions studied tends to decrease over time, in particular the emulsions based on demineralized whey in the presence of sodium caseinates (O/DSWs: from 92.85 after 24 h to 91.8% after 72 h) and in the absence of sodium caseinates (O/DSW: from 91.83 after 24 h to 84.69% after 72 h).

The emulsifying properties of whey proteins are due to the ability to reduce interfacial tensions between hydrophilic and hydrophobic components; they are often related to the solubility of the protein in water (Lorient and Linden 1994). The rheological behavior of an emulsion then often depends on the viscosity of its external phase (Allouche 2003). Ilia Anisa and Nour (2010) have shown that a negative correlation is evident between the viscosity of the external phase of the emulsion and the size of the dispersed fat globules.

There are various conflicting ideas and theories surrounding the effects of salt concentration on the characteristics of emulsions (Lim et al. 2015). According to Lim et al. (2015) the emulsion stability is influenced by many factors such as mixing speed and mixing duration, pH of solution, temperature of solution and salt concentration. Generally, higher mixing speed and longer mixing duration would produce smaller sizes of emulsion droplets which have higher interfacial area and droplet interaction, resulting in more stable emulsions (Briceno et al. 1997; Ashrafizadeh and Kamran 2010).

Sakka (2003) and Yang et al. (2007) suggested that higher pH (alkaline) would give more stability to the emulsions. According to Yang et al. (2007) for oil-in-water emulsions, higher pH would promote more affinity of surfactant molecules towards aggregation, resulting in more stable emulsions. Tambe and Sharma (1993) have found that low pH values of 4–6 favoured oil-in-water emulsions while high pH values of 8–10 favoured water-in-oil emulsions; they found out that for the case of oil-in-water emulsions, the stability was increased as the pH as increased from 4 to 6, but beyond pH 6 to pH 10, oil-in-water emulsions became less stable and on the flip side, water-in-oil emulsions were favoured.

The high or very low pH values would help in stabilizing the emulsions (Sjoblom et al. 1990). Higher temperature will first reduce the emulsions viscosity, eventually destabilizing and breaking the emulsions (Lim et al. 2015). Mat et al. (2006) highlighted that the rate of droplet film drainage is directly proportional to the temperature.

Ashrafizadeh and Kamran (2010) reported that at higher salinity of the water phase, the viscosity of oil-in-water emulsions was increased and the amount of separated water was reduced as well; and that the phenomenon could be caused by salt ions acting as barriers among the oil droplets and the continuous water phase, thus increasing the emulsions stability.

Figure 4 shows the microscopic aspect of the emulsions prepared from crude sweet whey (CSW) and demineralized (DSW).

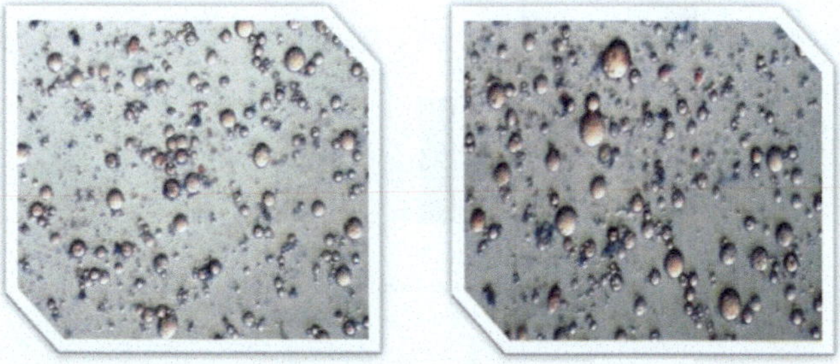

(a) (b)

Fig. 4. Microscopic aspect (Mg: 100x) of the emulsions studied at time 0 h showing the influence of demineralization of sweet whey on the size of fat globules: (a) emulsion O/DSWs and (b) emulsion O/CSW.

3.2.2 Foaming Power

Figure 5 shows the evolution of the mean diameter of the air bubbles formed from the crude sweet whey (CSW) and demineralized (DSW).

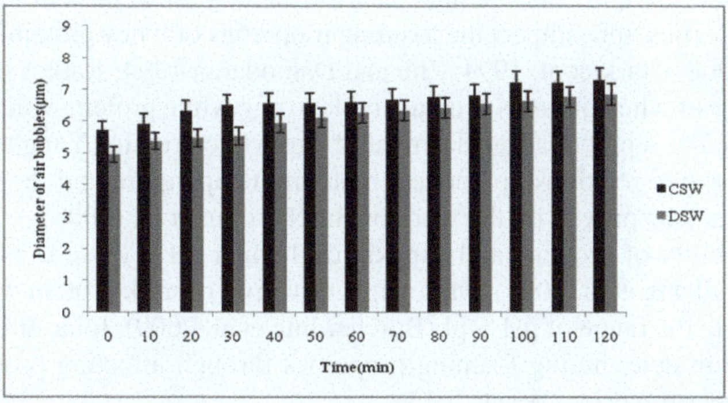

Fig. 5. Kinetics of the average diameter of the air bubbles of the crude sweet whey (CSW) and demineralized sweet whey (DSW).

The average diameter of the air bubbles tends to increase over time; for crude sweet whey (CSW) foam: the diameter varied from 5.72 at time 0 min to 7.31 µm after 120 min with respect to the foam prepared from demineralized sweet whey (DSW) which modulated from 4.96 in time 0 min to 6.85 µm after 120 min.

Figure 6 shows the evolution of the foaming capacity of the crude sweet whey (CSW) and demineralized (DSW).

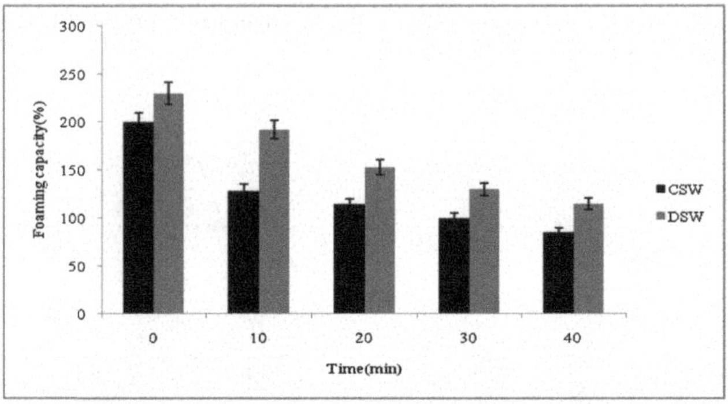

Fig. 6. Kinetics of the foaming capacity of the crude sweet whey (CSW) and demineralized sweet whey (DSW).

We observe that the curves of the foaming capacity are decreasing over time: for the crude sweet whey (CSW) foam: its foaming capacity has varied from 200 at time 0 min to 85.71% after 40 min and has modulated from 230 at time 0 min to 115% after 40 min for demineralized whey (DSW) foam.

The foaming properties or surface activities of proteins are closely related to the protein conformational properties, such as structure, size, shape, and amino-acid composition (Patino and Pilosof 2011) and external environment factors. External parameters, such as pH, ionic strength, and heating condition directly affect protein structure and its interfacial properties, thus impact the foaming properties of whey protein (Phillips et al. 1990, 2006, 2008; Clark et al. 1994; Zhu and Damodaran 1994; Raikos et al. 2007).

Foamability of whey proteins is comparable to egg white protein which allows them to be considered as a potential replacement of egg white protein (Yang and Foegeding 2011). The negative relationship between foaming temperature and the average foam bubble diameter was proved by the work of Borcherding et al. (2008).

The foamability of protein can be maximized when pH is close to pI (Hailing and Walstra 1981; Zhang et al. 2004), while higher overrun of milk protein was found with increasing pH in the range of pH > pI (Borcherding et al. 2009). Ionic strength plays an important role in determining foaming properties through affecting protein solubility and electrostatic properties (Wang 2013).

The overrun of 5% WPI (Whey protein isolate) had positive relation with NaCl concentration (from 0 to 1 M), and with the increasing of ionic strength, the stability showed a peak and then decreased (Mott et al. 1999). Zhu and Damodaran (1994) reported that foam stability reached the maximum value in the presence of low concentration of $CaCl_2$ (20 mM) and then decreased at higher $CaCl_2$ concentration.

The Fig. 7 shows the microscopic aspect of the foams prepared with the crude sweet whey (CSW) and demineralized (DSW).

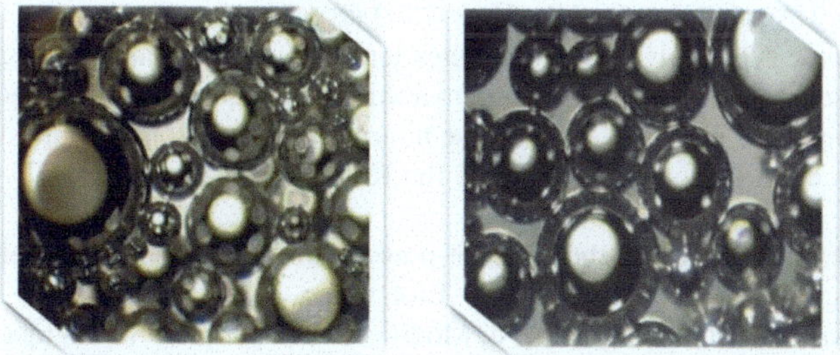

(c) (d)

Fig. 7. Microscopic aspect (Mg: 100x) of the foams studied of sweet whey at time 0 min: (c) CSW and (d) DSW

4 Conclusion

The demineralization of the crude sweet whey by the cation exchange resin modified on the one hand its physicochemical parameters, namely the ash, proteins, dry matter, viscosity, pH and acidity content, and on the other hand it improved interfacial properties, namely its emulsifying power (the average diameter of the fat globules and the index of creaming) and foaming (the average diameter of the air bubbles and the foaming capacity).

5 List of Abbreviations

CSW: Crude sweet whey.
DSW: Demineralized sweet whey.
O/CSW: Emulsion of oil in crude sweet whey without stabilizer.
O/CSWs: Emulsion of oil in crude sweet whey with stabilizer
O/DSW: Emulsion of oil in demineralized sweet whey without stabilizer.
O/DSWs: Emulsion of oil in demineralized sweet whey with stabilizer.
O/W/T80: Emulsion of oil in water in presence the tween 80 and without stabilizer.
O/W/T80/s: Emulsion of oil in water in presence the tween 80 and stabilizer.
Mg: Magnification.

References

Acem, K., Choukri, A.: Study of the emulsifying properties of whey proteins in crude and modified environments. J Appl. Sci. Res. **8**(7), 3293–3303 (2012a)

Acem, K., Choukri, A.: Analyse des propriétés tensioactives des lactosérums bruts dans une émulsion cosmétique. Revue écologie-environnement. Laboratoire de recherche d'agro-biotechnologie et de nutrition en zones semi-arides et laboratoire de physiologie végétale appliquée aux cultures hors sol, université Ibn Khaldoun, Tiaret, Algerie, No. 12, pp. 14–16 (2016)

Acem, K., Choukri, A.: Study of the effect of power's ultrasound on the protein's emulsifying properties of crudes wheys. Adv. Environ. Biol. **6**(10), 2689–2697 (2012b)

Acem, K., Choukri, A., Moulai Mostefa, N.: Modification physique et biologique du lactosérums en vue de sa valorisation dans le domaine des émulsions (cosmétique), revue de la recherche agronomique de l'INRA, actes du séminaire international sur la biotechnologie au service du secteur agroalimentaire, Tipaza, 17–18 juin 2008, pp. 9–14 (2008)

Acem, K., Choukri, A.: Etude du comportement interfacial des lactosérums bruts dans une émulsion alimentaire. Revue d'écologie-environnement. revue ecologie-environnement. Laboratoire de recherche d'agro-biotechnologie et de nutrition en zones semi-arides et laboratoire de physiologie végétale appliquée aux cultures hors sol, université Ibn Khaldoun, Tiaret, Algerie, No. 13, pp. 17–20 (2017)

Acem, K., Choukri, A., Moulai Mostefa, N., Mettai, K., Kadi, S., Fetouhi, B.: Analyse et valorisation des propriétés émulsifiantes des lactosérums déprotéinés. Conférence scientifique internationale sur l'environnement et l'agriculture (CSIEA), Hammamet. Tunisie. 24 au 25 Avril 2017 (2017a)

Acem, K., Mechouar, M., Ayad, I., Mettai, K., Kadi, S., Fetouhi, B.: Impact du lactosérum doux brut sur les propriétés physiques et chimiques des écosystèmes dulçaquicoles. Conférence scientifique internationale sur l'environnement et l'agriculture (CSIEA), à Hammamet. Tunisie. 24 au 25 Avril 2017 (2017b)

Acem, K., Mettai, K., Kadi, S., Fetouhi, B., Choukri, A., Moulai Mostefa, N.: Caractérisation et valorisation des propriétés polluantes du lactosérum brut. 1er colloque international en toxicologie et santé (ICTH). Faculté des Sciences de la Nature et de la Vie et Sciences de la Terre et l'Univers. Université 8Mai 1945, Guelma, 28–29 Nov 2016 (2016)

Adrian, J., Legrand, G., Franrue, R.: Dictionnaire de biochimie alimentaire et de nutrition, p. 233. Tec et Doc, Lavoisier, Paris (1981)

AFNOR: Méthodes d'analyses du lait et ses produits laitiers, recueil de normalisation française, 2ème édn., p. 580 (1980)

Allouche, J.: Developpement de nouvelles methodes pour l'elaboration d'emulsions multiples Eau/Huile/Eau, thèse, Institut National Polytechnique De Lorraine, Ecole National Supérieure des Industries Chimiques, p. 175 (2003)

Ashrafizadeh, S.N., Kamran, M.: Emulsification of heavy crude oil in water for pipeline transportation. J. Pet. Sci. Eng. **71**, 205–211 (2010)

Baumy, J.J., Brule, G.: Étude comparée de la solubilité et de la viscosité des solutions de caseinates et paracaseinates de sodium en présence de calcium. Le lait **66**, 65–77 (1986)

Borcherding, K., Lorenzen, P.C., Hoffmann, W., Schrader, K.: Effect of foaming temperature and varying time/temperature-conditions of pre-heating on the foaming properties of skimmed milk. Int. Dairy J. **18**(4), 349–358 (2008)

Borcherding, K., Lorenzen, P.C.H.R., Hoffmann, W.: Effect of protein content, casein-whey protein ratio and pH value on the foaming properties of skimmed milk. Int. J. Dairy Technol. **62**(2), 161–169 (2009)

Bottomley, R.C., Evans, M.T.A., Parhiason, C.J.: Whey proteins. In: Harris, P. (ed.) Food Gels. Applied Science, pp. 435–466. Elsevier, London, New York (1990)

Briceno, M., Isabel, R.M., Bullon, J., Salager, J.L.: Customizing drop size distribution to change emulsion viscosity. In: Proceedings of the 2nd World Congress on Emulsion, Bordeaux, France, 23–26 Sept 1997

Cheftel, J.C., Cuq, J.I., Lorient, D.: Protéines Alimentaires. Tec et Doc, p. 308 (1985)

Clark, D.C., Wilde, P.J., Marion, D.: The protection of beer foam against lipid-induced destabilization. J. Inst. Brew. **100**(1), 23–26 (1994)

Croguennec, T., Jeantet, R., Brule, G.: Fondements physicochimiques de la technologie laitière. Tec et Doc Lavoisier, p. 161 (2008)

Daufin, G., Rene, F., Aimar, P.: Les séparations par membrane dans les procédés de l'industrie alimentaire. Collection sciences & techniques agroalimentaires, p. 592. Technique & Documentation, France (1998)

FAO: Lait et produits laitiers dans la nutrition humaine, p. 262. FAO, Rome (1995)

Gana, S., Touzi, A.: Valorisation du lactosérum par la production de levures lactiques avec les procédés de fermentation discontinue et continue. Rev. Energ. Ren. **1**, 51–58 (2001)

Hailing, P.J., Walstra, P.: Protein-stabilized foams and emulsions. Crit. Rev. Food Sci. Nutr. **15**(2), 155–203 (1981)

Hoppe, G.K., Higgins, J.J.: Demineralization. In: Zadow, J.G. (ed.) Whey and Lactose Processing. Applied Science, pp. 91–131. Elsevier, London, New York (1992)

Ilia Anisa, A.N., Nour, H.A.: Affect of viscosity and droplet diameter on water-in-oil (w/o) emulsions: an experimental study. World Acad. Sci. Eng. Technol. **4**, 595–598 (2010)

Juillerat, M.-A., Badoud, R.: Acides aminés et protéines. In: Sciences et technologie des aliments: principes de chimie des constituants et de technologie des procédés. Presses polytechniques et universitaires romandes, CH-1015 Lausanne, pp. 47–104 (2010)

Kinsella, J.E., Whitehead, D.M.: Proteins in whey: chemical, physical, and functional properties. Adv. Food Nutr. Res. **33**, 343–438 (1989)

Klinkesorn, U., Sophanodora, P., Chinachoti, P., McClements, D.J.: Stability and rheology of corn oil in water emulsions containing maltodextrin. Food Res. Int. **37**, 851–859 (2004)

Lim, J.S., Wong, S.F., Law, M.C., Samyudia, Y., Dol, S.S.: A review on the effects of emulsions on flow behaviours and common factors affecting the stability of emulsions. J. Appl. Sci. **15**(2), 167–172 (2015)

Lorient, D., Linden, G.: Biochimie agro-industrielle: valorisation alimentaire de la production agricole, p. 367. Masson, Paris (1994)

Lorient, D., Luquet, J., Cheftel, J.C.: Protéines alimentaires, p. 309. Tech et Doc Lavoisier, Paris (1985)

Lowry, O.H., Rosebrough, N.J., Farr, A.L., Randall, R.J.: Protein measurement with the Folin phenol reagent. J. Biol. Chem. **193**, 265–275 (1951)

Martínez, K.D., Ganesan, V., Pilosof, A.M., Harte, F.M.: Effect of dynamic high-pressure treatment on the interfacial and foaming properties of soy protein isolate-hydroxypropylmethylcelluloses systems. Food Hydrocolloids **25**(6), 1640–1645 (2011)

Mathieu, J.: Initiation à la physicochimie du lait, p. 220. Tec et Doc, Lavoisier, Paris (1998)

Mollea, C., Marmo, L., Bosco, F. Valorization of cheese whey, a by-product from the dairy industry. In: Agricultural and Biological Sciences: Food Industry, Chap. 24, pp. 549–588 (2013)

Mott, C.L., Hettiarachchy, N.S., Qi, M.: Effect of xanthan gum on enhancing the foaming properties of whey protein isolate. JAOCS **76**(11), 1383–1386 (1999)

Patino, J.M.R., Pilosof, A.M.: Protein-polysaccharide interactions at fluid interfaces. Food Hydrocolloids **25**(8), 1925–1937 (2011)

Phillips, L., Yang, S., Kinsella, J.: Neutral salt effects on stability of whey protein isolate foams. J. Food Sci. **56**(2), 588–589 (2006)
Phillips, L., Yang, S., Schulman, W., Kinsella, J.: Effects of lysozyme, clupeine, and sucrose on the foaming properties of whey protein isolate and β-lactoglobulin. J. Food Sci. **54**(3), 743–747 (2008)
Phillips, L.G., Schulman, W., Kinsella, J.E.: pH and heat treatment effects on foaming of whey protein isolate. J. Food Sci. **5**, 1116–1119 (1990)
Raikos, V., Campbell, L., Euston, S.: Effects of sucrose and sodium chloride on foaming properties of egg white proteins. Food Res. Int. **40**, 347–354 (2007)
Sakka, S.: Sol-Gel science and technology: topics in fundamental research and applications. In: Sol-Gel Prepared Ferroelectrics and Related Materials, vol. 1, pp. 33–35. Springer, Boston (2003)
Sjoblom, J., Mingyuan, L., Hoiland, H., Johansen, E.J.: Water-in-crude oil emulsions from the Norwegian continental shelf part III. A comparative destabilization of model systems. Colloids Surf. **46**, 127–139 (1990)
Sottiez, P.: Produits dérivés des fabrications fromagères. In: Lait et produits laitiers, pp. 357–392. Tec et Doc, Lavoisier, Paris (1990)
Tambe, D.E., Sharma, M.M.: Factors controlling the stability of colloid-stabilized emulsions: an experimental investigation. J. Colloid Interface Sci. **157**, 244–253 (1993)
Wang, Z.: Foaming properties of whey protein isolate and λ-carrageenan mixed systems. Thesis, Master of Science, The Faculty of the Graduate School, The University of Missouri, Columbia, p. 73 (2013)
Yang, F., Niu, Q., Lan, Q., Sun, D.: Effect of dispersion pH on the formation and stability of Pickering emulsions stabilized by layered double hydroxides particles. J. Colloid Interface Sci. **306**, 285–295 (2007)
Yang, X., Foegeding, E.A.: The stability and physical properties of egg white and whey protein foams explained based on microstructure and interfacial properties. Food Hydrocolloids **25**(7), 1687–1701 (2011)
Zhang, Z., Dalgleish, D., Goff, H.: Effect of pH and ionic strength on competitive protein adsorption to air/water interfaces in aqueous foams made with mixed milk proteins. Colloids Surf., B **34**(2), 113–121 (2004)
Zhu, H., Damodaran, S.: Effects of calcium and magnesium ions on aggregation of whey protein isolate and its effect on foaming properties. J. Agric. Food Chem. **42**, 856–862 (1994)

Treatment of Medical Waste Using Electrocoagulation Process

K. Madi-Azegagh(✉), F. Aissani-Benissad, and I. Yahiaoui

Département de Génie des Procédés, Laboratoire de Génie de l'Environnement (LGE), Faculté de la Technologie, Université A. MIRA Bejaia, Bejaia, Algeria
madi.katiia@gmail.com

Abstract. The paper describes the removal of the anti-inflammatory "Ketoprofen (KTP)" by electrocoagulation process using Response Surface Methodology (RSM). pH, Current density, initial ketoprofen concentration were considered as input variables for RSM and the percentage removal (y%) of KTP was chosen as the response parameter. The disappearance of the molecule of KTP in synthetic wastewater and the appearance of by-products were followed by high performance liquid chromatography (HPLC) analysis. Second-order model was obtained and showed the classification of these parameters based on their level of significance. Pareto analysis result and the values of Student's distribution showed that current density had the most important effect on the response, which has a positive effect on the response, but the initial KTP concentration has a negative effect on the response. The analysis of variance (ANOVA) revealed an important coefficient of determination value (R^2 = 96.41%) between removal efficiency and predicted one by RSM developed model. Graphical response surface and counter plots were used to locate the optimum point. The electrocoagulation process was able to achieve 100% of KTP removal yield in 30 min.

Keywords: Electrocoagulation · Ketoprofen · Response surface methodology Optimization

1 Introduction

Water, a natural resource is essential for the existence of life. Although water is present in abundance, but only a limited amount of usable water is available. The level of ground water is decreasing day by day because of more industrialization, improved agriculture, inappropriate disposal techniques, excessive population growth and change in climatic conditions (Sushil and Amit 2014). Therefore, the optimum utilization of the limited water available becomes utmost priority and this could be done only by improving the quality of water and removing unwanted organic, inorganic and mineral substances from water streams (Amandeep et al. 2016). Pharmaceuticals have been identified as potential contaminants in aquatic environments among various water pollutants throughout the world and pollution by these drugs increased noticeably with their incessant use over the years (Sacher et al. 2008; Richardson and Lam 2005; Kolpin et al. 2002; Yang et al. 2010). Among the pharmaceuticals, non-steroidal

B. Abdelbaki et al. (Eds.): SMSD 2017, *Proceedings of the Third International Symposium on Materials and Sustainable Development*, pp. 527–539, 2018.
https://doi.org/10.1007/978-3-319-89707-3_57

anti-inflammatory drugs (NSAIDs) are most frequently found due to their widespread use (Liu et al. 2015; Tauxe-Wuersch et al. 2005; Nakada et al. 2006). Al though there is no evidence to suggest that NSAIDs are harmful to adults, they may be toxic to aquatic organisms and harmful to embryos, infants, children, and adults with weak constitutions and sensitivity to pharmaceuticals. Recent studies demonstrate that non-steroidal anti-inflammatory drugs (NSAIDs) are frequently found in surface water, sewer water, ground water and even in drinking water as pharmaceutically active compounds (Corcoll et al. 2014; Amandeep et al. 2016a). Among NSAIDs, ketoprofen (2-(3-benzoylphenyl)-propionic acid (KET)) is too consumed drug and is usually found in surface water (Tsvetkova and Peikova 2013). Accordingly, wastewater containing pharmaceuticals compounds should be treated prior discharging into the sewer networks and/or surface water. There are several processes for treating. Usually, the attention received by conventional biological treatment far exceeds that of other treatment methods, in part due to its low cost, high stability, and easy operating. However, the presence of toxic and bio-recalcitrant organics detracts the viability of biological treatment process (Kusi et al. 2006). During last two decades, different electrochemical technologies including electrochemical reduction, direct and indirect electrochemical oxidation and photo-assisted electrochemical methods have been utilized to remove pharmaceuticals compounds in wastewater (Ignasi and Enric 2012). One of the most interesting electrochemical technologies for the treatment of different wastewaters is electrocoagulation (EC) technique; it has a number of advantages such as low-cost, easy handling, and high efficiency in the removal of organic matter. On the other hand, the sludge generation rate, the use of sacrificial electrodes, and electrical energy consumption can be cited as drawbacks (Mollah et al. 2001). The EC method has been successfully applied to the remediation of some synthetic and industrial pharmaceutical wastewaters. Yehya et al. (2015) studied the removal of carbamazepine by electrocoagulation and found that 62% of carbamazepine was eliminated under slightly acidic initial conditions (pH 4) with a current density as high as 44 mA cm^{-2} (I = 4.5 A) using aluminum electrodes. Chou et al. (2011) reported total decontamination of synthetic salicylic acid solutions using aluminum electrodes at 1.2 mA cm^{-2} with a reasonable efficiency (Ignasi and Enric 2012). Deshpande et al. (2005) described the excellent decontamination of a pharmaceutical effluent using a cast iron anode with the removal of 72% of chemical oxygen demand (COD) and the improvement of the biochemical oxygen demand (BOD)/COD ratio from 0.18 to 0.3, pointing out the enhancement of the wastewater biodegradability (Ignasi and Enric 2012). Another study was reported by Daniel et al. (2013), this study is concerned with the removal of the anti-inflammatory dexamethasone from aqueous solution and hospital wastewater by electrocoagulation by using commercial aluminum electrodes (61 cm^2 effective area), the main effect of the electrocoagulation was that it removed colloids and reduced the organic load of the hospital wastewater. The results obtained show an increase in the removal of dexamethasone (up to 38.1%) with a rise of the current applied and a decrease of the inter-electrode distance, in aqueous solutions. The similar results were obtained with a hospital effluent for the removal of dexamethasone. According to the study of Ait Ouaissa et al. (2014), tetracycline was removal by electrocoagulation using aluminum as sacrificial anode of 28.5 cm^2 of submerged area, the removal efficiency of this compound was found at pH ranging from 3 to 10 and irrespective of the initial concentration. A more comprehensive work was further reported by Deshpande et al. (2010) for a high-strength wastewater from an industry manufacturing the drug ranitidine with 34 $g.L^{-1}$ COD, 16 $g.L^{-1}$ BOD, and 775 Pt–Co units color at pH 6.8–7.2. The EC experiments were performed with 150 mL of wastewaters in a stirred batch tank reactor equipped with 25 cm^2 Al/Al, mild steel/mild steel or stainless steel/stainless steel pairs. Comparative electrolyses at 8 mA cm^{-2} revealed that COD could be reduced by 24% at 25 min, 25% at 120 min, and 33% at 120 min as maximum, respectively, attaining final BOD/COD values of 0.546, 0.519, and 0.538 and final color of 235, 325, and 490 Pt–Co units. These results indicated a biodegradability enhancement of 9.3–13.8% and a better efficiency using the Al/Al pair during shorter time, with the lowest energy consumption of 95.83 $W.h.kg^{-1}$ COD (Ignasi and Enric 2012). In the EC process, many factors such as pH, applied electric current, the initial concentration and the application time influence the process efficiency. The process efficiency may be increased by the optimization of these factors. In conventional multifactor experiments, optimization is usually carried out by varying a single factor while keeping all the other factors fixed at a specific set of conditions (Wang et al. 2007; Tugba 2009). This method is time consuming and incapable of effective optimization. Recently, Response Surface Methodology (RSM) has been employed to optimize and understand the performance of complex systems (Wang et al. 2007; Tugba 2009; Sudarjanto et al. 2006; Korbahti 2007). The RSM is a statistical technique for designing experiments, building models, evaluating the effects of several factors, and searching optimum conditions for desirable responses. With RSM, the interactions of possible influencing parameters on treatment efficiency can be evaluated with a limited number of planned experiments. RSM is widely used in various fields (Wang et al. 2007; Tugba 2009; Sudarjanto et al. 2006; Korbahti 2007; Dos Santos et al. 2005).

In the present study, RSM, statistica version 10 including ANOVA was employed for the optimization of KTP removal, as a model pollutant, through EC process using iron electrodes. The percentage removal (y%) of KTP was chosen as the response parameter and pH (x_1), current density (x_2) and initial KTP concentration (x_3) were selected as process variables. The disappearance of the molecule of KTP in synthetic wastewater and the appearance of by-products were followed by high performance liquid chromatography (HPLC) analysis. The optimal conditions for complete KTP removal were also determined graphically from the model obtained.

2 Materials and Methods

2.1 Chemicals and Reagents

Ketoprofen ($\geq$98%, Formula: $C_{16}H_{14}O_3$ corresponds to a molecular weight 254.28 $g.mol^{-1}$), its molecular structure is shown in Fig. 1, acetonitrile ($\geq$99.9%) and acetic acid ($\geq$99.8%) were purchased from Sigma-Aldrich (France). Hydrochloric acid and sodium hydroxide supplied by prolabo (France).

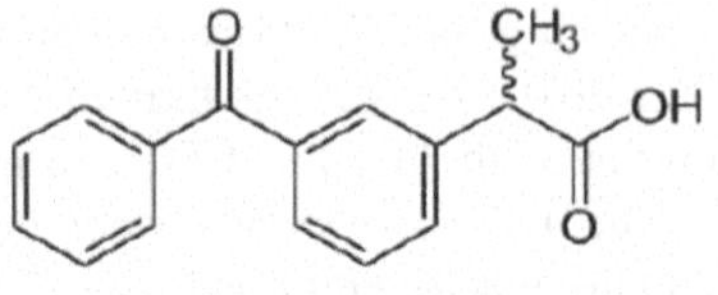

Fig. 1. Chemical structure of Ketoprofen

2.2 Experiments

Experiments were carried out in a batch electrochemical reactor of 1 L capacity with two rectangular iron electrodes which were used as the anode and the cathode, their dimensions were 80 mm × 65 mm × 1 mm with an inter-electrode distance of 10 mm, this distance was maintained constant during electrolysis (t = 30 min). They were connected to a precision DC power supply (VOLTCRAFT HPS-13015). Mixing was ensured by a magnetic stirring bar. Figure 2 summarizes the experimental setup. Before each run, the electrodes were treated with an HCl (0.2 $mol.L^{-1}$) aqueous solution and rinsed again with distilled water for cleaning prior use to avoid passivation. The experimental parameter ranges were selected as follows: pH (4–10), current density (9.61–38.46 $mA.cm^{-2}$) and initial KTP concentration (10–25 $mg.L^{-1}$), Synthetic wastewater was prepared by dissolving 40 $mg.L^{-1}$ into distilled water. All solutions were prepared with KTP, the composition of synthetic water includes also NaCl (3.5 $g.L^{-1}$) as a supporting electrolyte and pH is 8.2. Initial pH is then adjusted between 4 and 10 by the minute addition of either 0.1 M hydrochloric acid or sodium hydroxide solutions. After each electrocoagulation run, the liquid samples were filtered through a 0.45 μm size membrane filter. The injection volume was 20 μL and the retention time t_r of KTP is about 6.861 min.

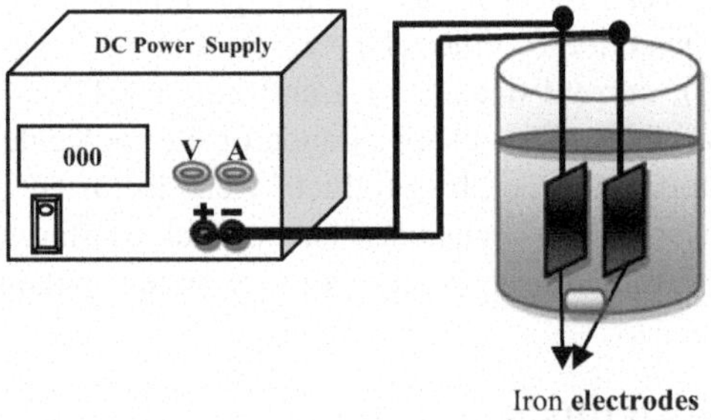

Fig. 2. Electrocoagulation set-up

2.3 Analytical Methods

In order to estimate the efficiency of KTP removal, samples were collected at the end of the experiments. The residual concentration of KTP in the synthetic wastewater was determined by HPLC analysis (Agilent 1100, UV, France) under isocratic mode, equipped a series auto sampler, a C_{18} column (diameter: 3.9 mm, length: 300 mm) and a spectra physics 2000 UV detector at a wave-length of 260 nm. The mobile phase consisted of ultra-pure water acidified (0.1%) and acetonitrile (Sigma-Aldrich, France) at 50:50 (v:v). It was filtered through a 0.45 μm size membrane filter, degassed, and delivered at 1 mL/min.

3 Response Surface Methodology (RSM)

Experimental design of the EC process for KTP removal was carried out by using the RSM. RSM is a collection of mathematical and statistical techniques that are beneficial for the modeling and analysis of problems in which a response of interest is influenced by several variables and the objective is to search the optimum conditions of variables to predict targeted responses (Tugba 2009; Myers et al. 2002). RSM consists of an empirical modeling technique devoted to the evaluation of relations existing between a group of controlled experimental factors and the observed results. RSM is an important branch of experimental design and a critical methodology in developing new processes, optimizing their performance and improving design and formulation of new products. The most extensive applications of RSM are in industrial research, particularly in situations where several input variables influence the process performance measure. This process performance measure is called the response and the input variables are called independent variables (Tugba 2009; Myers et al. 2002). After performing a full factorial design, response surface methodology (RSM) was applied according to central composite design experiments (CCD) in order to study the performance of elimination of KTP and to determine the optimum operational conditions, this type of design comprises two levels factorial design (+1, −1), superimposed by the centre points (coded 0), and the star points (+α, −α); α is the distance of the axial point from the centre domain (Kolz et al. 2005, Elmollaa et al. 2010, Hansen et al. 2007). The precise value of α depends on the desired properties for the design and on the number of factors involved (Yahiaoui et al. 2010, 2011). In this study, three operating factors were chosen as independent variables, namely: pH (x_1), current density (x_2), initial KTP concentration (x_3), their experimental ranges in coded and actual values are presented in Table 1. The response is expressed as the percentage removal (y%) of KTP, it is determined from the relation given below. Table 2 presents the responses obtained after EC.

Table 1. Values and levels of the operating parameters

Operating factors	Levels				
	$-\alpha = -2$	−1	0	+1	$+\alpha = +2$
Z_1: pH	4	5.50	7	8.50	10
Z_2: i ($mA.cm^{-2}$)	9.62	16.83	24.04	31.25	38.46
Z_3: $[KTP]_0$ ($mg.L^{-1}$)	5	10	15	20	25

Table 2. Response (Ketoprofen removal) values for different experimental conditions

Run	Natural values of parameters			Coded values of parameters			Observed value	Predicted value
N°	Z_1	Z_2	Z_3	X_1	X_2	X_3	y(%)	$\hat{y}$(%)
1	5.5	16.83	10	−1	−1	−1	42.4	40.87
2	5.5	16.83	20	−1	−1	1	26.71	23.55
3	5.5	31.25	10	−1	1	−1	99.22	98.63
4	5.5	31.25	20	−1	1	1	65.74	67.36
5	8.5	16.83	10	1	−1	−1	57.51	61.03
6	8.5	16.83	20	1	−1	1	21.51	27.24
7	8.5	31.25	10	1	1	−1	97.35	105.65
8	8.5	31.25	20	1	1	1	51.24	57.91
9	7	24.04	15	0	0	0	54.7	55.29
10	7	24.04	15	0	0	0	53.27	55.29
11	7	24.04	15	0	0	0	57.12	55.29
12	7	24.04	15	0	0	0	57.99	55.29
13	4	24.04	15	−1.68	0	0	44.49	50.79
14	10	24.04	15	1.68	0	0	70.06	59.78
15	7	9.62	15	0	−1.68	0	18.43	18.15
16	7	38.46	15	0	1.68	0	99.51	92.43
17	7	24.04	5	0	0	−1.68	100	96.70
18	7	24.04	25	0	0	1.68	46.04	42.05

$$y(\%) = \frac{[\mathrm{KTP}]_0 - [\mathrm{KTP}]_t}{[\mathrm{KTP}]_0} \times 100 \tag{1}$$

Where C_0 and C_t were the initial KTP concentration and its concentration at a given time t (mg.L^{-1}).

3.1 Data Analysis

For the evaluation of experimental data, the response variable was fitted by second-order model in the form of quadratic polynomial equation given below (Tugba 2009).

$$\hat{y} = b_0 + \sum b_i X_i + \sum b_{ii} X_i^2 + \sum b_{ij} X_i X_j \tag{2}$$

Where:

- $\hat{y}$ is the predicted response (KTP removal, %);
- x_i (i = 1, 2 and 3) are variables or independent factors;
- b_0, bi (i = 1, 2 and 3; j = 1,2 and 3) are the model coefficients.

Experimental data were analyzed using Statistica version 10 including ANOVA to obtain the interaction between the process variables and the response. Two-dimensional counter plots and three-dimensional curves of the response surface were developed using the same program.

4 Results and Discussion

Based on the experimental design results (Table 2) and after discarding the insignificant model terms (P > 0.05) which are shown in Table 3, an empirical relationship between the response and the independent variables (pH (x_1), current density (x_2) and initial KTP concentration (x_3)) was obtained and expressed by the following polynomial equation:

$$\hat{y} = 55.29 + 2.67x_1 + 22.11x_2 - 16.27x_3 + 4.99x_3^2 - 3.29x_1x_2 - 4.12x_1x_3 - 3.49x_2x_3 \quad (3)$$

Table 3. Estimated coefficients and corresponding T and P values

Coefficient	Coefficient estimate	T-value	P-value
b_0	56.06	51.72	0.00
b_1	2.67	4.55	0.02
b_2	22.11	37.60	0.00
b_3	−16.27	−27.67	0.00
b_{33}	4.99	7.92	0.00
b_{12}	−3.29	−4.28	0.02
b_{13}	−4.12	−5.36	0.01
b_{23}	−3.49	−4.54	0.02

The analysis of variance (ANOVA) of regression parameters of the predicted response surface quadratic model for KTP removal efficiency is required to test the significance and adequacy of the model (Yahiaoui et al. 2010). It can be seen from Table 4, the model F-value of 24.27 and a low probability value (P 0.000071) indicate that the model is significant for KTP removal. A good adjustment of Eq. (3) to experimental data was verified through the high correlation coefficient value obtained (R^2 = 96.41%), this value indicates that only 3.59% of the total variation could not be explained by the empirical model.

A comparison of the experimental and calculated responses (Fig. 3a) shows a clear agreement between the observed values and those predicted by the statistical model Eq. 3. According to the plot of the residuals in function of the predicted response

Table 4. Analysis of variance (ANOVA)

Source of variation	Sum of squares	Degree of freedom	Adjusted mean square	F-value	P-value
Regression	11040.16	9.00	1226.68	24.27	0.00
Residual error	404.37	8.00	50.55		
Total	11444.54				

R^2 = 96.41%. Adjusted R^2 = 93.89%

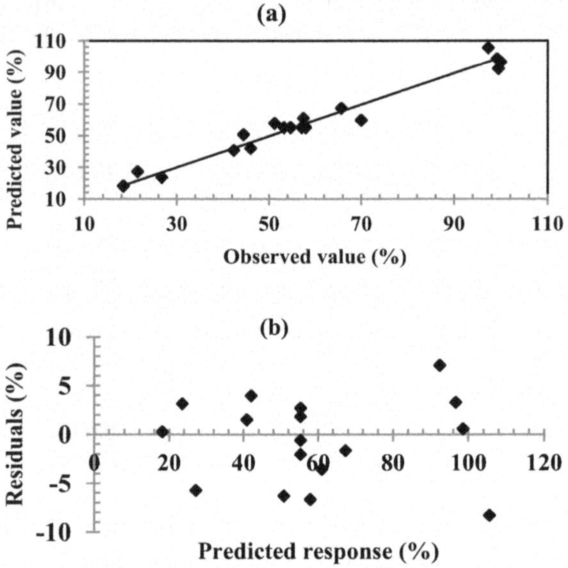

Fig. 3. Analysis of quality of model (a) Comparison of experimental and predicted responses; (b) Residual analysis for the estimated

(Fig. 3b) shows that the residuals have been distributed randomly, so the mathematical model was adequate and that there was no inconsistency between the experimental and calculated values of the response, to these plots.

The Pareto analysis informs us about the most significant parameters. In fact, this analysis calculates the percentage effect of each factor on the response, according to the following relation (Khataee et al. 2010; Amani-Ghadim et al. 2013).

$$P_i = \left(\frac{b_i^2}{\sum_i^n b_i^2}\right) \times 100 \qquad i \neq 0 \tag{4}$$

Figure 4 shows the result obtained from Pareto analysis. This figure suggest that current density had the most important effect on the response b_2 (59.21%) which has a positive effect on the response ($b_2 = +22.11$). These results are expectable because the dosage of iron ions released from anode was increased by increasing the current density and electrolysis time, according to the Faraday's law (Amani-Ghadim et al. 2013; Olad et al. 2010; Lakshmanan et al. 2009; Panizza and Cerisola 2010). After the current density, the initial KTP concentration was more significant than other factors (b_3 32.06%) with a negative effect on the response ($b_3 = -16.27$). The presumed reason can be deduced from Faraday's law (Daneshvar et al. 2007) which provides the same amount of iron dissolved at the same current density and time for all KTP concentrations. Consequently, the same amount of coagulant species would be produced in the liquid phase, which justifies the lower removal efficiency obtained at higher KTP concentrations. Indeed, the flocs produced at high KTP concentration were insufficient to complete pollutant removal. Similar results are reported by Daneshvar et al. (2007). Ait Ouaissa et al. (2014) also reported that the lower removal efficiency of tetracycline by electrocoagulation process was caused by a lower amount of coagulant $Al(OH)_3$ species, which became therefore limiting.

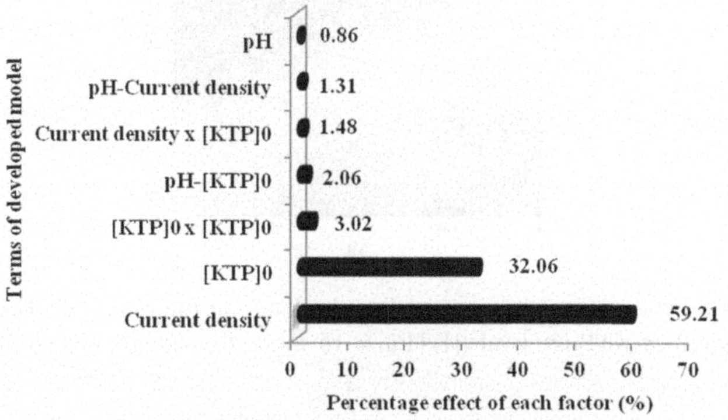

Fig. 4. Pareto graphic analysis

4.1 Optimization of the Electrocoagulation Process for KTP Removal

The main objective of the optimization is to determine the optimum value of variables for KTP removal with EC from the model developed by CCD. Figure 5 illustrates 3-dimensional response surface and 2-dimensional counter plot for investigating the effect of current density and pH on the removal of KTP, when the other parameters were kept constant at $[KTP]_0 = 15$ mg.L^{-1}. It is clear that the optimal conditions found for the highest removal efficiency of KTP were: pH = 7, i = 24.04 mA.cm^{-2} and $[KTP]_0 = 15$ mg.L$^{-1.}$

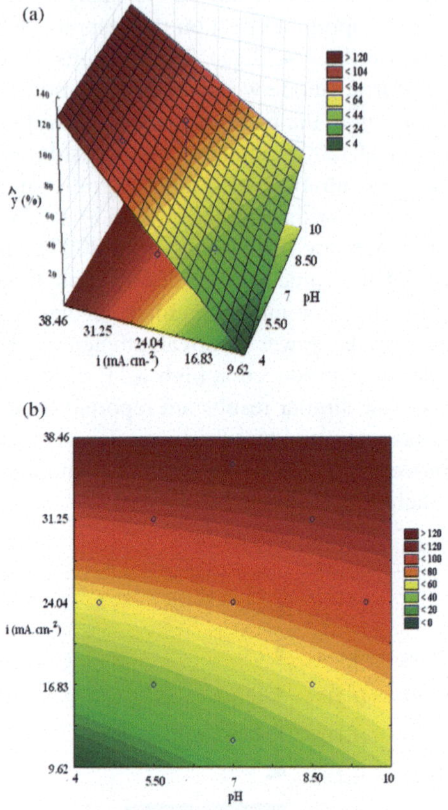

Fig. 5. Response surfaces (a) and counter plots (b) showing the effect of pH (x_1) and current density (x_2) on yield of KTP removal ($[KTP]_0 = 15$ mg.L^{-1})

5 Conclusion

A series of experiments was performed in order to find the effects of operating parameters for removal of the anti-inflammatory ketoprofen by electrocoagulation process. The effects of these important operational parameters including pH, Current density and initial KTP concentration were evaluated by the response surface and counter plots. Second-order model was obtained and showed the classification of these parameters (pH, current density and initial KTP concentration) based on their level of significance. Pareto analysis result and the values of Student's distribution showed that current density had the most important effect on the response P_2 (59.21%) which has a positive effect on the response ($b_2 = +22.11$) but the initial KTP concentration has a negative effect ($b_3 = -16.27$) on the response with a percentage P_3 (32.06%), as regards the pH, it has a negligible effect P_1 (0.86%) on the response comparing with the other parameters. The analysis of variance (ANOVA) revealed an important coefficient of determination value ($R^2 = 96.41\%$) between removal efficiency and predicted one by RSM developed model. Graphically, response surface and counter plots locate the optimal conditions; under some conditions, the electrocoagulation process was able to achieve 100% of KTP removal yield. The KTP removal mechanism between the current density and reaction time was based on flotation. To identify the by-products obtained at the end of EC and have a mechanism of reaction, GC-MS have been proposed as a technical analysis.

References

Sushil, K.K., Amit, K.: Potential of *M. oleifera* for the treatment of water and wastewater. Chem. Rev. **114**, 4993–5010 (2014)

Amandeep, K., Ahmad, U., Sushil, K.K.: Heterogeneous photocatalytic studies of analgesic and non-steroidal anti-inflammatory drugs. Appl. Catal. A **510**, 134–155 (2016)

Sacher, F., Gabriel, M., Graf, C., Brauch, H.J.: Pharmaceutical residues in river Rhine - results of a one-decade monitoring programme. J. Environ. Monit. **10**, 664–670 (2008)

Richardson, B.J., Lam, P.K.S.: Emerging chemicals of concern: pharmaceuticals and personal care products (PPCPs) in Asia, with particular reference to Southern China. Mar. Pollut. Bull. **50**, 913–920 (2005)

Kolpin, D.W., Furlong, E.T., Meyer, M.T., Thurman, E.M., Zaugg, S.D., Barber, L.B., Buxton, H.T.: Pharmaceuticals, hormones, and other organic wastewater contaminants in U.S. streams, 1999–2000: a national reconnaissance. Environ. Sci. Technol. **36**, 1202–1211 (2002)

Yang, H., Li, G., An, T., Gao, Y., Fu, J.: Photocatalytic degradation kinetics and mechanism of environmental pharmaceuticals in aqueous suspension of TiO_2: a case of sulfa drugs. Catal. Today **153**, 200–207 (2010)

Liu, Y.-J., Lo, S.-L., Liou, Y.-H., Hu, C.-Y.: Removal of nonsteroidal anti-inflammatory drugs (NSAIDs) by electrocoagulation–flotation with a cationic surfactant. Sep. Purif. Technol. **152**, 148–154 (2015)

Tauxe-Wuersch, A., De Alencastro, L.F., Grandjean, D., Tarradellas, J.: Occurrence of several acidic drugs in sewage treatment plants in Switzerland and risk assessment. Water Res. **39**, 1761–1772 (2005)

Nakada, N., Tanishima, T., Shinohara, H., Kiri, K., Takada, H.: Pharmaceutical chemicals and endocrine disrupters in municipal wastewater in Tokyo and their removal during activated sludge treatment. Water Res. **40**, 3297–3303 (2006)

Corcoll, N., Vicenç, A., Barceló, M., Casellas, M., Helena Guasch, B.H.: Pollution-induced community tolerance to non-steroidal anti-inflammatory drugs (NSAIDs) in fluvial biofilm communities affected by WWTP effluents. Chemosphere **112**, 185–193 (2014)

Tsvetkova, B., Peikova, L.: HPLC determination of ketoprofen in tablet dosage forms. Trakia J. Sci. **1**, 55–59 (2013)

Kusi, H., Koprivanac, N., Bozi, A.L., Selanec, I.: Photo-assisted Fenton type processes for the degradation of phenol: a kinetic study. J. Hazard. Mater. **136**, 632–644 (2006)

Ignasi, S., Enric, B.: Remediation of water pollution caused by pharmaceutical residues based on electrochemical separation and degradation technologies: a review. Environ. Int. **40**, 212–229 (2012)

Mollah, M.Y.A., Schennach, R., Parga, J.R., Cocke, D.L.: Electrocoagulation (EC)—science and applications. J. Hazard. Mater. **84**, 29–41 (2001)

Yehya, T., Favier, L., Kadmi, Y., Audonnet, F., Fayad, N., Gavrilescu, M., Vial, C.: Removal of carbamazepine by electrocoagulation: investigation of some key operational parameters. Environ. Eng. Manage. J. **3**, 639–645 (2015)

Chou, W.L., Wang, C.T., Huang, K.Y., Liu, T.C.: Electrochemical removal of salicylic acid from aqueous solutions using aluminium electrodes. Desalination **271**, 55–61 (2011)

Deshpande, A., Lokesh, K.S., Bejankiwar, R.S., Gowda, T.P.H.: Electrochemical oxidation of pharmaceutical effluent using cast iron electrode. J. Environ. Sci. Eng. **47**, 21–24 (2005)

Daniel, R.Arsand, Klaus, K., Ayrton, F.M.: Removal of dexamethasone from aqueous solution and hospital wastewater by electrocoagulation. Sci. Total Environ. **443**, 351–357 (2013)

Ait Ouaissa, Y., Chabani, M., Amrane, A., Bensmaili, A.: Removal of tetracycline by electrocoagulation: kinetic and isotherm modeling through adsorption. J. Environ. Chem. Eng. **2**, 177–184 (2014)

Deshpande, A.M., Satyanarayan, S., Ramakant, S.: Treatment of high-strength pharmaceutical wastewater by electrocoagulation combined with anaerobic process. Water Sci. Technol. **61**, 463–472 (2010)

Wang, J.P., Chen, Y.Z., Ge, X.W., Yu, H.Q.: Optimization of coagulation-flocculation process for a paper-recycling wastewater treatment using response surface methodology. Colloids Surf. A: Physicochem. Eng. Aspects **302**, 204–210 (2007)

Tugba, O.: The optimization of Cr(VI) reduction and removal by electrocoagulation using response surface methodology. J. Hazard. Mater. **162**, 1371–1378 (2009)

Sudarjanto, G., Keller-Lehmann, B., Keller, J.: Optimization of integrated chemical–biological degradation of a reactive azo dye using response surface methodology. J. Hazard. Mater. **38**, 160–168 (2006)

Korbahti, B.K.: Response surface optimization of electrochemical treatment of textile dye wastewater. J. Hazard. Mater. **145**, 277–286 (2007)

Dos Santos, K.S.C.R., Silva, H.S.R.C., Ferreira, E.I.: 32 factorial design and response surface analysis optimization of N-carboxybutylchitosan synthesis. Carbohyd. Polym. **59**, 37–42 (2005)

Myers, R.H., Montgomery, D.C., Anderson-Cook, C.M.: Response Surface Methodology: Process and Product Optimization Using Designed Experiments, 2nd edn. Wiley, New York (2002)

Kolz, A.C., Moorman, T.B., Ong, S.K., Scoggin, K.D., Douglass, E.A.: Degradation and metabolite production of tylosin in anaerobic and aerobic swine-manure lagoons. Water Environ. Res. **77**, 50–56 (2005)

Elmollaa, E.S., Chaudhuria, M., Eltoukhyb, M.M.: The use of artificial neural network (ANN) for modeling of COD removal from antibiotic aqueous solution by the Fenton process. J. Hazard. Mater. **179**, 127–134 (2010)

Hansen, H., Nunez, P., Raboy, D., Schippacasse, I., Grandon, R.: Electrocoagulation in wastewater containing arsenic: comparing different process designs. Electrochim. Acta **52**, 3464–3470 (2007)

Yahiaoui, I., Belattaf, A., Aissani-Benissad, F., Yahia Cherif, L.: Full factorial design applied to a biosorption of lead (II) ions from aqueous solution using a brewer's yeast (*Saccharomyces cerevisiae*). J. Chem. Eng. Data **56**, 3999–4005 (2011)

Yahiaoui, I., Aissani-Benissad, F., Ait-Amar, H.: Optimization of silver cementation yield in fixed bed reactor using factorial design and central composite design. Can. J. Chem. Eng. **88**, 1099–1106 (2010)

Khataee, A.R., Zarei, M., Khameneh Asl, S.: Photocatalytic treatment of a dye solution using immobilized TiO2 nanoparticles combined with photoelectro-Fenton process: Optimization of operational parameters. J. Electroanal. Chem. **648**, 143–150 (2010)

Amani-Ghadim, A.R., Abera, S., Olad, A., Ashassi-Sorkhabi, H.: Optimization of electrocoagulation process for removal of an azo dye using response surface methodology and investigation on the occurrence of destructive side reactions. Chem. Eng. Process. **64**, 68–78 (2013)

Olad, A., Amani-Ghadim, A.R., Dorraji, M.S.S., Rasoulifard, M.H.: Removal of the alphazurine FG dye from simulated solution by electrocoagulation. Clean **38**, 401–408 (2010)

Lakshmanan, D., Clifford, D.A., Samanta, G.: Ferrous and ferric ion generation during iron electrocoagulation. Environ. Sci. Technol. **43**, 853–3859 (2009)

Panizza, M., Cerisola, G.: Applicability of electrochemical methods to carwash wastewaters for reuse. Part 2: electrocoagulation and anodic oxidation integrated process. J. Electroanal. Chem. **638**, 236–240 (2010)

Daneshvar, N., Khataee, A.R., Amani Ghadim, A.R., Rasoulifard, M.H.: Decolorization of C.I. Acid Yellow 23 solution by electrocoagulation process: investigation of operational parameters and evaluation of specific electrical energy consumption (SEEC). J. Hazard. Mater. **148**, 566–572 (2007)

Biosorption of the Anionic Dye Direct Red 89 by the Aquatic Plant *Callitriche obtusangula*

Khaled Benabbas[1,2(✉)], Imene Hocini[1,2], and Nabila Khellaf[1,2]

[1] Department of Process Engineering, Badji Mokhtar University, P.O. Box 12, 23000 Annaba, Algeria
benabbaskhaled@gmail.com

[2] Laboratory of Organic Synthesis - Modeling and Optimization of Chemical Processes, Badji Mokhtar University, P.O. Box 12, 23000 Annaba, Algeria

Abstract. The use of aquatic plant biomass as sorbent is an economic way to mitigate water pollution because of abundance and low cost of this kind of biomass. In this study, we proposed the species *Callitriche obtusangula* (*C. obtusangula*) for the elimination of the azo dye Direct Red 89 (DR-89) which is a dangerous pollutant for humans, animals and their environments. Batch experiments were carried out to assess optimal parameters such as contact time, biomass dosage, agitation speed, pH and temperature. The results show that the maximum percentage of pollutant elimination was 46% after 90 min of treatment for an initial dye concentration of 50 mg/L, an agitation speed of 200 rpm, a biomass dose of 2 g/L, and a pH equal to 1. Thermodynamically, the study proved that the phenomenon was endothermic ($\Delta H = 34.84$ kJ/mol) and spontaneous ($\Delta G < 0$). The Infra-Red analysis (FT-IR) was used to determine the mechanism of DR-89 sorption by identifying the functional groups involved in the present phenomenon. The spectra showed that the *Callitriche* biomass contains Carboxylic Acid, Aromatic ring, Amines, Alcohol, Alkane which were responsible for the dye retention on the biomass surface.

Keywords: *C. obtusangula* · Organic pollutant · Optimization · Plant biomass

1 Introduction

The use of dyes by the textile industry for coloring several product is leading to produce and release colorful polluting water. These effluents have a great impact on the environment as well as on the humans and animals because most of these pollutants contain aromatic structures and azo groups which are carcinogenic, mutagenic even at low concentration. It is therefore necessary to find an appropriate strategy to eliminate these pollutants before their final discharge. In order to remedy this environmental problem, several treatment processes were used including coagulation-flocculation, oxidation processes, biological processes, membrane-based separation processes and adsorption (Nandi et al. 2009). Each of these methods has benefits and disadvantages. Recently, a particulate attention has been focused on the technology of adsorption on a solid surface

B. Abdelbaki et al. (Eds.): SMSD 2017, *Proceedings of the Third International Symposium on Materials and Sustainable Development*, pp. 540–548, 2018.
https://doi.org/10.1007/978-3-319-89707-3_58

such as activated carbon (AC) which is most often used because of its specific surface and high adsorption capacity. However, its elevated cost makes the process with AC a non-economical solution for an industrial application. Different research studies have demonstrated the effectiveness of non-conventional sorbent like rice husk, wheat bran, waste apricot, bagasse, fly ash, duckweed biomass and powdered peanut hull for eliminating organic and mineral pollutants (Guendouz et al. 2013). Additionally, these natural materials present in large quantities are mainly friendly to the environment.

The need to have a non-conventional sorbent with a high adsorption capacity is always at the request.

In this study, we proposed the elimination of the anionic dye, Direct Red 89 used in the textile industry of Eastern Algeria. *C. obtusangula*, a very abundant aquatic plant in the north-east Algeria was selected for this purpose. Only few scientific data exist in the field of pollution treatment using this species. Some operating parameters such as contact time, dose of biomass, temperature, pH and agitation speed were optimized.

2 Materials and Methods

C. Obtusangula was collected from natural pond located in the city of Annaba (Algeria), the raw biomass was dried in the open air before being washed with distilled water and dried again at 65 °C for 72 h. The dried biomass was stocked in sealed jar, for further utilization. The pollutant consisting of an anionic dye Direct Red 89 was provided by the textile industry of Constantine (Eastern Algeria). The physicochemical characteristics of the molecule are summarized in Table 1, its molecular structure is illustrated in Fig. 1 (Guendouz et al. 2013).

Table 1. Physico-chemical characteristics of the dye understudy

Molecular formula	$C_{44}H_{32}N_{10}Na_4O_{16}S_4$
Molecular weight	1372
λ_{max} (nm)	494
Dye nature	Direct dye

Biosorption essays were conducted in a large thermostated bath containing four batch reactors with 1 L capacity (Fig. 2). Each reactor was filled with 50 mg/L dye solution and stirred at 200 rpm, natural pH (pH = 7.1) and ambient temperature. Two control tests and two treatments were carried out in parallel. Controls were necessary to ensure that the reduction in pollutant concentration was due to the presence of *Callitriche* biomass and not to the reactor walls. Samples of 1 ml was taken at regular intervals (10 min), diluted with distilled water to a volume of 10 ml and centrifuged at 3000 rpm. The supernatant was then analyzed to determine the concentration of DR-89 by measuring the absorbance using a UV-vis spectrophotometer (SECOMAM Prim Light V9B S/N 2836). The optimization procedure was done by considering the following Intervals: 0.5–2.5 g/L for dose of biomass, 150–300 rpm for agitation speed, 1–13 for pH, 20–50 °C for temperature. The removal efficiency (Re) were calculated by the following Eq. (1):

$$Re = \frac{C_0 - C_e}{C_0} \times 100 \tag{1}$$

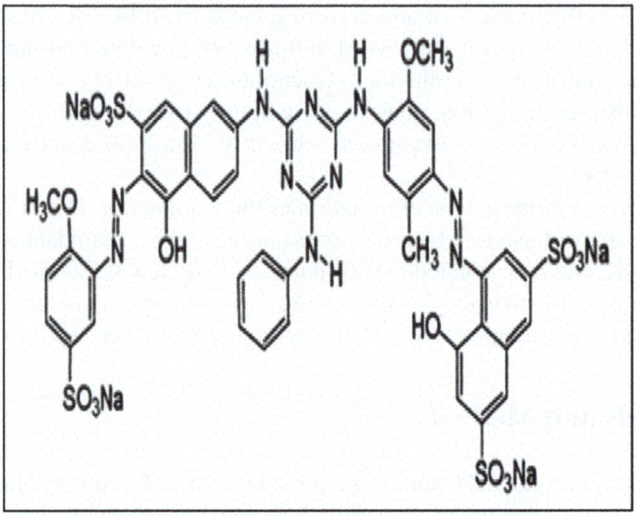

Fig. 1. Molecular structure of Direct Red-89.

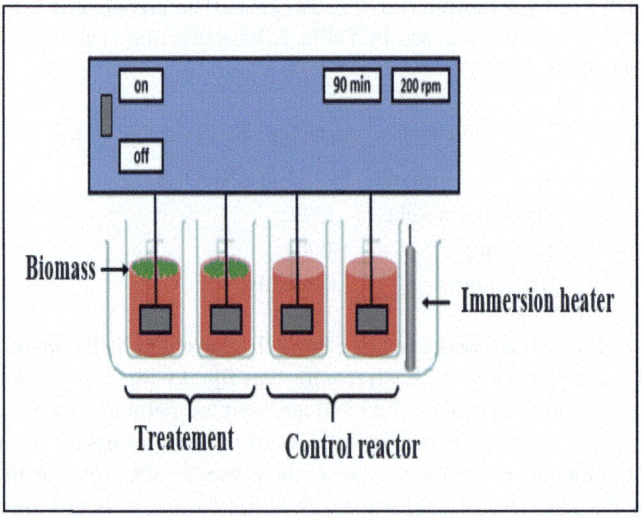

Fig. 2. Experimental device for the biosorption of DR-89 using *Callitriche* biomass.

Where C_0 and C_e are the initial and the equilibrium dye concentrations in the solution (mg/L), m (g/L) is the biomass dose and V (ml) is the volume of the aqueous solution. Fourier Transform Infrared (FT-IR) spectroscopy was performed on biosorbent samples before and after dye treatment. The samples were analyzed by a PerkinElmer FT-IR spectrophotometer (IRAffinity-1S model: Shimadzu) from 4000 to 500 cm^{-1}.

3 Results and Discussion

3.1 Adsorbent Characterization

FT-IR analysis was performed on *Callitriche* biomass before and after dye biosorption. As can be seen in Fig. 3 the results show that the peaks obtained at 835 and 1016 cm^{-1} are assigned respectively to the deformation of the aromatic C-H indicating the presence of aromatic ring (Tran et al. 2017) and stretching of C-OH assigned to a primary alcohol (Sun et al. 2013) these peaks have shifted toward the high frequencies at 837 and 1022 cm^{-1}, respectively. This suggests that these two functional groups could participate in dye retention on plant biomass.

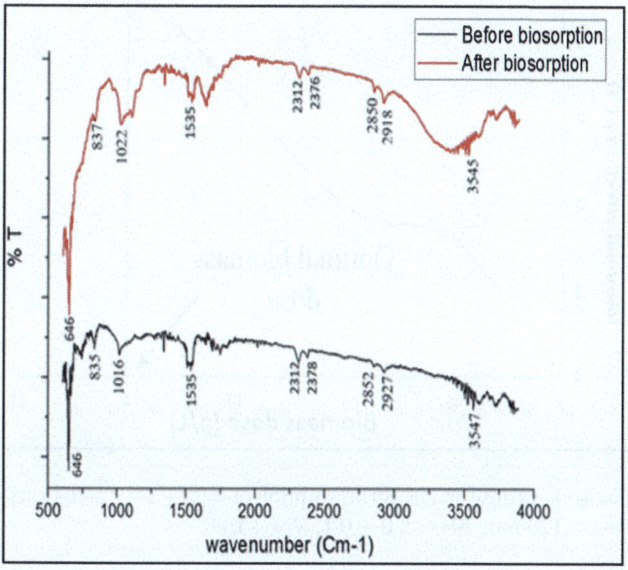

Fig. 3. FTIR spectra before and after DR-89 sorption by *Callitriche* biomass.

The peaks appeared at 646, 1535, 2378, 2852 2927 and 3547 cm^{-1} are assigned to the stretching of the S=O bond assigned to SO_3 function (Dizge et al. 2008), C=O of the ketone, $C{=}NH^+$ amine groups, the stretching of Alkane C-H bond (Tran et al. 2017) and the OH stretching (carboxylic acid) (Won et al. 2005). A little displacement of these peaks is observed in the red spectrum revealing a possible participation of NH^+, C-H and O-H in DR-89 biosorption using *Callitriche* biomass.

3.2 Adsorption Studies

3.2.1 Effect of Adsorbent Dose

The influence of the mass of the adsorbent was studied by varying it from 0.5–2.25 g. Figure 4 shows that the removal efficiency of DR-89 increased with biosorbent dose until a value of 2 g/L where a maximum dye removal (22%) was achieved. This is due to the rise of available adsorption sites on *Callitriche* biomass. Beyond this optimal value, the removal rate stabilized. This fact might be due to the inaccessibility of the organic compound through the biomass pores (Gupta et al. 2010). On the other hand, control treatment (biomass = 0 g/L) revealed no significant dye removal proving that the plant biomass was responsible for the DR-89 biosorption.

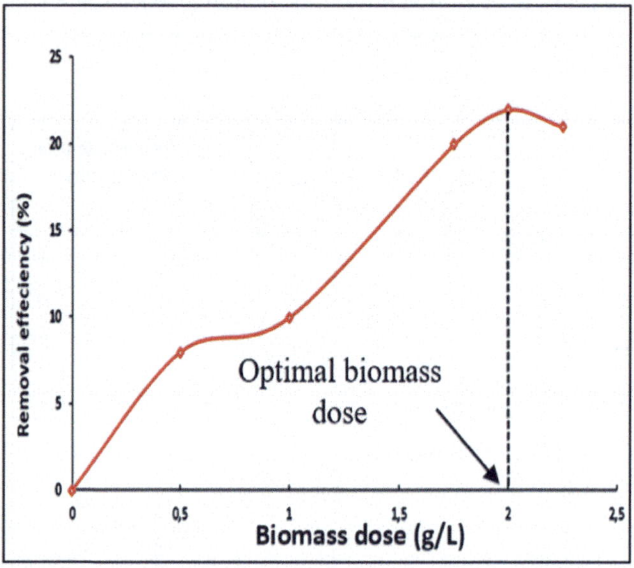

Fig. 4. Effect of biosorbent dose on DR-89 biosorption ($T = 20 \pm 2$ °C; agitation speed = 200 rpm; $C_0 = 50$ mg/L; time = 120 min; pH = 5.0 ± 0.1; V = 1 L).

3.2.2 Effect of pH Solution

The influence of initial solution pH of on the present process was examined. The pH was adjusted by using 1 M HCl or 1 M NaOH solutions. Figure 5 revels that the dye removal efficiency was at its maximum for Ph value between 1 and 3, at this acidic value the groups of amines (NH^{+3}), Carboxylic Acid (R-CO-OH^{+2}) and Alcohol (R-OH^{+2}) are protonated and they will have a positive charge which will interact with the negative sulfonate SO_3^- of the dye (Chiou et al. 2004; Sun et al. 2013). A decrease in DR-89 removal efficiency was observed at pH 7 (Data of PZC = 7.4 not shown) from which a further slight increase was recorded due to remaining positive charge in amine groups in the solution (Junxiong et al. 2009).

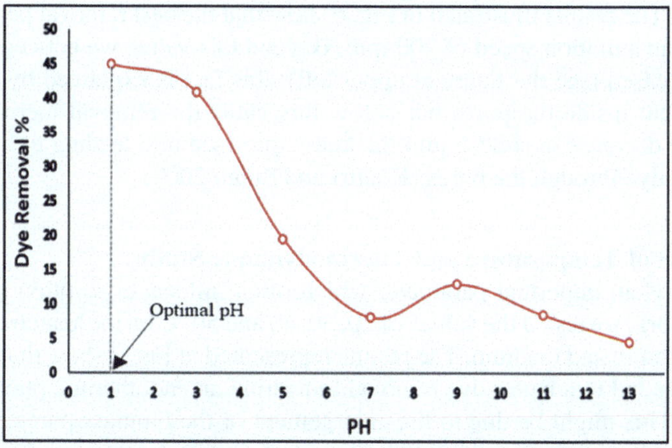

Fig. 5. Effect of pH solution on DR-89 (Volume = 1 L; T = 20 ± 2 °C; Agitation speed = 200 rpm; C_0 = 50 mg/L, m = 2 g/L).

3.2.3 Effect of Agitation Speed

Generally, agitation speed influences sorption phenomenon in a liquid-solid process. In the present work, we tested the effect the agitation speed for values ranged between 150

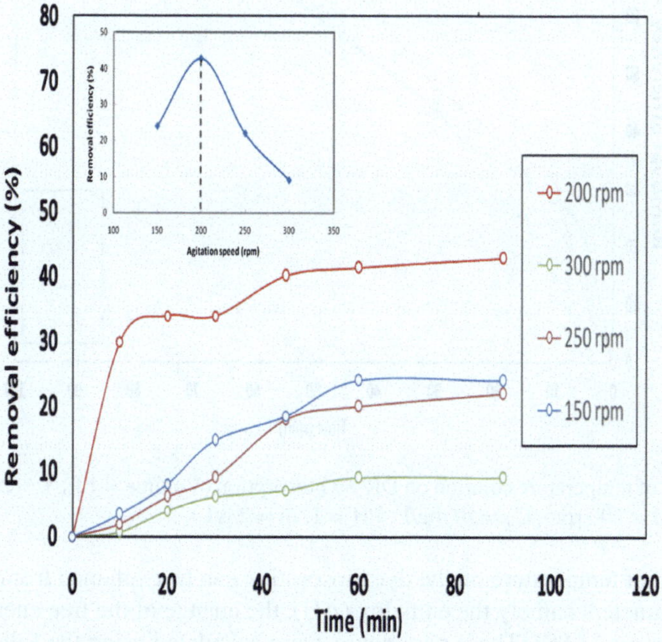

Fig. 6. Effect of agitation speed on DR-89 biosorption (Volume = 1 L; T = 20 ± 2 C_0 = 50 mg/L, pH = 1 ± 0.2, m = 2 g/L).

and 300 rpm. The results illustrated in Fig. 6 show that the best removal percentage was obtained for an agitation speed of 200 rpm. Beyond this value, we noticed that the dye removal decreased (see the figure at upper left); this fact is explained by a significant flow movement inside the pores but below this value the removal increase which is related in the decrease in solid-liquid thickness interface film leading to a better penetration of the dye through the pores (Khattri and Singh 2009).

3.2.4 Effect of Temperature and Thermodynamic Study

Temperature is an important parameter which often influences sorption processes. In the present work, we tested the values of 20, 30, 40 and 50 °C on the abatement of DR-89 from the contaminated medium. The results represented in Fig. 7 show that the increase in temperature led to a better dye removal indicating an endothermic phenomenon for biosorption. This might be due to the enlargement of the biomass pores creating new sites of adsorption (Dizge et al. 2008; Tan et al. 2007) Another explanation is that temperature could increase the kinetic energy of the organic molecule leading to a higher speed migration inside the pores (Son et al. 2005).

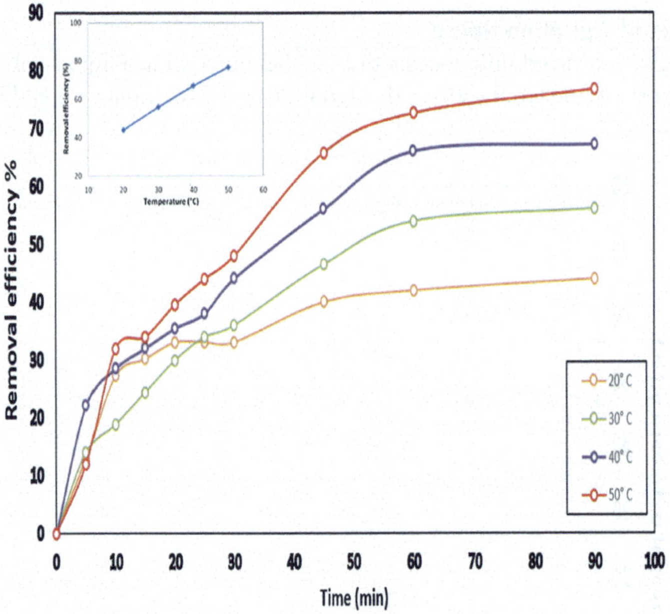

Fig. 7. Effect of temperature solution on DR-89 biosorption (Volume = 1 L; T = 20–50 ± 2oC;s agitation speed = 200 rpm; C_0 = 50 mg/L, PH = 1, m = 2 g/L).

The effect of temperature on the dye biosorption can be explained from the thermodynamic parameters namely the enthalpy (ΔH), the change of the free energy (ΔG), as well as the entropy (ΔS). These parameters were calculated using the following equation:

$$\text{Ln } K_L = -\frac{\Delta G}{RT} = \frac{\Delta H}{R} \times \frac{1}{T} + \frac{\Delta S}{R} \quad (2)$$

Where K_L (L mg^{-1}) is the equilibrium constant, R (J mol^{-1} K^{-1}) is the gas constant and T (K) the absolute temperature. The thermodynamic parameters for the adsorption for different temperatures are summarized in Table 2.

Table 2. Thermodynamic parameters

	ΔG (J/mol)	ΔS (J/mol)	ΔH (KJ/mol)
T = 30 °C	−457.92	116.48	34.84
T = 40 °C	−1622.72	-	-
T = 50 °C	−2787	-	-

The positive ΔH value confirms that the phenomenon is endothermic. The negative values of ΔG indicate that the Direct Red 89 adsorption on the biomass is spontaneous and the positive value of the entropy ΔS suggests an increase in the degree of disorder on the solid-liquid interface during the retention.

3.2.5 Proposal for Biosorption Mechanism of DR-89

According to the IR-FTIR analysis we can conclude that the amines grouping R-NH_2 alcohol R-OH and carboxylic acid RCOOH participate in dye retention because once protonated at low PH there will be load positively forming R-CO-OH^{+2}, R-NH^{+3}, R-OH^{+2} who can attract the negative dye charges thanks to the sulfonate groups $SO3^-$. On another side benzene could participate to the retention of the dye by electrophilic substitution with the sulfonate grouping dye - SO_3^- (Fig. 8).

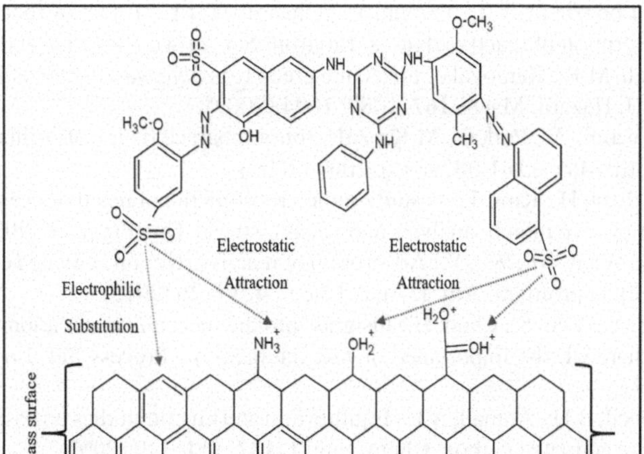

Fig. 8. Direct Red 89 adsorption mechanism

4 Conclusions

The results showed that *Callitriche* biomass had a significant biosorption capacity for Direct Red 89. It can be used as a biosorbent in fact of its abundance as well as its low cost. The results showed that the maximum of dye removed from the contaminated solution was 76.8% for a pH 1 and a temperature of 50 °C. Since the pollutant amount eliminated at ambient temperature (20 °C) was relatively low (44%), it is therefore necessary to carry out a physicochemical pretreatment of the plant biomass in order to improve the capacity of fixing the dye molecules on the biomass surface. This would avoid operating at high temperatures and would save energy.

References

Won, S.W., Choi, S.B., Yun, Y.-S.: Interaction between protonated waste biomass of corynebacterium glutamicum and anionic dye Reactive Red 4. Colloids Surf. Physicochem. Eng. Aspects **262**, 175–180 (2005)

Chiou, M., Ho, P., Li, H.: Adsorption of anionic dyes in acid solutions using chemically cross-linked chitosan beads. Dyes Pigm. **60**, 69–84 (2004)

Dizge, N., Aydiner, C., Demirbas, E., Kobya, M., Kara, S.: Adsorption of reactive dyes from aqueous solutions by fly ash: kinetic and equilibrium studies. J. Hazard. Mater. **150**, 737–746 (2008)

Guendouz, S., Khellaf, N., Zerdaoui, M., Ouchefoun, M.: Biosorption of synthetic dyes (Direct Red 89 and Reactive Green 12) as an ecological refining step in textile effluent treatment. Environ. Sci. Pollut. Res. **20**, 3822–3829 (2013)

Gupta, V.K., Jain, R., Malathi, S., Nayak, A.: Adsorption–desorption studies of indigocarmine from industrial effluents by using deoiled mustard and its comparison with charcoal. J. Colloid Interface Sci. **348**(2), 628–633 (2010)

Junxiong, C.A.I., Longzhe, C.U.I., Yanxin, W., Chengfu, L.I.U.: Effect of functional groups on sludge for biosorption of reactive dyes. J. Environ. Sci. **21**(4), 534–538 (2009)

Khattri, S.D., Singh, M.K.: Removal of malachite green from dye wastewater using neem sawdust by adsorption. J. Hazard. Mater. **167**, 1089–1094 (2009)

Nandi, B.K., Goswami, A., Purkait, M.K.: Adsorption characteristics of brilliant green dye on kaolin. J. Hazard. Mater. **161**(387–395), 2009 (2009)

Son, Y., Hong, J., Lim, H., Kim, T.: A study of heterobifunctional reactive dyes on nylon fibers: dyeing properties, dye moiety analysis and wash fastness. Dye Pigm. **66**, 2005 (2005)

Sun, D., Zhang, Z., Wang, M., Wu, Y.: Adsorption of reactive dyes on activated carbon developed from enteromorpha prolifera. Am. J. Anal. Chem. **4**, 17–26 (2013)

Tran, H.N., Wang, Y., You, S., Chao, H.: Insights into the mechanism of cationic dye adsorption on activated charcoal: the importance of π–π interactions. Process Saf. Environ. Prot. **107**, 168–180 (2017)

Tan, I.A.W., Hameed, B.H., Ahmad, A.L.: Equilibrium and kinetic studies on basic dye adsorption by oil palm fibre activated carbon. Chem. Eng. J. **127**, 111–119 (2007)

Extraction of Propionic Acid from Waste Water Using Rosmary

M. Timedjeghdine(✉), A. Djellouli, and A. Hasseine

Department of Chemical Engineering,
University of Mohamed Kheider, Biskra, Algeria
timedjeghdine@hotmail.fr

Abstract. Experimental liquid–liquid equilibrium (LLE) data for the extraction of propionic acid from water by rosmary at 291.15 K and at ambient pressure were investigated. Data for the binodal curve have been determined by cloud-point titration method and conjugate points on tie-line were obtained by correlating the refractive index of the binodal curves as a function of composition. The reliability of the tie lines generated from the composition data was tested and verified by two correlation models: Hand and Othmer, which were found to fit the data very well, producing high values of correlation coefficient (R^2). In addition, these experimental tie line data were also correlated with NRTL model and the binary interaction parameters of the model have been obtained by regression with experimental data. Finally, the distribution coefficients and the selectivity factor of the solvent used were calculated and presented.

Keywords: Liquid-liquid extraction · Propionic acid · Rosmary
Environnnement · NRTL

1 Introduction

Propionic acid is one of the important carboxylic acid widely used as mold inhibitor in baking and esterifying agent in the production of thermoplastics and in the manufacture of flavors and perfume bases (Wang et al. 2013; Lee 2008). The recovery of this acid present in aqueous solutions can be completed by extraction, which is considered as a kind of high efficiency and energy saving method. The successful implementation of the above processes is dependent on the accurate liquid–liquid equilibrium (LLE) data, which are also fundamental and important to simulate and design the extraction process. In recent years, several authors (Timedjeghdine et al. 2016, 2017; Merzougui et al. 2011, 2012, 2015; Laiadi et al. 2012; Bacha et al. 2016) have investigated LLE of ternary systems. The LLE data of the ternary system (water + propionic acid + rosmary) are reported at T = 291.15 K and 1 atm. Then, the solubility and tie line data for this system, determined by using the cloud-point method, are presented. This technique provides the concentration of all components in the mutually saturated liquid phases (Wongsawa et al. 2014). The distribution coefficients (D) and the selectivity factor (S) were determined to investigate the ability of the solvent to separate propionic acid from water in the recovery process (Ghanadzadeh Gilani et al. 2013). The quality of the experimental tie line data will be confirmed by the Othmer-Tobias (Othmer and Tobias 1942) and the

B. Abdelbaki et al. (Eds.): SMSD 2017, *Proceedings of the Third International Symposium on Materials and Sustainable Development*, pp. 549–555, 2018.
https://doi.org/10.1007/978-3-319-89707-3_59

Hand (Hand 1930) correlations on a mass-fraction basis. The tie-line data were correlated using the non-random two-liquid (NRTL) (Renon and Prausnitz 1968) model and the binary interaction parameters of the model have been obtained by regression with experimental data.

2 Experimental

2.1 Materials

The purity of the chemical species employed in this work and their origin were as follows: propoinic acid (Merck, >99%), rosmary, >95% (GC). These chemical compounds were used without further purification. Deionized and redistilled water was employed throughout all experiments.

2.2 Apparatus and Procedure

The solubility curve for the ternary system was determined by the cloud point method (Othmer et al. 1941) using a thermostated cell, equipped with a magnetic stirrer and an isothermal fluid jacket. The cell temperature was measured within an accuracy of ±0.2 K using a mercury-in-glass thermometer. The end-point was determined by observing the transition from a homogeneous to a heterogeneous mixture. For the water-rich and solvent-rich phases, (1) binary mixtures of (water + propionic acid) were titrated with rosmary, and (2) binary mixtures of (rosmary + propionic acid) were titrated with water, until the transition from homogeneity to cloudiness was observed. All these titrations involved the measurement of the refractive index using a Nahita Model 690/1 refractometer. All mixtures were prepared by weighing with an analytical balance (model Nahita 5034/200, accurate to ±0.0001 g).

The tie-line data were determined by a magnetically stirred equilibrium cell, which was kept in a constant-temperature bath maintained at (291.15 ± 0.2) K for 2 h. The prepared mixtures were left to settle for 24 h to allow them to separate into two clear liquids at equilibrium. After that, samples from both organic and aqueous phases had been taken using a syringe (2 ml) for analysis.

3 Results and Discussion

3.1 Experimental Solubility and Tie Line Data

The experimental solubility data from the binodal curve diagrams of ternary system (water + propionic acid + rosmary) at T = 291.15 K and atmospheric pressure were investigated. As can be seen in Fig. 1, only one liquid pair (water + rosmary) is heterogeneous and two liquid pairs (propionic acid + water or rosmary) are homogenous. This indicates that this investigated ternary system exhibit type-1 behaviour of LLE (Sandler, 1998). The tie-line data of this ternary system was obtained experimentally at the same temperature (T = 291.15 K) and are shown in Table 1.

Table 1. Experimental tie-line data for (water (1) +propionic acid (2) + rosmary (3)) ternary system at 291.15 K.

Water-rich phase (aqueous phase)			Solvent-rich phase (organic phase)		
x_1	x_2	x_3	x_1	x_2	x_3
0,0127	0,1755	0,8118	0,9639	0,0272	0,0089
0,0047	0,2904	0,7049	0,9533	0,0393	0,0074
0,0047	0,3762	0,6191	0,9459	0,0474	0,0067
0,0049	0,4507	0,5444	0,9403	0,0539	0,0058
0,0056	0,5255	0,4689	0,9351	0,0597	0,0052

3.2 LLE Correlation and Parameter Estimation

The NRTL model (Renon and Prausnitz 1968) was chosen to correlate the experimental data of the tie-lines. The non-randomness parameter in the NRTL model (α) is usually set equal to 0.2 in order to reduce the dimensions of the optimization search domain. In addition, Fig. 1 compares the results between the calculated and experimental tie lines. It is evident that the NRTL model can predict well the experimental behavior of the tie-lines.

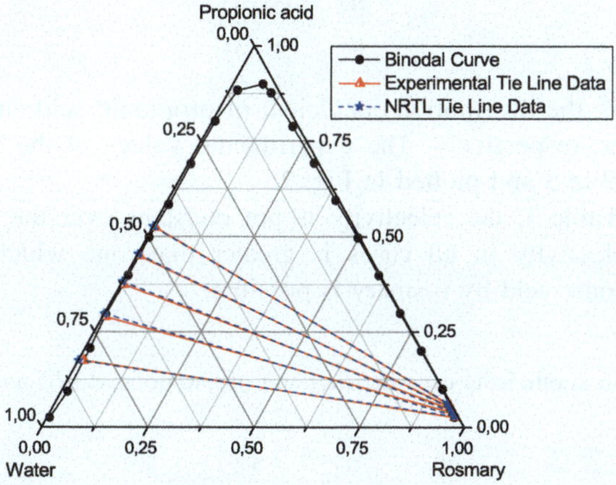

Fig. 1. LLE ternary diagram for water (1) + propionic acid (2) + rosmary (3) at 291.15 K.

In Table 2, the interaction binary parameters of the NRTL model are presented along with the model prediction error. This error is computed based on the root-mean square deviation (rmsd).

Table 2. NRTL binary interaction parameters (Aij and Aji) and root-mean square deviation (rmsd) values for the water (1) + propionic acid (2) + rosmary (3) ternary system.

i–j	Aij	Aji	rmsd (%)
water (1) + propionic acid (2) + rosmary (3)			
1-2	175.8	2656.6	
1-3	1200	2505.3	0.52
2-3	460.3	2981.5	

3.3 Distribution Coefficient and Selectivity Factor

The distribution coefficient, d_i, or the ratio of mass fraction of the ith component in rosmary rich-phase, x_{i3}, to that in water rich-phase, x_{i1}, is defined in the following form (Govindarajan and Sabarathinam 1995)

$$d_i = \frac{x_{i3}}{x_{i1}} \tag{1}$$

The effectiveness of extraction of propionic acid by rosmary from water is given by the selectivity (S) by the following equation (Cháfer et al. 2014)

$$S = \frac{d_2}{d_1} = \frac{x_{23}/x_{21}}{x_{13}/x_{11}} \tag{2}$$

where d_2 and d_1 are the distribution coefficient of propionic acid and the distribution coefficient of water, respectively. The experimental values of the selectivity factors (S) are given in Table 3 and plotted in Fig. 2.

According to Table 3, the selectivity is not constant over the whole two-phase region and the selectivity in all cases is greater than one which means that the extraction of propionic acid by rosmary is possible.

Table 3. Distribution coefficients of water (d_1) and propionic acid (d_2) and separation factors (S) at 291.15 K.

d_1	d_2	S
0.0109	0.1550	14.1368
0.0111	0.1353	12.2301
0.0108	0.1260	11.6425
0.0106	0.1196	11.2251
0.0111	0.1136	10.2442

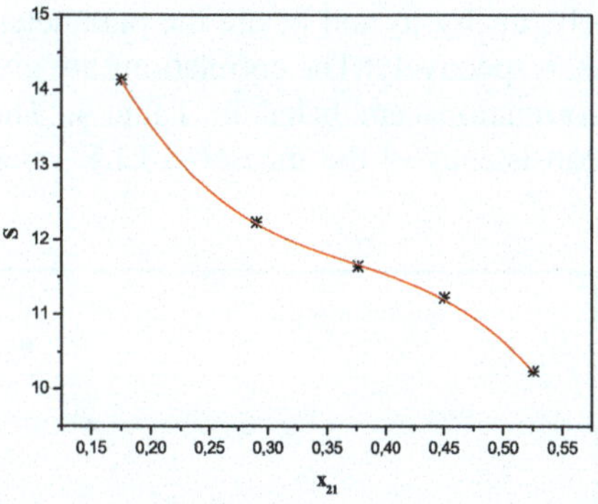

Fig. 2. Separation factors (*S*) of (water + propionic acid + rosmary) as a function of mass fraction of propionic acid in the aqueous phase x_{21} at 291.15 K.

3.4 Othmer-Tobias and Hand Correlations

For the studied system, the reliability of the experimental tie line data was verified by applying the Othmer-Tobias (Othmer and Tobias 1942) and the Hand (Hand 1930) correlations as shown in Eqs. (3) and (4) respectively:

$$ln\left[\frac{1-x_{33}}{x_{33}}\right] = a_1 + b_1 ln\left[\frac{1-x_{11}}{x_{11}}\right] \tag{3}$$

$$ln\left[\frac{x_{21}}{x_{11}}\right] = a_2 + b_2 ln\left[\frac{x_{23}}{x_{33}}\right] \tag{4}$$

Where x_{11} and x_{21} are the mass fractions of water and propionic acid in aqueous phase, respectively, x_{23} and x_{33}, are the mass fractions of propionic acid and rosmary in

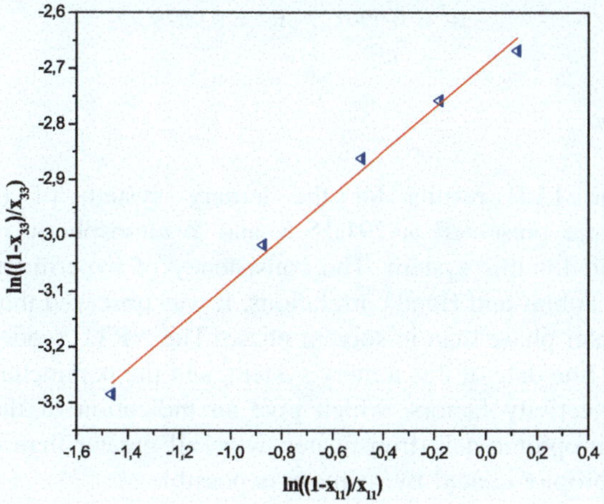

Fig. 3. Othmer–Tobias plots of water (1) +propionic acid (2) + rosmary (3) ternary system

organic phase, respectively, a_1, b_1, a_2 and b_2 are the parameters of the Othmer-Tobias and the Hand correlations, respectively. The correlations are given in Figs. 3 and 4, and the parameters of the correlations are listed in Table 4. The linearity of the plots indicates the degree of consistency of the measured LLE values in this work.

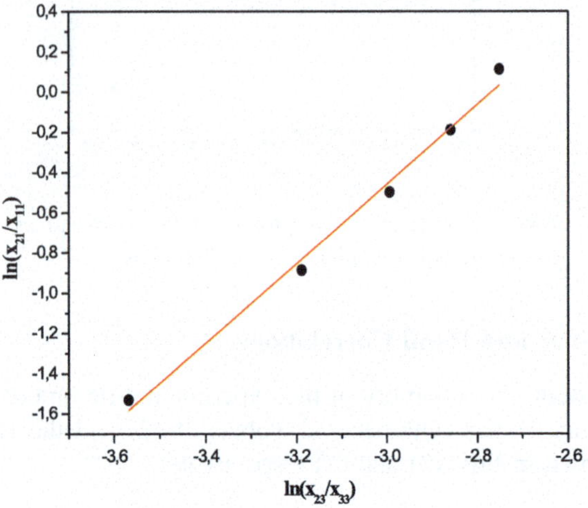

Fig. 4. Hand plots of water (1) +propionic acid (2) + rosmary (3) ternary system.

Table 4. Constants of othmer–tobias and hand correlations for the water (1) + propionic acid (2) + rosmary (3) ternary system at 291.15 K.

Correlations	
Othmer-Tobias	Hand
a_1 = −2.6926	a_2 = 5.4889
b_1= 0.3914	b_2 = 1.9815
R^2 = 0.9901	R^2 = 0.9879

4 Conclusions

In this work, the LLE results for the ternary system of (water + propionic acid + rosmary) were presented at 291.15 K and at atmospheric pressure. A type I behavior was found for this system. The consistency of experimental data was confirmed by Othmer-Tobias and Hand correlations. It was observed that propionic acid is more soluble in water phase than in solvent phase. The NRTL model used to correlate and analyze the tie line data of this ternary system, and the parameters of the model are calculated. The selectivity factors, which give an indication of the potential of the solvent to extract propionic acid from water, were all greater than unity. This means that extraction of propionic acid by rosmary is possible.

References

Wang, Y., Li, Y., Li, Y., Wang, L.Z., Dai, Y.: Extraction equilibria of monocarboxylic acids with trialkylphosphine oxide. J. Chem. Eng. Data **46**, 831–837 (2001)

Lee, L.L.: Molecular Thermodynamics of Electrolyte Solutions. World Scientific Publishing, USA (2008)

Timedjeghdine, M., Hasseine, A., Binous, H., Bacha, O., Attarakih, M.: Liquid-liquid equilibrium data for water + formic acid + solvent (butyl acetate, ethyl acetate, and isoamyl alcohol) at T = 291.15 K. Fluid Phase Equilib. **415**, 51–57 (2016)

Timedjeghdine, M., Hasseine, A., Binous, H., Bacha, O., Attarakih, M.: Liquid–liquid equilibrium data for water + acetic acid + solvent (dichloromethane + methyl isobutyl ketone) at T = 301.15 K. Desalin. Water Treat. **67**, 125–132 (2017)

Merzougui, A., Hasseine, A., Laiadi, D.: Liquid– liquid equilibria of n-heptane + toluene + aniline ternary system: experimental data and correlation. Fluid Phase Equilib. **308**, 142–146 (2011)

Merzougui, A., Hasseine, A., Laiadi, D.: Application of the harmony search algorithm to calculate the interaction parameters in liquid–liquid phase equilibrium modeling. Fluid Phase Equilib. **324**, 94–101 (2012)

Merzougui, A., Bonilla-Petriciolet, A., Hasseine, A., Laiadi, D., Labed, N.: Modeling of liquid–liquid equilibrium of systems relevant for biodiesel production using backtracking search optimization. Fluid Phase Equilib. **388**, 84–92 (2015)

Laiadi, D., Hasseine, A., Merzougui, A.: Homotopy method to predict liquid–liquid equilibria for ternary mixtures of (water + carboxylic acid + organic solvent). Fluid Phase Equilib. **313**, 114–120 (2012)

Bacha, O., Hasseine, A., Attarakih, M.: Measurement and correlation of liquid–liquid equilibria for water + ethanol + mixed solvents (dichloromethane or chloroform + diethyl ether) at *T* = 293.15 K. Phys. Chem. Liq. **54**, 245–257 (2016)

Wongsawa, T., Hronec, M., Soták, T., et al.: Ternary (liquid–liquid) equilibrium data of furfuryl alcohol with organic solvents at T = 298.2 K: experimental results and thermodynamic models. Fluid Phase Equilib. **365**, 88–96 (2014)

Ghanadzadeh Gilani, H., Ghanadzadeh Gilani, A., Shekarsaraee, S.: Experimental study of phase equilibria in aqueous mixtures of phosphoric acid with isoamyl acetate and methyl isoamyl ketone at T = (298.2, 308.2, and 318.2) K. Fluid Phase Equilib. **337**, 32–38 (2013)

Othmer, D.F., Tobias, P.E.: Ind. Eng. Chem. **346**, 90–692 (1942)

Hand, D.B.: Dineric distribution. J. Phys. Chem. **34**, 1961–2000 (1930)

Renon, H., Prausnitz, J.M.: Local compositions in thermodynamic excess functions for liquid Mixtures. AIChE J. **14**, 135–144 (1968)

Othmer, D.F., White, R.E., Trueger, E.: Liquid-liquid extraction data. Ind. Eng. Chem. **33**, 1240–1248 (1941)

Sandler, S.I.: Chemical and Engineering Thermodynamics. Wiley, New York (1998)

Govindarajan, M., Sabarathinam, P.L.: Fluid Phase Equilib. **108**, 269–292 (1995)

Cháfer, A., De La Torre, J., Lladosa, E., et al.: Liquid–liquid equilibria of 4-methyl-2-pentanone + 1-propanol or 2-propanol + water ternary systems: measurements and correlation at different temperatures. Fluid Phase Equilib. **361**, 23–29 (2014)

Investigate Method to Reduce Methane Emission During Maintenance Operations from Algerian Natural Gas Transportation

M. Louhibi–Bouiri(✉) and M. Hachemi

Research Unit Materials, Processes and Environment (UR-MPE), Faculty of Engineering Science, University M'Hamed Bougara of Boumerdes, 35000 Boumerdes, Algeria
bouiri_mounia@hotmail.com

Abstract. Methane concentration in the atmosphere has critically increased recently, its account for about 50% of the greenhouse gases (GHG) emissions. Increasing environmental concern and global warming have made it become an imminent consideration. The energy sector is the second largest contributor to methane emissions. The emissions of natural gas from gas transportation pipeline are an important factor for global warming. For this article we try find a way for preventing of waste this source because of economical aspect and environmental problems. First we calculate the volume of gas venting from pipeline and then get quantity of money value of this gas then we suggest a mobile compressor for saving this gas. Pipeline pump-down technique with portable compressor solution instead of venting will mainly allow 54.873 million m^3 gases saving, which costs about 11.628 million USD and benefits justified purchase of 4 portable compressors.

Keywords: Greenhouse gases · Methane emissions mitigation
Natural gas pipelines transportation · Gas venting · Portable compressor

1 Introduction

Methane (CH_4) is one of the six Greenhouse gases (GHGs) being mitigated under the Kyoto Protocol of the United Nations Framework Convention on Climate Change (UNFCCC). The others are carbon dioxide (CO_2), nitrous oxide (N_2O), hydrofluorocarbons (HFCs), perfluorocarbons (PFCs) and sulphur hexafluoride (SF_6). Atmospheric concentration and sources of CH_4 are not particularly well understood or well quantified, and they are highly disputable (Frankenberg et al. 2005; Miller et al. 2013). Also, official inventories underestimate actual CH_4 emissions (Brandt et al. 2014) despite, yet again, a rising global trend of CH_4 (Nisbet et al. 2014). According to the Intergovernmental Panel on Climate Change (IPCC) (IPCC 2007), the global atmospheric concentration of CH_4 rose from a pre-industrial value of 715 parts per billion (ppb) to 1732 ppb in the early 1990s and 1774 ppb in 2005, i.e. a rise of about 150% (Anifowose and Odubela 2015). This is of concern given that the lifetime of CH4 once released into the atmosphere is about 12 years (Xu et al. 2013), and it is 28 times more potent at trapping

B. Abdelbaki et al. (Eds.): SMSD 2017, *Proceedings of the Third International Symposium on Materials and Sustainable Development*, pp. 556–567, 2018.
https://doi.org/10.1007/978-3-319-89707-3_60

atmospheric heat than CO_2 over a 100 year timescale (IPCC 2006). Clearly, methane is crucial in the mitigation of global warming as its reduction will support an average global temperature rise of not greater than 2 °C (US EPA 2014). A significant fraction of GHG originates from the use of energy, primarily because fossil fuels (oil and natural gas) currently satisfy most of the global energy demands (Litto et al. 2007). Of the fossil fuels, natural gas has the lowest emissions of GHG per unit of energy, and there is much interest in switching from oil to gas. However, while natural gas is significantly cleaner than other fossil fuel (combustion GHG emission from natural gas per unit energy produced is 25% less than oil), non-combustion GHG emission (fugitive emissions) occurs during the process of extraction, production, processing, transmission, storage and distribution (Lechtenböhmer et al. 2007; Salomons et al. 2004).

Algeria's domestic natural gas pipeline system transports natural gas from Hassi R'Mel fields and processing facilities owned by Sonatrach, to export terminals and liquefaction plants along the Mediterranean Sea. It is divides into three main domestic pipeline systems:

❶The Hassi R'Mel to Arzew system is a collection of pipelines that move natural gas from HassiR'Mel to the export terminal and the LNG plant at Arzew. The system also includes an LPG pipeline. ❷The Hassi R'Mel to Skikda system transports natural gas from the Hassi R'Mel fields to the Skikda LNG plant. ❸The Alrar to Hassi R'Mel system transports natural gas produced in the Alrar and the southeastern region to the Hassi R'Mel processing facilities (Sonatrach 2014a, 2014b). The network includes also three transcontinental gas pipelines; one to Italy. The largest pipeline, Pipeline Enrico Mattei (GEM), came online in 1983 and runs 1.025 miles from Algeria to Italy via Tunisia. GEM's capacity is more than 1.3 Tcf/year and it is jointly owned by Sonatrach, the Tunisian government, and Eni. The Pedro Duran Farell (GPDF) pipeline started in 1996 and travels 325 miles to Spain via Morocco. GPDF's capacity is about 390 Bcf/year. The newest pipeline, MEDGAZ, came online in 2011 and is owned by Sonatrach, Cepsa, Endesa, Iberdrola, and GDF Suez. It stretches 125 miles onshore and offshore, from Algeria to Spain via the Mediterranean Sea (Sonatrach 2014a, 2014b).

These pipelines are important to both Algeria and Europe, because they transport nearly all the natural gas produced (70%), which provides about 30% of the European Union gas supply (Boukrif 2008) shown in Fig. 1. According to the BP Statistical Review of World Energy, pipelines exports of Algerian conventional natural gas to Europe was very high, 34.8 Gm^3 in 2012, compared to that of other African countries, characterized by a total amount of only 11.0 Gm^3 (Boudghene-Stambouli et al. 2012).

However, in the near future, the existing export gas pipeline from Algeria to Europe will not be able to cover the increasing gas demand. In 2035, the demand for natural gas in Europe is expected to exceed 630 Gm^3 (see Fig. 2) according to the International Energy Agency; 530 Gm^3 will be imported against 331 Gm^3 in 2010. This growth was mainly driven by the gas needs natural for generating electricity (GRTgaz 2013).

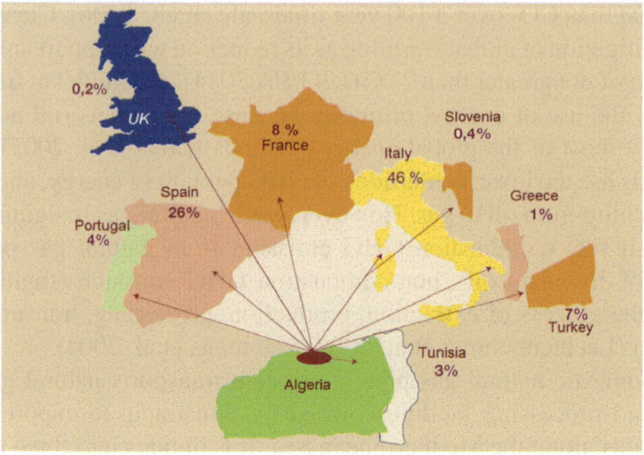

Fig. 1. Algerian exports of natural gas to Europe 2014

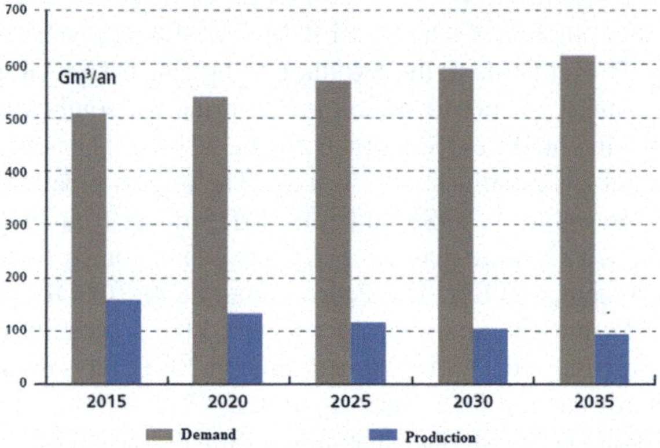

Fig. 2. Evolution of European natural gas demand and production in 2035 (GRTGaz 2013)

2 Backgrounds

The transmission of natural gas by pipeline is standard practice across Algeria. Natural gas is commonly transported over long distances from gas fields to customers and LNG units through a pipeline network. When the gas flows through the pipelines, it loses pressure owing to friction, and the gas is periodically recompressed in compressor stations. While there is no precise standard distance between compressor stations, they are located, on average every 100–150 km along a pipeline (Kirchgessner et al. 1997; Litto et al. 2007). These compressor stations typically contain between 1 and 15 compressors. The common types of compressors used are large slow speed reciprocating compressors driven by natural gas internal combustion (IC) engines, or centrifugal compressors powered by natural gas-fuelled turbines (Borraz-Sánchez 2010; Ríos-Mercado and Borraz-Sánchez 2015). To get an idea of how a compressor station is structured, a plan of a compressor station is shown in Fig. 3.

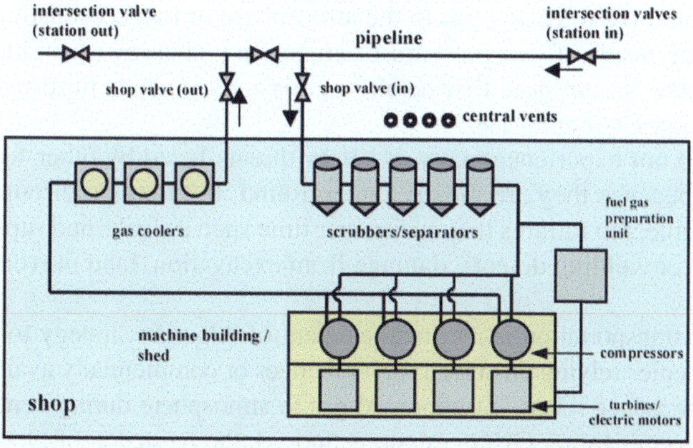

Fig. 3. Compressor station (Lechtenböhmer et al. 2007)

Readers interested in the design and arrangement of compressor stations in natural gas transmission systems are referred to the works of Kurz et al. (2012), Kopanos et al. (2015), Xenos et al. (2015). Valves and regulators are typical components installed for operational and safety reasons within a pipeline system. For example, by means of a valve, gas operators can restrict or direct the gas flow from one point to another. This is rather beneficial, among other circumstances, in order to perform scheduled maintenance or to satisfy demand requirements at specific points or to prevent loss of fluid by completely shutting down the gas flow through a specific pipeline section due to malfunctions. In real-practice, regulatory requirements force the transporter to install mainline block valves at certain fixed spacing along the transmission line.

Methane emissions in the natural gas transmission include leaks in natural gas pipelines and compressor stations can be grouped into three sources; ❶ Combustion emissions; ❷ Fugitive emissions; and ❸ Vented emissions. The first is emissions of methane resulting from incomplete combustion in the engine such as compressor engines, burners, and flares. The second source of emissions is fugitive emissions; these emissions are unintentional leaks emitted from sealed surfaces, such as packings and gaskets, or leaks from underground pipelines resulting from corrosion or faulty connections. Whilst the third is vented emissions am releases to the atmosphere by design or operational practice. Examples of vented emissions include: emissions from continuous process vents, such as dehydrator reboiler vents; maintenance practices, such as blowdowns; and small individual sources, such as gas-operated pneumatic device vents (Kirchgessner et al. 1997; Litto et al. 2007).

Causes of methane emissions are classified in different ways. Causes identified were: ❶ Start-up; ❷Normal operation; ❸Maintenance; ❹Upsets, and ❺ Mishaps.

Start-up operations, such as purging a newly constructed plant or pipeline, can involve purging natural gas directly to the atmosphere. Emissions associated with normal operation include emissions from process vents, fugitive emissions from packed or sealed surfaces or underground pipeline leaks, and emissions from gas-operated pneumatic devices. Maintenance operations involve blowing down equipment such as compressors, pipelines before maintenance begins (Kirchgessner et al. 1997). Process upsets usually involve releasing gas to the atmosphere or to a combustion device, such as a flare, as the result of overpressure or emergency shutdown conditions. Mishaps include accidental occurrences that result in emissions, such as third-party damage to pipelines (Kirchgessner et al. 1997).

Pipelines do not experience many of safety threats faced by other forms of freight transportation because they are mostly underground; but they go through a variety of areas and are subject to failures that occur over time such as leaks and ruptures resulting from corrosion or welding defects, damage from excavation, land movement, or incorrect operations.

Natural gas transportation is a critical element of Algerian strategy to manage GHG emissions. Strategies relying on current best practices or commercially available technologies, to avoid or reduce release of unburned gas to atmosphere during venting operations will be crucial to mitigating GHG emissions. Indeed, the reduction of pipeline emissions associated with CH_4 has thus become an issue of priority in the action plans and programs at Algerian national oil and gas company (Sonatrach). In the first place, since methane is the primary constituent of natural gas, the collection and utilization of methane provides a valuable, clean-burning energy source that improves quality of life in local communities and can generate revenue and improves living standards. Methane is both a potent GHG and has a short atmospheric lifetime; so reductions can also produce significant near-term results. In addition to all these facts, avoiding direct release of gas to atmosphere helps in reducing risks generated throughout the practice of this dangerous operation. Current practice creates an environmental Impact through the release of natural gas to atmosphere during maintenance repair since this typically requires removal of gas from the affected section of pipeline in order to ensure safe working conditions.

In recent years, Algeria has faced many problems in its gas pipeline network due to timeworn pipelines and to the obsolescence of equipment, especially for gas lines which were mainly installed between the 1961s and 2001s. For these reasons, considerable amounts of gases were vented directly to the atmosphere, causing huge disruption to meet the increasing natural gas demand, mainly by Europe. Gas venting is also responsible for greenhouse gas emissions, which contributes significantly to global warming and climate change of our planet. Algerian national oil and gas company (Sonatrach) had to improve the reliability of its natural gas transportation network; about 1210 km of pipelines (especially GZ0, GZ1, GZ2, GZ3 and GG1). Therefore, a huge rehabilitation plan of old pipelines has been prepared. The aim of this program is to comply with new regulations, enhance reliability, restore and improve the transport capacity.

3 Technologies and Practices

Natural gas companies have discovered many ways to reduce the amount of methane emissions from their systems. Table 1 outlines some of the major technologies and practices, summarizes the value of gas saved, and the expected payback period for the technologies and practices described (Ishkov et al. 2011).

Table 1. Emission reduction technologies and practices

Technology/Practice	Description	Value of gas saved	Payback (months)
Dry seals (Replacing Wet Seals)	Centrifugal compressors have seals along their shaft to keep gas from escaping. Wet, or oil-lubricated, seals are common and by design result in methane leakages, sometimes substantial ones. Dry seals operate mechanically without seal ring lubrication, which in its turn reduces gas leakage	324 000 USD per compressor	8 to 24 months
Directed inspection and maintenance (DI&M)	A technique to find, quantify and reduce leaks in a facility on a planned schedule based on frequency of leaks and cost-effectiveness. Quantification of leaks allows companies to prioritize repairs and evaluate cost-effectiveness	26 248 USD per station	Less than 12 months
Pipeline pump-down	Instead of venting gas from a section of pipeline that is being depressurized for maintenance, this technology uses a portable compressor to pump much of the gas in an adjacent operational pipeline segment	98 800 USD (assuming 4 pump-downs per month on 16 km segment)	2 months
Composite wrap for non-leaking pipeline defect	Composite wrap permanently repairs non-leaking pipeline defects. It is very cost-effective because it allows for pipeline repairs during pipeline operation, so gas does not need to be vented	5 600 USD per 8 m pipeline segment	6 months
Low-bleed pneumatics	Gas-operated pneumatic devices such as controllers vent methane by design. Because of the large number of pneumatic devices at compressor stations, the total emissions from such devices can be substantial at a facility. Low-bleed pneumatics can significantly reduce these leaks, through either replacement of high bleed devices with low bleeds or retrofitting existing high-bleed devices	675 USD to 1 850 USD	6–18 months

4 Methodology

Pipeline pump-down allows gas pipeline operators to reduce vented methane emissions when they take a section of pipeline offline for repairs. Normally, for safety reasons, the operator must vent the gas before repair work begins. This can involve the loss of substantial volumes of gas and money, as well as significant emissions of methane. Instead, with pipeline pump-down, the operator can use either a portable compressor, in combination with the in-line compressors, to move the majority of the gas to the next pipeline segment before depressurization.

Figure 4 illustrates the basic sequence of events for depressurizing a pipeline segment.

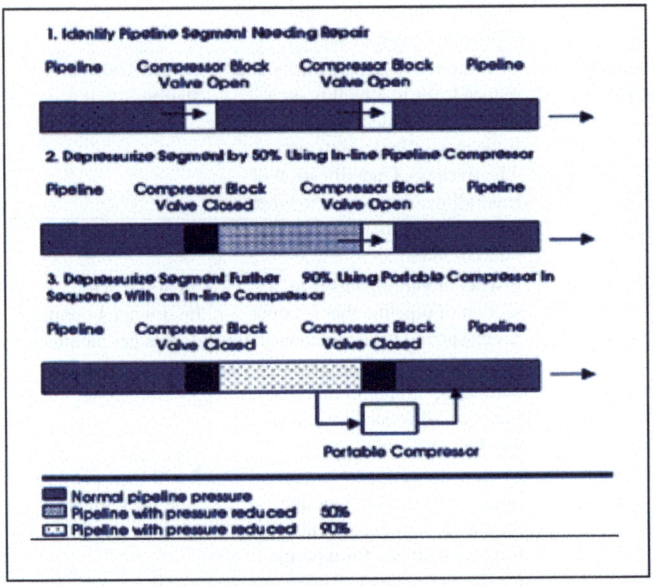

Fig. 4. Sequence of depressurizations events (US EPA 2006)

We can realize significant environmental and economic benefits by using downstream in-line compressors that may or may not be combined with portable compressors to lower gas-line pressure before performing maintenance and repair activities. Potential savings include:

- Significant reduction of fire/explosion risk of vented gas.
- Recovery and sale of natural gas that would have been vented to the atmosphere.
- Reduction of methane emissions.
- Reduction of nuisance (noise).

5 Results and Discussion

For our calculations, we take gas pipeline **GZ3** as an example; **GZ3** has maximum allowable operating pressure of 69.09 atmospheres (atm) and 20 km (km) distance between block valves. This pipeline has been chosen because it is the most leaking in the Algerian pipeline transportation network and it is concerned with rehabilitation program.

The Volume (Mcf) and value (USD) of the gas recovering from using in-line compressor and portable compressor are computed by the following equations:

$$V = \frac{\Pi D^2 L}{4RT}(P_i - P_f)MV_{\text{ideal gas}} \quad (1)$$

$$Ni = V - (V/R_i) \quad (2)$$

$$Vi = Ni \text{ x } 6 \text{ USD/MMBtu} \quad (3)$$

$$Np = Ni - (Ni/Rp) \quad (4)$$

$$V_g = N_p \text{ x } 6 \text{ USD/MMBtu} \quad (5)$$

$$Nt = Ni + Np \quad (6)$$

$$Vt = V_i + V_g \quad (7)$$

Where:

V:	Total amount of gas in the pipeline segment (m^3)
L:	Pipeline length between block valves(meter); 20×10^3 m
D:	GZ3 pipeline interior diameter(meter); 1.06 m
P_i:	GZ3 initial pipeline segment pressure(atm); 63.168 atm
P_f:	GZ3 final pipeline segment pressure(atm); 1 atm
R:	Gas constant R = 0.082 L.atm/mole.K
T:	Gas absolute temperature in pipeline segment(Kelvin); 294.25 °K
$MV_{\text{ideal gas}}$:	Molar volume of ideal gas22.4 x10^{-3} m^3/mole
Ni:	volume of gas recovered using an in-line compressor
Ri	is in-line compressor compression ratio
Vi:	Value of gas recovered using an in-line compressor
Np:	volume of gas recovered using a portable compressor
R_p	is the portable compressor compression ratio
Vg:	Value of gas recovered using a portable compressor
Nt:	Total gas recoverable with inline compressor and gas recoverable with portable compressor
Vt:	Total value of the recovered gas from (in-line compressor and portable compressor)

Gas price EU August 2017 is approximately 6 USD/MMBtu (Gas price source: Sonatrach-Trading activity)

With conversion factors:

1 km = 10^3 m
1 inch = 2.54 x 10^{-2}m (GZ3 is 42 inch)
° Kelvin = °Celsius + 273.15 (temperature in pipeline segment, assumed 21.1°C
1 litre = 10^{-3} m^3 ($MV_{ideal\ gas}$ = 22.4 litres/mole)
1 m^3 = 35.32 ft^3 and 1 Mcf = 10^3 ft^3
1 Mcf = 1 MMBtu

From these calculations, we can obtain that a total net value of **194 161.08 USD** can be achieved by recovering about **0.916 million m^3** of natural gas, through the use of portable compressor according to the same assumption.

Evaluate of economics of using a portable compressor in sequence with an in-line compressor are computed by the following equations:

$$Vc = Vcf + Vcl + Vcm \quad (8)$$

$$Vp = Vg - Vc \quad (9)$$

$$Vtotal = Vi + Vp \quad (10)$$

Where:

Vc: Total costs associated with a portable compressor
Vcf: Cost of fuel (fuel used by portable compressor to draw down pressure from 63.168 atm until suction lost)
Vcl: Cost of labor
Vcm: Cost of maintenance
Vp: Net value of recovered portable gas
Vg: Value of gas recovered using a portable compressor
Vtotal: Total value of recovered gas
Vi: Value of gas recovered using an in-line compressor

To calculate the cost of total investment, all expenses have to be included such as compressor purchase, freight, operating costs, etc..... The costs and gas savings from **GZ3** compressor pump-downs are summarized in Table 2.

Table 2. Total investments

Equipment/Practice	Costs USD
Portable compressor capital cost (69.09 atm, high flow)	3 to 6 million
Portable compressor O&M cost	5 000 to 30 000
Labor and transportation cost	5 000 to 20 000
Total investments	4 million
Total natural gas savings	11.628 million

From these calculations, it is clearly seen that use of portable compressor is technically and economically a feasible activity to reduce methane emissions from high pressure pipelines. Benefits justified purchase of **4 portable compressors**. Another consideration affects the beneficial use of portable pump-down compressors.

- Additional fuel use from downstream compressors can be large
- Fuel use by portable compressor
- Extra manpower and maintenance issues
- The time variable is very important, especially if service disruptions are a potential from the line being out of service
- The extra fuel consumed by inline compression due to increased friction loss is usually the critical variable

6 Conclusion

During this study, it has been demonstrated that avoiding natural gas venting directly to atmosphere is possible for the Algerian natural gas transportation network; furthermore, considerable cost saving and environmental benefits can be derived throughout the proposition of solution never used before at Algeria. Pipeline pump-down technique with portable compressor solution helps avoiding the outdated, dangerous and environmentally harmful practices.

The use of portable compressors instead of venting will mainly allow **54.873 million m^3** gases saving, which costs about **11.628 million USD** and benefits justified purchase of **4 portable compressors.**

As a developing country, Algeria has not any commitments for reducing GHG emissions under Kyoto Protocol. However, through launching CDM projects, Algeria could benefit from green investments which will contribute to sustainable development and to some extent bring additional economic benefits through carbon credits revenues and technology transfer from industrialized countries.

This work will demonstrate that when managing a complex asset like gas transportation network, it is possible to conciliate economics with safety and environment. So, it is time to change the current practices and to take actions to avoid gas venting.

References

Anifowose, B., Odubela, M.: Methane emissions from oil and gas transport facilities exploring innovative ways to mitigate environmental consequences. J. Clean. Prod. **92**, 121–133 (2015)

Borraz-Sánchez, C.: Optimization methods for pipeline transportation of natural gas, Ph.D. thesis, University of Bergen, Norway (2010)

Boukrif, N.: Le Gaz algérien dans l'équation de l'approvisionnement du marché Européen. Recherches Economiques et Managériale **4**, 2–25 (2008)

Brandt, A.R., Heath, G.A., Kort, E.A., O'Sullivan, F., Petron, G., Jordaan, S.M., Tans, P., Wilcox, J., Gopstein, A.M., Arent, D., Wofsy, S., Brown, N.J., Bradley, R., Stucky, G.D., Eardley, D., Harriss, R.: Methane leaks from north american natural gas systems. Sci. Energy Environ. **343**(6172), 733–735 (2014)

Cormack, J.: Overview of pipeline pumpdown and Natural Gas STAR Company TransCanada experience, Oil and Gas Systems Technology Transfer and Subcommittee Meeting, Lake Louise, Canada (2009)

Elshiekh, T.: Optimization of fuel consumption in compressor stations. Oil Gas Facil. **4**, 59–65 (2014)

Frankenberg, C., Meirink, J.F., Van Weele, M., Platt, U., Wagner, T.: Assessing methane emissions from global space-borne observations. Science **308**(5724), 1010–1014 (2005)

GRTGaz: Plan décennal de développement du réseau de transport de GRTgaz 2013–2022, France (2013)

IEA: Emission reductions in the natural gas sector through project-based mechanisms international energy agency, Paris, France (2003)

IPCC: Fugitive Emissions - Guidelines for National Green House Gas Inventories, Volume 2: Energy, Institute for Global Environmental Strategies, Japan (2006)

IPCC: Climate Change 2007: Synthesis Report, Contributions of Working Groups I, II, and III to the Fourth Assessment Report of the Intergovernmental Panel on Climate Change, Geneva (2007)

Ishkov, A., Akopova, G., Evans, M., Yulkin, G., Roshchanka, V., Waltzer, S., Romanov, K., Picard, D., Stepanenko, O., Neretin, D.: Understanding Methane Emissions Sources and Viable Mitigation Measures in the Natural Gas Transmission Systems: Russian and US Experience, International Gas Union Research Conference (2011)

Kirchgessner, D.A., Lott, R.A., Cowgill, R.M., Harrison, M.R., Shires, T.M.: Estimate of methane emissions from the US natural gas industry. Chemosphere **35**, 1365–1390 (1997)

Kopanos, G.M., Xenos, D.P., Cicciotti, M., Pistikopoulos, E.N., Thornhill, N.F.: Optimization of a network of compressors in parallel: operational and maintenance planning–The air separation plant case. Appl. Energy **146**, 453–470 (2015)

Kurz, R., Lubomirsky, M., Brun, K.: Gas compressor station economic optimization. Int. J. Rotating Mach. 2012, 9 p. (2012)

Lechtenböhmer, S., Dienst, C., Fischedick, M., Hanke, T., Fernandez, R., Robinson, D., Kantamaneni, R., Gillis, B.: Tapping the leakages: methane losses, mitigation options and policy issues for Russian long distance gas transmission pipelines. J. Greenhouse Gas Control **4**, 387–395 (2007)

Litto, R., Hayes, R.E., Liu, B.: Capturing fugitive methane emissions from natural gas compressor buildings. J. Environ. Manage. **84**, 347–361 (2007)

Miller, S.M., Wofsy, S., Michalak, A., Kort, E., Andrews, A., Biraud, S., Dlugokencky, E., Eluszkiewicz, J., Fischer, M., Janssens-Maenhout, G., Miller, B., Miller, J., Montzka, S., Nehrkorn, T., Sweeney, C.: Anthropogenic emissions of methane in the United States. Proc. Nat. Acad. Sci. U.S.A. **110**(50), 20018–20022 (2013)

Nisbet, E.G., Dlugokencky, E.J., Bousquet, P.: Methane on the rise-again. Science **343**(6170), 493–495 (2014)

Ríos-Mercado, R.Z., Borraz-Sánchez, C.: Optimization problems in natural gas transportation systems: a state-of-the-art review. Appl. Energy **147**, 536–555 (2015)

Salomons, S., Hayes, R.E., Poirier, M., Sapoundjiev, H.: Modeling a reverse flow reactor for the catalytic combustion of fugitive methane emissions. Comput. Chem. Eng. **28**, 1599–1610 (2004)

Sonatrach: Sonatrach une compagnie pétrolière et gazière intégrée (2014a). http://www.sonatrach.com/PDF/Presentation_Sonatrach_PetroliereGaziere.pdf

Sonatrach: Sonatrach une Dimension Gazière Internationale (2014b). http://www.sonatrach.com/PDF/Sonatrach_gaz.pdf

Stambouli, A.B., Khiat, Z., Flazi, S., Kitamura, Y.: A review on the renewable energy development in Algeria: current perspective, energy scenario and sustainability issues. Renew. Sustain. Energy Rev. **16**, 4445–4460 (2012)

US EPA: Directed Inspection and Maintenance at Compressor Stations, Lessons Learned from the Natural Gas STAR Program (2003)

US EPA: Using Pipeline Pump-Down Techniques to Lower Gas Line Pressure before Maintenance, Lessons Learned From Natural Gas STAR Partners (2006)

US EPA: Reducing methane emissions in pipeline maintenance and repair. Technology transfer workshop, Buenos Aires, Argentina (2008)

US EPA (2014): Inventory of U.S. Greenhouse Gas Emissions and Sinks: 1990–2012, Washington (2014)

Xenos, D.P., Cicciotti, M., Kopanos, G.M., Bouaswaig, A.E., Kahrs, O., Martinez-Botas, R., Thornhill, N.F.: Optimization of a network of compressors in parallel: Real Time Optimization (RTO) of compressors in chemical plants–an industrial case study. Appl. Energy **144**, 51–63 (2015)

Xu, Y., Zaelke, D., Velders, G.J.M., Ramanathan, V.: The role of HFCs in mitigating 21st century climate change. Atmos. Chem. Phys. **13**, 6083–6089 (2013)

Kinetics and Isotherms Adsorption of Reactive Dye by Thermally Treated Dolomite

Samira Ziane(✉) and Amine Khelifa

Laboratoire de Structure, Elaboration et Applications des Matériaux Moléculaires (S.E.A.2M.), Département de Génie des Procédés, Université de Mostaganem, B.P. 981, R.P., 27000 Mostaganem, Algeria
samirazianedz@gmali.com

Abstract. Kinetic and equilibrium data of the adsorption of reactive black 5 (RB5) from synthetic solutions on Algerian dolomite were determined experimentally. The obtained samples, i.e., raw dolomite and its form heated at 900 °C (D900), were characterized by XRD and SEM. The kinetic mechanism is enough complex, involving different models such as those of pseudo-second order, intraparticle diffusion, and Elovich. The adsorption of RB5 at equilibrium increases strongly with increasing adsorption temperature. The affinity sequence is D900 >> raw dolomite with amounts adsorbed of 125.9 and 38.20 mg g^{-1}, respectively. This difference would be correlated with the weight loss percentage and crystallographic properties. The kinetic mechanism involving different models such as those of pseudo-second-order, intraparticle diffusion, and Elovich. The isotherms are found to be suitably fitted by the Redlich-Peterson equation, a model including three adjustable parameters and requiring nonlinear least-square analysis. A close agreement exists between the evolution of kinetic, thermodynamic and equilibrium, parameters, indicate chemisorption process. The fact that maximum adsorption occurs at isoelectric point emphasizes the prevalence of the non-electrostatic interaction. The process reflects for D900 a weak chemical interaction via a mechanism of surface complexation. In view of its adsorbed amount, D900 appears very effective for removing reactive dyes from wastewaters.

Keywords: Dolomite · Reactive blac 5 · Adsorption · Redlich-Peterson model Mechanism

1 Introduction

The wastewater from many industries textile, rubber, paper, cosmetics, printing contain pollutants, pigment, dyes. …it is essential to treat this dye. In order to diminish the pollution. Many studies have been based on the toxicity of dyes and their impact on the environment [1, 2]. several methods have been employed to remove dyes from the aqueous phase such as photocatalysis, electrochemical, and adsorption etc. [3–9]. Adsorption is an economic and efficient process to remove dyes, pigments and other pollutants from wastewater. This study investigates the removal of reactive black 5 (RB5) by a raw dolomite and dolomite treated thermically at 900 °C (D900) from synthetic solutions of the azo textile dye. Dolomite is an important mineral formed by double carbonate of calcium and magnesium with ideal formula $CaMg(CO_3)_2$. thermal

B. Abdelbaki et al. (Eds.): SMSD 2017, *Proceedings of the Third International Symposium on Materials and Sustainable Development*, pp. 568–579, 2018.
https://doi.org/10.1007/978-3-319-89707-3_61

treatment of dolomite at 900 °C undergoes a modification of such kind to release CO2 and obtain two oxide CaO-MgO [10]. Before adsorption, raw Algerian dolomite and treated at 900 °C characterized by XRD and SEM.

The objective of this work is to examine the possibility of using this material for the RB5 removal from aqueous solutions. Before adsorption experiments, Algerian dolomite was treated at 900 °C, characterized by XRD and SEM. The effects of solution concentration, contact time, temperature and pH have been assessed; kinetic equations were applied to evaluate to study the mechanism of RB5 retention. Equilibrium data were examinated by Langmuir, Freundlich, Lagmuir Freundilch and Redlich Peterson models. A close agreement exists between the evolution of kinetic, equilibrium and thermodynamic parameters, suggesting chemisorption.

2 Materials and Methods

2.1 Materials

Raw dolomite used in this study was obtained from Tlemcen (Western region of Algeria). The grain size was 0.125–0.25 mm. Its chemical composition is principally constituted of CaO (31.2%) and MgO (21.1%). The raw dolomite was heated at 900 °C. This sample was named D900, which it was processed at the relevant temperature for 2 h, with a rate of 10 °C min^{-1}.

2.2 Adsorption Procedure

Commercial azo dye used in this procedure is Sakazol Black B named too reactive black 5 (RB5, chemical formula: $C_{26}H_{21}O_{19}N_5S_6Na_4$, Sigma-Aldrich, CAS: 17095-24-8, FW: 991.8 g mol^{-1}, Fig. 1), 20 mg of draw dolomite and/or D900 was mixed with 20 mL of aqueous RB5 solution. The solution was centrifuged, the analyse of the supernatant was carried by visible spectrophotometry, at 597 nm, using a Shimadzu 1240 UV-Vis spectrophotometer. To calculate the adsorbed quantity, it is enough to make the difference between the initial and final concentrations. The experimental conditions such as the effects of concentration, contact time, temperature and pH are regrouped in Table 1.

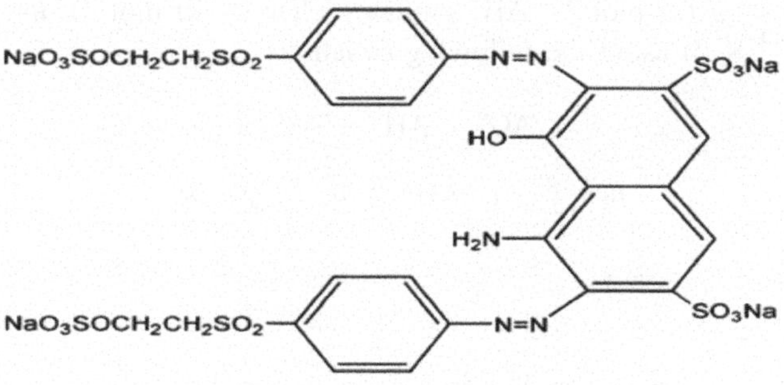

Fig. 1. Chemical structure of RB5.

Table 1. Experimental conditions during the adsorption of RB5

Contact time	1, 3, 5, 10, 20, 40, 60, 120, and 240 min; $C_{initial}$ = 80 mg L^{-1}; [solid]/[solution]: 1 g L^{-1}; T: 25 40 °C and 55 °C, pH = 6.9
Concentration	20, 40, 60, 80, 100, 150, 200, 300, and 400 mg L^{-1}; [solid]/[solution]: 1 g L^{-1}; contact time: 2 h

2.3 Theoretical Considerations

2.3.1 Adsorption Kinetics

In order to control the mechanism of the adsorption process, various kinetic equations were applied to examine the experimental data (Table 2).

Table 2. Kinetic adsorption models.

Kinetic models and equations	Parameters
Pseudo-first-order $\log(Q_e - Q_t) = \log Q_e - \frac{K_1 t}{2.303}$ [11] Pseudo-second-order $\frac{t}{Q_t} = \frac{1}{K_2 Q_e^2} + \frac{t}{Q_e}$ [12]	Q_t: the amount adsorbed at time t (mg g^{-1}) Q_e: adsorption capacity at equilibrium (mg g^{-1}) K_1: pseudo-first-order rate constant (min^{-1}) *t*: the contact time (min) K_2 (g mg^{-1} min^{-1}): the pseudo-second-order rate constant of adsorption *h*,: The initial adsorption rate, can be defined as h = $K_{2.}$ Q_e^2
Intraparticle diffusion $Q_t = K_{id} t^{1/2} + C$ [13]	K_{id} *(mg g^{-1} $min^{-1/2}$)*: the intraparticle diffusion rate constant C: constant
Elovich equation $Q_t = \frac{1}{\beta}\ln(\alpha.\beta) + \frac{1}{\beta}\ln t$ [14]	α: the initial adsorption rate (mg g^{-1} min^{-1}) β: the desorption constant (g mg^{-1}). The parameters α and β can be obtained from the linear plots of Q_t vs. $\ln t$

2.3.2 Adsorption Equilibrium Modeling

The equilibrium models of Langmuir, Freundlich, Langmuir-Freundlich (LF) and Redlich–Peterson (RP) were applied to fitting the results (Table 3).

2.3.3 Thermodynamic Study

The adsorption process was flowed also by thermodynamic parameters ΔG°, Gibb's free energy change (kJ mol^{-1}), ΔH° enthalpy change (kJ mol^{-1}) and ΔS°, entropy change (J mol^{-1} K^{-1}) using the following equations:

$$\Delta G^\circ = \Delta H^\circ - T\Delta S^\circ \tag{1}$$

$$\ln K_d = (-\Delta H^\circ/R.T) + (\Delta S^\circ/R) \tag{2}$$

Table 3. Isotherm equations applied to fit the experimental data.

Adsorption models	Equations	Parameters
Langmuir	$\frac{C_e}{Q_e} = \frac{1}{Q_m.K_L} + \frac{C_e}{Q_m}$ [15]	Q_e: equilibrium amount removed from solution (mg g^{-1}) C_e: equilibrium concentration (mg L^{-1}) Q_m: maximum amount per unit weight of adsorbent for complete monolayer coverage (mg g^{-1}) K_L: constant related to the affinity of binding sites (L mg^{-1}) K_F: constant taken as an indicator of adsorption capacity (L g^{-1}) *1/n*: constant indicative of the adsorption intensity
Freundlich	$\log Q_e = \log K_F + \frac{1}{n}\log C_e$ [16]	
Langmuir-Freundlich	$Q_e = \frac{K_{LF}C_e^{\beta}}{1+a_{LF}C_e^{\beta}}$ [17]	Q_e: amount adsorbed at equilibrium (mg g^{-1}) C_e: equilibrium solution concentration (mg L^{-1}) K: equilibrium constant (L mg^{-1}), β: heterogeneity factor that depends on surface properties of the adsorbent M: maximum amount adsorbed (mg g^{-1})
Redlich-Peterson	$Q_e = \frac{K_{RP}C_e}{1+a_{RP}C_e^{\beta}}$ [18]	

where K_d is the distribution coefficient (L g^{-1}), R is gas constant (8.314 J mol^{-1} K^{-1}) and T the absolute temperature (K), and. This coefficient reflects the overall RB5-material affinity and is evaluated by:

$$K_d = Q_e/C_e \tag{3}$$

3 Results and Discussion

3.1 Decomposition and Characterization

Decomposition of dolomite with the obtained phases, was discussed previously from XRD analysis [19, 20]. The D900 sample compared to draw dolomite contains mainly lime, periclase as well as a minor amount of calcite. According to the literature data [21], the decomposition of dolomite above 700 °C occurs in air in two steps, as follows:

first reaction:

$$CaMg(CO_3)_2 \rightarrow CaCO_3 + MgO + CO_2, \tag{4}$$

second reaction:

$$CaCO_3 \rightarrow CaO + CO_2 \quad (5)$$

About 800 °C the product of partial decomposition occurs and takes to calcium carbonate and magnesium oxide (first reaction). From 900 °C. The product was composed of a porous calcite (calcium carbonate) and the powdered magnesium oxide (second reaction) [22]. These considerations justified the choice of 900 °C.

SEM images of draw dolomite and D900 are depicted in Fig. 2. Dolomite evidences a compact structure with presence of bright area due to an inclusion of quartz. After thermal modification, the surface topography is profoundly affected. The micrograph of D900 reveals the grains undergo a disintegration with obtaining a crystallite multitude, of spherical form. That corresponds to the second reaction (5), which occurs starting from 900 °C. This reaction is associated with the CO_2 decarbonation of calcite.

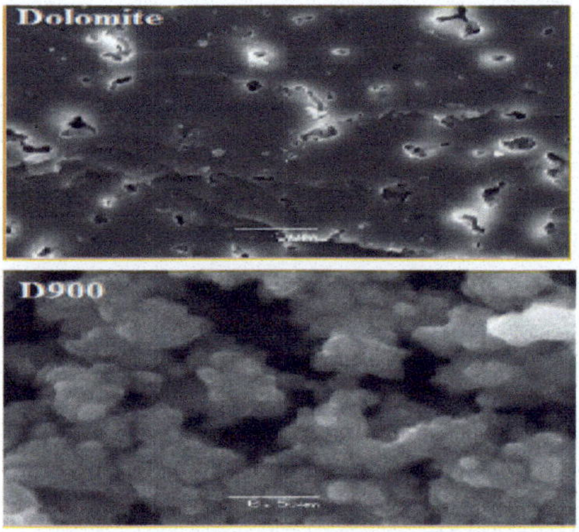

Fig. 2. SEM images of dolomite and dolomite treated at 900 °C.

3.2 Effect of Contact Time and Kinetic Study

The effects of contact time and temperature are demonstrated in Fig. 3. At 25 °C, it can be noticed that the removal rate is fast in the first 20 min, corresponding to the adsorption of 68% of the RB5 concentration for D900 and 41% for dolomite. After this initial stage, the rate gradually decreases, reaching equilibrium at about 2 h. rapid adsorption at the initial contact time is cause of to the availability of surface sites. So, the equilibrium was achieved after an agitation time of 2 h.

In order to study the adsorption kinetic, pseudo-second-order, intraparticle diffusion and Elovich's models were tested. Table 4

The fit of the experimental data was flowed the pseudo-second-order model were obtained, corresponding to high R^2 values, i.e. >0.96. The values of the calculated

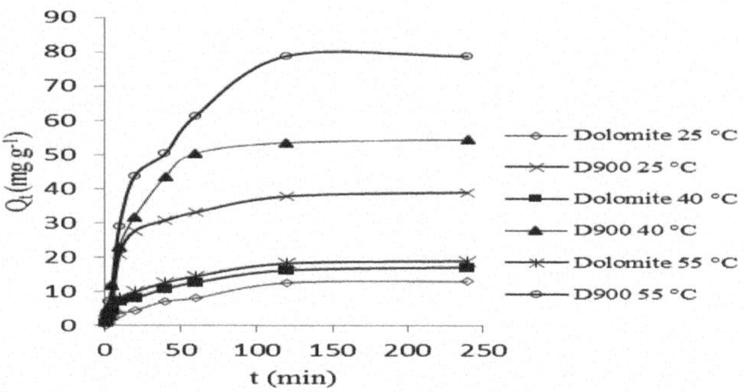

Fig. 3. Effect of contact time on the adsorption of RB5 by dolomite and D900 at Cinitial = 80 mg L^{-1}.

adsorbed quantity, Qe(cal), are approached by contribution in those experimental, Qe (exp), This model was also validated for RB5-activated carbon [23]. The initial sorption rate, *h*, increases with increasing adsorption temperature. For D900, the parameter *h* varies from 48.79 to 127.11 mg $g^{-1}min^{-1}$, when the temperature passes from 25 to 55 °C. Hence, the increase in temperature favors the initial adsorption process. The applicability of this equation shows that the possibility of chemisorption. Similar results have been found for the RB5- biocomposite system [24].

The kinetic parameters of intraparticle diffusion model are gathered in Table 4. High determination coefficients (0.93–0.99), corresponding to the second linear portion, were obtained. The intraparticle diffusion rate constant, K_{id}, increases with temperature, when one passes from 25 to 55 °C indicating an improvement of the pore diffusion into the dolomitic solides. Elovich's model was used to describe a chemisorption onto heterogeneous adsorbents. The obtained R^2 values (Table 4) confirm the applicability of this equation the results showed that Elovich's equation is appropriate, proving that RB5 is chemisorbed on the dolomite and D900. The α values increase but The β values parameter decrease with temperature. This would be explained that desorption of the molecule of RB5 is difficult. However, a close agreement exists between the evolution of values of *h* (pseudo-second-order), K_{id} (intraparticle diffusion) and α (Elovich), which increase with increasing adsorption temperature.

3.3 Adsorption Equilibrium

3.3.1 Isotherms

Figure 4 presents the equilibrium adsorption of RB5 onto dolomite, and D900 at 25, 40 and 55 °C. For two materials, the adsorption capacity strongly increases with increasing temperature. a initial concentration value of 400 mg L^{-1}, Dolomite adsorbed 26.36 and 38.18 but D900 adsorbed 51.8 and 125.9 mg g^{-1}, at 25 and 55 °C respectively, so that RB5 molecules become more reactive at higher temperatures. The affinity D900 >> raw dolomite would be correlated with the weight loss percentage and crystallographic properties [20]. The decomposition of dolomite (CaCO3.MgCO3) in MgO and CaO species, and a weight loss of 47% would explain the highest capacity

Table 4. Kinetic parameters of the RB5 adsorption by the thermally treated dolomites

Samples	T (°C)	Q_e(exp) (mg g^{-1})	Pseudo-second-order model				Intraparticle diffusion model			Elovich's model			
			Q_e(cal) (mg g^{-1})	K_2 (g $mg^{-1}min^{-1}$)	h (mg $g^{-1}min^{-1}$)	R^2	K_{id} (mg $g^{-1}min^{-1/2}$)	C (mg g^{-1})	R^2	Q_e (cal) (mg g^{-1})	α (mg g^{-1} min^{-1})	α β β (g mg^{-1})	R^2
Dolomite	25	12.72	14.92	0.0452	7.33	0.961	1.058	0.163	0.975	11.75	1.54	0.404	0.940
	40	16.36	18.86	0.0502	13.54	0.995	1.231	3.091	0.985	16.57	3.74	0.286	0.984
	55	18.40	20.40	0.0537	18.21	0.996	1.340	3.768	0.999	19.04	3.00	0.315	0.984
D900	25	37.95	43.47	0.0338	48.79	0.990	2.531	14.44	0.936	42.16	7.42	0.125	0.949
	40	53.63	62.50	0.0288	83.07	0.994	6.004	4.69	0.995	70.46	9.85	0.089	0.964
	55	78.95	90.90	0.0203	127.11	0.992	6.559	10.54	0.960	82.39	12.0	0.064	0.964

of D900. In return, the lowest capacity of raw dolomite could be due to a CaCO3. MgCO3 phase without decomposition and, thus, without CO2 removal, for which there is no weight loss.

At 55 °C, the maximum adsorption capacity of D900 is 3.3 times higher than that of raw dolomite. This ratio falls to 1.97 at 25 °C, so that the temperature effect on the retention capacity is more accentuated for a calcined dolomite. Using the classification of Giles et al. [25], the isotherms of dolomite is L-shaped. The initial curvature of the L-curve shows that the contaminant has a high affinity for the surface, while the slope falls steadily with a rise in concentration. For D900 the increase in the adsorption temperature from 40 to 55 °C transforms the curves from L- to H-shape. The H-type isotherms are indicative of adsorbate–substrate chemical forces rather than physical attraction. Our H-curves may be conformed to the H3 subgroup [26]. Curves of H3-type represent systems in which the adsorbate monolayer has been completed. Probably, it is about a chemisorbed monolayer, on the basis of the discussion above.

3.3.2 Modeling the Experimental Data

to describe experimental results the Langmuir [15] and Freundlich [16] models are widely used. The accuracy of model fit with the experimental data is given by the determination coefficient, R^2, and the average relative error, E%. The linearisation parameters of both Langmuir and Freundlich models are grouped in Table 5.

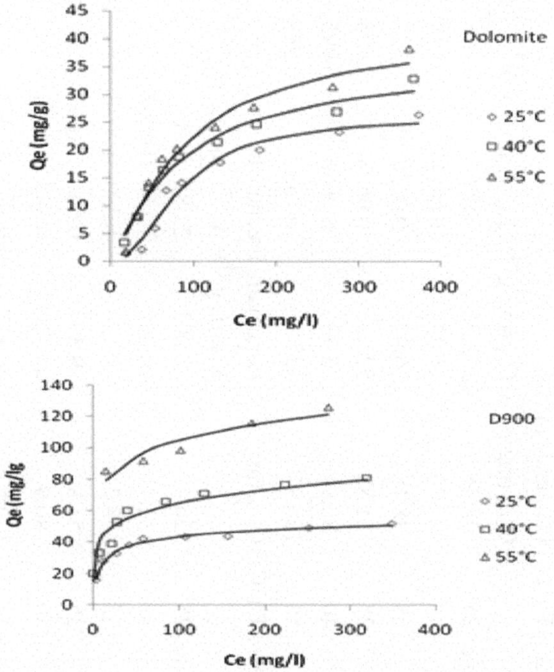

Fig. 4. Isotherms according to the LF model (—) and experimental data (…) for dolomite. Isotherms according to the RP model (—) and experimental data (…) for D900

Table 5. Adjustable parameters estimated from linear regression

Samples	T (°C)	Langmuir model					Freundlich model			
		Q_{exp} (mg g^{-1})	Q_m (mg g^{-1})	K_L (L mg^{-1})	R^2	E (%)	K_F (L $mg^{-1})^{1/n}$	n	R^2	E (%)
Dolomite	25	26.36	333.33	0.0003	0.008	33.97	0.078	0.947	0.854	35.85
	40	32.73	45.45	0.0069	0.935	11.16	0.820	1.524	0.885	19.06
	55	38.18	76.92	0.0030	0.322	29.34	0.655	1.194	0.797	36.38
D900	25	51.82	52.63	0.0603	0.995	15.73	9.57	2.75	0.738	19.79
	40	80.91	83.33	0.0666	0.995	42.09	4.35	6.02	0.917	10.86
	55	125.91	125	0.153	0.989	42.97	51.64	6.13	0.859	16.27

Samples	T (°C)	Q_{exp} (mg g^{-1})	Langmuir- Freundlich					Redlich-Peterson model				
			K_{LF} (L g^{-1})	β	a_{LF} (L $mg^{-1})^{\beta}$	R^2	E (%)	K_{RP} (L g^{-1})	β	a_{RP} (L $mg^{-1})^{\beta}$	R^2	E (%)
Dolomite	25	26.4	0.003	2.081	$9.71 * 10^{-5}$	0.969	20.16	0.156	1.691	$5.72 * 10^{-5}$	0.951	44.72
	40	32.7	0.206	1.166	0.0057	0.977	10.54	0.375	0.997	0.010	0.976	12.98
	55	38.2	0.136	1.280	0.0033	0.970	25.37	0.365	1.040	0.005	0.967	32.25
D900	25	51.8	9.524	0.661	0.1693	0.979	3.75	0.227	0.906	0.261	0.984	2.99
	40	80.9	49.520	0.384	0.5899	0.803	17.77	285.10	0.825	106.01	0.930	10.83
	55	125.9	58.480	0.172	0.1045	0.881	4.98	633.40	0.855	248.56	0.901	4.78

The application of the Freundlich model to the isotherms of untreated dolomite is inappropriate. A relatively better description is obtained especially for D900; the average relative error values being lower than those of Langmuir. This feature indicates that adsorption takes place onto energetically heterogeneous solids. However, the obtained results are found to be contradictory:

Although the E values are interesting those of R^2 are somewhat low (D900). The restricted validity of this model by considering a can be explained from its hypotheses: a physical process for RB5-dolomite, D900 systems is improbable, at least at high temperature.

The Langmuir-Freundlich [17] and Redlich-Peterson [18] models were used for the mathematical description of the adsorption equilibrium.. The related parameters were calculated and tabulated in Table 5. The isotherms of Redlich-Peterson (RP) describe efficiently the RB5 adsorption onto D900 (Fig. 4); the E values are globally < 10% with $R^2 > 0.90$. On the contrary, a poor agreement is found with dolomite. The K_{RP} values increase with the adsorption temperature and, the β exponent lies between 0 and 1, indicating a favorable adsorption onto heterogeneous adsorbents. The surface heterogeneity factor, β, depends on the surface properties, degree of crystallinity and distribution of active adsorption centers. Stefaniak et al. [27] found that many reasons are responsible for the energetic heterogeneity of the surfaces of dolomitic materials, namely the presence of Ca and Mg ions in the surface structure, the chemistry of adsorption sites, crystallographic defects, the influence of neighboring sites, impurities of surface, etc. As a complement of the RP model, that of Langmuir-Freundlich (Table 5) provides a satisfactory description of the dolomite isotherms (Fig. 4). The β_{LF} values for dolomite are >1, which is the sign of an unfavorable adsorption onto an energetically homogeneous material.

3.4 Thermodynamic Parameters

The values of enthalpy (ΔH), entropy (ΔS) and free energy (ΔG) are gathered in Table 6. The ΔG values are positive in the explored temperature range, indicate that the process is not spontaneous with a possibility of chemisorption. A better adsorption is found at high temperature because the calculated free energies decrease with temperature, for example the adsorption capacity D900 is higher at 55 °C, which confirms that the lowest (ΔG) values. The positive ΔH values shows that the RB5 adsorption is endothermic. Thereby, the process is favored by an increase in temperature through the activation of the adsorption sites. The positive values of ΔS proposed an increase in randomness at the materials/solution interface.

Table 6. Thermodynamic parameters for the RB5 adsorption onto thermally treated dolomites

Samples	ΔH (kJ mol^{-1})	ΔS (J mol^{-1}K^{-1})	ΔG (kJ mol^{-1})			R^2
			25 °C	40 °C	55 °C	
Dolomite	10.07	15.62	5.41	5.18	4.95	0.987
D900	33.87	103.08	3.15	1.61	0.058	0.999

4 Conclusion

The heating of dolomite at 900 °C products calcined partially decompose, leading to rigid porous calcite and fine powdered magnesium oxide., the SEM images evidence a more porous structure with newly created pores and slots so that the adsorption of RB5 by D900 is favored. Thermodynamic parameters $[\Delta G > 0, \Delta H > 0, \Delta S > 0]$ tend to chemisorption in close agreement with the evolution of isotherms (H-type), kinetics (applicability of the Elovich equation), pH (maximum adsorption at isoelectric point) and FTIR analysis (weak chemical interaction via a mechanism of inner-sphere complexation) [28].

The latter intervenes between nonbonding electrons of nitrogen atoms associated with amino groups and the surface oxides: MgO and/or CaO according to treatment temperature. The affinity of the dolomitic solids for RB5 is D900 > raw dolomite. Knowing that D900 adsorbs 125.9 mg g^{-1}, it has the potential to act as an adsorbent for the removal of reactive dyes from wastewaters

References

1. Shenai, V.A.: Azo dyes on textiles vs German ban: an objective assessment. Part III another study. Colourage **XLIII**(8), 41 (1996)
2. Greene, J.C., Baughman, G.L.: Effects of 46 dyes on population growth of fresh green alage Selenastrum capricornutum. Text. Chem. Colour. **28**, 4–23 (1996)
3. Bulut, Y., Aydin, H.: A kinetics and thermodynamics study of methylene blue adsorption on wheat shells. Desalination **194**, 259–267 (2006)
4. Demirbas, E., Kobya, M., Oncel, S., Sencan, S.: Removal of Ni(II) from aqueous solution by adsorption onto hazelnut shell activated carbon: equilibrium studies. Bioresour. Technol. **84**, 291–293 (2002)
5. Mahmoodi, N.M., Arami, M., Zhang, J.: Preparation and photocatalytic activity of immobilized composite photocatalyst (titania nanoparticle/activated carbon). J. Alloy. Compd. **509**, 4754–4764 (2011)
6. Mahmoodi, N.M., Arami, M., Bahrami, H., Khorramfar, S.: The effect of pH on the removal of anionic dyes from colored textile wastewater using a biosorbent. J. Appl. Polym. Sci. **120**, 2996–3006 (2011)
7. Maljaei, A., Arami, M., Mahmoodi, N.M.: Decolorization and aromatic ring degradation of colored textile wastewater using indirect electrochemical oxidation method. Desalination **249**, 1074–1078 (2009)
8. Sanghi, R., Bhattacharya, B.: Review on decolorisation of aqueous dye solutions by low cost adsorbents. Color. Technol. **118**, 256–269 (2002)
9. Malik, P.K.: Use of activated carbons prepared from sawdust and rice—husk for adsorption of acid dyes: a case study of Acid Yellow 36. Dyes Pigm. **56**, 239–249 (2003)
10. Wang, K., Yin, Z., Zhao, P., Han, D., Hu, X., Zhang, G.: Effect of chemical and physical treatments on the properties of a dolomite used in Ca looping. Energ. Fuels **29**, 4428–4435 (2015)
11. Lagergren, S.: Zur theorie der sogenannten adsorption elöster stoffe, Zur theorie der sogenannten adsorption gelöster stoffe (About the theory of so-called adsorption of soluble substances, K. Sven. Vetenskapsakad). Kungliga Svenska Vetenskapsademiens Handl. **24**, 1–39 (1898)

12. Ho, Y.S., McKay, G.: Pseudo-second order model for sorption process. Process Biochem. **34**, 451–465 (1999)
13. Weber, W.J., Morris, J.C.: Kinetics of adsorption on carbon from solution. J. Sanitary Eng. Div. Am. Soc. Civ. Eng. **89**, 31–59 (1963)
14. Chien, S.H., Clayton, W.R.: Application of Elovich equation to the kinetics of phosphate release and sorption in soils. Soil Sci. Soc. Am. J. **44**, 265–268 (1980)
15. Langmuir, I.: Adsorption of gases on plane surfaces of glass, mica and platinum. J. Am. Chem. Soc. **40**, 1361–1403 (1918)
16. Freundlich, H.M.F.: Over the adsorption in solution. J. Phys. Chem. **57**, 385–470 (1906)
17. Jaroniec, M., Derylo, A., Marczewski, A.W.: The Langmuir-Freundlich equation in adsorption from dilute solutions on solids. Mh. Chem. **114**, 393–397 (1983)
18. Redlich, O., Peterson, D.L.: A useful adsorption isotherm. J. Phys. Chem. **63**, 1024–1033 (1959)
19. Marouf, R., Khelifa, N., Marouf-Khelifa, K., Schott, J., Khelifa, A.: Removal of pentachlorophenol from aqueous solutions by dolomitic sorbents. J. Colloid Interface Sci. **297**, 45–53 (2006)
20. Boucif, F., Marouf-Khelifa, K., Batonneau-Gener, I., Schott, J., Khelifa, A.: Preparation, characterisation of thermally treated Algerian dolomite powders and application to azo- dye adsorption. Powder Technol. **201**, 277–282 (2010)
21. Wiedemann, H.G., Bayer, G.: Note on the thermal decomposition of dolomite. Thermochim. Acta **121**, 479–485 (1987)
22. Karaca, S., Gürses, A., Ejder, M., Açikyildiz, M.: Adsorptive removal of phosphate from aqueous solutions using raw and calcinated dolomite. J. Hazard. Mater. **128**, 273–279 (2006)
23. Ahmad, A.A., Idris, A., Hameed, B.H.: Organic dye adsorption on activated carbon derived from solid waste. Desalin. Water Treat. **51**, 2554–2563 (2013)
24. Umpuch, C., Sakaew, S.: Adsorption characteristics of reactive black 5 onto chitosan-intercalated montmorillonite. Desalin. Water Treat. (2013). https://doi.org/10.1080/19443994.2013.867541
25. Giles, C.H., Mc Ewan, T.H., Nakhwa, S.N., Smith, D.: Studies in adsorption. Part XI. A system of classification of solution adsorption isotherms, and its use in diagnosis of adsorption mechanisms and in measurement of specific surface areas of solids. J. Chem. Soc. **60**, 3973–3993 (1960)
26. Giles, C.H., Smith, D., Huitson, A.: A general treatment and classification of the solute adsorption isotherm: I. Theoretical. J. Colloid Interface Sci. **47**, 755–765 (1974)
27. Stefaniak, E., Bilinski, B., Dobrowolski, R., Staszczuk, P., Wojcik, J.: The influence of preparation conditions on adsorption properties and porosity of dolomite-based sorbents. Colloids Surf. A **208**, 337–345 (2002)
28. Ziane, S., Khelifa, K.M., Benmekki, H., Schott, J., Khelifa, A.: Removal of a reactive textile azo dye by dolomitic solids: kinetic, equilibrium, thermodynamic, and FTIR studies. J. Desalin. Water Treat. **56**, 695–708 (2015)

Degradation of a Pharmaceutical Pollutant by Coupling Photo-Fenton and Adsorption Processes

S. Cherif(✉), M. R. Boudraa, A. Moussa, R. Maachi, N. Nasrallah, and N. Guendouz

Laboratoire de Génie de la Réaction, Faculté de Génie Mécanique et Génie des Procédés, BP 32 El-Alia, 16000 Bab-Ezzouar, Algeria
Cherif14sonia@hotmail.com

Abstract. This study concerns a novel process for the treatment of industrial aqueous pharmaceutical effluents. In our case, it is diclofenac sodium loaded with an organic matter. The former is difficult to be treated biologically. This hybrid process couples photofenton and adsorption on commercial bentonite within a single reactor. The efficiency of an advanced oxidation process, which is photo-fenton on diclofenac, has been tested. We attempt to increase the elimination efficiency by acting on pH, H_2O_2/Fe^{2+} ratio, and the initial pollutant concentration. The maximum yield achieved is 65%. In order to complete the elimination of our pollutant we have used adsorption on commercial bentonite, the various degradation parameters have been optimized: contact time, mass of the adsorbent, and concentration of the pollutant. The combination of the two processes photo-fenton/adsorption allowed us to increase the removal efficiency, which reached 87%.

Keywords: Photo-fenton · Diclofenac · Advanced oxidation · Adsorption Bentonite

1 Introduction

The use and misuse of medicines by the world's population has led to the appearance of these in trace form in various aquatic environments and in different treatment plants around the world. Since these pharmaceuticals are highly chemically stable, considered as emerging and recalcitrant pollutants, their removal requires the application of new, more efficient and less expensive technologies such as advanced oxidation processes (POAs) coupled with another process to arrive at the total mineralization of the pollutant.

B. Abdelbaki et al. (Eds.): SMSD 2017, *Proceedings of the Third International Symposium on Materials and Sustainable Development*, pp. 580–592, 2018.
https://doi.org/10.1007/978-3-319-89707-3_62

2 Material and Method

2.1 Generality of the Pollutant Studied: Diclofenac Sodium

In this study, we are interested in a single pollutant that is diclofenac sodium: it is an anti-inflammatory active substance of the non-steroidal family (NSAIDs). A popular pharmaceutical drug. Diclofenac, mostly used as the sodium salt (diclofenac-Na), is used in human medical care as an analgesic, antiarthritic, antirheumatic compound belonging to the group of the nonsteroidal anti-inflammatory drugs. It represents an important drug in ambulatory care. It is used worldwide and has a production volume estimated to be in the hundreds of tons annually. It is used in the form of tablets, capsules, suppositories, intravenous solutions, and in ointments and gels for dermal application. It is readily metabolized after oral use, but assimilation is lower after dermal application. Schwaiger et al. (2004) [1].

The general characteristics of this substance and the physico-chemical properties are presented in Table 1 below.

Table 1. Physico-chemical characteristics of diclofenac.

Diclofenac		References
Chemical formula	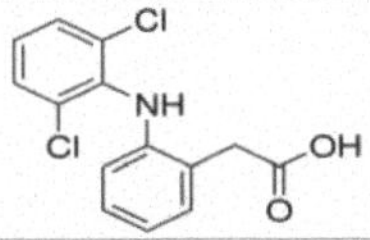	Scheytt et al. (2005) [2]
Molecular weight	318,15 g//mol	Rainsford et al. (2008) [3]
Solubility in water	2,37 mg/l	Scheytt et al. (2005) [2]
pKa	4,15	
Log Kow	4,51	
Log D	1,90	
Log koc	2,20–3,42	Carballa et al. (2008) [4]
Log kd	1,26–2,18	

Experimental procedure for the Fenton Photo Process

Reagents: The degradation experiments of our pollutants according to the photo-Fenton method are carried out using products of analytical qualities, which are the following:

Sulfate of iron ($FeSO_4$, $7H_2O$−278.02 g/mol), hydrogen peroxide (H_2O_2), at 10 V, sulfuric acid (H_2SO_4) concentrated at 95–97%, and the stock solutions were prepared with distilled water.

2.2 Experimental Design

For the advanced oxidation processes in homogeneous phase (Photo Fenton process), the procedure comprises the following steps:

A solution containing the diclofenac is placed in the photoreactor, to which is added a desired concentration of the iron sulphate and the oxygen peroxide, the whole is irradiated by the UV lamp.

The samples are taken at regular intervals and analyzed by UV-Visible spectroscopy.

The experimental device of the Photo-Fenton process is given in Fig. 1.

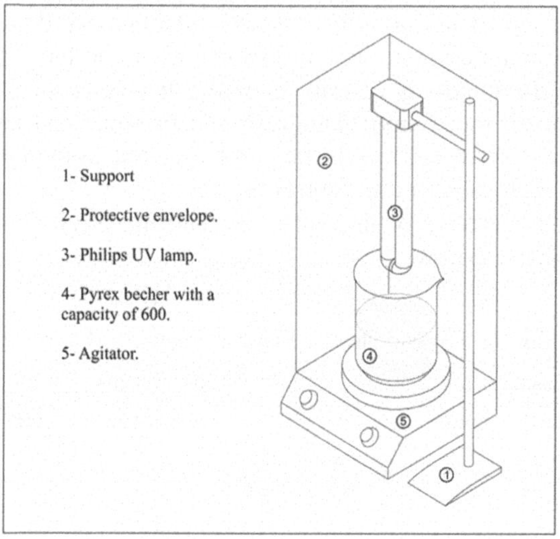

Fig. 1. Experimental device of the Photo - Fenton process

In addition to all the reagents used, the photo-Fenton process uses TUV lamps PL-L 36 W/4P (UVC 36 W) compact discharge lamps with single-ended low-pressure mercury vapor.

2.3 Preparation of the Diclofenac Solution for the Adsorption Process

Bentonite in Algeria: In Algeria, the most economically important bentonite deposits are in western Algeria. In particular, the career of Maghnia (Hammam Boughrara), whose reserves are estimated at one million tons and Mostaganem (M'zila) with reserves of two million tons. The bentonite used in our experiments is bentonite from Maghnia.

Experimental design: A well-defined mass of bentonite is added to the reactor containing the diclofenac solution for degradation; the whole is placed under stirring with taking samples at regular intervals of time until reaching equilibrium. The experimental setup is given in the following Fig. 2:

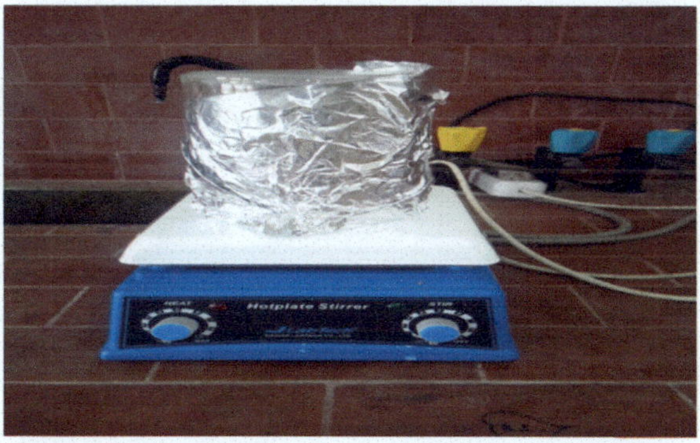

Fig. 2. Experimental device for an adsorption process

3 Results and Discussion

3.1 Direct Photolysis

A solution of diclofenac with 50 ppm concentrations is subjected to irradiation by the UV lamp used; we followed the evolution of the reduced concentration for 120 min.

The results obtained (Fig. 3) show that diclofenac undergoes very little degradation, for a 120-min irradiation the decrease is 3%. This confirms the theory that the anti-inflammatory studied is of high chemical and refractory stability, and that irradiation alone has no influence on degradation, Méndez-Arriaga et al. (2010) [5], Cardoso da Silva et al. (2013) [6].

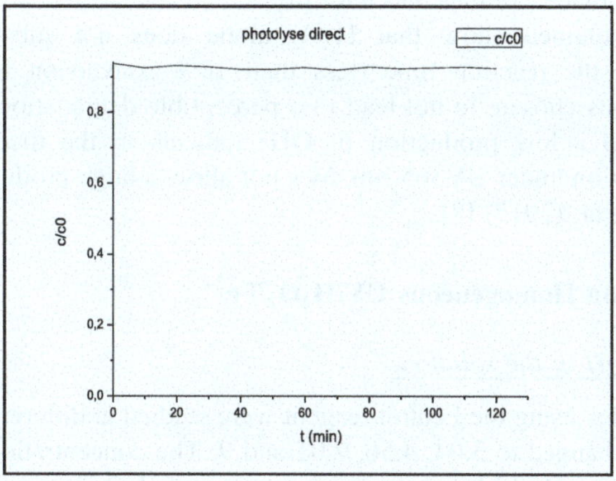

Fig. 3. Direct photolysis of diclofenac;[diclo] = 50 ppm, pH = 4.56, T = 20 °C, V = 500 ml.

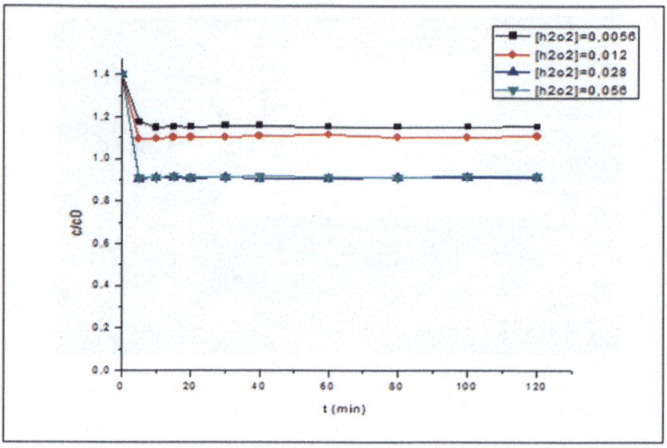

Fig. 4. The reduced concentration of Diclo as a function of the initial concentration of H_2O_2, *[Diclofenac]* = *50* ppm, pH = *4.56 and T* = 20 °C, *V* = *500* ml.

3.2 Photolysis in the Presence of H_2O_2

To accelerate the photolysis process, we have added a clean, non-toxic H_2O_2 oxidizer. The amount of the added H_2O_2 is taken from the theoretical value:

$$C_{14}H_{11}Cl_2NO_2 + 33H_2O_2 \rightarrow 14CO_2 + 2HCl + HNO_3 + 37H_2O \quad (1)$$

Then a complete oxidation of one mole of diclofenac theoretically requires 33 mol of H_2O_2, which corresponds to a concentration of 0.0056 M, then we have increased this concentration twice, 5 times and ten times.

The results obtained show that H_2O_2 alone does not substantially degrade diclofenac during the reaction time (less than 10% conversion in 120 min). The operating conditions chosen do not lead to a perceptible degradation of the pollutant.

This is due to a low production of OH• radicals in the medium Kiwi et al. (2000) [7], irradiation under UV365 nm does not allow a large production of hydroxyl radicals. Huang et al. (2015) [8].

3.3 Photo-Fenton Homogeneous $UV/H_2O_2/Fe^{2+}$

(a) *Effect of the pH of the solution:*

Degradation kinetics using the Fenton reagent were studied at different pHs. The pH of the medium was changed to 3.04, 4.56, 7.01 and 9. The concentration of $FeSO_4$ (Fe^{2+} ions) being set at $0.1.10^{-3}$ M and the concentration of H_2O_2 set at 0.006 M.

The degradation curves of diclofenac at different pHs are shown in Fig. 5.

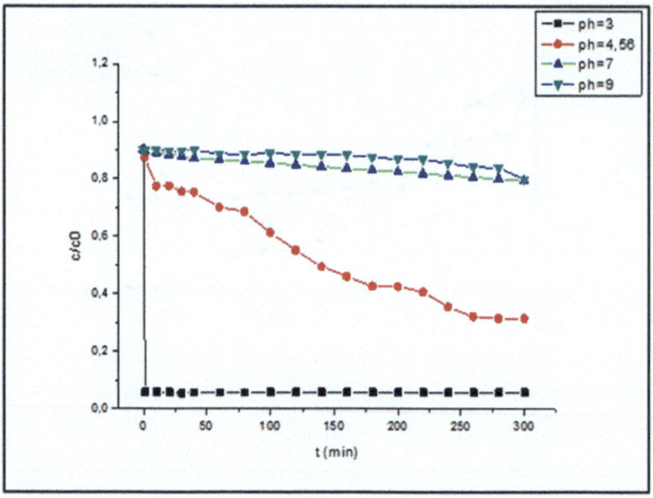

Fig. 5. The degradation curve of diclofenac at different pH

The increase in pH values causes the degradation of diclofenac to decrease. This result could be attributed to a decrease in the production of OH radicals. Souza et al. (2014) [9]. Indeed, the pH of the solution controls the production of these reactive oxygen species. As the pH of the solution increases, precipitation of iron ions in the form of hydroxide complexes is promoted, thereby decreasing the catalytic decomposition of H_2O_2 to form the OH• groups. Goi et al. (2008) [10].

In an acid medium, at a pH of less than 4, diclofenac becomes insoluble and precipitates, the precipitation of diclofenac should not be confused with decomposition. Pérez-Estrada et al. (2005) [11],

(b) *Effect of the concentration of H_2O_2 on the Photo-Fenton process (presence of catalyst)*

The concentration of hydrogen peroxide is an important factor in the Photo-Fenton process. Thus, we tried to determine the optimal concentration of hydrogen peroxide to improve the efficiency of the Photo-Fenton process Fig. 6 illustrates the variation of the C/C0 ratio as a function of time for 3 concentrations of H_2O_2.

The addition of hydrogen peroxide H_2O_2 improves the degradation rates compared to the direct photolysis and UV/H_2O_2 processes, reaching 65% after 300 min of UV irradiation, this is due to the additional production of hydroxyl radicals Ahmed et al. (2011) [12].

From this Fig. 4 it appears that the concentration 0.006 M of H_2O_2 improves the efficiency of the Photo-Fenton process. After 300 min of treatment, the degradation has increased, the concentration reduced and decreased. Indeed the gradual increase in the concentration of H_2O_2 allows the production of a higher quantity of hydroxyl radicals necessary for the oxidation of diclofenac, Selvam et al. (2010) [13], Alalm et al. (2015) [14].

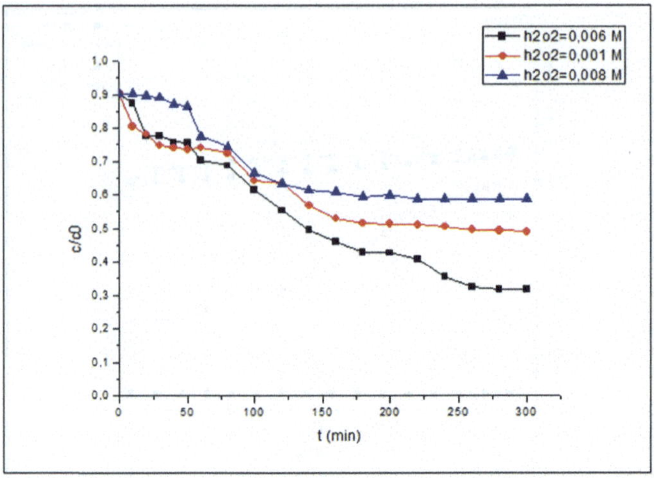

Fig. 6. The evolution of the reduced concentration of Diclofenac as a function of the concentration of H_2O_2, [Diclofenac] = 50 ppm, $[Fe^{2+}] = 0.0001$ M, pH = 4.56 and T = 20 °C.

However, for the 0.008 concentration the degradation yield decreases, this phenomenon can be explained by the acceleration of the parasitic reactions by the trapping of the OH• groups by H_2O_2 that leads to the formation of the hydroxyl radicals Ravina et al. (2002) [15].

Since the concentration of hydroxyl radicals is constant, the oxidation rate of the organic compound can be described by an apparent kinetic law of order 1 with respect to the concentration of organic compound, it follows that the degradation kinetics of the organic substrate may be described by the equation below:

$$\ln \frac{C_0}{C} = -K_{app}.t \tag{2}$$

The monitoring of the disappearance of diclofenac as a function of time by varying the concentration of H_2O_2 from 0.001 to 0.008 M is given in Table 2. It gives for each

Table 2. Synthesis of the apparent constants and initial degradation velocities in depending on the concentration of H_2O_2 and the $[H_2O_2]/[Fe^{2+}]$ ratio.

[Diclo] = 50 ppm pH = 4,56 T = 20°					
$[Fe^{2+}]$ M	$[H_2O_2]$ M	K_{app} (min^{-1})	$R = [H_2O_2]/[Fe^{2+}]$	V_0 (g/l.minn)	efficiency%
0,0001	0,001	0,00196	10	0,098	44
0,0001	0,006	0,0038	60	0,19	65
0,0001	0,008	0,00172	80	0,086	32

ratio the values of the apparent kinetic constants of disappearance, the initial velocity and the yield after 300 min of treatment.

(c) *Effect of the initial concentration of the pollutant:*

The pollutant concentration is one of the most important parameters influencing the Photo Fenton process according Pignatello et al. (2006) [16].

The effect of the initial concentration of diclofenac on the degradation yields in this process is shown in Fig. 7.

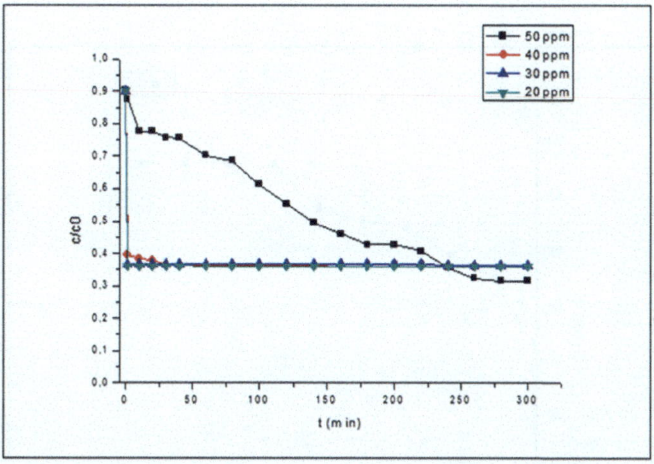

Fig. 7. Effect of the initial concentration of the pollutant

The curves clearly show that concentrations less than 50 ppm do not require coupling with UV, the fenton process is sufficient for them.

As soon as the two reactants are injected into the reactor, the yield reaches 57%. For the concentrations 40.30 and 20 ppm, the hydroxyl radical is the main element responsible for the degradation of the pollutant. Elmolla et al. (2010) [17].

3.4 Parametric Study of Adsorption

In the present research, we were interested in the study of the adsorption of diclofenac on bentonite, which is a natural adsorbent. This adsorbent was used to adsorb our active ingredient (diclofenac); the preparations of the sample and the adsorption procedure have already been described previously. In order to optimize the elimination efficiency of our pollutant, different parameters were studied: contact time, pH, mass of adsorbent and initial concentration of active principle.

(a) *Effect of contact time*

In order to optimize the elimination efficiency of diclofenac by adsorption, it is necessary to start the study by the influence of the contact time on the degradation of the pollutant. The dose of pollutant being fixed at 50 g/l.

The following Fig. 7 shows the influence of contact time on the degradation of the pollutant.

The results presented in Fig. 8 show that the reduction of diclofenac by adsorption on bentonite is constant from 120 min. It can be deduced that the contact time or adequate equilibrium time is 120 min, since the equilibrium is reached from 120 min.

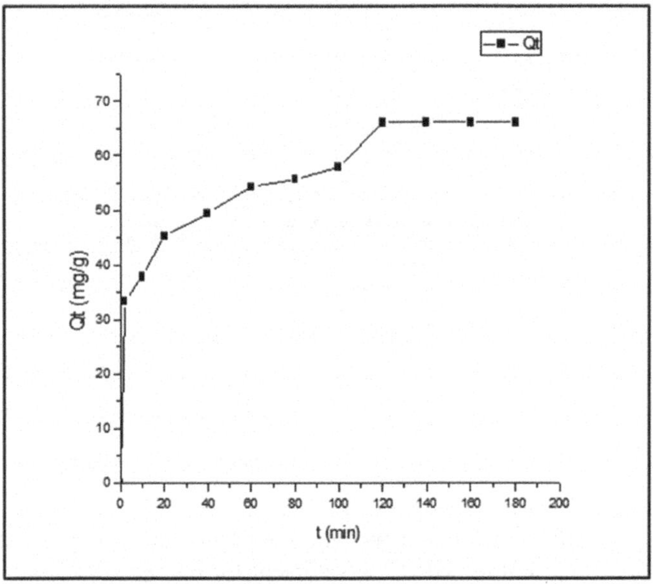

Fig. 8. Influence of contact time on adsorption on bentonite

(b) *Effect of adsorbent mass*

The search for the optimum mass of adsorbent is an essential step in any study on adsorption. Zhang et al. (2012) [18]. This is why we have begun studying the influence of the bentonite mass on the adsorption of the active ingredient (diclofenac).

Figure 9 shows that there is an increase in the adsorption of diclofenac as a function of the mass of bentonite added in the reaction medium. The optimum mass of bentonite is 0.3 g for a concentration of the pollutant 50 ppm (the yield reaches 90%). The increase in the rate of diclofenac reduction with the increase in the adsorbent mass is due to the availability of a large active surface area. This deduction is reported by various authors Zha et al. (2013) [19].

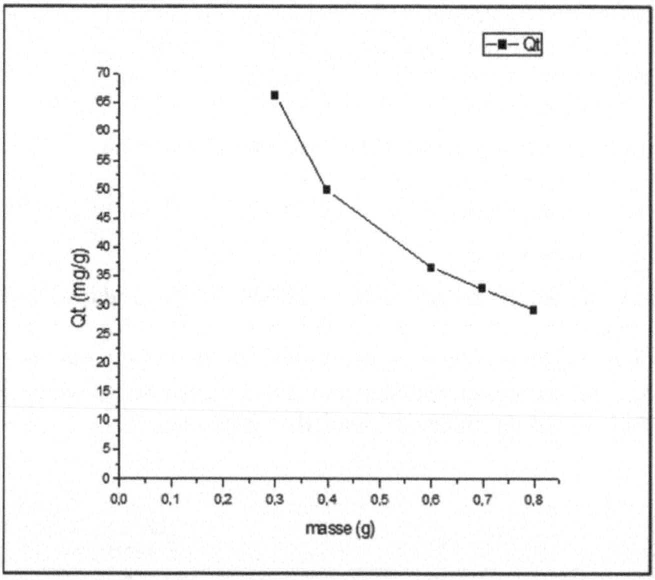

Fig. 9. Influence of the mass on the adsorption of diclofenac

(c) *Effect of the initial concentration of the active ingredient*

The initial concentration of diclofenac is a determining parameter for the adsorption process on diclofenac (Fig. 10).

Since the agitation rate is constant, the diffusion of the molecules towards the active surface of the adsorbent material is affected by the initial concentration. The increase in concentration accelerates the diffusion of this concentration due to the increase in

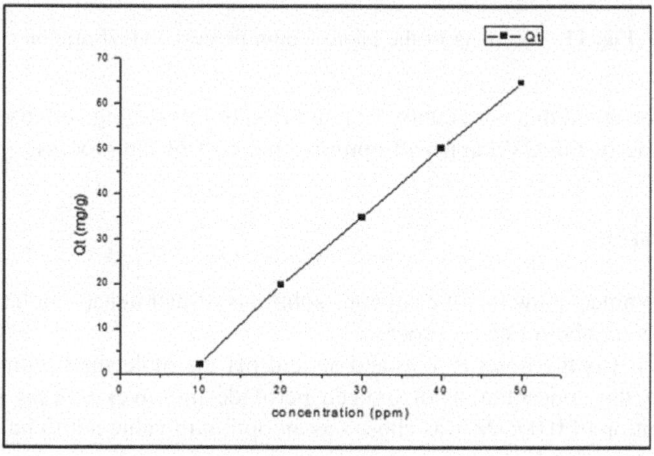

Fig. 10. Effect of the initial concentration of the active ingredient

attraction forces of the concentration gradient. The elimination efficiency is between 10% and 80% for the different concentrations. Padmavathy et al. (2015), [20], Banat et al. (2016) [21].

3.5 Coupling of the Photo Fenton Process and Adsorption

Since the optimum removal efficiency of our pollutant obtained in the optimization of the various parameters by the photo-Fenton process is 65%, we need to supplement the degradation rate by another process as efficient as the first one, which is adsorption. The curve below shows the results of the removal efficiency during coupling photo-fenton adsorption.

For economic reasons (energy consumption of the UV lamp) we stopped the process at 50% and we completed the removal by adsorption, which increased the elimination efficiency of the molecule studied to 87% (Fig. 11).

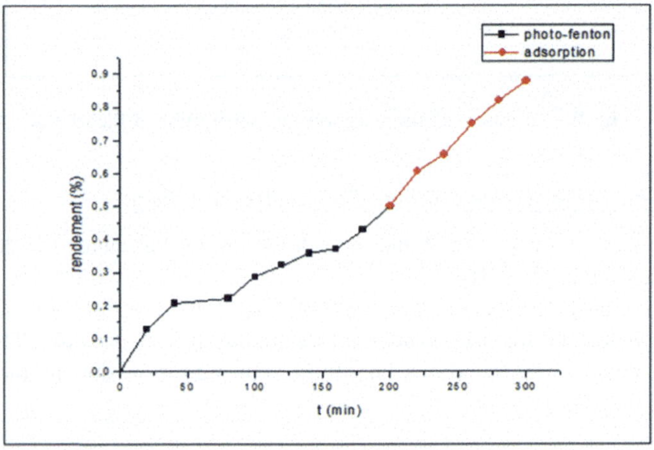

Fig. 11. Coupling of the photo fenton process and adsorption

It should be noted that our country is a highly sunny emerging country so the use of sunlight instead of the UV lamp will optimize the cost of the process.

4 Conclusion

The results obtained show that the aqueous solutions of diclofenac can be degraded by the homogeneous photo-Fenton process.

The ideal pH is the free pH 4.56 and at acid pH the molecule becomes insoluble.

Increasing the concentration of oxygen peroxide improves process performance. The concentration of 0.006 M was chosen as an optimum value not to be exceeded for the hydrogen peroxide added to the solution at the beginning of irradiation in order to ensure a better degradation of the organic matter.

The yield is 65% under the best conditions $[H_2O_2] = 0.006$ M, a temperature of 20 ± 3 °C and a concentration $[FeSO_4]$ 0.0001 M, a free pH of 4.56.

We have had a very good yield with the adsorption process, which is close to 90%.

To complete the degradation of our pollutant we coupled the photo-fenton reaction to adsorption, which increased process efficiency to 87%.

References

1. Schwaiger, J., Ferling, H., Mallowa, U., Wintermayr, H., Negele, R.D.: Toxic effects of the non-steroidal anti-inflammatory drug diclofenac Part I: histopathological alterations and bioaccumulation in rainbow trout. Aquat. Toxicol. **68**, 141–150 (2004)
2. Scheytt, T., Mersmann, P., Lindstadt, R., Heberer, T.: Octanol/water partition coefficients of 5 pharmaceuticals from human medical care: carbamazepine, clofibric acid, diclofenac, ibuprofen, and propyphenazone. Water Air Soil Pollut. **165**, 3–11 (2005)
3. Rainsford, K.D., Kean, W.F., Ehrlich, G.E.: Review of the pharmaceutical properties and clinical effects of the topical NSAID formulation, diclofenac epolamine. Curr. Med. Res. Opin. 0300-7995 **24**(10), 2967–2992 (2008)
4. Carballa, M., Fink, G., Omil, F., Lema, J.M., Ternes, T.: Determination of the solid–water distribution coefficient (Kd) for pharmaceuticals, estrogens and musk fragrances in digested sludge. Water Res. **42**, 287–295 (2008)
5. Méndez-Arriaga, F., Esplugas, S., Giménez, J.: Degradation of the emerging contaminant ibuprofen in water by photo-Fenton. Water Res. **44**, 589–595 (2010)
6. Cardoso da Silva, J.C., Teodoro, J.A.R., Franco Afonso, R.J.C., Aquino, S.F., Rodinei Augustia, R.: Photolysis and photocatalysis of ibuprofen in aqueous medium: characterization of byproducts via liquid chromatography coupled to high-resolution mass spectrometry and assessment of their toxicities against Artemia Salina. J. Mass Spectrom. (2013). https://doi.org/10.1002/jms.3320
7. Kiwi, J., Lopez, A., Nadtochenko, V.: Mechanism and kinetics of the OH• radical intervention during Fenton oxidation in the presence of a significant amount of radical scavenger (Cl^-). Env. Sci. Techn. **34**, 2162–2168 (2000)
8. Huang, Y.-H., Huang, Y.-J., Tsai, H.-Ch., Chen, H.-T.: Degradation of phenol using low concentration of ferric ions by the photo-Fenton process. J. Taiwan Inst. Chem. Eng. **41**, 699–704 (2010)
9. Souza, B.M., Dezotti, M.W.C., Boaventura, R.A.R., Vilar, V.J.P.: Intensification of a solar photoFenton reaction at near neutral pH with ferrioxalate complexes: a case study on diclofenac removal from aqueous solutions. Chem. Eng. J. **256**, 448–457 (2014)
10. Goi, A., Veressinina, Y., Trapido, M.: Degradation of salicyic acid by Fenton and modified Fenton treatment. Chem. Eng. J. **143**, 1–9 (2008)
11. Pérez-Estrada, L.A., Maldonado, M.I., Gernjak, W., Agùera, A., Fernandez-Alba, A.R., Ballesteros, M.M., Malato, S.: Decomposition of diclofenac by solar driven photocatalysis at pilot plant scale. Catal. Today **101**, 219–226 (2005)
12. Ahmed, B., Limem, E., Ahmed, A.W., Nasi, B.: PhotoFenton treatment of actual agroindustrial wastewaters. Ind. Eng. Chem. Res. **50**(11), 6673–6680 (2011)
13. Alalm, M.G., Tawfik, A., Ookawara, S.: Degradation of four pharmaceuticals by solar photoFenton process: Kinetics and costs estimation. J. Environ. Chem. Eng. **3**, 46–51 (2015)

14. Selvam, K., Muruganandham, M., Swaminathan, M.: Enhanced heterogeneous ferrioxalate photo-Fenton degradation of reactive orange 4 by solar light. Sol. Energy Mater. Sol. Cells **89**, 61–74 (2010)
15. Ravinaa, M., Campanellab, L., Kiwia, J.: Accelerated mineralization of the drug Diclofenac via Fenton reactions in a concentric photo-reactor. Water Res. **36**, 3553–3560 (2002)
16. Pignatello, J.J., Oliveros, E., MacKay, A.: Advanced oxidation processes for organic contaminant destruction based on the Fenton reaction and related chemistry. Crit. Rev. Environ. Sci. Technol. **36**, 1–84 (2006). Erratum, 37, 273-275 (2007)
17. Elmolla, E.S., Chaudhuri, M.: Photo-Fenton Treatment of Antibiotic Wastewater. Nat. Environ. Pollut. Technol. **9**, 65–370 (2010)
18. Zhang, Y.X., Zhao, Y., Zhu, Y., Wu, H.Y., Wang, H.T., Lu, W.J.: Adsorption of mixed cationic-nonionic surfactant and its effect on bentonite structure. J. Environ. Sci. **24**, 1525–1532 (2012)
19. Zha, S.X., Zhou, Y., Jin, X.Y., Chen, Z.L.: The removal of amoxicillin from wastewater using organobentonite. J. Environ. Manag. **129**, 569–576 (2013)
20. Padmavathy, K.S., Madhub, G., Haseena, P.V.: (ICETEST- 2015) study on effects of pH, adsorbent dosage, time, initial concentration and adsorption isotherm study for the removal of hexavalent chromium (Cr (VI)) from wastewater by magnetite nanoparticles. Procedia Technol. **24**, 585–594 (2016). International Conference on Emerging Trends in Engineering, Science and Technology
21. Banat, F.A., Al-Bashir, B., Al-Asheh, S., Hayajneh, O.: Adsorption of phenol by bentonite. Environ. Pollut. **107**, 391–398 (2000)

Synthesis and Structural Characterization of Glass Foam Composites for Electromagnetic Application

Younes Lamri[1,2(✉)], Ratiba Benzerga[2], Azzedine Ayadi[1], Laurent Le Gendre[2], and Fayrouz Benhaoua[1]

[1] Unité de Recherché Matériaux, Procédés et Environnement (UR-MPE), Université M'hamed Bouguera – Boumerdes, Boumerdès, Algérie
lamri90younes@gmail.com

[2] Institut d'Electronique et des Télécommunications de Rennes (IETR), 18 Rue Henri Wallon, 22004 St-Brieuc, France

Abstract. This article deals with the synthesis and structural characterization of glass foams made from cullet and loaded with carbon fibers or used tires for electromagnetic absorption application. All the raw materials used for this work come from recycling or reconditioning. Two carbonaceous loads are tested; the carbon fiber scraps are used because they show an interest in the field of absorption and parallel to this, another load type (tire waste) is used for its high carbon content. This novel later load is used, to our knowledge, for a first time. Samples with different weight contents of the loads are achieved and characterized. The used percentages of loads, are respectively, between 0 and 10 wt.% and between 0 and 20 wt.% for carbon fibers and tire waste. In this article, we present and discuss the elaboration method, the morphological and structural results of the loaded glass foam composites. A low density was obtained from both composite, with a density between 0.28 and 0.53 g/cm^3. An excess content of tire waste load leads to "collapse" of the gas bubbles during the heat treatment of the sample which induces an increase in the density of the composite. This is due to the decomposition of this load during the elaboration process. This result is confirmed by the scanning electron microscopy observation and also by gravimetric thermal analysis of the tire waste. Contrary to this, composites loaded with carbon fibers show better structure homogeneity and non-decomposition of carbon fibers during a heat treatment.

Keywords: Foam glass · Composites · Cullet · Tires · Carbon fibers
Electromagnetic wave

1 Introduction

The increasing use of electromagnetic (EM) waves, for communication (mobile phone, WiFi, etc.) and electronic equipment's, causes an EM pollution problem. Today, both the population and the producers of these devices, that are using the EM waves, become sensitive to this pollution. For this, development of EM absorbing materials becomes a

B. Abdelbaki et al. (Eds.): SMSD 2017, *Proceedings of the Third International Symposium on Materials and Sustainable Development*, pp. 593–601, 2018.
https://doi.org/10.1007/978-3-319-89707-3_63

concern of researchers to protect human health and devices from this increased of EM pollution (Chen *et al.* 2011).

Alongside that, the use of waste to produce new material becomes very frequent and even very advisable in order to meet the demand to reduce our waste but also to meet the growing demand of raw materials of our society.

In this context of recycling wastes and protecting our environment from EM waves, glass foams loaded with carbon fibers and used tires have been studied for absorbing application.

Today, two categories of electromagnetic (EM) absorbers exist. Indeed, as the EM waves are composed of electrical and magnetic fields, absorber materials must absorb one of these components. For this, materials with magnetic or dielectric losses are often studied as absorbers (Chin and Lee 2007; Dosoudil *et al.* 2010). Some absorbing materials combine dielectric and magnetic losses (Meng *et al.* 2011; Liu *et al.* 2009; Al-Ghamdi *et al.* 2016). The carbonaceous materials are the most used loads for the production of dielectric absorber materials.

Likewise, glass foam made from cullet (glass waste) are used for thermal and acoustic insulation (for example: insulation of roofs, walls, floors, ceilings) (Ayadi *et al.* 2011; Bernardo and Albertini 2006). Other studies showed that glass foams could be also used as photo-catalyst for hydrocarbons degradation (Lebullenger *et al.* 2010). In recent years, some studies have shown the potentiality of using glass foams for EM applications (Benzerga *et al.* 2015; Laur *et al.* 2017; Benhaoua *et al.* 2015). Otherwise, millions of tires are thrown away each year. These tires are put in inappropriate places or are burned leading to serious dangers where their destruction is the source of toxic and harmful emissions. In this context, glass foams loaded with carbon used tires and waste of carbon fibers have been studied for absorbing application.

In this article, the elaboration method is presented. The morphological and structural results of the glass foam composites loaded with carbon fibers and tire waste are presented and discussed.

2 Elaboration and Characterization Techniques

2.1 Glass Foam Composites Synthesis

The glass foams were prepared from glass waste soda-lime (Mediterranean Float Glass, Algeria). The composition of this glass waste is shown in Table 1. This waste is mainly composed of SiO_2 (~72 wt.%) and Na_2O (~13.5 wt.%), as well as other oxides in lower percentages. After grinding and sieving the glass waste to a size particles of 100 μm, the glass powder is mixed with 1 wt.% of the foaming agent (calcium carbonate $CaCO_3$) and with the desired percentage of the used tire granulates or carbon fibers. The tire granulates have a diameter between 1 mm and 2.5 mm and the carbon fiber length is 3 mm.

Table 1. Composition of glass waste

Element	SiO_2	Na_2O	CaO	MgO	Al_2O_3	K_2O	Fe_2O_3
wt.%	71.73	13.49	8.69	4.33	0.98	0.60	0.07

The mixture of the different compounds (cullet, foaming agent and the different loads of used tires or carbon fibers) is then humidified with 15 wt.% of water. The mixture is then placed in a refractory steel mold and heated at 800 °C for 30 min. A gas emission, produced by the reaction of the foaming agent ($CaCO_3 \rightarrow CaO + CO_2$), associated with a glass softening (at this temperature of 800 °C), traps the produced gas in the softened matrix and allows the production of the foamed glass. The latter is a porous material with a density that depends on the nature and percentage of foaming agent used, the glass composition, and the heating temperature of the mixture (Benzerga *et al.* 2015). The Fig. 1 presents the thermal regime used to achieve our samples.

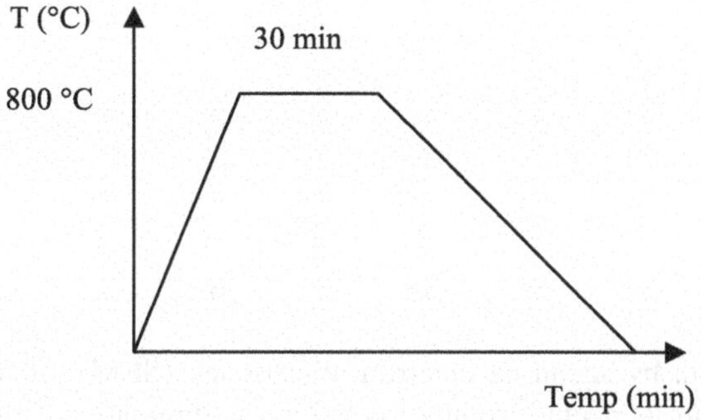

Fig. 1. Thermal regime for the production of glass foams.

2.2 Morphological and Thermal Characterization

Synthesized samples are cut for morphological characterization. The structure of the samples was observed by a scanning electron microscope (SEM) JEOL 5600. The chemical composition was measured by an EDS OXFORD instruments 51-ADD 0076. The apparent density of the glass foams was calculated by measuring the mass and the volume of each sample. A measurement uncertainty of ±0.01 was estimated and was taken into account for the presented results.

The differential thermal analysis (DTA) and the gravimetric thermal analysis (GTA) were carried out with SDT Q600 equipment.

3 Results of Structure and Morphology of Glass Foams

Various samples of glass foams containing different weight percentages (between 0 wt.% and 20 wt.% of tire pellets and between 0 wt.% and 10 wt.% of carbon fibers) were prepared. The same percentage of foaming agent (1 wt.% of $CaCO_3$) was used for these different samples. Table 2 summarizes the composition of these samples and their measured apparent densities. It can be noted that the measured densities are between 0.28 g/cm^3 and 0.43 g/cm^3 for tire-filled foams and between 0.28 g/cm^3 and 0.52 g/cm^3 for carbon-filled foams; with an increasing trend as a function of the weight percentage of the filler. The sample produced with a load percentage of 10% by weight of carbon fibers has the highest density.

Table 2. The different weight percentages of the components and the apparent densities of the different samples made with aggregates of tires and the carbon fibers.

Sample	wt.% cullet	wt.% $CaCO_3$	wt.% used tires	wt.% Carbon fiber	Apparent density (g/cm^3)
Smp-01	99	1	0	-	0.28
Smp-02	95	1	4	-	0.22
Smp-03	91	1	8	-	0.30
Smp-04	87	1	12	-	0.27
Smp-05	83	1	16	-	0.37
Smp-06	79	1	20	-	0.43
Smp-07	93	1	-	2	0.21
Smp-08	95	1	-	4	0.27
Smp-09	93	1	-	6	0.35
Smp-10	91	1	-	8	0.32
Smp-11	89	1	-	10	0.52

The observation, by scanning electron microscopy (SEM), of the samples of the unloaded glass foam and glass foams loaded with aggregates of tires and of carbon fibers, is shown in Fig. 2.

Figure 2 shows, first, a conventional porous structure (Fig. 2-a) of a unloaded glass foams as well as macroporosities and microporosities in the walls of this structure. Figure 2-b shows foam loaded with tires having open and closed porosities and an example of an area where the tire load is located in the glass foam structure. The tire load (Fig. 2-b) has a different appearance from tire aggregates that have not undergone heat treatment. This is probably due to the decomposition of the tire during heat treatment. Figure 2-c shows glass foam loaded with carbon fibers. This figure shows a porous structure with the presence of carbon fibers. The fibers located in the walls of the pores, in the form of packages, present a very similar appearance to the carbon fibers which are not heat-treated. The low degradation of carbon fibers under heating may be due to their high temperature resistance.

Figure 3 shows the pictures of the foams elaborated with different contents of the tire granulate. This figure shows, first of all, that the unloaded sample (0 wt.%) has a relatively homogeneous structure. In fact, a gradient of porosity dimensions is to be noted between the bottom and the top of this sample. This behavior during foaming is often observed in the literature (Fernandes *et al.* 2009; Bernardo and Albertini 2006). A presence of large porosities (with diameter larger than 3 mm) should be noted on these images for samples with high load amounts. The quantity of these large porosities increase with the increase of this load amount, except for the most heavily loaded sample.

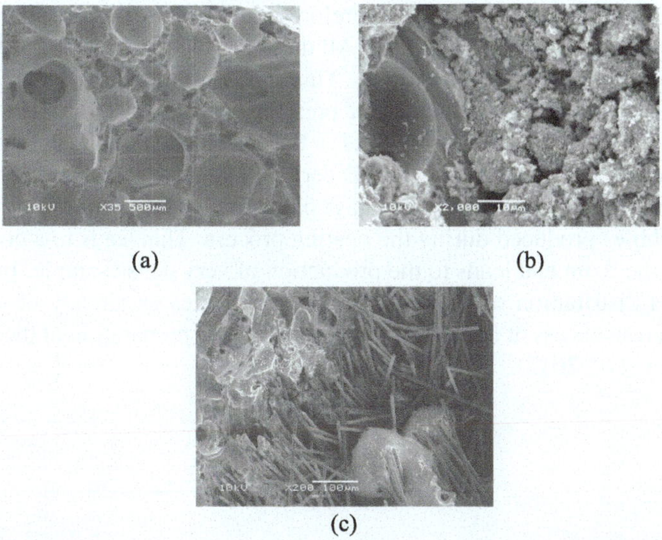

Fig. 2. Scanning electron microscope of (a) unloaded foam, (b) foam loaded with tires and (c) loaded with carbon fibers.

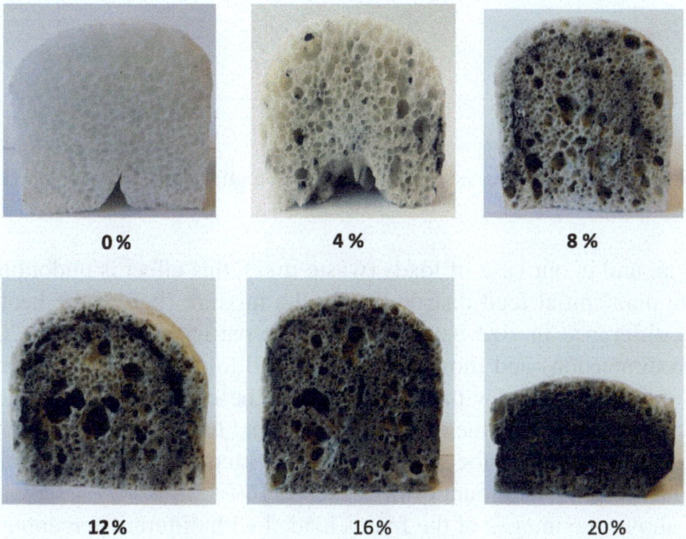

Fig. 3. Glass foams with different aggregate contents of used tires.

The result of the DTA-GTA analysis of the tire granulates, achieved between room temperature and 800 °C and shown in Fig. 4, shows a significant weight loss (greater than 60 wt.%) starting at 200 °C. This decomposition, studied by (de Sousa *et al.* 2017), showed that the increase in temperature causes weight loss, which corresponds mainly to the decomposition of natural rubber (from 200 °C to 400 °C). The gases produced by heating the tires would therefore consist of CO_2, CO, H_2, CH_4, C3H6, C_3H_8, C_4H_6 (Williams and Besler 1995). All these produced gases induced the production of large porosities in final samples. This gases production, increasing with the increase of load amount, consolidate our previous remark on the aspect of the glass foams us a function of the load percentage. For the most loaded samples (20% by weight), a greater apparent density is noted. This is certainly due to excessive degassing during the heat treatment, due to the high percentage of the tire load, which leads to a "collapse" of the gas bubbles produced during the heating process. This leads to a collapse of the porosities of the foam and leads to the production of very dense sample. In fact, a non-homogeneous distribution of the dimension of porosities as already observed in the literature when an excess of foaming agent is used for the preparation of the foams (Laur *et al.* 2017; Li *et al.* 2012).

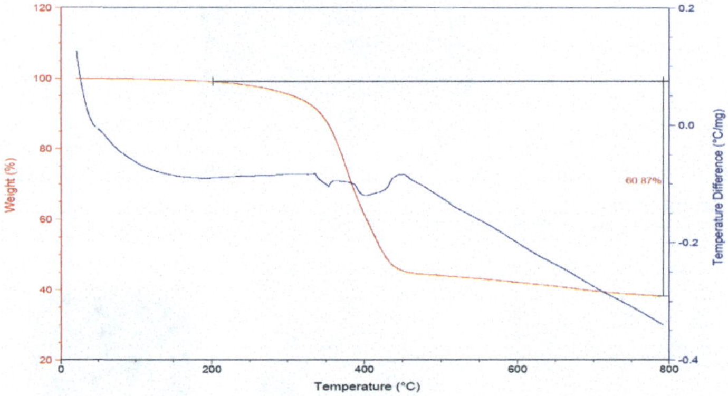

Fig. 4. Gravimetric thermal analysis (GTA) and differential thermal analyzes (DTA) of tire granulates.

In addition, and in our case of loads (waste tires), this effect is undoubtedly accentuated by the poor initial feed distribution in the mixture (before the heat treatment) caused by a difference in size between the cullet particles and the $CaCO_3$ powders (micrometric dimension) and those of the tires (millimetric dimension). So, the poor distribution of tire, associated with the degassing of these tires during the heat treatment, may justify the appearance of the macroporosities and load distribution inhomogeneity which can be observed on all the samples loaded with tires, and explains also the high density obtained when the amount of tires exceed 16%.

Figure 5 shows the images of the foams loaded with different percentage of carbon fibers. These foams have a more homogeneous structure, in comparison with foam composites loaded with tires. These figures confirm also the pervious result related to the density increase with the increase of carbon fiber amount (Table 2). This is due the fact that the fibers, by their presence in the mixture, prevent the formation of gas bubbles during the heat treatment.

Fig. 5. Glass foams with different carbon fiber contents

Figure 6 shows the DTA-GTA analysis result of the carbon fibers, carried out between room temperature and 1000 °C, showing a low loss of mass (4.8 wt.%) at 600 °C. This weak decomposition was explained by the strong resistance of carbon fibers to high temperature. This non-decomposition of the carbon load induces the good homogeneity (distribution of porosities) observed on our fiber loaded foams, which is better than those of tire loaded foams.

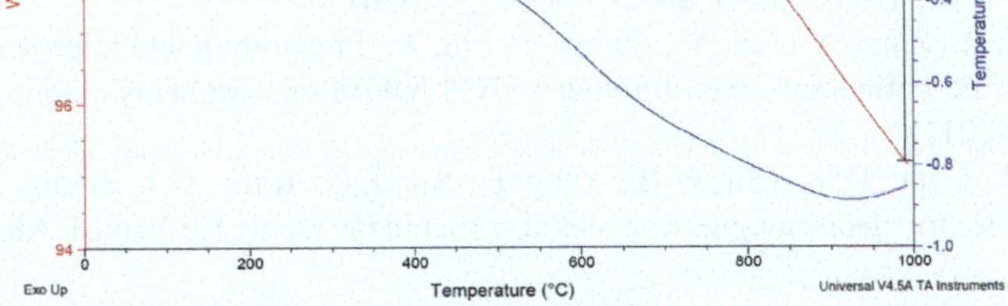

Fig. 6. Gravimetric thermal analysis (GTA) and differential thermal analyzes (DTA) of carbon fibers.

4 Conclusion

This paper deals with the synthesis and structural characterization of glass foam composites for electromagnetic absorbing application. Two carbonaceous loads are used (waste of tires and carbon fibers) with different weight percentage between 0 wt.% and 20 wt. %. A $CaCO_3$ foaming agent was used to obtain the foam structure. This foaming agent induces the emission of CO_2 gas which will subsequently be trapped by the molten glass matrix and form pores at a temperature corresponding to its decomposition. The obtained samples show non-homogeneity of the structure (dimensions of porosities) and of the distribution of load for tire loaded foams. This observation has been explained, first, by the poor initial homogeneity of mixture (before heat treatment) induced by a difference in size between the cullet particles and the $CaCO_3$ powders (micrometric dimension) and those of the tires (millimetric dimension) and secondly, by the decomposition (more than 60 wt.%) of the tire granulates. The latter induces degassing during the preparation of the samples which could explain the inhomogeneity in structure of these samples leading to the presence of porosities with diameter of several millimeters.

The homogeneous appearance of the carbon fiber-loaded foams, compared to that loaded with tires, is explained by the low decomposition of the fiber load (4.8 wt.%) which does not subsequently induce the presence of large porosities. On the other hand, our results show that the increase of the fiber content leads to an increase of the density of the produced composites. These results show that a compromise between the morphology (homogeneity), the density of the obtained materials and their magnetic absorption capacity exists and will be taken in account for the targeted application. The latter has not been presented in this paper and will be the subject of a future publication.

References

Chen, K., Li, X., Lu, D., Yu, F., Yin, Z., Wu, T.: Study on microwave absorption properties of metal-containing foam glass. Mater. Sci. Eng. B **176**, 1239–1242 (2011)

Chin, W.S., Lee, D.G.: Development of the composite RAS (radar absorbing structure) for the X-band frequency range. Compos. Struct. **77**, 457–465 (2007)

Dosoudil, R., Ušáková, M., Franek, J., Sláma, J., Grusková, A.: Particle size and concentration effect on permeability and EM-Wave absorption properties of hybrid ferrite polymer composites. IEEE Trans. Magn. **46**(2), 436–439 (2010)

Meng, F., Zhao, R., Zhan, Y., Lei, Y., Zhong, J., Liu, X.: Preparation and microwave absorption properties of Fe-phthalocyanine oligomer/Fe3O4 hybrid microspheres. Appl. Surf. Sci. **257**, 5000–5006 (2011)

Liu, X.G., Li, B., Geng, D.Y., Cui, W.B., Yang, F., Xie, Z.G., Kang, D.J., Zhang, Z.D.: (Fe, Ni)/C nanocapsules for electromagnetic-wave-absorber in the whole Ku-band. CARBON **47**, 470–474 (2009)

Al-Ghamdi, A.A., Al-Hartomy, O.A., Al-Solamy, F.R., Dishovsky, N., Malinova, P., Atanasova, G., Atanasov, N.: Conductive carbon black/magnetite hybrid fillers in microwave absorbing composites based on natural rubber. Comp. Part B **96**, 231–241 (2016)

Ayadi, A., Stiti, N., Boumchedda, K., Rennai, H., Lerari, Y.: Elaboration and characterization of porous granules based on waste glass. Powder Technol. **208**, 423–426 (2011)

Bernardo, E., Albertini, F.: Glass foams from dismantled cathode ray tubes. Ceram. Int. **32**, 603–608 (2006)

Lebullenger, R., Chenu, S., Rocherullé, J., Merdrignac-Conanec, O., Cheviré, F., Tessier, F., Bouzaza, A., Brosillon, S.: Glass foams for environmental applications. J. Non-Cryst. Solids **356**, 2562–2568 (2010)

Benzerga, R., Laur, V., Lebullenger, R., Le Gendre, L., Genty, S., Sharaiha, A., Queffelec, P.: Waste-glass recycling: a step toward microwave applications. Mater. Res. Bull. **67**, 261–265 (2015)

Laur, V., Benzerga, R., Lebullenger, R., Le Gendre, L., Lanoë, G., Sharaiha, A., Queffelec, P.: Green foams for microwave absorbing applications: synthesis and characterization. Mater. Res. Bull. **96**, 100–106 (2017)

Benhaoua, F., Ayadi, A., Stiti, N., Lamri, Y., Ratni, H.: Elaboration and characterization of acellular glass based cullet loaded with carbon fibers. Mater. Lett. **160**, 278–281 (2015)

Fernandes, H.R., Tulyaganov, D.U., Ferreira, J.M.F.: Preparation and characterization of foams from sheet glass and fly ash using carbonates as foaming agents. Ceram. Int. **35**, 229–235 (2009)

de Sousa, F.D.B., Scuracchio, C.H., Hu, G.H., Hoppe, S.: Devulcanization of waste tire rubber by microwaves. Polym. Degrad. Stab. **138**, 169–181 (2017)

Li, X., Lv, D., Chen, K.: Effects of graphite additive on dielectric properties and microwave absorption properties of zinc-containing foam glass. J. Non-Cryst. Solids **358**, 2917–2921 (2012)

Williams, P.T., Besler, S.: Pyrolysis-thermogravimetric Analysisof Tyres and Tyre Components. Elsevier Science Ltd. (1995). https://doi.org/10.1016/0016-2361(95)00083-H

Materials Processing Technology

Bonded Composite Repairs Analysis in Pipes Under Internal Pressure Using Finite Element Technics

A. Achour[1(✉)], B. Bachir Bouiadjra[2], and D. Ouinas[1,2]

[1] LMNEPM Laboratory, Civil Engineering Department, University of Mostaganem, 27000 Mostaganem, Algeria
ar.aida@yahoo.fr

[2] LMPM Laboratory, Department of Mechanical Engineering, University of Sidi Bel Abbes, BP 89, Cité Ben M'hidi, Sidi Bel Abbes, Algeria

Abstract. The use of composite systems as a repair methodology in the pipeline industry has grown in recent years. However, there are still no widely accepted standards governing the design and installation of such systems but the potential cost savings can be significant. Whilst composite repairs are gaining acceptance and approvals they do not as yet have full regulatory approval and this should be considered during the repair assessment process. In this study, the analysis of the behaviour of circumferential through cracks in repaired pipe with bonded composite patch subjected to internal pressure is performed using three dimensional finite element methods. The stress intensity factor is used as a fracture criteria which describes the elastic behavior of the structure. The obtained results show that the presence of the bonded composite patch reduces significantly the stress intensity factor, what can improve the lifespan of the pipe.

Keywords: Pipe · Crack · Bonded composite repair · Stress intensity factor Finite element method

1 Introduction

This trend is expected to continue with the aging international infrastructure, coupled with greater demands for pipeline systems to meet the needs of a global energy market [1].

For almost two decades, composite materials have been used to repair and reinforce transmission pipelines. This effort has been accompanied by an extensive array of engineering analysis and testing programs by pipeline companies, research organizations, and composite manufacturers [1–3].

The history of onshore pipeline repairs using composite materials is presented by Alexander and Francini [4]. Even though the long-term performance of composites incorporating physical and chemical degradation is of great importance in pipeline applications, yet the data reflecting relevant environmental conditions is still sparse [5, 6].

B. Abdelbaki et al. (Eds.): SMSD 2017, *Proceedings of the Third International Symposium on Materials and Sustainable Development*, pp. 605–612, 2018.
https://doi.org/10.1007/978-3-319-89707-3_64

Toutanji and Dempsey [2] worked on establishing fiber reinforced polymer composites as an effective mean for the repair and rehabilitation of pipelines. They developed a theoretical model with stress expressions and circumferential stress curves. This study focused on the application of fiber reinforced polymer on steel pipes.

The main aim of this work is to assess the composite patch repair of pipes under different loading condition using finite element analysis. A complete finite element model and analysis of the crack repair with composite patch of circumferential through cracks in pipes subjected to internal pressure is presented. Also, the evaluation of the long-term performance of composite repair systems will be addressed. Finally, an attempt will be made to provide industry with an overview of the current state of the art in composite repair technology and how the integrity of pipeline systems is being restored using composite materials.

2 Geometrical and Materials Models

2.1 Review and Final Stage

This study presents a three dimensional finite element analyses by the code Abaqus for semi circular complete cracks in pipes repaired with Glass/epoxy patch. The pipes were subjected to internal pressure of amplitude $p = 100$ MPa. (Fig. 1). Figure 2 presents the geometrical model of the pipe and the cracks. The geometrical characteristics of the cracked pipe are:

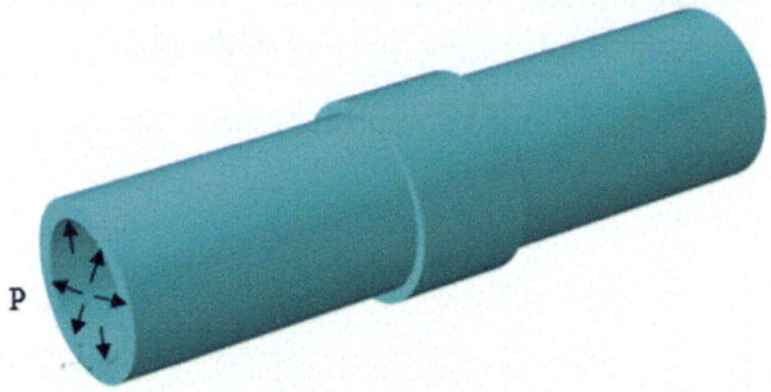

Fig. 1. Repaired pipe under pressure

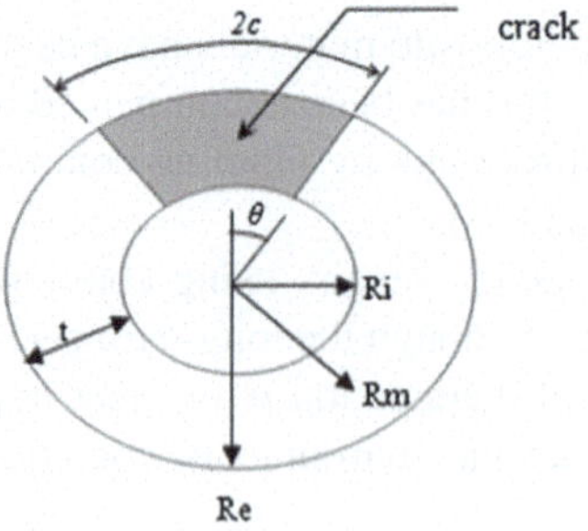

Fig. 2. Geometrical model of the pipe and the crack

- External diameter $R_0 = 31$ mm
- Internal diameter $R_i = 24$ mm
- Equivalent diameter $R_m = 27{,}5$ mm

The size of the crack is represented with the angleθ. The pipe material is the SA312 Type 304 stainless steel. The patch repair is bonded to the pipe with the structural adhesive FM73.

Table 1 gives the elastic properties of the pipe, the patch and the adhesive.

Table 1. Materials properties

Materials	E (GPa)	υ	G (GPa)
Patch	150	0.21	5.5
Glass/epoxy Adhesive FM73	2.55		
Pipe SA312 Type 304 stainless steel	204		

2.2 Finite Element Models

The analysis involved a three-dimensional finite element method to supplement and analyze the experiments by using a commercially available finite element code ABAQUS [7]. The finite element model consisted of three subsections to model the cracked pipe, the adhesive, and the composite patch.

Figure 3 *shows the overall mesh of the specimen and* Fig. 4 *present the mesh refinement in the crack tip region. The Stress intensity factor (SIF° at the crack front was extracted using the virtual crack closure technique (VCCT), This technique was originally proposed in 1977 by Rybicki and Kanninen* [8], *is a very attractive SIF extraction technique because of its good accuracy, a relatively easy algorithm of application capability to calculate SIF for all three fracture modes. The proposed technique has been extended by other authors* [9, 10]. *Currently, the three-dimensional virtual crack closure technique (3D VCCT) is often chosen as a tool for SIF calculations* [9].

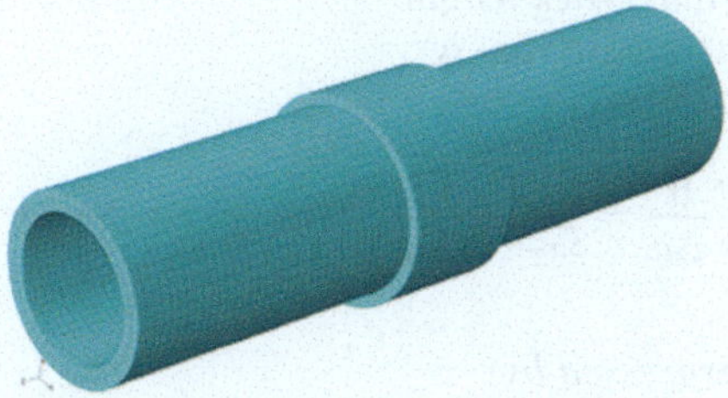

Fig. 3. Typical mesh model of the global structure

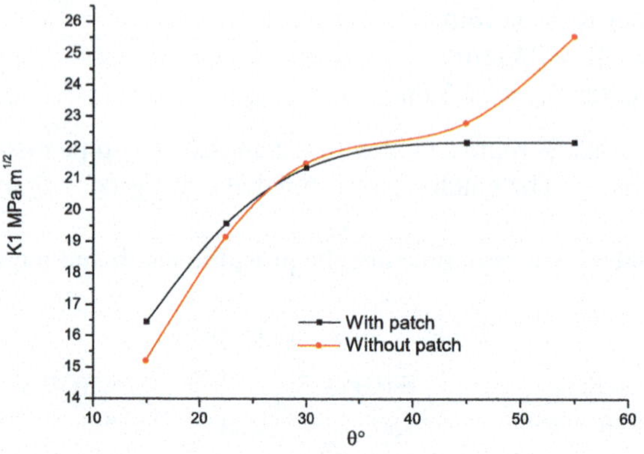

Fig. 4. SIF vs crack angle for repaired and un-repaired crack at the internal position of the crack

The VCCT is based on the energy balance proposed by Irwin. In this technique, SIF are obtained for three fracture modes from the equation:

$$G_i = \frac{K_i^2}{E} \tag{1}$$

where G_i is the energy release rate for mode i, K_i the stress intensity factor for mode i, E the elastic modulus.

The idea presented in [8] *is based on the calculation of the energy release rate, using Irwin assumption that the energy released in the process of crack expansion is equal to work required to close the crack to its original state as the crack extends by a small amount Δa. Irwin computed this work as:*

$$W = \int_0^{\Delta a} u(r).\sigma.(\Delta a - r)dr \tag{2}$$

Where u is the relative displacement, σ the stress, r the distance from the crack tip, and Δa the change in virtual crack length.

Therefore, the energy release rate is:

$$G = \lim_{\Delta a \to 0} \frac{W}{\Delta a} = \lim_{\Delta a \to 0} \frac{1}{2.\Delta a} \int_0^{\Delta a} u(r).\sigma(r - \Delta a).dr \tag{3}$$

The work W can be expressed by:

$$W = \frac{1}{2} F.u \tag{4}$$

Equation 3 *allows the calculation of the energy release rate for 2D FE models (mode I and II). The application of the VCCT in 3D FE models is commonly called the 3D VCCT. The extension of the method from 2D to 3D requires replacing* Eq. (3) *with* Eq. (5)

$$G = \frac{1}{2h\Delta a}\int_0^h\int_0^{\Delta a} u(r,s).\sigma(r-\Delta a,s).ds.dr \tag{5}$$

Where the distance from the crack tip in he third direction and h is the element size in the third direction.

To apply Eq. 5 *to FE models comprising 8-node brick elements, the integrals 5 are replaced with the sum:*

$$G = \frac{1}{2h\Delta a}\sum_{k=1}^{2} F_{ki}.u_{ki} \tag{6}$$

Where index i controls direction and index k controls the node b number.

3 Results and Discussions

In this study, the performance of bonded composite repair for pipe under pressure is analysed by computing the stress intensity factor at the crack front. A comparison between the behaviour of repaired and unrepaired crack in pipe under pressure is presented and then effect of the crack size on the SIF variation is discussed.

3.1 Comparison Between Repaired and Unrepaired Crack

Figure 4 presents the distribution of the stress intensity factor along crack front for patched and unpatched cracks. The crack angle θ is taken equal to (45°). It can be seen that, for unrepaired crack, the difference in the stress intensity factor between the external and the internal crack tips is not sensitive enough. This difference is about 5%. The presence of the bonded composite patch reduces significantly the stress intensity factor at the crack front particularly at the external crack tips. The reduction of the SIF at the internal crack tip is almost negligible. This behaviour is because the external crack tip is directly in contact with the adhesive layer and the composite patch. The stress Transfer from the cracked pipe to the composite patch throughout the adhesive layer is maximal at the external crack tip, what explains the significant reduction of the SIF at this tip. Quantitatively, the SIF id reduced at the external crack tip from 19 MPa $(m)^{1/2}$ to 7 MPa $(m)^{1/2}$, the relative reduction is about 63%. The rate of reduction of the SIF decreases proportionally to the distance along crack front.

The highest difference in the SIF between the external internal crack tips presents a major disadvantage for repairing pipe and make the prediction of the path of crack propagation very difficult. In the aeronautical application, this problem is solved by the technique of double-sided composite patch. This technique allows the equilibrium of stresses transfer between the faces of the repaired plates. The application of this technique for damaged pipe is practically impossible. One can confirm that the repair of external cracks in pipe more efficient than the internal cracks. In addition, the detection of external cracks is simpler. These observations can be generalised to the case of corroded pipes. Indeed, the repair of corroded pipes with bonded composite patch is more efficient for external corrosion under stresses. The composite patch reduces the stresses intensity in the corroded region of the pipe what reduce the risk of crack initiation and propagation. According to the precedent observation, one can conclude that the performances of the pipe repair are better for thin walled pipes.

It can be also noted according to the results of Fig. 5 that the variation of the SIF along the crack front for patched crack is approximately linear, the variation of the SIF according to the position along the crack front can be expressed by:

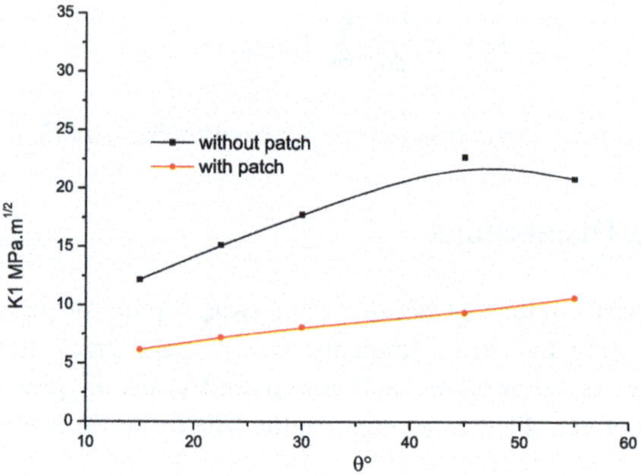

Fig. 5. SIF vs crack angle for repaired and unrepaired crack at the external position of the crack

$$K = A.x + B$$

Where x is the position along the crack front. A and B are coefficients. These coefficients depend on several parameters of which:

- The geometrical and mechanical properties of the pipe
- The geometrical and mechanical properties of the composite patch
- The geometrical and mechanical properties of the adhesive
- The applied pressure.

It is thus necessary to perform parametrical studies in order to establish analytical equations of the SIF for repaired crack. The authors will achieve this objective in future works.

3.2 Effect of the Cracks Size

It is known that the durability of bonded composite repairs in damaged structures depends essentially on the nature of the defects and its size. The application of this technique will be no effective if the size of the defect to be repaired exceeds a critical value. In this section, the effect of the crack size represented by the crack angle θ on the SIF variation is analysed.

Figure 6 presents the variation of the stress intensity factor at internal crack tip according to the crack angle θ for repaired and unrepaired cracks. It can be seen that the stress intensity factor for patched crack exhibits an asymptotic behaviour as the crack, length increases. It can be also noted that the reduction of the SIF by composite patch is weak except when the crack angle exceeds the value of 40°. This is because the stress transfer between the pipe and the patch increases significantly as the crack angle increases too. This stress transfer can affect the internal crack tip.

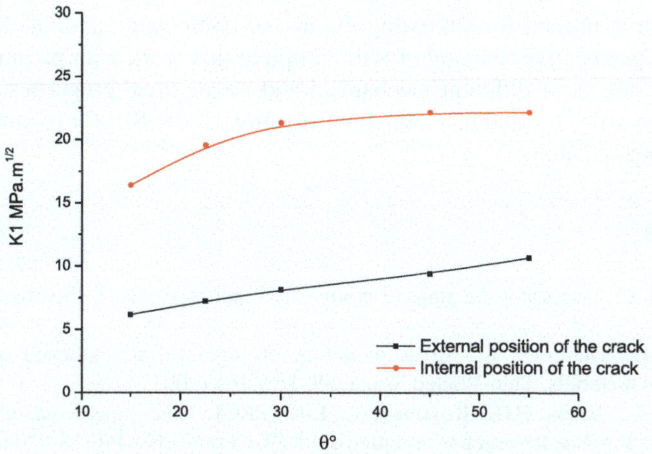

Fig. 6. SIF vs crack angle for patched crack

Figure 6 presents the variation of the SIF at the external crack tip according to the crack angle for patched and un-patched cracks. The reduction of the SIF by the patch is very significant for this crack tip,what confirm the results of Fig. 5.

This reduction is more significant when the crack angle increases. The rate of reduction of the SIF is about 55% when the crack angle is less than 25° and it is about 67% when the crack angle exceeds the value of 45°.

According to the results of Fig. 6, On cane note that the variation of the SIF for patched crack according to the crack angle is approximately linear with weak slope. This behaviour differs from the case of repaired plate in tension where the SIF exhibits an asymptotic behaviour as the crack length increases [11, 12]. The difference is due to the fact that for repaired plate subjected to tension loading there is pure mode I of crack propagation and the SIF at the crack front is very significant, what allows a high stress transfer from the cracked plate to the composite patch.

4 Conclusions

Composite repair systems are going to continue to play an important role in rehabilitating the aging pipeline systems around the world. Prior research, as well as ongoing efforts, has focused on demonstrating how composite materials can restore the serviceability of pipelines using engineering analysis and testing methods.

It was shown in this study that the reduction of the stress intensity by the composition patch repair is very significant at the external position of the crack, what can improve the lifespan of repairs. The main disadvantage of the technique of bonded composite repair in pipe is the impossibility to bond double sided composite patch in order to equilibrate the stress transfer between the internal and external positions of the cracks. More research is needed for validating the use of composite materials for long-term service. As a perspective, the author will complete this work with parametrical study including the effects of different mechanical and geometrical properties of the repair components in order to establish analytical function of the SIF for repaired pipe with bonded composite patch.

References

1. Alexander, C.: Assessing the state of composite repair systems, J. Pipeline Gas Technol. (2009)
2. Toutanji, H., Dempsey, S.: Stress modeling of pipelines strengthened with advanced composites materials. Thin-Walled Struct. **39**, 153–165 (2001)
3. Kiefner, J.F., Vieth, P.H., Roytman, I.: Line PRCI, Line Pipe Research Supervisory Committee, Pipeline Research Committee of PRCI, Contract No. PR218-9304, 20 December 1996
4. Alexander, C.R., Francini, R.: State of the art assessment of composite systems used to repair transmission pipelines. In: Proceedings of the 16th International Pipeline Conference, Calgary, Canada, Paper No. IPC2006-10484, 25–29 September 2006
5. Offshore Standard DNV-OS-C501, Composite Components; 2003
6. Ochoa, O.O.: Composite riser experience and design guidance, final project report for minerals management service-DOI, OTRC Library No. 10/06A176 (2006). http://www.mms.gov/tarprojects/490.htm
7. ABAQUS/CAE Ver 6.9 User's Manual. Hibbitt, Karlsson & Sorensen, Inc (2007)
8. Rybicki, E.F., Kanninen M.F.: A finite element calculation of stress intensity factors by a modified crack closure integral. Eng. Fract. Mech. **9**, 931–938 (1977)
9. Krueger, R.: The virtual crack closure technique: history, approach and applications. ICASE Report No. 2002-10, Hampton, VA, USA (2002)
10. Leski, A.: Implementation of the virtual crack closure technique in engineering FE calculations. Finite Elem. Anal. Des. **43**, 261–268 (2007)
11. Bachir Bouiadjra, B., Belhouari, M., Serier, B.: Computation of the stress intensity factor for repaired crack with bonded composite patch in mode I and mixed mode. Compos. Struct. **56**, 401–406 (2002)
12. Belhouari, M., BachirBouiadjra, B., Serier, B.: Comparison of double and single bonded repairs to symmetric composite structures: a numerical analysis. Compos. Struct. **65**, 47–53 (2004)

Elaboration and Characterization of Ceramic Materials from Kaolin-Phosphate Mixture

Chouia Fateh(✉) and Belhouchet Houcine

Laboratoire des Matériaux Non Métalliques, Institut d'Optique et Mécanique de Précision, Université Sétif 1, 19000 Sétif, Algeria
fatehchouia@gmail.com

Abstract. This work focuses on the reaction between kaolin and natural phosphate. Four mixtures were prepared by varying the kaolin content 20–80% by weight with a step of 20. All powders were ground and dried and then pressed in cylindrical form. Finally, the samples were sintered at 1300 °C for 2 h under static air. The reactions and the phase transformations of the powders were studied by X-ray diffraction (X-ray diffraction), infrared spectroscopy (FT-IR) and SEM. X-ray diffraction analysis showed that the mixing of two raw materials produces several composites with different amounts (HAp, anorthite, β-TCP and mullite). The presence of the different composites was confirmed by micrographic observations. The results showed that the 20K samples had a high degree of densification and a low open porosity.

Keywords: DD2 · Natural phosphate · Reaction-sintering · HAp · Anorthite Composites

1 Introduction

Algeria is a country that is full of cheap natural minerals (kaolin, phosphate, dolomite, quartz,…), but unfortunately not used to their true value until now, especially in the field of ceramics. Kaolin is natural clayey raw materials which are generally heterogeneous mixtures of minerals accompanying kaolinite, which is the majority mineral (Mecif et al. 2013). Natural Phosphate (NP) is a general term that describes assemblages of natural minerals containing a high concentration of phosphate minerals. About 80% of the world's production of natural phosphate is derived from sedimentary marine deposits mainly composed of apatite (Rida et al. 2013; Bezzi et al. 2001).

Anorthite is an economical ceramic that has good physical proper ties such as low thermal expansion coefficient, high thermal shock resistance and low dielectric constant. With these excellent characteristics, the anorthite ceramics are applied in electronics industry, heat exchangers and biomedical materials (Mucahit and Sedat 2010; Han et al. 2013).

Hydroxyapatite (Hap) bioceramics as bone substitute materials have the advantages of abundant supply, low cost and absence of immunogenicity (Sumit et al. 2007). Therefore, the preparation of fine and well sinter able Hap powder is the most important step

B. Abdelbaki et al. (Eds.): SMSD 2017, *Proceedings of the Third International Symposium on Materials and Sustainable Development*, pp. 613–619, 2018.
https://doi.org/10.1007/978-3-319-89707-3_65

in the fabrication of ceramic Hap with high sintered density and desirable microstructure (Mezahi et al. 2009).

The objective of this work initially is the elaboration of ceramic materials from kaolin and local phosphate. This study focuses on the reaction between kaolin and natural phosphate.

2 Materials and Methods

The samples studied were prepared from the mixture of two raw materials: the kaolin (DD2) which extracted from the quarry of Djebel Debbagh in Guelma and is white in color and rich in alumina and the natural phosphate (NP) from the mine of Djebel El-Onk (East of Algeria). Their chemical compositions are given in Table 1.

Table 1. The chemical composition

Elements	SiO_2	Al_2O_3	CaO	P_2O_5	Fe_2O_3	Na_2O	SO_3	K_2O	MgO	L.O.I
DD2	45,52	38,75	0,18	–	0,04	0,05	–	0,03	–	15,44
NP	02,30	0,41	53,16	26,74	0,69	0,85	3,26	0,10	0,75	11,74

Seven compositions were prepared, while varying the percentage of the kaolin from 20 to 80 mass% with a step of 20%. These are named 20K, 40K, 60K and 80K, respectively.

The powders were milled through planetary ball milling (Fritsch P6) f or 5 h. The slurry was dried at 110 1C, powdered and sieve through a 63 μm mesh, then compacted at a pressure of 75 MPa using a cold uniaxial press. Cylindrical specimens of 13 mm diameter were produced. The mixtures powders were sintered under normal conditions during 2 h at temperatures between 1000 and 1400 °C. XRD analyses were carried out using a Bruker D8 diffractometer. The XRD tests conditions w ere N i-filtered Cu kα X radiation (35 kV–30 mA) with a scanning speed of 37° (2θ) per minute and at an increment of 0.05 1. The chemical functional groups were investigated by Fourier transform infrared spectroscopy (FT-I R) Shimadzu Corporation type FT-IR 8300 E within the wave number range of 4000–400 cm^{-1}. The morphology of the samples was observed by SEM scanning electron microscopy (Hitachi S3500-N).

3 Results and Discussion

3.1 Analysis by X-Ray Diffraction (XRD)

X-ray diffraction patterns obtained from the samples heated at 1300 °C for duration of two hours are shown in Fig. 1. The XRD spectra of the samples show the presence of the three phases of HAp, anorthite and mullite.

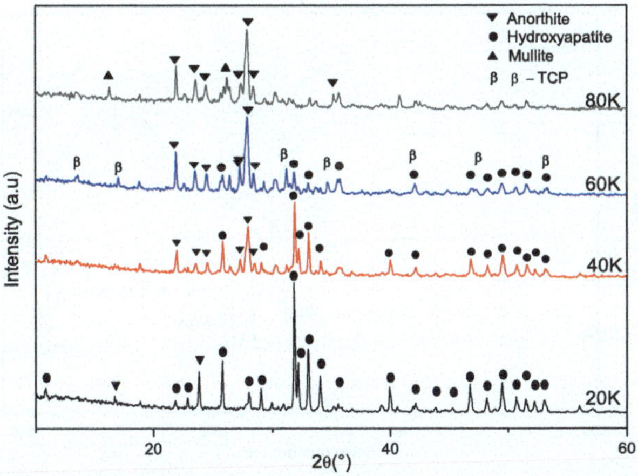

Fig. 1. XRD patterns of the samples sintered at 1300 °C for 2 h

The intensity of the anorthite increased progressively with the percentage increase of kaolin in the mixture. The formation of this anorthite resulting from the reaction between kaolin and CaO, comes from NP dissociation, while the intensity of the peaks corresponding to the HAp decreases. This reduction is justified by the decrease in the amount of phosphate in the mixtures.

In the sample 60K, there is a clear modification of the formed phases. The decomposition of HAp and the formation of β-TCP can be described according to the equation and the following molar proportions (Chu et al. 2013; Sobczak-Kupiec and Wzorek 2012):

$$Ca_{10}(PO_4)_6(OH)_2 \rightarrow 3\beta - Ca_3(P_2O_8) + CaO + H_2O \quad (1)$$

For the 80K sample, the peaks corresponding to the HAp disappeared completely, while the intensity of the anorthite increases. In these samples a new phase (mullite) appears.

3.2 FT-IR Analyzes

The functional groups of different mixtures synthesized at different temperatures were evaluated by FT-IR (Fig. 2). The spectra corresponding to 20K and 40K samples reveal the presence of two bands at 424 and 686 cm^{-1} relating to the vibrations of bond Si–O (Sahnoun and Bouaziz 2012; Yang et al. 2012). A new band was appeared at 470 cm^{-1} related to the deformation of Si–O band. Other peaks observed at 1045 cm^{-1} that demonstrates the presence of Hap: $\nu_3PO_4^{3-}$ (Cho et al. 2014). In addition, two sharp peaks corresponding to ν_4 O–P–O bond at 570–600 cm^{-1} and ν_4 non-degenerate P–O symmetric stretching at about 925 cm^{-1} were appeared (Ahn et al. 2013; Mukherjeea et al. 2014). The last broad band observed at 1153 cm^{-1} corresponds to the ν_{as} PO_4^{3-} vibrations. The majority of the bands disappeared except the two bands at 600 and 1045 cm^{-1} (ν_3 PO_4^{3-}) which becomes broader with a reduction in their intensity.

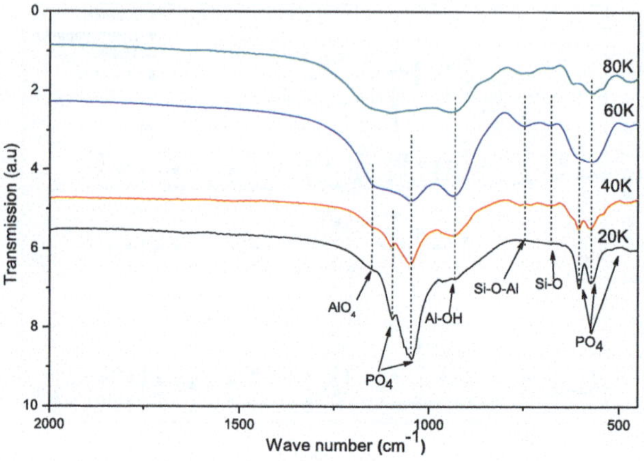

Fig. 2. FT-IR spectra of the samples sintered at 1300 °C for 2 h.

With increasing of kaolin to 60 and 80%, the evolution of FT-IR spectra also denotes the gradual disappearance of PO_4^{3-} groups. This is indicated by the decrease in the sharpness of the corresponding peaks in the region around 1045 and 570 cm^{-1}. For the 80K sample, the appearance of a new intense band at 1097 cm^{-1} is related to the vibrations of AlO_4 hybrids (Li et al. 2009).

3.3 Apparent Density and Open Porosity

Density is a very interesting parameter used to characterize a sintered product. Figure 3 shows the apparent density of the sintered samples at 1300 °C for 2 h as a function of percentage of kaolin. The change in the degree of densification with the reaction develops between kaolin and natural phosphate in all mixtures.

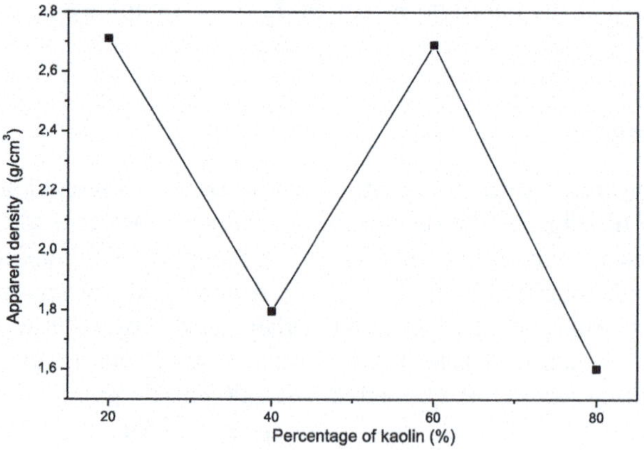

Fig. 3. Apparent density variation of samples as a function of percentage of kaolin.

The apparent density reached its maximum value (2.78 g/cm^3) in the case of the 20K sample. The decrease in apparent density in the 40K sample may be due to the release of some gases (CO_2 and H_2O) enclosed within the densified material early. The viscosity of the vitreous phase is greatly reduced by CaO from the dissociation of HAp. The low viscosity facilitates the release of entrapped CO_2 and H_2O in the pores (Wawrzyn et al. 2013). In the case of the 60K sample, the apparent density increases slightly and reaches a value of 2.32 g/cm^3, this being justified by the formation of tricalcium phosphate (β-TCP), which was confirmed by analysis of X-ray diffraction (Fig. 4).

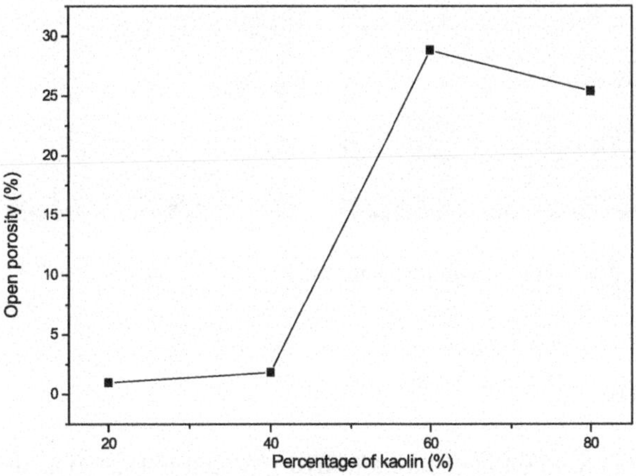

Fig. 4. Open porosity variation of samples as a function of percentage of kaolin.

3.4 SEM Analysis

The microstructures shown in Fig. 5(a–d) correspond to the sintered mixtures at 1300 °C for 2 h.

The SEM image corresponding to sample 20K (Fig. 5-a) shows the presence of HAp grains. It may be noted that the surface of the sintered sample has shown quite different morphology with discontinuous pores. We also note the presence of small size of inter-granular type. For the sample 40K (Fig. 5-b) it is observed that the growth of the grains occurs, and the grains have become large.

The microstructures of samples 60K and 80K show the appearance of irregularly shaped grains and large grains of rounded anorthite and mullite distributed in the anorthite matrix, in sample 80 K, that the mullite is essentially in the form of needles.

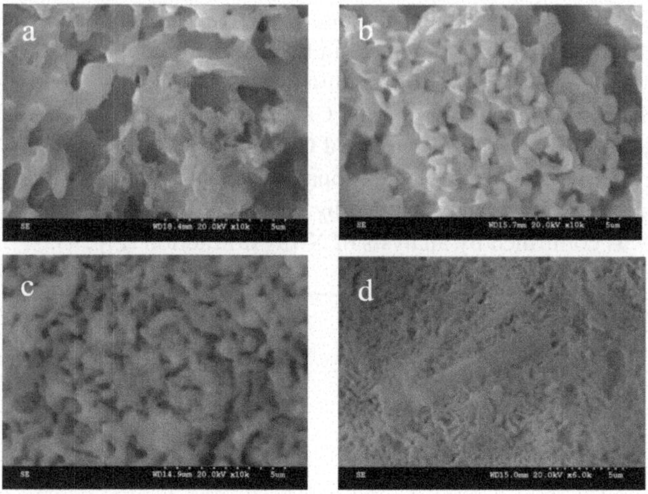

Fig. 5. SEM micrographs of the samples sintered at 1300 °C for 2 h: (a) 20K, (b) 40K, (c) 60K, (d) 80K

4 Conclusion

Ceramic materials based on anorthite and hydroxyapatite (HAp) as the majority phase were prepared from local raw material mixtures, kaolin (DD2) from Djebel Debagh in Guelma and natural phosphate (PN) from Djebel el Onk in Tebessa. The X-ray diffraction patterns obtained show the formation of anorthite-HAp-β-TCP and anorthite-HAp composites in samples containing 40 and 60 kaolin. For lower percentages, the HAp-anorthite composites are obtained. Anorthite and anorthite-mullite composite are formed for higher levels of kaolin percentage. In addition, the β-TCP phase was formed only in the 60K sample. The 20K sample has a high degree of densification. Reactive sintering of the Kaolin/Natural Phosphate mixture is a feasible route to obtain anorthite, mullite and hydroxyapatite which can be used in different fields.

References

Mecif, A., Soro, J., Harabi, A.: Preparation of mullite- and zircon-based ceramics using kaolinite and zirconium oxide: a sintering study. Am. Ceram. Soc. **93**, 1306–1312 (2010)

Rida, K., Chemmal, B., Boukhemkhem, A.: Removal of methylene blue from aqueous solution by natural phosphate. Optoelectron. Adv. Mater. **13**, 493–500 (2013)

Bezzi, N., Merabet, D., Benabdeslem, N., Arkoub, H.: Caractérisation physico-chimique du minerai de phosphate de bled el hadba-Tebessa. Annales de Chimie Science des Matériaux **26**, 5–23 (2001)

Mucahit, S., Sedat, A.: Utilization of recycled paper processing residues and clay of different sources for the production of porous anorthite ceramics. Eur. Ceram. Soc. **30**, 1785–1793 (2010)

Han, Y., Li, C., Bian, C., Li, S., Wang, C.A.: Porous anorthite ceramics with ultra-low thermal conductivity. Eur. Ceram. Soc. **33**, 2573–2578 (2013)

Sumit, P., Agarwal, A.K., Rai, K.N., Garg, A.: Development of high strength hydroxyapatite by solid-state-sintering processs. Ceram. Int. **33**, 419–426 (2007)

Mezahi, F.Z., Oudadesse, H., Harabi, A., Lucas-Girot, A., Le-Gal, Y., Chaair, H., Cathelineau, G.: Dissolution kinetics and structural behavior of natural hydroxyapatite vs. thermal treatment: comparison to synthetic hydroxyapatite. J. Therm. Anal. Calorim. **95**, 21–29 (2009)

Chu, K.T., Ou, S.F., Chen, S.Y., Chiou, S.Y., Chou, H.H., Ou, K.L.: Research of phase transformation induced biodegradable properties on hdroxyapatite and tricalcium phosphate based bioceramic. Ceram. Int. **39**, 1455–1462 (2013)

Sobczak-Kupiec, A., Wzorek, Z.: The influence of calcination parameters on free calcium oxide content in natural hydroxyapatite. Ceram. Int. **38**, 641–647 (2012)

Sahnoun, R., Bouaziz, J.: Sintering characteristics of kaolin in the presence of phosphoric acid binder. Ceram. Int. **38**, 1–7 (2012)

Yang, S.Q., Yuan, P., He, H.P., Qin, Z., Zhou, Q., Zhu, J., Liu, D.: Effect of reaction temperature on grafting of γ-aminopropyl triethoxysilane (APTES) onto kaolinite. Appl. Clay Sci. **62**, 8–14 (2012)

Cho, J.S., Lee, J.C., Chung, S.H., Seo, J.K., Rhee, S.H.: Effect of grain size and density of spray-pyrolyzed hydroxyapatite particles on the sinterability of hydroxyapatite disk. Ceram. Int. **40**, 6691–6697 (2014)

Ahn, G., Lee, J.Y., Seol, D.W., Pyo, S.G., Lee, D.: The effect of calcium phosphate cement–silica composite materials on proliferation and differentiation of pre-osteoblast cells. Mater. Lett. **109**, 302–305 (2013)

Mukherjeea, S., Kundu, B., Sen, S., Chanda, A.: Improved properties of hydroxyapatite–carbon nanotube biocomposite: mechanical, in vitro bioactivity and biological studies. Ceram. Int. **40**, 5635–5643 (2014)

Li, J., Lin, H., Li, J., Wu, J.: Effects of different potassium salts on the formation of mullite as the only crystal phase in kaolinite. Eur. Ceram. Soc. **29**, 2929–2936 (2009)

Wawrzyn, E.K., Małolepszy, J., Murzyn, P.: Sintering behavior of kaolin with calcite. Procedia Eng. **57**, 572–582 (2013)

Optimization of Tool Geometry Parameters for Rotary Peeling Veneer Process Based on TLBO Algorithm

Bourab Fatma-zohra(✉), Aknouche Hamid, and Hamrani Abderrachid

Unité de Recherche Matériaux-Procédé et Environnement (UMPRE), Université M'hamed Bougura Boumerdés (UMBB), Boumerdès, Algeria
fz.bouarab@gmail.com, h.aknouche@yahoo.fr

Abstract. Rotary peeling veneer is a very specific machining process, where the chip is the final product. The fact that works related to this manufacturing process are rare, our objective is to investigate on the optimal cutting parameters, tool edge geometry, through the use of Teaching-Learning based optimization (TLBO) algorithm in order to obtain the best quality with the desired thickness of the veneer product. A study is carried out to identify the objective function that best characterize the machining parameters to be optimized. The challenge is to maintain the best possible quality of peeled veneer with the control of the pre-splitting condition and the veneer thickness variation. The developed algorithm, implemented in Matlab, used in this study is described through two pseudo-codes: main algorithm and the TLBO algorithm. In the main algorithm, the whole resolution procedure is prescribed. The second algorithm is dedicated to the description of all steps of the TLBO technique. Preliminary numerical results obtained from TLBO algorithm are consistent with the experimental ones. The proposed numerical model allows us to predict the characteristic tool angles for different chip thicknesses and friction coefficient. The need to use a pressure bar to produce a quality veneer is numerically proved.

Keywords: Peeling veneer machining · Teaching-learning based optimization Veneer thickness

1 Introduction

Wood is the most widely used natural material in the manufacturing industry; it can be defined as an orthotropic material with specific chemical and physical properties [1]. The machining of wood products has acquired great importance in recent years, identified as being different from that for homogeneous metal removal in the need to avoid creating splintering, delaminating or burning. The various properties of the wood fiber and their orientations have a significant effect on the machining process [2]. The study of the wood machining process is around the chip formation, tool wear, work piece surface quality, crack initiation and propagation into the wood, with taking into account several factors such as humidity, temperature, and vibrations that can affect the wood during the machining process. There are three methods of cutting veneers,

B. Abdelbaki et al. (Eds.): SMSD 2017, *Proceedings of the Third International Symposium on Materials and Sustainable Development*, pp. 620–636, 2018.
https://doi.org/10.1007/978-3-319-89707-3_66

sawing, slicing/half round slicing, and rotary peeling is the most common [3]. The veneer is created by spinning the log and peeling off a continuous sheet (Fig. 1). This is the least-expensive way to produce the veneer. Therefore, one of the big challenges is to maintain the best possible quality of peeled veneer, and this can be achieved with the control of the veneer thickness variation and the surface roughness. Optimum choice of the peeling veneer process parameters is primordial for the economic, efficient, and effective utilization of this particular wood machining process. Peeling veneer machining parameters are generally selected either based on the experience, and expertise of the operator. This way of working in most of the cases is unreliable, and selected machining parameters are far from the optimum. In recent years, soft computing and numerical techniques such as meta-heuristics optimization, fuzzy logic, experimental design method, artificial neural networks and other methods are gaining more attention from researchers dealing with manufacturing and machining processes. However, it is noticed that the lion's share of the researcher's attention is targeted towards metal cutting; very few works have been dedicated to wood cutting and even less on the optimization of the corresponding cutting parameters. The fact that works related to the peeling veneer process are extremely rare, our objective is to investigate on the optimal tool edge geometry of this process through the use of the meta-heuristic optimization algorithm: Teaching-Learning based Optimization (TLBO) algorithm, with the aim of obtaining the best quality with the desired thickness of the veneer product. A detailed study on the mechanics of the wood cutting process is carried out through the definition of the geometry of the cutting tool and the corresponding cutting forces experienced during cutting, and the most important part about the chip deformation. The aim of this preliminary study is the Identification of the objective function that best characterize the machining parameters to be optimized. Numerical results obtained from TLBO algorithm are coherent with the experimental ones. The proposed numerical model allows us to predict the characteristic tool angles for different chip thicknesses and friction coefficients.

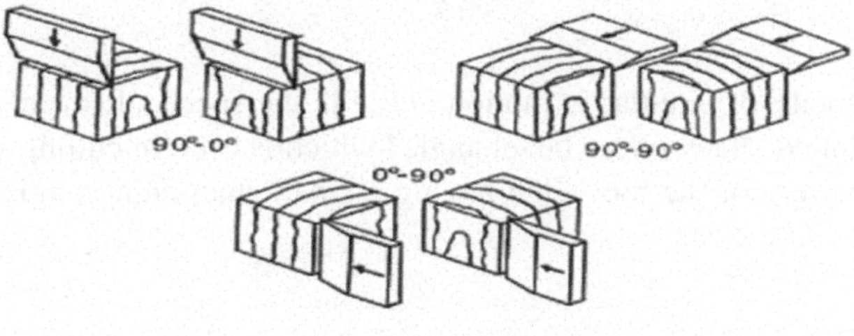

Fig. 1. Main cutting directions and cutting corresponding to the anisotropy [5]

2 Mechanics of the Wood Cutting Process

On the mechanics of cutting process, we are interested, in general, on defining the interactions between the cutting tool and the chip using the laws of engineering mechanics [4]. The mechanical properties of the material have to be taken into account:

the anisotropy nature, the viscoelasticity and plastic properties of the wood. By the determination of the relationships between the forces and strains (deformations), then the cutting force depending on the properties of the material and the cutting tool, we can provide scientific bases to the phenomena of wood cutting.

2.1 Characteristic Angles and Notations

The geometry of the tool (angles of the edges) is one of the most important factors that effect the quality of machining products. In addition, the anisotropy and other material characteristics of wood have also a direct effect on the machining operation. The main cutting directions corresponding to the anisotropy are presented in Fig. (1), [5].

The characteristic angles of the wood machining tool are shown in Fig. 2 [5], we can quote:

α: Usually called "the clearance angle", in veneer cutting tool this angle is only around a 1°.

β: Usually called "the sharpening or bevel angle"; in veneer cutting knives, the angle values are around 20°.

γ: Usually called "the rake angle", for veneer cutting tools are around 70°.

δ': Which is an oblique angle lowers the true cutting angle according to the following equation:

$$\tan \delta' = \tan \delta . \cos \alpha \tag{1}$$

$$\delta = \alpha + \beta \tag{2}$$

where δ is "the cutting angle"

α': this angle is called "the moving clearance angle" occurs especially in peeling veneer and drilling due to the combination of the two effects of a circular and a linear feed motion:

$$\tan \alpha' = \frac{e}{\upsilon} \tag{3}$$

where υ the velocity of circular is motion and e is the speed of linear motion.

β_1: Usually called "back micro bevel angle" which is used in cutting veneer to avoid excessive wear of the tool edge having a small sharpening angle.

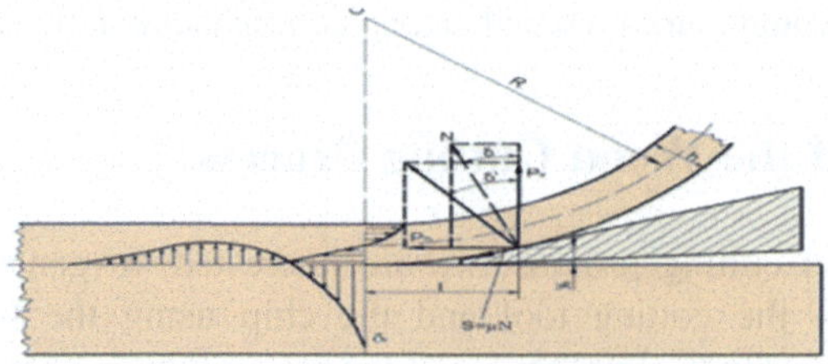

Fig. 3. Force and stress relations of the chip [5]

2.2 Cutting Forces

Forces experienced by a tool during cutting are detrimental in design of mechanical structure of cutting machine, predicting power consumption, determining the tool life and increasing the productivity. The experimental measurement of these cutting forces is a primordial tool allowing the development of mechanical cutting models for a better understanding of the phenomena observed during cutting. These models enable us to design or optimize processes, machines, tools and wood preparation [6]. Concerning the study of forces development during wood cutting, we can find several works treated by many authors [7–12]. The basic concepts related to the wood cutting forces are presented in the following.

Firstly, we have to suppose that the chip deforms on the knife's surface on a radius *R*, (Fig. 4). The normal force N operates orthogonally on the tool's front surface, this force vector can be divided into two components: horizontal and vertical. When the chip slides on the knife's front surface, a frictional force $S = \mu N$ arise (μ is the coefficient of friction at the tool-chip-work piece interface).

The resultant force from the sum of the components shows an inclination towards the direction of the movement [1]. Its angle from the vertical will be given as follows:

$$P_h = N \sin\delta + \mu N \cos\delta \tag{4}$$

$$P_v = N \cos\delta - \mu N \sin\delta \tag{5}$$

From which:

$$\tan\delta' = \frac{P_h}{P_v} = \frac{\tan\delta + \mu}{1 - \mu.\tan\delta} \tag{6}$$

Where P_h and P_v are the horizontal and vertical cutting forces respectively. In the Fig. 3 (on the left side) the internal stress distributions are given. The stress distribution due to bending is asymmetric; the compression stress is near constant on the compressed side where the load on the material equals its strength. The result of the asymmetric stress distribution is that the neutral axis shifts to the tension side [1]. The stress distribution in front of the cutting edge can be evaluated according to the following equation:

$$\sigma_{zr} = k.y \tag{7}$$

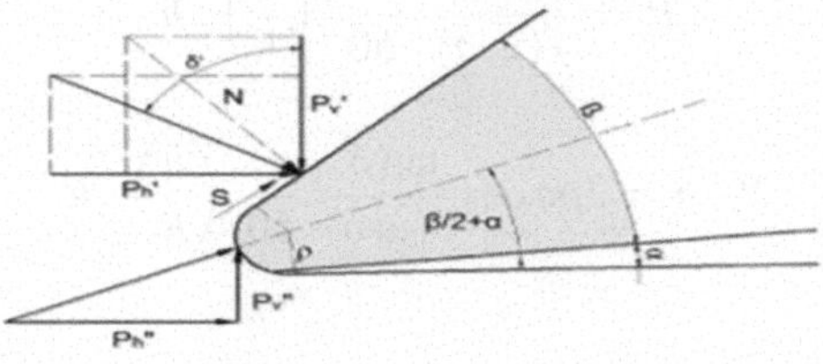

Fig. 6. The position of the force acting on the edge [5]

Where: k is the deformation modulus of the wood (determined approximately with an indenter), y is the deformation occurs in a radial direction. The following correlation is given by the Boussinesq-problem [4]:

$$p = \frac{4E}{\pi(1-\upsilon^2)} \cdot \frac{y}{d} = k.y \tag{8}$$

Where: p is the pressure, d the diameter of the indenter, E the modulus of elasticity and υ Poisson's ratio. For more details about Boussinesq-problem we refer the interested reader to the following references [13–15] the k value depends on the diameter of the indenter according to Eq. 8. The y deformation in a distance x from the edge of the knife can be calculated as:

$$y = -\frac{P_v.e^{-\beta.x}}{2\beta E.I}[\cos\beta x + \beta.l(\cos\beta x - \sin\beta x)] \tag{9}$$

$$\beta = \sqrt{4}\frac{k}{4EI} \tag{10}$$

Where:

P_v The vertical force component.
I The moment of inertia of the chip.
l The distance of the attack point of the P_v force.

Using the above equations, the maximum tensile stress value is given from Eqs. 8 and 9 as:

$$\sigma_{zr\max} = \frac{kP_v(1+\beta.1)}{2\beta^3.E.I} \tag{11}$$

The changing of radial deformation in the plane in front of the edge is illustrated on Fig. 5, we can observe that deformation in a given distance x changes from tensile to compressive field (Fig. 6).

Finally, we determine the two components of the cutting force $(P_v\, and\, P_h)$ as follow [1]:

$$P_v = \frac{1}{f(\mu,\delta)} \cdot \frac{E.b}{50} \cdot \left(\frac{h}{R}\right)^2 .h \tag{12}$$

$$P_v/b = 2\rho\sigma_c + \frac{\tan\delta'}{f(\mu,\delta)} \cdot \frac{E}{50} \cdot \left(\frac{h}{R}\right)^2 .h \tag{13}$$

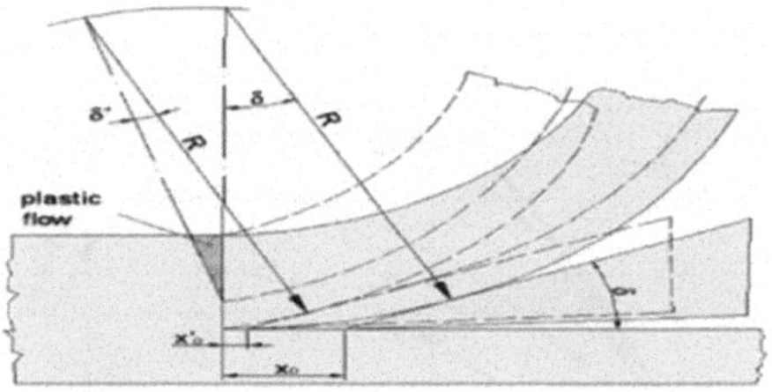

Fig. 7. Plastic chip deformation [5]

Where:

$$f(\mu, \delta) = \sin\delta(1 - \mu.\tan\delta) + (\sin\delta + x_0/R).\tan\delta.(\tan\delta + \mu) \quad (14)$$

x_0 : *Geometrical pre – splitting* (Fig. 8).
b : *Width of the chip,*
h : *Thichness of the chip,*
ρ : *The radius of the edge.*

We can write the vertical component of the whole resultant force acting on the tool as follows:

$$P_{vedge}/b = 2\rho\sigma_c\sin(\beta/2 + \alpha) - \frac{B.h}{\tan\delta'} \quad (15)$$

The horizontal component has the form:

$$P_{hedge}/b = 2\rho\sigma_c[\cos(\beta/2 + \alpha) + \mu\sin(\beta/2 + \alpha)] + \mathrm{B.h} \quad (16)$$

$$B = \frac{\tan\delta'}{f(\mu, \delta)} \cdot \frac{E}{50} \cdot \left(\frac{h}{R}\right) \quad (17)$$

The plot of the trigonometric function occurring in the second term of equation [13]: $\left(\tan\left(\delta'\right)/f(\mu, \delta)\right)$ *is* is illustrated in Fig. 9. In the interval of the angle δ, we find that the trigonometric function have a parabolic shape with a minimum at about $\delta = 40^\circ$, and that means that the cutting force is minimal in the given range. In the case of veneer cutting, δ is around 21°.

2.3 Chip Deformation

As mentioned previously, the ratio h/R is constant and known at a given δ. But in reality, that is not totally true, because chip deformation has more than one degree of freedom, depending on the combined action of internal and external forces, and therefore, the h/R ratio cannot be calculated from static equilibrium equations [1].

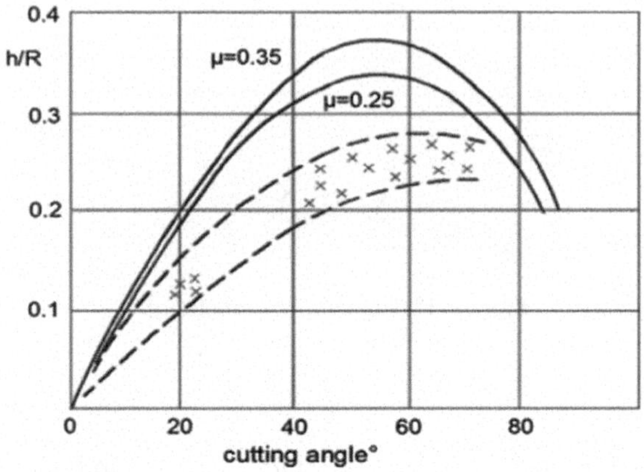

Fig. 9. h = R ratio as a function of cutting angle for two μ friction coefficients. (Dotted line is the boundary for measurement data) [5].

In order to determine the h/R ratio, the minimum of potential energy principle may be used. To solve this problem, the virtual work produced by internal/external forces is described as follows:

$$\Delta(U_i + U_e) = 0 \tag{18}$$

And

$$U_e = -\sum F_j \delta_{yj} \tag{19}$$

Where: U_i, U_e are the work done by internal and external forces respectively, Δ symbol of variation, F_j external forces, δ_{yj} virtual displacements.

The work done by internal forces in a bent beam is given by the following equation (Ritz method):

$$U_i = \frac{1}{2}\int E.I^2_{y''}dz = \frac{1}{4}\left(\frac{\pi}{2l}\right)^4.E.I.y^2.l \tag{20}$$

Where $I^2_{y''}$ means the curvature of the neutral axis. The work done by external forces is expressed in the following form:

$$U_e = \frac{p_h \Delta l}{2} + \frac{p_v.y_0}{2} \tag{21}$$

In which the longitudinal deformation of the chip is given by

$$\Delta l = \frac{P_h \Delta l}{E.b.h} \tag{22}$$

Keeping in mind equation [6], the tangential and radial force components can be expressed using the bent beam theory:

$$p_h = \frac{3EIy_0}{l^3}(\tan\delta + \mu) \tag{23}$$

$$p_v = \frac{3EIy_0}{l^3}(1 - \mu\tan\delta) \tag{24}$$

Taking the functional $F = U_i - U_e$ and its derivative with respect to y, setting the functional equal to zero, we get

$$\frac{dF}{dy} = 0 \tag{25}$$

Solving the above equation and using the $l \cong R \sin\delta$ approximation, the h/R is given by the following equation:

$$\frac{h}{R} = \frac{2\sin\delta}{\sqrt{3}(\tan\delta + \mu)}\sqrt{\mu\tan\delta} \tag{26}$$

In the above development we have supposed that the total bending work occurs in the elastic field and no plastic deformation occurs, but that is not the case. Beyond the yield limit, the bending forces remain constant and do no increase linearly with the deformation. Then we can add the following correction:

$$\frac{h}{R} = \frac{2\sin\delta}{\sqrt{3}(\tan\delta + \mu)}\sqrt{\mu\tan\delta} \tag{27}$$

Then

$$h = R\frac{2\sin\delta}{\sqrt{3}(\tan\delta + \mu)}\sqrt{(\mu\tan\delta)} \tag{28}$$

In Fig. 10, we observe that the values of the measuring results are systematically smaller than those obtained theoretically. This can be explained by the fact that the upper side of the chip yields plastically, and this leads to the rotation of the bent cross section, thus reducing the apparent value of the δ angle [1] (Fig. 7).

The values of material parameters for the wood specimen used in this study are listed in the following Table 1:

Table 1. Material parameters of the wood specimen [34].

Wood species	Moisture	Young's modulus	Deformation modulus	Compre strength	Tensile strength
Red oak	12%	E = 11,900 MPa	k = 85000 MPa = mm	σ_c = 47000 KPa	σ_t = 7200 KPa

3 Optimization Problem

In this section, the optimization model of rotary peeling veneer process is formulated based on the above analysis. The optimization algorithm is carried out using a relatively recent heuristic search method, whose mechanics are inspired by works on the philosophy of teaching and learning, namely Teaching–Learning-Based Optimization (TLBO). A short presentation of the basics of this algorithm is described below

3.1 Teaching-Learning Based Optimization (TLBO) Algorithm

Teaching–learning-based optimization algorithm (TLBO) is a teaching–learning process inspired algorithm recently proposed by [29] based on the effect of influence of a teacher on the output of learners in a class. Teacher and learners are the two vital components of the algorithm and describes two basic modes of the learning, through teacher (known as teacher phase) and interacting with the other learners (known as learner phase). The output in TLBO algorithm is considered in terms of results or grades of the learners which depend on the quality of teacher. A high quality teacher is usually considered as a highly learned person who trains learners so that they can have better results in terms of their marks or grades. Moreover, learners also learn from the interaction among themselves which also helps in improving their results. The working of TLBO is divided into two parts, 'Teacher phase' and 'Learner phase'. Working of both the phase is explained below [32].

3.1.1 Teacher Phase

It is the first part of the algorithm where learners learn through the teacher. During this phase a teacher tries to increase the mean result of the class in the subject taught by him or her depending on his or her capability. At any iteration i, assume that there are 'm' number of subjects (i.e., design parameters), 'n' number of learners (i.e., population size, $k = 1, 2, \ldots\ldots\ldots, n$) and $M_{j,i}$ be the mean result of the learners in a particular subject j ($j = 1, 2, \ldots, m$) the best overall result $X_{total-kbest,i}$, obtained in the entire population of learners considering all the subjects together can be considered as the result of best learner K_{best}.

However, as the teacher is usually considered as a highly learned person who trains learners so that they can have better results, the best learner identified is considered as the teacher.

The difference between the existing mean result of each subject and the corresponding result of the teacher for each subject is given by:

$$difference\_mean_{j,k,i} = r_i(X_{j,kbest,i} - T_F M_{j,i}) \qquad (29)$$

Where, $X_{j,kbest,i}$ is the result of the best learner (i.e., teacher) in subject j, T_F is the teaching factor which decides the value of mean to be changed, and r_i is the random number in the range [0, 1]. The value of T_F is decided randomly with equal probability as:

$$T_F = round[1 + \text{rand}(0, 1)2 - 1] \quad (30)$$

T_F is not a parameter of the TLBO algorithm. The value of T_F is not given as an input to the algorithm and its value is randomly decided by the algorithm using Eq. (30). After conducting a number of experiments on many benchmark functions it is concluded that the algorithm performs better if its value is between 1 and 2, however, the algorithm is found to perform much better if the value of T_F is either 1 or 2 and hence to simplify the algorithm, the teaching factor is suggested to take either 1 or 2 depending on the rounding up criteria given by Eq. (30). However, one can take any value of T_F in between 1 and 2.

Based on the $\text{Difference\_Mean}_{j,k,i}$ the existing solution is updated in the teacher phase according to the following expression.

$$X'_{j,k,i} = X_{j,k,i} + \text{Difference\_Mean}_{j,k,i} \quad (31)$$

Where $X'_{j,k,i}$ is the updated value of $X_{j,k,i}$. Accept $X'_{j,k,i}$ if it gives better function value.

All the accepted function values at the end of the teacher phase are maintained and these values become the input to the learner phase [33].

3.1.2 Learner Phase

It is the second part of the algorithm where learners increase their knowledge by interaction among themselves. A learner interacts randomly with other learners for enhancing his or her knowledge. A learner learns new things if the other learner has more knowledge than him or her. Considering a population size of 'n', the learning phenomenon of this phase is expressed below. Randomly select two learners P and Q such that $X'_{total-P,i} \neq X'_{total-Q,i}$ (where, $X'_{total-P,i}$ and $X'_{total-Q,i}$ are the updated values of $X_{total-P,i}$ and $X_{total-Q,i}$, respectively at the end of teacher phase).

$$X''_{j,P,i} = X'_{j,P,i} + r_i(X'_{j,P,i} - X'_{j,Q,i}) \text{ if } X'_{total-P,i} \prec X'_{total-Q,i} \quad (32)$$

$$X''_{j,P,i} = X'_{j,P,i} + r_i * (X'_{j,Q,i} - X'_{j,P,i}) \text{ if } X'_{total-Q,i} \prec X'_{total-P,i} \quad (33)$$

Accept $X''_{j,P,i}$, if it gives a better function value. All the accepted function values at the end of the learner phase are maintained and these values become the input to the teacher phase of the next iteration. The values of r_i used in Eqs. (29), (32) and (33) can be different. The flow chart of TLBO algorithm is given in Fig. 11 [33].

3.2 Optimization Model Formulation

3.2.1 Objective Function

The mono-objective constrained optimization problem can be formulated, from Eq. (28), as follows:

$$\begin{cases} \min\left|\left(R\frac{2\sin\delta}{\sqrt{3}(\tan\delta+\mu)}\sqrt{\mu\tan\delta}\right)-h\right| \\ subjectto : geometrocalpre - splittingx_0 = 0(equalityconstraint) \end{cases} \tag{34}$$

In this study, we envisage optimizing the cutting angle with fixing the veneer thickness between the ranges of (0.1–3 mm) and give at each iteration a value of friction coefficient (Table 2)

Table 2. Adopted search space parameters.

Cutting angle δ	Chip thickness h	Friction coefficients μ	Radius R
$0^\circ \leq \delta \leq 85^\circ$	$0.5\,\text{mm} \leq h \leq 3\,\text{mm}$	$0.20 \leq \mu \leq 0.45$	$5\,\text{mm} \leq R \leq 200\,\text{mm}$

4 Developed Algorithm

The developed algorithm, implemented in Matlab, used in this study is described through two flowcharts: main algorithm and the TLBO algorithm. In the main algorithm (Fig. 11), the whole procedure of the resolution is prescribed. The second algorithm (Fig. 12) is dedicated to the description of all steps of the TLBO technique.

The steps of the proposed optimizing process parameters of rotary veneer cutting operation using TLBO is given as follows:

STEP 1: Define the optimization problem and initialize the optimization parameters:
Initialize population size N = 1000
Number of generations = 100
Number of design variables = fixed h and μ + random δ, R
Limits of design variables taken from Table 2
Define optimization problem from Eq. 34
STEP 2: Generate the population Initialize random population according to the population size (number of learners) and the number of design variables (the subjects).
STEP 3: Teacher phase Compute the mean of the population to give the mean for the current subject.
STEP 4: Learner phase Mutual interactions between learners enhance and increase their knowledge
STEP 5: Termination criterion, Stop if the maximum generation number is achieved; otherwise repeat from **STEP3.**

5 Results and Discussion

Figure 12 shows the fitness value of the best particle obtained whilst iterations runs. At the beginning, the fitness value is observed at the rate of 1. When the iterative process reaches approximately 10 iterations, it is found that the value of fitness stabilize at zero. This observation confirm the convergence of the adopted TLBO algorithm

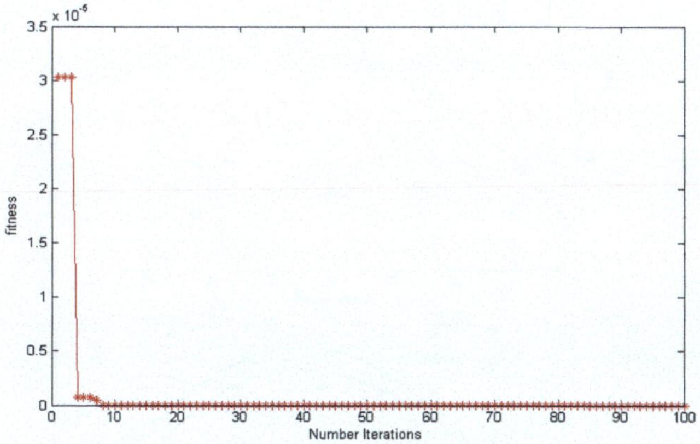

Fig. 12. Performance of TLBO algorithm.

Figure 13 presents the values of h/R ratio obtained from our TLBO code, as a function of the optimal parameter δ, for different values of μ friction coefficients at the tool-chip-workpiece interface. Compared with the results obtained in Fig. 9, our results seem in good accordance and perform a prediction of the cutting angles δ and radius R for different values of friction coefficients μ and veneer thickness h. This observation confirms the fact that the adopted optimization model is valid.

After validating our numerical model, consideration has been given to the prediction of tool angle δ for different values of friction coefficients μ and veneer thickness h, in the particular case of rotary peeling veneer process. The assigned intervals for machining parameters are listed in Table 3.

Table 3. Adopted search space parameters for rotary peeling veneer process

Cutting angle δ	Chip thickness h	Friction efficient μ	Radius R
$17^{\circ} \leq \delta \leq 27^{\circ}$	$0.5\,\text{mm} \leq h \leq 3\,\text{mm}$	$0.20 \leq \mu \leq 0.45$	$5\,\text{mm} \leq R \leq 200\,\text{mm}$

Figure 14 of the adopted TLBO algorithm, this is due to the fact that the objective function chosen in our study (Eq. 34) is conditioned by the pre-splitting restriction ($x_0 = 0$). In practice, the use of small cutting angles is more inclined to cause pre-splitting, especially for production of chips (veneer) thicker than 2 mm [1]. Therefore, the use of a pressure bar is always essential to hamper pre-splitting.

The final section is dedicated to the results of forces development during wood cutting in rotary peeling veneer process. The same observation can be made concerning the cutting forces, where the pre-splitting phenomenon is clearly apparent.

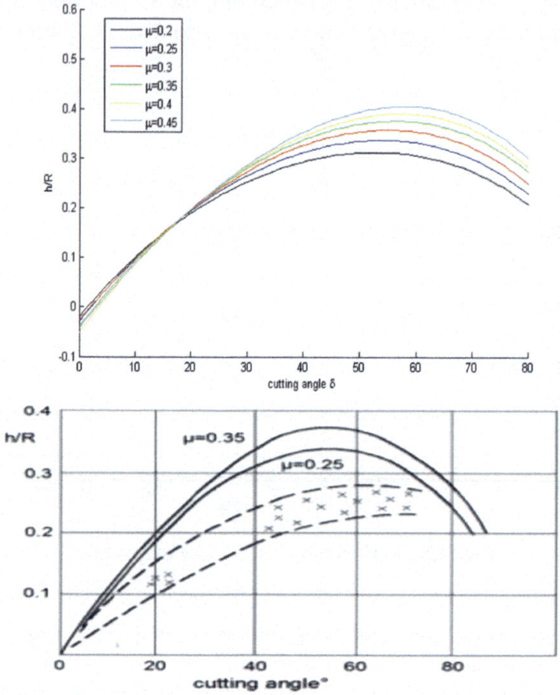

Fig. 13. h/R ratio as a function of cutting angle δ for different values of μ friction coefficients obtained from our TLBO code

In Fig. 15, the condition of pre-splitting is verified for values of h greater than 2.5 mm, where the maximum tensile stress value $\sigma_{zr\max}$ is larger than the tensile strength σ_t of the wood. The variation of friction coefficient μ does not appear to have a significant effect on the maximum tensile stress value $\sigma_{zr\max}$.

The force components $P_{h_{edge}}$ and $P_{v_{edge}}$ for each cutting direction are shown in Figs. 16 and 17, respectively.

The tangential force $P_{h_{edge}}$ has positive values and increase with the increasing of veneer thickness h. The radial force $P_{v_{edge}}$ has negative values and decrease with the increasing of veneer thickness h. The variation of the two force components is intensified when the chip become thicker than 2 mm. The effect of friction coefficient is more visible on the tangential force component compared to the radial one. This is due to the fact that friction force opposes motion between the two surfaces of tool and wood, and the most affected force component will be the one located along the tangent of tool feed direction.

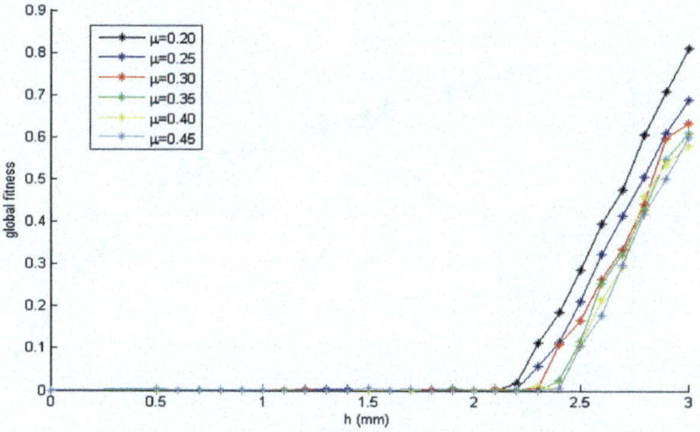

Fig. 14. Global fitness as a function of thickness h for different values of μ.

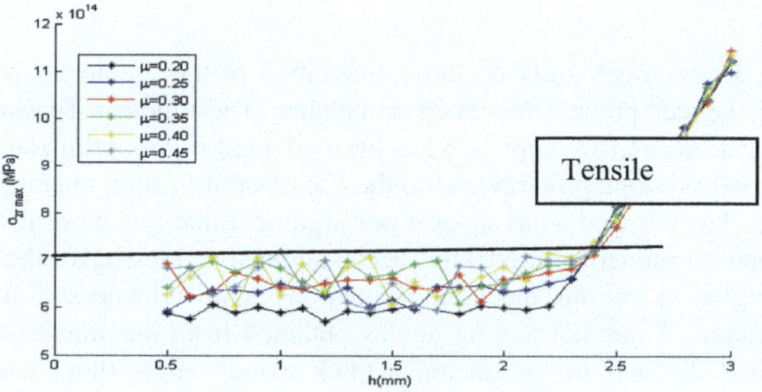

Fig. 15. $\sigma_{zr\max}$ as a function of h for different values of friction coefficients μ.

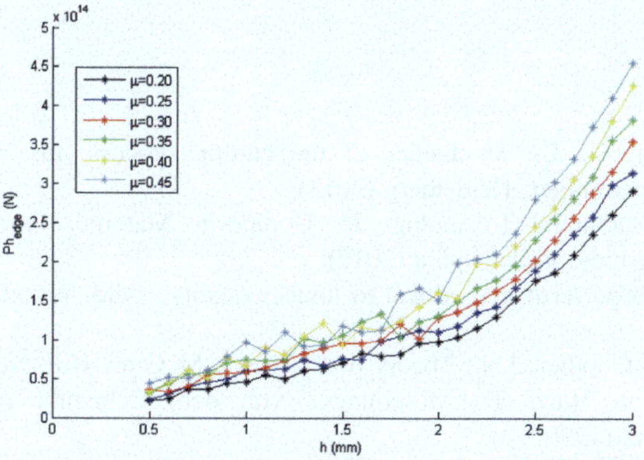

Fig. 16. Variation of horizontal (tangential) cutting force $P_{h_{edge}}$ with varying chip thickness h for different values of μ.

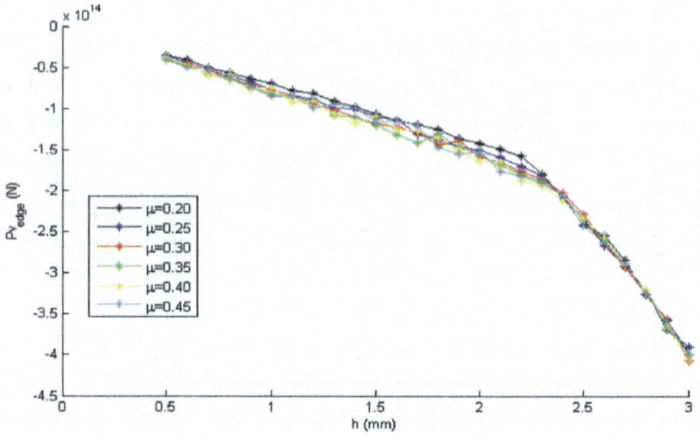

Fig. 17. Variation of vertical (radial) cutting force $P_{v_{edge}}$ with varying chip thickness h for

6 Conclusion

In this paper, a numerical study on the optimization of tool geometry parameters for rotary peeling veneer process has been conducted. The efficient Teaching–Learning Based Optimization (TLBO) approach has been adopted as an optimization technique. A theoretical and bibliographic review on the use of optimization techniques for wood cutting process has allowed us to support our argument that this work is the first of its kind. The proposed numerical model has been validated and tested for the prediction of the cutting angles in various machining situations (chip thicknesses, friction coefficients). The range of optimal cutting angles obtained from our numerical analysis is between 19 and 23, and for producing a thick wood veneer (thickness larger than 2 mm) the use of a pressure bar is primordial to avoid pre-splitting and thus have a quality veneer product. This study deal with optimization on wood cutting process, and generates useful data for our future works (adding mechanics of pressure bar, application to wood slicing and milling processes, etc.).

References

1. Csandy, E., Magoss, E.: Mechanics of the cutting process. In: Mechanics of Wood Machining, p. 1. Springer, Heidelberg (2013)
2. Hocheng, H.: Machining Technology for Composite Materials Principles and Practice. Woodhead Publishing, Philadelphia (2012)
3. Kirbach, E.: Saw performance critical to lumber quality, yield. Wood Technol. **39**, 22–24 (1995)
4. Timoshenko, S., Goodier, J.N.: Theory of Elasticity. McGraw-Hill, New York (1951)
5. des Spanens von Holz, T.: Mitteilungen von dem Lehrstuhl fur Holzbearbeitung. Fortschrittberichte (1990)
6. Marchal, R., Mothe, F., Denaud, L.-E., Thibaut, B., Bleron, L.: Cutting forces in wood machining basics and applications in industrial processes, wood machining micromechanics and fracture. Holzforschung (2009)

7. Koch, P., et al.: Wood machining processes. Wood Machining Processes (1964)
8. Eyma, F., Méausoone, P.-J., Martin, P.: Study of the properties of thirteen tropical wood species to improve the prediction of cutting forces in mode B. Ann. Forest Sci. **61**, 55–64 (2004)
9. Eyma, F., Méausoone, P.-J., Larricq, P., Marchal, R.: Utilization of a dynamometric pendulum to estimate cutting forces involved during routing. Comparison with actual calculated values. Ann. Forest Sci. **62**, 441–447 (2005)
10. Fischer, R., Gottlober, Ch.: Basics in the optimization of wood cutting in the example of peripheral milling. In: Proceedings of the 16th International Symposium on Wood Machining (2003)
11. Fischer, R.: Micro processes at cutting edge, some basics of machining wood, Vienna. In: Proceedings of the 2nd International Symposium on Wood Machining (2004)
12. Palmqvist, J., Lenner, M., Gustafsson, S.I.: Cutting-forces when up-milling in beech. Wood Sci. Technol. **39**, 674–684 (2005)
13. Sneddon, I.N.: The relation between load and penetration in the axisymmetric Boussinesq problem for a punch of arbitrary profile. Int. J. Eng. Sci. **3**, 47–57 (1965)
14. Rostovtsev, N.A.: On the theory of elasticity of a non homogeneous medium. J. Appl. Math. Mech. **28**, 745–757 (1964)
15. Giannakopoulos, A.E., Suresh, S.: Indentation of solids with gradients in elastic properties: Part I. Point force. Int. J. Solids Struct. **34**, 2357–2392 (1997)
16. Palanisamy, P., Rajendran, I., Shanmugasundaram, S.: Optimization of machining parameters using genetic algorithm and experimental validation for end-milling operations. Int. J. Adv. Manuf. Technol. **32**, 644–655 (2007)
17. Saravanan, R., Janakiraman, V.:. Study on reduction of machining time in CNC turning centre by genetic algorithm. In: International Conference on Computational Intelligence and Multimedia Applications (2007)
18. Armarego, E.J.A., Smith, A.J.R., Wang, J.: Constrained optimization strategies and CAM software for single-pass peripheral milling. Int. J. Prod. Res. **31**, 2139–2160 (1993)
19. Raju, K.V.M.K., Janardhana, G.R., Kumar, P.N., Rao, V.D.P.: Optimization of cutting conditions for surface roughness in CNC end milling. Int. J. Precis. Eng. Manuf. **12**, 383–391 (2011)
20. Tandon, V., El-Mounayri, H., Kishawy, H.: NC end milling optimization using evolutionary computation. Int. J. Mach. Tools Manuf. **42**, 595–605 (2002)
21. Wang, J.: Computer-aided economic optimization of end-milling operations. Int. J. Prod. Econ. **54**, 307–320 (1998)
22. Wang, J., Armarego, E.J.A.: Computer-aided optimization of multiple constraint single pass face milling operations. Mach. Sci. Technol. **5**, 77–99 (2001)
23. Wang, Z.G., Wong, Y.S., Rahman, M.: Optimisation of multi-pass milling using genetic algorithm and genetic simulated annealing. Int. J. Adv. Manuf. Technol. **24**, 727–732 (2004)
24. Rao, R.V., Kalyankar, V.D.: Parameter optimization of modern machining processes using teaching–learning-based optimization algorithm. Eng. Appl. Artif. Intell. **26**, 524–531 (2013)
25. Yildiz, A.R.: Optimization of multi-pass turning operations using hybrid teaching learning-based approach. Int. J. Adv. Manuf. Technol. **66**, 1319–1326 (2013)
26. Rao, R.V., Kalyankar, V.D.: Parameter optimization of modern machining processes using teaching–learning-based optimization algorithm. Eng. Appl. Artif. Intell. **26**, 524–531 (2013)
27. Togan, V.: Design of planar steel frames using teaching-learning based optimization. Eng. Struct. **34**, 225–232 (2012)

28. Yu, K., Wang, X., Wang, Z.: An improved teaching-learning-based optimization algorithm for numerical and engineering optimization problems (2014)
29. Rao, R.V., Patel, V.: An elitist teaching-learning-based optimization algorithm for solving complex constrained optimization problem. Int. J. Ind. Eng. Comput. **3**, 535–560 (2012)
30. Tiryaki, S., Malkoçoğlu, A., Özşahin, Ş.: Using artificial neural networks for modeling surface roughness of wood in machining process. Constr. Building Mater. **66**, 329–335 (2014)
31. Gawronski, T.: Optimization of CNC routing operations of wood en furniture parts. Int. J. Adv. Manuf. Technol. **67**, 2259–2267 (2013)
32. Venkata Rao, R., Kalyankar, V.D.: Multi-pass turning process parameter optimization using teaching–learning-based optimization algorithm (2012)
33. Venkata Rao, R., Savsani, V.J.: Mechanical Design Optimization Using Advanced Optimization Techniques. Springer Series in Advanced Manufacturing. Springer, London (2012)
34. Forest Products Laboratory: Wood handbook—Wood as an engineering material. General Technical report FPL-GTR-190. Department of Agriculture, Forest Service, Forest Products Laboratory, Madison (2010)

Light Scattering Applied to the Study of Biological Tissues: To an Optical Biopsy

H. Chorfi[1(✉)], K. Ayadi[1], and L. Boufendi[2]

[1] Applied Optics Laboratory, Institute of Optics and Precision Mechanics, Setif-1-University, 19000 Setif, Algeria
chorfihichem@univ-setif.dz,
hichem.chorfi@univ-orleans.fr
[2] GREMI, Orléans University,
BP6744, 14 Rue d'Issoudun, 45067 Orléans Cedex 2, France

Abstract. This work is devoted to the study of light scattering in biological tissues. It aims to determine indicators that permit to differentiate between cancerous and normal tissues of a human organ and to seek a mean of therapy transport. For this, we analyzed the intensity and the spectral variation of the scattered light as a function of its scattering direction.

This work gives some main results of the experiment and the simulation. We stepped impressive to discriminate between cancerous and normal tissues of a human organ, such as: Vesicle, Breast.

Keywords: Biological materials · Light scattering · Optical analysis
Mie theory

1 Introduction

The light scattering is a well-developed technique and perfectly adapted to the study of the properties of organized molecular systems. It is used successfully for the study and characterization of heterogeneous media. In these media, light scattering is due to spatial fluctuations in the refractive index, concentration and random positioning of the different diffusers.

The Rayleigh scattering is a limiting case of Mie scattering. Nevertheless, it differs in several perceptible aspects when compared for particles of very different sizes.

The Mie scattering designates the elastic scattering by particles whose radius oscillates between 0.1 and 10 times the wavelength. This theory is complex and these quantitative results are obtained with particles of pseudo symmetric geometries [1].

The elastic scattering of light is the type the most dominant of the interaction light – Tissues. Statically, the diffusion of the light comes from the spatial heterogeneity of the Optical refractive index. In turn, the refraction index depends on the concentration and type of tissues [2]. Various structures of tissues such as cellular organelles and the extracellular matrix that give rise to the heterogeneous spatially distribution in the refraction index and can affect the propagation of light in the tissues, including the spectroscopic, or angular characteristics of the light diffused from emerging tissues.

B. Abdelbaki et al. (Eds.): SMSD 2017, *Proceedings of the Third International Symposium on Materials and Sustainable Development*, pp. 637–643, 2018.
https://doi.org/10.1007/978-3-319-89707-3_67

In principle, a structure of tissue of any size may give rise to the diffusion of the light (Fig. 1).

Fig. 1. Observation of the scattering phenomenon as a function of the particle size; -a- Rayleigh scattering, -b- Mie scattering.

2 Mie Theory

The first to solve the problem of light scattering by a spherical particle using Maxwell's theory of electromagnetism is Gustav Mie.

The diffusion theory of Mie concerns the diffusion of an electromagnetic wave by a particle of homogeneous spherical shape whose radius is of the same order of magnitude as the incident wave [3].

This theory shows that the intensity of scattered light as a function of wavelength decreases strongly from a wavelength approximately equal to the size of the particle. Above this value, it does not depend substantially on the wavelength of the radiation [4].

Mie's theory is much more complex than that of Rayleigh scattering. Most of the notions introduced in the theory of incident light and the size of the particle are two parameters that determine the parameter "q" which is one of the basic parameters of the theory of Mie which is given *by:*

$$m = \pi R/\lambda \tag{1}$$

Where: R is the particle size.

λ: Wavelength of incident light.

This important parameter can be obtained from the relations of the scattering and extinction sections which present as infinite sums of polynomials [5, 6] associated with the electric and magnetic multipolar contributions of the response of the particle to the wave electromagnetic field.

$$C_{ext} = \frac{2\pi}{m^2}\sum_{i=1}^{\infty}(2i+1)\mathrm{Re}(a_1 + b_1) \tag{2}$$

$$C_{sca} = \frac{2\pi}{m^2}\sum_{i=1}^{\infty}(2i+1)(|a_1|^2 + |b_1|^2) \tag{3}$$

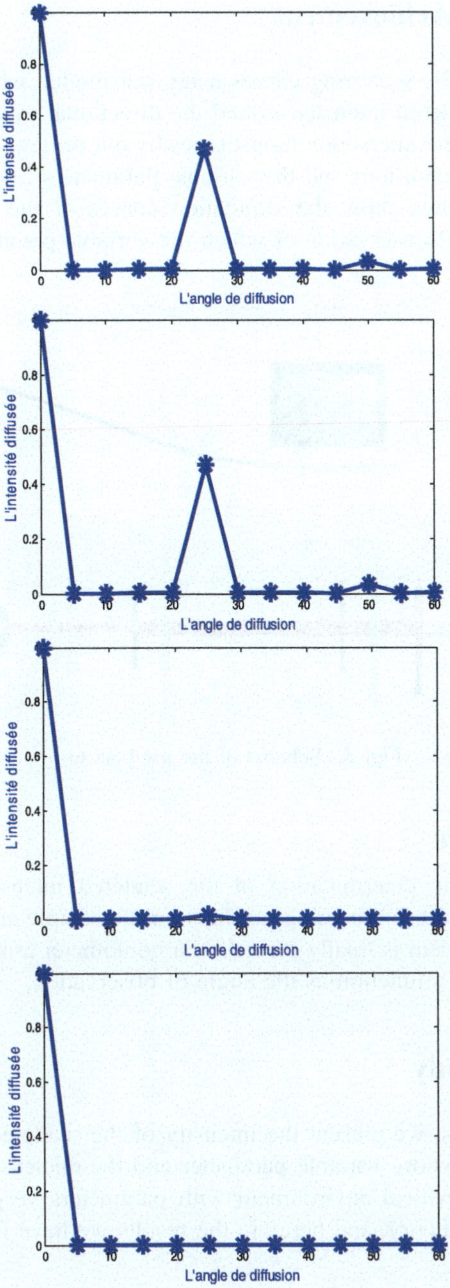

Fig. 2. Curves of scattered intensities as a function of scattering angle for different sizes: a - 2.3 μm, b - 2.9 μm, c - 6.3 μm, d - 9.1 μm.

3 Experimental Achievement

First, we plotted the Mie scattering curves using our model, which allows to plot the distribution of the scattered intensity around the directional transmission.

The simulation of the interaction light-tissues by our program was made taking into account the working conditions and the variable parameters in our assembly.

The following figures show the simulation curves of the scattered intensity as function as the angle of observation of which the variable parameter is the size of the micro-diffuser (Fig. 2):

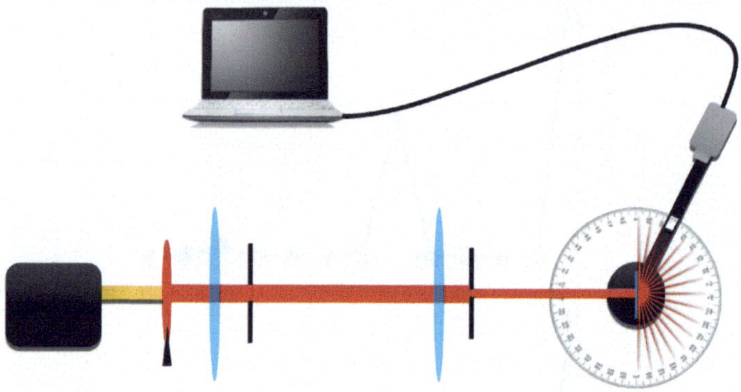

Fig. 3. Scheme of the used set-up.

3.1 The Used Set-Up

The used setup for the determination of the scattered intensity is represented by (Fig. 3). The collimated beam coming out falls on the sample and it will diffuse in all directions. A CCD camera is finally placed on a goniometer arm to capture images of scattered intensities as a function as the angle of observation.

4 Simulation Study

In the following curves, we present the intensity of the scattered light simulated as a function as the angle whose variable parameter and the nucleus size, for this part we have created a mathematical environment with parameters very close to the experimental using the Mie theory, and here are the results we have found (Fig. 4):

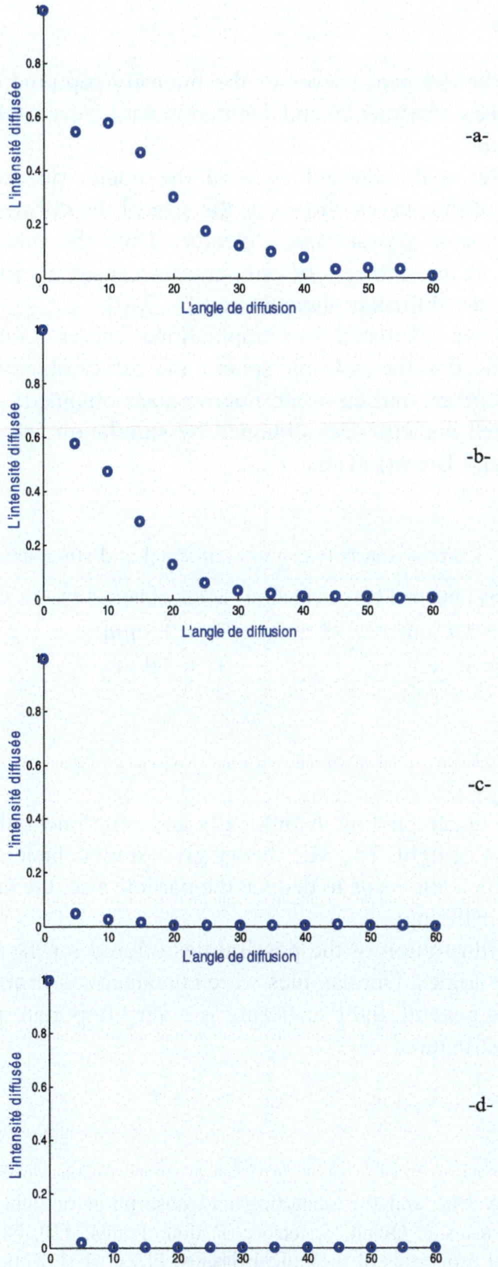

Fig. 4. The simulation curves of scattered intensities as a function of the scattering angle for different particle sizes: a - 2.3 μm, b - 2.9 μm, c - 6.3 μm, d - 9.1 μm.

5 Observations

The observation of the different curves of the intensity captured as a function of the scattering angle for the experimental and simulation study reveals the same pace for the different cases studied.

If we use samples with different sizes of the nuclei (in descending order, for example), the shape of the curves widens as the size of the diffusers decreases and the phenomenon of diffusion appears more clearly. Thus the inhomogeneity and the variation in the size of the diffusers of our samples causes a variation in the density inside which reveals the diffusion phenomenon [4, 7–9].

Using our model, we calculated the computational curves assuming the parameters of anisotropy normalized to the isotropic sphere, the curves obtained are approximated to the experimental curves, and these are micron sizes obtained:

Comparison of cell nucleus sizes obtained by simulation and experimental set-up (study of a Vesicle and Breast) (Table 1).

Table 1. Comparison between experimental and simulation results

	Sizes obtained by calculation	Sizes obtained by the experimental
Vesicle	$T_1 = 3.4$ μm	$T_1 = 3.8$ μm
Breast	$T_1 = 4.3$ μm	$T_1 = 5.0$ μm

6 Conclusion

Our work consists in understanding theoretically and experimentally the measurements based on the diffusion of light. The Mie theory gives a modelable representation of the optical properties, which allows us to deduce the particle size, the nature of the particles and the presence of pathology.

This study is an illustration of the possibilities offered for the study of cytological and histological pathologies. Our samples were characterized from the variation of the scattered intensity. In general, light scattering is a very important phenomenon for [8] characterizing microstructures.

References

1. Horvath, H.: Gustav Mie and the scattering and absorption of light by particles: historic developments and basics. J. Quant. Spectrosc. Radiat. Transf. **110**, 787–799 (2009)
2. Steven, L.J.: Optical properties of biological tissues. Phys. Med. Biol. **58**, 39 (2013)
3. Brunsting, A., Mullaney, P.F.: Light scattering from coated spheres: model for biological cells. Appl. Opt. **11**, 675–680 (1972)
4. Chorfi, H., et al.: Morphological characterization of particles by the intensity and polarization of the scattered radiation, Optik Int. J. Light ElectronOptics. https://doi.org/10.1016/j.ijleo.2017.10.056

5. Yang, P., et al.: Investigation of scattering coefficients and anisotropy factors of human cancerous and normal prostate tissues using Mie theory. In: Proceedings of SPIE, vol. 8941, pp. 1–2 (2014)
6. Huiling, H., et al.: Light scattering by a spheroidal bubble with geometrical optics approximation. J. Quant. Spectrosc. Radiat. Transf. **113**, 1467–1475 (2012)
7. Lemta: Rayonnement dans les tissus biologiques (2013). http://lemta.univ-lorraine.fr/rayonmt_bio.html
8. Oncoprof: Techniques standards d'anatomie pathologique (2015). http://www.oncoprof.net/Generale2000/g04_Diagnostic/Histologie/Technique-texte/dg_ap_tech01.html
9. Zerrari, N.: Caractérisation des tissus biologiques mous par Diffusion multiple de la Lumière, Thèse de Doctorat, université Claude Bernard, Lyon (2014)

The Effect of Doping with N and Cu Atoms on the Hydrogen Sensing Properties of the ZnO$(10\bar{1}0)$ Surface and ZnO Nanowires: A First-Principles Study

M. A. Lahmer[1,2(✉)]

[1] Department of Physic, University of Boumerdes, 35000 Boumerdes, Algeria
malahmer@univ-boumerdes.dz
[2] Laboratoire Revetement, Materiaux et Environementul, University of Boumerdes, 35000 Boumerdes, Algeria

Abstract. The effect of doping with acceptor atoms such as copper and nitrogen on the electronic structure and hydrogen sensing properties of the ZnO$(10\bar{1}0)$ surface and the ZnO nanowires was investigated using the first principles method. Our results show that the doping process can greatly affect the electronic structure of ZnO surfaces and nanowires, and consequently their responses to hydrogen adsorption can be significantly enhanced after doping with acceptors. We found that hydrogen adsorption is more favored on the doped ZnO$(10\bar{1}0)$ surface than onto the clean surface, and this makes the Cu-doped and the N-doped surfaces more efficient for H gas sensing applications than clean ZnO surface. Our results show also that the Cu-doped surface is more efficient for the H_2 sensing applications than the N-doped surface however the N-doped surface is better for the atomic hydrogen sensing applications. We have also examined the effect of doping with Cu and N on the hydrogen sensing properties of the ZnO nanowires. The obtained results show that the H gas sensing properties of the undoped ZnO nanowires are many times better than those of the ZnO$(10\bar{1}0)$ surface. We found also that, the doping process can greatly enhance the atomic hydrogen sensing properties of the ZnO nanowires.

Keywords: ZnO · Adsorption · First-principles · Hydrogen · Doping

1 Introduction

Zinc oxide (ZnO) is a wide-band-gap semiconductor with a band gap of 3.3 eV at room temperature and with many potential applications in optoelectronic and transparent devices (Özgür et al. 2005). Because of its gas sensing properties, this material has become a very attractive candidate for gas sensor construction. Today, the ZnO gas sensors are believed to be the most promising due to their low cost, low power consumption, high response, and rapid recovery (Gu et al. 2012). In the last years, many theoretical and experimental works have focused on the high performance of ZnO based gas

B. Abdelbaki et al. (Eds.): SMSD 2017, *Proceedings of the Third International Symposium on Materials and Sustainable Development*, pp. 644–656, 2018.
https://doi.org/10.1007/978-3-319-89707-3_68

sensors employed in detecting a wide range of gases (Bai et al. 2013; Wongchoosuk et al. 2009; Wright et al. 2010).

It is very known that the most stable structure of ZnO is the wurtzite structure. From structural point of view, the wurtzite structure has different types of surfaces which can be classified into two main categories; polar and non-polar surfaces (Wöll 2007). The non-polar surfaces can be predominant in ZnO nanostructures such as; nanowires and nanotubes, and they are widely used in surface functionalization applications. Among the ZnO low index surfaces, the ZnO$(10\bar{1}0)$ surface has attracted much attention and has been extensively studied (Meyer and Marx. 2003; D'Amico et al. 2012).

Generally, the use of an undoped semiconductor material in its crude state of growth does not lead to high performances. For this reason, the doping step is necessary to achieve the desired properties.

The doping of a semiconductor surface can significantly change its electronic and optical properties. It can also change its response to adsorbed gases or atoms. However, the performances of the doped surface, especially in gas sensing applications, depend on the choice of the dopant. In the case of ZnO, several studies have showed that the doping of the surface by suitable dopant atoms can significantly improve the performance of gas sensors based on ZnO (Han et al. 2011).

In this paper, we use density-functional theory (DFT) to investigate the effect of doping with acceptor atoms like Cu and N on the hydrogen sensing properties of the ZnO$(10\bar{1}0)$ surface as well as on those of the ZnO nanowires.

2 Computational Details

Our calculations are performed by using the first-principles pseudopotential method based on density functional theory (DFT) as implemented in the Quantum Espresso package (Giannozzi et al. 2009). The exchange-correlation interaction was treated by using the GGA-PBE exchange-correlation functional (Perdew et al. 1996). Ultrasoft pseudopotentials are used to deal with the interaction between the ion cores and valence electrons, and the cutoff energy for the plane wave basis was set to 40 Ry. The BFGS method was used to do structural relaxation and to find the most stable geometries.

In this work, the ZnO$(10\bar{1}0)$ surface was modeled by using a slab model with a (2×2) surface unit cell consisting of 8 layers and a vacuum layer of 14 Å separating the slabs (Fig. 1). The top three Zn–O bilayers are allowed to relax while the atoms of the bottom ZnO bilayer are fixed at their bulk positions to mimic the bulk substrate.

To study the interaction of hydrogen with ZnO nanowires, a triangular nanowire was used. The choice of this geometry is related to the limitation of computational resources only. Each nanowire is surrounded by a vacuum layer of at least 12 Å separating the NWs.

The $(4 \times 4 \times 1)$ k point's grid was used for the Brillouin zone integration in the case of the ZnO surface, while a grid of $(1 \times 1 \times 8)$ k points was used to study the properties of ZnO nanowires. The Cu doping was realized by replacing one of the Zn atoms with a copper atom, while the N doping was done by replacing one oxygen atom with a nitrogen atom.

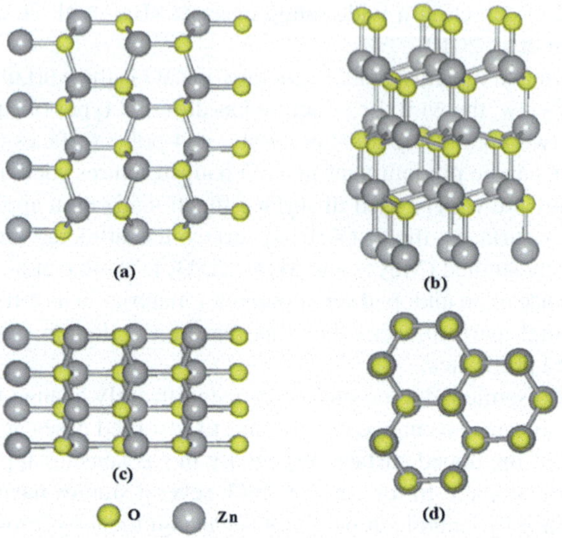

Fig. 1. (a)+(b) Side view of the ZnO$(10\bar{1}0)$ surface and the triangular ZnO NW structures used in this study, (c)+(d) Top view of the same structures.

The formation energy of the X dopant (X = Cu or N) in a given structure (surface or NW) was calculated according to:

$$E_f = E_{\mathrm{ZnO:X}} - E_{\mathrm{ZnO}} - \mu_{\mathrm{X}} + \mu_{\mathrm{V}} \tag{1}$$

where $E_{\mathrm{ZnO:X}}$ is the total energy of the X-doped structure. E_{ZnO} is the total energy of the clean structure. μ_{V} is the chemical potential of the removed atom (Zn or O) and μ_{X} is the chemical potential of the dopant atom.

The hydrogen adsorption energy was calculated according to;

$$E_A = E_{slab+nH} - E_{slab} - n_H\mu_H \tag{2}$$

Where E_{slab} and $E_{slab+nH}$ are the total energies of the slab without and with n hydrogen atoms adsorbed, respectively. μ_H is the chemical potential of the hydrogen atom.

The chemical potentials of Zn and O atoms (μ_{Zn} and μ_O) are not independent, but related one to the other by the bulk thermodynamic equilibrium condition:

$$\Delta\mu_{Zn} + \Delta\mu_O = \Delta H^f_{ZnO} \tag{3}$$

Which gives the allowed range for Zn and O chemical potentials to be:

$$\frac{1}{2}E(O_2) + \Delta H^f_{ZnO} \leq \mu_O \leq \frac{1}{2}E(O_2) \tag{4}$$

$$\frac{1}{2}E_{Zn}(hcp) + \Delta H^f_{ZnO} \leq \mu_{Zn} \leq \frac{1}{2}E_{Zn}(hcp) \tag{5}$$

Where $E_{Zn}(hcp)$ is the total energy of the Zn unit cell and $E(O_2)$ is the total energy of the O_2 molecule.

The DFT-PBE formation enthalpies (per atom or unit formula) for ZnO, Zn(hcp), Cu(FCC), N_2 and O_2 are −3.43 eV, −1.13, −2.65 eV, −16.56 eV and −8.65 eV versus experimental values of −3.60 eV, −1.30 eV, −2.95 eV, −9.91 eV and −5.12 eV, respectively. We note here that the spin polarization calculations give a binding energy of −4.98 eV for the oxygen molecule.

3 Results and Discussions

3.1 The Adsorption of Hydrogen on the Clean, Cu-, and N-Doped ZnO$(10\bar{1}0)$ Surfaces

First of all, we will study the effect of doping with acceptor atoms like Cu and N on the properties of the ZnO$(10\bar{1}0)$ surface. The formation energy is a key quantity that can give us useful information's about the preferred site, the solubility, and the stability of a given defect in the host of a semiconductor.

To find the most stable configuration of the doped ZnO$(10\bar{1}0)$ surface, we have calculated the formation energy of each dopant atom in all possible sites as shown in Fig. 2 and the obtained results are listed in Table 1. As seen, these two atoms have different behaviors. While the Cu atom prefers to be in the top layer of the ZnO$(10\bar{1}0)$ surface, the nitrogen atom prefers to be in the second layer. Moreover, the obtained results show that the Cu_{Zn} defect has the lowest formation energy under O-rich conditions, while under Zn-rich conditions the N_O defect has the lowest formation energy. This means that doping with Cu atoms is very easy under O-rich conditions and can lead to a high solubility of copper in ZnO. On the other hand, the formation energy value of Cu_{Zn} under O-rich conditions is very similar to that obtained by Gallino et al. for the neutral Cu_{Zn} in bulk ZnO using the functional B3LYP (Gallino and Di Valentin 2011).

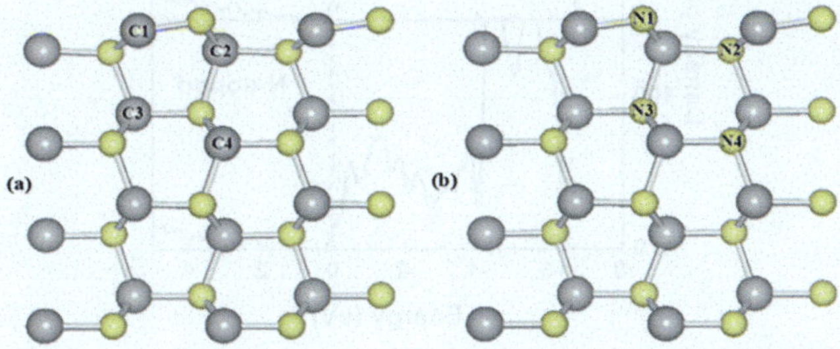

Fig. 2. The ZnO$(10\bar{1}0)$ surface with a dopant atom in different configurations; (a) the Cu-doped surface, (b) the N-doped surface.

Table 1. The Calculated Formation Energy E_f (eV) for Cu_{Zn} and Cu_O atoms for the doping sites shown in Fig. 2.

Surface	Doping site	O-rich	Zn-rich
N-doped	N1	5.31	1.88
	N2	4.80	1.37
	N3	5.09	1.66
	N4	5.06	1.63
Cu-doped	C1	−1.30	2.13
	C2	−0.87	2.55
	C3	−0.87	2.55
	C4	−0.81	2.62

Now we will discuss the effect of doping with acceptor atoms (Cu and N) on the electronic properties of ZnO$\left(10\bar{1}0\right)$ surface. Figure 3 show the total density of states (DOS) as obtained in the case of the clean, Cu-doped, and N-doped surfaces. The zero value in the energy scale indicates the position of the Fermi level.

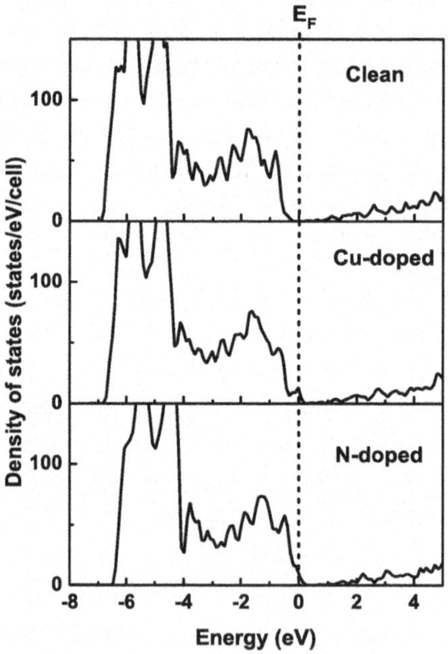

Fig. 3. The calculated density of states (DOS) of the clean, the Cu-doped, and the N-doped ZnO $\left(10\bar{1}0\right)$ surface, respectively.

In the case of the clean surface, there is no states in the band gap and the Fermi level is located in the top the valence band maximum (VBM), which means that the clean surface has a semiconductor behavior. The surface doping with Cu or N atoms will lead to the formation of acceptor states above the valence band maximum (VBM). The Fermi level which was initially located in the forbidden gap of the clean surface is now shifted downwards after doping with acceptor atoms (Cu and N), in good agreement with the previous studies (Xia et al. 2014; Li et al. 2010). These results mean that surface doping with Cu or N will lead to the formation of a surface with a p-type conductivity. Moreover, it appear that the N-related states are shallower than the Cu-related states, which means that the p-type conductivity of the N-doped surface is better than that of the Cu-doped surface.

On the other hand, our results show that the doping process can greatly affect the work function of the ZnO$(10\bar{1}0)$ surface. The calculated work function values in the case of the clean, the Cu-doped, and the N-doped surface are 3.88 eV, 4.39 eV and 4.46 eV, respectively. This means that the doping of the surface with copper or nitrogen increases the work function by at least 0.5 eV.

Now we will investigate the effect of doping with acceptor atoms (Cu and N) on the response of the ZnO$(10\bar{1}0)$ surface to hydrogen adsorption. To do this, we will try to explore the changes in the adsorption energies of atomic and molecular hydrogen as induced by the presence of acceptor dopants. For more accuracy, we have calculated the adsorption energies of atomic and molecular hydrogen on all possible adsorption sites (Fig. 4) and the obtained results are showed in Table 2. These results show clearly that the adsorption energies of atomic and molecular hydrogen on the ZnO$(10\bar{1}0)$ surface are greatly affected by the doping process. We find that the adsorption energy of the hydrogen molecule which it is around −0.04 eV/molecule in the case of the clean surface becomes two times lower after doping with copper or nitrogen atoms and this is true for all possible adsorption sites. This means that the surface doping with acceptor atoms will enhance the hydrogen molecules adsorption

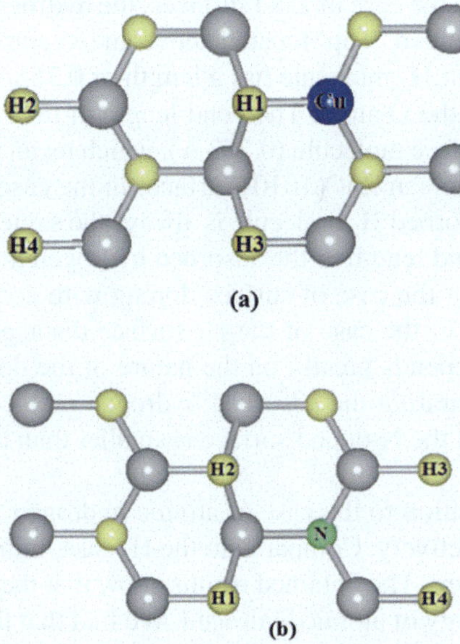

Fig. 4. Possible hydrogen adsorption sites on the doped ZnO$(10\bar{1}0)$ surface; (a) the Cu-doped surface, and (b) the N-doped surface.

on this surface, and therefore the H_2 sensing properties of the doped surface are better than those of the clean surface. We found also that the H_2 adsorption energy on the Cu-doped ZnO surface is lower than that of the N-doped surface, which mean that the H_2 sensing properties of the Cu-doped surface are better than those of the N-doped surface. These results appear to be in good agreement with the recent experimental results. Although there are no available results on the effect of N doping on the hydrogen gas sensing properties of the ZnO surface, a recent study of Chow et al. showed that the doping of ZnO nanostructures with Cu atoms enhances their responses to the H_2 adsorption (Chow et al. 2013).

Table 2. The H_2 adsorption energy E_a(eV), the H_2 bond length (Å), the O-H_2 distance (Å), The atomic H adsorption energy E_a(H) in eV, and the OH bond length (Å) for the adsorption configurations shown in Fig. 4.

Surface	Ads site	$E_a(H_2)$	r(H-H)	O-H_2	E_a(H)	r(O-H)
N-doped	H1	−0.088	0.75	2.50	−1.62	0.97
	H2	−0.087	0.75	2.50	−1.62	0.97
	H3	−0.085	0.75	2.44	−1.70	0.97
	H4	−0.088	0.75	2.52	−1.70	0.97
Cu-doped	H1	−0.096	0.75	2.58	−1.51	0.97
	H2	−0.092	0.75	2.52	−1.10	0.97
	H3	−0.090	0.75	2.53	−1.09	0.97
	H4	−0.094	0.75	2.58	−0.98	0.97
Clean	Any	−0.049	0.75	2.50	−0.27	0.97

It is very known that in the case of ZnO surfaces, the hydrogen molecule is adsorbed preferentially on the site called 'Top' located above an oxygen atom. In the case of the clean surface, the adsorbed H_2 molecule has a length of 0.755Å and it is adsorbed at a distance of 2.50 Å above the O atom. The bond length of the H_2 molecule in this case is very close to that of the free molecule (0.752 Å), which mean that hydrogen molecule interacts weakly with the clean ZnO$(10\bar{1}0)$ surface. In the case of the doped surfaces, the bond length of the adsorbed H_2 molecule is always the same and it is equal to 0.755 Å. This means that the bond length of the adsorbed hydrogen molecule is insensitive to the nature of the dopant in the case of surface doping with acceptor atoms. However, this statement is not true in the case of the H_2-surface distance. The obtained results show that this distance depends greatly on the nature of the dopant. As we can see in Table 2, the distance separating the adsorbed hydrogen molecule from the top of the ZnO surface in the case of the N-doped surface is smaller than that obtained in the case of the Cu-doped surface.

Now we turn our attention to the case of atomic hydrogen adsorbed on both clean and doped surfaces, respectively. Compared to the H_2 case, things are quite different in the case of atomic hydrogen. The obtained results show that the doping process greatly affects the adsorption energy of atomic hydrogen. We find that the adsorption of atomic hydrogen is more favored on the Cu-doped or the N-doped surfaces than onto the clean surface. The difference in the adsorption energy between undoped and doped surfaces is higher than 0.7 eV. These results allow us to suggest that the doping with acceptor atoms is an efficient mechanism to enhance the hydrogen sensing properties of the ZnO $(10\bar{1}0)$ surface.

On the other hand, our result showed that the adsorption energies of atomic H on the N-doped surface are lower than those of the Cu-doped surface, which means that the hydrogen sensing properties of the N-doped surface are better than those of the Cu-doped surface.

To make this study more comprehensive, we will try to estimate the change in the hydrogen gas sensing properties of the ZnO$(10\bar{1}0)$ surface induced by the doping process. To do this step, we will use a simple model in which we will calculate the variation of the number of the adsorbed hydrogen atoms and molecules onto the ZnO $(10\bar{1}0)$ surface as function of the dopant concentration and the temperature.

In this model, we will consider that the surface is doped in a homogeneous way with N_D acceptor atoms. We have showed previously that the presence of an acceptor atom will affect the adsorption energy of H (or H_2) not only on the nearest O atom but also on other oxygen atoms which are located around the dopant atom at a given distance. To take this effect into account, the modeled surface is divided in two regions. The first region is constituted of atoms located around the dopant atom at a distance (r) lower or equal to 5 Å, and it is characterized by an adsorption energy $\overline{E}_a$ (equal to the average value of the adsorption energy values in Table 2). The second region is the rest of the surface and it is characterized by an adsorption energy E_{a0} equal to that of the clean surface.

For a given temperature, the number of adsorbed hydrogen atom on the clean surface is given by:

$$N_H(T) = N_{sites} e^{-E_{a0}/kT} \tag{6}$$

N_{sites} is the number of the adsorption sites on the surface.

The number of adsorbed hydrogen atom on the X-doped surface is given by:

$$N_{HD}(T) = N_{sites} e^{-E_{av}/kT} \tag{7}$$

where E_{av} is the averaged value of the hydrogen adsorption energy on the doped surface and it is given by:

$$E_{av} = N_D \overline{E}_a \left(\pi r^2\right) + \left(1 - \pi r^2 N_D\right) E_{a0} \tag{8}$$

where r is the rayon of the first region.

The enhancement in the number of adsorbed hydrogen atoms (molecules) is given in this case by:

$$\eta(T) = \frac{N_{HD}(T) - N_H(T)}{N_H(T)} \tag{9}$$

Figure 5 shows the variation of η as function of the temperature (T) and the dopant concentration (N_S) for both Cu-doped and N-doped surfaces, respectively.

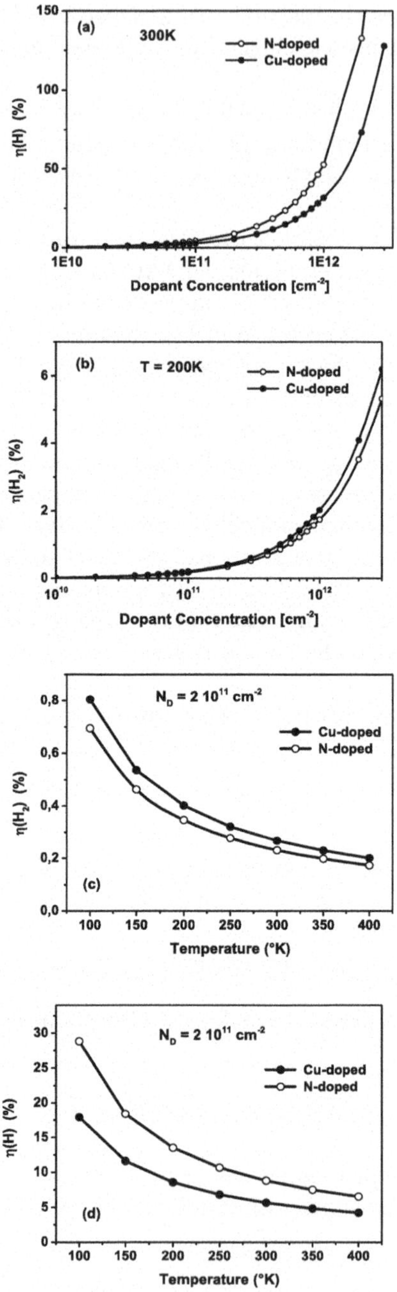

Fig. 5. The enhancement in hydrogen gas sensing η vs. dopant concentration (a+b) and temperature (c+d) for the Cu-doped and the N-doped $ZnO(10\bar{1}0)$ surfaces.

Let us start by discussing the effect of the dopant concentration on the calculated values of $\eta(H)$ and $\eta(H_2)$. The obtained results show that, for both Cu-doped and N-doped surfaces, η is always greater than zero. However, it appears that η depends greatly on the concentration of the dopant. We can see that for a low dopant concentration ($N_s < 2\ 10^{12}\ cm^{-2}$), there was no significant difference between the response of Cu-doped and N-doped surfaces to adsorption of atomic and molecular hydrogen. However, for a dopant concentration value of $10^{12}\ cm^{-2}$, the enhancement of the hydrogen sensing properties of both doped surfaces at room temperature is higher than 25%. This means that the surface doping with acceptor atoms can be regarded as a good solution to fabricate room-temperature H gas sensors. Nevertheless, we should note here that, in practice, the performances of the gas sensors can be affected by many factors such as the presence of point defects which has been ignored in this work.

3.2 The Adsorption of Hydrogen on the Undoped, Cu-Doped, and N-Doped ZnO Nanowires

Now, we turn our attention to the effect of doping with acceptor atoms like copper and nitrogen on the response of the ZnO nanowires to the hydrogen adsorption. First of all, we will discuss the formation energies of the Cu_{Zn} and N_O dopants in their most stable sites. In similarity with the ZnO$(10\bar{1}0)$ surface case, we find that the most stable site for the Cu_{Zn} dopant is a 3-fold site located in the external side of the nanowire while the most stable site of N_O is a 4-fold site located just below the external layer, in good agreement with previous studies (Haffad et al. 2012). Our calculations give formation energy values of −1.89 eV and 4.32 eV for Cu_{Zn} and N_O under the O-rich conditions. Under Zn-rich conditions, the calculated values are 1.54 eV and 0.89 eV. It is very clear that the formation energies of both Cu and N atoms in the ZnO nanowire are smaller than those of the ZnO$(10\bar{1}0)$ surface, which means that the solubility of these atoms in the ZnO NW can be higher than their solubility in the ZnO surface.

Figure 6 shows the total density of states as calculated for clean, Cu-doped, and N-doped ZnO nanowires. In the case of the clean nanowire, the corresponding electronic structure show the presence of a direct band gap of 1.63 eV located at the Γ point. This means that the perfect NW has a semiconductor behavior. After doping with copper atoms, new levels appear above the valence band maximum (VBM).

These levels are located at $E_v + 0.3$ eV and $E_v + 0.83$ eV, respectively. Our analysis shows that the first one is a fully occupied level while the second is a half-occupied level. On the other hand, the obtained results in the case of the N-doped ZnO nanowire show the presence of an empty level located at $E_v + 0.15$ eV. The Fermi level in this case is shifted downwards in the valence band and the N-doped NW can show a p-type conductivity.

To study the effect of doping with acceptors like Cu and N on the H sensing properties of the triangular ZnO NW, we have calculated the adsorption energies of atomic and molecular hydrogen on the nearest oxygen atoms to the dopant atom. Table 3 shows the calculated adsorption energies for both clean and doped nanowires, respectively.

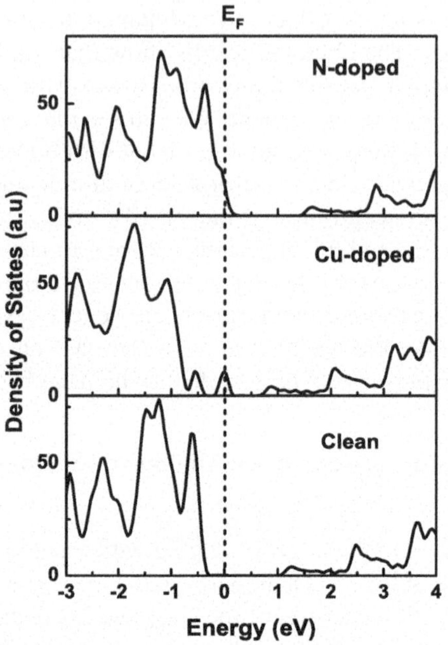

Fig. 6. The calculated density of states (DOS) near the valence band maximum of the clean, the cu-doped, and the N-doped ZnO nanowires, respectively. The Fermi level position is indicated by the dashed line.

Table 3. The H_2 adsorption energy E_a(eV), the H_2 bond length (Å), the O-H_2 distance (Å), The atomic H adsorption energy E_a(H) in eV, and the OH bond length (Å) for adsorbed hydrogen on clean and doped ZnO NW's.

ZnO NW	$E_a(H_2)$	r(H-H)	r(O-H_2)	E_a(H)	r(O-H)
Undoped	−0.615	0.75	2.44	−0.563	0.98
Cu-doped	−0.085	0.75	2.53	−0.893	0.95
N-doped	−0.078	0.75	2.50	−1.685	0.97

Compared to the ZnO$(10\bar{1}0)$ surface case, the behavior of the adsorbed hydrogen on the ZnO nanowires is very different. A big difference in the obtained H adsorption energies exist between the two cases. Let us discuss firstly the case of the undoped ZnO nanowire. Surprising, the obtained adsorption energy for the H_2 molecule in this case is not only smaller than its value in the case of the ZnO$(10\bar{1}0)$ surface, but it is also smaller than that of atomic hydrogen adsorbed on the same nanowire. This means that the adsorption of molecular hydrogen on the ZnO nanowire is more favored than the adsorption of atomic H. On the other hand, our results show that the adsorption energy values of both atomic and molecular hydrogen onto the ZnO NW's are many times smaller than those of the ZnO$(10\bar{1}0)$ surface case. This means that the hydrogen sensing performance of ZnO nanowires are better than those of the ZnO surface. This result appears to be in good agreement with the recent work of Tonezzer et al. (2015). In their work, Tonezzer et al. have showed that 1D ZnO nanowires have a gas-sensing performance better than that of 2D ZnO nanostructures (Tonezzer et al. 2015). All these results allow us to suggest that the ZnO nanowires are many times better than the ZnO surfaces for H gas sensing applications.

After doping with acceptor atoms, unexpected behavior has been observed. The H_2 adsorption energies are increased, thing completely different from the case of the ZnO $(10\bar{1}0)$ surface discussed previously. On the other hand, our results show that doping with acceptor atoms will enhance the stability of atomic hydrogen on the ZnO nanowires, and therefore, the atomic H sensing properties of these nanowires will be improved.

4 Conclusions

In summary, the effect of doping with acceptor atoms on the electronic structure and hydrogen gas sensing properties of the ZnO$(10\bar{1}0)$ surface as well as those of the ZnO nanowires was investigated using the first principles method. The obtained results show that the doping process affects greatly the properties of the surface as well as its response to adsorbed hydrogen. We find that a high solubility of Cu can be obtained if the doping process is done under O-rich conditions. On the other hand, our results show that the doping of the ZnO$(10\bar{1}0)$ surface with acceptor atoms like Cu and N can greatly enhances the adsorption of both molecular and atomic hydrogen on this surface. However, The H_2 adsorption energy value on the Cu-doped surface is lower than its value on the N-doped surface and this make the Cu-doped surface more efficient for H_2 sensing applications than the N-doped surface. In the case of atomic hydrogen, the situation is quite different. Our results show that the N-doped surface is better than the Cu-doped surface.

We have also examined the effect of doping with copper and nitrogen on the hydrogen sensing properties of triangular ZnO nanowires. The obtained results show that the adsorption energy values of atomic and molecular hydrogen on ZnO NW are many times smaller than those of the ZnO$(10\bar{1}0)$ surface, which means that the H sensing properties of the ZnO NW are better than those of the surface. On the other hand, our results show that the H_2 adsorption energy on the doped ZnO NW is higher than that calculated in the case of the clean nanowire, which is completely different from the case of the ZnO$(10\bar{1}0)$ surface. We find also that the doping process can enhance greatly the atomic hydrogen sensing properties of the ZnO nanowires.

References

Özgür, Ü., Alivov, Y.I., Liu, C., Teke, A., Reshchikov, M.A., Doğan, S., Avrutin, V., Cho, S.-J., Morkoç, H.: A comprehensive review of ZnO materials and devices. J. Appl. Phys. **98**, 041301 (2005)

Gu, H., Wang, Z., Hu, Y.: Hydrogen gas sensors based on semiconductor oxide nanostructures. Sensors **12**, 5517–5550 (2012)

Bai, S., Guo, T., Zhao, Y., Luo, R., Li, D., Chen, A., Liu, C.C.: Mechanism enhancing gas sensing and first-principle calculations of Al-doped ZnO nanostructures. J. Mater. Chem. A **1**, 11335–11342 (2013)

Wongchoosuk, C., Choopun, S., Tuantranont, A., Kerdcharoen, T.: Au-doped zinc oxide nanostructure sensors for detection and discrimination of volatile organic compounds. Mater. Res. Innov **13**, 185–188 (2009)

Wright, J.S., Lim, W., Norton, D.P., Pearton, S.J., Ren, F., Johnson, J.L., Ural, A.: Nitride and oxide semiconductor nanostructured hydrogen gas sensors. Semicond. Sci. Technol. **25**, 024002 (2010)

Wöll, C.: The chemistry and physics of zinc oxide surfaces. Prog. Surf. Sci. **82**, 55–120 (2007)

Meyer, B., Marx, D.: Density-functional study of the structure and stability of ZnO surfaces. Phys. Rev. B **67**, 035403 (2003)

D'Amico, N.R., Cantele, G., Ninno, D.: First-principles calculations of clean and defected ZnO surfaces. J. Phys. Chem. C **116**, 21391–21400 (2012)

Han, N., Liu, H., Wu, X., Li, D., Chai, L., Chen, Y.: Pure and Sn-, Ga- and Mn-doped ZnO gas sensors working at different temperatures for formaldehyde, humidity, NH_3, toluene and CO. Appl. Phys. A **104**, 627–633 (2011)

Giannozzi, P., et al.: J. Phys. Condens. Matter **21**, 395502 (2009)

Perdew, J.P., Burke, K., Ernzerhof, M.: Generalized gradient approximation made simple. Phys. Rev. Lett. **77**, 3865–3868 (1996)

Gallino, F., Di Valentin, C.: Copper impurities in bulk ZnO: a hybrid density functional study. J. Chem. Phys. **134**, 144506 (2011)

Xia, C., Wang, F., Hu, C.: Theoretical and experimental studies on electronic structure and optical properties of Cu-doped ZnO. J. Alloy Comp. **589**, 604–608 (2014)

Li, P., Deng, S.H, Zhang, L., Li, Y.B., Zhang, X.Y., Xu, J.R.: First-principle studies on the conductive behaviors of Ga, N single-doped and Ga–N codoped ZnO. Compt. Mater. Sci. **50**, 153–157 (2010)

Chow, L., Lupan, O., Chai, G., Khallaf, H., Ono, L.K., Cuenya, B.R., Tiginyanu, I.M., Ursakif, V.V., Sonteac, V., Schulte, A.: Synthesis and characterization of Cu-doped ZnO one-dimensional structures for miniaturized sensor applications with faster response. Sens. Actuators A **189**, 399–408 (2013)

Haffad, S., Samah, M., Cicero, G.: Effect of nitrogen impurities on the physical properties of ZnO nanowires: first-principles study. Phys. Rev. B **85**, 165207 (2012)

Tonezzer, M., Le Dang, T.T., Bazzanella, N., Nguyen, V.H., Iannotta, S.: Comparative gas-sensing performance of 1D and 2D ZnO nanostructures. Sens. Actuators, B **220**, 1152–1160 (2015)

Macromolecular Complex Architectures: Synthesis and Characterization

N. Cherifi[1(✉)], A. Benaboura[2], and L. Billon[3]

[1] Centre de Recherche Scientifique et technique en Analyses physicochimique, BP 384, 42004 Bou-Ismail, Tipaza , Algeria
ncherifiouradi@yahoo.fr

[2] Faculté de Chimie, Laboratoire de Synthèse Macromoléculaire et Thio-organique Macromoléculaire, Université des Sciences et de la technologie Houari Boumediene, B.P. 32, 16111 El-Alia, Algiers, Algeria

[3] Université de Pau et des Pays de l'Adour, CNRS, IPREM-Equipe de Physique et Chimie des Polymères, UMR 5254 Hélioparc, 2 avenue du président Angot, 64053 Pau, France

Abstract. The use of polymer materials becomes very important in our everyday life and their applications include several industries. The majority of such materials were synthesized for a long time via free radical polymerization FRP process. However, the obtained architectures were generally not controlled. In the mid 1990s, new controlled polymerization mechanisms were developed in the objective to synthesis a new controlled polymers able to satisfy our various needs in different sectors.

In this contribution, we are particularly interested in the synthesis and characterization of two controlled macromolecular families: gradient and block copolymers. As it will be presented, the nitroxyde mediated polymerization NMP was successfully applied in the synthesis of those complex architectures, using Styrene, Butyl acrylate and methyl methacrylate as monomers at different temperatures. Many experimental techniques were preferentially chosen to confirm the obtaining of required architectures. As example, let us quote SEC, H-RMN, RMN-HMBC and DSC techniques.

Keywords: Controlled radical polymerization CRP · NMP · Gradient polymers Block polymers · Dispersity

1 Introduction

It is well known in macromolecular field that although the broad application of Free Radical Polymerization FRP in the synthesis of polymer materials, it often induces the formation of matrices with non controlled microstructures, and very broad mass distributions. Such results are generally explained by the development of spontaneous transfer reactions and irreversible bimolecular terminations susceptible to exist during FRP technique. In order to eliminate these spontaneous disadvantages, many other mechanisms were developed, in the mid 1990s, to define the so called controlled radical polymerization CRP. The three most successful techniques quoted in bibliography are: Atom

B. Abdelbaki et al. (Eds.): SMSD 2017, *Proceedings of the Third International Symposium on Materials and Sustainable Development*, pp. 657–664, 2018.
https://doi.org/10.1007/978-3-319-89707-3_69

Transfer Radical Polymerization (ATRP) (Braunecker 1993), (Coessens et al. 2001), Radical Addition Fragmentation and Transfer (RAFT) (Moad et al. 2005), (Moad et al. 2006) and Nitroxyde Mediated Polymerization (NMP) (Hawker et al. 2001), (Charleux and Nicolas 2007). These techniques are based on the development of dynamic equilibration between dormant and active species, with decrease of the instantaneous radical concentration. They permit the obtaining of complex polymeric architectures able to satisfy our various needs to polymer materials in different sectors: pharmacy, cosmetic, textile, aeronautics, coatings…

In this work, we were particularly interested by the synthesis of two controlled families, which were not synthetically possible prior the development of controlled radical polymerization, and which present various applications in nanotechnology, medical and pharmaceutical field. The first family is block copolymers defined by abrupt changement in monomer compositions of considered polymers (Singh et al. 2015). The second family is gradient copolymers characterized by a gradual change in monomer compositions along the chains (Pakula and Matyjaszewski 1996), (Cherifi et al. 2011). Depending on the nature of the monomers used, we can distinguish two types of gradient materials (Ziegler and Matyjaszewski 2001), (Shinoda et al. 2003), (Shinoda 2004): (a) spontaneous gradients copolymers obtained by batch process, with monomers exhibiting different reactivity ratios. (b) Forced gradient copolymers obtained by so called semi batch process, when monomers present close reactivity ratios.

In our synthesis, we have preferentially chosen to use the nitroxyde mediated polymerization NMP technique in order to synthesize poly (butyl acrylate-*b*-methyles methacrylate/styrene) copolymer, poly (butyl acrylate-*b*-styrene) and spontaneous and forced poly (butyl acrylate-grad-methyl methacrylate) copolymers by applying respectively batch and semi-batch processes as it will be explained in experimental part.

2 Experimental Part

2.1 Materials

n-Butyl acrylate (BA, Aldrich, 99%), methyl methacrylate (MMA, Aldrich, 99%), (Sty, Aldrich, 99%) and 2, 2-azobis (amidinopropane) dihydrochloride (AIBA, Aldrich) were used as received. The alcoxyamine BlocBuilder® (2-methylaminoxypropionic-SG1) (MAMA) and the nitroxyde SG1 (88%) were provided by Arkema Chemicals. All solvents were used without further purification.

2.2 Copolymer Synthesis

Diblock copolymer Synthesis were realized in two steps. The first one concern the synthesis of the first block PBA, generally called macro initiator, by mixing BA monomer with AIBA initiator and SG1 (5 mol% vs initiator) at 120 °C.

In the second step, two diblock copolymers were synthesized: PBA-b-P (MMA-co-S) and PBA-b-PS.

Gradient copolymer synthesis, realized by both batch and semi batch processes.

1. Spontaneous gradient copolymer (Batch 50/50). In a round flask, we mixed BA with MMA in different compositions in presence of SG1and MAMA as initiator. The mixture was degassed for 30 min then placed into an oil bath thermostaté at 115 °C. After 6 h and 30 min, the reaction was stopped by cooling down the polymeric solution to room temperature.
2. Forced gradient copolymer, obtained via semi-batch process. Contrary to batch process, in these syntheses, the MMA monomer was degassed, and then transferred into a 25 ml Hamilton syringe. In the other hand, BA monomer was mixed with BlocBuilder MAMA in the reaction flask equipped with a septum. The mixture was purged under nitrogen flow for 30 min and the round bottom flask was immersed into an oil bath at 115 °C. The PBA was first polymerized alone for 60 min, and then the addition of methyl methacrylate, using a pump, was initiated at a constant addition rate equal to 2.2 mL h^{-1} for the semi-batch 2.2 and equal to 4.3 mL h^{-1} for the semi-batch 4.3.

For all the polymerizations, samples were withdrawn during the polymerization for further NMR, SEC and DSC analyses.

The final recovered copolymers were precipitated twice into a methanol/water (90/10) mixture for HMBC characterization.

2.3 Analytical Techniques

The crude polymer solutions were analyzed by size exclusion chromatography (SEC). Number-average molar mass (Mn) and dispersity (Mw/Mn) were calculated with a calibration derived from PS standards using OmniSEC®Software.

The proton nuclear magnetic resonance (1H NMR) spectra of the crude polymer solutions were recorded at 25 °C using a Bruker Advanced AM400 spectrometer (400 MHz), in CDC13 as solvent, in order to determine the monomer conversions and the copolymer compositions.

The copolymer microstructures were also analyzed by carbon nuclear magnetic resonance (13C NMR) and hetero-nuclear multiple bond correlation (HMBC) NMR sequences in CDC13.

Differential scanning calorimetry technique (DSC) was carried out on Q100 from TA instruments, to measure the copolymer glass transition temperature (Tg) and to prove the gradient profile of synthesized materials.

3 Results and Discussion

In the present study, we have been interested by the copolymerization of n-butyl acrylate BA with methyl methacrylate MMA and/or Styrene, in different experimental conditions, applying NMP technique.

First, the H-NMR and SEC techniques were applied to studying the variation of the logarithmic monomer concentration versus time and the evolution of Mn and the dispersity (Mw/Mn) as a function of conversion for all synthesized materials. It has been found that all realized copolymers exhibit features of controlled systems. As example, Fig. 1 exhibits obtained results with gradient copolymers.

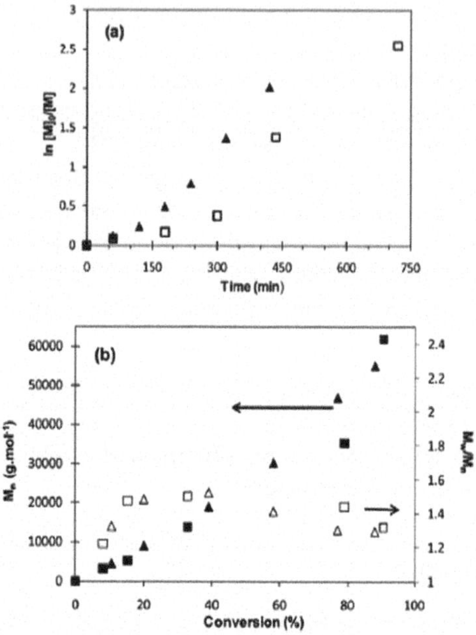

Fig. 1. Semi-batch copolymerization of BA and MMA. (a) Variation of ln [M]0/[M] versus time. (b) Evolution of molar mass (Mn) and dispersity (Mw/Mn) versus conversion. The squares correspond to the Semibatch-2.2 experiment and the triangles correspond to the Semibatch-4.3 experiment.

In the second step, the formation of targeted architectures was checked by different techniques. As first example, Fig. 2 exhibits H-RMN obtained results of Poly (BA-*b*-MMA/Sty), proving the successful formation of diblock copolymers.

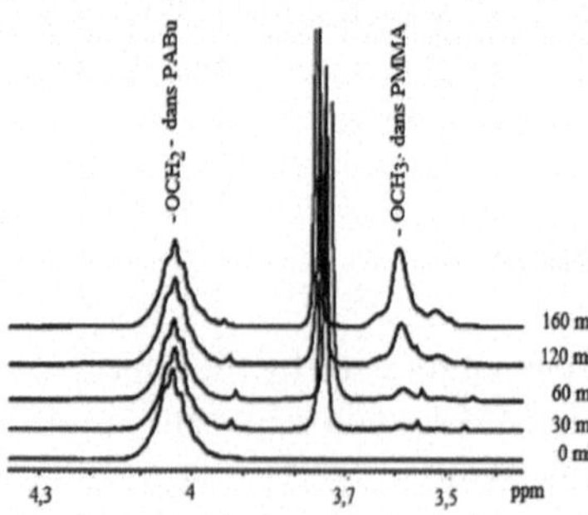

Fig. 2. H-RMN Spectra diblocs copolymers Poly (BA-*b*-MMA/Sty) at (4–3) ppm corresponding to O-C$\underline{H}_2$ of PBA and O-C$\underline{H}_3$ of PMMA.

The same observation was recorded with the second diblock P (BA-*b*-S) as exhibited in Fig. 3.

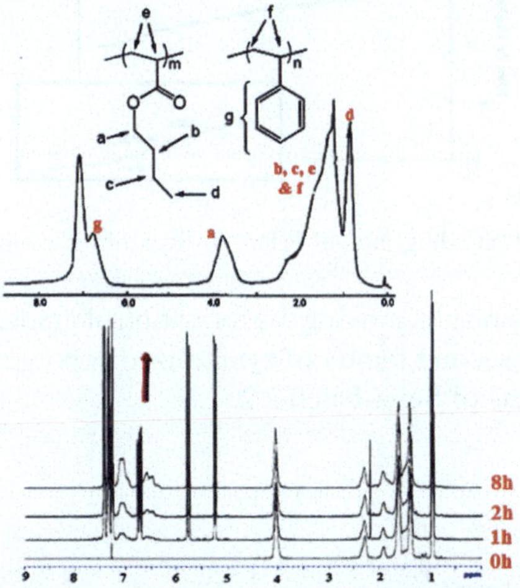

Fig. 3. H-RMN Spectra diblocs copolymers of Poly (BA-*b*-Sty)

Figure 4 shows SEC traces of these two synthesized diblocs.

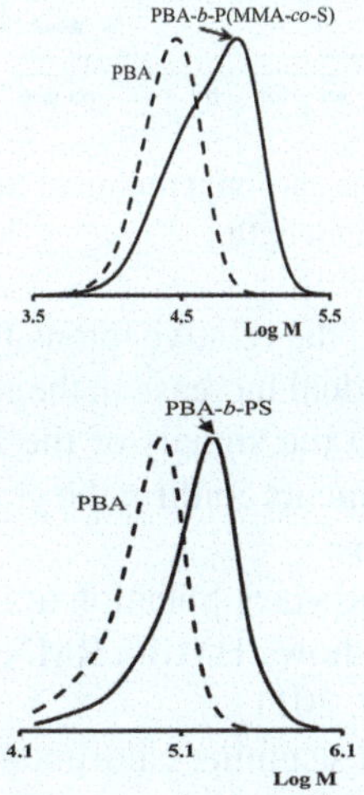

Fig. 4. SEC traces profiles for final diblock materials

The examination of DSC diagrams of all synthesized diblocs shows the presence of two transition temperatures corresponding to the homopolymers constituting each copolymer. Figure 5 exhibits the case of P(BA-*b*-Sty).

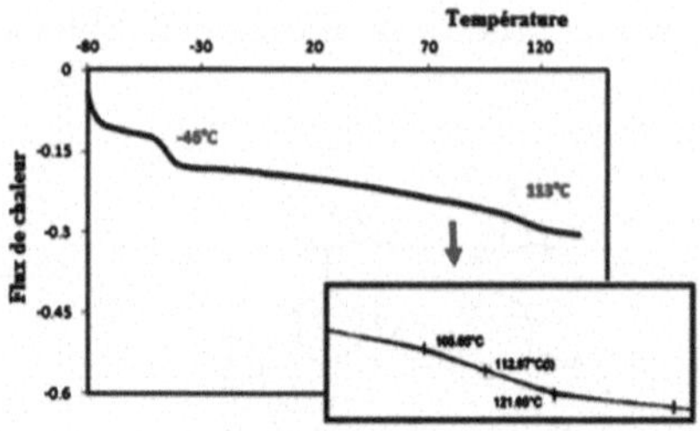

Fig. 5. DSC diagrams of P(BA -*b*-Sty) diblocs copolymers

The same satisfactory results proving the formation of gradient profile were obtained when characterizing the second family of synthesized copolymers. As example, Fig. 6 exhibits 13C-NMR results of Semi-batch 2,2.

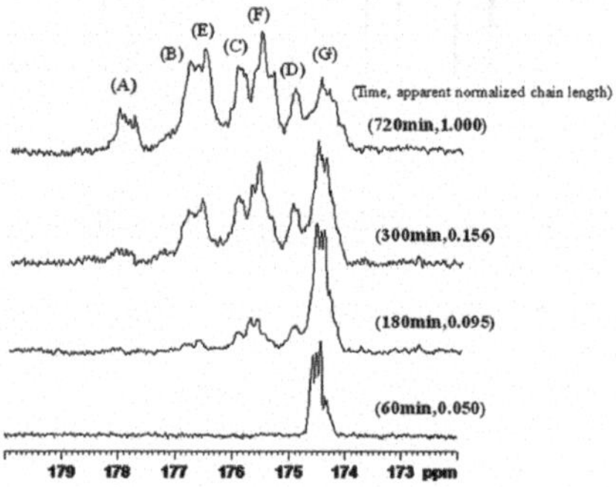

Fig. 6. 13C NMR spectra showing the carbonyl region of Semibatch-2.2 copolymers as function of time and apparent normalized chain length.

As it can be seen, a decrease in the relative intensity of BBB triad signal versus time was observed, together with a gradual increase in the relative intensities of (A), (B), (C) and (E) signals, corresponding to the signals of the MMA or the BA centered triads (Cherifi et al. 2011). Such changements confirm the gradual change in monomer compositions along the copolymer chains.

In the hand, the results obtained via application of HMBC RMN technique were also satisfactory. As example, Fig. 7 shows HMBC RMN corresponding to Semibatch-4.3 gradient copolymer. (Cherifi et al. 2011).

The application of differential scanning calorimetry technique reveals the presence of one broad Tg for batch gradient copolymer as illustrated on Fig. 8, whereas the two semi-batch gradient copolymers exhibit a microphase separation expressed by the presence of two glass transition temperatures. Such behaviors were alreay reported in bibliography of various gradient copolymers (Kim et al. 2006). (karaky et al. 2007), (Karaky et al. 2008).

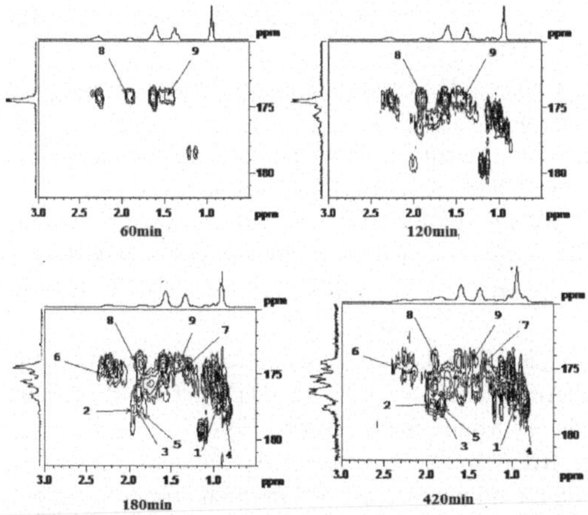

Fig. 7. 2D HMBC NMR spectra for Semibatch-4.3 as function of time

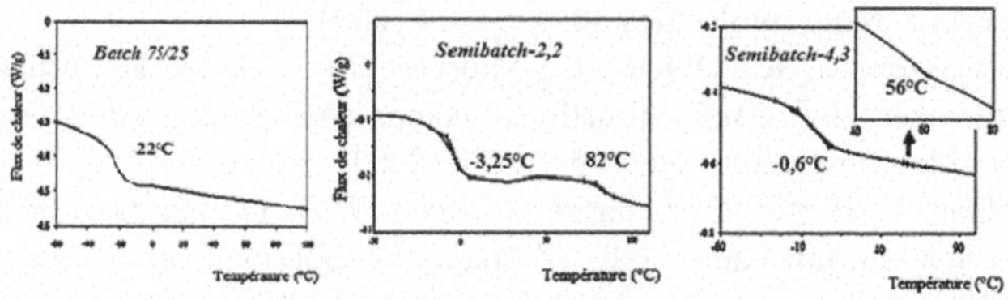

Fig. 8. Glass transitions temperature values of synthesized copolymers.

4 Conclusions

In this work, attention was given to the synthesis of poly (n-butyl acrylate-co-methyl methacrylate) copolymers by the nitroxide mediated polymerization (NMP) technique. Two macromolecular complex architetures were chosen, diblock and gradient copolymers.

First, the criteria of controlled radical polymerization were cheked using 1HRMN and SEC techniques. After, various techniques were also applied, 13C-RMN, RMN-HMBC, DSC and DSC/SEC techniques, in order to verify diblock and gradient profiles. It has been found that all made syntheses were successfully controlled, and the examination of the obtained results confirms the successful formation of required profiles.

References

Braunecker, W.A., Matyjaszewski, K.: Controlled living radical: features, developments and perspectives. Prog. Polym. Sci. **32**, 93–146 (2007)

Coessens, V., Pintauer, T., Matyjaszeski, K.: Functional polymers by atom radical polymerisation. Prog. Polym. Sci. **2001**(26), 337–377 (2001)

Moad, G., Rizzardo, E., Thang, S.H.: Living radical polymerization by the RAFT process. Aust. J. Chem. **58**, 379–410 (2005)
Moad, G., Rizzardo, E., Thang, S.H.: Living radical polymerization by the RAFT process. Aust. J. Chem. **2006**(59), 669–692 (2006)
Hawker, J.C., Bosman, A.W., Harth, E.: New polymer synthesis by nitroxide mediated living radical polymerizations. Chem. Rev. **101**, 3661–3688 (2001)
Charleux, B., Nicolas, J.: Water-soluble SG1-based alkoxyamines: a breakthrough in controlled/living free-radical polymerization in aqueous dispersed media. Polymer **48**, 5813–5833 (2007)
Singh, A.N., Rupesh, D., Thakre, J.C., More, P.K., Sharma, Y.K.A.: Polymer-Plastics Technology and Engineering, vol. 54, pp. 1077–1095 (2015)
Pakula, T., Matyjaszewski, K.: Copolymers with controlled distribution of comonomers along the chain, 1. Structure, thermodynamics and dynamic properties of gradient copolymers. Comput. Simul. Macromol. Theory Simul. **5**, 987–1006 (1996)
CherifI, N., Issoulie, A., Khoukh, A., Benaboura, A., Save, M., Derail, C., Billon, L.: Synthetic methodology effect on the microstructure and thermal properties of poly (n-butyl acrylate-co-methyl methacrylate) synthesized by nitroxide mediated polymerization. Polym. Chem. **2011**(2), 1769–1777 (2011)
Ziegler, M.J., Matyjaszewski, K.: Atom transfer radical copolymerization of methyl methacrylate and n-butyl acrylate. Macromolecules **34**, 415–424 (2001)
Shinoda, H., Matyjaszewski, K., Okrasa, L., Mierzwa, M., Pakula, T.: Structural control of poly(methyl methacrylate)-g-poly(dimethylsiloxane) copolymers using controlled radical polymerization. Macromolecules **36**, 4772–4778 (2003)
Mignard, E., Leblanc, T., Bertin, D., Guerret, O., Reed, W.F.: Online monitoring of controlled radical polymerization: nitroxide-mediated gradient copolymerization. Macromolecules **37**, 966–975 (2004)
Kim, J., Mok, M.M., Sandoval, R.W., Woo, D.J., Torkelson, J.M.: Uniquely broad glass transition temperatures of gradient copolymers relative to random and block copolymers containing repulsive comonomers. Macromolecules **39**, 6152–6160 (2006)
Karaky, K., Pouchan, C., Desbrieres, J., Billon, L.: Amphiphilic gradient copolymers shape composition influence on the surface/bulk properties. Macromolecules **40**, 458–464 (2007)
Karaky, K., Derail, C., Reiter, G., Billon, L.: Tuning the surface/bulk properties by the control of the amphiphilic profile in gradient copolymer. Macromol. Symp. **267**, 31 (2008)

Study of COD Adsorption on Deferent Activated Carbon Prepared from the Date Stones of the South of Algeria

Ouahiba Belaid[1(✉)], Ahmed A. Bebba[1], Zineb Barhoum[2], and Imane Raache[3]

[1] Laboratory of Valorization and Promotion of Sahara Resources (VPRS), University Kasdi Merbah Ouargla, BP 511, Ouargla, Algeria
hiba.belaid@yahoo.com

[2] Laboratory LAR-CHYDE, Chemical Engineering Department, University of Biskra, Biskra, Algeria

[3] Process Engineering Department, Faculty of Applied Sciences, University Kasdi Merbah Ouargla, Ouargla, Algeria

Abstract. For the aim of valorization of local lignocelluloses natural residue from different activated carbon with a high surface areas were prepared by chemical activation using the phosphoric acid H_3PO_4 from the date stones of two locals varieties (Ghars and Mech Degla) of the regions of Biskra and El Oued in the south of Algeria and characterized by different techniques, the textural properties determined by N2 adsorption at 77 k and scanning electronic microscopy SEM. The surface function was found by the Boehm method and the elemental analysis of activated carbons was carried out.

The adsorption of the organic pollutant (COD) of wastewater from the common Sidi Khouiled (Ouargla) by the activated carbons CGB1, CGE1, CMB1 has been studied. Under the process conditions there is a maximum absorption capacity of this pollutant 62.5 mg/g for CGB1. The experimental data were well described by the isothermal equilibrium equation Langmuir type II and the calculated adsorption capacity was 71.43 mg/g comparable by models Freundlich and Temkin. The kinetics of COD adsorption obeys the intra cellular model.

Keywords: Date stones · Activated carbon · Adsorption · COD Isotherm

1 Introduction

The adsorption processes using activated carbons are widely used to remove the various organic pollutants from urban wastewaters. However, commercially available activated carbon is becoming too expensive. In the last few years, special emphasis has been placed on the preparation of activated carbons from several agricultural by-products due to the growing interest in low-cost activated carbons from renewable sources, especially for application in wastewater treatment. Researchers have studied the production of activated carbon from cherry stones (Dilek 2014), waste tea

B. Abdelbaki et al. (Eds.): SMSD 2017, *Proceedings of the Third International Symposium on Materials and Sustainable Development*, pp. 665–675, 2018.
https://doi.org/10.1007/978-3-319-89707-3_70

(Auta and Hameed 2011), apricot stones (Djilani et al. 2015), olive stones (Bohli et al. 2015; Hazzaa and Hussein 2015), oil-palm stones (Hoseinzadeh Hesas et al. 2013), walnut shells (Jung et al. 2014), cattail (Shi et al. 2010), coconut shells (Isah et al. 2015), tobacco residues (Kilic et al. 2011), rice husks (Liu et al. 2012), date stone (Foo and Hameed 2012; Danish et al. 2014). Date stones are abundant by-products of date palm industry in the south of Algeria. In an attempt to valorize this well know material, we have chosen three varieties of date stones coming from two different regions: Biskra and El-Oued. The selected varieties are called Ghers Biskra (CGB1), Mech degla Biskra (CMB1) and Ghers El-Oued (CGE1),this activated carbon purpose of eliminating a quantity of organic pollutants COD in the urban wastewater of the region of sidi Khouiled wilayat Ouargla from south of Algeria.

2 Materials and Methods

2.1 Preparation and Characterization of Activated Carbon Prepared

The precursors were crushed and then sieved to reduce their particle size of 2 mm after washing with distilled water to eliminate the impurities (dust and water soluble substances. The granular date stones were impregnated with a weighed amount of H_3PO_4 aqueous solution and further heat treated at 600 °C during 3 h in air atmosphere.

Textural characterization of the ACs was carried out by N_2 adsorption at 77 K using Quantachrome Instruments, Model NOVAWIN2 by Brunauer–Emmett–Teller (BET) method, Surface morphology and the presence of porosity of the activated carbon prepared in this work were studied using Quanta 600 W environmental scanning electron microscope (ESEM) operated at accelerating volt-ages of 2 kV with magnification of 15,000X. Bohm titration method (Boehm 1994) was applied in order to determination of the quantification of acidic and basic surface oxide groups of the adsorbents. The neutralization realized with different solutions: 0.1M of HCl solution (for basic groups) and 0.1M of NaOH solution (for acidic groups). The elemental analysis was performed by the spectroscopy X-rays with photoemission (XPS).

All these procedure of preparation and characterization of activated carbons details were presented in reference (Belaid et al. 2017).

2.2 Batch Mode Treatment of Wastewater Samples

Sampling is carried out using an automatic sampler wastewater. The samples are taken at the entrance to the station after the screening and sandblasting. The sewage bottles are closed and transported to the laboratory at Temperature of 4 °C in a cooler. Sampling and conservation Samples are taken according to Rodier (Rodier et al. 2009).

All the experiments were carried out at ambient temperature (25 °C) in batch mode. Adsorption experiments were conducted in different batches for all the experimental conditions like adsorbent treatment time, adsorbent dose, PH of the solution, agitation speed. The influence of various operating parameters were studied by varying one parameter and keeping others constant. To determine the contribution of the adsorbent dose to COD reduction, 100 mL of sample was treated with different doses of adsorbent ranging between 0.10 and 1.0 g, the other conditions includes treatment time of 240 min, pH of 6.92 and the agitation speed is 200 rpm. The pH studies were performed with constant initial concentration, adsorbent dose (0.60 g) and contact time 4 h but with varying pH from 2 to 12 using dilute NaOH or HNO_3 solutions for the adjustment. Effect of contact time of the adsorbent with urban wastewater was investigated by agitating 100 mL sample, pH = 5, constant initial concentrations and adding 0.60 g activated carbon for different time periods varying between 5 and 240 min. the treated wastewater sample was filtered through 0.45 μm filter papers before analyses. The final concentration of the polluant organic COD was measured using UV Spectrophotometer (DR-2800, Germany) by the cuve test LANGE of two different range LCK 114 (15–150 mg/l) and LCK 314 (150–1000 mg/l). The amount of adsorption at equilibrium, qe and the removal percentage (R%) were calculated from Equations:

$$Q_e = \frac{(C_i - C_f) * m}{V} \tag{1}$$

$$R(\%) = \frac{(C_i - C_f)}{C_i} \tag{2}$$

Where, Qe is the uptake (mg adsorbate/g adsorbent), C_i is the initial COD concentration, C_f is the final COD (mg/L), m is the adsorbent amount (g) and V is the wastewater volume (1).

3 Results and Discussion

3.1 Textural Properties of the Activated Carbons

The SEM micrographs of all the activated carbons CMB1, CGB1 and CGE1 in Fig. 1. Show clearly well-developed pore structures.

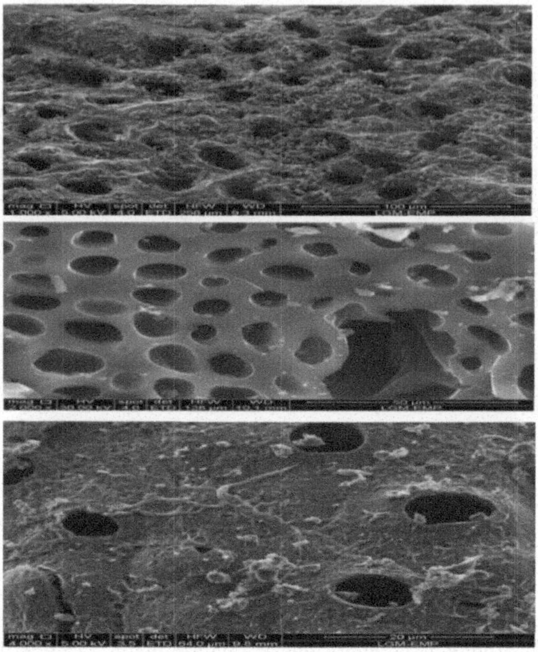

(a)
(b)
(c)

Fig. 1. Scanning electron micrographs for the three varieties of activated carbon: (a) CGE1, (b) CMB1 and (c) CGB1.

Table 1. Textural and chemical properties for the three varieties of Activated carbon

Methods		Propretés	CGB1	CGE1	CMB1
Porous structure	BET	S_{BET} (m^2/g)	1,118	950	794
	DFT	Vp (cm^3/g)	0.50	0.43	0.36
	Single point	V_T (cm^3/g)	0.56	0.48	0.40
	t-plot	S_{ext} (m^2/g)	21	13	26
		S_{micro} (m^2/g)	1,097	937	767
		V_{mico} (cm^3/g)	0.52	0.46	0.37
	V_T–V_{micro}	$V_{méso}$ (cm^3/g)	0.04	0.02	0.03
		S_{mico}/S_{BET} (%)	98	99	97
		$Sext/S_{BET}$ (%)	2	1	3
		V_{micro}/V_T (%)	93	96	92
	4Vp/SBET	D_{moy} (nm)	0.89	0.9	0.91
	D–R plot	V_{micro} (cm^3/g)	0.43	0.37	0.31
		L (nm)	1.85	1.85	1.85
		E_0 (kJ/mol)	4.28	4.28	4.28
Elemental analyzed		C %	86.33	84.18	84.80
		O %	8.51	13.02	12.40
		N %	1.81	0.19	1.43
		P %	2.24	2.21	1.48
Functional groups	Boehm titration	$V_{carboxylic}$ (meq/g)	1.40	1.90	2.55
		$V_{lactone}$ (meq/g)	0.30	0.60	0.05
		$V_{phénolic}$ (meq/g)	0.45	0.13	0.20
		V_{acidic} (meq/g)	2.15	2.63	2.80
		V_{basic} (meq/g)	0.50	1.00	0.25

3.2 Effect of Activated Carbon Dose

Figure 2 gives the variation of percentage reduction for COD removal versus ACS dose that increased with increase in the quantity of ACs up to 0.60 g. The maximum percentage COD removed 83.88%, 79.74%, 70.8% for CGB1, CGE1 and CMB1 respectively, the reason for stabilization of the adsorbed amount is the reversibility of the interaction between the adsorbate and the adsorbent, this step expresses the equilibrium step between coal and organic pollutants (Table 1).

3.3 Effect of Solution PH

The result of this effect is as shown in Fig. 3. The maximum COD reduction obtained at pH = 5 corresponds 85.35% to the CGB1.

3.4 Effect of Contact Time

The evolution of the adsorption process at pH = 5 as a function of time shows Fig. 4 that the equilibrium is established after three (03) hours, from this contact time the quantity of the adsorbed COD oxygen demand remains constant, the high capacity of the maximum adsorption is 39.55 mg/g for the CGB1 coal corresponding to a maximum removal rate of 95.81% greater than the adsorption yield for the other active carbons (CGE1 and CMB1).

3.5 Effect of Agitation Speed

The results of the adsorption tests are presented in Fig. 5. that the adsorption yield increases with stirring speed up to 500 rpm, the yield is higher to 88% for the three carbons CGB1, CGE1 and CMB1, after this speed this yield of adsorption was decreased to 37%.

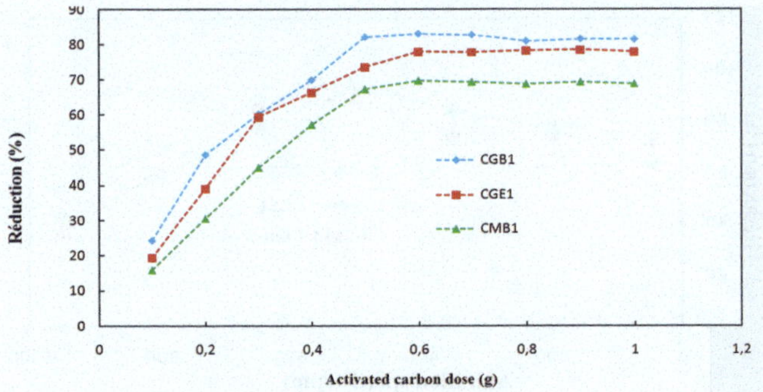

Fig. 2. Effect of adsorbent dosage on the adsorption of COD.

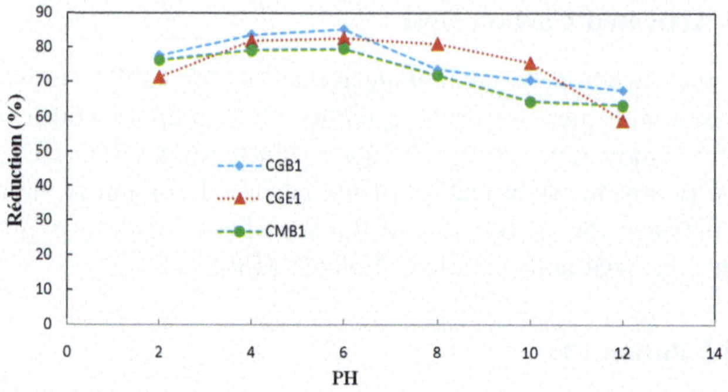

Fig. 3. Effect of PH on the adsorption of COD.

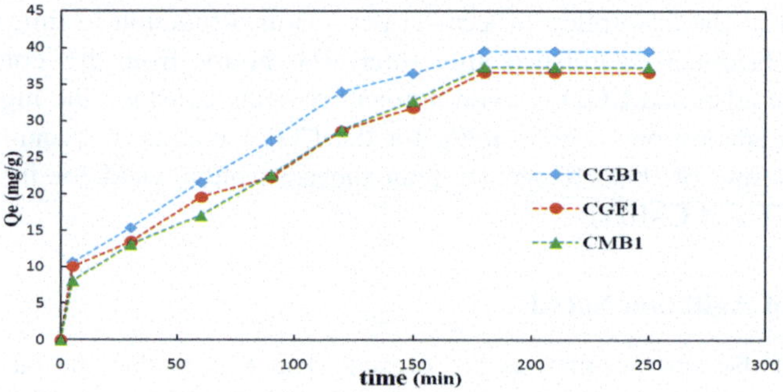

Fig. 4. Effect of contact time on the adsorption of COD

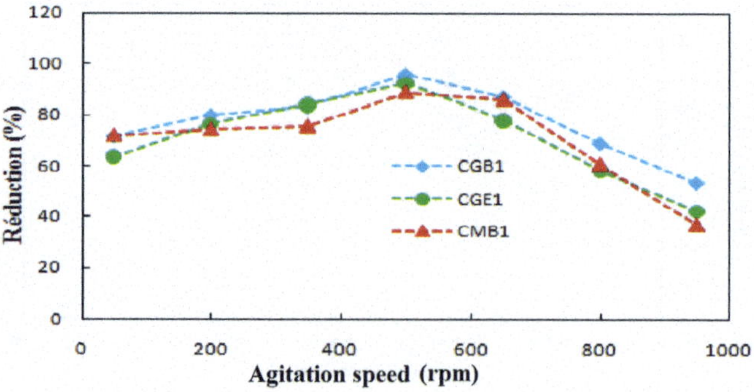

Fig. 5. Effect of agitation speed on the adsorption of COD

3.6 Modeling the Adsorption Isotherms

3.6.1 Langmuir Isotherm

The adsorption capacity data for the removal of the organic pollutant COD were calculated using the linear Langmuir isotherm equation (Langmuir 1916; Allali-Hassani et al. 1990), which is mathematically represented by the following equation:

$$Q_e = Q_{max} + \frac{K_L.C_e}{1 + K_L.C_e}$$

Were Ce = the equilibrium concentration; K_L = Langmuir constant related to adsorption capacity and energy of adsorption; Q_{max} the maximum adsorption capacities.

The saturation monolayer can be represented by plotting of Ce/qe versus Ce (type II). in a linear graphical relation indicating the applicability of the model for different adsorbent doses. The maximum adsorption capacities of COD were 71.43 mg/g, 62.5 mg/g, 50 mg/g for, CGB1 and CGE1 CMB1, respectively with R^2 greater than 0.99.

3.6.2 Freundlich Isotherm

The Freundlich isotherm (Freundlich 1907) is commonly used for a heterogeneous surface energy system. The well-known logarithmic form of Freundlich isotherm is given by the following equation:

$$\log q_e = \log K_F + \frac{1}{n} \log C_e$$

where Ce is the equilibrium concentration of the adsorbate, qe is the amount of adsorbate adsorbed per unit mass of adsorbent, K_F and n are Freundlich constants with n giving an indication of how favorable is the adsorption process and K_F is the adsorption capacity of the adsorbent. The Freundlich isotherm has been illustrated to be a special case of heterogeneous surface energies but it can be easily extended to this case. The slope of 1/n ranging between 0 and 1 is a measure of adsorption intensity or surface heterogeneity, where the surface becomes more heterogeneous as its value gets closer to zero (Fytianos *et al.* 2000). A value for 1/n below one indicates a normal Langmuir isotherm while 1/n above one is indicative of cooperative adsorption (Ahmad and Hameed 2009). The plot of log qe versus log Ce gave a straight line with slope of 1/n; whereas K_F was calculated from the intercept value. The values of 1/n for CGB1, CGE1 and CMB1 were 0.187, 0.176, 0.123 and the values of K_F were 24.63, 23.71 and 24.75, respectively with R^2 greater than 0.93.

3.6.3 Temkin Isotherm

The Temkin isotherm (Temkin *et al.* 1940) has been used in the form linear as follows:

$$q_e = B \ln K_T + B \ln C_e$$

Where K_T is the equilibrium binding constant corresponding to the maximum binding energy and constant B is related to the heat of adsorption. A plot of qe versus ln Ce enables the determination of the isotherm constants K_T and B. The values of constant B for CGB1, CGE1 and CMB1 respectively were 9.437 8.416 5.121 and the values of K_T were 5 5.83 51.35 with R_2 greater than 0.93.

The three isotherms models have been tested and the equilibrium data fits well to all adsorption isotherms. The Langmuir type II model was found to be the best fit with R^2 which were higher than 0.99 at the range of adsorbent doses considered.

3.7 Adsorption Kinetic Studies

3.7.1 Lagergren Pseudo-First Order Model

The Lagergren pseudo-first order model (Lagergren 1898) is most commonly used to describe the adsorption of solute from a liquid solution. The linearized form of this model is given by:

$$\ln(q_e - q_t) = \ln(q_e) - \frac{k.t}{2.303}$$

where k is the kinetics constant of pseudo-first order adsorption (min^{-1}), qe and qt (mg/g) are the amounts adsorbed at equilibrium and at time t (min), respectively.

3.7.2 Pseudo-Second Order Model

In this model, the rate limiting step is the surface adsorption that involves chemisorption, where the adsorbate removal from a solution is due to physicochemical interactions between the two phases (Ho and McKay 1999). The model is usually represented by its linearized form as follows:

$$\frac{1}{q_t} = \frac{1}{k.q_e^2} + \frac{1}{q_e}t$$

where, K^2 (g/mg/min) is the pseudo-second order rate constant of adsorption.

3.7.3 Elovich's Model

In recent years, Elovich's model has been successfully used to describe the adsorption of pollutants from aqueous solutions (Low 1960). The linearized form of this model is given by:

$$q_t = \frac{1}{\ln(\alpha + \beta)} + \frac{1}{\beta}\ln t$$

where, α is the initial adsorption rate (mg/g/min), and $1/\beta$ (mg/g) is a parameter related to the number of sites available for adsorption.

3.7.4 Intra Particle Diffusion Model

The intraparticle diffusion model (Weber and Morris 1963) describes adsorption processes, where the rate of adsorption depends on the speed at which adsorbate diffuses towards adsorbent (i.e., the process is diffusion controlled), which is presented by:

$$q_t = k_d.\sqrt{t} + C$$

where, kd is the rate constant of the intraparticle transport (mg/g/min$^{1/2}$) and q_t (mg/g) is a constant related to the thickness C of the boundary layer. High values of C indicate greater boundary layer effect.

The validity of the kinetic models is examined from the R^2 value of the coefficient of linear regression, reported in the Tables 2. For the model of pseudo first order and pseudo second order, Elovich, the correlation coefficient is always less than or equal to 0.96 and 0.95, indicating a poor correlation for the three activated carbons CGB1, CGE1 and CMB1. On the other hand, the application of the intra particle diffusion model leads to coefficients much better than or equal to 0, 971.

Table 2. Kinetic parameters for the adsorption of COD.

	CGB1	CGE1	CMB1
Pseudo first-order model	ln(Qe − Qt) = 3.66 − 0.015 t	ln(Qe − Qt) = 3.523 − 0.012 t	ln(Qe − Qt) = 3.608 − 0.012 t
R^2	0.965	0.962	0.955
K_1 (min^{-1})	0.035	0.028	0.028
Qe (mg/g)	38.90	33.89	36.90
Pseudo second-order model	t/Qt = 1.064 + 0.02 t	t/Qt = 0.973 + 0.024 t	t/Qt = 1.109 + 0.023 t
R^2	0.96	0.926	0.902
K_2 g.mg^{-1}.min^{-1}	3.76*10 − 4	5.92*10 − 4	4.77*10 − 4
Qe (mg/g)	50	41.67	43.48
Elovich's model	Qt = 10.85 lnt − 22.62	Qt = 9.676 lnt − 20.27	Qt = 10.03 lnt − 22.04
R^2	0.91	0.895	0.866
Kp	0.0923	0.103	0.1
Intraparticle diffusion model	Qt = 2.616√t + 2.381	Qt = 2.380√t + 1.699	Qt = 2.52√t + 0.331
R^2	0.971	0.975	0.977
K_d (mg/g/min$^{1/2}$)	2.616	2.380	2.52
C	2.381	1.699	0.331

4 Conclusions

The characterization physicochemical and textural of the activated carbons prepared from the date's stone of the variety Ghars and Mech degla of Biskra and El-Oued regions of southern Algeria by chemical activation was determined by different techniques, surface functions by the method of Boehm as well as elemental chemical analysis by XPS. The specific surface area, microporous volume and diameter of the pores by scanning electron microscopy (SEM) and by nitrogen adsorption N_2 at 77 K (BET method). These characterizations have shown that the activated carbons prepared have a microporosity well developed (96.69%) and large specific areas (569.509 to 1118.097 m^2/g) and volumes of microporous (from 0.256 to 0.521 cm^3/g), the variety Ghars of the locality of Biskra is the one that gives the best result. The adsorption capacity of the three coals prepared was tested by elimination of organic pollutant COD from urban wastewater in the Sidi Khouiled region. The effects of adsorbent dose of pH, contact time and agitation speed on COD have been studied, the maximum capacity of adsorption (62.5 mg/g) was obtained for an optimal conditions: contact time of 3 h, a mass of adsorbent (0.1 to 0.6 g), a pH(5), a stirring speed of 500 rpm when working at room temperature.

The equilibrium data fitted to Langmuir, Freundlich and Temkin isotherms. However, the equilibrium data were best described by Langmuir isotherm model. kinetic data were best fitted by the intra cellular diffusion model.

References

Dilek, A.: Production and characterization of activated carbon from sour cherry stones by zinc chloride. Fuel **115**, 804–811 (2014)

Ahmad, A.A., Hameed, B.H.: Reduction of COD and color of dyeing effluent from a cotton textile mill by adsorption onto bamboo-based activated carbon. J. Hazard. Mater. **172**, 1538–1543 (2009)

Allali-Hassani, M., Dusart, O., Mazet, M.: Elimination de micropolluants par adsorption sur charbon actif dans l'eau: cas d'un tensioactif non ionique; influence de co-adsorbats removal of non ionic surfactant from water onto activated carbon, influence of organic compounds. Water Res. **24**, 699–708 (1990)

Auta, M., Hameed, B.H.: Preparation of waste tea activated carbon using potassium acetate as an activating agent for adsorption of Acid Blue 25 dye. Chem. Eng. J. **171**, 502–509 (2011)

Belaid, O., Bebba, A.A., Sekirifa, M.L., Baameur, L., Al-Dujaili, A.H.: Preparation and characterization of chemically activated carbons from different varieties of date stones. Desalin. Water Treat. **65**, 267–273 (2017)

Boehm, H.: Some aspects of the surface chemistry of carbon blacks and other carbons. Carbon **32**, 759–769 (1994)

Bohli, T., Ouederni, A., Fiol, N., Villaescusa, I.: Evaluation of an activated carbon from olive stones used as an adsorbent for heavy metal removal from aqueous phases. C. R. Chim. **18**, 88–99 (2015)

Danish, M., Hashim, R., Ibrahim, M.N.M., Sulaiman, O.: Optimized preparation for large surface area activated carbon from date (Phoenix dactylifera L.) stone biomass. Biomass Bioenergy **61**, 167–178 (2014)

Djilani, C., Zaghdoudi, R., Djazi, F., Bouchekima, B., Lallam, A., Modarressi, A., Rogalski, M.: Adsorption of dyes on activated carbon prepared from apricot stones and commercial activated carbon. J. Taiwan Inst. Chem. Eng. **53**, 112–121 (2015)

Foo, K.Y., Hameed, B.H.: An overview of dye removal via activated carbon adsorption process. Desalin. Water Treat. **19**, 255–274 (2012)

Freundlich, H.: Über die adsorption in lösungen. Zeitschrift fur Physikalische Chemie **57**, 385–470 (1907)

Fytianos, K., Voudrias, E., Kokkalis, E.: Sorption–desorption behaviour of 2, 4-dichlorophenol by marine sediments. Chemosphere **40**, 3–6 (2000)

Hazzaa, R., Hussein, M.: Adsorption of cationic dye from aqueous solution onto activated carbon prepared from olive stones. Environ. Technol. Innov. **4**, 36–51 (2015)

Ho, Y.-S., McKay, G.: Pseudo-second order model for sorption processes. Process Biochem. **34**, 451–465 (1999)

Hoseinzadeh, H.R., Arami-Niya, A., Wan Daud, W.M.A., Sahu, J.N.: Comparison of oil palm shell-based activated carbons produced by microwave and conventional heating methods using zinc chloride activation. J. Anal. Appl. Pyrol. **104**, 176–184 (2013)

Isah, A.U., Abdulraheem, G., Bala, S., Muhammad, S., Abdullahi, M.: Kinetics, equilibrium and thermodynamics studies of C.I. Reactive Blue 19 dye adsorption on coconut shell based activated carbon. Int. Biodeterior. Biodegrad. **102**, 265–273 (2015)

Jung, S.-H., Oh, S.-J., Choi, G.-G., Kim, J.-S.: Production and characterization of microporous activated carbons and metallurgical bio-coke from waste shell biomass. J. Anal. Appl. Pyrol. **109**, 123–131 (2014)

Kilic, M., Apaydin-Varol, E., Putun, A.E.: Adsorptive removal of phenol from aqueous solutions on activated carbon prepared from tobacco residues: equilibrium, kinetics and thermodynamics. J. Hazard. Mater. **189**, 397–403 (2011)

Lagergren, S.: About the theory of so-called adsorption of soluble substances (1898)

Langmuir, I.: The constitution and fundamental properties of solids and liquids. Part I. solids. J. Am. Chem. Soc. **38**, 2221–2295 (1916)

Liu, Y., Guo, Y., Gao, W., Wang, Z., Ma, Y., Wang, Z.: Simultaneous preparation of silica and activated carbon from rice husk ash. J. Clean. Prod. **32**, 204–209 (2012)

Low, M.: Kinetics of chemisorption of gases on solids. Chem. Rev. **60**, 267–312 (1960)

Rodier, J., Legube, B., Merlet, N.: Coll 2009: L'analyse de l'eau, 9 ème édn. DUNOD, Paris 1217 (2009)

Shi, Q., Zhang, J., Zhang, C., Li, C., Zhang, B., Hu, W., Xu, J., Zhao, R.: Preparation of activated carbon from cattail and its application for dyes removal. J. Environ. Sci. **22**, 91–97 (2010)

Temkin, M., Pyzhev, V.: Recent modifications to Langmuir isotherms (1940)

Weber, W.J., Morris, J.C.: Kinetics of adsorption on carbon from solution. J. Sanit. Eng. Div. **89**, 31–60 (1963)

Study of Structural, Morphological and Optical, Properties of Fe Doped SnO_2 Semiconductor Thin Films Prepared by Sol-Gel Technique

S. Benkara[1(✉)], H. Ghamri[2], and M. Zaabat[3]

[1] Electrical Engineering Department, Larbi Ben M'hidi University, Oum El Bouaghi, Algeria
salimabenkara@gmail.com

[2] Physic Department Hadj Lakhder University, Batna, Algeria

[3] Laboratory of Active Components and Materials, Oum El Bouaghi University, Oum El Bouaghi, Algeria

Abstract. The undoped and iron doped tin oxide thin films were deposited on glass substrates by sol-gel dip coating technique, using tin chloride and iron III chloride as the starting materials. The effect of method conditions and Fe doping on the structural, morphological and optical properties of applied thin films have been studied by X-ray diffraction (XRD), and UV–vis spectroscopy. Surface topography of thin films was examined by atomic force microscopy (AFM).

XRD patterns showed an increase in peak intensities of the rutile and cubic crystalline phases of SnO_2 by increasing the Fe content. SnO_2 nanoparticles in the range of 9–20 nm size were obtained by the Scherrer equation using FWHM (Full Width at Half Maximum) values of the main peaks in the XRD diffraction pattern when calcined at 500 °C.

The UV–Vis–spectroscopy analyses have shown that all the thin films were transparent in the visible region with an average transmittance ranging from 70% to 87%. And showed a direct band gap reducing with increase in Fe^{3+} doping from 3.82 to 3.72 eV.

AFM images shown thin films with different contents of Fe. The effect of Fe concentration has been discussed.

Keywords: Thin films · SnO_2 oxide · Fe doped Sn · Optical analysis
AFM analysis

1 Introduction

Transparent conductive oxides (TCO) materials have been widely studied in recent years due to their optical and electrical properties such as: In_2O_3, SnO_2, ZnO, Cd_2SnO_4, $CdSnO_2$.

To improve the electrical properties of these transparent oxides, scientists have resorted to incorporate the dopants in these latter.

Obtaining such materials having a good compromise between transparency to visible light and good electrical conductivity, is a major industrial challenge.

B. Abdelbaki et al. (Eds.): SMSD 2017, *Proceedings of the Third International Symposium on Materials and Sustainable Development*, pp. 676–682, 2018.
https://doi.org/10.1007/978-3-319-89707-3_71

Among these TCO materials tin dioxide (SnO_2) is the first transparent conductor commercialized abundantly, and an important semiconductor which has wider band gap of 3.6 eV and much higher exciton binding energy of 130 meV (Demichelis et al. 1984).

Due to its properties such as low electrical resistivity, high optical transmittance and high reflectivity in the infrared region, it can be used as anode in flat panel displays (Kotz et al. 1991). It is also one of the most metal oxides used in the control of air pollution and toxic gas detection (Takahata et al. 1988), as a conductive electrode in the thin film photovoltaic cells (Shang et al. 2012) and as a heat reflector in solar applications (Fukano et al. 2005). Sol-gel dip coating is employed to prepare thin films; this method is suitable to prepare homogenous layers with different doping.

In the present work we report the process for preparation of pure and Fe doped SnO_2 thin films from $SnCl_2{\cdot}2H_2O$ and $FeCl_3$ solutions by sol-gel dip coating method. The effect of Fe dopant on structural, morphological and optical properties of SnO_2 thin films has been studied.

2 Experimental

The samples were prepared by sol-gel technique using dip coating apparatus. As a starting material and dopant source, tin (II) chloride dihydrate ($SnCl_2.2H_2O$) and iron III chloride ($FeCl_3{\cdot}6H_2O$). Firstly $SnCl_2.2H_2O$ (2.5 g) was dissolved into 30 ml of absolute ethanol, then the solution, which was heated to 70 °C previously, was stirred by a magnetic stirring apparatus for 2 h.

Fe doped SnO_2 films have been prepared by adding an appropriate amount (1%, 2%, 3%, 5% and 10%) of $FeCl_3{\cdot}6H_2O$ to $SnCl_2{\cdot}2H_2O$ and dissolved by ethanol.

The obtained solution was continuously stirred for 2 h. Then the two solutions were stirred and heated at 70 °C for another 2 h. Finally, both mixtures were separately aged for 24 h at room temperature, a clear and homogeneous solutions was obtained.

The cleaned glass substrates were immersed in these solutions and withdrawn at the rate of 4 cm/min. The deposited films was then dried at 200 °C and finally annealed at temperature of 500 °C for 60 min. The procedures from coating to drying were repeated 10 times.

The Xray- diffraction patterns of samples were recorded by an X-ray diffractometer D6 Advance Bruker system using Cu Kα radiation (Cu Kα = 0.15406). The optical properties of the films were studied using A Jasko V-630 visible spectrophotometer. Atomic force microscope (AFM) measurements were carried out using A100-AFM microscopy.

3 Results and Discussion

3.1 X-Ray Diffraction Studies

To verify the crystallinity, we present the diffraction pattern of Fe doped SnO_2 thin layers after annealing of 500 °C.

Figure 1 shows peaks at 2θ = 26.7°, 34.8°, 38.14°. 51.55°, 53.25° and 66.96° respectively correspond to the crystalline planes of SnO_2 rutile phase (110), (101), (200), (211), (220) and (301). Peak positions obtained are in good agreement with the tetragonal rutile structure of SnO_2 (Patil et al. 2012).

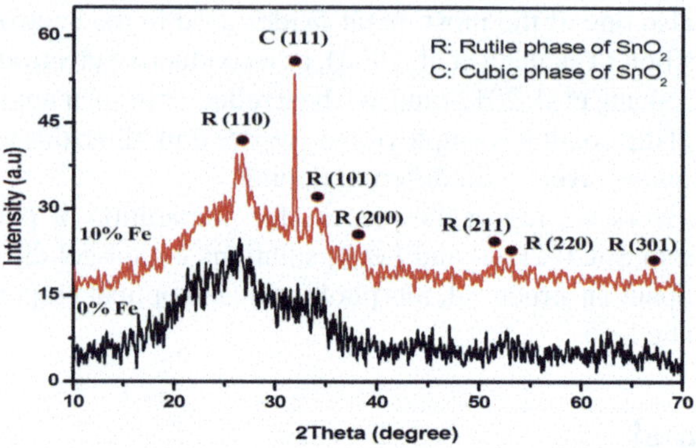

Fig. 1. XRD pattern of undoped and 10% Fe doped SnO_2 thin films.

The spectrum also shows a peak located at 2θ = 31.88° assigned to the cubic phase (111) of SnO_2. The appearance of the cubic phase in the literature (Haines and Leger 1997) confirms our result and the reason to get it is due to the method followed and the materials used.

Moreover, no metallic iron phase, iron oxide or other phases were detected, which means a good dilution of the iron atoms in the SnO_2 structure in the XRD detection limits (Junbo et al. 2005).

The grain size (D) is calculated using the Debye-Scherrer formula (Patil et al. 2012):

$$D = \frac{0 \cdot 9\lambda}{\beta \cos \theta} \tag{1}$$

Where λ = 0.154056 nm is X-ray wavelength, β is FWHM, θ is Bragg's angle in degrees.

Figure 2 shows the width at half maximum (β) of 10% Fe doped SnO_2 and diffraction angle attributed to (211) peak.

The grain sizes are: 33.00; 48.39; 37.46; 38.74; 54.03 and 35.58 nm, respectively corresponding to the planes (110), (101), (200), (211), (220) and (301).

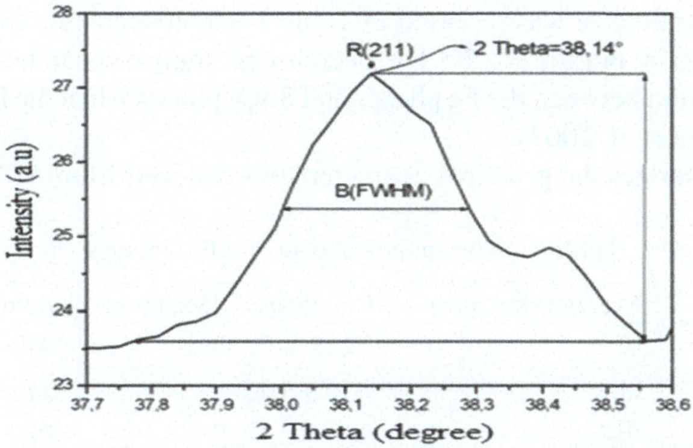

Fig. 2. Determination of the grain size from the measurement of FWHM (β) by the Debye Scherrer.

3.2 Surface Morphology Studies

Figure 3 shows AFM images in 2D and measurements of the roughness of pure and iron doped SnO_2 doped oxide layers deposited on glass substrates.

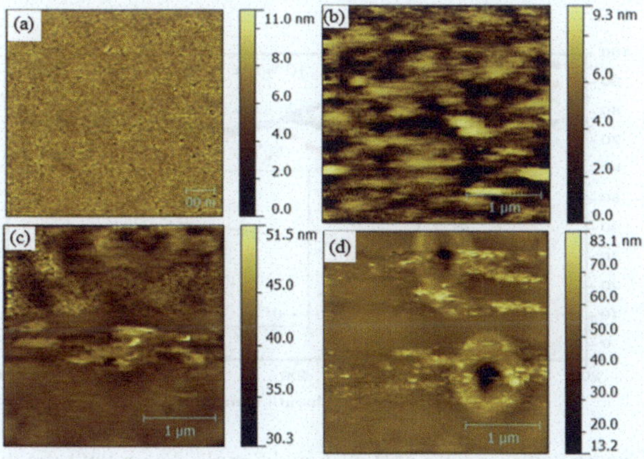

Fig. 3. AFM images in 2D of SnO_2 thin films with different concentrations of Fe: (a) pure SnO_2, (b) 1% Fe-SnO_2, (c) 3% Fe-SnO_2 and (d) 5% Fe-SnO_2.

These images reveals that the gains are aleatory distributed throughout the sample surface, these surfaces are not rough enough. We can also be observed a small change of the grain size as a function of doping.

Pure SnO_2 is a spherical-shaped structure; the average grain size is 19 nm, whereas the root mean square (RMS) roughness is 27.28 nm. However, its topography changed following the introduction of Fe.

The average grain size was reported in Table 1 which decreases for 1% Fe and 5% Fe but increases to 20 nm for 3%. Fe The decrease in grain size can be attributed to the growing competition between the Fe phases and SnO_2 phases when the Fe concentration increases (Sakuma et al. 2007).

Table 1 summarizes the principal characteristics deduced from AFM images.

Table 1. The analytical data of AFM images

	Average size (nm)	Roughness (RMS) (nm)	Grain number
Pure SnO_2	19	27.28 ± 11.32	141
1% Fe	14.2	2.15 ± 1.12	77
3% Fe	20.5	4.83 ± 4.39	57
5% Fe	15.7	7.5 ± 2.21	71

3.3 Optical Properties

Figure 4 shows the optical transmission spectra of Fe doped SnO_2 films elaborated with different concentrations. It can be seen in this figure the presence of a good transparency region between 200 and 800 nm. The transmission is estimated at 60 to 90% in the visible which is strongly influenced by the doping, leading to a decrease in transmittance for 1% and 3% of Fe, and then rises again for 5% and 10% Fe reaching almost 90%.

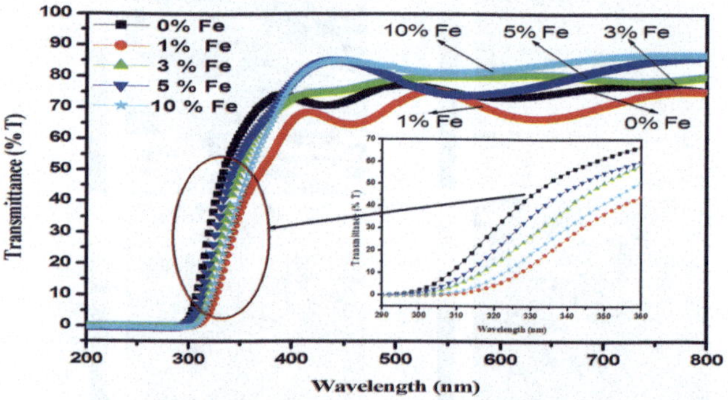

Fig. 4. Transmittance spectra of 0, 1, 3, and 5 mol% Fe, in SnO_2:Fe thin films.

We also note that the doped layer by 1% Fe has the smallest transparency in the visible. In fact, we observed a small shift of the absorption threshold towards higher energy with increasing doping.

Figure 4 also indicates that the doped films with 1% and 5% have more oscillations interference with respect to the other, indicating an increase in the film thickness (Khodja et al. 2014; Martinez et al. 2012).

The energy of the band gap is derived by the Tauc formula (Pankove 1971):

$$(\alpha h\upsilon)^2 = A\left(h\upsilon - E_g\right) \tag{2}$$

Where α, Eg and hν are: the absorption coefficient, the energy gap and the optical beam energy, respectively and A is a constant.

On Fig. 5 we represented the curves $(\alpha h\upsilon)^2$ for SnO_2 doped with different concentrations of iron.

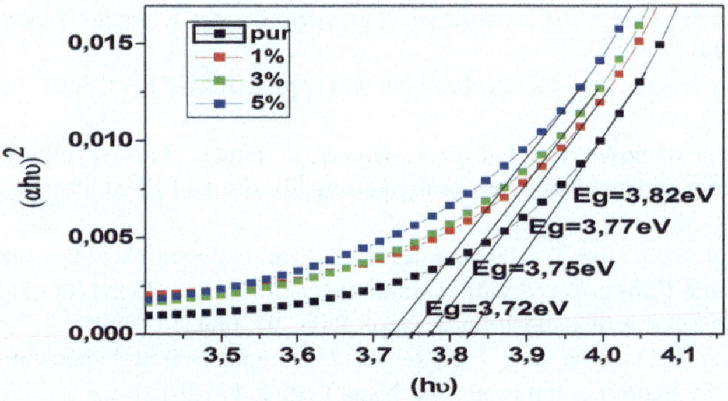

Fig. 5. Plot of $(\alpha h\upsilon)^2$ vs. photon energy of undoped and Fe-doped SnO_2 films.

As observed, the optical gap of the films is decreased when the concentration of Fe increases. The variation in the optical gap can be explained by the distortions in the crystal lattice of SnO_2 after the introduction of impurities that lead to the formation of the tail band in the band gap (Shajira et al. 2014).

4 Conclusions

The structural, morphological, optical and electrical thin layers of pure tin oxide doped with iron according to the synthesis conditions were studied in this work.

The X-ray spectrum of the SnO_2 thin films shows the formation of the crystalline structure after thermal annealing of 500 °C with great predominance of rutile nanoparticles. The grain size calculated from the Scherrer formula, vary between 33 and 54 nm according to the conditions of elaboration.

The morphological properties were examined by the AFM images, which show a aleatory distribution of gains on some samples. The surfaces are not sufficiently rough. The average grain size ranges between 14.2 and 20.5 nm and the roughness factor (RMS) varies between 7.5 and 27.28 nm.

Characterization of thin films developed by UV-Visible spectrophotometer showed a visible transmission is over 90% for the samples doped with 10% of Fe. Optical gap values vary between 3.72 and 3.82 eV.

References

Demichelis, F., Minetti-Mezzetti, E., Tagliaferro, A., Tresso, E.: Determination of optical properties of SnO_2 films. In: II Nuovo Cimento D, vol. 4, pp. 68–78 (1984)

Kotz, R., Stucki, S., Carcer, B.: Electrochemical waste water treatment using high overvoltage anodes. Part1: Physical and electrochemical properties of SnO_2 anodes. J. Appl. Electrochem. **21**, 14–20 (1991)

Takahata, K.: Tin dioxide sensors development and applications. Chem. Sens. Technol. **1**, 39–55 (1988)

Shang, G., Wu, J., Huang, M., Lin, J., Lan, Z., Huang, Y., Fan, L.: Facile synthesis of mesoporous tin oxide spheres and their applications in dye-sensitized solar cells. J. Phys. Chem. C **116**(38), 20140–20145 (2012)

Fukano, T., Motohiro, T., Ida, T., Hashizume, H.: Ionization potentials of transparent conductive indium tin oxide films covered with a single layer of fluorine-doped tin oxide nanoparticles grown by spray pyrolysis deposition. J. Appl. Phys. **97**, 084314 (2005)

Patil, G.E., Kajale, D.D., Gaikwad, V.B., Jain, G.H.: Preparation and characterization of SnO_2 nanoparticles by hydrothermal route. Int. Nano Lett. **2**, 17 (2012)

Haines, J., Leger, J.M.: X-ray diffraction study of the phase transitions and structural evolution of tin dioxide at high pressure: relationships between structure types and implications for other rutile-type dioxides. Phys. Rev. B **55**, 11144 (1997)

Junbo, W., Ming, Y., Yingmin, L., Licheng, C., Yan, Z., Bingjun, D.: Synthesis of Fe-doped nanosized SnO_2 powders by chemical co-precipitation method. J. Non-Cryst. Solids **351**, 228–232 (2005)

Sakuma, J., Nomura, K., Barrero, C., Takeda, M.: Mössbauer studies and magnetic properties of SnO_2 doped with 57Fe. Thin Solid Films **515**, 8653–8655 (2007)

Khodja, S., Touam, T., Chelouche, A., Boudjouan, F.: Effects of stabilizer ratio on structural, morphological, optical and waveguide properties of ZnO nano-structured thin films by a sol–gel process. Superlattices Microstruct. **75**, 485–495 (2014)

Martinez, D.Y.T., Perez, R.C., Delgado, G.T., Angel, O.Z.: Structural, morphological, optical and photocatalytic characterization of ZnO–SnO_2 thin films prepared by the sol–gel technique. J. Photochem. Photobiol. A Chem. **235**, 49–55 (2012)

Pankove, J.I.: Optical Processes in Semiconductors. Prentice-Hall Inc., Englewoord Cliffs (1971)

Shajira, P.S., Junaid Bushiri, M., Nair, B.B., Ganesh Chandra, P.V.: Energy band structure investigation of blue and green light emitting Mg doped SnO_2 nanostructures synthesized by combustion method. J. Lumin. **145**, 425–429 (2014)

Effect of a Thermal Degradation on the Mechanical Behavior of a 5-Harness Satin Weave Carbon–PEEK

T. Saoudi[1(✉)] and M. A. Belouchrani[1,2]

[1] Laboratoire Génie des Matériaux, Ecole Militaire Polytechnique, BP 17C, 16046 Bordj El Bahri, Algiers, Algeria
saoudi.toufik@gmail.com

[2] Ecole Nationale Supérieure de Technologie, Dergana, Algiers, Algeria

Abstract. This paper presents an experimental analysis of a 5-harness satin weave carbon–PEEK (Polyether-ether-ketone) composite. In order to understand the thermal behavior of composites, when exposed to moderate and high temperatures, a thermal degradation study is carried out. To this end, a preliminary characterization of the composite before thermal exposure were performed by Differential Scanning Calorimetry (DSC), thermogravimetric analysis (TGA) and Dynamic mechanical analysis (DMA). The results obtained by tensile and three-point bending test showed the effect of thermal heating on the mechanical behavior of the resin and the composite.

Keywords: Carbon-PEEK · High temperature properties · Mechanical behavior Thermal degradation

1 Introduction

Manufacturers in the aeronautics sector are increasingly using organic matrix composites in order to reduce structures as much as possible. These materials combine excellent mechanical properties at low density.

Composites with a thermosetting matrix, such as epoxy, have been widely used since the Boeing 707 was commissioned in the 1950s. Although they have excellent mechanical properties, they suffer from some notable disadvantages, such as the need for storage at low temperature, the presence of irreversible defects after forming, etc. Thus, the high thermoplastic resins performances such as polyphenylene sulfide (PPS) or polyether-ether-ketone (PEEK) are a promising alternative. These resins have many qualities, namely good mechanical properties over a wide temperature range, less moisture absorption and as well as good resistance to chemical attack, damage and shock, see for example (Kemmish and Hay 2010), (Jonas and Legras 1991), (Kemmish and Hay 1985), and Stober *et al.* (1984). All these characteristics predispose them to aeronautical applications.

B. Abdelbaki et al. (Eds.): SMSD 2017, *Proceedings of the Third International Symposium on Materials and Sustainable Development*, pp. 683–697, 2018.
https://doi.org/10.1007/978-3-319-89707-3_72

Generally, the reinforcements used in structural composite materials are fibers, the most common being glass fibers and carbon fibers. Nevertheless, the need to lighten the structures favors the use of carbon fibers; see (Chung 1994).

The structures studied are made of textile composite material. The name of textile composite comes from the process of obtaining the reinforcement (weaving, braiding, knitting or sewing). A fairly exhaustive description of these types of composites is carried out by some authors, including (El Hage 2006) and (Byun and Chou 2000).

The introduction of composite materials, however, is progressive and conservative because of their long life which remains difficult to predict. Their long-term behavior is all the more random because they are subjected to external mechanical stresses or exposed to high temperatures. They are, however, exposed in service to numerous thermal, chemical and mechanical stresses, which can lead to significant losses properties and compromise the integrity of structures.

Composites result from the association of two constituents, fibers and matrix, whose very diverse properties create heterogeneity of behavior. When these composites are subjected to temperature variations, the differences in coefficients of expansion between the fibers and the matrix and between layers of different orientations of the stack are such that residual stresses are generated in the material.

Functional or accidental thermal exposure may also contribute to degrading the material long before it is subjected to mechanical stress, see Lee *et al.* (1993), and (Daniel 1980). This degradation affects the matrix and, more particularly, its interface with the fibers. The parameters, such as volume fraction, fiber orientation and thickness, play a very important role in the propagation of damage (Saadaoui 1996). It will be all the more important if the thermal exposure is beyond the glass transition temperature Tg, see (Chang 1987), Griffis *et al.* (1981), and (Collings and Mead 1988).

In shipbuilding applications, in order to solve fire problems, regulation requires that the composite material be resistant to high temperatures for at least 30 to 60 min. This is the time it takes for the crew to respond to the main source of fire, Asokendu *et al.* (2004). For this, knowledge of their residual states becomes more than necessary. Indeed, the thermal stability is a determining factor in the choice of the composite material. The importance of this aspect in the safety of structures has favored research in this field.

Thermogravimetric analysis (TGA) is a very efficient way to study the thermal stability of polymers. It makes it possible to follow the evolution of the loss of mass as a function of the temperature. Using this analysis, the degradation temperature corresponds to a sudden drop in the weight loss curve. This temperature is an index which gives a good characterization of the thermal stability of a polymer.

According to (Verdu 1992), the majority of the processes of thermal degradation is accompanied by gas emissions, hence the possibility of using gravity criteria to achieve a classification of the stabilities. On the other hand, for Dyakonov *et al.* (1996), if the degradation products formed are non-volatile, numerous chain cuts can occur without significant loss of mass. It is also the opinion of (Hilaire 1991) who pointed out that there are many mechanisms of degradation which, by splitting chains without producing volatiles, lead to a decrease in mechanical properties without, however, a loss of mass. For this author, the degradation products will either be volatile or dissolved in the matrix.

The aim of the present paper is to investigate the effect of a thermal degradation on the mechanical properties of 5-Harness Satin (5HS) weave composite consists of T300 3 k carbon fibers and a high performance thermoplastic matrix PEEK. Also, parametric studies were performed to analyze the influence of the temperature range and the exposure time on the mechanical behavior of these composite.

2 Materials and Methods

2.1 Raw Materials

The materials used in this study were PEEK 450G matrix provided by Victrex and T300 3 k carbon fibers fabric with 5 Harness Satin (5HS) architecture provided by Soficar, since the volume fraction of the fibers is 50%. The composite are hot pressed according to an angle-ply [0°/45°/0°/45°/0°/45°/0°] stacking sequence. The laminates with the dimension of $(300 \times 300 \times 2.2)$ mm^3, see Fig. 1, are obtained by thermo-compression (consolidation) of the pre-impregnated fabrics, with a pressure of 10 bars and a temperature of 390 °C.

Fig. 1. Geometric model for 5-Harness satin weave and the plate of the composite carbon/PEEK

Temperature and duration of application are the main factors to be taken into account. Even if the melting temperature of PEEK is 343 °C it is necessary to heat at least to 390 °C to obtain a viscosity sufficient to allow easy shaping, see Fig. 2. This temperature makes it possible to completely erase the thermal history of the PEEK and thus to more easily control the crystallinity of the final material.

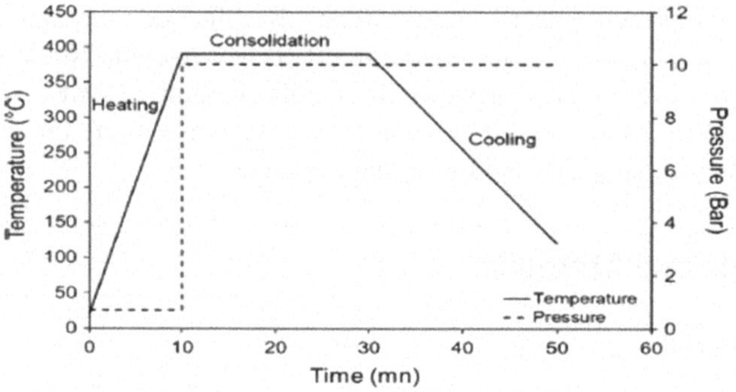

Fig. 2. Cycle of thermos-compression

2.2 Thermal Degradation

To investigate the effect of thermal degradation on the properties of the composite and the matrix, the samples are subjected to a thermal source. The oven is first preheated to the exposure temperature, before each operation. Samples (resin and composites) are exposed to temperatures of 150 °C, 250 °C, 300 °C and 340 °C for 15 min or 1 h. After cooling at room temperature, various characterizations are carried out.

2.3 Differential Scanning Calorimetry (DSC)

DSC is used widely for examining polymeric materials to determine their thermal transitions. The scanning thermal analysis is carried out using a DSC apparatus of the type NETZSCH 204 F1. A 9.3 mg mass heated to 400 °C at a heating rate of 10 °C/min, under an inert nitrogen atmosphere.

2.4 Dynamic Mechanical Analysis (DMA)

Dynamic mechanical analysis (DMA) is a widely used technique for characterizing the properties of a material as a function of temperature, time, frequency, stress, atmosphere or a combination of these parameters. Since a small deformation is applied to a sample in a cyclic manner.

The tests are carried out in three-point bending with a DMA NETZSCH 242C apparatus. A frequency of 1 Hz and a heating rate of 5° C/min are used. The sample having a size of $(50 \times 10 \times 2.2)\ mm^3$ is subjected to an oscillation in constant frequency (1 Hz) and amplitude bending (1N), while the temperature of the sample is increased linearly up to 500 °C.

2.5 Thermogravimetric Analysis (TGA)

Thermogravimetric analysis is a thermal technique used to visualize the mass variations of a sample as a function of temperature or time according to a temperature program and a controlled atmosphere. The thermogravimetric analyzer is a TGA Q50 V6.1, under argon. Mass sample of 7 mg is deposited in an alumina nacelle. The temperature program applied for the thermal analysis is from 0 °C to 750 °C at a heating rate of 10 °C/min.

2.6 Mechanical Tests

The specimens of the composite are cut from the previously described plates by means of a water jet cutting method according to the dimensions recommended by the ISO 527 standards for tensile test and NF T 51-001 for three-point bending test. The both tensile and flexural tests are carried out on a Zwick/Roell Z050 machine.

3 Results and Discussions

In this part, a preliminary characterization of the composites before the thermal exposure is presented. On the other hand, the effect of thermal degradation on the mechanical properties was performed.

3.1 Differential Scanning Calorimetry (DSC)

As shown in Fig. 3 the evolutions of the heat flux as a function of the temperature obtained by DSC of a PEEK.

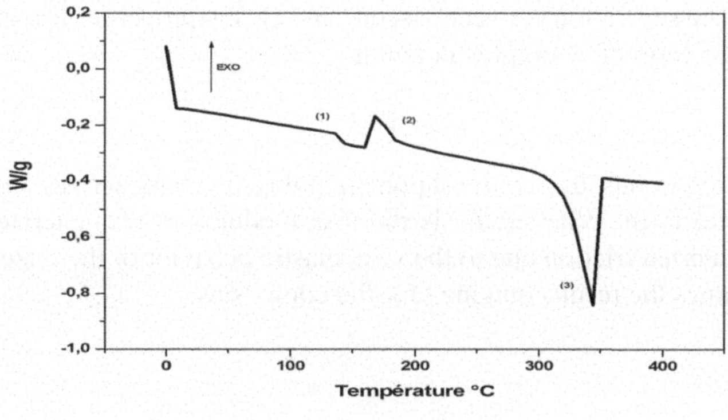

Fig. 3. Thermogram of PEEK obtained by DSC

In the diagram of Fig. 3, the following transitions are observed: (1) Deflection of the heat flux and non-return to alignment with the baseline, associated with the glass transition, transition from vitreous to rubbery state, which is of the order of 143 °C. (2) Exothermic peak just after the glass transition corresponding to the so-called "cold crystallization". (3) High-temperature endothermic peak corresponding to the melting of the crystalline phase, which is of the order of 343 °C. These results are consistent with supplier data.

Figure 4 illustrates an example of a thermogram trace obtained by (DCS) for the composite carbon/PEEK, while respecting the operating conditions described above.

In this case, they have a glass transition and a melting point. The composite does not exhibit cold crystallization (due to the absence of the exothermic peak just after the glass transition), which indicates that these materials are at their maximum crystallinity rate.

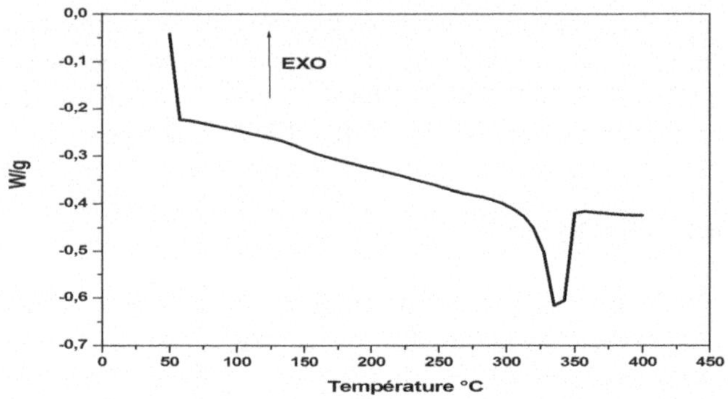

Fig. 4. Thermogram of composite carbon/PEEK obtained by DSC

3.2 Dynamic Mechanical Analysis (DMA)

DMA is a thermomechanical test that allows the viscoelastic characteristics of a material to be studied. The separation between viscous and elastic properties is expressed mathematically in the form of a complex notation:

$$E^{*} = E' + i.E'' \tag{1}$$

Where E' represents the conservation modulus. It characterizes the cumulative energy in elastic form. Whereas E'' is the loss modulus, it characterizes the energy dissipated by internal friction due to the viscoelastic behavior of the material.

Figure 5 shows the results obtained for the composite.

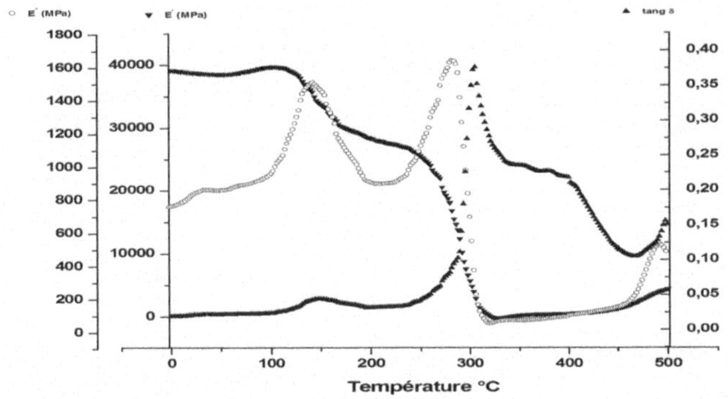

Fig. 5. Evolution of the conservation modulus (E'), the loss modulus (E'') and the tangent of the loss angle (tan (δ)) as a function of temperature

The plot of the real module (E') as a function of the temperature makes it possible to distinguish four zones;

(a) A glassy zone (temperatures between 0 °C and 130 °C) where the modulus remains practically constant.
(b) A transition zone in which the module drops (temperatures between 130 °C and 200 °C). The modulus drop is associated with a peak of the tangent of the damping factor (tan (δ)), which is of the order of 149 °C. It is the relaxation of the PEEK matrix which corresponds to generalized motions of the macromolecular chains relative to one another and is associated with the glass transition.
(c) A rubber zone (temperatures between 200 °C and up to 290 °C softening and melting of the crystalline part), where the movements of the chains can be carried out very rapidly, corresponding to the lowest level of the actual module until the material melts.
(d) The PEEK is a semi-crystalline thermoplastic material, in the rubber zone the modulus evolution as a function of temperature does not take the form of a stable plateau as in the case of thermosetting materials. The module continues to decrease as the temperature increases.

3.3 Thermogravimetric Analysis (TGA)

The results obtained are shown in Fig. 6, which represent the rate of degradation and the residual mass. The degradation temperature corresponds to the peaks of the derivative of the residual mass curve (DTG). It shows that the behavior of the composite has two steps of degradation. The first degradation step is beginning just before 450 °C. In the second step, the residual mass of composite at 700 °C is 5%. These results show that the degradation temperature of the composite is higher than the maximum operating temperature (240 °C). They also show that thermal degradation can occur only at accidental extreme temperatures.

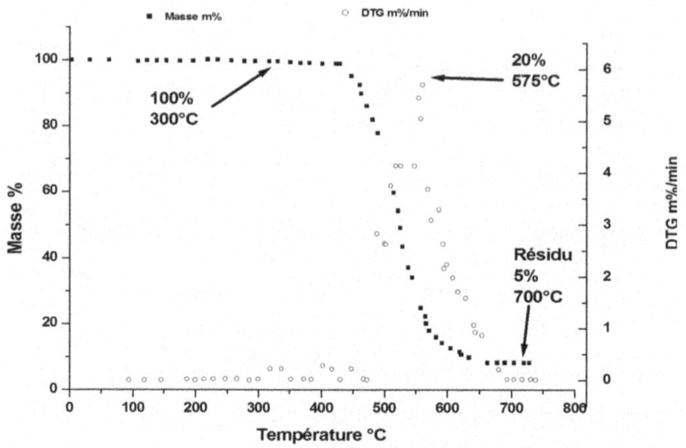

Fig. 6. Evolution of mass as a function of temperature

The work of Dyakonov *et al.* (1996) confirmed the importance of coupling the exposure temperature and the processing time to the physical properties of the polymer material. The evolution of the mass of the composite as a function of the duration of exposure at temperatures of 200 °C, 300 °C and 500 °C is illustrated in Fig. 7, the isothermal ATG were carried out by a rapid rise in heating temperature (30 °C/min) and then an isotherm of 90 min at the desired temperature. If the mass loss is not significant (about 6%) for 200 °C exposure, this is not the case for exposure to temperatures of 300 °C and 500 °C, a loss of mass of 10% at 300 °C and 47% at 500 °C.

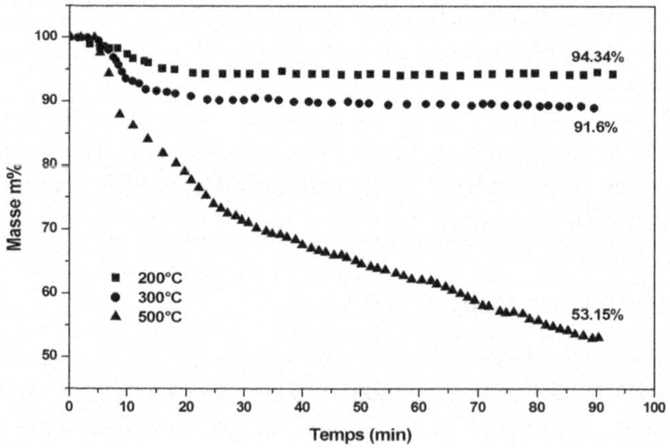

Fig. 7. Evolution of mass as a function of exposure time

3.4 Mechanical Tests

The tensile behavior of the resin only and of the composite before and after thermal exposure is shown in Figs. 8 and 9.

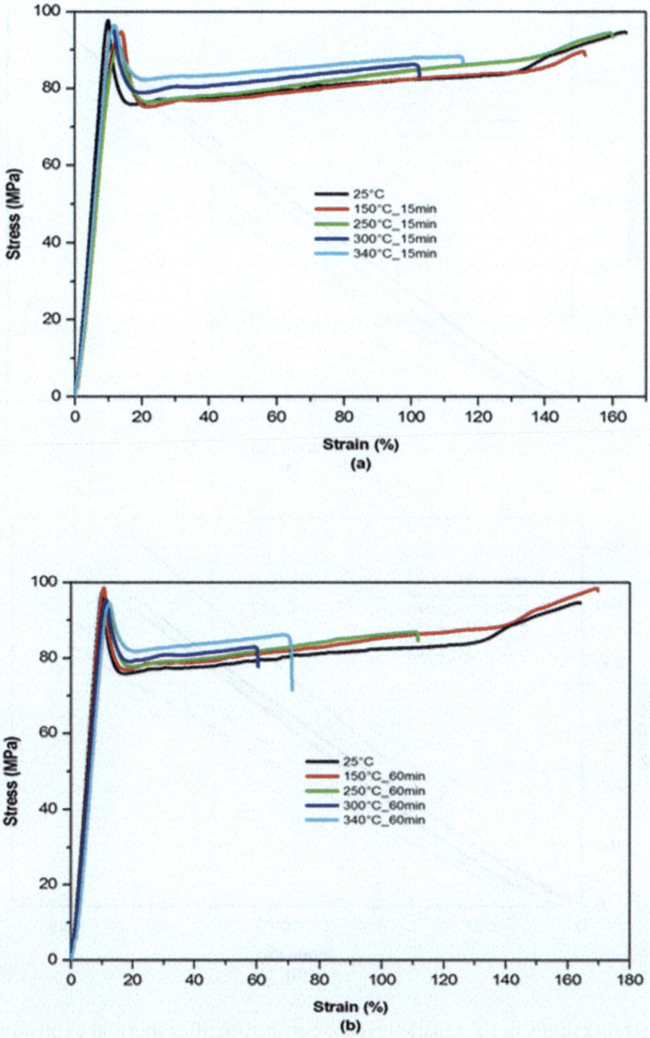

Fig. 8. Stress-strain curves in the tensile test for resin after thermal exposure at: (a) 15 min (b) 60 min

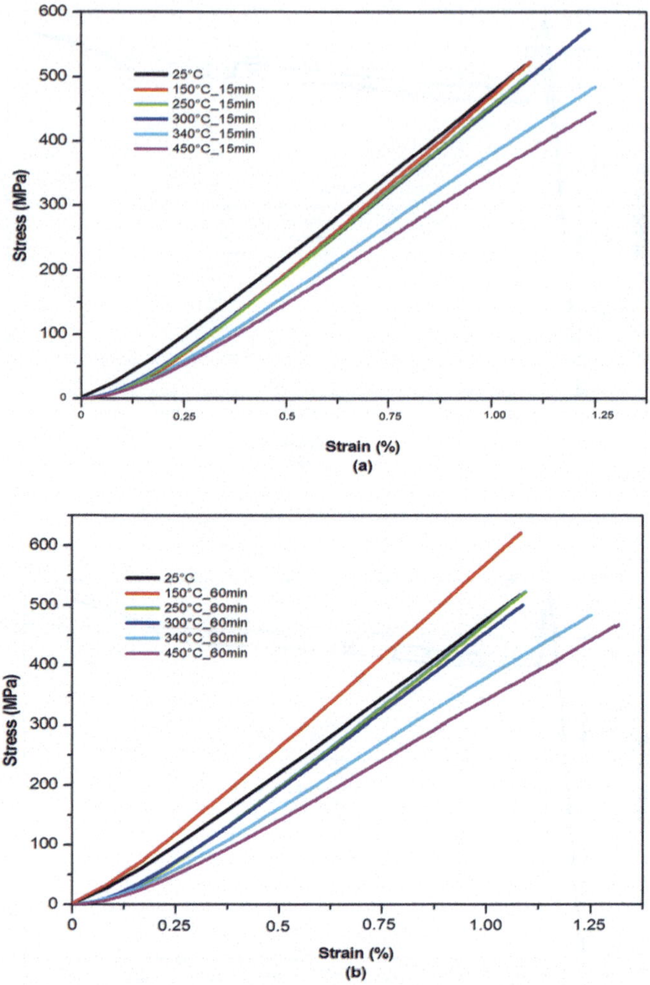

Fig. 9. Stress-strain curves in the tensile test for composite after thermal exposure at: (a) 15 min (b) 60 min

Figure 8 shows that the stress-strain curves of the resin alone, not exposed thermally and exposed to 150 °C, 250 °C, 300 °C and 340 °C, have the same evolution. The response of the resin is first linear elastic, before reaching a yield point where the phenomenon of striction begins. The length of the striction areas then increases with the load. A stationary plastic flow is then observed, followed by a progressive increase of the resistance (hardening).

These behavioral curves were used to determine the mechanical characteristics of the resin before and after thermal degradation. Table 1 shows the values of the modulus of elasticity $\boldsymbol{E}$, yield strain ε_y, yield strength S_y, ultimate strength S_u and ultimate strain ε_u.

Table 1. Evolution of the tensile properties of the resin alone at different temperatures

Properties	E (GPa)	ε_y (%)	S_y (MPa)	ε_u (%)	S_u (MPa)
25 °C	3,12	10,23	97,79	164,10	94,76
150 °C_15 min	2,93	14,16	94,71	151,86	89,85
150 °C_60 min	3,10	10,54	98,67	169,73	98,55
250 °C_15 min	2,78	13,87	91,03	160,04	94,50
250 °C_60 min	2,70	12,02	93,89	111,73	86,91
300 °C_15 min	3,06	11,69	95,31	102,61	86,36
300 °C_60 min	2,90	11,41	95,43	60,37	83,00
340 °C_15 min	2,95	12,02	96,36	115,77	88,30
340 °C_60 min	2,68	12,38	94,64	71,35	86,07

Table 1 shows an increase in the yield strain of the resin, at the exposure time of 15 min, about 38% at 150 °C and 35% at 250 °C. At the same temperatures (150 °C and 250 °C), the ultimate strain decreased by 7.5% and 2.5%, respectively. For temperatures above 300 °C, the ultimate strain recorded a drop which reached 63% at 300 °C and 56.5% at a temperature of 340 °C and exposure time of 60 min. On the other hand, the modulus of elasticity and the ultimate strength seems less affected than the previous parameters. This drop has reached 12% at 300 °C for ultimate strength and 14% at 340 °C for the modulus of elasticity.

The following figure (see Fig. 9) illustrates the evolution of the stress-strain curves of the composite as a function of the temperature and the exposure time.

In terms of the overall behavior of the laminate, the influence of temperature on the tensile response depends little on the state of the material. The oriented fibers at 0° support the load and the response of the laminate is dominated by the behavior of the fibers. The materials then have an elastic-brittle behavior.

These behavioral curves were used to determine the mechanical characteristics of the composite before and after thermal degradation. Table 2 shows the values of the modulus of elasticity $\boldsymbol{E}$, ultimate strength $\boldsymbol{S_u}$ and ultimate strain $\boldsymbol{\varepsilon_u}$.

Table 2 a shows that the ultimate strength has a small variation (increase) at temperatures below 300 °C and then decreases by 14% to 450 °C. On the other hand, Table 2 shows that ultimate strain is gradually increasing with increasing temperature. The increase in ultimate strain reached 21.3% at 450 °C. An increase in the modulus of elasticity of the composite to about 20% at 150 °C then decreased by 25% at 450 °C.

The three-point bending curves of the composite are illustrated in Fig. 10. The load-displacement curve consists of two phases:

- The first phase reflects the behavior of the material at the beginning of the stress. It is a reversible elastic linear behavior of a material that is still strong (without damage).
- The second phase, situated beyond the peak of the curve, corresponds to a ruin in the form of a staircase (sudden fall of the load). In fact, the sudden fall of the load represents the rupture of the first layer. It should be understood that at the rupture of the first fold, there will still remain other folds which will constitute a certain resistance. After the rupture of the remaining folds a rapid fall of the load ending in the ruin of the material.

Draw a load-displacement curve of the bending test. The modulus of elasticity is determined from the tangent at the origin of the load-displacement curve, using the following formula:

$$E = \frac{L^3}{4bh^3} * \frac{F}{d} \tag{2}$$

Where ***F***: applied load (N); ***L***: distance between the two supports (mm); ***b***: width of the specimen (mm); ***h***: height of the specimen (mm); ***d***: displacement (mm) and ***E*** is the modulus of elasticity (MPa).

These behavioral curves were used to determine the mechanical characteristics of the composite before and after thermal degradation. Table 3 shows the values of the modulus of elasticity ***E***, ultimate load $\boldsymbol{F_u}$ and failure displacement $\boldsymbol{D_f}$.

Table 3 a shows that the failure displacement has a small variation at temperatures below 300 °C and then increases by 64% at 340 °C. On the other hand, except at 150 °C

Table 2. Evolution of the tensile properties of the composite at different temperatures

Properties	E (GPa)	ε_u (%)	S_u (MPa)
25 °C	48,44	1,08	518
150 °C_15 min	49,25	1,09	523
150 °C_60 min	58,12	1,10	621
250 °C_15 min	48,19	1,08	501
250 °C_60 min	49,12	1,07	520
300 °C_15 min	47,88	1,24	574
300 °C_60 min	48,19	1,08	501
340 °C_15 min	40,75	1,25	484
340 °C_60 min	40,62	1,22	482
450 °C_15 min	37,42	1,25	445
450 °C_60 min	36,21	1,31	468

Table 3. Evolution of the three point bending properties of the composite at different temperatures

Properties	E (GPa)	F_u (N)	D_f (mm)
25 °C	33,61	948	1,31
150 °C_15 min	30,86	933	1,40
150 °C_60 min	32,56	1010	1,33
250 °C_15 min	32,70	845	1,23
250 °C_60 min	28,53	925	1,63
300 °C_15 min	25,40	882	2,06
300 °C_60 min	27,26	878	1,60
340 °C_15 min	26,02	834	2,16
340 °C_60 min	24,70	873	1,95

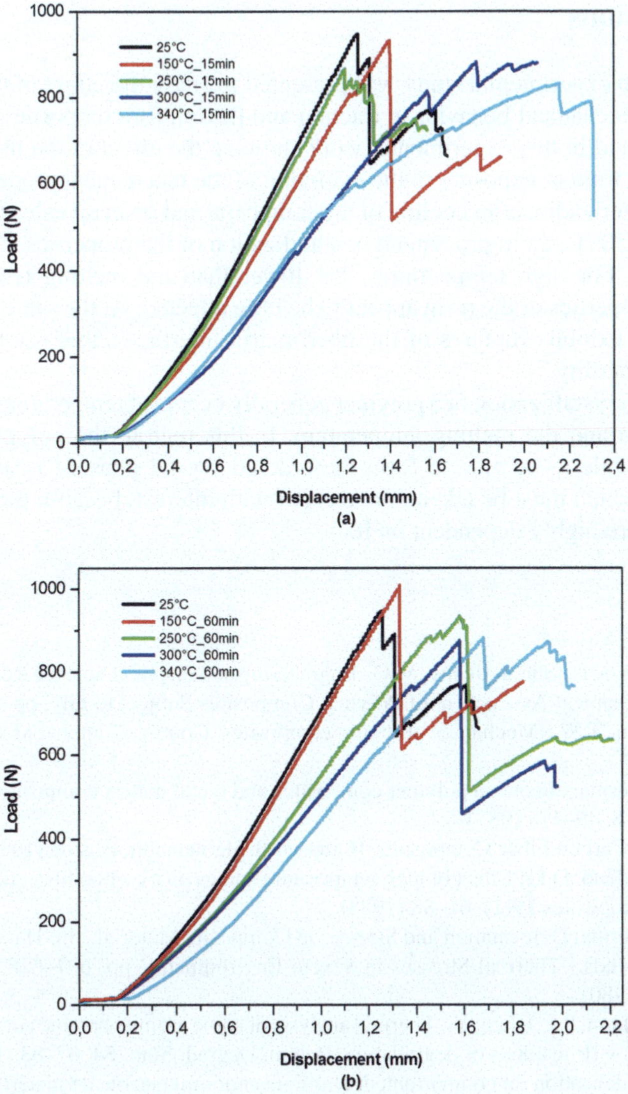

Fig. 10. Load-displacement curves in the bending test for composite after thermal exposure at: (a) 15 min (b) 60 min

and exposure time of 60 min, the ultimate load is gradually decreasing with increasing temperature. The increase in ultimate load reached 12% at 340 °C. An decrease in the modulus of elasticity of the composite to about 15% at 250 °C and 25% at 340 °C.

4 Conclusions

In this paper an experimental study was presented to study the effect of thermal degradation in the mechanical behavior in traction and flexion for composite or resin alone. The results found in this experimental work showing the effect of the thermal overexposure and the time of exposure in the evolution of the mechanical properties.

The characterization after cooling of the composite and resin revealed that at temperatures below 150 °C, an improvement or stabilization of the properties of the materials was observed. For high temperatures, but lower than the melting temperature, the mechanical properties of the resin appear to be little affected. At the same temperatures, the composite exhibits ruptures of the fiber/matrix interfaces, this resulted in loss of strength and ductility.

Finally the crystallization of a polymer generally occurs at temperatures between the glass transition and the melting temperature. In this region, the polymer chains are sufficiently mobile to be able to form the ordered crystal phase. Crystallization is a phenomenon which must be taken into account with interest, because the properties of the polymers are highly dependent on it.

References

Asokendu, S., Reza, L., Sha, J., Jenny, M.C.: Engineering and Physical Science Research Council. Thermo-Mechanical Assessment of Polymer Composites Subject to Fire, pp. 20–60 (2004)

Byun, J.H., Chou, T.W.: Mechanics of textile composites. Compr. Compos. Mater. **1**, 719–761 (2000)

Chang, C.L.: Thermal effects on polymer composites and metal matrix composites. Theor. Appl. Fract. Mech. **8**, 49–57 (1987)

Chung, D.D.L.: Carbon Fiber Composites. Butterworth Heinemann, Boston (1994)

Collings, T.A., Mead, D.L.: Effect of high temperature spikes on a carbon fibre- reinforced epoxy laminate. Composites **19**(1), 61–88 (1988)

Daniel, I.M.: Thermal Deformation and Stresses in Composite Materials. In: Hasselman, D.P.H., Heiier, R.A. (eds.) Thermal Stresses in Severe Environments, pp. 607–628. Plenum Press, New York (1980)

Dyakonov, T., Mann, P.J., Chen, Y.: Thermal analysis of some aromatic amine cured model epoxy resin systems – II: residues of degradation. Polym. Degrad. Stab. **54**, 67–83 (1996)

El Hage, C.: Modélisation du comportement élastique endommageable de matériaux composites à renfort tridimensionnel. Thèse Université de Technologie de Compiègne (2006)

Griffis, C.A., Masumura, R.A., Chang, C.L.: Thermal response of graphite epoxy composite subjected to rapid heating. J. Compos. Mater. **15**, 427–442 (1981)

Hilaire, B.: Etude du vieillissement thermique d'une résine polyamide thermostable. Thèse de Doctorat de l'ENSAM, Paris (1991)

Jonas, A., kegras, R.: Thermal stability and crystallization of poly(aryl ether ether ketone). Polymer **32**(5), 2691–2706 (1991)

Kemmish, D., Hay, J.N.: The effect of physical ageing on the properties of amorphous PEEK. Polymer **26**, 905–912 (1985)

Kemmish, D.: Update on the Technology and Applications of Polyaryletherketones. Smithers Rapra, London (2010)

Lee, S., Plamondon, M., Gaudert, P.C.: The effect of elevated temperature spikes on the mechanical properties of carbon fibre epoxy composites. In: 38th International SAMPE Symposium, Anaheim, California, pp. 1582–1593 (1993)
Saadaoui, H.: Évaluation des endommagements thermiques et mécaniques du composite graphite/époxyde par émission acoustique et acousto-ultrasonique. Thèse Ecole de génie, Université de Moncton, Canada (1996)
Stober, E.J., Seferis, J.C., Keenan, J.D.: Polymer **24**, 1845–1852 (1984)
Verdu, J.: Matériaux polymères, relations structure-propriétés. Cours ENSAM, Technique de l'ingénieur, Paris, pp. 2–20 (1992)

Author Index

A
Abdelaziz, I., 465
Abderrachid, Hamrani, 620
Abderrahim, Benabbas, 306
Aberkane, A., 3
Aboutaleb, Djamila, 219
Acem, K., 516
Achour, A., 605
Adkhis, A., 337
Adour, L., 492
Aili, D., 492
Aissani-Benissad, F., 527
Ait-kaci, H., 116
Ait-Kaci, Hocine, 15
Ali, M., 465
Amara, Mohamed, 254
Ammar, Mohammed Hadj, 219
Amrane, Meryem Imene, 52
Arbia, W., 492
Ayadi, Azzedine, 593
Ayadi, K., 637
Azibi, M., 353
Aziz, Z., 75
Azizi, C., 3

B
Baa, N., 166
Bachir Bouiadjra, B., 605
Bahoussi, Rawia Imene, 52
Baira, M., 10, 134, 365
Baït, N., 372, 378
Bakhti-Siad, A., 10
Bakora, Ahmed, 34
Barhoum, Zineb, 665
Bebba, Ahmed A., 665
Beggah, Y., 189
Bekhemas, K., 330
Bekhti-Siad, A., 420
Belaid, Ouahiba, 665
Belfar, Abbas, 15
Belfedal, A. E. K., 365
Belouchrani, M. A., 683
Benabbas, Khaled, 540
Benaboura, A., 126, 358, 372, 378, 657
Benbelaid, S., 408
Benchikh, Imene, 52
Bengueddach, Abdelkader, 52
Benhalima, Z., 420
Benhaoua, Fayrouz, 593
Benkara, S., 676
Benmesmoudi, S., 385
Bennegadi, L. M., 288
Bennini, L., 345
Bensouici, F., 67
Benturki, Asma, 228
Benturki, Oumessaâd, 228
Benzerga, Ratiba, 593
Benzidane, R., 288
Berkak, H., 451
Bessi, A., 296
Bezzar, A., 385
Bezzazi, B., 408
Billon, L., 657
Boudaren, C., 296
Boudine, B., 296
Boudraa, M. R., 580
Boufendi, L., 637
Bouhani-Benzian, H., 134

B. Abdelbaki et al. (Eds.): SMSD 2017, *Proceedings of the Third International Symposium on Materials and Sustainable Development*, pp. 699–701, 2018.
https://doi.org/10.1007/978-3-319-89707-3

Bouhani-Benziane, H., 10, 365
Bouiadjra, Mohamed Bachir, 95
Boulassel, Fatma Zohra, 306
Bouloufa, Abdeslam, 181
Boumchedda, K., 141
Boumerdassi, Karima, 313
Boumezoued, A., 3
Bounabi, L., 126, 166
Bourada, Fouad, 34, 147
Bourada, Mohamed, 147
Bouremana, Mohamed, 147
Bouslah Mokhnachi, N., 126, 166
Boutouchent-Guerfi, N., 358
Bouttout, Abdelouahab, 254
Brahimi, Nawal, 25
Brixi, N., 385

C
Chedad, A., 241
Chellil, A., 189, 279, 398
Chellil, S., 279, 398
Chelouche, A., 81
Chenine, D., 75
Cheraitia, Abdallah, 25
Cherif, S., 580
Cherifi, N., 657
Chorfi, H., 637
Choukri, A., 516
Choumane, Fatima Zohra, 506

D
Dakhli, A., 288
Daoud, Mounir, 228
Debbakh, S., 141
Dehbi, Atallah, 52
Demdoum, Abdellah, 451
Deratani, André, 481
Dib, A., 358
Djamil, Rechem, 197
Djellouli, A., 549
Djouadi, D., 81

E
El Abbas, Adda Bedia, 34
Elmeiche, N., 241

F
Fateh, Chouia, 613
Fatma-zohra, Bourab, 620
Ferhat, M., 157
Ferkhi, Mosbah, 301

G
Gassara, Sana, 481
Ghamri, H., 676
Goual, I., 451
Grassl, B., 372, 378
Gronheid, R., 157
Gueddouda, M. K., 451
Guendouz, N., 580
Guergouri, K., 3

H
Hachemaoui, A., 330
Hachemi, M., 556
Haddad, H., 44
Haddadine, N., 126, 166
Hadj Mostefa, Adda, 267
Halouane, M., 345
Hamid, Aknouche, 620
Hammiche, L., 81
Hammouda, Ahmed, 219
Hamouche, Houria, 105
Hamzi, Souad, 206
Haouanoh, D., 67
Hasseine, A., 549
Hebali, K., 10
Hocini, Imene, 540
Houcine, Belhouchet, 613

K
Kachi, M., 439
Kaci, Abdelhakim, 147
Kadi, Tarik, 15
Kakaa, Faiza, 301
Kandouci, Fatma, 506
Kecira, Zoubida, 228
Kerour, Safia, 181
Khalfi, R., 44
Khelifa, Amine, 568
Khellaf, Nabila, 540
Kheloui, M., 141
Khial, A., 3
Khiar, C., 345
Krour, Baghdad, 95

L
Lahmer, M. A., 644
Lamri, Younes, 593
Lasladj, Meryem, 181
Le Gendre, Laurent, 593
Lecheb, S., 189, 279, 398
Loucif Seiad, M., 157

Louhibi–Bouiri, M., 556
Lounici, Y., 189
Lounis, Zoubida, 52

M
Maachi, R., 580
Maachou, Bouhana, 506
Madi-Azegagh, K., 527
Mahi, K., 116
Mahmoudi, Noureddine, 267
Malek, O. H., 465
Mazari, T., 345
Mebsout, Hanane, 95
Mechab, Ismail, 206
Mechakra, H., 189, 279, 398
Melbouci, Bachir, 427
Menad, S., 345
Merdaci, Slimane, 267
Merzouk, H., 44
Merzouk, L. Abane, 337
Messani, B., 116
MKuppuswamy, V. K., 157
Mohamed, M., 465
Mouna, Rekaik, 306
Moussa, A., 580
Mouzaia, F., 81

N
Nadjem, A., 439
Nasrallah, N., 580
Nour, A., 398

O
Okba, F., 89
Ouaaz, Oum Essaad, 427
Ouazib, F., 166
Ouinas, D., 605
Ourahmoun, O., 175

R
Raache, Imane, 665
Rabehi, B., 141
Rabia, C., 345
Ramdani, Amina, 481
Reguig, Farouk Hamza, 52
Remadnia, M., 439
Rouagdia, K., 439

S
Safi, B., 189, 279
Safi, Brahim, 219
Sahnoun, M., 10, 134, 365, 420
Sahnoun, O., 10
Sail, L., 385
Saoudet, N., 516
Saoudi, T., 683
Saoula, N., 353
Sebouai, O., 385
Selma, Kaour, 197
Senhadji, Yassine, 206
Sereir, Z., 288
Serier, Aicha, 313
Settet, A. T., 398
Shabat, Mohammed M., 105
Si Ziani, N., 365
Siad-Bekhti, A., 134
Siziani, N., 134

T
Tabti, Affaf, 52
Talalghil, R., 67
Talantikite-Touati, D., 44
Taleb, Safia, 481
Tergou, S., 516
Timedjeghdine, M., 549
Toubane, M., 67
Toumi, Nadia, 52
Tounsi, A., 44
Tounsi, Abdelouahed, 147
Tounsi, Abdelouhed, 34
Trari, H., 516

Y
Yahiaoui, A., 330
Yahiaoui, I., 527

Z
Zaabat, M., 676
Zehaf, R., 288
Ziane, Samira, 568

The manufacturer's authorised representative in the EU is Springer Nature Customer Service Centre GmbH, Europaplatz 3, 69115 Heidelberg, Germany. If you have any concerns regarding our products, please contact ProductSafety@springernature.com

Printed and bound by CPI Group (UK) Ltd, Croydon, CR0 4YY
05/10/2026
02228141-0017